SOME OTHER BOOKS BY

Isaac Asimov

ASIMOV'S BIOGRAPHICAL ENCYCLOPEDIA
OF SCIENCE AND TECHNOLOGY

THE HUMAN BODY

THE HUMAN BRAIN

UNDERSTANDING PHYSICS (3 VOLUMES)

THE UNIVERSE

ASIMOV'S GUIDE TO THE BIBLE (2 VOLUMES)

ASIMOV'S GUIDE TO SHAKESPEARE (2 VOLUMES)

ISAAC ASIMOV'S TREASURY OF HUMOR

THE NOBLE GASES

PHOTOSYNTHESIS

THE SEARCH FOR THE ELEMENTS

Asimov's

GUIDE TO

Science

Asimov's
GUIDE TO
Science

BY

Isaac Asimov

BASIC BOOKS, INC., PUBLISHERS

New York   London

LIBRARY OF CONGRESS CATALOG CARD NUMBER: 72–76720
SBN 465-00472-5
MANUFACTURED IN THE UNITED STATES OF AMERICA
10 9 8 7

TO

Janet Jeppson

who shares my interest in science

Preface

THE RAPID ADVANCE of science is exciting and exhilarating to anyone who is fascinated by the unconquerability of the human spirit and by the continuing efficacy of the scientific method as a tool for penetrating the complexities of the Universe.

But what if one is also dedicated to keeping up with every phase of scientific advance for the deliberate purpose of interpreting that advance for the general public? For him, the excitement and exhilaration is tempered by a kind of despair.

Science will not stand still. It is a panorama that subtly dissolves and changes even while we watch. It cannot be caught in its every detail at any one moment of time without being left behind at once.

In 1960, *The Intelligent Man's Guide to Science* was published, and at once the advance of science flowed past it. In order to consider quasars and lasers, for instance (which were unknown in 1960 and household words a couple of years later), *The New Intelligent Man's Guide to Science* was published in 1965.

But still science drove on inexorably. Now there came the question of pulsars, of black holes, of continental drift, men on the moon, REM sleep, gravitational waves, holography, cyclic-AMP, and so on, and so on, and so on—all post-1965.

So it is time for a new edition, the third. But what do we call it now? *The New New Intelligent Man's Guide to Science?* Obviously not.

Since 1965, however, I have done, with my own name in the title, a two volume guide to the Bible and also a two volume guide to Shakespeare. Why not use the same system here? Enter, then, the 1972 edition of my guide to science, entitled, straightforwardly, *Asimov's Guide to Science.*

ISAAC ASIMOV

New York
1972

Contents

PART TWO

The Biological Sciences

Part One

THE
Physical Sciences

CHAPTER 1

What Is Science?

ALMOST in the beginning was curiosity.

Curiosity, the overwhelming desire to know, is not characteristic of dead matter. It is also not characteristic of some forms of living organism, which, for that very reason, we can scarcely bring ourselves to consider alive.

A tree does not display curiosity about its environment in any way we can recognize; nor does a sponge or an oyster. The wind, the rain, the ocean currents bring them what is needful, and from it they take what they can. If the chance of events is such as to bring them fire, poison, predators, or parasites, they die as stoically and as undemonstratively as they lived.

Early in the scheme of life, however, independent motion was developed by some organisms. It meant a tremendous advance in their control of the environment. A moving organism no longer had to wait in stolid rigidity for food to come its way; it went out after it.

This meant that adventure had entered the world—and curiosity. The individual that hesitated in the competitive hunt for food, that was overly conservative in its investigation, starved. As early as that, curiosity concerning the environment was enforced as the price of survival.

The one-celled paramecium, moving about in a searching way, cannot have conscious volitions and desires in the sense that we do, but it has a drive, even if only a "simple" physical-chemical one, which causes it to behave as if it were investigating its surroundings for food. And this "act of curiosity" is what we most easily recognize as being inseparable from the kind of life that is most akin to ours.

As organisms grew more intricate, their sense organs multiplied and became both more complex and more delicate. More messages of greater variety were received from and about the external environment. Along with that (whether as cause or effect we cannot tell), there developed an increasing complexity of the nervous system, the living instrument that interpreted and stored the data collected by the sense organs.

There comes a point where the capacity to receive, store, and interpret messages from the outside world may outrun sheer necessity. An organism may for the moment be sated with food, and there may, at the moment, be no danger in sight. What does it do then?

It might lapse into an oysterlike stupor. But the higher organisms, at least, still show a strong instinct to explore the environment. Idle curiosity, we may call it. Yet, though we may sneer at it, we judge intelligence by it. The dog, in moments of leisure, will sniff idly here and there, pricking up its ears at sounds we cannot hear; and so we judge it to be more intelligent than the cat, which in its moments of leisure grooms itself or quietly and luxuriously stretches out and falls asleep. The more advanced the brain, the greater the drive to explore, the greater the "curiosity surplus." The monkey is a byword for curiosity. Its busy little brain must and will be kept going on whatever is handy. And in this respect, as in many others, man is but a supermonkey.

The human brain is the most magnificently organized lump of matter in the known universe, and its capacity to receive, organize, and store data is far in excess of the ordinary requirements of life. It has been estimated that in a lifetime a human being can learn up to 15 trillion items of information.

It is to this excess that we owe our ability to be afflicted by that supremely painful disease, boredom. A human being forced into a situation where he has no opportunity to utilize his brain except for minimal survival will gradually experience a variety of unpleasant symptoms, up to and including serious mental disorganization.

What it amounts to, then, is that the normal human being has an intense and overwhelming curiosity. If he lacks the opportunity to satisfy it in ways immediately useful to him, he will satisfy it in other ways—even regrettable ways to which we have attached admonitions such as: "Curiosity killed the cat," "Mind your own business."

The overriding power of curiosity, even with harm as the penalty, is reflected in the myths and legends of the human race. The Greeks had the tale of Pandora and her box. Pandora, the first woman, was given a box that she was forbidden to open. Quickly and naturally enough she opened it and found it full of the spirits of disease, famine, hate, and all kinds of evil—which escaped and have plagued the world ever since.

In the Biblical story of the temptation of Eve, it seems fairly certain (to me, at any rate) that the serpent had the world's easiest job. He might have saved his tempting words: Eve's curiosity would have driven her to taste the forbidden fruit even without temptation. If you are of a mind to interpret the Bible allegorically, you may think of the serpent as simply the representation of this inner compulsion; in the conventional cartoon picturing Eve standing under the tree with the forbidden fruit in her hand, the serpent coiled around the branch might be labeled "Curiosity."

If curiosity, like any other human drive, can be put to ignoble use —the prying invasion of privacy that has given the word its cheap and unpleasant connotation—it nevertheless remains one of the noblest properties of the human mind. For its simplest definition is "the desire to know."

This desire finds its first expression in answers to the practical needs of human life—how best to plant and cultivate crops, how best to fashion bows and arrows, how best to weave clothing—in short, the "applied arts." But after these comparatively limited skills have been mastered, or the practical needs fulfilled, what then? Inevitably the desire to know leads on to less limited and more complex activities.

It seems clear that the "fine arts" (designed to satisfy inchoate and boundless and spiritual needs) were born in the agony of boredom. To be sure, one can easily find more mundane uses and excuses for the fine arts. Paintings and statuettes were used as fertility charms and as religious symbols, for instance. But one cannot help suspecting that the objects existed first and the use second.

To say that the fine arts arose out of a sense of the beautiful may also be putting the cart before the horse. Once the fine arts were developed, their extension and refinement in the direction of beauty would have followed inevitably, but even if this had not happened, the fine arts would have developed nevertheless. Surely the fine arts antedate any possible need or use for them, other than the elementary need to occupy the mind as fully as possible.

Not only does the production of a work of fine art occupy the mind satisfactorily; the contemplation or appreciation of the work supplies a similar service to the audience. A great work of art is great precisely because it offers a kind of stimulation that cannot readily be found elsewhere. It contains enough data of sufficient complexity to cajole the brain into exerting itself past the usual needs, and, unless a person is hopelessly ruined by routine or stultification, that exertion is pleasant.

But if the practice of the fine arts is a satisfactory solution to the problem of leisure, it has this disadvantage: it requires, in addition to an active and creative mind, a physical dexterity. It is just as interesting to pursue mental activities that involve only the mind, without the supplement of manual skill. And, of course, such an activity is available. It is the pursuit of knowledge itself, not in order to do something with it but for its own sake.

Thus the desire to know seems to lead into successive realms of greater etherealization and more efficient occupation of the mind—from knowledge of accomplishing the useful, to knowledge of accomplishing the esthetic, to "pure" knowledge.

Knowledge for itself alone seeks answers to such questions as "How high is the sky?" or "Why does a stone fall?" This is sheer curiosity—curiosity at its idlest and therefore perhaps at its most peremptory. After all, it serves no apparent purpose to know how high the sky is or why the stone falls. The lofty sky does not interfere with the ordinary business of life, and, as for the stone, knowing why it falls does not help us to dodge it more skillfully or soften the blow if it happens to hit us. Yet there have always been people who ask such apparently useless questions and try to answer them out of the sheer desire to know—out of the absolute necessity of keeping the brain working.

The obvious method of dealing with such questions is to make up an esthetically satisfying answer: one that has sufficient analogies to what is already known to be comprehensible and plausible. The expression "to make up" is rather bald and unromantic. The ancients liked to think of the process of discovery as the inspiration of the muses or a revelation from heaven. In any case, whether it was inspiration, revelation or the kind of creative thinking that goes into storytelling, their explanations depended heavily on analogy. The lightning bolt is destructive and terrifying, but it appears, after all, to be hurled like a weapon and does the damage of a hurled weapon—a fantastically violent one. Such a weapon must have a wielder similarly enlarged in scale, and so the thunderbolt becomes the hammer of Thor or the flashing spear of Zeus. The more-than-normal weapon is wielded by a more-than-normal man.

Thus a myth is born. The forces of nature are personified and become gods. The myths react on one another, are built up and improved by generations of mythtellers until the original point may be obscured. Some may degenerate into pretty stories (or ribald ones), whereas others may gain an ethical content important enough to make them meaningful within the framework of a major religion.

Just as art may be fine or applied, so may mythology. Myths may be maintained for their esthetic charm, or they may be bent to the physical uses of mankind. For instance, the earliest farmers would be intensely concerned with the phenomenon of rain and why it fell so capriciously. The fertilizing rain falling from the heavens on the earth presented an obvious analogy to the sex act, and, by personifying both heaven and earth, man found an easy explanation of the release or withholding of the rains. The earth-goddess, or the sky-god, was either pleased or offended, as the case might be. Once this myth was accepted, farmers had a plausible basis for bringing rain; namely, appeasing the god by appropriate rites. These rites might well be orgiastic in nature—an attempt to influence heaven and earth by example.

6

The Greek myths are among the prettiest and most sophisticated in our literary and cultural heritage. But it was the Greeks also who, in due course, introduced the opposite way of looking at the universe—that is, as something impersonal and inanimate. To the mythmakers, every aspect of nature was essentially human in its unpredictability. However mighty and majestic the personification, however superhuman Zeus or Marduk or Odin might be in powers, they were also—like mere men—frivolous, whimsical, emotional, capable of outrageous behavior for petty reasons, susceptible to childish bribes. As long as the universe was in the control of such arbitrary and unpredictable deities, there was no hope of understanding it, only the shallow hope of appeasing it. But in the new view of the later Greek thinkers, the universe was a machine governed by inflexible laws. The Greek philosophers now devoted themselves to the exciting intellectual exercise of trying to discover just what the laws of nature might be.

The first to do so, according to Greek tradition, was Thales of Miletus, about 600 B.C. He was saddled with an almost impossible number of discoveries by later Greek writers, and it may be that he first brought the gathered Babylonian knowledge to the Greek world. His most spectacular achievement was that of predicting an eclipse for 585 B.C.—and having it take place.

In engaging in this intellectual exercise, the Greeks assumed, of course, that nature would play fair; that, if attacked in the proper manner, it would yield its secrets and would not change position or attitude in mid-play. (Thousands of years later Albert Einstein expressed this feeling when he said, "God may be subtle, but he is not malicious.") There was also the feeling that the natural laws, when found, would be comprehensible. This Greek optimism has never entirely left the human race.

With confidence in the fair play of nature, man needed to work out an orderly system for learning how to determine the underlying laws from the observed data. To progress from one point to another by established rules of argument is to use "reason." A reasoner may use "intuition" to guide his search for answers, but he must rely on sound logic to test his theories. To take a simple example: if brandy and water, whiskey and water, vodka and water, and rum and water are all intoxicating beverages, one may jump to the conclusion that the intoxicating factor must be the ingredient these drinks hold in common—namely, water. There is something wrong with this reasoning, but the fault in the logic is not immediately obvious, and in more subtle cases the error may be hard indeed to discover.

The tracking-down of errors or fallacies in reasoning has amused thinkers from Greek times to the present. And of course we owe the earliest foundations of systematic logic to Aristotle of Stagira who in the fourth century B.C. first summarized the rules of rigorous reasoning.

7

The essentials of the intellectual game of man-against-nature are three. First, you must collect observations about some facet of nature. Second, you must organize these observations into an orderly array. (The organization does not alter them but merely makes them easier to handle. This is plain in the game of bridge, for instance, where arranging the hand in suits and order of value does not change the cards or show the best course of play, but makes it easier to arrive at the logical plays.) Third, you must derive from your orderly array of observations some principle that summarizes the observations.

For instance, we may observe that marble sinks in water, wood floats, iron sinks, a feather floats, mercury sinks, olive oil floats, and so on. If we put all the sinkable objects in one list and all the floatable ones in another and look for a characteristic that differentiates all the objects in one group from all in the other, we will conclude: Heavy objects sink in water and light objects float.

The Greeks named their new manner of studying the universe *philosophia* ("philosophy"), meaning "love of knowledge" or, in free translation, "the desire to know."

The Greeks achieved their most brilliant successes in geometry. These successes can be attributed mainly to their development of two techniques: abstraction and generalization.

Here is an example. Egyptian land surveyors had found a practical way to form a right angle: they divided a rope into twelve equal parts and made a triangle in which three parts formed one side, four parts another, and five parts the third side—the right angle lay where the three-unit side joined the four-unit side. There is no record of how the Egyptians discovered this method, and apparently their interest went no further than to make use of it. But the curious Greeks went on to investigate why such a triangle should contain a right angle. In the course of their analysis, they grasped the point that the physical construction itself was only incidental; it did not matter whether the triangle was made of rope or linen or wooden slats. It was simply a property of "straight lines" meeting at angles. In conceiving of ideal straight lines, which were independent of any physical visualization and could exist only in imagination, they originated the method called abstraction—stripping away nonessentials and considering only those properties necessary to the solution of the problem.

The Greek geometers made another advance by seeking general solutions for classes of problems, instead of treating individual problems separately. For instance, one might discover by trial that a right angle appeared in triangles, not only with sides 3, 4, and 5 feet long, but also in those of 5, 12, and 13 feet and of 7, 24, and 25 feet. But these were merely numbers without meaning. Could some common property be found that would describe all right triangles? By careful reasoning the

Greeks showed that a triangle was a right triangle if, and only if, the lengths of the sides had the relation $x^2 + y^2 = z^2$, z being the length of the longest side. The right angle lay where the sides of length x and y met. Thus for the triangle with sides of 3, 4, and 5 feet, squaring the sides gives $9 + 16 = 25$; similarly, squaring the sides of 5, 12, and 13 gives $25 + 144 = 169$; and squaring 7, 24, and 25 gives $49 + 576 = 625$. These are only three cases out of an infinity of possible ones and, as such, trivial. What intrigued the Greeks was the discovery of a proof that the relation must hold in all cases. And they pursued geometry as an elegant means of discovering and formulating generalizations.

Various Greek mathematicians contributed proofs of relationships existing among the lines and points of geometric figures. The one involving the right triangle was reputedly worked out by Pythagoras of Samos about 525 B.C. and is still called the Pythagorean theorem in his honor.

About 300 B.C., Euclid gathered the mathematical theorems known in his time and arranged them in a reasonable order, such that each theorem could be proved through the use of theorems proved previously. Naturally, this system eventually worked back to something unprovable: If each theorem had to be proved with the help of one already proved, how could one prove theorem No. 1? The solution was to begin with a statement of truths so obvious and acceptable to all as to need no proof. Such a statement is called an "axiom." Euclid managed to reduce the accepted axioms of the day to a few simple statements. From these axioms alone, he built an intricate and majestic system of "Euclidean geometry." Never was so much constructed so well from so little, and Euclid's reward is that his textbook has remained in use, with but minor modification, for more than 2,000 years.

Working out a body of knowledge as the inevitable consequence of a set of axioms ("deduction") is an attractive game. The Greeks fell in love with it, thanks to the success of their geometry—sufficiently in love with it to commit two serious errors.

First, they came to consider deduction as the only respectable means of attaining knowledge. They were well aware that for some kinds of knowledge deduction was inadequate; for instance, the distance from Corinth to Athens could not be deduced from abstract principles but had to be measured. The Greeks were willing to look at nature when necessary; however, they were always ashamed of the necessity and considered that the highest type of knowledge was that arrived at by cerebration. They tended to undervalue knowledge which was too directly involved with everyday life. There is a story that a student of Plato, receiving mathematical instruction from the master, finally asked impatiently: "But what is the use of all this?" Plato, deeply offended, called a slave and ordered him to give the student a coin. "Now," he said, "you need not feel your instruction has been entirely to no purpose." With that, the student was expelled.

There is a well-worn belief that this lofty view arose from the Greek's slave-based culture, in which all practical matters were relegated to the slaves. Perhaps so, but I incline to the view that the Greeks felt that philosophy was a sport, an intellectual game. We regard the amateur in sports as a gentleman socially superior to the professional who makes his living at it. In line with this concept of purity, we take almost ridiculous precautions to make sure that the contestants in the Olympic games are free of any taint of professionalism. The Greek rationalization for the "cult of uselessness" may similarly have been based on a feeling that to allow mundane knowledge (such as the distance from Athens to Corinth) to intrude on abstract thought was to allow imperfection to enter the Eden of true philosophy. Whatever the rationalization, the Greek thinkers were severely limited by their attitude. Greece was not barren of practical contributions to civilization, but even its great engineer, Archimedes of Syracuse, refused to write about his practical inventions and discoveries; to maintain his amateur status, he broadcast only his achievements in pure mathematics. And lack of interest in earthly things—in invention, in experiment, in the study of nature—was but one of the factors that put bounds on Greek thought. The Greeks' emphasis on purely abstract and formal study—indeed, their very success in geometry—led them into a second great error and, eventually, to a dead end.

Seduced by the success of the axioms in developing a system of geometry, the Greeks came to think of the axioms as "absolute truths" and to suppose that other branches of knowledge could be developed from similar "absolute truths." Thus in astronomy they eventually took as self-evident axioms the notions that (1) the earth was motionless and the center of the universe, and (2) whereas the earth was corrupt and imperfect, the heavens were eternal, changeless, and perfect. Since the Greeks considered the circle the perfect curve and since the heavens were perfect, it followed that all the heavenly bodies must move in circles around the earth. In time their observations (arising from navigation and calendar making) showed that the planets did not move in perfectly simple circles, and so they were forced to allow planets to move in ever more complicated combinations of circles, which, about 150 A.D., were formulated as an uncomfortably complex system by Claudius Ptolemaeus (Ptolemy) at Alexandria. Similarly, Aristotle worked up fanciful theories of motion from "self-evident" axioms, such as the proposition that the speed of an object's fall was proportional to its weight. (Anyone could see that a stone fell faster than a feather.)

Now this worship of deduction from self-evident axioms was bound to wind up at the edge of a precipice, with no place to go. After the Greeks had worked out all the implications of the axioms, further important discoveries in mathematics or astronomy seemed out of the question. Philosophic knowledge appeared complete and perfect, and, for nearly 2,000 years after the Golden Age of Greece, when questions involving the material universe arose, there was a tendency to settle matters to the satisfaction of all by saying, "Aristotle says . . . ," or, "Euclid says. . . ."

Having solved the problems of mathematics and astronomy, the Greeks turned to more subtle and challenging fields of knowledge. One was the field of the human soul.

Plato was far more interested in such questions as "What is justice?" or "What is virtue?" than in why rain fell or how the planets moved. As the supreme moral philosopher of Greece, he superseded Aristotle, the supreme natural philosopher. The Greek thinkers of the Roman period found themselves drawn more and more to the subtle delights of moral philosophy and away from the apparent sterility of natural philosophy. The last development in ancient philosophy was an exceedingly mystical "neo-Platonism" formulated by Plotinus about 250 A.D.

Christianity, with its emphasis on the nature of God and His relation to man, introduced an entirely new dimension into the subject matter of moral philosophy and increased its superiority as an intellectual pursuit over natural philosophy. From 200 A.D. to 1200 A.D., Europeans concerned themselves almost exclusively with moral philosophy, in particular with theology. Natural philosophy was nearly forgotten.

The Arabs, however, managed to preserve Aristotle and Ptolemy through the Middle Ages, and, from them, Greek natural philosophy eventually filtered back to Western Europe. By 1200, Aristotle had been rediscovered. Further infusions came from the dying Byzantine Empire, which was the last area in Europe that maintained a continuous cultural tradition from the great days of Greece.

The first and most natural consequence of the rediscovery of Aristotle was the application of his system of logic and reason to theology. About 1250, the Italian theologian Thomas Aquinas established the system called "Thomism," based on Aristotelian principles, which still represents the basic theology of the Roman Catholic Church. But men soon began to apply the revival of Greek thought to secular fields as well.

Because the leaders of the Renaissance shifted emphasis from matters concerning God to the works of humanity, they were called "humanists," and the study of literature, art, and history is still referred to as "the humanities."

To the Greek natural philosophy, the Renaissance thinkers brought a fresh outlook, for the old views no longer entirely satisfied. In 1543 the Polish astronomer Nicolaus Copernicus published a book that went so far as to reject a basic axiom of astronomy: he proposed that the sun, not the earth, be considered the center of the universe. (He retained the notion of circular orbits for the earth and other planets, however.) This new axiom allowed a much simpler explanation of the observed motions of heavenly bodies. Yet the Copernican axiom of a moving earth was far less "self-evident" than the Greek axiom of a motionless earth, and so it is not surprising that it took nearly a century for the Copernican theory to be accepted.

In a sense, the Copernican system itself was not a crucial change. Copernicus had merely switched axioms; and Aristarchus of Samos had already anticipated this switch to the sun as the center 2,000 years earlier. This is not to say that the changing of an axiom is a minor matter. When mathematicians of the nineteenth century challenged Euclid's axioms and developed "non-Euclidean geometries" based on other assumptions, they influenced thought on many matters in a most profound way: today the very history and form of the universe are thought to conform to a non-Euclidean (Riemannian) geometry rather than the "commonsense" geometry of Euclid. But the revolution initiated by Copernicus entailed not just a shift in axioms but eventually involved a whole new approach to nature. This revolution was carried through in the person of the Italian Galileo Galilei.

The Greeks, by and large, had been satisfied to accept the "obvious" facts of nature as starting points for their reasoning. It is not on record that Aristotle ever dropped two stones of different weight to test his assumption that the speed of fall was proportional to an object's weight. To the Greeks, experimentation seemed irrelevant. It interfered with and detracted from the beauty of pure deduction. Besides, if an experiment disagreed with a deduction, could one be certain that the experiment was correct? Was it likely that the imperfect world of reality would agree completely with the perfect world of abstract ideas, and, if it did not, ought one to adjust the perfect to the demands of the imperfect? To test a perfect theory with imperfect instruments did not impress the Greek philosophers as a valid way to gain knowledge.

Experimentation began to become philosophically respectable in Europe with the support of such philosophers as Roger Bacon (a contemporary of Thomas Aquinas) and his later namesake Francis Bacon. But it was Galileo who overthrew the Greek view and effected the revolution. He was a convincing logician and a genius as a publicist. He described his experiments and his point of view so clearly and so dramatically that he won over the European learned community. And they accepted his methods along with his results.

According to the best-known story about him, Galileo tested Aristotle's theories of falling bodies by asking the question of nature in such a way that all Europe could hear the answer. He is supposed to have climbed to the top of the Leaning Tower of Pisa and dropped a ten-pound sphere and a one-pound sphere simultaneously; the thump of the two balls hitting the ground in the same split second killed Aristotelian physics.

Actually Galileo probably did not perform this particular experiment, but the story is so typical of his dramatic methods that it is no wonder it has been widely believed through the centuries..

Galileo undeniably did roll balls down inclined planes and measured the distance that they traveled in given times. He was the first to conduct time experiments, the first to use measurement in a systematic way.

His revolution consisted in elevating "induction" above deduction as the logical method of science. Instead of building conclusions on an assumed set of generalizations, the inductive method starts with observations and derives generalizations (axioms, if you will) from them. Of course, even the Greeks obtained their axioms from observation; Euclid's axiom that a straight line is the shortest distance between two points was an intuitive judgment based on experience. But whereas the Greek philosopher minimized the role played by induction, the modern scientist looks on induction as the essential process of gaining knowledge, the only way of justifying generalizations. Moreover, he realizes that no generalization can be allowed to stand unless it is repeatedly tested by newer and still newer experiments—unless it withstands the continuing test of further induction.

The present general viewpoint is just the reverse of the Greeks. Far from considering the real world an imperfect representation of ideal truth, we consider generalizations to be only imperfect representatives of the real world. No amount of inductive testing can render a generalization completely and absolutely valid. Even though billions of observers tend to bear out a generalization, a single observation that contradicts or is inconsistent with it must force its modification. And no matter how many times a theory meets its tests successfully, there can be no certainty that it will not be overthrown by the next observation.

This, then, is a cornerstone of modern natural philosophy. It makes no claim of attaining ultimate truth. In fact, the phrase "ultimate truth" becomes meaningless, because there is no way in which enough observations can be made to make truth certain, and therefore "ultimate." The Greek philosophers recognized no such limitation. Moreover, they saw no difficulty in applying exactly the same method of reasoning to the question "What is justice?" as to the question "What is matter?" Modern science, on the other hand, makes a sharp distinction between the two types of question. The inductive method cannot make generalizations about what it cannot observe, and, since the nature of the human soul, for example, is not observable by any direct means yet known, this subject lies outside the realm of the inductive method.

The victory of modern science did not become complete until it established one more essential principle—namely, free and cooperative communication among all scientists. Although this necessity seems obvious to us now, it was not obvious to the philosophers of ancient and medieval times. The Pythagoreans of ancient Greece were a secret society who kept their mathematical discoveries to themselves. The alchemists of the Middle Ages deliberately obscured their writings to keep their so-called findings within as small an inner circle as possible. In the sixteenth century, the Italian mathematician Niccolo Tartaglia, who discovered a method of solving cubic equations, saw nothing wrong in attempting to keep it a secret. When Geronimo Cardano, a fellow mathematician, wormed the secret out of Tartaglia and published it as his own, Tartaglia naturally was outraged, but aside from Cardano's trickery in claiming the credit, he was certainly correct in his reply that such a discovery had to be published.

Nowadays no scientific discovery is reckoned a discovery if it is kept secret. The English chemist Robert Boyle, a century after Tartaglia and Cardano, stressed the importance of publishing all scientific observations in full detail. A new observation or discovery, moreover, is no longer considered valid, even after publication, until at least one other investigator has repeated the observation and "confirmed" it. Science is the product not of individuals but of a "scientific community."

One of the first groups (and certainly the most famous) to represent such a scientific community was the Royal Society of London for Improving Natural Knowledge, usually called simply the "Royal Society." It grew out of informal meetings, beginning about 1645, of a group of gentlemen interested in the new scientific methods originated by Galileo. In 1660, the Society was formally chartered by King Charles II.

The members of the Royal Society met and discussed their findings openly, wrote letters describing them in English rather than Latin, and pursued their experiments with vigor and vivacity. Nevertheless, through most of the seventeenth century they remained in a defensive position. The attitude of many of their learned contemporaries might be expressed by a cartoon, after the modern fashion, showing the lofty shades of Pythagoras, Euclid, and Aristotle staring down haughtily at children playing with marbles, labeled "Royal Society."

All this was changed by the work of Isaac Newton, who became a member of the society. From the observations and conclusions of Galileo, of the Danish astronomer Tycho Brahe, and of the German astronomer Johannes Kepler, who figured out the elliptical nature of the orbits of the planets, Newton arrived by induction at his three simple laws of motion and his great fundamental generalization—the law of universal gravitation. The educated world was so impressed with this discovery that Newton was idolized, almost deified, in his own lifetime. This majestic new universe, built upon a few simple assumptions, now made the Greek philosophers look like boys playing with marbles. The revolution that Galileo had initiated at the beginning of the seventeenth century was triumphantly completed by Newton at the century's end.

It would be pleasant to be able to say that science and man have lived happily ever since. But the truth is that the real difficulties of both were only beginning. As long as science had remained deductive, natural philosophy could be part of the general culture of all educated men. But inductive science became an immense labor—of observation, learning, and analysis. It was no longer a game for amateurs. And the complexity of science grew with each decade. During the century after Newton, it was still possible for a man of unusual attainments to master all fields of scientific knowledge. But, by 1800, this had become entirely impracticable. As time went on, it was increasingly necessary for a scientist to limit himself to a portion of the field if he intended an intensive concern with it. Specialization was forced on science by its own inexorable growth. And with each generation of scientists, specialization has grown more and more intense.

The publications of scientists concerning their individual work have never been so copious—and so unreadable for anyone but their fellow specialists. This has been a great handicap to science itself, for basic advances in scientific knowledge often spring from the cross-fertilization of knowledge from different specialties. What is even more ominous is is that science has increasingly lost touch with nonscientists. Under such circumstances scientists come to be regarded almost as magicians— feared rather than admired. And the impression that science is incomprehensible magic, to be understood only by a chosen few who are suspiciously different from ordinary mankind, is bound to turn many youngsters away from science.

In the 1960's, indeed, strong feelings of outright hostility toward science were to be found among the young—even among the educated young in the colleges. Our industrialized society is based on the scientific discoveries of the last two centuries, and our society finds it is plagued by undesirable side-effects of its very success.

Improved medical techniques have brought about a runaway increase in population; chemical industries and the internal-combustion engine are fouling our water and our air; the demand for materials and for energy is depleting and destroying the Earth's crust. And this is all too easily blamed on "science" and "scientists" by those who do not quite understand that if knowledge can create problems, it is not through ignorance that we can solve them.

Yet modern science need not be so complete a mystery to nonscientists. Much could be accomplished toward bridging the gap if scientists accepted the responsibility of communication—explaining their own fields of work as simply and to as many as possible—and if non-scientists, for their part, accepted the responsibility of listening. To gain a satisfactory appreciation of the developments in a field of science, it is not essential to have a total understanding of the science. After all, no one feels that he must be capable of writing a great work of literature in order to appreciate Shakespeare. To listen to a Beethoven symphony with pleasure does not require the listener to be capable of composing an equivalent symphony of his own. By the same token, one can appreciate and take pleasure in the achievements of science even though he does not himself have a bent for creative work in science.

But what, you may ask, would be accomplished? The first answer is that no one can really feel at home in the modern world and judge the nature of its problems—and the possible solutions to those problems—unless he has some intelligent notion of what science is up to. But beyond this, initiation into the magnificent world of science brings great esthetic satisfaction, inspiration to youth, fulfillment of the desire to know, and a deeper appreciation of the wonderful potentialities and achievements of the human mind.

It is with this in mind that I have undertaken to write this book.

CHAPTER 2

The Universe

The Size of the Universe

There is nothing about the sky that makes it look particularly distant to a casual observer. Young children have no great trouble in accepting the fantasy that "the cow jumped over the moon"—or "he jumped so high, he touched the sky." The ancient Greeks, in their myth-telling stage, saw nothing ludicrous in allowing the sky to rest on the shoulders of Atlas. Of course, Atlas might have been astronomically tall, but another myth suggests otherwise. Atlas was enlisted by Hercules to help him with the eleventh of his famous twelve labors—fetching the golden apples (oranges?) of the Hesperides ("the far west"—Spain?). While Atlas went off to fetch the apples, Hercules stood on a mountain and held up the sky. Granted that Hercules was a large specimen, he was nevertheless not a giant. It follows then that the early Greeks took quite calmly to the notion that the sky cleared the mountaintops by only a few feet.

It is natural to suppose, to begin with, that the sky is simply a hard canopy in which the shining heavenly bodies are set like diamonds. (Thus the Bible refers to the sky as the "firmament," from the same Latin root as the word "firm.") As early as the sixth to fourth centuries B.C., Greek astronomers realized that there must be more than one canopy. For while the "fixed" stars moved around the Earth in a body, apparently without changing their relative positions, this was not true of the Sun, Moon, and five bright starlike objects (Mercury, Venus, Mars, Jupiter, and Saturn)—in fact, each moved in a separate path. These seven bodies were called planets (from a Greek word meaning "wanderer"), and it seemed obvious that they could not be attached to the vault of the stars.

17

The Greeks assumed that each planet was set in an invisible vault of its own and that the vaults were nested one above the other, the nearest belonging to the planet that moved fastest. The quickest motion belonged to the Moon, which circled the sky in about twenty-nine and a half days. Beyond it lay in order (so thought the Greeks) Mercury, Venus, the Sun, Mars, Jupiter, and Saturn.

The first scientific measurement of any cosmic distance came about 240 B.C. Eratosthenes of Cyrene, the head of the Library at Alexandria, then the most advanced scientific institution in the world, pondered the fact that on June 21, when the noonday sun was exactly overhead at the city of Syene in Egypt, it was not quite at the zenith at noon in Alexandria, 500 miles north of Syene. Eratosthenes decided that the explanation must be that the surface of the earth curved away from the sun. From the length of the shadow in Alexandria at noon on the solstice, straightforward geometry could yield an answer as to the amount by which the earth's surface curved in the 500-mile distance from Syene to Alexandria. From that one could calculate the circumference and the diameter of the earth, assuming that the earth was spherical in shape—a fact which Greek astronomers of the day were ready to accept.

Eratosthenes worked out the answer (in Greek units), and, as nearly as we can judge, his figures in our units came out at about 8,000 miles

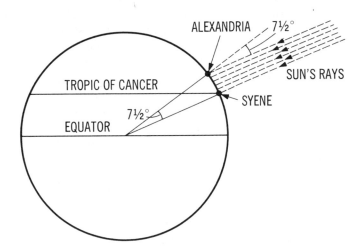

Eratosthenes measured the size of the earth from its curvature. At noon, June 21, the sun is directly overhead at Syene, which lies on the Tropic of Cancer. But, at this same time, the sun's rays, seen from farther north in Alexandria, fall at an angle of 7.5° to the vertical and therefore cast a shadow. Knowing the distance between the two cities and the length of the shadow in Alexandria, Eratosthenes made his calculations.

for the diameter and 25,000 miles for the circumference of the earth. This, as it happens, is just about right. Unfortunately, this accurate value for the size of the earth did not prevail. About 100 B.C. another Greek astronomer, Posidonius of Apamea, repeated Eratosthenes' work, but reached the conclusion that the earth was but 18,000 miles in circumference.

It was the smaller figure that was accepted by Ptolemy and, therefore, throughout medieval times. Columbus accepted the smaller figure and thought that a 3,000-mile westward voyage would take him to Asia. Had he known the earth's true size, he might not have ventured. It was not until 1521–23, when Magellan's fleet (or rather the one remaining ship of the fleet) finally circumnavigated the earth, that Eratosthenes' correct value was finally established.

In terms of the earth's diameter, Hipparchus of Nicaea, about 150 B.C., worked out the distance to the moon. He used a method that had been suggested a century earlier by Aristarchus of Samos, the most daring of all Greek astronomers. The Greeks had already surmised that eclipses of the moon were caused by the earth coming between the sun and the moon. Aristarchus saw that the curve of the earth's shadow as it crossed the moon should indicate the relative sizes of the earth and the moon. From this, geometric methods offered a way to calculate how far distant the moon was in terms of the diameter of the earth. Hipparchus, repeating this work, calculated that the moon's distance from the earth was thirty times the earth's diameter. Taking Eratosthenes' figure of 8,000 miles for the earth's diameter, that meant the moon must be about 240,000 miles from the earth. This again happens to be about correct.

But finding the moon's distance was as far as Greek astronomy managed to carry the problem of the size of the universe—at least correctly. Aristarchus had made a heroic attempt to determine the distance to the sun. The geometric method he used was absolutely correct in theory, but it involved measuring such small differences in angles that, without the use of modern instruments, he was unable to get a good value. He decided that the sun was about twenty times as far as the moon (actually it is about 400 times). Although his figures were wrong, Aristarchus nevertheless did deduce from them that the sun must be at least seven times larger than the earth. Pointing out that it was illogical to suppose that the large sun circled the small earth, he decided that the earth must be revolving around the sun.

Unfortunately, no one listened to him. Later astronomers, beginning with Hipparchus and ending with Claudius Ptolemy, worked out all the heavenly movements on the basis of a motionless earth at the center of the universe, with the moon 240,000 miles away and other objects an undetermined distance farther. This scheme held sway until 1543, when

Nicolaus Copernicus published his book, which returned to the viewpoint of Aristarchus and forever dethroned earth's position as the center of the universe.

The mere fact that the sun was placed at the center of the solar system did not in itself help determine the distance of the planets. Copernicus adopted the Greek value for the distance of the moon, but he had no notion of the distance of the sun. It was not until 1650 that a Belgian astronomer, Godefroy Wendelin, repeated Aristarchus' observations with improved instruments and decided that the sun was not twenty times the moon's distance (5 million miles) but 240 times (60 million miles). The value was still too small, but it was much better than before.

In 1609, meanwhile, the German astronomer Johannes Kepler opened the way to accurate distance determinations with his discovery that the orbits of the planets were ellipses, not circles. For the first time, it became possible to calculate planetary orbits accurately and, furthermore, to plot a scale map of the solar system. That is, the relative distances and orbit-shapes of all the known bodies in the system could be plotted. This meant that if the distance between any two bodies in the system could be determined in miles, all the other distances could be calculated at once. The distance to the sun, therefore, need not be calculated directly, as Aristarchus and Wendelin had attempted to do. The determination of the distance of a nearer body, such as Mars or Venus, outside the earth–moon system would do.

One method by which cosmic distances can be calculated involves the use of parallax. It is easy to illustrate what this term means. Hold your finger about three inches before your eyes and look at it first with just the left eye and then with just the right. Your finger will shift position against the background, because you have changed your point of view. Now if you repeat this procedure with your finger farther away, say at arm's length, the finger again will shift against the background, but this time not so much. Thus the amount of shift can be used to determine the distance of the finger from your eye.

Of course, for an object fifty feet away the shift in position from one eye to the other begins to be too small to measure; we need a wider "baseline" than just the distance between our two eyes. But all we have to do to widen the change in point of view is to look at the object from one spot, then move twenty feet to the right and look at it again. Now the parallax is large enough to be measured easily and the distance can be determined. Surveyors make use of just this method for determining the distance across a stream or ravine.

The same method, precisely, can be used to measure the distance to the moon, with the stars playing the role of background. Viewed from an observatory in California, for instance, the moon will be in one position against the stars. Viewed at the same instant from an observatory in England, it will be in a slightly different position. From this change in position, and the known distance between the two observatories (in a straight line through the earth), the distance of the moon can be calculated. Of course, we can, in theory, enlarge the baseline by making observations from observatories at directly opposite sides of the earth; the length of the baseline is then 8,000 miles. The resulting angle of parallax, divided by two, is called the "geocentric parallax."

The shift in position of a heavenly body is measured in degrees or subunits of a degree—minutes and seconds. One degree is 1/360 of the circuit around the sky; each degree is split into sixty minutes of arc, and each minute into sixty seconds of arc. A minute of arc is therefore 1/(360 × 60) or 1/21,600 of the circuit of the sky, while a second of arc is 1/(21,600 × 60) or 1/1,296,000 of the circuit of the sky.

By trigonometry, Claudius Ptolemy was able to measure the distance of the moon from its parallax, and his result agreed with the earlier figure of Hipparchus. It turned out that the geocentric parallax of the moon is fifty-seven minutes of arc (nearly a full degree). The shift is about equal to the width of a twenty-five-cent piece as seen at a distance of five feet. This is easy enough to measure even with the naked eye. But when it came to measuring the parallax of the sun or a planet, the angles involved were too small. The only conclusion that could be reached was that the other bodies were much farther than the moon. How much farther, no one could tell.

Trigonometry alone, in spite of its refinement by the Arabs during the Middle Ages and by European mathematicians of the sixteenth century, could not give the answer. But measurement of small angles of parallax became possible with the invention of the telescope (which Galileo first built and turned to the sky in 1609, after hearing of a magnifying tube that had been made some months earlier by a Dutch spectaclemaker).

The method of parallax passed beyond the moon in 1673, when the Italian-born French astronomer Jean Dominique Cassini determined the parallax of Mars. He determined the position of Mars against the stars while, on the same evenings, the French astronomer Jean Richer, in French Guiana, was making the same observation. Combining the two, Cassini obtained his parallax and calculated the scale of the solar system. He arrived at a figure of 86 million miles for the distance of the sun from the earth, a figure that was only 7 per cent less than the actual value.

Since then, various parallaxes in the solar system have been measured with increasing accuracy. In 1931, a vast international project was made out of the determination of the parallax of a small planetoid named Eros, which happened at that time to approach the earth more closely than any heavenly body except the moon. Eros on this occasion showed a large parallax that could be measured with considerable precision, and the scale of the solar system was determined more accurately than ever before. From these calculations and by the use of methods still more accurate than those involving parallax, the distance of the sun from the earth is now known to average approximately 92,965,000 miles, give or take a thousand miles or so. (Because the earth's orbit is elliptical, the actual distance varies from 91.4 million to 94.6 million miles.)

This average distance is called an "astronomical unit" (A. U.), and other distances in the solar system are given in this unit. Saturn, for instance, turned out to be, on the average, 887 million miles from the sun, or 9.54 A. U. As the outer planets—Uranus, Neptune, and Pluto—were discovered, the boundaries of the solar system were successively enlarged. The extreme diameter of Pluto's orbit is 7,300 million miles, or 79 A. U. And some comets are known to recede to even greater distances from the sun.

By 1830, the solar system was known to stretch across billions of miles of space, but obviously this was by no means the full size of the universe. There were still the stars.

Astronomers felt certain that the stars were spread throughout space and that some were closer than others, if only because some were so much brighter than others. This should mean that the nearer stars would show a parallax when compared with the more distant ones. However, no such parallax could be detected. Even when the astronomers used as their baseline the full diameter of the earth's orbit around the sun (186 million miles), looking at the stars from the opposite ends of the orbit at half-year intervals, they still could observe no parallax. This meant, of course, that even the nearest stars must be extremely distant. As better and better telescopes failed to show a stellar parallax, the estimated distance of the stars had to be increased more and more. That they were visible at all at the vast distances to which they had to be pushed made it quite plain that they must be tremendous balls of flame like our own sun.

But telescopes and other instruments continued to improve. In the 1830's, the German astronomer Friedrich Wilhelm Bessel made use of a newly invented device, called the "heliometer" ("sun measure") because it was originally intended to measure the diameter of the sun with great precision. It could, of course, be used equally well to measure other distances in the heavens, and Bessel used it to measure the distance between two stars. By noticing the change in this distance from month to month, he finally succeeded in measuring the parallax of a star. He chose a small star in the constellation Cygnus, called 61 Cygni. His reason for choosing it was that it showed an unusually large shift in position from year to year against the background of the other stars, which could only mean that it was nearer than the others. (This steady,

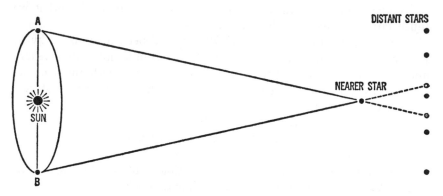

Parallax of a star measured from opposite points on the earth's orbit around the sun.

but very slow, motion across the sky, called "proper motion," should not be confused with the back-and-forth shift against the background that indicates parallax.) Bessel pinpointed the successive positions of 61 Cygni against the "fixed" neighboring stars (presumably much more distant) and continued his observations for more than a year. Then, in 1838, he reported that 61 Cygni had a parallax of 0.31 second of arc—the width of a twenty-five-cent piece as seen from a distance of 10 miles! This parallax, observed with the diameter of the earth's orbit as the baseline, meant that 61 Cygni was about 64 trillion (64,000,000,000,-000) miles away. That is 9,000 times the width of our solar system. Thus, compared to the distance of even the nearest stars, the solar system shrinks to an insignificant dot in space.

Because distances in the trillions of miles are inconvenient to handle, astronomers shrink the numbers by giving the distances in terms of the speed of light—186,282 miles per second. In a year, light travels 5,880,000,000,000 (nearly 6 trillion) miles. That distance is therefore called a "light-year." In terms of this unit, 61 Cygni is about 11 light-years away.

Two months after Bessel's success (so narrow a margin by which to lose the honor of being the first!), the British astronomer Thomas Henderson reported the distance of the star Alpha Centauri. This star, located low in the southern skies and not visible from the United States or Europe, is the third brightest in the heavens. It turned out that Alpha Centauri had a parallax of .75 second of arc, more than twice that of 61 Cygni. Alpha Centauri was therefore correspondingly closer. In fact, it is only 4.3 light-years from the solar system and is our nearest stellar neighbor. Actually it is not a single star, but a cluster of three.

In 1840, the German-born Russian astronomer, Friedrich Wilhelm von Struve announced the parallax of Vega, the fourth brightest star in the sky. He was a little off in his determination, as it turned out, but this was understandable, because Vega's parallax was very small and it was much farther away—27 light-years.

By 1900, the distances of about 70 stars had been determined by the parallax method (and by 1950, nearly 6,000). One hundred light-years is about the limit of the distance that can be measured with any accuracy, even with the best instruments. And beyond this are countless stars at much greater distances.

With the naked eye we can see about 6,000 stars. The invention of the telescope at once made plain that this was only a fragment of the universe. When Galileo raised his telescope to the heavens in 1609, he not only found new stars previously invisible, but, on turning to the Milky Way, received an even more profound shock. To the naked eye, the Milky Way is merely a luminous band of foggy light. Galileo's telescope broke down this foggy light into myriads of stars, as numerous as the grains in talcum powder.

The first man to try to make sense out of this was the German-born English astronomer William Herschel. In 1785, Herschel suggested that the stars of the heavens were arranged in a lens-shape. If we look toward the Milky Way, we see a vast number of stars, but when we look out to the sky at right angles to this wheel, we see relatively few stars. Herschel deduced that the heavenly bodies formed a flattened system, with the long axis in the direction of the Milky Way. We now know that, within limits, this picture is correct, and we call our star system the Galaxy, which is actually another term for Milky Way, because galaxy comes from the Greek word for "milk."

Herschel tried to estimate the size of the Galaxy. He assumed that all the stars had about the same intrinsic brightness, so that one could tell the relative distance of a star by its brightness. (By a well-known law, brightness decreases as the square of the distance, so if star A is one-ninth the brightness of star B, it should be three times as far as star B.)

By counting samples of stars in various spots of the Milky Way, Herschel estimated that there were about 100 million stars in the Galaxy altogether. From the levels of their brightness, he decided that the diameter of the Galaxy was 850 times the distance to the bright star Sirius and that its thickness was 155 times that distance.

We now know that the distance to Sirius is 8.8 light-years, so Herschel's estimate was equivalent to a Galaxy about 7,500 light-years in diameter and 1,300 light-years thick. This turned out to be far too conservative. But like Aristarchus' overconservative measure of the distance to the sun, it was a step in the right direction. (Furthermore, Herschel used his statistics to show that the sun was moving at 12 miles per second in the direction of the constellation Hercules. The sun did move after all, but not in the fashion the Greeks had thought.)

Beginning in 1906, the Dutch astronomer Jacobus Cornelis Kapteyn conducted another survey of the Milky Way. He had photography at his disposal and knew the true distance of the nearer stars, so he was able to make a better estimate than Herschel had. Kapteyn decided that the dimensions of the Galaxy were 23,000 light-years by 6,000. Thus Kapteyn's model of the Galaxy was four times as wide and five times as thick as Herschel's; but it was still overconservative.

To sum up, by 1900 the situation with respect to stellar distances was the same as that with respect to planetary distances in 1700. In 1700, the moon's distance was known but the distance of the farther planets could only be guessed at. In 1900, the distance of the nearer stars was known, but that of the more distant stars could only be guessed at.

The next major step forward was the discovery of a new measuring rod—certain variable stars which fluctuated in brightness. This part of the story begins with a fairly bright star called Delta Cephei, in the constellation Cepheus. On close study, the star was found to have a cycle of varying brightness: from its dimmest stage it rather quickly doubled in brightness, then slowly faded to its dim point again. It did this over and over with great regularity. Astronomers found a number of other stars that varied in the same regular way, and in honor of Delta Cephei all were named "Cepheid variables," or simply "Cepheids."

The Cepheids' periods (the time from dim point to dim point) vary from less than a day to as long as nearly two months. Those nearest our sun seem to have a period in the neighborhood of a week. The period of Delta Cephei itself is 5.3 days, while the nearest Cepheid of all (the Pole Star, no less) has a period of 4 days. (The Pole Star, however, varies only slightly in luminosity; not enough to be noticeable to the unaided eye.)

The importance of the Cepheids to astronomers involves their brightness, which is a subject that requires a small digression.

Ever since Hipparchus, the brightness of stars has been measured by the term "magnitude." The brighter the star, the lower the magnitude. The twenty brightest stars he called "first magnitude." Somewhat dimmer stars are "second magnitude." Then, third, fourth, and fifth, until the dimmest, those just barely visible, are of the "sixth magnitude."

In modern times—1856, to be exact—Hipparchus' notion was made quantitative by the English astronomer Norman Robert Pogson. He showed that the average first-magnitude star was about 100 times brighter than the average sixth-magnitude star. Allowing this interval of five magnitudes to represent a ratio of one hundred in brightness, the ratio for one magnitude must be 2.512. A star of magnitude 4 is 2.512 times as bright as a star of magnitude 5, and 2.512×2.512, or about 6.3 times as bright as a star of magnitude 6.

Among the stars, 61 Cygni is a dim star with a magnitude of 5.0

(modern astronomical methods allow magnitudes to be fixed to the nearest tenth and even to the nearest hundreth in some cases). Capella is a bright star, with a magnitude of 0.9; Alpha Centauri still brighter, with a magnitude of 0.1. And the measure goes on to still greater brightnesses which are designated by magnitude 0 and beyond this by negative numbers. Sirius, the brightest star in the sky, has a magnitude of – 1.6. The planet Venus attains a magnitude of – 4; the full moon, – 12; the sun, – 26.

These are the "apparent magnitudes" of the stars as we see them—not their absolute luminosities independent of distance. But if we know the distance of a star and its apparent magnitude, we can calculate its actual luminosity. Astronomers base the scale of "absolute magnitudes" on the brightness at a standard distance, which has been established at ten "parsecs," or 32.6 light-years. (The "parsec" is the distance at which a star would show a parallax of one second of arc; it is equal to a little more than 19 trillion miles, or 3.26 light-years.)

Although Capella looks dimmer than Alpha Centauri and Sirius, actually it is a far more powerful emitter of light than either of them. It merely happens to be a great deal farther away. If all were at the standard distance, Capella would be much the brightest of the three. Capella has an absolute magnitude of – 0.1, Sirius 1.3, and Alpha Centauri 4.8. Our own sun is just about as bright as Alpha Centauri, with an absolute magnitude of 4.86. It is an ordinary, medium-sized star.

Now to get back to the Cepheids. In 1912, Miss Henrietta Leavitt, an astronomer at the Harvard Observatory, was studying the smaller of the Magellanic Clouds—two huge star systems in the Southern Hemisphere named after Ferdinand Magellan, because they were first observed during his voyage around the globe. Among the stars of the Small Magellanic Cloud, Miss Leavitt detected twenty-five Cepheids. She recorded the period of variation of each and to her surprise found that the longer the period, the brighter the star.

This was not true of the Cepheid variables in our own neighborhood; why should it be true of the Small Magellanic Cloud? In our own neighborhood, we know only the apparent magnitudes of the Cepheids; not knowing their distances or absolute brightnesses, we have no scale for relating the period of a star to its brightness. But in the Small Magellanic Cloud, all the stars are effectively at about the same distance from us, because the cloud itself is so far away. It is as though a person in New York were trying to calculate his distance from each person in Chicago. He would conclude that all the Chicagoans were about equally distant from himself—what is a difference of a few miles in a total distance of a thousand? Similarly, a star at the far end of the Cloud is not significantly farther away than one at the near end.

With the stars in the Small Magellanic Cloud at about the same distance from us, their apparent magnitude could be taken as a measure of their comparative absolute magnitude. So Miss Leavitt could consider the relationship she saw a true one: that is, the period of the Cepheid variables increased smoothly with the absolute magnitude. She was thus able to establish a "period-luminosity curve"—a graph which showed what period a Cepheid of any absolute magnitude must have, and conversely what absolute magnitude a Cepheid of a given period must have.

If Cepheids everywhere in the universe behaved as they did in the Small Magellanic Cloud (a reasonable assumption), then astronomers had a *relative* scale for measuring distances, as far out as Cepheids could be detected in the best telescopes. If they spotted two Cepheids with equal periods, they could assume that both were equal in absolute magnitude. If Cepheid A seemed four times as bright as Cepheid B, Cepheid B must be twice as distant from us. In this way, the relative distances of all the observable Cepheids could be plotted on a scale map. Now if the actual distance of just one of the Cepheids could be determined, so could the distances of all the rest.

Unfortunately, even the nearest Cepheid, the Pole Star, is hundreds of light-years away, much too far to measure its distance by parallax. Astronomers had to use less direct methods. One usable clue was proper motion: on the average, the more distant a star is, the smaller its proper motion. (Recall that Bessel decided 61 Cygni was relatively close because it had a large proper motion.) A number of devices were used to determine the proper motions of groups of stars, and statistical methods were brought to bear. The procedure was complicated, but the results gave the approximate distances of various groups of stars which contained Cepheids. From the distances and the apparent magnitudes of those Cepheids, their absolute magnitudes could be determined, and these could be compared with the periods.

In 1913, the Danish astronomer Ejnar Hertzsprung found that a Cepheid of absolute magnitude − 2.3 had a period of 6.6 days. From that, and using Miss Leavitt's period-luminosity curve, he could determine the absolute magnitude of any Cepheid. (It turned out, incidentally, that Cepheids generally were large, bright stars, much more luminous than our sun. Their variations in brightness are probably the result of pulsations. The stars seem to expand and contract steadily, as though they are ponderously breathing in and out.)

A few years later, the American astronomer Harlow Shapley repeated the work and decided that a Cepheid of absolute magnitude − 2.3 had a period of 5.96 days. The agreement was close enough to allow astronomers to go ahead. They had their yardstick.

In 1918, Shapley began observing the Cepheids of our own Galaxy in an attempt to determine the Galaxy's size by this new method. He concentrated on the Cepheids found in groups of stars called "globular clusters"—densely packed spherical aggregates of tens of thousands to tens of millions of stars, with diameters of the order of 100 light-years.

These clusters (the nature of which was first observed by Herschel a century earlier) present an astronomical environment quite different from that prevailing in our own neighborhood in space. At the center of the larger clusters, stars are packed together with a density of 500 per ten cubic parsecs, as compared with one star per ten cubic parsecs in our own neighborhood. Starlight under such conditions would be far brighter than moonlight on earth, and a planet situated at the center of such a cluster would know no night.

There are about one hundred known globular clusters in our Galaxy and probably as many again that have not yet been detected. Shapley calculated the distance of the various globular clusters at from 20,000 to 200,000 light-years from us. (The nearest cluster, like the nearest star, is in the constellation Centaurus. It is visible to the naked eye as a starlike object, "Omega Centauri." The most distant, NGC 2419, is so far off as scarcely to be considered a member of the Galaxy.)

Shapley found the clusters were distributed in a large sphere that the plane of the Milky Way cut in half; they surrounded a portion of the main body of the Galaxy like a halo. Shapley made the natural assumption that they encircled the center of the Galaxy. His calculations placed the central point of this halo of globular clusters within the Milky Way in the direction of the constellation Sagittarius and about 50,000 light-years from us. This meant that our solar system, far from being at the center of the Galaxy, as Herschel and Kapteyn had thought, was far out toward one edge.

Shapley's model pictured the Galaxy as a giant lens about 300,000 light-years in diameter. This time its size was overestimated, as another method of measurement soon showed.

From the fact that the Galaxy had a disk shape, astronomers from William Herschel onward assumed that it was rotating in space. In 1926, the Dutch astronomer Jan Oort set out to measure this rotation. Since the Galaxy is not a solid object, but is composed of numerous individual stars, it is not to be expected that it rotates in one piece, as a wheel does. Instead, stars close to the gravitational center of the disk must revolve around it faster than those farther away (just as the planets closest to the sun travel fastest in their orbits). This means that the stars toward the center of the Galaxy (i.e., in the direction of Sagittarius) should tend to drift ahead of our sun, whereas those farther from the center (in the direction of the constellation Gemini) should tend to lag behind us in their revolution. And the farther a star is from us, the greater this difference in speed should be.

On these assumptions, it became possible to calculate the rate of rotation around the galactic center from the relative motions of the stars. The sun and nearby stars, it turned out, travel at about 150 miles a second relative to the galactic center and make a complete revolution around the center in approximately 200 million years. (The sun travels in a nearly circular orbit, but the orbit of some stars, such as Arcturus, are quite elliptical. The fact that the various stars do not rotate in perfectly parallel orbits accounts for the sun's relative motion toward the constellation Hercules.)

Having estimated a value for the rate of rotation, astronomers were then able to calculate the strength of the gravitational field of the galactic center and, therefore, its mass. The galactic center (which contains most of the mass of the Galaxy) turns out to be well over 100 billion times as massive as our sun. Since our sun is a star of average mass, our Galaxy therefore contains perhaps 100 to 200 billion stars—up to 2,000 times the number estimated by Herschel.

From the curve of the orbits of the revolving stars, it is also possible to locate the center around which they are revolving. The center of the Galaxy in this way has been confirmed to be in the direction of Sagittarius, as Shapley found, but only 27,000 light-years from us, and the total diameter of the Galaxy comes to 100,000 light-years instead of 300,000. In this new model, now believed to be correct, the thickness of the disk is some 20,000 light-years at the center and falls off toward the edge: at the location of our sun, which is two thirds of the way out toward the extreme edge, the disk is perhaps 3,000 light-years thick. But these are only rough figures, because the Galaxy has no sharply definite boundaries.

If the sun is so close to the edge of the Galaxy, why is not the Milky Way much brighter in the direction toward the center than in the opposite direction, where we look toward the edge? Looking toward Sagittarius, we face the main body of the Galaxy with nearly 100 billion stars, whereas out toward the edge there is only a scattering of some millions. Yet in each direction the band of the Milky Way seems of about the same brightness. The answer appears to be that huge clouds of obscuring dust hide much of the center of the Galaxy from us. As much as half the mass of the galactic outskirts may be composed of such clouds of dust and gas. Probably we see no more than 1/10,000 (at most) of the light of the galactic center.

This explains why Herschel and other early students of the Galaxy thought our solar system was at the center, and also, it seems, why Shapley originally overestimated the size of the Galaxy. Some of the clusters he studied were dimmed by the intervening dust, so that the Cepheids in them seemed dimmer and therefore more distant than they really were.

Even before the size and mass of the Galaxy itself had been determined, the Cepheid variables of the Magellanic Clouds (where Miss Leavitt had made the crucial discovery of the period-luminosity curve) were used to determine the distance of the clouds. They proved to be more than 100,000 light-years away. The best modern figures place the Large Magellanic Cloud at about 150,000 light-years from us and the Small Magellanic Cloud at 170,000 light-years. The Large Cloud is no more than half the size of our Galaxy in diameter; the Small Cloud, no more than a fifth. Besides, they seem to be less densely packed with stars. The Large Magellanic Cloud contains 5 billion stars (only 1/20 or less the number in our Galaxy), while the Small Magellanic Cloud has only 1.5 billion.

That was the situation as it stood in the early 1920's. The known universe was less than 200,000 light-years in diameter and consisted of our Galaxy and its two neighbors. The question then arose as to whether anything existed outside that.

Suspicion rested upon certain small patches of luminous fog, called nebulae (from the Greek word for "cloud"), which astronomers had long noted. The French astronomer Charles Messier had catalogued 103 of them about 1800. (Many are still known by the numbers he gave them, preceded by the letter "M" for Messier.)

Were these nebulosities merely clouds as they seemed to be? Some, such as the Orion Nebula (first discovered in 1656 by the Dutch astronomer Christian Huygens) seemed to be just that. It was a cloud of gas and dust, equal in mass to about 500 suns like ours, and illuminated by hot stars contained within them. Others, on the other hand, turned out to be globular clusters—huge assemblages of stars.

But there remained patches of luminous cloud that seemed to contain no stars at all. Why, then, were they luminous? In 1845, the British astronomer William Parsons (third Earl of Rosse) using a 72-inch telescope he had spent his life building, had ascertained that some of these patches had a spiral structure; thus they came to be called "spiral nebulae"; but that did not help explain the source of the luminosity.

The most spectacular of these patches, known as M-31, or the Andromeda Nebula (because it is in the constellation Andromeda) was first studied in 1612 by the German astronomer Simon Marius. It is an elongated oval of dim light about half the size of the full moon. Could it be composed of stars so distant that they could not be made out separately even in large telescopes? If so, the Andromeda Nebula must be incredibly far away and incredibly large to be visible at all at such a distance. (As long ago as 1755, the German philosopher Immanuel Kant had speculated on the existence of such far-distant star collections. "Island universes," he called them.)

In 1924, the American astronomer Edwin Powell Hubble turned the new 100-inch telescope at Mount Wilson in California on the Andromeda Nebula. The powerful new instrument resolved portions of the Nebula's outer edge into individual stars. This showed at once that the Andromeda Nebula, or at least parts of it, resembled the Milky Way and that there might be something to this "island universe" notion.

Among the stars at the edge of the Andromeda Nebula were Cepheid variables. With these measuring rods it was found that the Nebula was nearly a million light-years away! So the Andromeda Nebula was far, far outside our Galaxy. Allowing for its distance, its apparent size showed that it must be a huge conglomeration of stars, almost rivaling our own Galaxy.

Other nebulosities, too, turned out to be conglomerations of stars, even farther away than the Andromeda Nebula. These "extra-galactic nebulae" all had to be recognized as galaxies—new universes which reduced our own to just one of many in space. Once again the universe had expanded. It was larger than ever—not merely hundreds of thousands of light-years across, but perhaps hundreds of millions.

Through the 1930's, astronomers wrestled with several nagging puzzles about these galaxies. For one thing, on the basis of their assumed distances, all of them turned out to be much smaller than our own. It seemed an odd coincidence that we should be inhabiting by far the largest galaxy in existence. For another thing, globular clusters surrounding the Andromeda galaxy seemed to be only one half or one third as luminous as those of our own Galaxy. (Andromeda is about as rich in globular clusters as our own Galaxy, and its clusters are spherically arranged about Andromeda's center. This seems to show that Shapley's assumption that our own clusters were so arranged was a reasonable one. Some galaxies are amazingly rich in globular clusters. The galaxy M-87, in Virgo, possesses at least a thousand.)

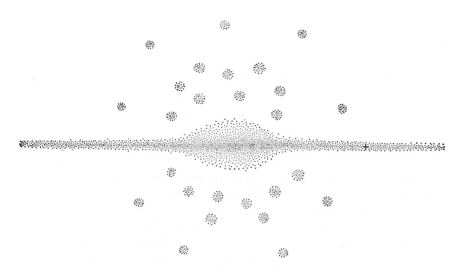

A model of our Galaxy seen edgewise. Globular clusters are arrayed around the central portion of the Galaxy. The position of our sun is indicated by +.

The most serious puzzle is that the distances of the galaxies seemed to imply that the universe was only about 2 billion years old (for reasons I shall discuss later in this chapter). This was puzzling, for the earth itself was considered by geologists to be older than that, on what was thought to be the very best kind of evidence.

The beginning of an answer came during World War II, when the German-born American astronomer Walter Baade discovered that the yardstick by which the galaxies' distances had been measured was wrong.

In 1942 Baade took advantage of the wartime blackout of Los Angeles, which cleared the night sky at Mount Wilson, to make a detailed study of the Andromeda galaxy with the 100-inch Hooker telescope (named for John B. Hooker, who had provided the funds for its construction). With the improved seeing, he was able to resolve some of the stars in the inner regions of the galaxy. He immediately noted some striking differences between these stars and those in the outskirts of the galaxy. The brightest stars in the interior were reddish, whereas those of the outskirts were bluish. Moreover, the red giants of the interior were not nearly so bright as the blue giants of the outskirts: the latter had up to 100,000 times the luminosity of our sun, whereas the internal red giants had only up to 1,000 times that luminosity. Finally, the outskirts of the Andromeda galaxy, where the bright blue stars were found, was loaded with dust, whereas the interior, with its somewhat less bright red stars, was free of dust.

To Baade, it seemed that here were two sets of stars with different structure and history. He called the bluish stars of the outskirts Population I and the reddish stars of the interior, Population II. Population I stars, it turns out, are relatively young, with high metal content, and follow nearly circular orbits about the galactic center in the median plane of the galaxy. Population II stars are relatively old, with low metal content, with orbits that are markedly elliptical, and with considerable inclination to the median plane of the galaxy. Both populations have been broken down into finer subgroups since Baade's discovery.

When the new 200-inch Hale telescope (named for the American astronomer, George Ellery Hale, who supervised its construction) was set up on Palomar Mountain after the war, Baade continued his investigations. He found certain regularities in the distribution of the two populations, and these depended on the nature of the galaxies involved. Galaxies of the class called "elliptical" (systems with the shape of an ellipse and rather uniform internal structure) apparently were made up mainly of Population II stars, as were globular clusters in any galaxy. On the other hand, in "spiral galaxies" (that is, with arms which make them look like a pinwheel) the spiral arms were composed of Population I, set against a Population II background.

It is estimated that only about 2 per cent of the stars in the universe are of the Population I type. But our own sun and the familiar stars in

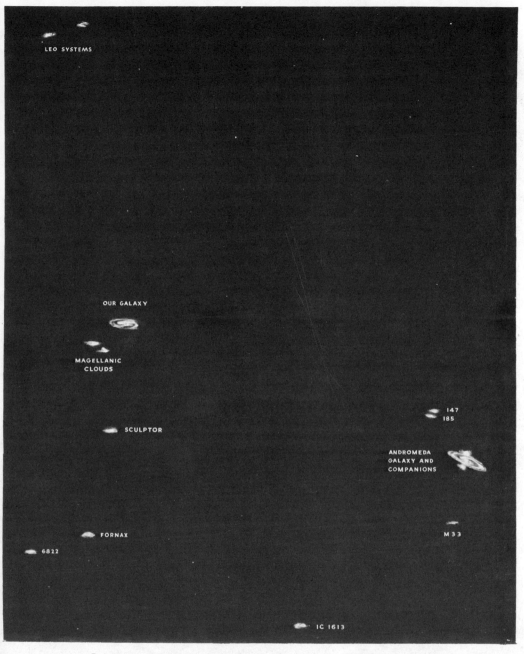

Our region of the universe—a drawing showing the other galaxies
in our neighborhood.

The spiral galaxy in Andromeda.

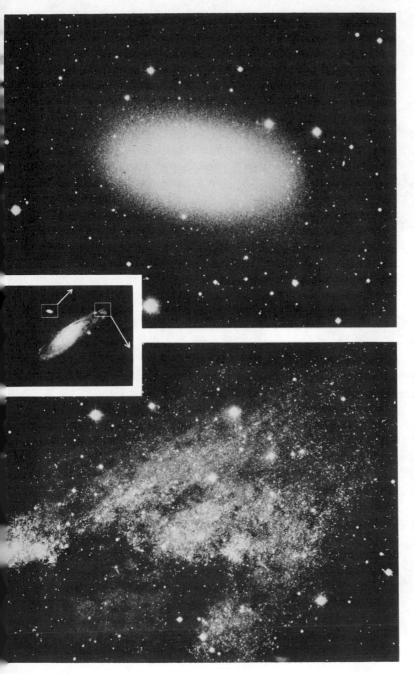

Stellar populations I and II. At the left a spiral arm of the Andromeda galaxy, photographed in blue light, shows giant and supergiant stars of Population I; at the right the galaxy NGC 205, a companion of the Andromeda galaxy, photographed in yellow light, shows stars of Population II, the brightest of which are red stars one-hundredth as bright as the blue giants of Population I. (The large stars in both pictures are foreground stars belonging to our own Milky Way.)

A spiral galaxy in Coma Berenices, seen edge on.

The Crab Nebula, the remains of a supernova, photographed in red light.

A spiral galaxy in broadside view—the "whirlpool nebula" in Canes Venatici.

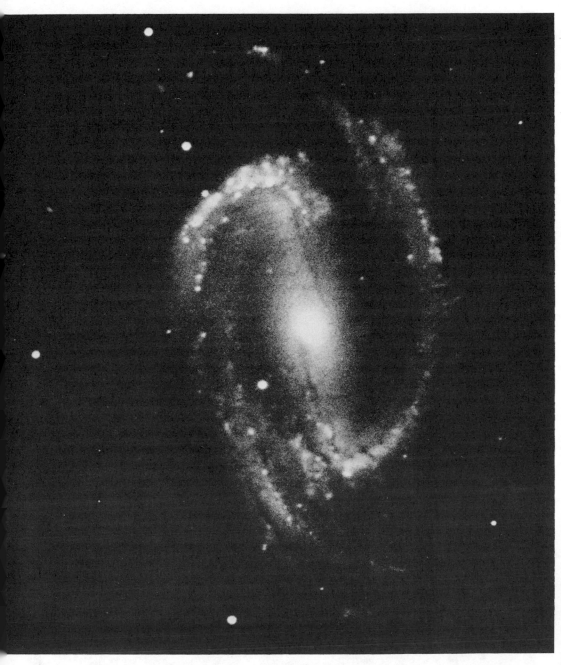

A barred spiral galaxy in Eridanus.

Two colliding galaxies—NGC 4038 and 4039.

The Horsehead Nebula in Orion, south of Zeta Orionis, photographed in red light.

our neighborhood fall into this class. From this fact alone, we can deduce that ours is a spiral galaxy and that we lie in one of the spiral arms. (This explains why there are so many dust clouds, both light and dark, in our neighborhood: the spiral arms of a galaxy are clogged with dust.) Photographs show that the Andromeda galaxy also is of the spiral type.

Now to get back to the yardstick. Baade began to compare the Cepheid stars found in globular clusters (Population II) with those found in our spiral arm (Population I). It turned out that the Cepheids in the two populations were really of two different types, as far as the relation between period and luminosity was concerned. Cepheids of Population II followed the period-luminosity curve set up by Leavitt and Shapley. With this yardstick, Shapley had accurately measured the distances to the globular clusters and the size of our Galaxy. But the Cepheids of Population I, it now developed, were a different yardstick altogether! A Population I Cepheid was four or five times as luminous as a Population II Cepheid of the same period. This meant that use of the Leavitt scale would result in miscalculation of the absolute magnitude of a Population I Cepheid from its period. And if the absolute magnitude was wrong, the calculation of distance must be wrong; the star would actually be much farther away than the calculation indicated.

Hubble had gauged the distance of the Andromeda galaxy from the Cepheids (of Population I) in its outskirts—the only ones that could be resolved at the time. Now, with the revised yardstick, the galaxy proved to be about 2.5 million light-years away, instead of less than a million. And other galaxies had to be moved out in proportion. (The Andromeda galaxy is still a close neighbor, however. The average distance between galaxies is estimated to be something like 20 million light-years.)

At one stroke, the size of the known universe was more than doubled. This instantly solved the problems that had plagued the 1930's. Our Galaxy was no longer larger than all the others; the Andromeda galaxy, for instance, was definitely more massive than ours. Second, it now appeared that the Andromeda galaxy's globular clusters were as luminous as ours; they had seemed less bright only because of the misjudgment of their distance. Finally, for reasons I will explain later, the new scale of distances allowed the universe to be considered much older—at least 5 billion years old and very likely considerably more than that—which brought it into line with the geologists' estimates of the age of the earth.

Doubling the distance of the galaxies does not end the problem of size. We must now consider the possibility of still larger systems—of clusters of galaxies and supergalaxies.

Actually, modern telescopes have shown that clusters of galaxies do exist. For instance, in the constellation of Coma Berenices there is a large, ellipsoidal cluster of galaxies about 8 million light-years in diameter. The "Coma Cluster" contains about 11,000 galaxies, separated by an average distance of only 300,000 light-years (as compared with an average of something like 3 million light-years between galaxies in our own vicinity).

Our own Galaxy seems to be part of a "local cluster" that includes the Magellanic Clouds, the Andromeda galaxy, and three small "satellite galaxies" near it, plus some other galaxies for a total of nineteen members altogether. Two of these, called "Maffei One" and "Maffei Two" (for Paolo Maffei, the Italian astronomer, who first reported them), were discovered only in 1971. The lateness of the discovery was owing to the fact that they can only be detected through dust clouds that lie between them and ourselves.

Of the local cluster, only our own Galaxy, Andromeda, and the two Maffeis are giants, whereas the rest are dwarfs. One of the dwarfs, IC 1613, may contain only 60 million stars; hence it is scarcely more than a large globular cluster. Among galaxies, as among stars, dwarfs far outnumber giants.

If galaxies do form clusters and clusters of clusters, does that mean that the universe goes on forever and that space is infinite? Or is there some end, both to the universe and to space? Well, astronomers can make out objects up to an estimated 9 billion light-years away, and there is no sign of an end of the universe—yet. At the theoretical level, there are arguments both for an end of space and for no end, for a beginning in time and for no beginning. Having considered space, let us consider time next.

The Birth of the Universe

Mythmakers have invented many fanciful creations of the universe (usually concentrating on the earth itself, with all the rest dismissed quickly as "the sky" or the "heavens"). Generally, the time of creation is set not very far in the past (although we should remember that to people in the preliterate stage, a time of a thousand years was even more impressive than a billion years is today).

The creation story with which we are most familiar is, of course, that given in the first chapters of Genesis, which, some people hold, is an adaptation of Babylonian myths, intensified in poetic beauty and elevated in moral grandeur.

Various attempts have been made to work out the date of the Creation on the basis of the data given in the Bible (the reigns of the various kings, the time from the Exodus to the dedication of Solomon's temple, the ages of the patriarchs both before and after the flood). Medieval Jewish scholars put the date of the Creation at 3760 B.C. and the Jewish calendar still counts its years from that date. In 1658 A.D., Archbishop James Ussher of the Anglican Church calculated the date

43

of the Creation to be 4004 B.C., at 8 P.M. of October 22 of that year, to be exact. Some theologians of the Greek Orthodox Church put Creation as far back as 5508 B.C.

Even as late as the eighteenth century, the Biblical version was accepted by the learned world, and the age of the universe was considered to be only six or seven thousand years at most. This view received its first major blow in 1785 in the form of a book entitled *Theory of the Earth*, by a Scotch naturalist named James Hutton. Hutton started with the proposition that the slow natural processes working on the surface of the earth (mountain building and erosion, the cutting of river channels, and so on) had been working at about the same rate throughout the earth's history. This "uniformitarian principle" implied that the processes must have been working for a stupendously long time to produce the observed phenomena. Therefore the earth must be not thousands but many millions of years old.

Hutton's views were immediately derided. But the ferment worked. In the early 1830's, the British geologist Charles Lyell reaffirmed Hutton's views, and in a three-volume work entitled *Principles of Geology* presented the evidence with such clarity and force that the world of science was won over. The modern science of geology can be dated from that work.

Attempts were made to calculate the age of the earth on the basis of the uniformitarian principle. For instance, if one knew the amount of sediment laid down by the action of water each year (a modern estimate is one foot in 880 years), one could calculate the age of a layer of sedimentary rock from its thickness. It soon became obvious that this approach could not determine the earth's age accurately, because the record of the rocks was obscured by processes of erosion, crumbling, upheavals, and other forces. Nevertheless, even the incomplete evidence indicated that the earth must be at least 500 million years old.

Another way of measuring the age of the earth was to estimate the rate of accumulation of salt by the oceans, a suggestion first advanced by the English astronomer Edmund Halley as long ago as 1715. Rivers steadily washed salt into the sea; since only fresh water left it by evaporation, the salt concentration rose. Assuming that the ocean had started as fresh water, the time necessary for the rivers to have endowed the oceans with their salt content of over 3 per cent could have been as long as a billion years.

This great age was very agreeable to the biologists, who during the latter half of the nineteenth century were trying to trace the slow development of living organisms from primitive one-celled creatures to the complex higher animals. They needed long eons for the development to take place, and a billion years gave them sufficient time.

However, by the mid-nineteenth century astronomical considerations brought in sudden complications. For instance, the principle of the "conservation of energy" raised an interesting problem with respect to the sun. The sun was pouring out energy in colossal quantities and had been doing so throughout recorded history. If the earth had existed for countless eons, where had all this energy come from? It could not have come from the usual sources familiar to mankind. If the sun had started as solid coal burning in an atmosphere of oxygen, it would have been reduced to a cinder (at the rate it was delivering energy) in the space of about 2,500 years.

The German physicist Hermann Ludwig Ferdinand von Helmholtz, one of the first to enunciate the law of conservation of energy, was particularly interested in the problem of the sun. In 1854, he pointed out that if the sun were contracting, its mass would gain energy as it fell toward its center of gravity, just as a rock gains energy when it falls. This energy could be converted into radiation. Helmholtz calculated that a contraction of the sun by a mere ten thousandth of its radius could provide it with a 2,000-year supply of energy.

The British physicist William Thomson (later Lord Kelvin) did more work on the subject and decided that on this basis the earth could not be more than 50 million years old, for at the rate the sun had spent energy, it must have contracted from a gigantic size, originally as large as the earth's orbit around the sun. (This meant, of course, that Venus must be younger than the earth and Mercury still younger.) Lord Kelvin went on to estimate that if the earth itself had started as a molten mass, the time needed to cool to its present temperature, and therefore its age, would be about 20 million years.

By the 1890's the battlelines seemed drawn between two invincible armies. The physicists seemed to have shown conclusively that the earth could not have been solid for more than a few million years, while geologists and biologists seemed to have proved just as conclusively that the earth must have been solid for not less than a billion years.

And then something new and completely unexpected turned up, and the physicists found themselves with their case crumbling.

In 1896, the discovery of radioactivity made it clear that the earth's uranium and other radioactive substances were liberating large quantities of energy and had been doing so for a very long time. This finding made Kelvin's calculations meaningless, as was pointed out first, in 1904, by the New Zealand-born British physicist, Ernest Rutherford, in a lecture —with the aged (and disapproving) Kelvin, himself, in the audience.

There is no point in trying to decide how long it would take the earth to cool if you don't take into account the fact that heat is being constantly supplied by radioactive substances. With this new factor, it might take the earth billions of years, rather than millions, to cool from a molten mass to its present temperature. The earth might even be warming with time.

Actually, radioactivity itself eventually gave the most conclusive evidence of the earth's age, for it allowed geologists and geochemists to calculate the age of rocks directly from the quantity of uranium and lead they contain. By the clock of radioactivity, some of the earth's rocks are now known to be nearly 4 billion years old, and there is every reason to think that the earth is somewhat older than that. An age of 4.7 billion years for the earth in its present solid form is now accepted as likely. And, indeed, some of the rocks brought back from our neighbor world, the moon, have proven to be just about that old.

And what of the sun? Radioactivity, together with discoveries concerning the atomic nucleus, introduced a new source of energy, much larger than any previously known. In 1930, the British physicist Sir Arthur Eddington set a train of thought working when he suggested that the temperature and pressure at the center of the sun must be outrageously high: the temperature might be as high as 15 million degrees. At such temperatures and pressures, the nuclei of atoms could undergo reactions which could not take place in the bland mildness of the earth's environment. The sun is known to consist largely of hydrogen. If four hydrogen nuclei combined (forming a helium atom), they would liberate large amounts of energy.

Then, in 1938, the German-born American physicist Hans Albrecht Bethe worked out the possible ways in which this combination of hydrogen to helium could take place. There were two processes by which this could occur under the conditions at the center of stars like the sun. One involved the direct conversion of hydrogen to helium; the other involved a carbon atom as an intermediate in the process. Either set of reactions can occur in stars; in our own sun, the direct hydrogen conversion seems to be the dominant mechanism. Either brought about the conversion of mass to energy. (Einstein, in his Special Theory of Relativity, had shown that mass and energy were different aspects of the same thing and could be interconverted; furthermore, that a great deal of energy could be liberated by the conversion of a small amount of mass.)

The rate of radiation of energy by the sun requires the disappearance of solar mass at the rate of 4.2 million tons per second. At first blush this seems a frightening loss, but the total mass of the sun is 2,200,000,000,000,000,000,000,000,000 tons, so the sun loses only 0.00000000000000000002 per cent of its mass each second. Assuming the sun to have been in existence for 6 billion years, as astronomers now believe, and if it has been radiating at its present rate all that time, it would have expended only 1/40,000 of its mass. It is easy to see, then, that the sun can continue to radiate energy at its present rate for billions of years to come.

By 1940, then, an age of 6 billion years for the solar system as a whole seemed reasonable. The whole matter of the age of the universe might have been settled, but astronomers had thrown another monkey wrench into the machinery. Now the universe as a whole seemed too youthful to account for the age of the solar system. The trouble arose from an examination of the distant galaxies by the astronomers and from a phenomenon first discovered in 1842 by an Austrian physicist named Christian Johann Doppler.

The "Doppler effect" is familiar enough; it is most commonly illustrated by the whistle of a passing locomotive, which rises in pitch as it approaches and drops in pitch as it recedes. The change in pitch is due simply to the fact that the number of sound waves striking the eardrum per second changes because of the source's motion.

As Doppler suggested, the Doppler effect applies to light waves as well as to sound. When light from a moving source reaches the eye, there is a shift in frequency—that is, color—when the source is moving fast enough. For instance, if the source is traveling toward us, more light waves are crowded into each second and the light perceived shifts toward the high-frequency violet end of the visible spectrum. On the other hand, if the source is moving away, fewer waves arrive per second and the light shifts toward the low-frequency red end of the spectrum.

Astronomers have been studying the spectra of stars for a long time, and they are well acquainted with the normal picture—a pattern of bright lines against a dark background or dark lines against a bright background showing the emission or absorption of light by atoms at certain wavelengths, or colors. They have been able to calculate the velocity of stars moving toward or away from us (i.e., radial velocity) by measuring the displacement of the usual spectral lines toward the violet or red end of the spectrum.

It was the French physicist Armand Hippolyte Louis Fizeau who, in 1848, pointed out that the Doppler effect in light could best be observed by noting the position of the spectral lines. For that reason, the Doppler effect is called the "Doppler-Fizeau effect" where light is concerned.

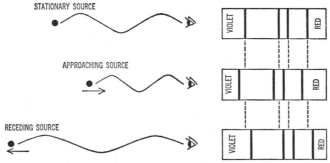

The Doppler-Fizeau effect. The lines in the spectrum shift toward the violet end (*left*) when the light source is approaching. When the source recedes, the spectral lines shift toward the red end (*right*).

The Doppler-Fizeau effect has been used in a variety of ways. Within our solar system, it could be used to demonstrate the rotation of the sun in a new way. The spectral lines originating from that limb of the sun being carried toward us in the course of its vibration would be shifted toward the violet (a "violet shift"). The lines from the other limb would show a "red shift" since that limb was receding.

To be sure, the motion of sunspots is a better and more obvious way of detecting and measuring solar rotation (which turns out to have a period of about 25 days, relative to the stars). However, the effect can also be used to determine the rotation of featureless objects, such as the rings of Saturn.

The Doppler-Fizeau effect can be used for objects at any distance, so long as those objects can be made to produce a spectrum for study. Its most dramatic victories, therefore, were in connection with the stars.

In 1868, the British astronomer Sir William Huggins measured the radial velocity of Sirius and announced that it was moving away from us at 29 miles per second. (We have better figures now, but he came reasonably close for a first try.) By 1890, the American astronomer James Edward Keeler, using better instruments, was producing reliable results in quantity; he showed, for instance, that Arcturus was approaching us at a rate of 3¾ miles per second.

The effect could even be used to determine the existence of star systems, the details of which could not be made out by telescope. In 1782, for instance, an English astronomer, John Goodricke (a deaf-mute who died at twenty-two; a first-rate brain in a tragically defective body), studied the star Algol, which increased and decreased regularly in brightness. Goodricke explained it by supposing that a dark companion circled Algol. Periodically the dark companion passed in front of Algol, eclipsing it and dimming its light.

A century passed before this plausible hypothesis was supported by additional evidence. In 1889, the German astronomer Hermann Carl Vogel showed that the lines of Algol's spectrum underwent alternate red and violet shifts that matched its brightening and dimming. First it receded while the dark companion approached and then approached while the dark companion receded. Algol was an "eclipsing binary star."

In 1890, Vogel made a similar and more general discovery. He found that some stars were both advancing and receding. That is, the spectral lines showed both a red shift and a violet shift, appearing to have doubled. Vogel interpreted this as indicating that the star was an eclipsing binary with the two stars (both bright) so close together that they appeared as a single star even in the best telescopes. Such stars are "spectroscopic binaries."

But there was no need to restrict the Doppler-Fizeau effect to the stars of our Galaxy. Objects beyond the Milky Way could be studied in this way, too. in 1912, the American astronomer Vesto Melvin Slipher found, on measuring the radial velocity of the Andromeda galaxy, that it was moving toward us at approximately 125 miles per second. But when he went on to examine other galaxies, he discovered that most of them were moving away from us. By 1914, Slipher had figures on 15 galaxies; of these, 13 were receding, all at the healthy clip of several hundred miles per second.

As research along these lines continued, the situation grew more remarkable. Except for a few of the nearest galaxies, all were fleeing from us. Furthermore, as techniques improved so that fainter, more distant galaxies could be tested, the red shift increased.

In 1929, Hubble at Mount Wilson suggested that there was a regular increase in these velocities of recession in proportion to the distance of the galaxy involved. If galaxy A was twice as far from us as galaxy B, then galaxy A receded at twice the velocity of galaxy B. This is sometimes known as "Hubble's law."

Hubble's law certainly continued to be borne out by observations. Beginning in 1929, Milton La Salle Humason at Mount Wilson used the 100-inch telescope to obtain spectra of dimmer and dimmer galaxies. The most distant galaxies he could test were receding at 25,000 miles per second. When the 200-inch telescope came into use, still more distant galaxies could be studied, and by the 1960's, objects were detected so distant that their recession velocities were as high as 150,000 miles per second.

Why should this be? Well, imagine a balloon with small dots painted on it. When the balloon is inflated, the dots move apart. To a manikin standing on any one of the dots, all the other dots would seem to be receding, and the farther away from him a particular dot was, the faster it would be receding. It would not matter on which particular dot he was standing; the effect would be the same.

The galaxies behave as though the universe were expanding like a balloon. Astronomers have now generally accepted the fact of this expansion, and Einstein's "field equations" in his General Theory of Relativity can be construed to fit an expanding universe.

But this raises profound questions. Does the visible universe have a limit? The farthest objects we can now see (about 9 billion light-years away) are receding from us at four-fifths the speed of light. If Hubble's law of the increase in recession velocity holds, at about 11 billion light-years from us the galaxies are receding with the speed of light. But the speed of light is the maximum possible velocity, according to Einstein's theory. Does that mean there can be no visible galaxies more distant?

There is also the age question. If the universe has been expanding constantly, it is logical to suppose that it was smaller in the past than it is now, and that at some time in the distant past it began as a dense core of matter. And that is where the conflict over the age of the universe lay in the 1940's. From its rate of expansion and the present distance of the galaxies, it appeared that the universe could not be more than 2 billion years old. But the geologists, thanks to radioactivity, were now certain that the earth must be nearly 4 billion years old, at least.

Fortunately, the revision of the Cepheid yardstick in 1952 saved the situation. By doubling, possibly tripling, the size of the universe, it doubled or tripled its age, and so the rocks and the red shift now agreed that both the solar system and the galaxies were 5 or 6 billion years old.

By the 1960's, the situation was thrown into some confusion again. The British astronomer Fred Hoyle, after analyzing the probable composition of Population I and Population II stars, decided that, of the two processes by which stars burn hydrogen to form helium, the slower one was predominant. On that basis, he estimated that some stars must be at least 10, perhaps 15, billion years old. Then the American astronomer Allen Sandage found that stars in a cluster called NGC 188 appeared to be at least 24 billion years old, while the Swiss-American astronomer Fritz Zwicky speculated on ages as great as a million billion years. Such ages would not conflict with the rocks' evidence on the age of the earth, for the earth could certainly be younger than the universe, but if the universe has been expanding at the present rate for 24 billion years or more, it would seem that it should be more spread out than it is. So the astronomers have a new problem to resolve.

Assuming that the universe expands and that Einstein's field equations agree with that interpretation, the question still arises inexorably: Why? The easiest, and almost inevitable, explanation is that the expansion is the result of an explosion at the beginning. In 1927, the Belgian mathematician Abbé Georges Edouard Lemâitre suggested that all matter came originally from a tremendously dense "cosmic egg," which exploded and so gave birth to the universe as we know it. Fragments of the original sphere of matter formed galaxies, which are still rushing outward in all directions as a result of that unimaginably powerful multi-billion-year-old explosion.

The Russian-American physicist George Gamow has elaborated this notion. His calculations led him to believe that the various elements as we know them were formed in the first half-hour after the explosion. For 250 million years after the explosion, radiation predominated over matter, and the universe's matter, as a consequence, remained dispersed as a thin gas. After a critical point in the expansion, however, matter came into predominance and began to condense into the beginnings of galaxies. Gamow believes that the expansion will probably continue until all the galaxies, except for those in our own local cluster, have receded beyond the reach of our most powerful instruments. We will then be alone in the universe.

Where did the matter in the "cosmic egg" come from? Some

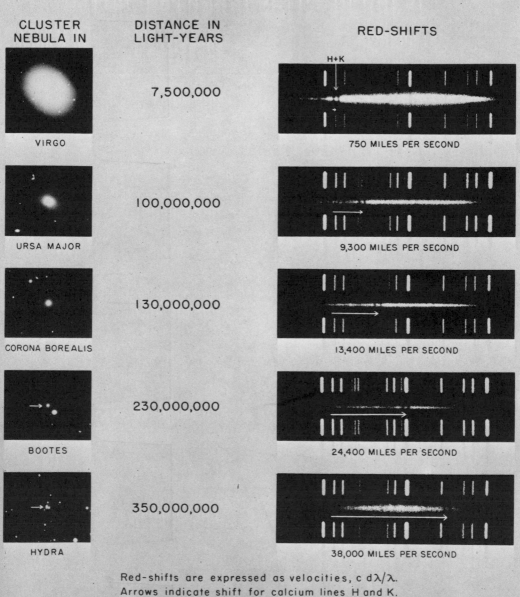

RELATION BETWEEN RED-SHIFT AND DISTANCE
FOR EXTRAGALACTIC NEBULAE

CLUSTER NEBULA IN	DISTANCE IN LIGHT-YEARS	RED-SHIFTS
VIRGO	7,500,000	750 MILES PER SECOND
URSA MAJOR	100,000,000	9,300 MILES PER SECOND
CORONA BOREALIS	130,000,000	13,400 MILES PER SECOND
BOOTES	230,000,000	24,400 MILES PER SECOND
HYDRA	350,000,000	38,000 MILES PER SECOND

Red-shifts are expressed as velocities, $c \, d\lambda/\lambda$.
Arrows indicate shift for calcium lines H and K.
One light-year equals about 6 trillion miles,
or 6×10^{12} miles

Red shifts of distant galaxies.

A globular cluster in Canes Venatici.

The Small
Magellanic Cloud.

The Large
Magellanic Cloud.

Cornell University's new radio telescope. The reflector of this radio-radar telescope at Arecibo, Puerto Rico, is 1,000 feet in diameter and is suspended in a natural bowl.

Panoramic view of the "radio sky," made with the Ohio State University radio telescope, shows how the sky would look if our eyes were sensitive to radio waves instead of light. The broad arch of radiation comes from the plane of our Galaxy. The bright dots represent radio sources. The large bright dot near the top center is Cassiopeia A, the remnant of an exploded star. The brighter of the two large dots, slightly lower and to the right, is Cygnus A, a tremendously powerful radio galaxy at a distance of 600 million light-years from earth.

The Jodrell Bank radio telescope. Its dish is 250 feet in diameter.

astronomers suggest that the universe started as an extremely thin gas, gradually contracted under the force of gravitation to a superdense mass, and then exploded. In other words, it began an eternity ago in the form of almost complete emptiness, went through a contracting stage to the "cosmic egg," exploded, and is going through an expanding stage back to an eternity of almost complete emptiness. We just happen to be living during the very temporary period (an instant in eternity) of the fullness of the universe.

Other astronomers, notably W. B. Bonnor of England, argue that the universe has gone through an unending series of such cycles, each lasting perhaps tens of billions of years—in other words, an "oscillating universe." In 1965, Sandage suggested a period of 82 billion years for each oscillation.

Whether the universe is simply expanding, or contracting and then expanding, or oscillating, all these theories have in mind an "evolutionary universe."

In 1948, the British astronomers Hermann Bondi and Thomas Gold put forward a theory, since extended and popularized by another British astronomer, Fred Hoyle, that forbade evolution. Their universe is called the "steady-state universe" or "continuous creation universe." They agree that the galaxies are receding and that the universe is expanding. When the farthest galaxies reach the speed of light, so that no light from them can reach us, they may be said to have left our universe. However, while the galaxies and clusters of galaxies of our universe move apart, new galaxies are continually forming among the old ones. For every galaxy that disappears over the speed-of-light edge of the universe, another joins our midst. Therefore the universe remains in a steady state, with galaxies always at the same density in space.

Of course, new matter has to be created continually to replace the galaxies that leave, and no such creation of matter has been detected. This is not surprising, however. In order to supply new matter to form galaxies at the necessary rate, only one atom of hydrogen need be formed per year in a billion liters of space. This is creation at far too slow a rate to be detected by the instruments now at our disposal.

If we are to suppose matter to be created continuously, at however slow a rate, we must ask, "Where does this new matter come from?" What happens to the law of conservation of mass-energy? Surely matter cannot be made out of nothing. Hoyle replies that the energy for the creation of new matter may be siphoned from the energy of the expansion. In other words, the universe may be expanding a bit more slowly than it would be if matter were not being formed, and the matter being formed could be manufactured at the expense of the energy being pumped into expansion.

The argument between the proponents of the evolutionary and steady-state views has been hot. The best way of deciding between the two theories would be to study the far-distant reaches of the universe, billions of light-years away.

If the steady-state theory is correct, then the universe is much the same everywhere, and its appearance billions of light-years away ought to be equivalent to its appearance in our own neighborhood. By the evolutionary theory, however, the universe billions of light-years away would be seen by light that had been formed billions of years ago. That light had been formed when the universe was young, and not long after the "big bang." What we see in the young universe ought to be quite different from what we see in our own neighborhood where the universe is old.

Unfortunately, it is hard to make out clearly what we see by telescope in the very distant galaxies, and until the 1960's, the information gathered was insufficient. When the evidence finally began to come in, it involved (as we shall see) radiation other than that of ordinary light.

The Death of the Sun

Whether the universe is evolutionary or steady state is a point that does not affect individual galaxies or clusters of galaxies directly. Even if all the distant galaxies recede and recede until they are out of range of the best possible instruments, our own Galaxy will remain intact, its component stars held firmly within its gravitational field. Nor will the other galaxies of the local cluster leave us. But changes within our Galaxy, possibly disastrous to our planet and its life, are by no means excluded.

The whole conception of changes in heavenly bodies is a modern one. The ancient Greek philosophers, Aristotle in particular, believed that the heavens were perfect and unchangeable. All change, corruption, and decay were confined to the imperfect regions that lay below the nethermost sphere—the moon. This seemed only common sense, for certainly from generation to generation and century to century there was no important change in the heavens. To be sure, there were the mysterious comets that occasionally materialized out of nowhere—erratic in their comings and goings, ghostlike as they shrouded stars with a thin veil, baleful in appearance, for the filmy tail looked like the streaming hair of a distraught creature prophesying evil (in fact, the word "comet" comes from the Latin word for "hair"). About twenty-five of these objects are visible to the naked eye each century.

Aristotle tried to reconcile these apparitions with the perfection of the heavens by insisting that they belonged to the atmosphere of the corrupt and changing earth. This view prevailed until late in the sixteenth century. But, in 1577, the Danish astronomer Tycho Brahe attempted to measure the parallax of a bright comet and discovered that it could not be measured (this was before the days of the telescope). Since the moon's parallax *was* measurable, Tycho Brahe was forced to conclude that the comet lay far beyond the moon and that there was change and imperfection in the heavens. (The Roman philosopher Seneca had suspected this in the first century A.D.)

Actually, changes even in the stars had been noticed much earlier, but apparently they had aroused no great curiosity. For instance, there are the variable stars that change noticeably in brightness from night to night, even to the naked eye. No Greek astronomer made any reference to variations in the brightness of any star. It may be that we have lost the records of such references; on the other hand, perhaps the Greek astronomers simply chose not to see these phenomena. One interesting case in point is Algol, the second brightest star in the constellation Perseus, which suddenly loses two thirds of its brightness, then regains it, and does this regularly every 69 hours. (We know now, thanks to Goodricke and Vogel, that Algol has a dim companion star that eclipses it and diminishes its light at 69-hour intervals.) The Greek astronomers made no mention of the dimming of Algol, nor did the Arab astronomers of the Middle Ages. Nevertheless, the Greeks placed the star in the head of Medusa, the demon who turned men to stone, and the very name "Algol," which is Arabic, means "the ghoul." Clearly, the ancients felt uneasy about this strange star.

A star in the constellation Cetus, called Omicron Ceti, varies irregularly. Sometimes it is as bright as the Pole Star; sometimes it vanishes from sight. Neither the Greeks nor the Arabs said a word about it, and the first man to report it was a Dutch astronomer, David Fabricius, in 1596. It was later named Mira (Latin for "wonderful"), astronomers having grown less frightened of heavenly change by then.

Even more remarkable was the sudden appearance of "new stars" in the heavens. This the Greeks could not altogether ignore. Hipparchus is said to have been so impressed by the sighting of such a new star in the constellation Scorpio in 134 B.C. that he designed the first star map, in order that new stars in the future might be more easily detected.

In 1054 A.D., in the constellation Taurus, another new star was sighted—a phenomenally bright one, in fact. It surpassed Venus in brightness, and for weeks it was visible in broad daylight. Chinese and Japanese astronomers recorded its position accurately, and their records have come down to us. In the Western world, however, the state of astronomy was so low at the time that no European record of this remarkable occurrence has survived, probably because none was kept.

It was different in 1572, when a new star as bright as that of 1054 appeared in the constellation Cassiopeia. European astronomy was reviving from its long sleep. The young Tycho Brahe carefully observed the new star and wrote a book entitled *De Nova Stella*. It is from the title of that book that the word "nova" was adopted for any new star.

In 1604, still another remarkable nova appeared, in the constellation Serpens. It was not quite as bright as that of 1572, but it was bright enough to outshine Mars. Johannes Kepler observed this one, and he too wrote a book about the subject.

After the invention of the telescope, novae became less mysterious. They were not new stars at all, of course; merely faint stars that had suddenly brightened to visibility.

Increasing numbers of novae were discovered with time. They would brighten many thousandfold, sometimes within the space of a few days, and then dim slowly over a period of months to their previous obscurity. Novae showed up at the average rate of 20 per year per galaxy (including our own).

From an investigation of the Doppler-Fizeau shifts that took place during nova formation and from certain other fine details of their spectra, it became plain that the novae were exploding stars. In some cases the star material blown into space could be seen as a shell of expanding gas, illuminated by the remains of the star. Such stars are called "planetary nebulae."

This sort of nova formation does not necessarily signify the death of a star. It is a tremendous catastrophe, of course, for the luminosity of such a star may increase a millionfold in less than a day. (If our sun were to become a nova, it would destroy all life on earth and possibly vaporize the planet.) But the explosion apparently ejects only 1 or 2 per cent of the star's mass, and afterward the star settles back to a reasonably normal life. In fact, some stars seem to undergo such explosions periodically and still survive.

The most remarkable nova that appeared after the invention of the telescope was one that was discovered by the German astronomer Ernst Harwig in the Andromeda Galaxy in 1885 and was given the name "S Andromedae." It was just below visibility to the naked eye; in the telescope it looked one tenth as bright as the entire Andromeda galaxy. At the time, no one realized how distant the Andromeda galaxy was, or how large, and so the brightness of its nova occasioned no particular excitement. But after Hubble worked out the distance of the Andromeda galaxy, the brilliance of that nova of 1885 suddenly staggered astronomers. Hubble eventually discovered a number of novae in the Andromeda galaxy, but none even approached the 1885 nova in brightness. The nova of 1885 must have been 10,000 times as bright as ordinary novae. It was a "supernova." Looking back now, we realize that the novae of 1054, 1572, and 1604 also were supernovae. What is more, they must have been in our own Galaxy, which would account for their extreme brightness. In 1965, Bernard Goldstein of Yale presented evidence to the effect that a fourth supernova in our Galaxy may have flared in 1006, if an obscure note by an Egyptian astrologer of the period is to be accepted.

Supernovae apparently are quite different in physical behavior from ordinary novae, and astronomers are eager to study their spectra in detail.

The main difficulty is that they are so rare. About 3 per thousand years is the average for any one galaxy, according to Zwicky. Although astronomers have managed to spot about 50 so far, all these are in distant galaxies and cannot be studied in detail. The 1885 supernova of Andromeda, the closest to us in the last 350 years, appeared a couple of decades before photography in astronomy had been fully developed; consequently, no permanent record of its spectrum exists. (However, the distribution of supernovae in time is random. In one galaxy recently, three supernovae were detected in just seventeen years. Astronomers on earth may yet prove lucky.)

The brightness of a supernova (absolute magnitudes range from -14 to an occasional -17) could only come about as a result of a complete explosion—a star literally tearing itself to pieces. What would happen to such a star? Well, let us go back a little. . . .

As early as 1834, Bessel (the astronomer who was later to be the first to measure the parallax of a star) noticed that Sirius and Procyon shifted position very slightly from year to year in a manner which did not seem related to the motion of the earth. Their motions were not in a straight line but wavy, and Bessel decided that each must actually be moving in an orbit around something.

From the manner in which Sirius and Procyon were moving in these orbits, the "something" in each case had to have a powerful gravitional attraction that could belong to nothing less than a star. Sirius' companion, in particular, had to be as massive as our own sun to account for the bright star's motions. So the companions were judged to be stars, but since they were invisible in telescopes of the time, they were referred to as "dark companions." They were believed to be old stars growing dim with time.

Then in 1862 the American instrument maker, Alvan Clark, testing a new telescope, sighted a dim star near Sirius and, sure enough, on further observation this turned out to be the companion. Sirius and the dim star circled about a mutual center of gravity in a period of about 50 years. The companion of Sirius ("Sirius B" it is now called, with Sirius itself as "Sirius A") has an absolute magnitude of only 11.2, and so it is only about 1/400 as bright as our sun, though it is just as massive.

This seemed to check with the notion of a dying star. But in 1914 the American astronomer Walter Sydney Adams, after studying the spectrum of Sirius B, decided that the star had to be as hot as Sirius A itself and hotter than our sun. The atomic vibrations that gave rise to the particular absorption lines found in its spectrum could only be taking place at very high temperatures. But if Sirius B was so hot, why was its light so faint? The only possible answer was that it was considerably smaller than our sun. Being hotter, it radiated more light per unit of surface, but to account for the small total amount of light, its total surface had to be small. In fact, the star could not be more than 16,000 miles in diameter—only twice the earth's diameter. Yet Sirius B had a mass equal to that of our sun! Adams found himself trying to imagine this mass mashed down into a volume as small as that of Sirius B. The star's density would have to be nearly 3,000 times that of platinum.

This represented nothing less than a completely new state of matter. Fortunately, by this time physicists had no trouble in suggesting the answer. They knew that in ordinary matter the atoms are composed of very tiny particles, so tiny that most of the volume of an atom is "empty" space. Under extreme pressure the subatomic particles could be forced together into a superdense mass. Yet even in superdense Sirius B, the subatomic particles are far enough apart to move about freely so that the far-denser-than-platinum substance still acts as a gas. The English physicist Ralph Howard Fowler suggested in 1925 that this be called a "degenerate gas," and the Soviet physicist Lev Davidovich Landau pointed out in the 1930's that even ordinary stars such as our own Sun ought to consist of degenerate gas at the center.

The companion of Procyon ("Procyon B"), first detected in 1896 by J. M. Schaberle at Lick Observatory, was also found to be a superdense star although only five-eighths as massive as Sirius B and, as the years passed, more examples were found. These stars are called "white dwarfs," because they combine small size with high temperature and white light. White dwarfs are probably quite numerous and may make up as much as 3 per cent of all stars. However, because of their small size, only those in our own neighborhood are likely to be discovered in the foreseeable future. (There are also "red dwarfs," considerably smaller than our sun, but not as small as white dwarfs. Red dwarfs are cool and of ordinary density. They may be the most common of all stars but, because of their dimness, are as difficult to detect as are white dwarfs. A pair of red dwarfs, a mere 6 light-years distant from us, was only discovered in 1948. Of the 36 stars known to be within 14 light-years of the sun, 21 are red dwarfs, and 3 are white dwarfs. There are no giants among them, and only two, Sirius and Procyon, are distinctly brighter than our sun.)

The year after Sirius B was found to have its astonishing properties, Albert Einstein presented his General Theory of Relativity, which was mainly concerned with new ways of looking at gravity. Einstein's views of gravity led to the prediction that light emitted by a source possessing a very strong gravitational field should be displaced toward the red (the "Einstein shift"). Adams, fascinated by the white dwarfs he had discovered, carried out careful studies of the spectrum of Sirius B and found that there was indeed the red shift predicted by Einstein. This was not only a point in favor of Einstein's theory but also a point in favor of the superdensity of Sirius B, for in an ordinary star such as our sun the red-shift effect would be only one thirtieth as great. Nevertheless, in the early 1960's this very small Einstein shift produced by our sun was detected, and the General Theory of Relativity was further confirmed.

But what have white dwarfs to do with supernovae, the subject that prompted this discussion? To answer that, let us go back to the supernova of 1054. In 1844 the Earl of Rosse, investigating the location in Taurus where the oriental astronomers had reported finding the 1054 supernova, studied a small cloudy object. Because of its irregularity and its clawlike projections, he named the object the "Crab Nebula." Continued observation over decades showed that the patch of gas was slowly expanding. The actual rate of expansion could be calculated from the Doppler-Fizeau effect, and this, combined with the apparent rate of expansion, made it possible to compute the distance of the Crab Nebula as 3,500 light-years from us. From the expansion rate it was also determined that the gas had started its expansion from a central explosion point nearly 900 years ago, which agrees well with the date 1054. So there can be little doubt that the Crab Nebula, which now spreads over a volume of space some 5 light-years in diameter, represents the remnants of the 1054 supernova.

No similar region of turbulent gas has been observed at the reported sites of the supernovae of Tycho and Kepler, although small spots of nebulosity have been observed close to each site. There are some 150 planetary nebulae, however, in which doughnut-shaped rings of gas may represent large stellar explosions. A particularly extended and thin gas cloud, the Veil Nebula in Cygnus, may be what is left of a supernova explosion 30,000 years ago. When it took place it must have been even closer and brighter than the supernova of 1054—but no civilization existed on earth to record the spectacle.

There are even suggestions that a very faint nebulosity enveloping the constellation Orion may be what is left of a still older supernova.

In all these cases, though, what happened to the star that exploded? The difficulty, or impossibility, of locating it points to exceeding dimness and that, in turn, suggests a white dwarf. If so, are all white dwarfs the remnants of stars that have exploded? In that case, why do some white dwarfs, such as Sirius B, lack an enclosing envelope of gas? Will our own sun some day explode and become a white dwarf? These queries lead us into the problem of the evolution of stars.

Of the stars near us, the bright ones seem to be hot and the dim ones cooler, according to a fairly regular brightness-temperature scale. If the surface temperatures of various stars are plotted against their absolute magnitudes, most of the familiar stars fall within a narrow band, increasing steadily from dim coolness to bright hotness. This band is called the "main sequence." It was first plotted in 1913 by the American astronomer Henry Norris Russell, following work along similar lines by Hertzsprung (the astronomer who first determined the absolute magnitudes of the Cepheids). A graph showing the main sequence is therefore called a Hertzsprung-Russell diagram, or H-R diagram.

Not all stars belong in the main sequence. There are some red stars that, despite their rather low temperature, have large absolute magnitudes, because their substance is spread out in rarefied fashion into tremendous size. Among these "red giants," the best-known are Betelgeuse and Antares. They are so cool (it was discovered in 1964) that many have atmospheres rich in water vapor, which would decompose to hydrogen and oxygen at the higher temperatures of our own sun. The high-temperature white dwarfs also fall outside the main sequence.

In 1924, Eddington pointed out that the interior of any star must be very hot. Because of a star's great mass, its gravitational force is immense. If the star is not to collapse, this huge force must be balanced by an equal internal pressure—from radiation energy. The more massive the star, the higher the central temperature required to balance the gravitational force. To maintain this high temperature and radiation pressure, the more massive stars must be burning energy faster, and they must be brighter than less massive ones. This is the "mass-luminosity law." The relationship is a drastic one, for luminosity varies as the sixth or seventh power of the mass. If the mass is increased by 3 times, then the luminosity increases by a factor of six or seven 3's multiplied together, say 750-fold.

It follows that the massive stars are spendthrift with their hydrogen fuel and have a shorter life. Our sun has enough hydrogen to last it at its present radiation rate for many billions of years. A bright star such as Capella must burn out in about 20 million years, and some of the brightest stars—for example, Rigel—cannot possibly last more than one or two million years. This means that the very brightest stars must be very youthful. New stars are perhaps even now being formed in regions where space is dusty enough to supply the raw material.

Indeed, the American astronomer George Herbig detected two stars in the dust of the Orion Nebula, in 1955, that were not visible in photographs of the region taken some years back. They may represent stars that were actually born as we watched.

By 1965, hundreds of stars were located that were so cool, they didn't quite shine. They were detected by their infrared radiation and are therefore called "infrared giants" because they are made up of large quantities of rarefied matter. Presumably, these are quantities of dust and gas, gathering together and gradually growing hotter. Eventually, they will become hot enough to shine, and whether they will join the main sequence at some point then depends on the total mass of the gathered-together matter.

The next advance in the study of the evolution of stars came from analysis of the stars in globular clusters. The stars in a cluster are all about the same distance from us, so their apparent magnitude is pro-

portional to their absolute magnitude (as in the case of the Cepheids in
the Magellanic Clouds). Therefore, with their magnitude known, an
H-R diagram of these stars can be prepared. It is found that the cooler
stars (burning their hydrogen slowly) are on the main sequence, but the
hotter ones tend to depart from it. In accordance with their high rate
of burning, and with their rapid aging, they follow a definite line showing various stages of evolution, first toward the red giants and then back,
across the main sequence again, and down toward the white dwarfs.

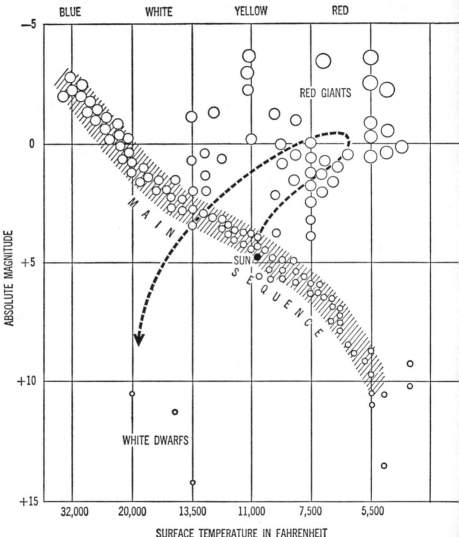

The Hertzsprung-Russell diagram. The dotted line indicates
the evolution of a star. The relative size of the stars are
given only schematically, not according to scale.

From this and from certain theoretical considerations as to the manner in which subatomic particles can combine at certain high temperatures and pressures, Fred Hoyle has drawn a detailed picture of the course of a star's evolution. According to Hoyle, in its early stages a star changes very little in size or temperature. (This is the position our sun is in now and will continue to be in for a long time.) As it converts its hydrogen in the extremely hot interior into helium, the helium accumulates at the center of the star. When this helium core reaches a certain size, the star starts to change its size and temperature dramatically. It becomes cooler and expands enormously. In other words, it leaves the main sequence and moves in the red-giant direction. The more massive the star, the more quickly it reaches this point. In the globular clusters, the more massive ones have already progressed varying lengths along the road.

The expanded giant releases more heat, despite its lower temperature, because of its larger surface area. In the far distant future, when the sun leaves the main sequence, or even somewhat before, it will have heated to the point where life will be impossible on the earth. That point, however, is still billions of years in the future.

Until recently the hydrogen-to-helium conversion was the only source of energy recognized in stars, and that raised a problem with respect to the red giants. By the time a star has reached the red-giant stage, most of its hydrogen is gone. How, then, can it go on radiating energy in such large quantities? Hoyle suggested that the helium core itself contracts, and as a result it rises to a temperature at which the helium nuclei can fuse to form carbon, with the liberation of additional energy. In 1959, the American physicist David Elmer Alburger showed in the laboratory that this reaction actually can take place. It is a very rare and unlikely sort of reaction, but there are so many helium atoms in a red giant that enough such fusions can occur to supply the necessary quantities of energy.

Hoyle goes further. The new carbon core heats up still more, and still more complicated atoms, such as those of oxygen and neon begin to form. While this is happening, the star is contracting and getting hotter again; it moves back toward the main sequence. By now the star has begun to acquire a series of layers, like an onion. It has an oxygen-neon core, then a layer of carbon, then one of helium, and the whole is enveloped in a skin of still-unconverted hydrogen.

As the temperature at the center continues to increase, more and more complex types of reactions can go on. The neon in the new core can combine further to magnesium, which can combine in turn to form silicon, and then, in turn, iron. At a late stage in its life, the star may be built up of more than half a dozen concentric shells, in each of which a different fuel is being consumed. The central temperature may have reached 3 to 4 billion degrees by then.

However, in comparison with its long life as a hydrogen-consumer, the star is on a quick toboggan slide through the remaining fuels. Its life off the main sequence is a merry one, but short. Once the star begins to form iron, it has reached a dead end, for iron atoms represent the point of maximum stability and minimum energy content. To alter iron atoms in the direction of more complex atoms, or of less complex atoms, requires an input of energy.

Furthermore, as central temperatures rise with age, radiation pressure rises, too, and in proportion to the fourth power of the temperature. When the temperature doubles, the radiation pressure increases 16-fold, and the balance between it and gravitation becomes ever more delicate. A temporary imbalance will have more and more drastic results, and, if radiation pressure shoots up a little too quickly, the explosion of a nova can result. The loss of some mass probably relieves the situation, at least temporarily, and the star may then continue to age without further catastrophe for another million years or so.

It may be, though, that the balance is maintained and the star does not relieve the situation by a minor explosion. In that case, the central temperatures may rise so high, according to Hoyle's suggestion, that the iron atoms are driven apart into helium. But for this to happen, as I have just said, energy must be poured into the atoms. The only place the star can get this energy from is its gravitational field. When the star shrinks, the energy it gains can be used to convert iron to helium. The amount of energy needed is so great, however, that the star must shrink drastically to a tiny fraction of its former volume, and this must happen, according to Hoyle, "in about a second."

In the blink of an eye, then, the ordinary star is gone, and a white dwarf takes its place. That is the fate the far, far future holds in store for our sun, and stars currently brighter than the sun will reach that stage sooner, perhaps within 5 billion years.

All this purports to explain the formation of a white dwarf without an explosion. It may be the story of dwarfs such as Sirius B and Procyon B. But where do supernovae come in?

The Indian astronomer Subrahmanyan Chandrasekhar, working at Yerkes Observatory, calculated that no star more than 1.4 times the mass of our sun (now called "Chandrasekhar's limit") could become a white dwarf by the "normal" process Hoyle described. And in fact all the white dwarfs so far observed turn out to be below Chandrasekhar's limit in mass. It also turns out, though, that the Crab Nebula, which is accepted as the remnant of a supernova explosion, and which, it seems certain, has a white dwarf at its center, posesses more than 1.4 times the mass of our sun, if we count the mass of the ejected gas.

So we now have to explain how the original over-the-limit star could have become a white dwarf. The reason for Chandrasekhar's limit is that, the more massive the star, the more it has to shrink (i.e., the denser it has to become) to provide the energy necessary to reconvert its iron to helium, and there is a limit to the possible shrinkage, so to speak. However, a very massive star can get around that limit. When such a star starts to collapse, its iron core is still surrounded with a voluminous outer mantle of atoms not yet built up to a maximum stability. As the outer regions collapse and their temperature rises, these still combinable substances "take fire" all at once. The result is an explosion which blasts the outer material away from the body of the star. The white dwarf left at the conclusion of such an explosion is then below Chandrasekhar's limit, although the original star was above it.

This may be the explanation not only of the Crab Nebula but also of all supernovae. Our sun, by the way, being below Chandrasekhar's limit, may become a white dwarf some day but apparently will never become a supernova.

Hoyle suggests that the matter blasted into space by a supernova may spread through the galaxies and serve as raw material for the formation of new, "second-generation" stars, rich in iron and other metallic elements. Our own sun is probably a second-generation star, much younger than the old stars of some of the dust-free globular clusters. Those "first-generation" stars are low in metals and rich in hydrogen. The earth, formed out of the same debris of which the sun was born, is extraordinarily rich in iron—iron which once may have existed at the center of a star that exploded many billions of years ago.

As for white dwarfs, though they are dying, it seems that their death will be indefinitely prolonged. Their only source of energy is their gravitational contraction, but this force is so immense that it can supply the charily radiating white dwarfs with enough energy to last tens of billions of years before they dim out altogether and become "black dwarfs."

Or, perhaps, as we shall see later in the chapter, it may be that even the white dwarf is not the most extreme case of stellar evolution, but that stars may exist even more shrunken—where the subatomic particles making them up approach until they are in virtual contact and the mass of an entire star is compressed into a globe perhaps no more than ten miles across.

The detection of such extremes had to await new methods of probing the universe, taking advantage of radiations other than those of visible light.

The Windows to the Universe

Man's greatest weapons in his conquest of knowledge are his understanding mind and the inexorable curiosity that drives it on. And his resourceful mind has continually invented new instruments which have opened up horizons beyond the reach of his unaided sense organs.

The best-known example is the vast surge of new knowledge that followed the invention of the telescope in 1609. The telescope, essentially, is simply an oversized eye. In contrast to the quarter-inch pupil of the human eye, the 200-inch telescope on Palomar Mountain has more than 31,000 square inches of light-gathering area. Its light-collecting power intensifies the brightness of a star about a million

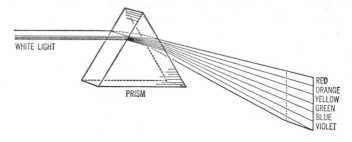

Newton's experiment splitting the spectrum of white light.

times, as compared with what the naked eye can see. This telescope, first put into use in 1948, is the largest in use today, but the Soviet Union, whose largest telescope now is a 102-incher, is constructing two more of that size and a 236-inch telescope which will be the largest in the world when it is done. Meanwhile, during the 1950's Merle A. Ture developed an image tube which electronically magnified the faint light gathered by a telescope, tripling its power. Nevertheless, the law of diminishing returns is setting in. To make still bigger telescopes will be useless, for the light absorption and temperature variations of the earth's atmosphere are what now limits the ability to see fine detail. If bigger telescopes are to be built, it will have to be for use in an airless observatory, perhaps an observatory on the moon.

But mere magnification and light-intensification are not the full measure of the telescope's gift to man. The first step toward making it something more than a mere light collector came in 1666 when Isaac Newton discovered that light could be separated into what he called a "spectrum" of colors. He passed a beam of sunlight through a triangularly shaped prism of glass and found that the beam spread out into a band made up of red, orange, yellow, green, blue, and violet light, each color fading gently into the next. (The phenomenon itself, of course, has always been familiar in the form of the rainbow, the result of sunlight passing through water droplets, which act like tiny prisms.)

What Newton showed was that sunlight, or "white light," is a mixture of many specific radiations (that we now recognize as wave forms of varying wavelengths) which impress the eye as so many different colors. A prism separates the colors because, on passing from air into glass, and from glass into air, light is bent, or "refracted," and each wavelength undergoes a different amount of refraction—the shorter the wavelength, the greater the refraction. The short wavelengths of violet light are refracted most; the long wavelengths of red, least.

Among other things, this explains an important flaw in the very earliest telescopes, which was that objects viewed through them were surrounded by obscuring rings of color, because the lenses through which light passed dispersed that light into spectra.

Newton despaired of correcting this as long as lenses of any sort were used. He therefore designed and built a "reflecting telescope" in which a parabolic mirror, rather than a lens, was used to magnify an image. Light of all wavelengths was reflected alike so that no spectra were formed on reflection and rings of color ("chromatic aberration") were not to be found.

In 1757, the English optician John Dollond prepared lenses of two different kinds of glass; one kind canceling out the spectrum-forming tendency of the other. In this way, "achromatic" ("no color") lenses could be built. Using such lenses, "refracting telescopes" became popular again. The largest such telescope, with a 40-inch lens, is at Yerkes Observatory near Williams Bay, Wisconsin, and was built in 1897. No larger refracting telescopes have been built since or are likely to be built, for still larger lenses would absorb so much light as to cancel their superior magnifying powers. The giant telescopes of today are all of the reflecting variety, in consequence, since the reflecting surface of a mirror absorbs very little light.

In 1814, a German optician, Joseph von Fraunhofer, went beyond Newton. He passed a beam of sunlight through a narrow slit before allowing it to be refracted by a prism. The spectrum that resulted was actually a series of images of the slit in light of every possible wavelength. There were so many slit images that they melted together to form the spectrum. Fraunhofer's prisms were so excellently made and produced such sharp slit images that it was possible to see that some of the slit images were missing. If particular wavelengths of light were missing in sunlight, no slit image would be formed at that wavelength and the sun's spectrum would be crossed by dark lines.

Fraunhofer mapped the location of the dark lines he detected, recording over 700. They have been known as "Fraunhofer lines" ever since. In 1842, the lines of the solar spectrum were first photographed by the French physicist Alexandre Edmond Becquerel. Such photography greatly facilitated spectral studies, and, with the use of modern instruments, more than 30,000 dark lines have been detected in the solar spectrum and their wavelengths measured.

In the 1850's, a number of scientists toyed with the notion that the lines were characteristic of the various elements present in the sun. The dark lines would represent absorption of light at the wavelengths in question by certain elements; bright lines would represent characteristic emissions of light by elements. About 1859, the German chemists Robert

Wilhelm Bunsen and Gustav Robert Kirchhoff worked out a system for identifying elements in this way. They heated various substances to incandescence, spread out their glow into spectra, measured the location of the lines (in this case, bright lines of emission, against a dark background) on a background scale, and matched up each line with a particular element. Their "spectroscope" was quickly applied to discovering new elemetns by means of new spectral lines not identifiable with known elements. Within a couple of years Bunsen and Kirchhoff discovered cesium and rubidium in this manner.

The spectroscope was also applied to the light of the sun and the stars and soon turned up an amazing quantity of new information, chemical and otherwise. In 1862, the Swedish astronomer Anders Jonas Angstrom identified hydrogen in the sun by the presence of spectral lines characteristic of that element.

Hydrogen could also be detected in the stars, although, by and large, the spectra of the stars varied among themselves because of differences in their chemical constitution (and other properties, too). In fact, stars could be classified according to the general nature of their spectral line pattern. Such a classification was first worked out by the Italian astronomer Pietro Angelo Secchi in the mid-nineteenth century, on the basis of a few scattered spectra. By the 1890's, the American astronomer Edward Charles Pickering was studying stellar spectra by the tens of thousands, and the spectral classification could be made finer.

Originally, the classification was by capital letters in alphabetical order, but as more and more was learned about the stars, it became necessary to alter that order to put the spectral classes into a logical arrangement. If the letters are arranged in order of stars of decreasing temperature, we have O, B, A, F, G, K, M, R, N, and S. Each classification can be further subdivided by numbers from 1 to 10. The sun is a star of intermediate temperature with a spectral class of G-0, while Alpha Centauri is G-2. The somewhat hotter Procyon is F-5, while the considerably hotter Sirius is A-0.

Just as the spectroscope could locate new elements on earth, so it could locate them in the heavens. In 1868, the French astronomer Pierre Jules César Janssen was observing a total eclipse of the sun in India, and reported sighting a spectral line he could not identify with any produced by any known element. The English astronomer Sir Norman Lockyer, sure that the line represented a new element, named it "helium," from the Greek word for "sun." Not until nearly thirty years later was helium found on the earth.

The spectroscope eventually became a tool for measuring the radial velocity of stars, as we saw earlier in this chapter, and for exploring many other matters—the magnetic characteristics of a star, its temperature, whether the star is single or double, and so on.

Moreover, the spectral lines were a veritable encyclopedia of information about atomic structure, which, however, could not properly be utilized until after the 1890's, when the subatomic particles within the atom were first discovered. For instance, in 1885, the German physicist Johann Jakob Balmer showed that hydrogen produced a whole series of lines that were regularly spaced according to a rather simple formula. This was used, a generation later, to deduce an important picture of the structure of the hydrogen atom (see Chapter 7).

Lockyer himself showed that the spectral lines produced by a given element altered at high temperatures. This indicated some change in the atoms. Again, this was not appreciated until it was later found that an atom consisted of smaller particles, some of which were driven off at high temperatures, altering the atomic structure and the nature of the lines the atom produced. (Such altered lines were sometimes mistaken for indications of new elements but, alas, helium remained the only new element ever discovered in the heavens.)

When, in 1830, the French artist, Louis Jacques Mandé Daguerre produced the first "daguerreotypes" and thus introduced photography, this, too, soon became an invaluable instrument for astronomy. Through the 1840's, various American astronomers photographed the moon, and one picture, by the American astronomer George Phillips Bond, was a sensation at the Great Exhibition of 1851 in London. They also photographed the sun. In 1860, Secchi made the first photograph of a total eclipse of the sun. By 1870, photographs of such eclipses had proved that the corona and prominences were part of the sun and not of the moon.

Meanwhile, beginning in the 1850's, astronomers were also making pictures of the distant stars. By 1887, the Scottish astronomer David Gill was making stellar photography routine. Photography was well on its way to becoming more important than the human eye in observing the universe.

The technique of photography with telescopes steadily improved. A major stumbling block was the fact that a large telescope can cover only a very small field. If an attempt is made to enlarge the field, distortion creeps in at the edges. In 1930, the Russian-German optician Bernard Schmidt designed a method for introducing a correcting lens that would prevent such distortion. With such a lens, a wide swatch of sky can be photographed at one swoop and studied for interesting objects that can then be studied intensely by an ordinary telescope. Since such telescopes are almost invariably used for photographic work, they are called "Schmidt cameras."

The largest Schmidt cameras now in use are a 53-inch instrument, first put to use in 1960 in Tautenberg, East Germany, and a 48-inch instrument used in conjunction with the 200-inch Hale telescope on Mount Palomar. The third largest is a 39-inch instrument put into use at an observatory in Soviet Armenia in 1961.

About 1800, William Herschel (the astronomer who first guessed the shape of our galaxy) performed a very simple but interesting experiment. In a beam of sunlight transmitted through a prism, he held a thermometer beyond the red end of the spectrum. The mercury climbed! Plainly some form of invisible radiation existed at wavelengths below the visible spectrum. The radiation Herschel had discovered became known as "infrared"—below the red—and, as we now know, fully 60 per cent of the sun's radiation is in the infrared.

At about the same time the German physicist Johann Wilhelm Ritter was exploring the other end of the spectrum. He found that silver nitrate, which breaks down to metallic silver and darkens when it is

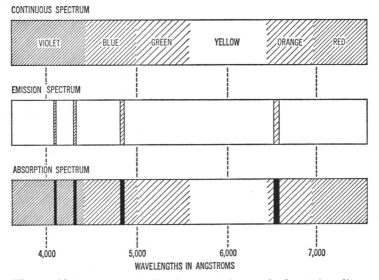

The visible spectrum, indicating emission and absorption lines.

exposed to blue or violet light, would break down even more rapidly if it were placed beyond the point in the spectrum where violet faded out. Thus Ritter discovered the "light" now called "ultraviolet" (beyond the violet). Between them, Herschel and Ritter had widened the time-honored spectrum and crossed into new realms of radiation.

These new realms bear promise of yielding much information. The ultraviolet portion of the solar spectrum, invisible to the eye, shows up in nice detail by way of photography. In fact, if a quartz prism is used (quartz transmits ultraviolet light, whereas ordinary glass absorbs most of it), quite a complicated ultraviolet spectrum can be recorded, as was first demonstrated in 1852 by the British physicist George Gabriel Stokes. Unfortunately, the atmosphere transmits only the "near ultraviolet"— that part with wavelength almost as long as violet light. The "far ultraviolet," with its particularly short wavelengths, is absorbed in the upper atmosphere.

In 1860, the Scottish physicist James Clerk Maxwell worked out a theory which predicted a whole family of radiation associated with electric and magnetic phenomena ("electromagnetic radiation")—a family of which ordinary light was only one small portion. The first definite evidence bearing out his prediction came a quarter of a century later, seven years after Maxwell's premature death through cancer. In 1887, the German physicist Heinrich Rudolf Hertz, generating an oscillating current from the spark of an induction coil, produced and detected radiation of extremely long wavelengths—much longer than those of ordinary infrared. These came to be called "radio waves."

The wavelengths of visible light can be measured in microns (millionths of a meter). They range from 0.39 micron (extreme violet) to 0.78 micron (extreme red). Next come the "near infrared" (0.78 to 3 microns), the "middle infrared" (3 to 30 microns), and then the "far infrared" (30 to 1,000 microns). It is here that radio waves begin: the so-called "microwaves" run from 1,000 to 160,000 microns and long-wave radio goes as high as many billions of microns.

Radiation can be characterized not only by wavelength, but also by "frequency," the number of waves of radiation produced in each second. This value is so high for visible light and the infrared that it is not commonly used in these cases. For the radio waves, however, frequency reaches down into lower figures and comes into its own. One thousand waves per second is a "kilocycle," while a million waves per second is a "megacycle." The microwave region runs from 300,000 megacycles down to 1,000 megacycles. The much longer radio waves used in ordinary radio stations are down in the kilocycle range.

Within a decade after Hertz's discovery, the other end of the spectrum opened up similarly. In 1895, the German physicist Wilhelm Konrad Roentgen accidentally discovered a mysterious radiation which he called "X-rays." Their wavelengths turned out to be shorter than ultraviolet. Later, "gamma rays," associated with radioactivity, were shown by Rutherford to have wavelengths even smaller than those of X-rays.

The short-wave half of the spectrum is now divided roughly as follows: The wavelengths from 0.39 down to 0.17 micron belong to the "near ultraviolet," from 0.17 down to 0.01 micron to the "far ultraviolet," from 0.01 to 0.00001 micron to X-rays, and gamma rays range from this down to less than a billionth of a micron.

Newton's original spectrum was thus expanded enormously. If we consider each doubling of wavelength as equivalent to one octave (as is the case in sound), the electromagnetic spectrum over the full range studied amounts to almost 60 octaves. Visible light occupies just one octave near the center of the spectrum.

With a wider spectrum, of course, we can get a fuller view of the stars. We know, for instance, that sunshine is rich in ultraviolet and in

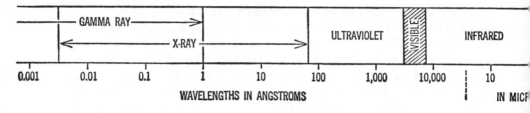

The spectrum of electromagnetic radiation.

infrared. Our atmosphere cuts off most of these radiations; but in 1931, quite by accident, a radio window to the universe was discovered.

Karl Jansky, a young radio engineer at the Bell Telephone Laboratories, was studying the static that always accompanies radio reception. He came across a very faint, very steady noise which could not be coming from any of the usual sources. He finally decided that the static was caused by radio waves from outer space.

At first the radio signals from space seemed strongest in the direction of the sun, but day by day the direction of strongest reception slowly drifted away from the sun and made a circuit of the sky. By 1933, Jansky decided the radio waves were coming from the Milky Way and, in particular, from the direction of Sagittarius, toward the center of the Galaxy.

Thus was born "radio astronomy." Astronomers did not take to it immediately, for it had serious drawbacks. It gave no neat pictures—only wiggles on a chart which were not easy to interpret. More important, radio waves are much too long to resolve a source as small as a star. The radio signals from space had wavelengths hundreds of thousands and even millions of times the wavelength of light, and no ordinary radio receiver could give anything more than a general idea of the direction they were coming from.

These difficulties obscured the importance of the new discovery, but a young radio ham named Grote Reber carried on, for no reason other than personal curiosity. Through 1937 he spent time and money building in his backyard a small "radio telescope" with a parabolic "dish" about 30 feet in diameter to receive and concentrate the radio waves. Beginning in 1938, he found a number of sources of radio waves other than the one in Sagittarius—one in the constellation Cygnus, for instance, and another in Cassiopeia. (Such sources of radiation were at first called "radio stars," whether the sources were actually stars or not, but are now usually called "radio sources.")

During World War II, while British scientists were developing radar, they discovered that the sun was interfering by sending out signals in the microwave region. This aroused their interest in radio astronomy, and after the war the British pursued their tuning-in on the sun. In 1950, the found that much of the sun's radio signals were associated with sun-

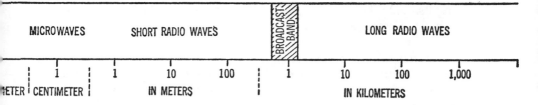

| MICROWAVES | SHORT RADIO WAVES | BROADCAST BAND | LONG RADIO WAVES |

| | 1 | | 1 | 10 | 100 | | 1 | 10 | 100 | 1,000 |
| ETER | CENTIMETER | | IN METERS | | | | | IN KILOMETERS | | |

spots. (Jansky had conducted his experiments during a period of minimal sunspot activity, which is why he detected the galactic radiation rather than that of the sun.)

The British pioneered in building large antennae and arrays of widely separated receivers (a technique first used in Australia) to sharpen reception and pinpoint radio stars. Their 250-foot dish at Jodrell Bank in England, built under the supervision of Sir Bernard Lovell, was the first really large radio telescope.

In 1947 the Australian astronomer John C. Bolton narrowed down the third strongest radio source in the sky, and it proved to be none other than the Crab Nebula. Of the 2,000 or so radio sources detected here and there in the sky, this was the first to be pinned down to an actual visible object. It seemed unlikely that a white dwarf was giving rise to the radiation, since other white dwarfs did not. The source was much more likely to be the cloud of expanding gas in the nebula.

This strengthened other evidence that cosmic radio signals arise primarily from turbulent gas. The turbulent gas of the outer atmosphere of the sun gives rise to radio waves, so that what is called the "radio sun" is much larger than the visible sun. Then, too, Jupiter, Saturn, and Venus, each with a turbulent atmosphere, have been found to be radio emitters. In the case of Jupiter, however, the radiation first detected in 1955, in records going back to 1950, seems somehow to be associated with a particular area, which moves so regularly that it can be used to determine Jupiter's period of rotation within a hundredth of a second. Does this mark an association with a portion of Jupiter's solid surface— a surface never seen under the obscuring clouds of a giant atmosphere? If so, why? And in 1964, it was reported that Jupiter's period of rotation had altered quite suddenly, though, to be sure, only very slightly. Again, why? It was also shown, in 1965, that Jupiter is most likely to emit a strong burst of radio waves when its satellite Io is in the first or third quarter (to the left or right of Jupiter as seen from Earth). Still again, why? So far, radio studies have raised more questions about the planets than they have answered, but there is nothing so stimulating to science and scientists as a good unanswered question.

Jansky, who started it all, was largely unappreciated in his lifetime and died in 1950 at the age of forty-four, just as radio astronomy was hitting its stride. He received posthumous recognition in that the strength of radio emission is now measured in "janskies."

Radio astronomy probed far out into space. Within our Galaxy, there is a strong radio source (the strongest outside the solar system) which is called "Cass" because it is located in Cassiopeia. Walter Baade and Rudolph Minkowski at Palomar trained the 200-inch telescope on the spot where this source was pinpointed by British radio telescopes, and they found streaks of turbulent gas. It is possible that these may be remnants of the supernova of 1604, which Kepler observed in Cassiopeia.

A still more distant discovery was made in 1951. The second strongest radio source lies in the constellation Cygnus. Reber first reported it in 1944. As radio telescopes later narrowed down its location, it began to appear that this radio source was outside our Galaxy—the first to be pinpointed beyond the Milky Way. Then, in 1951, Baade, studying the indicated portion of the sky with the 200-inch telescope, found an odd galaxy in the center of the field. It had a double center and seemed to be distorted. Baade at once suspected that this odd, distorted, double-centered galaxy was not one galaxy but two, joined broadside-to like a pair of clashing cymbals. Baade thought they were two colliding galaxies—a possibility which he had already discussed with other astronomers.

It took another year to settle the matter. The spectroscope showed absorption lines which could only be explained by supposing the dust and gas of the two galaxies to be coming into collision. The collision is now accepted to be a fact. Moreover, it seems likely that galactic collisions are fairly common, especially in dense clusters, where galaxies may be separated by distances not much more than their own diameters.

When two galaxies collide, the stars themselves are not likely to encounter one another: they are so widely spaced that one galaxy could pass through the other without any stars coming even close. But dust and gas clouds are stirred into vast turbulence and thereby generate very powerful radio radiation. The colliding galaxies in Cygnus are 260 million light-years away, yet their radio signals reaching us are stronger than those of the Crab Nebula, only 3,500 light-years away. By this token, we should be able to detect colliding galaxies at far greater distances than we can see with the optical telescope. The 250-foot Jodrell Bank radio telescope, for instance, should outrange the 200-inch Hale telescope.

And yet as the number of radio sources found among the distant galaxies increased and passed the hundred mark, astronomers grew uneasy. Surely they could not all be brought about by colliding galaxies. That would be overdoing a good thing.

In fact, the whole notion of galactic collisions in the sky grew shaky. The Soviet astrophysicist, Victor Amazaspovich Ambartsumian, advanced theoretical reasons in 1955 for supposing that radio galaxies were exploding rather than colliding. By the early 1960's, Fred Hoyle was backing this view and suggesting that radio galaxies might be subjected to whole series of supernovae. In the crowded center of a galactic nucleus, a supernova may explode and may heat a nearby star to just the point where it, too, lets go in a supernova explosion. The second explosion sets off a third and that sets off a fourth, and so on, domino fashion. In a sense, the whole center of a galaxy is exploding.

The possibility that this may be so has been greatly strengthened by the discovery, in 1963, that the galaxy M-82, in the constellation of Ursa Major (a strong radio source about 10 million light-years away), is such an "exploding galaxy."

Investigation of M-82 with the 200-inch Hale telescope, making use of the light of a particular wavelength, showed great jets of matter up to a thousand light-years long emerging from the galactic center. From the amount of matter exploding outward, the distance it had traveled, and its rate of travel, it seems likely that 1.5 million years ago the light of some 5 million stars exploding in the nucleus, almost simultaneously, first reached us.

The New Objects

By the time astronomers had entered the 1960's, it might have been easy for them to suppose that there were few surprises left among the physical objects in the heavens. New theories, new insights, yes; but surely little in the way of startling new varieties of stars, galaxies, or anything else could remain after three centuries of observation with steadily more sophisticated instruments.

If any astronomers thought this, they were due for an enormous shock—the first coming as a result of the investigation of certain radio sources that looked unusual but not surprising.

The radio sources first studied in deep space seemed to exist in connection with extended bodies of turbulent gas: the Crab Nebula, distant galaxies, and so on. There did exist a few radio sources, however, that seemed unusually small. As radio telescopes grew more refined and as the view of the radio sources was sharpened, it began to seem possible that radio waves were being emitted by individual stars.

Among these compact radio sources were several known as 3C48, 3C147, 3C196, 3C273, and 3C286. The "3C" is short for "Third Cambridge Catalog of Radio Stars," a listing compiled by the English astronomer Martin Ryle and his coworkers, while the remaining numbers represent the placing of the source on that list.

In 1960, the areas containing these compact radio sources were combed by Sandage with the 200-inch telescope, and in each case a star did indeed seem to be the source. The first star to be detected was that associated with 3C48. In the case of 3C273, the brightest of the objects, the precise position was obtained by Cyril Hazard, in Australia, who recorded the moment of radio blackout as the moon passed before it.

The stars involved had been recorded on previous photographic sweeps of the sky and had always been taken to be nothing more than faint members of our own galaxy. Painstaking photographing, spurred by their unusual radio-emission, now showed, however, that that was *not* all there was to it. Faint nebulosities proved to be associated with some of the objects, and 3C273 showed signs of a tiny jet of matter emerging from it. In fact, there were two radio sources in connection with 3C273: one from the star and one from the jet. Another point of interest that arose after close inspection was that these stars were unusually rich in ultraviolet light.

It would seem then that the compact radio-sources, although they looked like stars, might not be ordinary stars after all. They eventually came to be called "quasi-stellar sources" ("quasi-stellar" means "star-resembling"). As the term became more and more important to astronomers, quasi-stellar radio sources became too inconvenient a mouthful and in 1964 it was shortened by the Chinese-American physicist Hong Yee Chiu to "quasar" ("*quasi*-stell*ar*"), an uneuphonious word that is now firmly embedded in astronomic terminology.

Clearly, the quasars were interesting enough to warrant investigation with the full battery of astronomic techniques, and that meant spectroscopy. Such astronomers as Allen Sandage, Jesse L. Greenstein, and Maarten Schmidt, labored to obtain the spectra. When they accomplished the task in 1960, they found themselves with strange lines they could not identify. Furthermore, the lines in the spectra of one quasar did not match those in any other.

In 1963, Schmidt returned to the spectrum of 3C273, which, as the brightest of these puzzling objects, showed the clearest spectrum. Six lines were present, of which four were spaced in such a way as to seem to resemble a series of hydrogen lines—except that no such series ought to exist in the place in which they were found. What, though, if those lines were located elsewhere but were found where they were because they had been displaced toward the red end of the spectrum? If so, it was a large displacement, one that indicated a recession at the velocity of over 25,000 miles per second. This seemed unbelievable, and yet, if such a displacement existed, the other two lines could also be identified: one represented oxygen minus two electrons, the other magnesium minus two electrons.

Schmidt and Greenstein turned to the other quasar spectra and found that the lines there could also be identified, provided huge red shifts were assumed.

Such enormous red shifts could be brought about by the general expansion of the universe; but if the red shift was equated with distance in accordance with Hubble's law, it turned out that the quasars could not be ordinary stars of our own galaxy at all. They had to be among the most distant objects known—billions of light-years away.

By the end of the 1960's, a concentrated search had uncovered 150 quasars. The spectra of about 110 of them were studied. Every single one of these showed a large red shift—larger ones, indeed, than that of 3C273. The distance of a couple of them is estimated to be about 9 billion light-years.

If the quasars are indeed as far away as the red shift makes them seem, astronomers are faced with some puzzling and difficult points. For one thing, they must be extraordinarily luminous to appear as bright as they do at such a distance; they must be anywhere from thirty to a hundred times as luminous as an entire ordinary galaxy.

Yet if this is so, and if the quasars had the form and appearance of a galaxy, they ought to contain up to a hundred times as many stars as an ordinary galaxy and be up to five or six times as large in each dimension. Even at their enormous distances they ought to show up as distinct oval blotches of light in large telescopes. Yet they don't. They remain starlike points in even the largest telescope, which seems to indicate that, despite their unusual luminosity, they were far smaller in size than ordinary galaxies.

The smallness in size was accentuated by another phenomenon, for as early as 1963 the quasars were found to be variable in the energy they emitted, both in the visible-light region and in the radio-wave region. Increases and decreases of as much as three magnitudes were recorded over the space of a few years.

For radiation to vary so markedly in so short a time, a body must be small. Small variations might result from brightenings and dimmings in restricted regions of a body, but large variations must involve the body as a whole. If the body is involved as a whole, some effect must make itself felt across the full width of the body within the time of variation. But no effect can travel faster than light, so that if a quasar varies markedly over a period of a few years, it cannot be more than a light-year or so in diameter. Actually, some calculations indicate quasars may be as little as a light-week (500 billion miles) in diameter.

Bodies which are at once so small and so luminous must be expending energy at a rate so great that the reserves cannot last long (unless there is some energy-source as yet undreamed-of, which is not impossible, of course). Some calculations indicate that a quasar can only deliver energy at this enormous rate for a million years or so. In that case, the quasars we see only became quasars a short time ago, cosmically speaking, and there must be a number of objects that were once quasars but are quasars no longer.

Sandage, in 1965, announced the discovery of objects that may indeed be aged quasars. They seemed like ordinary bluish stars, but they possessed huge red shifts as quasars do. They were as distant, as luminous, as small as quasars, but they lacked the radio-wave emission. Sandage called them "blue stellar objects," which can be abbreviated to BSO's.

The BSO's seem to be more numerous than quasars: a 1967 estimate places the total number of BSO's within reach of our telescopes at 100,000. There are so many more BSO's than quasars because the bodies last so much longer in BSO form than in quasar form.

The chief interest in the quasars (aside from the knotty puzzle of what they actually are) lies in the fact that they are at once so unusual and so distant. Perhaps they represent a kind of body that existed only in the youth of the universe. (After all, a body that is 9 billion light-years away is seen only by light that left it 9 billion years ago, and that is a time that may have been only shortly after the explosion of the "cosmic egg.") If so, then it is clear that the appearance of the universe was radically different billions of years ago. The universe, in that case, evolved, as the proponents of the "big bang" theory maintain, and is not eternally changeless-on-the-average, as the proponents of "continuous creation" insist.

The use of the quasars as evidence in favor of the "big bang" did not go unchallenged. A number of astronomers advance evidence to the effect that the quasars are not really very distant and therefore cannot be taken to represent objects characteristic of the youth of the universe. Astronomers holding this view must then explain the enormous red-shifts in the quasar spectra by some effect other than vast distance, and this is not easy to do. On the whole, though the question is far from settled and there are enormous difficulties involved in both views, the weight of opinion seems to be on the side of quasars as very distant objects.

Even if they are, there seems to be reason to question whether they are entirely characteristic only of the youth of the universe. Are they distributed fairly evenly, so that they are to be found in the universe at all ages?

Thus, back in 1943, the American astronomer Carl Seyfert observed an odd galaxy, one with a very bright and very small nucleus. Others of the sort have since been observed, and the entire group is now referred to as "Seyfert galaxies." Though only a dozen were known by the end of the 1960's, there is reason to suspect that as many as 1 per cent of all galaxies may be of the Seyfert type.

Can it be that Seyfert galaxies are objects intermediate between ordinary galaxies and quasars? Their bright centers show light-variations that would make those centers almost as small as quasars. If the centers were further intensified and the rest of the galaxy further dimmed, they would become indistinguishable from a quasar and one Seyfert galaxy, 3C120, is almost quasarlike in appearance.

The Seyfert galaxies have only moderate red-shifts and are not enormously distant. Can it be that the quasars are very distant Seyfert galaxies, so distant that we can only see the luminous and small centers; and so distant that we can only see the largest, so that we get the impression that quasars are extraordinarily luminous, whereas we should rightly suspect that only a few very large Seyfert galaxies make up the quasars we can see despite their distance?

But if we consider the Seyfert galaxies near us to be either small quasars or, perhaps, large quasars in the course of development, then it may be that the quasar distribution is not characteristic only of the youth of the universe and that their existence is not strong evidence, after all, of the "big bang" theory.

As it happened, however, the "big bang" received strong support in another direction. In 1949, Gamow had calculated that the radiation associated with the "big bang" should have died down with the expansion of the universe to the point where it would now consist of radio-wave radiation coming equally from all parts of the sky as a kind of radio-background. He suggested that the radiation would be that to be expected of objects at a temperature of 5° K. (that is, 5° above absolute zero or – 268° C.).

In 1965, just such a background radio-wave radiation was detected and reported by A. A. Penzias and R. W. Wilson of Bell Telephone Laboratories in New Jersey. The temperature associated with the radiation was 3° K., which was in not-too-bad agreement with Gamow's prediction. No explanation, other than that of the "big bang," has as yet been offered for the existence of this background radiation so that for the moment the "big bang" theory of the evolutionary universe beginning with the explosion of a large volume of condensed matter, seems to hold the field.

If radio-wave radiation had given rise to that peculiar and puzzling astronomical body, the quasar, research at the other end of the spectrum suggested another body—just as peculiar, if not quite as puzzling.

In 1958, the American astrophysicist Herbert Friedman discovered that the sun produced a considerable quantity of X-rays. These could not be detected from the earth's surface for the atmosphere absorbed them; but rockets, shooting beyond the atmosphere and carrying appropriate instruments, could detect the radiation with ease.

For a while, the source of solar X-rays was a puzzle. The temperature of the sun's surface is only 6,000° C.—high enough to vaporize any form of matter, but not high enough to produce X-rays. The source had to lie in the sun's corona, a tenuous halo of gases stretching outward from the sun in all directions for many millions of miles. Although the corona delivers fully half as much light as the full moon, it is completely masked by the light of the sun itself and is visible only during eclipses, at least under ordinary circumstances. In 1930, the French astronomer Bernard Ferdinand Lyot invented a telescope which, at high altitudes and on clear days, could observe the inner corona even in the absence of an eclipse.

The corona was felt to be the X-ray source because, even before the rocket studies of X-rays, it had been suspected of possessing unusually high temperatures. Studies of the spectrum of the corona (during eclipses) had revealed lines that could not be associated with any known element. A new element was suspected and named "coronium." In 1941, however, it was found that the lines of coronium could be produced by iron atoms that had had many subatomic particles broken away from them. To break off all those particles, however, required a temperature of something like a million degrees, and such a temperature would certainly be enough to produce X-rays.

X-ray radiation increases sharply when a solar flare erupts into the corona. The X-ray intensity at that time implies a temperature as high as 100 million degrees in the corona above the flare. The reason for such enormous temperatures in the thin gas of the corona is still a matter of controversy. (Temperature here has to be distinguished from heat. The temperature is a measure of the kinetic energy of the atoms or particles in the gas, but since the particles are few, the actual heat content per unit of volume is low. The X-rays are produced by collisions between the extremely energetic particles.)

X-rays came from beyond the solar system, too. In 1963, rocket-borne instruments were launched by Bruno Rossi and others to see if solar X-rays were reflected from the moon's surface. They detected, instead, two particularly concentrated X-ray sources elsewhere in the sky. The weaker ("Tau X-1" because it was in the constellation Taurus) was quickly associated with the Crab Nebula. In 1966, the stronger, in the constellation Scorpio ("Sco X-1") was found to be associated with an optical object which seemed the remnant (like the Crab Nebula) of an old nova. Since then several dozen other, and weaker, X-ray sources have been detected in the sky.

To be giving off energetic X-rays with an intensity sufficient to be detected across an interstellar gap required a source of extremely high temperature and large mass. The concentration of X-rays emitted by the sun's corona would not do at all.

To be at once massive and have a temperature of a million degrees suggested a kind of "superwhite dwarf." As long ago as 1934, Zwicky had suggested that the subatomic particles of a white dwarf might, under certain conditions, combine into unchanged particles called "neutrons." These could then be forced together until actual contact was made. The result would be a sphere no more than ten miles across which would yet retain the mass of a full-sized star. In 1939, the properties of such a "neutron star" were worked out in some detail by the American physicist J. Robert Oppenheimer. Such an object would attain so high a surface temperature, at least in the initial stages after its formation, as to emit X-rays in profusion.

The search by Friedman for actual evidence of the existence of such "neutron stars" centered on the Crab Nebula, where it was felt that the terrific explosion that had formed it might have left behind, not a condensed white dwarf, but a supercondensed neutron star. In July 1964, the moon passed across the Crab Nebula and a rocket was sent beyond the atmosphere to record the X-ray emission. If it were coming from a neutron star, then the X-ray emission would be cut off entirely and at once as the moon passed before the tiny object. If the X-ray emission were from the Crab Nebula generally, then it would drop off gradually as the moon eclipsed the nebula bit by bit. The latter proved to be the case, and the Crab Nebula seemed to be but a larger and much more intense corona, about a light-year in diameter.

For a moment, the possibility that neutron stars might actually exist and be detectable dwindled, but in the same year that the Crab Nebula failed its test, a new discovery was made in another direction. The radio waves from certain sources seemed to indicate a very rapid fluctuation in intensity. It was as though there were "radio twinkles" here and there.

Astronomers quickly designed instruments capable of catching very short bursts of radio-wave radiation. They felt this would make it possible to study these fast changes in greater detail. One astronomer making use of such a radio telescope was Anthony Hewish at Cambridge University Observatory.

He had hardly begun operating the telescope with its new detector when he detected bursts of radio-wave energy from a place midway between Vega and Altair. It was not difficult to detect and would have been found years earlier if astronomers had expected to find quite such short bursts and had developed the equipment to detect them. The bursts were, as it happened, astonishingly brief, lasting only 1/30 of a second. What was even more astonishing, the bursts followed one another with remarkable regularity at intervals of 1 1/3 seconds. The intervals were so regular, in fact, that the period could be worked out to a hundred-millionth of a second: it was 1.33730109 seconds.

Naturally, there was no way of telling what these pulses represented, at least not at first. Hewish could only think of it as a "pulsating star," each pulsation giving out a burst of energy. This was shortened almost at once to "pulsar," and it is by that name that the new object came to be known.

One should speak of the new objects in the plural, for once Hewish found the first, he searched for others. By February 1968, when he announced the discovery, he had located four. Other astronomers avidly began searching, and more were quickly discovered. In two more years, nearly forty more pulsars were located.

Two-thirds of them are located very close to the galactic equator, which is a good sign that pulsars generally are part of our own galaxy. Some may be as close as a hundred light-years or so. (There is no reason to suppose they don't exist in other galaxies, too, but at the distance of other galaxies they are probably too faint to detect.)

All the pulsars are characterized by extreme regularity of pulsation, but the exact period varies from pulsar to pulsar. One had a period as long as 3.7 seconds. In November 1968, astronomers at Green Bank, West Virginia, detected a pulsar in the Crab Nebula that had a period of only 0.033089 seconds. It was pulsing thirty times a second.

Naturally, the question is, What can produce such short flashes with such fantastic regularity? Some astronomical body must be undergoing some very regular change at intervals rapid enough to produce the pulses. Could it be a planet that was circling a star in such a way that once each revolution it moved beyond the star (as seen from the direction of earth) and, as it emerged, emitted a powerful flash of radio waves? Or else could a planet be rotating and, each time it did so, would some particular spot on its surface, which leaked radio waves in vast quantity, sweep past our direction?

To do this, however, a planet must revolve about a star or rotate about its axis in a period of seconds or fractions of a second, and this was unthinkable. For pulses as rapid as those of pulsars, some object must be rotating or revolving at enormous velocities. That requires very small size combined with huge temperatures, or huge gravitational fields, or both.

This instantly brought white dwarfs to mind, but even white dwarfs could not revolve about each other, or rotate on their axes, or pulsate, with a period short enough to account for pulsars. White dwarfs were still too large, and their gravitational fields were still too weak.

Thomas Gold at once suggested that a neutron star was involved. He pointed out that a neutron star was small enough and dense enough to be able to rotate about its axis in four seconds or less. What's more, it had already been theorized that a neutron star would have an enormously intense magnetic field, with magnetic poles that need not be at the pole of rotation. Electrons would be held so tightly by the neutron star's gravity that they could emerge only at the magnetic poles. As they were thrown off, they would lose energy, in the form of radio waves. This would mean that there would be a steady sheaf of radio waves emerging from two opposite points on the neutron star's surface.

If, as the neutron star rotates, one or both of those sheafs of radio waves sweeps past our direction, then we will detect a short burst of radio-wave energy once or twice each revolution. If this is so, we would detect only pulsars which happen to rotate in such a way as to sweep at least one of the magnetic poles in our direction. Some astronomers estimate that only one neutron star out of a hundred would do so. They guess that there might be as many as 100,000 neutron stars in the galaxy, but that only 1000 would be detectable from earth.

Gold went on to point out that if his theory were correct, the neutron star would be leaking energy at the magnetic poles and its rate of rotation would be slowing down. This means that the shorter the period of a pulsar, the younger it is and the more rapidly it would be losing energy and slowing down.

The most rapid pulsar known is in the Crab Nebula. It might well be the youngest, since the supernova explosion that would have left the neutron star behind took place only a thousand years ago.

The period of the Crab Nebula pulsar was studied carefully, and it was indeed found to be slowing, just as Gold had predicted. The period was increasing by 36.48 billionths of a second each day. The same phenomenon was discovered in other pulsars as well, and as the 1970's opened, the neutron-star hypothesis was widely accepted.

Sometimes a pulsar will suddenly speed up its period very slightly, then resume the slowing trend. Some astronomers suspect this may be the result of a "starquake," a shifting of mass distribution within the neutron star. Or perhaps it might be the result of some sizable body plunging into the neutron star and adding its own momentum to that of the star.

There was, of course, no reason why the electrons emerging from the neutron star should lose energy only as microwaves. This phenomenon should produce waves all along the spectrum. It should produce visible light, too.

Keen attention was focused on the sections of the Crab Nebula where visible remnants of the old explosion might exist. Sure enough, in January 1969, it was noted that the light of a dim star within the Nebula *did* flash on and off in precise time with the microwave pulses. It would have been detected earlier if astronomers had had the slightest idea that they ought to search for such rapid alternations of light and darkness. The Crab Nebula pulsar was the first optical pulsar discovered —the first visible neutron-star.

The Crab Nebula pulsar released X-rays, too. About 5 per cent of all the X-rays from the Crab Nebula emerged from that tiny flickering light. The connection between X-rays and neutron stars, which seemed extinguished in 1964, thus came triumphantly back to life.

Nor is even the neutron star the limit. When Oppenheimer worked out the properties of the neutron star in 1939, he predicted also that it was possible for a star that was massive enough and cool enough, to collapse altogether to nothingness. When such collapse proceeded past the neutron-star stage, the gravitational field would become so intense that no matter, no light could escape from it. Nothing could be seen of it; it would simply be a "black hole" in space.

Will it be possible at some time in the future to detect such black holes—surely the ultimate in strange new objects in the universe? That remains to be seen.

Are quasars large clusters of neutron stars? Are they single neutron-stars of galactic mass? Are they phenomena associated with black-hole formation? That, too, remains to be seen.

But if there are objects in the universe that surprise us, there are also surprises in the vast not-so-empty spaces between the stars. The non-emptiness of "empty space" has proven to be a matter of difficulty for astronomers in observations relatively close to home.

In a sense, the galaxy hardest for us to see is our own. For one thing, we are imprisoned within it, while the other galaxies can be viewed as a whole from outside. It is like the difference between trying to view a city from the roof of a low building and seeing it from an airplane. Furthermore, we are far out from the center and, to make matters worse, we lie in a spiral arm clogged with dust. In other words, we are on a low roof on the outskirts of the city on a foggy day.

The space between stars, generally speaking, is not a perfect vacuum under the best of conditions. There is a thin gas spread generally through interstellar space within galaxies. Spectral absorption lines due to such "interstellar gas" were first detected in 1904 by the German astronomer Johannes Franz Hartmann. That would be supportable. The trouble is, however, that in the outskirts of a galaxy, the concentration of gas and dust becomes much thicker. We can see such dark fogs of dust rimming the nearer galaxies.

We can actually "see" the dust clouds, in a negative way, within our own Galaxy as dark areas in the Milky Way. Examples are the dark Horsehead Nebula, outlined starkly against the surrounding brilliance of millions of stars, and the even more dramatically named Coal Sack in the Southern Cross, a region of scattered dust particles 30 light-years in diameter and about 400 light-years away from us.

Although the gas and dust clouds hide the spiral arms of the Galaxy from direct vision, they do not hide the structure of the arms from the spectroscope. Hydrogen atoms in the clouds are ionized (broken up into electrically charged subatomic particles) by the energetic radiation from the bright Population I stars in the arms. Beginning in 1951, streaks of ionized hydrogen were found by the American astronomer William Wilson Morgan, marking out the lines of the blue giants, i.e., the spiral arms. Their spectra were similar to the spectra shown by the spiral arms of the Andromeda galaxy.

The nearest such streak of ionized hydrogen includes the blue giants in the constellation of Orion, and this streak is therefore called the

The sun's corona.

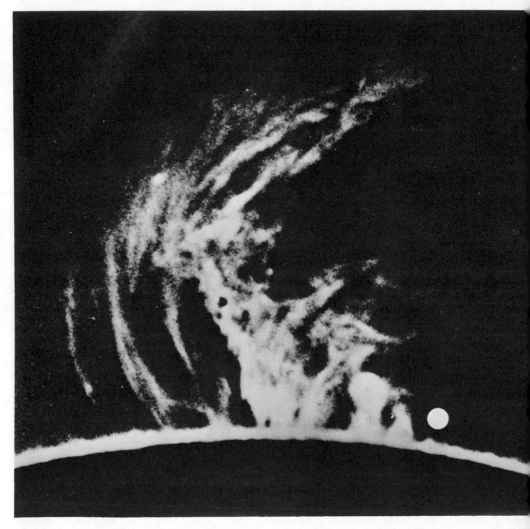

A solar flare, stretching 140,000 miles from the sun, photographed in the light of calcium. The white circle represents the earth on a similar scale.

Daniel's Comet, photographed July 17, 1907. The streaks are stars.

The Coal Sack, a huge cloud of dust and gas in the Southern Cross.

"Orion Arm." Our solar system is in that arm. Two other arms were located in the same way. One lies farther out from the galactic center than our own and includes giant stars in the constellation Perseus (the "Perseus Arm"). The other lies closer to the galactic center and contains bright clouds in the constellation Sagittarius (the "Sagittarius Arm"). Each arm seems to be about 10,000 light-years long.

Then radio came along as a still more powerful tool. Not only could it pierce through the obscuring clouds, but it made the clouds themselves tell their story—through their own voice. This came about as a result of the work of the Dutch astronomer H. C. Van de Hulst. In 1944, the Netherlands was ground under the heavy boot of the Nazi army, and astronomic observation was nearly impossible. Van de Hulst, confining himself to pen and paper work, studied the characteristics of ordinary nonionized hydrogen atoms, of which most of the interstellar gas is composed.

He suggested that every once in a while such atoms, on colliding, might change their energy state and, in so doing, emit a weak radiation in the radio part of the spectrum. A particular hydrogen atom might do so only once in 11 million years, but among the vast numbers present in intergalactic space, enough would be radiating each moment to produce a continuously detectable emission.

Van de Hulst calculated that the wavelength of the radiation should be 21 centimeters. Sure enough, with the development of new radio techniques after the War, this "song of hydrogen" was detected in 1951 by Edward Mills Purcell and Harold Irving Ewen at Harvard University.

By tuning in on the 21-centimeter radiation of collections of hydrogen, astronomers were able to trace out the spiral arms and follow them for long distances—in most cases nearly all the way around the Galaxy. More arms were found, and maps of the concentration of hydrogen show half a dozen or more streaks.

What is more, the song of hydrogen told something about its movements. Like all waves, this radiation is subject to the Doppler-Fizeau effect. It allows astronomers to measure the velocity of the moving hydrogen clouds, and thereby to explore, among other things, the rotation of our Galaxy. This new technique confirmed that the Galaxy rotates in a period (at our distance from the center) of 200 million years.

In science, each new discovery unlocks doors leading to new mysteries. And the greatest progress comes from the unexpected—the discovery that overthrows previous notions. An interesting example at the moment is a puzzling phenomenon brought to light by radio study of a concentration of hydrogen at the center of our Galaxy. The hydrogen seems to be expanding, yet is confined to the equatorial plane of the Galaxy. The expansion itself is surprising, because there is no theory to account for it. And if the hydrogen is expanding, why has it not all dissipated away during the long lifetime of the Galaxy? Is it a sign perhaps that, some 10 million years ago, as Oort suspects, its center exploded, as that of M-82 did much more recently? Then too, the plane of hydrogen is not perfectly flat. It bends downward on one end of the Galaxy and upward on the other. Why? No good explanation has yet been offered.

Hydrogen is not, or should not, be unique as far as radio waves are concerned. Every different atom, or combination of atoms, is capable of emitting characteristic radio-wave radiation, or of absorbing characteristic radio-wave radiation from a general background. Naturally, then, astronomers sought to find the telltale fingerprints of atoms other than the supremely common hydrogen.

Almost all the hydrogen that occurs in nature is of a particularly simple variety called "hydrogen-1." There is a more complex form, which is "deuterium" or "hydrogen-2." The radio-wave radiation from various spots in the sky were combed for the wavelengths that theory predicted. In 1966, it was detected, and the indications are that the quantity of hydrogen-2 in the universe is about 5 per cent that of hydrogen-1.

Next to the varieties of hydrogen as common components of the universe are helium and oxygen. An oxygen atom can combine with a hydrogen atom to form a "hydroxyl group." This combination would not be stable on earth, for the hydroxyl group is very active and would combine with almost any other atom or molecule it encountered. It would, notably, combine with a second hydrogen atom to form a molecule of water. In interstellar space, however, where the atoms are spread so thin that collisions are few and far between, a hydroxyl group, once formed, would persist undisturbed for long periods of time. This was pointed out in 1953 by the Soviet astronomer I. S. Shklovskii.

Such a hydroxyl group would, calculations showed, emit or absorb four particular wavelengths of radio waves. In October 1963, two of them were detected by a team of radio engineers at Lincoln Laboratory of M.I.T.

Since the hydroxyl group is some seventeen times as massive as the hydrogen atom alone, it is more sluggish and moves at only one-fourth the velocity of the hydrogen atom at any given temperature. In general, movement blurs the wavelengths so that the hydroxyl wavelengths are sharper than those of hydrogen. Its shifts are easier to determine, and it is easier to tell whether a gas cloud, containing hydroxyl, is approaching or receding.

Astronomers were pleased, but not entirely astonished, at finding evidence of a two-atom combination in the vast reaches between the stars. Automatically, they began to search for other combinations, but not with a great deal of hope. Atoms are spread out so thin in interstellar space that the chance of more than two atoms coming together long enough to form a combination seemed remote. The chance that atoms less common than oxygen (such as those of carbon and nitrogen, which are next most common of those that are able to form combinations) would be involved seemed out of the question.

But then, beginning in 1968, came the real surprises. In November of that year, they discovered the telltale radio-wave fingerprints of water molecules (H_2O). Those molecules were made up of two hydrogen atoms and an oxygen atom—three atoms altogether. In the same month, even more astonishingly, ammonia molecules (NH_3) were detected. These were composed of four-atom combinations: three atoms of hydrogen and one of nitrogen.

In 1969, another four-atom combination, including a carbon atom, was detected. This was formaldehyde (H_2CO).

In 1970, a number of new discoveries were made, including the presence of a five-atom molecule, cyanoacetylene, which contained a chain of three carbon atoms ($HCCCN$). And then, as a climax (at least for that year), came methyl alcohol, a molecule of six atoms (CH_3OH).

Astronomers found themselves with a totally new, and quite unexpected, subdivision of the science before them: "astrochemistry."

How those atoms come together to form molecules so complicated, and how such molecules manage to remain in being despite the flood of hard radiation from the stars, which ordinarily might be expected to smash them apart, astronomers can't say. Presumably, these molecules are formed under conditions that are not quite as empty as we assumed interstellar space to be—perhaps in regions where dust clouds are thickening toward star-formation.

If so, still more complicated molecules may be detected, and their presence may revolutionize our views on the formation of planets and on the development of life on those planets. Astronomers are combing the radio-wave radiation bands avidly for additional and different molecular traces.

CHAPTER 3

The Earth

Birth of the Solar System

However glorious the unimaginable depths of the universe and however puny the earth in comparison, it is on the earth that we live and to the earth that we must return.

By the time of Newton, it had become possible to speculate intelligently about the creation of the earth and the solar system as a separate problem from the creation of the universe as a whole. The picture of the solar system showed it to be a structure with certain unifying characteristics.

1. All the major planets circle the sun in approximately the plane of the sun's equator. In other words, if you were to prepare a three-dimensional model of the sun and its planets, you would find it could be made to fit into a very shallow cakepan.

2. All the major planets circle the sun in the same direction—counterclockwise if you were to look down on the solar system from the direction of the North Star.

3. Each major planet (with some exceptions) rotates around its axis in the same counterclockwise sense as its revolution around the sun, and the sun itself also rotates counterclockwise.

4. The planets are spaced at smoothly increasing distances from the sun and have nearly circular orbits.

5. All the satellites, with minor exceptions, revolve about their respective planets in nearly circular orbits in the plane of the planetary equator and in a counterclockwise direction.

The general regularity of this picture naturally suggested that some single process had created the whole system.

What, then, is the process that produced the solar system? All the theories so far proposed fall into two classes: catastrophic and evolutionary. The catastrophic view is that the sun was created in single blessedness and gained a family as the result of some violent event. The evolutionary ideas hold that the whole system came into being in an orderly way.

In the eighteenth century, when scientists were still under the spell of the Biblical stories of such great events as the Flood, it was fashionable to assume that the history of the earth was full of violent catastrophes. Why not one supercatastrophe to start the whole thing going? One popular theory was the proposal of the French naturalist Georges Louis Leclerc de Buffon that the solar system had been created out of the debris resulting from a collision between the sun and a comet. Buffon's theory collapsed, however, when it was discovered that comets were only wisps of extremely thin dust.

In the nineteenth century, as such concepts of long-drawn-out natural processes as Hutton's uniformitarian principle (see page 000) won favor, catastrophes went out of fashion. Instead, scientists turned more and more to theories involving evolutionary processes, following Newton rather than the Bible.

Newton himself had suggested that the solar system might have been formed from a thin cloud of gas and dust that slowly condensed under gravitational attraction. As the particles came together, the gravitational field would become more intense, the condensation would be hastened, and finally the whole mass would collapse into a dense body (the sun), made incandescent by the energy of the contraction.

In essence, this is the basis of the most popular theories of the origin of the solar system today. But a great many thorny problems had to be solved to answer specific questions. How, for instance, could a highly dispersed gas be brought together by the extremely weak force of gravitation? In recent years, scientists have proposed another plausible mechanism—the pressure of light. Now particles in space are bombarded by radiation from all sides, but, if two particles come close enough together to shade each other, they will be under less radiation pressure on the shaded than on the unshaded sides. The difference in pressure will tend to push them toward each other. As they come closer, gravitational attraction will accelerate their meeting.

If this is the way the sun was created, what about the planets? Where did they come from? The first attempts to answer this were put forward by Immanuel Kant in 1755 and independently by the French astronomer and mathematician Pierre Simon de Laplace in 1796. Laplace's picture was the more detailed.

As Laplace described it, the vast, contracting cloud of matter was rotating to start with. As it contracted, the speed of its rotation increased, just as a skater spins faster when he pulls in his arms. (This is due to the

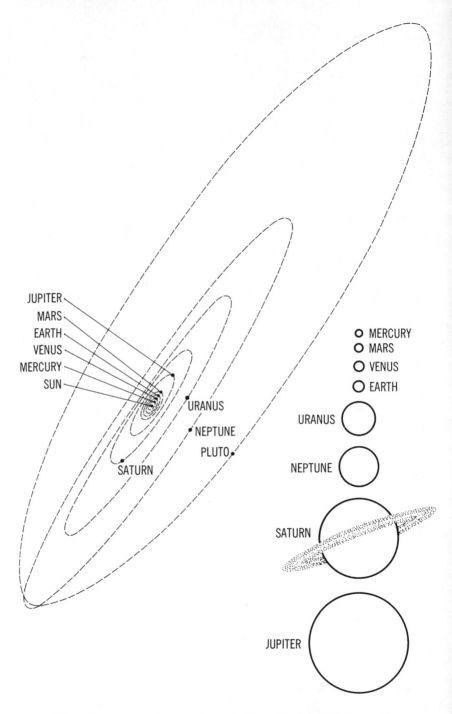

The solar system, drawn schematically, with an indication of
the hierarchy of planets according to relative size.

"conservation of angular momentum." Since angular momentum is equal to the speed of motion times the distance from the center of rotation, when the distance from the center decreases the speed of motion increases in compensation.) And as the rotating cloud speeded up, according to Laplace, it began to throw off a ring of material from its rapidly rotating equator. This removed some of the angular momentum, so that the remaining cloud slowed down, but, as it contracted further, it again reached a speed at which it threw off another ring of matter. So the coalescing sun left behind a series of rings—doughnut-shaped clouds of matter. These rings, Laplace suggested, slowly condensed to form the planets, and along the way they themselves threw off small rings that formed their satellites.

Laplace's "nebular hypothesis" seemed to fit the main features of the solar system very well—and even some of its details. For instance, the rings of Saturn might be satellite rings that had failed to coagulate. (Put all together, they would indeed form a satellite of respectable size.) Similarly, the asteroids, circling around the sun in a belt between Mars and Jupiter, might be products of sections of a ring which had not united to form a planet. And when Helmholtz and Kelvin worked up theories attributing the sun's energy to its slow contraction (see page 45), that, too, seemed to fit right in with Laplace's picture.

The nebular hypothesis held the field through most of the nineteenth century. But fatal flaws began to appear well before its end. In 1859, James Clerk Maxwell, analyzing Saturn's rings mathematically, showed that a ring of gaseous matter thrown off by any body could only condense to a collection of small particles like the rings of Saturn; it would never form a solid body, because gravitational forces would pull the ring apart before such a condensation materialized.

The problem of angular momentum also arose. It turned out that the planets, making up only a little more than 0.1 per cent of the mass of the whole solar system, carried 98 per cent of its total angular momentum! Jupiter alone possesses 60 per cent of all the angular momentum of the solar system. The sun, then, retained only a tiny fraction of the angular momentum of the original cloud. How did almost all of the angular momentum get shoved into the small rings split off the nebula? The problem is all the more puzzling since, in the case of Jupiter and Saturn which have satellite systems that seem like miniature solar systems and have, presumably, been formed in the same way, the central planetary body retains most of the angular momentum.

By 1900, the nebular hypothesis was so dead that the idea of any evolutionary process at all seemed discredited. The stage was set for the revival of a catastrophic theory. In 1905, two American scientists, Thomas Chrowder Chamberlin and Forest Ray Moulton, proposed a new one, this time explaining the planets as the result of a near collision between our sun and another star. The encounter pulled gaseous matter out of both suns, and the clouds of material left in the vicinity of our sun afterward condensed into small "planetesimals" and these into planets. This is the "planetesimal hypothesis." As for the problem of angular momentum, the British scientists James Hopwood Jeans and Harold Jeffreys proposed, in 1918, a "tidal hypothesis," suggesting that the passing sun's gravitational attraction had given the dragged-out masses of gas a kind of sidewise yank (put "English" on them, so to speak) and thus imparted angular momentum to them. If such a catastrophic theory were true, then planetary systems would have to be extremely scarce. Stars are so widely spaced that stellar collisions are 10,000 times less common than are supernovae, which are themselves not common. It is estimated that in the lifetime of the galaxy, there has been time for only ten encounters of the type that would produce solar systems by this theory.

However, these initial attempts at designing catastrophes failed when put to the test of mathematical analysis. Russell showed that in any such near collision the planets would have to end up thousands of times as far from the sun as they actually are. Furthermore, attempts to patch up the theory by imagining a variety of actual collisions, rather than near misses, had little success. During the 1930's, Lyttleton speculated about the possibility of a three-star collision, and later Hoyle had suggested that the sun had had a companion that had gone supernova and left planets as a legacy. In 1939, however, the American astronomer Lyman Spitzer showed that any material ejected from the sun under any circumstances would be so hot that it would not condense into planetesimals, but would merely expand into a thin gas. That seemed to end all thought of catastrophe (although, in 1965, a British astronomer, M. M. Woolfson, got around this by suggesting that the sun may have drawn its planetary material from a very diffuse, cool star, so that extreme temperatures need not be involved).

And so, after the planetesimal theory had come to a dead end, astronomers returned to the evolutionary idea and took another look at Laplace's nebular hypothesis.

By that time, their view of the universe had expanded enormously. They now had to account for the formation of galaxies. This called, of course, for much bigger clouds of gas and dust than Laplace had envisaged as the parent of the solar system. And it now appeared that such vast collections of matter would experience turbulence and would break up into eddies, each of which could condense into a separate system.

In 1944, the German astronomer Carl F. von Weizsäcker made a thorough analysis of this idea. He calculated that the largest eddies would contain enough matter to form galaxies. During the turbulent contraction of such an eddy, subeddies would develop. Each subeddy would be large enough to give birth to a solar system (with one or more suns).

On the outskirts of the solar eddy itself, subsubeddies might give rise to planets. This would take place at junctions where subsubeddies met, moving against each other like meshing gears; at such places dust particles would collide and coalesce. As a result of these collisions, first planetesimals and then planets would form.

The Weizsäcker theory, in itself, did not solve the matter of the angular momentum of the planets any more than the much simpler Laplacian version did. The Swedish astrophysicist Hannes Alfven took into account the magnetic field of the sun. As the young sun whirled rapidly, its magnetic field acted as a brake, slowing it up, and the angular momentum was passed on to the planets. Hoyle elaborated on this notion so that the Weizsäcker theory, modified to include magnetic as well as gravitational forces, seems the best one yet to account for the origin of the solar system.

There remain irregularities in the solar system, however, that no overall theory for the general formation can easily account for, and which would probably require subtheories, so to speak. For instance, there are the comets: small bodies in vastly elongated orbits that circle the sun in periods of dozens, hundreds, even thousands of years. Their orbits are completely unplanetlike; they enter the inner solar system from all angles; and they are made up in part of light, low-melting substances, which vaporize and stream away as the temperature goes up when they approach the neighborhood of the sun.

In 1950, Oort suggested the existence of a vast shell of small, icy bodies, slowly circling the sun at the distance of a light-year or more. There might be as many as 100 billion of them altogether—material left over from the original cloud of dust and gas that condensed to form the solar system, material too far out to be effectively captured by gravitational forces and remaining as an outermost shell, not drawn inward.

On the whole, they would remain undisturbed in their orbit. But every once in a while, some chance combination of gravitational pulls from nearby stars, might slow up one body or another sufficiently to allow it to fall toward the inner solar system, move around the sun, and go shooting out to the cloud. In doing so, they approach from all possible directions. If they pass near one of the large outer planets, the planetary gravitational pull may further alter their orbits to keep them permanently within the bounds of the planetary system. Once within the planetary system, the heating and vaporizing effect of the sun breaks up their substance in a very short time, geologically speaking. However, there are many more where they come from, since Oort estimates that only 20 per cent of the total supply of cometary bodies have been sent hurtling in toward the sun in all the billions of years of the existence of the solar system.

A second irregularity is that represented by the planetoids, a group of tens of thousands of tiny planetary bodies (the largest is less than five hundred miles in diameter; the smallest, under a mile) that, for the most part, are to be found between the orbits of Mars and Jupiter. If the spacing of the planets were absolutely regular, astronomers would expect to find a planet about where the largest of the planetoids is. Did a planet once actually exist there? Did it explode for some reason, sending fragments scattering? Were there subexplosions, which would account for some planetoids' having elongated orbits, while others have unusually tilted ones (though all rotate more or less counterclockwise)? Or is it that, thanks to the overriding effect of the gravitational field of nearby giant Jupiter, that the cloud in the region between the orbits of Mars and Jupiter, coalesced into planetesimals but never into a single planet? The problem of the origin of the planetoids remains open.

Pluto, the outermost planet, first discovered in 1930 by the American astronomer Clyde William Tombaugh, is another problem. The other outer planets—Jupiter, Saturn, Uranus, and Neptune—are large, gaseous, speedily rotating giants; Pluto is small, dense, and rotates once in 6.4 days. Furthermore, its orbit is elongated more than any other planet, and it is tipped at a greater angle to the general plane of revolution than any other planet. Its orbit is so elongated that when it approaches that part which is closest to the sun, it is actually closer, for twenty years or so, than Neptune can ever approach.

Some astronomers wonder if Pluto might once have been a satellite of Neptune. It is a little large for the role, but that hypothesis would account for its slow rotation, since 6.4 days might have been the time of its revolution about Neptune—a revolution equal to rotation, as in the case of our moon. Perhaps a vast cosmic accident freed Pluto of Neptune's grip and sent it hurtling into an elongated orbit. The same accident may have forced Triton, Neptune's large satellite, to circle in an orbit far removed from that of Neptune's equator, and moved Neptune closer to the sun, for its orbit ought to be distinctly farther out if the evenly increasing separation of successive planets were to be obeyed. Unfortunately, astronomers haven't the faintest notion of the kind of cosmic accident that could have resulted in all this.

The rotation of the planets also offers its problems. Ideally, all the planets ought to rotate in a counterclockwise direction (as viewed from a point high above the earth's North Pole) with their axes of rotation perpendicular to the plane of their revolution about the sun. This is reasonably true for the sun itself and for Jupiter, the two major bodies of the solar system, but there is a puzzling variation in the others whose plane of rotation we can measure.

The earth's axis is tipped about 23½° to the vertical, while the axes of Mars, Saturn, and Neptune, are tipped by 25°, 27°, and 29° respectively. Uranus represents an even more extreme case, for its axis is tipped by 98°—or a little more than a right angle—so that in effect its axis is lined up with its plane of rotation and it rolls along its orbit like a top rolling on its side, instead of standing upright (or leaning a little) on its peg. Uranus has five small satellites whose orbits are all tipped with the planet's axis so they remain in Uranus' equatorial plane.

What has tipped so many of the planets, and what has tipped Uranus so drastically, is still a puzzle, yet not nearly as much a puzzle as that posed by the planet Venus. For a long time now, astronomers have understood that when a small body circles a larger one, tidal pulls slow the rotation of the small body until it faces one side constantly to the large one, rotating exactly once every revolution. This is true of the moon with respect to the earth, for instance, so that it rotates about its axis (and revolves about the Earth) once in 29½ days.

It was thought very likely that Mercury and Venus, so close to the sun, were likewise slowed, and that both planets rotated once per revolution—Mercury in 88 days and Venus in 225 days. Mercury is hard to observe because it is small, distant, and extraordinarily close to the masking light of the sun. However, as long ago as the 1880's, the Italian astronomer Giovanni Virginio Schiaparelli noted vague markings on Mercury's surface and used them to measure its period of rotation. He decided Mercury *did* rotate in 88 days, once per revolution.

The case of Venus was much more difficult. An eternal layer of clouds totally and permanently obscured the surface of Venus as far as ordinary vision was concerned. No markings could be seen and, down to the 1960's, no direct determination of the period of rotation of the nearest planet was possible (though that of distant Pluto was known.)

In the 1960's, however, it became possible to "see" astronomical bodies by something other than the reflection of light waves. Tight beams of short radio-waves could be sent out in the direction of such a body, and the reflected beam could be detected on earth. In 1946, this had been done in the case of the moon. The short radio-waves were of the type used in radar, and so "radar astronomy" was born.

Radar reflections from the moon were of minor importance, however, since the moon's surface could be seen very well by the reflection of ordinary sunlight. What about Venus, however? Radar waves could slip past the cloud layer and touch the actual surface before being reflected. In 1961, this was accomplished by groups of scientists in the United States, in Great Britain, and in the Soviet Union. From the time it took the radar waves to reach Venus and return, a more accurate measure could be made of Venus' distance at that moment (and therefore of all the distances in the solar system). It was not long afterward that radar contact was made with Mercury, too; a Soviet team was first to succeed, in 1962.

The nature of the reflected radar-beam varies according to whether the surface it has touched is rough or smooth, and whether it is rotating or not. Roughness tends to broaden the reflected beam, while rotation tends to expand the wavelength range. The extent of change depends upon the degree of roughness or the rapidity of rotation.

In 1965, the nature of the reflected radar-beam from Mercury made it clear to the Americans Rolf Buchanan Dyce and G. H. Pettengill that Mercury had to be rotating faster than had been thought. The period of rotation was not 88 days but 59 days! This discovery, which was optically confirmed in 1968, was a considerable surprise, but astronomers quickly recovered. The period of Mercury's rotation was just two-thirds its period of revolution, which meant that it presented alternate faces to the sun at each perihelion (point of closest approach). Tidal effects were studied, and it was found that this situation was stable and could be accounted for.

Venus was something else again. There was considerable satisfaction over the fact that the radar beam could bring back information concerning the planet's solid surface—something that could not be done by light waves.

The surface was, for instance, rough. Late in 1965 it was decided that there were at least two huge mountain ranges on Venus. One of them runs from north to south for about 2,000 miles and is several hundred miles wide. The other, even larger, runs east to west. The two ranges are named for the first two letters of the Greek alphabet and are the "Alpha Mountains" and the "Beta Mountains."

But earlier than that, in 1964, it turned out that Venus was rotating slowly. So far, so good, for the period of rotation was thought to be (a matter of pure speculation) 225 days. The period turned out to be 243 days, and the axis of rotation was just about perpendicular to the plane of revolution. It was disappointing that the period of rotation was not exactly 225 days (equal to the period of revolution), for that would have been expected and easily explained. What really astonished the astronomers was that the rotation was in the "wrong" direction. Venus rotated clockwise (as viewed from high above Earth's north pole), east to west, instead of west to east as did all the other planets, except Uranus. It was as though Venus were standing on its head, with its North Pole pointing downward and its South Pole pointing upward.

Why? Nobody can yet offer any explanation.

Furthermore, the time of rotation is such that every time Venus makes its closest approach to Earth, it presents the same (cloud-hidden) side to us. Can there be some gravitational influence of Earth over Venus? But how can small and distant Earth compete with the more distant but enormously larger sun? This is puzzling, too.

In short, in the latter part of the 1960's, Venus emerged as the puzzle-planet of the solar system.

And yet there are puzzles even closer to home. The moon is extraordinarily large in some ways. It is 1/81 as massive as the Earth. No other satellite in the solar system is nearly as large in comparison to the planet it circles. What's more, the moon does not circle Earth in the plane of Earth's equator, but has an orbit markedly tipped to that plane—an orbit that is more nearly in the plane within which the planets generally revolve about the sun. Is it possible that the Moon originally was not a satellite of the Earth, but was an independent planet somehow captured by Earth? Are Earth and the moon twin planets?

Burning curiosity about the origin of the moon and the past history of the Earth-Moon system, is one of the motives that have led scientists to such an excited study of the moon's surface, up to and including manned landings on our satellite.

Of Shape and Size

One of the major inspirations of the ancient Greeks was their decision that the earth has the shape of a sphere. They conceived this idea originally (tradition credits Pythagoras of Samos with being the first to suggest it about 525 b.c.) on philosophical grounds—e.g., that a sphere was the perfect shape. But the Greeks also verified it with observations. Around 350 b.c., Aristotle marshaled conclusive evidence that the earth was not flat but round. His most telling argument was that as one traveled north or south, new stars appeared over the horizon ahead and visible ones disappeared below the horizon behind. Then, too, ships sailing out to sea vanished hull first in whatever direction they traveled, while the cross section of the earth's shadow on the moon, during a lunar eclipse, was always a circle, regardless of the position of the moon. Both these latter facts could be true only if the earth were a sphere.

Among scholars, at least, the notion of the spherical earth never entirely died out, even during the Dark Ages. The Italian poet Dante Alighieri assumed a spherical earth in that epitome of the medieval view, *The Divine Comedy.*

It was another thing entirely when the question of a *rotating* sphere arose. As long ago as 350 b.c., the Greek philosopher Heraclides of Pontus suggested that it was far easier to suppose that the earth rotated on its axis than that the entire vault of the heavens revolved around the earth. This, however, most ancient and medieval scholars refused to accept, and, as late as 1632, Galileo was condemned by the Inquisition at Rome and forced to recant his belief in a moving earth.

Nevertheless, the Copernican theory made a stationary earth completely illogical, and slowly its rotation was accepted by everyone. It was only in 1851, however, that this rotation was actually demonstrated experimentally. In that year, the French physicist Jean Bernard Léon Foucault set a huge pendulum swinging from the dome of a Parisian church. According to the conclusions of physicists, such a pendulum ought to maintain its swing in a fixed plane, regardless of the rotation of the earth. At the North Pole, for instance, the pendulum would swing in a fixed plane, while the earth rotated under it, counterclockwise, in twenty four hours. Since a person watching the pendulum would be carried with the earth (which would seem motionless to him), it would seem to that person that the pendulum's plane of swing was turning clockwise through one full revolution every twenty four hours. At the South Pole, the same thing would happen except that the pendulum's plane of swing would turn counterclockwise.

At latitudes below the poles, the plane of the pendulum would still turn (clockwise in the Northern Hemisphere and counterclockwise in the Southern), but in longer and longer periods as one moved farther

Carl F. von Weizsäcker's model of the origin of the solar system. His theory holds that the great cloud from which it was formed broke up into eddies and subeddies that then coalesced into the sun, the planets, and their satellites.

and farther from the poles. At the equator, the pendulum's plane of swing would not alter at all.

During Foucault's experiment, the pendulum's plane of swing turned in the proper direction and at just the proper rate. The observer, so to speak, could see with his own eyes the earth turn under the pendulum.

The rotation of the earth brings with it many consequences. The surface moves fastest at the Equator, where it must make a circle of 25,000 miles in twenty-four hours, at a speed of just over 1,000 miles an hour. As one travels north (or south) from the Equator, a spot on the earth's surface need travel more slowly, since it must make a smaller circle in the same twenty-four hours. Near the poles, the circle is small indeed, and, at the poles, the surface is motionless.

The air partakes of the motion of the surface of the earth over which it hovers. If an air mass moves northward from the Equator, its own speed (matching that of the Equator) is faster than that of the surface it travels toward. It overtakes the surface in the west-to-east journey and drifts eastward. This drift is an example of a "Coriolis effect," named for the French mathematician Gaspard Gustave de Coriolis, who first studied it in 1835.

The effect of such Coriolis effects on air masses is to set them to turning with a clockwise twist in the Northern Hemisphere. In the Southern Hemisphere, the effect is reversed, and a counterclockwise twist is produced. In either case, "cyclonic disturbances" are set up. Massive storms of this type are called "hurricanes" in the North Atlantic and "typhoons" in the North Pacific. Smaller but more intense storms of this sort are "cyclones" or "tornadoes." Over the sea, such violent twisters set up dramatic "sea spouts."

However, the most exciting deduction obtained from the earth's rotation was made two centuries before Foucault's experiment, in Isaac Newton's time. At that time, the notion of the earth as a perfect sphere had already held sway for nearly 2,000 years, but then Newton took a careful look at what happens to such a sphere when it rotates. He noted the difference in the rate of motion of the earth's surface at different latitudes and considered what it must mean.

The faster the rotation, the stronger the centrifugal effect—that is, the tendency to push material away from the center of rotation. It follows, therefore, that the centrifugal effect increases steadily from zero at the stationary poles to a maximum at the rapidly whirling equatorial belt. This means that the earth should be pushed out most around its middle. In other words, it should be an "oblate spheroid," with an "equatorial bulge" and flattened poles. It must have roughly the shape of a tangerine rather than that of a golf ball. Newton even calculated that the polar flattening should be about 1/230 of the total diameter, which is surprisingly close to the truth.

The earth rotates so slowly that the flattening and bulging are too slight to be readily detected. But at least two astronomical observations supported Newton's reasoning, even in his own day. First, Jupiter and Saturn were clearly seen to be markedly flattened at the poles, as was first pointed out by the Italian-born, French astronomer Giovanni Domenico Cassini in 1687. Both planets are much larger than the earth and rotate much faster, so that Jupiter's surface, for instance, is speeding at 27,000 miles per hour at its equator. With centrifugal effects born of such speeds, no wonder it is flattened.

Second, if the earth really bulges at the Equator, the varying gravitational pull on the bulge by the moon, which most of the time is either north or south of the Equator in its circuit around the earth, should cause the earth's axis of rotation to mark out a double cone, so that each pole points to a steadily changing point in the sky. The points mark out a circle about which the pole makes a complete revolution every 26,000 years. In fact, Hipparchus of Nicaea had noted this shift about 150 B.C. when he compared the position of the stars in his day with those recorded a century and a half earlier. The shift of the earth's axis has the effect of causing the sun to reach the point of equinox about fifty seconds of arc eastward each year (that is, in the direction of morning). Since the equinox thus comes to a preceding (i.e., earlier) point each year, Hipparchus named this shift the "precession of the equinoxes," and it is still known by that name.

Naturally scientists set out in search of more direct proof of the earth's distortion. They resorted to a standard device for solving geometrical problems—trigonometry. On a curved surface, the angles of a triangle add up to more than 180 degrees. The greater the curvature, the greater the excess over 180 degrees. Now if the earth was an oblate spheroid, as Newton had said, the excess should be greater on the more sharply curved surface of the equatorial bulge than on the less curved surface toward the poles. In the 1730's, French scientists made the first test by doing some large-scale surveying at separate sites in the north and the south of France. On the basis of these measurements, the French astronomer Jacques Cassini (son of Giovanni Domenico, who had pointed out the flattening of Jupiter and Saturn) decided that the earth bulged at the poles, not at the Equator! To use an exaggerated analogy, its shape was more like that of a cucumber than of a tangerine.

But the difference in curvature between the north and the south of France obviously was too small to give conclusive results. Consequently, in 1735 and 1736 a pair of French expeditions went forth to more widely separated regions—one to Peru, near the Equator, and the other to Lapland, approaching the Arctic. By 1744, their surveys had given a clear answer: the earth was distinctly more curved in Peru than in Lapland.

Today the best measurements show that the diameter of the earth is 26.68 miles longer through the Equator than along the axis through the poles (i.e., 7,926.36 miles against 7,899.78 miles).

Perhaps the most important scientific result of the eighteenth-century inquiry into the shape of the earth was that it made the scientific community dissatisfied with the state of the art of measurement. No decent standards for precise measurement existed. This dissatisfaction was partly responsible for the adoption, during the French Revolution half a century later, of the logical and scientifically worked-out "metric" system based on the meter. The metric system now is used by scientists all over the world, to their great satisfaction, and it is the system in general public use in every civilized country except the English-speaking nations, chiefly Great Britain and the United States.

The importance of accurate standards of measure cannot be overestimated. A good percentage of scientific effort is continually being devoted to improvement in such standards. The standard meter and standard kilogram were made of platinum-iridium alloy (virtually immune to chemical change) and were kept in a Paris suburb under conditions of great care; in particular, under constant temperature to prevent expansion or contraction.

New alloys such as "Invar" (short for invariable), composed of nickel and iron in certain proportions, were discovered to be almost unaffected by temperature change. These could be used in forming better standards of length and the Swiss-born, French physicist Charles Edouard Guillaume, who developed Invar, received the Nobel Prize for physics in 1920 for this discovery.

In 1960, however, the scientific community abandoned material standards of length. The General Conference of Weights and Measures adopted as standard the length of a tiny wave of light produced by the rare gas krypton. Exactly 1,650,763.73 of these waves (far more unchanging than anything man-made could be) equal one meter, a length which is now a thousand times as exact as it had been before.

The smoothed-out, sea-level shape of the earth is called the "geoid." Of course, the earth's surface is pocked with irregularities—mountains, ravines, and so on. Even before Newton raised the question of the planet's over-all shape, scientists had tried to measure the magnitude of these minor deviations from a perfect sphere (as they thought). They resorted to the device of a swinging pendulum. Galileo, in 1581, as a seventeen-year-old boy, had discovered that a pendulum of a given length always completed its swing in just about the same time, whether the swing was short or long; he is supposed to have made the discovery while watching the swinging chandeliers in the cathedral of Pisa during services. There is a lamp in the cathedral still called "Galileo's lamp," but it was not hung until 1584. (Huygens hooked a pendulum to the gears of a clock and used the constancy of its motion to keep the clock going with even accuracy. In 1656, he devised the first modern clock in this way—the "grandfather clock"—and at once increased the accuracy of timekeeping tenfold.)

The period of the pendulum depends both on its length and on the gravitational force. At sea level, a pendulum with a length of 39.1 inches makes a complete swing in just one second, a fact worked out in 1644 by Galileo's pupil, the French mathematician, Marin Mersenne. The investigators of the earth's irregularities made use of the fact that the period of a pendulum's swing depends on the strength of gravity at any given point. A pendulum that swings perfect seconds at sea level, for instance, will take slightly longer than a second to complete a swing on a mountain top, where gravity is slightly weaker because the mountain top is farther from the center of the earth.

In 1673, a French expedition to the north coast of South America (near the Equator) found that at that location the pendulum was slowed even at sea level. Newton later took this as evidence for the existence of the equatorial bulge, since that lifted the camp farther from the earth's center, and weakened the force of gravity. After the expedition to Peru and Lapland had proved his theory, a member of the Lapland expedition, the French mathematician Alexis Claude Clairault, worked out methods of calculating the oblateness of the earth from pendulum swings. Thus the geoid, or sea-level shape of the earth, can be determined, and it turns out to vary from the perfect oblate spheroid by less than 300 feet at all points. Nowadays, gravitational force is also measured by a "gravimeter," a weight suspended from a very sensitive spring. The position of the weight against a scale in the background indicates the force with which it is pulled downward and hence measures variations in gravity with great delicacy.

Gravity at sea level varies by about 0.6 per cent, being least at the Equator, of course. The difference is not noticeable in ordinary life, but it can affect sports records. Achievements at the Olympic Games depend to some extent on the latitude (and altitude) of the city in which they are conducted.

A knowledge of the exact shape of the geoid is essential for accurate map-making, and only 7 per cent of the earth's land surface can really be said to be accurately mapped. As late as the 1950's, the distance between New York and London, for instance, was not known to better than a mile or so, and the locations of some islands in the Pacific were known only within a possible error of several miles. In these days of air travel and (alas!) potential missile-aiming, this is inconvenient. But truly accurate mapping has now been made possible—oddly enough, not by surveys of the earth's surface, but by astronomical measurements of a new kind. The first instrument of these new measurements was the man-made satellite called Vanguard I, launched by the United States on March 17, 1958. Vanguard I makes a revolution around the earth in two and a half hours, and in the first couple of years of its lifetime it had already made more revolutions than the moon had in all the centuries it has been observed with the telescope. By observations of Vanguard I's position at specific times from specific points of the earth, the distances between these observing points can be calculated precisely. In this way, positions and distances not known to within a matter of miles were, in 1959, determined to within a hundred yards or so. (Another satellite named Transit I-B, launched by the United States on April 13, 1960, was the first of a series specifically intended to extend this into a system for the accurate location of position on the earth's surface, something which could greatly improve and simplify air and sea navigation.)

Like the moon, Vanguard I circles the earth in an ellipse which is not in the earth's equatorial plane. As in the case of the moon, the perigee (closest approach) of Vanguard I shifts because of the attraction of the equatorial bulge. Because Vanguard I is far closer to the bulge and far smaller than the moon, it is affected to a greater extent, and because of its many revolutions, the effect of the bulge can be well studied. By 1959, it was certain that the perigee shift of Vanguard I was not the same in the Northern Hemisphere as in the Southern. This showed that the bulge was not quite symmetrical with respect to the Equator. The bulge seemed to be twenty-five feet higher (that is, twenty-five feet more distant from the earth's center) at spots south of the Equator than at spots north of it. Further calculations showed that the South Pole was fifty feet closer to the center of the earth (counting from sea level) than was the North Pole.

Further information, obtained in 1961, based on the orbits of Vanguard I and Vanguard II (the latter having been launched on February 17, 1959) indicates that the sea-level Equator is not a perfect circle. The equatorial diameter is 1,400 feet (nearly a quarter of a mile) longer in some places than in others.

Newspaper stories have described the earth as "pear-shaped" and the Equator as "egg-shaped." Actually, these deviations from the perfectly smooth curve are perceptible only to the most refined measurements. No one looking at the earth from space would see anything resembling a pear or an egg; he would see only what would seem a perfect sphere. Besides, detailed studies of the geoid have shown so many regions of very slight flattening and very slight humping that, if the earth must be described dramatically, it had better be called "lumpy shaped."

A knowledge of the exact size and shape of the earth makes it possible to calculate its volume, about 260 billion cubic miles. Calculating the earth's mass, however, is a more complex matter, but Newton's law of gravitation gives us something to begin with. According to Newton, the gravitational force (f) between any two objects in the universe can be expressed as follows:

$$f = \frac{gm_1m_2}{d^2}$$

where m_1 and m_2 are the masses of the two bodies concerned and d is the distance between them, center to center. As for g, that represents the "gravitational constant."

What the value of the constant was, Newton could not say. If we can learn the values of the other factors in the equation, however, we can find g, for by transposing the terms we get:

$$g = \frac{fd^2}{m_1m_2}$$

To find the value of g, therefore, all we need to do is to measure the gravitational force between two bodies of known mass at the separation of a known distance. The trouble is that gravitational force is the weakest force we know, and the gravitational attraction between two masses of any ordinary size that we can handle is almost impossible to measure.

Nevertheless, in 1798 the English physicist Henry Cavendish, a wealthy, neurotic genius who lived and died in almost complete seclusion but performed some of the most astute experiments in the history of science, managed to make the measurement. Cavendish attached a

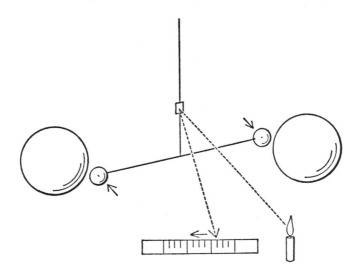

Henry Cavendish's apparatus for measuring gravity. The two small balls are attracted by the larger ones, causing the thread on which they are suspended to twist. The mirror shows the amount of this slight twist by the deflection of reflected light on the scale.

ball of known mass to each end of a long rod and suspended this dumbbell-like contraption on a fine thread. Then he placed a larger ball, also of known mass, close to each ball on the rod—on opposite sides, so that gravitational attraction between the fixed large balls and the suspended small balls would cause the horizontally hung dumbbell to turn, thus twisting the thread. The dumbbell did indeed turn slightly. Cavendish now measured how much force was needed to produce this amount of twist of the thread. This told him the value of f. He also knew m_1 and m_2, the masses of the balls, and d, the distance between the attracted balls. So he was able to compute the value of g. Once he had that, he could calculate the mass of the earth, because the earth's gravitational pull (f) on any given body can be measured. Thus Cavendish "weighed" the earth for the first time.

The measurements have since been greatly refined. In 1928, the American physicist Paul R. Heyl at the United States Bureau of Standards determined the value of g to be 0.00000006673 dyne centimeter squared per gram squared. You need not be concerned about those units, but note the smallness of the figure. It is a measure of the weakness of gravitational force. Two one-pound weights placed a foot apart attract each other with a force of only one half of one billionth of an ounce.

The fact that the earth itself attracts such a weight with the force of one pound even at a distance of 4,000 miles from its center emphasizes how massive the earth is. In fact, the mass of the earth turns out to be 6,595,000,000,000,000,000,000 tons or, in metric units, 5,983,000,000,000,000,000,000,000 kilograms.

From the mass and volume of the earth, its average density is easily calculated. In metric units, the answer comes out to 5.522 grams per cubic centimeter (5.522 times the density of water). The density of the earth's surface rocks averages only about 2.8 grams per cubic centimeter, so the density of the interior must be much greater. Does it increase smoothly all the way down to the center? The first proof that it does not—that the earth is made up of a series of different layers—came from the study of earthquakes.

The Layers of the Planet

On November 1, 1755, a great earthquake, possibly the most violent of modern times, struck the city of Lisbon, demolishing every house in the lower part of the city. Then a tidal wave swept in from the ocean. Sixty thousand people were killed, and the city was left a scene of devastation.

The shock was felt over an area of one and a half million square miles, doing substantial damage in Morocco as well as in Portugal. Because it was All Soul's Day, people were in church, and it is said that all over southern Europe those in the cathedrals saw the chandeliers dance and sway.

The Lisbon disaster made a great impression on the scholars of the day. It was an optimistic time when many thinkers felt that the new science of Galileo and Newton would place in man's hands the means of making the earth a human paradise. This blow showed that there were still giant, unpredictable, and apparently malicious forces beyond man's control. The earthquake inspired Voltaire, the great literary figure of the time, to write his famous pessimistic satire *Candide*, with its ironical refrain that all was for the best in this best of all possible worlds.

We are accustomed to thinking of dry land as shaking with the effect of an earthquake, but the earth beneath the ocean floor may be set to quivering, too, with even more devastating effects. The vibration sets up long, gentle swells in the ocean which, on reaching the shallow shelves in the neighborhood of land, pile up into towers of water, sometimes fifty to one hundred feet high. If the waves hit with no warning, thousands are drowned. The popular name for such earthquake-generated waves are "tidal waves," but this is a misnomer. They may resemble monstrous tides, but they have entirely different causes. Nowadays, they are referred to by the Japanese name "tsunami." Japan's coastline is particularly vulnerable to such waves, so this nomenclature is justified.

After the Lisbon disaster, to which a tsunami had added its share of destruction, scientists began turning their thoughts earnestly to what the causes of earthquakes might be. The best the ancient Greeks were able to do had been Aristotle's suggestion that it was caused by masses of air, imprisoned underground and trying to escape. Modern scientists, however, suspected that it might be the effect of earth's internal heat on stresses within the solid rock itself.

The English geologist John Michell (who had studied the forces involved in "torsion," or twisting, later used by Cavendish to measure the mass of the earth) suggested in 1760 that earthquakes were waves set up by the shifting of masses of rock miles below the surface. To study earthquakes properly, an instrument for detecting and measuring these waves had to be developed, and this did not come to pass until one hundred years after the Lisbon quake. In 1855, the Italian physicist Luigi Palmieri devised the first "seismograph" (from Greek words meaning "earthquake-writing").

In its simplest form, the seismograph consists of a massive block suspended by a comparatively weak spring from a support firmly fixed in bedrock. When the earth moves, the suspended block remains still, because of its inertia. However, the spring attached to the bedrock stretches or contracts a little with the earth's motion. This motion is recorded on a slowly rotating drum by means of a pen attached to the stationary block, writing on smoked paper. Actually, two blocks are used, one oriented to record the earthquake waves traveling north and south, the other, east and west. Nowadays, the most delicate seismographs, such as the one at Fordham University, use a ray of light in place of a pen, to avoid the frictional drag of the pen on the paper. This ray shines on sensitized paper, making tracings that are developed as a photograph.

The English engineer John Milne, using seismographs of his own design, showed conclusively in the 1890's that Michell's description of earthquakes as waves propagated through the body of the earth was correct. Milne was instrumental in setting up stations for the study of earthquakes and related phenomena in various parts of the world, particularly in Japan. By 1900, thirteen seismograph stations were in existence, and today there are over 500, spread over every continent including Antarctica.

The earth suffers a million quakes a year, including at least ten disastrous ones and a hundred serious ones. Some 15,000 people are killed by these tremors each year. The most murderous quake is supposed to have taken place in northern China in 1556, with an estimated 830,000 dead. As recently as 1923, a quake that shook Tokyo devastated the city and left 143,000 dead.

The largest earthquakes are estimated to release a total energy equal to 100,000 ordinary atomic bombs or, if you prefer, one hundred large H-bombs. It is only because their energies are dissipated over a large area that they are not more destructive than they are. They can make the earth vibrate as though it were a gigantic tuning fork. The Chilean earthquake of 1960 caused our planet to vibrate at a frequency of just under once an hour (20 octaves below middle C and quite inaudible).

Earthquake intensity is measured on a scale from 0 up through 9, where each number represents an energy release ten times that of the number below. (No quake of intensity greater than 9 has ever been recorded, but the Good Friday quake in Alaska in 1964 recorded an intensity of 8.5.) This is called the "Richter scale" because it was introduced in 1935 by the American seismologist Charles Francis Richter.

About 80 per cent of earthquake energy is released in the areas bordering the vast Pacific Ocean. Another 15 per cent is released in an east–west band sweeping across the Mediterranean. These earthquake zones (see map on page 118) are closely associated with volcanic areas, which is one reason why the effect of internal heat was associated with earthquakes.

Volcanoes are a natural phenomenon that are as frightening as earthquakes and longer-lasting, though, of course, in most cases their effects are confined to a smaller area. About 500 volcanoes are known to have been active in historical times, two-thirds of them along the rim of the Pacific.

On rare occasions, when a volcano traps and overheats huge quantities of water, appalling catastrophes can take place. On August 26–27, 1883, the small East Indian volcanic island Krakatoa, situated in the strait between Sumatra and Java, exploded with a roar that has been described as the loudest sound ever formed on earth during historic times. The sound was heard by human ears as far away as 3,000 miles and could be picked up by instruments all over the globe. The sound waves traveled several times completely about the planet. Five cubic miles of rock were fragmented and hurled into the air; ash fell over an area of 300,000 square miles. Ashes darkened the sky over hundreds of square miles, leaving in the stratosphere dust that brightened sunsets for years. Tsunamis a hundred feet in height killed 36,000 people on the shores of Java and Sumatra, and their waves could be detected easily in all parts of the world.

A similar event, with even greater consequences, may have taken place over three thousand years before in the Mediterranean Sea. In 1967, American archaeologists discovered the ash-covered remains of a city on the small island of Thera, 80 miles north of Crete. About 1400 B.C., apparently, it exploded as Krakatoa did. The tsunami that resulted struck the island of Crete, then the home of a long-developed and admirable civilization, a crippling blow from which that civilization never recovered. The Cretan control of the seas vanished, and a period of turmoil and darkness eventually followed; recovery took many centuries. The dramatic disappearance of Thera lived on in the minds of survivors, and its tale passed down the line of generations with embellishments. It may very well have given rise to Plato's tale of Atlantis, which was told about eleven centuries after the death of Thera and of Cretan civilization.

And yet perhaps the most famous single volcanic eruption in the history of the world was a minute one, compared to Krakatoa or Thera. It was the eruption of Vesuvius in 79 (at that time it had been considered a dead volcano), which buried the Roman resort cities of Pompeii and Herculaneum. The famous encyclopedist Gaius Plinius Secundus (better known as Pliny) died in that catastrophe, which was described by his nephew, Pliny the Younger, an eyewitness.

Excavations of the buried cities began in serious fashion after 1763. These offered an unusual opportunity to study relatively complete remains of a city that had existed during the most prosperous period of ancient times.

Another unusual phenomenon is the actual birth of a new volcano. Such an awesome event was witnessed in Mexico on February 20, 1943, when in the village of Paricutin, 200 miles west of Mexico City, a volcano began to appear in what had been a quiet cornfield. In eight months, it had built itself up to an ashy cone 1,500 feet high. The village had to be abandoned, of course.

Modern research in volcanoes and their role in forming much of the earth's crust began with the French geologist Jean Etienne Guettard in the mid-eighteenth century. For a while, in the late eighteenth century, the single-handed efforts of the German geologist Abraham Gottlob Werner popularized the false notion that most rocks were of sedimentary origin, from an ocean that had once been world-wide ("neptunism"). The weight of the evidence, particularly that presented by Hutton, made it quite certain, however, that most rocks were formed through volcanic action ("plutonism"). Both volcanoes and earthquakes would seem the expression of the earth's internal energy, originating for the most part from radioactivity (see Chapter 6).

Once seismographs allowed the detailed study of earthquake waves, it was found that those most easily studied came in two general varieties: "surface waves" and "bodily waves." The surface waves follow the curve of the earth; the bodily waves go through the interior—and by virtue of this short cut usually are the first to arrive at the seismograph. These bodily waves in turn are of two types: primary ("P waves") and secondary ("S waves"). The primary waves, like sound waves, travel by alternate compression and expansion of the medium (to visualize them,

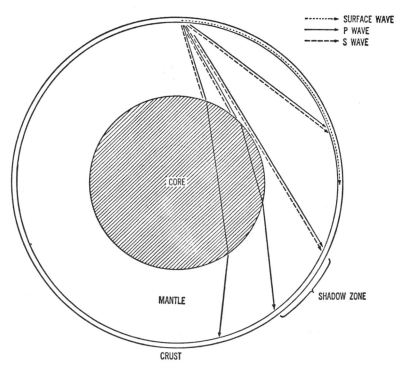

Earthquake waves' routes in the earth. Surface waves travel along the crust. The earth's liquid core refracts the P-type bodily waves. S waves cannot travel through the core.

think of the pushing together and pulling apart of an accordion). Such waves can pass through any medium—solid or fluid. The secondary waves, on the other hand, have the familiar form of snakelike wiggles at right angles to the direction of travel, and they cannot travel through liquids or gases.

The primary waves move faster than secondary waves and consequently reach a seismograph station sooner. From the time lag of the secondaries, it is possible to estimate the distance of the earthquake. And its location or "epicenter" (the spot on the earth's surface directly above the rock disturbance) can be pinpointed by getting distance bearings at three or more stations: the three radii trace out three circles that will intersect at a single point.

The speed of both the P and S types of wave is affected by the kind of rock, the temperature, and the pressure, as laboratory studies have shown. Therefore earthquake waves can be used as probes to investigate conditions deep under the earth's surface.

A primary wave near the surface travels at 5 miles per second; 1,000 miles below the surface, judging from the arrival times, its velocity must be nearly 8 miles per second. Similarly, a secondary wave has a velocity of less than 3 miles per second near the surface and 4 miles per second at a depth of 1,000 miles. Since increase in velocity is a measure of increase in density, we can estimate the density of the rock beneath the surface. At the surface of the earth, as I have mentioned, the average density is 2.8 grams per cubic centimeter; 1,000 miles down it amounts to 5 grams per cubic centimeter; 1,800 miles down, nearly 6 grams per cubic centimeter.

At the depth of 1,800 miles, there is an abrupt change. Secondary waves are stopped cold. The British geologist R. D. Oldham maintained, in 1906, that this must mean that the region below is liquid: the waves have reached the boundary of the earth's "liquid core." And primary waves on reaching this level change direction sharply; apparently they are refracted by entering the liquid core.

The boundary of the liquid core is called the "Gutenberg discontinuity," after the American geologist Beno Gutenberg, who in 1914 defined the boundary and showed that the core extended to 2,160 miles from the earth's center. The density of the various deep layers of the earth were worked out in 1936 from earthquake data by the Australian mathematician Keith Edward Bullen. His results were confirmed by the data yielded by the huge Chilean earthquake of 1960. We can therefore say that at the Gutenberg discontinuity, the density of the material jumps from 6 to 9, and thereafter it increases smoothly to 11.5 grams per cubic centimeter at the center.

What is the nature of the liquid core? It must be composed of a substance which has a density of from 9 to 11.5 grams per cubic centimeter under the conditions of temperature and pressure in the core. The pressure is estimated to range from 10,000 tons per square inch at the top of the liquid core to 25,000 tons per square inch at the center of the earth. The temperature is less certain. On the basis of the rate at which temperature is known to increase with depth in deep mines and of the rate at which rocks can conduct heat, geologists estimate (rather roughly) that temperatures in the liquid core must be as high as 5,000° C. (The center of the much larger planet Jupiter may be as high as 500,000° C.)

The substance of the core must be some common element—common enough to be able to make up a sphere half the diameter of the earth and one third its mass. The only heavy element that is at all common in the universe is iron. At the earth's surface its density is only 7.86 grams per cubic centimeter, but under the enormous pressures of the core, it would have a density in the correct range—9 to 12 grams per cubic centimeter. What is more, under center-of-the-earth conditions it would be liquid.

If more evidence is needed, meteorites supply it. These fall into two broad classes: "stony" meteorites, composed chiefly of silicates, and "iron" meteorites, made up of about 90 per cent iron, 9 per cent nickel, and 1 per cent other elements. Many scientists believe that the meteorites are remnants of a shattered planet; if so, the iron meteorites may be pieces from the liquid core of that planet and the stony meteorites fragments of its mantle. (Indeed, in 1866, long before seismologists had probed the earth's core, the composition of the iron meteorites suggested to the French geologist Gabriel Auguste Daubrée that the core of our planet was made of iron.)

Today most geologists accept the liquid nickel-iron core as one of the facts of life as far as the earth's structure is concerned. One major refinement, however, has been introduced. In 1936, the Danish geologist Inge Lehmann, seeking to explain the puzzling fact that some primary waves show up in a "shadow zone" on the surface from which most such waves are excluded, proposed that a discontinuity within the core about 800 miles from the center introduced another bend in the waves and sent a few careening into the shadow zone. Gutenberg supported this view, and now many geologists differentiate between an "outer core" that is liquid nickel-iron and an "inner core" that differs from the outer core in some way, perhaps in being solid or slightly different chemically. As a result of the great Chilean earthquakes of 1960, the entire globe was set into slow vibrations at rates matching those predicted by taking the inner core into account. This is strong evidence in favor of its existence.

The portion of the earth surrounding the nickel-iron core is called the "mantle." It seems to be composed of silicates, but judging from the velocity of earthquake waves passing through them, these silicates are different from the typical rocks of the earth's surface—something first shown in 1919 by the American physical chemist Leason Heberling Adams. Their properties suggest that they are rocks of the so-called "olivine" type (olive-green in color), which are comparatively rich in magnesium and iron and poor in aluminum.

The mantle does not quite extend to the surface of the earth. A

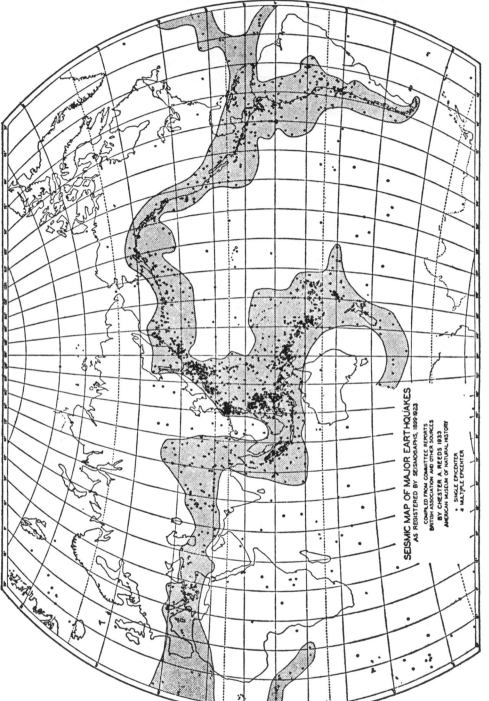

SEISMIC MAP OF MAJOR EARTHQUAKES
AS REGISTERED BY SEISMOGRAPHS, 1899-1923

COMPILED FROM COMMITTEE REPORTS
BRITISH ASSOCIATION AND OTHER SOURCES
BY CHESTER A. REEDS 1933
AMERICAN MUSEUM OF NATURAL HISTORY

• SINGLE EPICENTER
▲ MULTIPLE EPICENTER

The earth's earthquake belts. They follow the principal zones of new mountain building.

Croatian geologist named Andrija Mohorovicic, while studying the waves produced by a Balkan earthquake in 1909, decided that there was a sharp increase in wave velocity at a point about twenty miles beneath the surface. This "Mohorovicic discontinuity" (known as "Moho" for short) is now accepted to be the boundary of the earth's "crust."

The nature of the crust and of the upper mantle is best explored by means of the "surface waves" I mentioned earlier. Like the "bodily waves," the surface waves come in two varieties. One kind are called "Love waves" (after their discoverer A. E. H. Love). The Love waves are horizontal ripples, like the shape of a snake moving over the ground. The other variety is the "Rayleigh waves" (named after the English physicist John William Strutt, Lord Rayleigh); these ripples are vertical, like the path of a sea serpent moving through the water.

Analysis of these surface waves (notably by Maurice Ewing of Columbia University) shows that the crust is of varying thickness. It is thinnest under the ocean basins, where the Moho discontinuity in some places is only 8 to 10 miles below sea level. Since the oceans themselves are 5 to 7 miles deep in spots, the solid crust may be as thin as 3 miles under the ocean deeps. Under the continents, on the other hand, the Moho discontinuity lies at an average depth of about 20 miles below sea level (it is about 22 miles under New York City, for instance), and it plunges to a depth of nearly 40 miles beneath mountain ranges. This fact, combined with evidence from gravity measurements, shows that the rock in mountain ranges is less dense than the average.

The general picture of the crust is that of a structure composed of two main types of rock—basalt and granite—with the less dense granite riding buoyantly on the basalt, forming continents and, in places where the granite is particularly thick, mountains (just as a large iceberg rises higher out of the water than a small one). Young mountains thrust their granite roots deep into the basalt, but, as the mountains are worn down by erosion, they adjust by floating slowly upward (to maintain the equilibrium of mass called "isostasy," a name suggested in 1889 by the American geologist Clarence Edward Dutton). In the Appalachians, a very ancient mountain chain, the root is about gone.

The basalt beneath the oceans is covered with a quarter to a half mile of sedimentary rock, but little or no granite—the Pacific basin is completely free of granite. The thinness of the crust under the oceans has suggested a dramatic project: Why not drill a hole through the crust down to the Moho discontinuity and tap the mantle to see what it is made of? It will not be an easy task, for it will mean anchoring a ship over an abyssal section of the ocean, lowering drilling gear through miles of water, and then drilling through a greater thickness of rock than anyone has yet drilled. Early enthusiasm for the project evaporated, however, and the matter now lies in abeyance.

The "floating" of the granite in the basalt inevitably suggests the possibility of "continental drift." In 1912, the German geologist Alfred Lothar Wegener suggested that the continents were originally a single piece of granite, which he called "Pangaea" ("All-Earth"). At some early stage of the earth's history this fractured and the continents drifted apart. He argued that they were still drifting—Greenland, for instance, moving away from Europe at the rate of a yard a year. What gave him (and others, dating back to Francis Bacon about 1620) the idea was mainly the fact that the eastern coastline of South America seemed to fit like a jigsaw piece into the shape of the western coast of Africa.

For a half-century, Wegener's theory was looked upon with hard disfavor. As late as 1960, when the first edition of this book was published, I felt justified, in view of the state of geophysical opinion at that time, in categorically dismissing it. The most telling argument against it was that the basalt underlying both oceans and continents was simply too stiff to allow the continental granite to drift sideways.

And yet evidence in favor of the supposition that the Atlantic Ocean once did not exist and that the separate continents once formed a single land-mass, grew massively impressive. If the continents were matched, not by their actual shore line (an accident of the current sea-level) but by the central point of the continental slope (the shallow floor of the ocean neighboring the continents which is exposed during ages of low sea-level), then the fit is excellent all along the Atlantic, in the north as well as the south. Then, too, rock formations in parts of western Africa, match the formations in parts of eastern South America in fine detail. Past wanderings of the magnetic poles look less startling if one considers that the continents, not the poles, wandered.

Perhaps the most devastating piece of evidence arrived in 1968 when a 2½-inch fossilized bone from an extinct amphibian was found in Antarctica. Such a creature could not possibly have lived so close to the South Pole, so Antarctica must once have been farther from the pole, or at least milder in temperature. The amphibian could not have crossed even a narrow stretch of salt water, so Antarctica must have been part of a larger body of land, containing warmer areas.

That still leaves the question of what it was that caused the original supercontinent to break up and drift apart. About 1960, the American geologist Harry Hammond Hess suggested that molten mantle material might be welling up—along certain fracture-lines running the length of the Atlantic Ocean, for instance—be forced sideways near the top of the mantle, cool, and harden. The ocean floor is, in this way, pulled apart and stretched. It is not, then, that the continents drift, but that they are pushed apart by a spreading sea-floor.

As the story seems now, Pangaea did exist, after all, and was intact as recently as 225 million years ago, when the dinosaurs were coming into prominence. Judging from the evolution and distribution of plants and animals, the breakup must have become pronounced about 200 million years ago. Pangaea then broke into three parts. The northern part (North America, Europe, and Asia) is called "Laurasia"; the southern part (South America, Africa, and India) is called "Gondwana," from an Indian province. Antarctica plus Australia formed a third part.

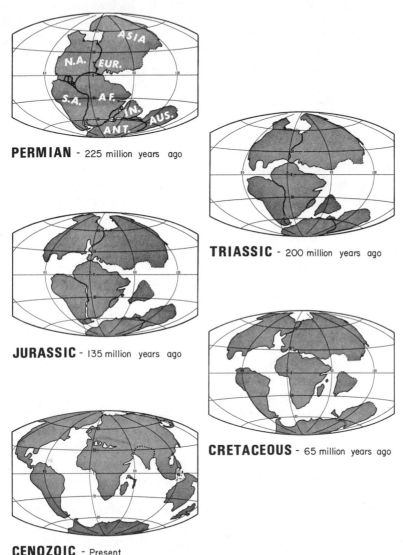

PERMIAN - 225 million years ago

TRIASSIC - 200 million years ago

JURASSIC - 135 million years ago

CRETACEOUS - 65 million years ago

CENOZOIC - Present

Some 65 million years ago, with the dinosaurs already extinct and the mammals ruling earth, South America separated from Africa on the west, and India on the east separated and moved up toward southern Asia. Finally, North America split off from Europe, India crunched up into Asia (with the Himalayan Mountains folding up at the junction

121

line), Australia moved away from its connection with Antarctica, and the continental arrangement as we have it at present was seen.

An even more startling suggestion as to the changes that may have taken place on the earth over geologic periods, dates back to 1879, when the British astronomer George Howard Darwin (a son of Charles Darwin) suggested that the moon was a piece of the earth that had broken loose in early times, leaving the Pacific Ocean as the scar of the separation.

This is an attractive thought, since the moon makes up only a little over 1 per cent of the combined earth-moon mass and is small enough for its width to lie within the stretch of the Pacific. If the Moon were made up of the outer layers of the earth, it would account for the moon's having no iron core and being much less dense than the earth, and for the Pacific floor's being free of continental granite.

The possibility of an earth-moon breakup seems unlikely on various grounds, however, and virtually no astronomer or geologist now thinks that it can have taken place (however, remember the fate of the continental-drift theory). Nevertheless, the moon seems certainly to have been closer in the past than it is today.

The moon's gravitational pull produces tides both in the ocean and in the earth's solid crust. As the earth rotates, ocean water is dragged across sections of shallow floor, while layers of rock rub together as they rise and fall. The friction represents a slow conversion of the earth's energy of rotation into heat, so that its rotational period gradually increases. The effect is not great in human terms, for the day lengthens by one second in about a hundred thousand years. As the earth loses rotational energy, the angular momentum must be conserved. What the earth loses, the moon gains. Its speed increases as it revolves about the earth, which means it drifts farther away very slowly.

If one works backward in time toward the far geologic past, we see that the earth's rotation must speed up, the day be significantly shorter, the moon significantly closer, and the whole effect more rapid. Darwin calculated backward to find out when the moon was close enough to earth to form a single body, but even if we don't go that far we ought to find evidence of a shorter day in the past. For instance, about 570 million years ago—the time of the oldest fossils—the day may have been only a little over 20 hours long, and there may have been 428 of them in a year.

Nor is this only theory now. Certain corals lay down bands of calcium carbonate more actively at some seasons than others, so that you can count annual bands just as in tree trunks. It is also suggested that some lay down calcium carbonate more actively by day than by night, so that there are very fine daily bands. In 1963, the American paleontologist John West Wells counted the fine bands in fossil corals and reported there were, on the average, 400 daily bands per annual bands in corals dating back 400 million years and 380 daily bands per annual band in corals dating back only 320 million years.

Of course, the question is, If the moon was much closer to the earth then, and the earth rotated more rapidly, what happened in still earlier periods? If Darwin's theory of an earth-moon separation is not so, what *is* so?

One suggestion is that the moon was captured at some time in the past. If it were captured 600 million years ago, for instance, that might account for the fact that at about that time, we begin to find numerous fossils in rocks, whereas earlier rocks have nothing but uncertain traces of carbon. Perhaps the earlier rocks were washed clean by the vast tides that accompanied the capture of the moon. (There was no land life at the time; if there had been, it would have been destroyed.) If the moon were captured, it would have been closer then than now, and there would be a lunar recession and a lengthening of the day since, but nothing of the sort before.

Another suggestion is that the moon was formed in the neighborhood of the earth, out of the same gathering dust-cloud, and has been receding ever since, but that it never was actually part of the earth. The truth is that astronomers still don't know, but they hope to find out through a continued exploration of the Moon's surface either by men or machines landed on our companion.

The fact that the earth consists of two chief portions—the silicate mantle and the nickel-iron core (in about the same proportions as the white and yolk of an egg)—has persuaded most geologists that the earth must have been liquid at some time in its early history. It might then have consisted of two mutually insoluble liquids. The silicate liquid, being the lighter, would float to the top and cool by radiating its heat into space. The underlying iron liquid, insulated from direct exposure to space, would give up its heat far more slowly and would thus remain liquid to the present day.

There are at least three ways in which the earth could have become hot enough to melt, even from a completely cold start as a collection of planetesimals. These bodies, on colliding and coalescing, would give up their energy of motion ("kinetic energy") in the form of heat. Then, as the growing planet was compressed by gravitational force, still more energy would be liberated as heat. Third, the radioactive substances of the earth—uranium, thorium, and potassium—have delivered large quantities of heat over the ages as they have broken down; in the early stages, when there was a great deal more radioactive material than now, radioactivity itself might have supplied enough heat to liquefy the earth.

Not all scientists are willing to accept a liquid stage as an absolute necessity. The American chemist Harold Clayton Urey, in particular, believes that most of the earth was always solid. He argues that in a largely solid earth an iron core could still be formed by a slow separation of iron; even now, he suggests, iron may be migrating from the mantle into the core at the rate of 50,000 tons a second.

Cooling of the earth from an original molten or near-molten state would help to explain its wrinkled exterior. As the cooling earth shrank, its crust would occasionally buckle. Minor buckling would give rise to earthquakes. Larger buckling, or a steady accumulation of smaller adjustments, would eventually produce mountain ranges. The mountain-building eras, however, would be relatively brief. After mountains were formed, they would be worn down by erosion in fairly short order (on the geological time scale), and then would come a long period of stability before compressional forces built up great enough strains to start a new crust-buckling stage. Thus during most of its lifetime the earth would be a rather drab and featureless planet, with low continents and shallow seas.

The trouble with this view is that the earth does not seem really to be cooling off. The thought that it must be doing so arises from the natural assumption that a hot body *must* cool down if there is no source of continuing heat. True! But in Earth's case, there *is* a source of continuing heat, one that was not understood prior to the twentieth century. This new source became apparent with the discovery of radioactivity in 1896 when it appeared that a hitherto utterly unsuspected form of energy lay hidden deep within the recesses of the atom.

It appears that over the last several hundred million years radioactivity has been generating enough heat in the crust and mantle at least to keep the earth's internal temperature from falling; if anything, the earth may be very slowly heating up. Yet, despite that, we are now living at the tag end of a mountain-building era (fortunately for those of us who are fond of rugged scenery). If the earth has not been cooling and shrinking during that period, how were our present mountains built?

A couple of decades ago a theory was put forward by the Israeli physicist Chaim L. Perkeris and elaborated by the American geologist D. T. Briggs. This theory, which resembles the later notion of the spreading of the ocean floor, begins by supposing that heat coming from the core periodically sets up a series of vertical eddies in the mantle. The eddies of heated material rise toward the crust and sink again after they cool there. Since the mantle is not liquid, merely plastic, this motion is very slow—perhaps not more than two inches a year.

Now, where two neighboring eddies move downward, a portion of crust is sucked downward too, forming a root of light crustal material in the heavier mantle. This root is converted by the mantle's heat into granite. Afterward, isostasy causes the root and its overlay of light material to rise and form a mountain chain. The period of mountain building, lasting perhaps 60 million years, is followed by a quiescent period of 500 million years during which enough heat accumulates in the mantle to start a new cycle. It may be then that mountain-building and continental drift are interrelated.

The Ocean

The earth is unusual among the planets of the solar system in possessing a surface temperature that permits water to exist in all three states: liquid, solid, and gas. The earth is also the only body in the solar system, as far as we know, to have oceans. Actually I should say "ocean," because the Pacific, Atlantic, Indian, Arctic, and Antarctic oceans all comprise one connected body of salt water in which the Europe-Asia-Africa mass, the American continents, and smaller bodies such as Antarctica and Australia can be considered islands.

The statistics of this ocean are impressive. It has a total area of 140 million square miles and covers 71 per cent of the earth's surface. Its volume, reckoning the average depth of the oceans as two and one third miles, is about 326 million cubic miles—0.15 per cent of the total volume of our planet. It contains 97.2 per cent of all the H_2O on the earth and is the source of the earth's fresh water supply as well, for 80,000 cubic miles of it are evaporated each year to fall again as rain or snow. As a result of such precipitation, there is some 200,000 cubic miles of fresh water under the continents' surface and about 30,000 cubic miles of fresh water gathered into the open as lakes and rivers.

The ocean is of peculiar importance to life. Almost certainly the first forms of life originated there, and, from the standpoint of sheer quantity, the oceans still contain most of our planet's life. On land, life is confined to within a few feet of the surface (though birds and airplanes do make temporary sorties from this base); in the oceans, life permanently occupies the whole of a realm as deep as seven miles or more in some places.

And yet, until recent years mankind has been as ignorant of the oceans, and particularly of the ocean floor, as of another planet. Even today, astronomers know more about the surface of the moon than geologists know about the surface of the earth under the oceans.

The founder of modern oceanography was an American naval officer named Matthew Fontaine Maury. In his early thirties, he was lamed in an accident that, however unfortunate for himself, brought benefits to humanity. Placed in charge of the depot of charts and instruments (undoubtedly intended as a sinecure), he threw himself into the task of charting ocean currents. In particular, he studied the course of the Gulf Stream, which had first been investigated as early as 1769 by the American scholar Benjamin Franklin. Maury gave it a description that has become a classic remark in oceanography: "There is a river in the ocean." It is certainly a much larger river than any on land. It transports a thousand times as much water each second as does the Mississippi. It is 50 miles wide at the start, nearly ½ mile deep, and moves at speeds of up to 4 miles an hour. Its warming effect is felt even in the far northern island of Spitzbergen.

Maury also initiated international cooperation in studying the ocean; he was the moving figure behind a historic international conference held in Brussels in 1853. In 1855, he published the first textbook in oceanography, entitled *Physical Geography of the Sea.* The Naval Academy at Annapolis honored his achievements by naming Maury Hall after him.

Since Maury's time, the ocean currents have been thoroughly mapped. They move in large clockwise circles in the oceans of the Northern Hemisphere and in large counterclockwise circles in those of the Southern, thanks to the Coriolis effect (see page 105). A current moving directly along the Equator is not subjected to a Coriolis effect and may move in a straight line. Such a thin, straight current was located in the Pacific Ocean, moving due east for several thousand miles along the Equator. It is called the "Cromwell current" after its discoverer, the American oceanographer Townsend Cromwell. A similar current, somewhat slower, was discovered in the Atlantic, in 1961, by the American oceanographer Arthur D. Voorhis.

Furthermore, oceanographers have even begun to explore the more sluggish circulation of the ocean depths. That the deeps cannot maintain a dead calm is clear from several indirect forms of evidence. For one thing, the life at the top of the sea is continually consuming its mineral nutrients—phosphate and nitrate—and carrying this material down to the depths with itself after death; if there were no circulation to bring it up again, the surface would become depleted of these minerals. For another thing, the oxygen supplied to the oceans by absorption from the air would not percolate down to the depths at a sufficient rate to support life there if there were no conveying circulation. Actually oxygen is found in adequate concentration down to the very floor of the abyss. This can be explained only by supposing that there are regions in the ocean where oxygen-rich surface waters sink.

The engine that drives this vertical circulation is temperature difference. The ocean's surface water is cooled in arctic regions, and it therefore sinks. This continual flow of sinking water spreads out all along the ocean floor, so that even in the tropics the bottom water is very cold —near the freezing point. Eventually the cold water of the depths wells up toward the surface, for it has no other place to go. After rising to the surface, the water warms and drifts off toward the Arctic or the Antarctic, there to sink again. The resulting circulation, it is estimated, would

126

bring about complete mixing of the Atlantic Ocean, if something new were added to part of it, in about 1,000 years. The larger Pacific Ocean would undergo complete mixing in perhaps 2,000 years.

The continental barriers complicate this general picture. To follow the actual circulations, oceanographers have resorted to oxygen as a tracer. Cold water absorbs more oxygen than warm water can. The arctic surface water, therefore, is particularly rich in oxygen. After it sinks, it steadily loses oxygen to organisms feeding in it. So by sampling the oxygen concentration in deep water at various locations, it is possible to plot the direction of the deep-sea currents.

Such mapping has shown that one major current flows from the Arctic Ocean down the Atlantic under the Gulf Stream and in the opposite direction, another from the Antarctic up the south Atlantic. The Pacific Ocean gets no direct flow from the Arctic to speak of because the only outlet into it is the narrow and shallow Bering Strait. This is why it is the end of the line for the deep-sea flow. That the north Pacific is the dead end of the global flow is shown by the fact that its deep waters are poor in oxygen. Large parts of this largest ocean are therefore sparsely populated with life forms and are the equivalent of desert areas on land. The same may be said of nearly land-locked seas like the Mediterranean, where full circulation of oxygen and nutrients is partly choked off.

More direct evidence for this picture of the deep-sea currents was obtained in 1957 during a joint British-American oceanographic expedition. The investigators used a special float, invented by the British oceanographer John C. Swallow, which is designed to keep its level at a depth of a mile or more and is equipped with a device for sending out short-wave sound waves. By means of these signals the float can be tracked as it moves with the deep-sea current. The expedition thus traced the deep-sea current down the Atlantic along its western edge.

All this information will acquire practical importance when the world's expanding population turns to the ocean for more food. Scientific "farming of the sea" will require knowledge of these fertilizing currents, just as land farming requires knowledge of river courses, ground water, and rainfall. The present harvest of seafood—some 55 million tons per year—can, with careful and efficient management, be increased (it is estimated) to something over 200 million tons per year, while leaving sea life enough leeway to maintain itself adequately. (This, of course, presupposes that we do not continue our present course of heedlessly damaging and polluting the ocean, particularly those portions of the ocean—nearest the continental shores—that contain and offer man the major portion of sea organisms. So far, we are not only failing to rationalize a more efficient use of the sea for food, but are decreasing its ability to yield us the quantity of food we harvest now.)

Food is not the only important resource of the ocean. Sea water contains in solution vast quantities of almost every element. As much as 4 billion tons of uranium, 300 million tons of silver, and 4 million tons of gold are contained in the oceans, but in dilution too great for practical extraction. However, both magnesium and bromine are now obtained from sea water on a commercial scale. By the end of the 1960's, the value of the magnesium obtained from the ocean was $70 million per year, while 75 per cent of all the bromine produced in the world came from the sea. Moreover, an important source of iodine is dried seaweed, the living plants having previously concentrated the element out of sea water to an extent that man cannot yet profitably duplicate.

Much more prosaic material is dredged up from the sea. From the relatively shallow waters bordering the United States, some 20 million tons of oyster shells are obtained each year to serve as a valuable source of limestone. In addition, 50 million cubic yards of sand and gravel are obtained in similar fashion.

Scattered over the deeper portions of the ocean floor are metallic nodules that have precipitated out about some nucleus that may be a pebble or a shark tooth. (It is the oceanic analog of the formation of a pearl about a sandgrain inside an oyster.) These are usually referred to as manganese nodules because they are richest in that metal. It is estimated that there are 31,000 tons of these nodules per square mile of the Pacific floor. Obtaining these in quantity would be difficult indeed and the manganese content alone would not make it worthwhile under present conditions. However, the nodules contain 1 per cent nickel, 0.5 per cent of copper, and 0.5 per cent cobalt. These minor constituents make the nodules far more attractive than they would otherwise be.

Even the 97 per cent of the ocean substance that is actually water is important. Mankind presses ever harder on the limited fresh-water supplies of the planet; eventually more and more use will have to be made of ocean water from which the salts have been removed, a process known as "desalination." Already some 700 desalination plants, with a capacity of up to 30,000 gallons of fresh water per day, exist throughout the world. On the whole, such sea-borne fresh-water cannot yet compete with rain-borne fresh-water in most parts of the world, but the technology involved is as yet young.

It is only within the last century that man has plumbed the great deeps of the ocean. The sea bottom first became a matter of practical interest to mankind (rather than one of intellectual curiosity to a few scientists) when it was decided to lay a telegraph cable across the Atlantic. In 1850, Maury had worked up a chart of the Atlantic sea-bottom for purposes of cable-laying. It took fifteen years, punctuated by many breaks and failures, before the Atlantic cable was finally laid—under the incredibly persevering drive of the United States financier Cyrus West Field, who lost a fortune in the process. (More than twenty cables now span the Atlantic.)

Systematic exploration of the sea bottom began with the famous around-the-world expedition of the British "Challenger" in the 1870's. To measure the depth of the oceans the "Challenger" had no better device than the time-honored method of paying out four miles of cable with a weight on the end until it reached the bottom. Over 360 soundings were made in this fashion. This procedure is not only fantastically laborious (for deep sounding) but also of low accuracy. Ocean-bottom exploration was revolutionized in 1922 with the introduction of echo-sounding by means of sound waves; in order to explain how this works, a digression on sound is in order.

Mechanical vibrations set up longitudinal waves in matter (in air, for instance), and we can detect some of these as sound. We hear different wavelengths as sounds of different pitch. The deepest sound we hear has a wavelength of 22 meters and a frequency of 15 cycles per second. The shrillest sound a normal adult can hear has a wavelength of 2.2 centimeters and a frequency of 15,000 cycles per second. (Children can hear somewhat shriller sounds.)

The absorption of sound by the atmosphere depends on the wavelength. The longer the wavelength, the less sound is absorbed by a given thickness of air. For this reason, foghorn blasts are far in the bass register so that they can penetrate as great a distance as possible. The foghorn of the "Queen Mary" sounds at twenty-seven vibrations per second, about that of the lowest note on the piano. It can be heard at a distance of 10 miles, and instruments can pick up the sound at a distance of 100 to 150 miles.

Sounds deeper in pitch than the deepest we can hear also exist. Some of the sounds set up by earthquakes or volcanoes are in this "infrasonic" range. Such vibrations can encircle the earth, sometimes several times, before being completely absorbed.

The efficiency with which sound is reflected depends on the wavelength in the opposite way. The shorter the wavelength, the more efficient the reflection. Sound waves with frequencies higher than those of the shrillest sounds we hear are even more efficiently reflected. Some animals can hear shriller sounds than we can and make use of this. Bats squeak to emit soundwaves with "ultrasonic" frequencies as high as 130,000 cycles per second and listen for the reflections. From the direction in which reflections are loudest and from the time lag between squeak and echo they can judge the location of insects to be caught and twigs to be avoided. They can thus fly with perfect efficiency if they are blinded, but not if they are deafened. (The Italian biologist Lazzaro Spallanzani, who first observed this in 1793, wondered if bats could see with their ears, and, of course, in a sense, they do.)

Porpoises, as well as guacharos (cave-dwelling birds of Venezuela), also use sounds for "echo-location" purposes. Since they are interested in locating larger objects, they can use the less efficient sound waves in the audible region for the purpose. (The complex sounds emitted by the large-brained porpoises and dolphins may even, it is beginning to be suspected, be used for purposes of general communication—for talking, to put it bluntly. The American biologist John C. Lilly has been investigating this possibility exhaustively.)

To make use of the properties of ultrasonic sound waves, men must first produce them. Small-scale production and use are exemplified by the "dog whistle" (first constructed in 1883). It produces sound in the near ultrasonic range that can be heard by dogs, but not by humans.

A route whereby much more could be done was opened by the French chemist Pierre Curie and his brother, Jacques, who in 1880 discovered that pressures on certain crystals produced an electric potential ("piezoelectricity"). The reverse was also true. Applying an electric potential to a crystal of this sort produced a slight constriction as though pressure were being applied ("electrostriction"). When the technique for producing a very rapidly fluctuating potential was developed, crystals could be made to vibrate quickly enough to form ultrasonic waves. This was first done in 1917 by the French physicist Paul Langevin, who immediately applied the excellent reflective powers of this short-wave sound to the detection of submarines. During World War II, this method was perfected and became "sonar" ("sound navigation and ranging," "ranging" meaning "determining distance").

The determination of the distance of the sea bottom by the reflection of ultrasonic sound waves was what replaced the sounding line. The time interval from the sending of the signal (a sharp pulse) and the return of its echo measures the distance to the bottom. The only thing the operator has to worry about is whether the reading signals a false echo from a school of fish or some other obstruction. (Obviously the instrument is useful to fishing fleets.)

The echo-sounding method, not only is swift and convenient, but also makes it possible to trace a continuous profile of the bottom over which the vessel moves, so that oceanographers are obtaining a picture of the topography of the ocean bottom. It turns out to be more rugged than the land surface, and its features have a grander scale. There are plains of continental size and mountain ranges longer and higher than any on land. The island of Hawaii is the top of an underwater mountain 33,000 feet high—higher than anything in the Himalayas—so that Hawaii may fairly be called the tallest mountain on the earth. There are also numerous flat-topped cones, called "seamounts" or "guyots." The latter name honors the Swiss-American geographer Arnold Henry Guyot, who brought scientific geography to the United States when he emigrated to America in 1848. Seamounts were first discovered during World War II by the American geologist Harry Hammond Hess, who located nineteen in quick succession. At least 10,000 exist, mostly in the Pacific. One of these, discovered in 1964 just south of Wake Island, is over 14,000 feet high.

Moreover, there are deep abysses (trenches) in which the Grand Canyon would be lost. The trenches, all located alongside island archipelagoes, have a total area amounting to nearly 1 per cent of the ocean bottom. This may not seem much, but it is actually equal to one half the area of the United States, and the trenches contain fifteen times as much water as all the rivers and lakes in the world. The deepest of them are in the Pacific; they are found there alongside the Philippines, the Marianas, the Kuriles, the Solomons, and the Aleutians. There are other great abysses in the Atlantic off the West Indies and the South Sandwich Islands, and there is one in the Indian Ocean off the East Indies.

Besides the trenches, oceanographers have traced on the ocean bottom canyons, sometimes thousands of miles long, which look like river channels. Some of them actually seem to be extensions of rivers on land, notably a canyon extending from the Hudson River into the Atlantic. At least twenty such huge gouges have been located in the Bay of Bengal alone, as a result of oceanographic studies of the Indian Ocean during the 1960's. It is tempting to suppose that these were once river beds on land, when the ocean was lower than now. But some of the undersea channels are so far below the present sea level that it seems altogether unlikely they could ever have been above the ocean. In recent years, various oceanographers, notably Maurice Ewing and Bruce C. Heezen, have developed another theory: that the undersea canyons were gouged out by turbulent flows ("turbidity currents") of soil-laden water in an avalanche down the off-shore continental slopes at speeds of up to sixty miles an hour. One turbidity current, which focused scientific attention on the problem, took place in 1929 after an earthquake off Newfoundland. The current snapped a number of cables, one after the other, and made a great nuisance of itself.

The most dramatic find concerning the sea bottom, though, was foreshadowed as 1853, when the Atlantic-cable project was in progress. Soundings of the ocean depth were taken, and it was reported that there seemed signs of an undersea plateau in the middle of the ocean. The Atlantic seemed shallower in the middle than on either side.

Naturally, it was only practical to make a few soundings by actual line, but in 1922, the German oceanographic vessel *Meteor* began to make soundings in the Atlantic with ultrasonic devices. By 1925, they were able to report a vast undersea mountain-range winding down the Atlantic. The highest peak broke through the water surface and appeared as islands such as the Azores, Ascension, and Tristan da Cunha.

Later soundings elsewhere showed that the mountain range was not confined to the Atlantic. At its southern end it curves around Africa and moves up the western Indian Ocean to Arabia. In mid-Indian Ocean, it branches so that the range continues south of Australia and New Zealand and then works northward in a vast circle all around the Pacific Ocean. What began (in men's minds) as the Mid-Atlantic Ridge became the Mid-Oceanic Ridge. And in one rather basic fashion, the Mid-Oceanic Ridge is not like the mountain ranges on the continent. The continental highlands are of folded sedimentary rocks, while the vast oceanic ridge is of basalt squeezed up from the hot lower depths.

After World War II, the details of the ocean floor were probed with new energy by Ewing and Heezen. Detailed soundings in 1953 showed, rather to their astonishment, that a deep canyon ran the length of the Ridge and right along its center. This was eventually found to exist in all portions of the Mid-Oceanic Ridge, so that sometimes it is called the "Great Global Rift." There are places where the Rift comes quite close to land: it runs up the Red Sea between Africa and Arabia, and it skims the borders of the Pacific through the Gulf of California and up the coast of the state of California.

At first it seemed that the Rift might be continuous, a 40,000-mile crack in the earth's crust. Closer examination, however, showed that it consisted of short, straight sections that were set off from each other as though earthquake shocks had displaced one section from the next. And, indeed, it was along the Rift that the earth's quakes and volcanoes tended to occur.

The Rift was a weak spot up through which heated molten rock, "magma," welled slowly from the interior—cooling, piling up to form the Ridge, and spreading out farther still. The spreading can be as rapid as 16 centimeters per year, and the entire Pacific Ocean floor could be covered with a new layer in 100 million years. Indeed, sediment drawn up from the ocean floor is rarely found to be older, which would be remarkable in a planetary life 45 times as long, were it not for the concept of "sea-floor spreading."

The Rift and its branches seem to divide the earth's crust into six large plates and some smaller ones. As a result of the activity along the Rift, these plates move, but as units; there is no motion to speak of among the surface features of a given plate. It is the movement of these plates that accounts for the breakup of Pangaea and the continental drifting since. There is nothing to show that the drifting may not eventually bring the continents together again, perhaps in a new arrangement. There may have been many Pangaeas formed and broken up in the earth's lifetime, with the latest breakup most clearly seen in the records only because it is the latest.

This concept of the motion of the plates may serve to explain many features of earth's crust whose origin was obscure earlier. When two plates come together slowly, the crust buckles and bulges both up and down, forming mountains and their "roots." Thus, the Himalayan Mountains seem to have been formed when the plate bearing India made slow contact with the plate bearing the rest of Asia.

On the other hand, when two plates come together too rapidly to allow buckling, the surface of one plate may gouge its way under the other, forming a trench, a line of islands, and a disposition toward volcanic activity. Such trenches and islands are found in the western Pacific, for instance.

Plates pull apart under the influence of sea-floor spreading, as well as come together. The Rift passes right through western Iceland, which is (very slowly) pulling apart. Another place of division is at the Red Sea, which is rather young and exists only because Africa and Arabia have already pulled apart somewhat. (The opposite shores of the Red Sea fit closely if put together.) This process is continuing, so that the Red Sea is, in a sense, a new ocean in the process of formation. Active upwelling in the Red Sea is indicated by the fact that at the bottom of that body of water there are, as discovered in 1965, sections with a temperature of 56° C. and a salt concentration at least five times normal.

The existence of the Rift is of greatest immediate importance, naturally, to those people who live on those parts of earth's land surface that happen to be in its neighborhood. The San Andreas Fault in California is actually part of the Rift, for instance, and it was the yielding of that fault which caused the San Francisco earthquake of 1906 and the Good Friday earthquake in Alaska in 1964.

The deep sea, surprisingly enough, contains life. Until nearly a century ago, life in the ocean was thought to be confined to the surface region. The Mediterranean, long the principal center of civilization, is indeed rather barren of life in its lower levels. But though this sea is a semidesert—warm and low in oxygen—the English naturalist Edward Forbes dredged up living starfish from a depth of a quarter of a mile in the 1840's. Then, in 1860, a telegraph cable was brought up from the Mediterranean bottom, a mile deep, and was found to be encrusted with corals and other forms of life.

In 1872, the "Challenger," under the direction of the British naturalist Charles Wyville Thomson, in a voyage spanning 69,000 miles, made the first systematic attempt to dredge up life forms from the ocean bottom; he found plenty. Nor is the world of underwater life a region of eerie silence by any means. An underwater-listening device, the "hydrophone," has, in recent years, shown that sea creatures click, grunt, snap, moan, and, in general, make the ocean depths as maddeningly noisy as ever the land is.

Since World War II, numerous expeditions have explored the abyss. A new "Challenger" probed the Marianas Trench in the western Pacific in 1951 and found that it (and not one off the Philippine Islands) was the deepest gash in the earth's crust. The deepest portion is now called the "Challenger Deep." It is over 36,000 feet deep. If Mount

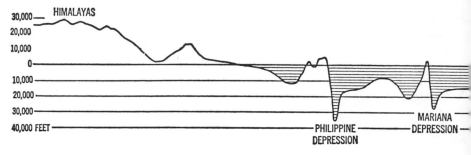

Profile of the Pacific bottom. The great trenches in the sea
floor go deeper below sea level than the height of the
Himalayas, and the Hawaiian peak stands higher from the
bottom than the tallest land mountain.

Everest were placed in it, a mile of water would roll over its topmost
peak. Yet the "Challenger" brought up bacteria from the floor of the
abyss. They look much like bacteria of the surface, but cannot live at a
pressure of less than a thousand atmospheres!

The creatures of the trenches are so adapted to the great pressures
of these bottoms that they are unable to rise out of their trench; in
effect, they are imprisoned in an island. They have experienced a segre-
gated evolution. Yet they are in many respects related to other organisms closely enough so that it seems their evolution in the abyss has not
gone on for a very long time. One can visualize some groups of ocean
creatures being forced into lower and lower depths by the pressure of
competition, just as other groups were forced higher and higher up the
continental shelf until they emerged onto the land. The first group had
to become adjusted to higher pressures, the second to the absence of
water. On the whole, the latter adjustment was probably the more
difficult, so we should not be amazed that life exists in the abyss.

To be sure, life is not as rich in the depths as nearer the surface. The
mass of living matter below four and one half miles is only a tenth as
great per unit volume of ocean as it is estimated to be at two miles. Furthermore, there are few, if any, carnivores below four and one half
miles, since there are insufficient prey to support them. They are scav-
engers instead, eating anything organic that they can find. The recent-
ness with which the abyss has been colonized is brought out by the
disclosure that no species of creature found there has been developed
earlier than 200 million years ago, and most have histories of no more
than 50 million years. It is only at the beginning of the age of the dinosaurs that the deep sea, hitherto bare of organisms, was finally invaded
by life.

Nevertheless, some of the organisms that invaded the deep survived
there, whereas their relatives nearer the surface died out. This was
demonstrated, most dramatically, in the late 1930's. On December 25,

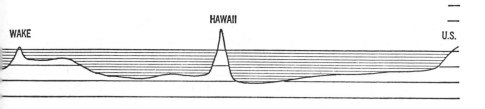

HAWAII

WAKE

U.S.

1938, a trawler fishing off South Africa brought up an odd fish about five feet long. What was odd about it was that its fins were attached to fleshy lobes rather than directly to the body. A South-African zoologist, J. L. B. Smith, who had the chance of examining it, recognized it as a matchless Christmas present. It was a coelacanth, a primitive fish that zoologists had thought extinct for 70 million years. Here was a living specimen of an animal that was supposed to have disappeared from the earth before the dinosaurs reached their prime.

World War II halted the hunt for more coelacanths, but in 1952 another of a different genus was fished up off Madagascar. By now numbers have been found. Because it is adapted to fairly deep waters, the coelacanth dies soon after being brought to the surface.

Evolutionists have been particularly interested in studying the coelacanth specimens because it was from this fish that the first amphibians developed; in other words, the coelacanth is a rather direct descendant of our fishy ancestors.

Just as the ideal way to study outer space is to send men out there, so the ideal way to study the ocean deeps is to send men down there. The first practical diving-suit to make this possible was designed in 1830 by Augustus Siebe. A diver in a modern diving suit can go down about 300 feet. In 1934, Charles William Beebe managed to get down to about 3,000 feet in his "bathysphere," a small, thick-walled craft equipped with oxygen and with chemicals to absorb carbon dioxide. His co-worker, Otis Barton, plumbed to a depth of 4,500 feet in 1948, using a modified bathysphere called a "benthoscope."

The bathysphere was an inert object suspended from a surface vessel by a cable (a snapped cable meant the end). What was needed was a maneuverable ship of the abyss. Such a ship, the "bathyscaphe," was invented in 1947 by the Swiss physicist Auguste Piccard. Built to withstand great pressures, it used a heavy ballast of iron pellets (which are automatically jettisoned in case of emergency) to take it down and a "balloon" containing gasoline (which is lighter than water) to provide buoyancy and stability. In its first test off Dakar, West Africa, in 1948, the bathyscaphe (unmanned) descended 4,500 feet.

Later, Piccard and his son Jacques built an improved version of the bathyscaphe and named the new vessel "Trieste," because the then Free

City of Trieste had helped finance its construction. In 1953, Piccard plunged two and a half miles into the depths of the Mediterranean.

The "Trieste" was bought by the United States Navy for research. On January 14, 1960, Jacques Piccard and a Navy man, Don Walsh, took it to the bottom of the Marianas Trench, plumbing seven miles to the deepest part of any abyss. There, at the ultimate ocean depth, where the pressure was 1,100 atmospheres, they found water currents and living creatures. In fact, the first creature seen was a vertebrate, a one-foot-long, flounderlike fish, with eyes.

In 1964, the French-owned bathyscaphe "Archimède" made ten trips to the bottom of the Puerto Rico Trench, which, with a depth of five and one quarter miles, is the deepest abyss in the Atlantic. There, too, every square foot of the ocean floor had its life form. Oddly enough, the bottom did not descend smoothly into the abyss; rather it seemed terraced, like a giant, spread-out staircase.

The Icecaps

The extremities of our planet have always fascinated mankind, and one of the most adventurous chapters in the history of science has been the exploration of the polar regions. Those regions are charged with romance, spectacular phenomena, and elements of man's destiny—the strange auroras in the sky, the extreme cold, and especially the immense ice-caps, or glaciers, which hold the key to the world climate and man's way of life.

The actual push to the poles came rather late in human history. It began during the great age of exploration following the discovery of the Americas by Christopher Columbus. The first Arctic explorers went chiefly to find a sea route around the top of North America. Pursuing this will-o'-wisp, the English navigator Henry Hudson (in the employ of Holland) found Hudson Bay and his death in 1610. Six years later, another English navigator, William Baffin, discovered what came to be called Baffin Bay and penetrated to within 800 miles of the North Pole. Eventually, in the years 1846 to 1848, the British explorer John Franklin worked his way over the northern coast of Canada and discovered the "Northwest Passage" (and a most impractical passage for ships it then was). He died on the voyage.

There followed a half century of efforts to reach the North Pole, motivated in large part by sheer adventure and the desire to be the first to get there. In 1873, the Austrian explorers Julius Payer and Carl Weyprecht reached within 600 miles of the Pole and named a group of islands they found Franz Josef Land, after the Austrian emperor. In 1896, the Norwegian explorer Fridtjof Nansen drifted on the Arctic ice to within 300 miles of the Pole. At length, on April 6, 1909, the American explorer Robert Edwin Peary arrived at the Pole itself.

Saturn and its rings, photographed with the 100-inch telescope on Mount Wilson.

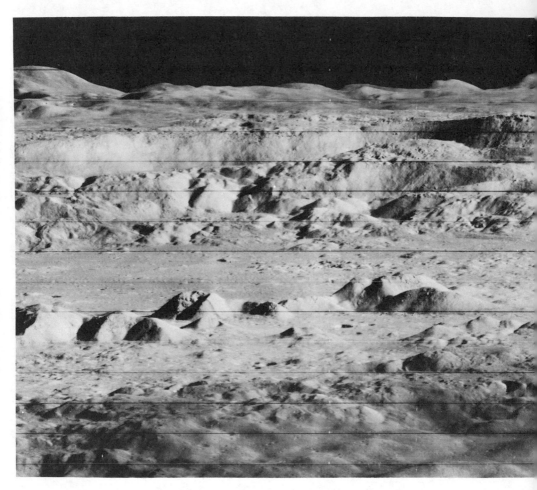

This photograph of the crater Copernicus was taken from 28.4 miles above the surface of the moon by Lunar Orbiter II.

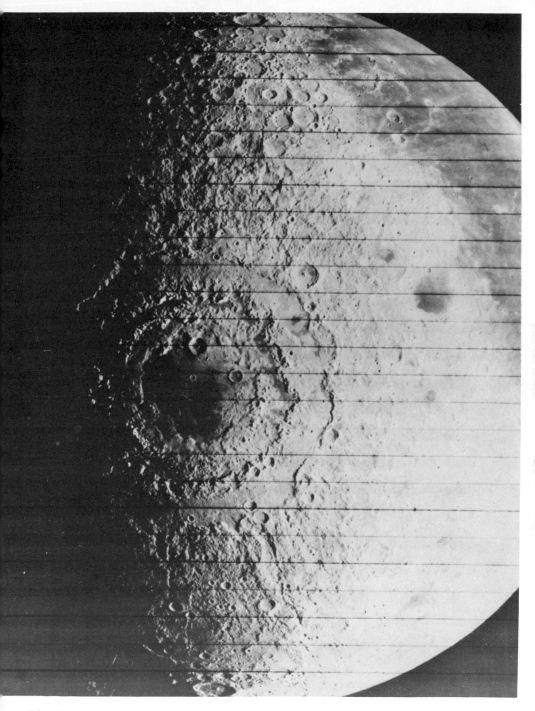

The Orientale Basin photographed from 1690 miles above the
moon's surface by Lunar Orbiter IV.

Foucault's famous experiment in Paris in 1851, which showed the rotation of the earth on its axis by means of the swing of a pendulum; the plane of its swings turned clockwise.

The Ranger VII spacecraft, which took close-up photographs of the moon before impacting on the lunar surface on July 31, 1964. The Ranger VII mission terminated with the acquisition of some 4,000 television records of a preselected area of the lunar surface. Six television cameras transmitted pictures during the last seventeen minutes of flight, the last being taken at an altitude of approximately 480 meters above the moon's surface. The Ranger project is part of the National Aeronautics and Space Administration's Lunar and Planetary Programs.

Astronaut Edwin E. Aldrin Jr., lunar module pilot, is photographed walking near the lunar module during the Apollo II extravehicular activity.

Apollo II astronaut Edwin E. Aldrin deploys Solar Wind Composition experiment on moon's surface.

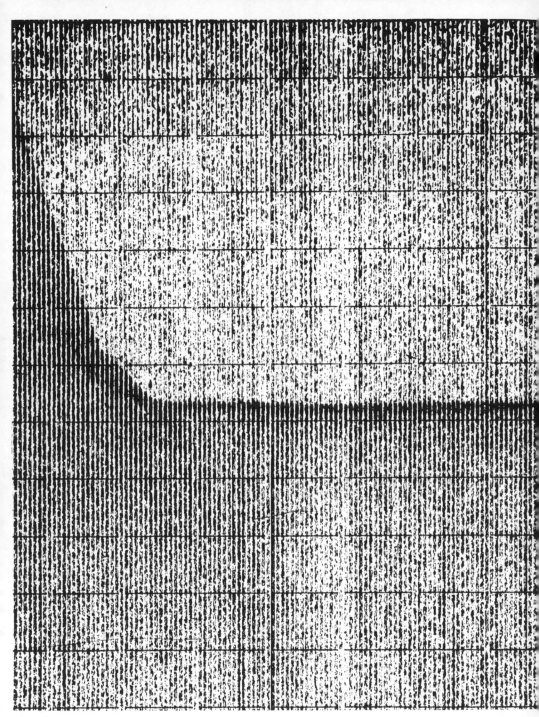

Foot of a continent. Trace made with precision depth recorder of the foot of the North American continent east of Eleuthera Island, Bahamas. The vertical exaggeration is 12 to 1. The sloping surface (*left*), with an inclination of 16°, is the foot of the continental slope. The flat surface is a portion of the Abyssal Plain, which occurs here at a depth of 4,825 meters.

Concretions on the rim of the Kharga Oasis depression in the western desert of Egypt. Formed by ground water in Eocene limestone, they remained after the crumbling rock in which they were embedded weathered away. Their surfaces are deeply etched by sandblast.

Glacial strata exposed in the Franz Josef Fiord of northeastern Greenland.

Glacial formation atop Mount Kilimanjaro in central Africa.

Sedimentary formation under water in Moriches Inlet off Long Island, photographed from an airplane.

Grand Canyon from the air, clearly showing the erosion of the rock by the Colorado River.

The heart of Antarctica. An aerial photograph of the Sentinel Mountains made during the IGY.

By now, the North Pole has lost much of its mystery. It has been explored on the ice, from the air, and under water. Richard Evelyn Byrd and Floyd Bennett were the first to fly over it, in 1926, and submarines have traversed its waters.

Meanwhile, the largest northern icecap, which is centered in Greenland, has drawn a number of scientific expeditions. The Greenland glacier has been found to cover about 640,000 of that island's 840,000 square miles, and its ice is known to reach a thickness of a mile in some places.

As the ice accumulates, it is pushed down to the sea, where the edges break off or "calve" to form icebergs. Some 16,000 icebergs are thus formed in the Northern Hemisphere each year, 90 per cent of them breaking off the Greenland icecap. The icebergs work slowly southward, particularly down the west Atlantic. About 400 icebergs per year pass Newfoundland and threaten shipping lanes; between 1870 and 1890, fourteen ships were sunk and forty damaged by collision with icebergs.

The climax came in 1912, when the luxury liner "Titanic" collided with an iceberg and sank on her maiden voyage. An international watch over the positions of these inanimate monsters has been maintained ever since. During the years since this Ice Patrol has come into existence, not one ship has been sunk by an iceberg.

Far larger than Greenland is the South Pole's great continental glacier. The Antarctic icecap covers seven times the area of the Greenland glacier and has an average thickness of one and one half miles, with three-mile depths in spots. This is due to the great size of the Antarctic continent—some 5 million square miles, though how much of this is land and how much ice-covered sea is still uncertain. Some explorers believe that Antarctica is a group of large islands bound together by ice, but at the moment the continent theory seems to have the upper hand.

The famous English explorer James Cook (better known as Captain Cook) was the first European to cross the Antarctic Circle. In 1773, he circumnavigated the Antarctic regions. (It was perhaps this voyage that inspired Samuel Taylor Coleridge's *The Rime of the Ancient Mariner*, published in 1798, which described a voyage from the Atlantic to the Pacific by way of the icy regions of Antarctica.)

In 1819, the British explorer Williams Smith discovered the South Shetland Islands, just fifty miles off the coast of Antarctica; in 1821, a Russian expedition sighted a small island ("Peter I Island") within the Antarctic Circle; and, in the same year, the Englishman George Powell and the American Nathaniel B. Palmer first laid eyes on a peninsula of the Antarctic continent itself—now called Palmer Peninsula.

In the following decades, explorers inched toward the South Pole. By 1840, the American naval officer Charles Wilkes announced that the land strikes added up to a continental mass, and, subsequently, he was proved right. The Englishman James Weddell penetrated an ocean inlet east of Palmer Peninsula (now called Weddell Sea) to within 900 miles of the Pole. Another British explorer, James Clark Ross, discovered the other major ocean inlet into Antarctica (now called the Ross Sea) and got within 710 miles of the Pole. In 1902–04, a third Briton, Robert Falcon Scott, traveled over the Ross ice shelf (a section of ice-covered ocean as large as the state of Texas) to within 500 miles of it. And, in 1909, still another Englishman, Ernest Shackleton, crossed the ice to within about 100 miles of the Pole.

On December 16, 1911, the goal was finally reached by the Norwegian explorer Roald Amundsen. Scott, making a second dash of his own, got to the South Pole just three weeks later, only to find Amundsen's flag already planted there. Scott and his men perished on the ice on their way back.

In the late 1920's, the airplane helped to make good the conquest of Antarctica. The Australian explorer George Hubert Wilkins flew over 1,200 miles of its coastline, and Richard Evelyn Byrd, in 1929, flew over the South Pole. By that time the first base, Little America I, had been established in the Antarctic.

The North and South polar regions became focal points of the greatest international project in science of modern times. This had its origin in 1882–83, when a number of nations joined in an "International Polar Year" of exploration and scientific investigation of phenomena such as the aurorae, the earth's magnetism, etc. The project was so successful that, in 1932–33, it was repeated with a second International Polar Year. In 1950, the United States geophysicist Lloyd Berkner (who had been a member of the first Byrd Antarctic Expedition) proposed a third such year. The proposal was enthusiastically adopted by the International Council of Scientific Unions. This time scientists were prepared with powerful new research instruments and bristling with new questions—about cosmic rays, about the upper atmosphere, about the ocean depths, even about the possibility of the exploration of space. An ambitious "International Geophysical Year" (IGY) was arranged, and the time selected was July 1, 1957, to December 31, 1958 (a period of maximum sunspot activity). The enterprise enlisted heart-warming international cooperation; even the cold-war antagonists, the Soviet Union and the United States, managed to bury the hatchet for the sake of science.

Although the most spectacular achievement of the IGY, from the standpoint of public interest, was the successful launching of man-made satellites by the Soviet Union and the United States, science reaped many other fruits which were no less important. Outstanding among these was a vast international exploration of Antarctica. The United States alone set up seven stations, probing the depth of the ice and

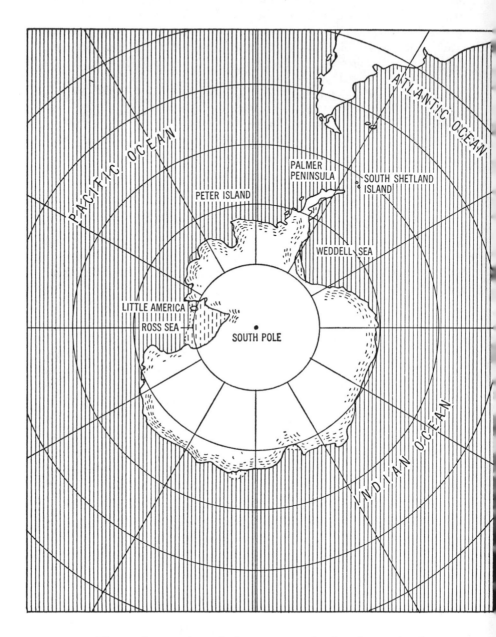

The major continental glaciers are today largely restricted to Greenland and Antarctica. At the height of the last ice age, the glaciers extended over most of northern and western Europe and south of the Great Lakes on the North American continent.

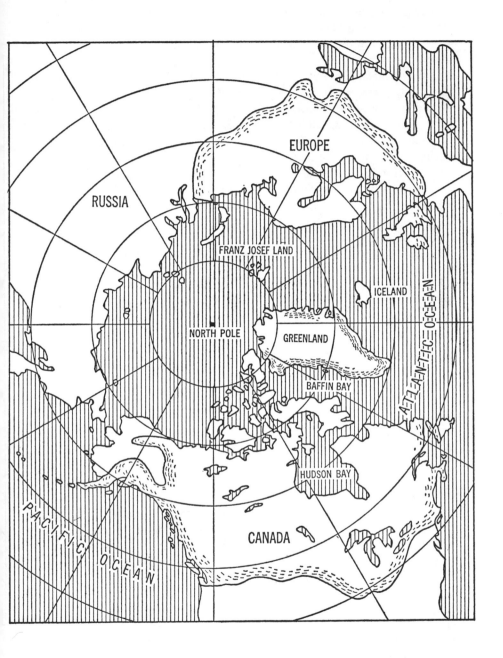

bringing up from miles down samples of the air trapped in it (which must date back millions of years) and of bacterial remnants. Some bacteria, frozen one hundred feet below the ice surface and perhaps a century old, were revived and grew normally. In January, 1958, the Soviet group established a base at the "Pole of Inaccessibility"—the spot in Antarctica farthest inland—and there, 600 miles from the South Pole, recorded new lows in temperature. In August 1960—the Antarctic midwinter—a temperature of $-127°$ F., cold enough to freeze carbon dioxide, was recorded. In the course of the following decade, dozens of year-round stations were operating in Antarctica.

In the most dramatic Antarctic feat, a British exploring team headed by Vivian Ernest Fuchs and Edmund Percival Hillary crossed the continent by land for the first time in history (with special vehicles and all the resources of modern science at their disposal, to be sure). Hillary, by the way, had also been the first, along with the Sherpa mountaineer Tenzing Norgay, to climb Mount Everest, the highest mountain on earth, in 1953.

The success of the IGY and the warmth generated by this demonstration of cooperation in the midst of the cold war led to an agreement in 1959 among twelve nations to bar all military activities (including nuclear explosions and the dumping of radioactive wastes) from the Antarctic. Thus Antarctica will be reserved for scientific activities.

The earth's load of ice, amounting to nearly 9 million cubic miles, covers about 10 per cent of its land area. About 86 per cent of the ice is piled up in the Antarctic continental glacier and 10 per cent in the Greenland glacier. The remaining 4 per cent makes up the small glaciers in Iceland, Alaska, the Himalayas, the Alps, and a few other locations.

The Alpine glaciers have been under study for a long time. In the 1820's, two Swiss geologists, J. Venetz and Jean de Charpentier, noticed that rocks characteristic of the central Alps were scattered over the plains to the north. How had they got there? The geologists speculated that the mountain glaciers had once covered a much larger area and had left boulders and piles of debris behind when they retreated.

A Swiss zoologist, Jean Louis Rodolphe Agassiz, looked into this notion. He drove lines of stakes into the glaciers and waited to see whether they moved. By 1840, he had proved beyond doubt that glaciers flowed like very slow rivers at a rate of about 225 feet per year. Meanwhile, he had traveled over Europe and found marks of glaciers in France and England. He found boulders foreign to their surroundings in other areas and scoured marks on rock that could only have been made by the grinding of glaciers, carrying pebbles encrusted along their bottoms.

Agassiz went to the United States in 1846 and became a Harvard professor. He found signs of glaciation in New England and the Midwest.

By 1850, it seemed quite obvious that there must have been a time when a large part of the Northern Hemisphere was under a large continental glacier. The deposits left by the glacier have been studied in detail since Agassiz' time. These studies have shown that the ice advanced and retreated four times. They were as far south as Cincinnati a mere 18,000 years ago. When they advanced, the climate to the south was wetter and colder; when they retreated (leaving lakes behind, of which the largest still in existence are the Canadian-American Great Lakes), the climate to the south grew warmer and drier.

The last retreat of the ice took place between 8,000 and 12,000 years ago. Before the ice ages, there was a period of mild climate on the earth lasting at least 100 million years. There were no continental glaciers, even at the poles. Coal beds in Spitzbergen and signs of coal even in Antarctica testify to this, because coal marks the site of ancient lush forests.

The coming and going of glaciers leaves its mark, not only on the climate of the rest of the earth, but on the very shape of the continents. For instance, if the now-shrinking glaciers of Greenland and Antarctica were to melt completely, the ocean level would rise nearly 200 feet. It would drown the coastal areas of all the continents, including many of the world's largest cities, with the water level reaching the twentieth story of the Manhattan skyscrapers. On the other hand, Alaska, Canada, Siberia, Greenland, and even Antarctica would become more habitable.

The reverse situation takes place at the height of an ice age. So much water is tied up in the form of land-based icecaps (up to three or four times the present amount) that the sea-level mark is as much as 440 feet lower than it now is. When this is so, the continental shelves are exposed.

The continental shelves are relatively shallow portions of the ocean adjoining the continents. The sea floor slopes more or less gradually until a depth of about 130 meters is achieved. After this the slope is much steeper, and considerably greater depths are achieved rapidly. The continental shelves are, structurally, part of the continents they adjoin: it is the edge of the shelf that is the true boundary of the continent. What it amounts to is that at the present moment, there is enough water in the ocean basins to flood the borders of the continent.

Nor is the continental shelf small in area. It is much broader in some places than others; there is considerable shelf area off the east coast of the United States, but little off the west coast (which is at the edge of a crustal plate). On the whole, though, the continental shelf is some fifty miles wide on the average and makes up a total area of 10 million square miles. In other words, a potential continental area rather greater than the Soviet Union in size, is drowned under ocean waters.

It is this area that is exposed during periods of maximum glaciation and was indeed exposed in the last great Ice Ages. Fossils of land animals (such as the teeth of elephants) have been dredged up from the continental shelves, miles from land and under yards of water. What's more, with the northern continental sections ice-covered, rain was more common than now, farther south, so that the Sahara Desert was then grassland. The drying of the Sahara as the icecaps receded took place not long before the beginning of historic times.

There is thus a pendulum of habitability. As the sea level drops, large continental areas become deserts of ice, but the continental shelves become habitable, as do present-day deserts. As the sea-level rises, there is further flooding of the lowlands, but the polar regions become habitable, and again deserts retreat.

The major question regarding the Ice Ages involves their cause. What makes the ice advance and retreat, and why is it that the glaciations have been relatively brief, the present one having occupied only 1 million of the last 100 million years?

It takes only a small change in temperature to bring on or to terminate an ice age—just enough fall in temperature to accumulate a little more snow in the winter than melts in the summer or enough rise to melt a little more snow in the summer than falls in the winter. It is estimated that a drop in the earth's average annual temperature of only 3.5° C. is sufficient to make glaciers grow, whereas a rise of the same amount would melt Antarctica and Greenland to bare rock in a matter of centuries.

Such changes in the temperature of the earth have indeed taken place in the past. A method has now been evolved by which primeval temperatures can be measured with amazing accuracy. The American

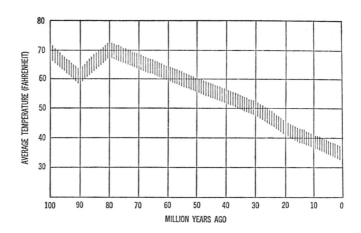

The record of the ocean temperatures during the last 100 million years.

chemist Jacob Bigeleisen, working with H. C. Urey, showed in 1947, that the ratio of the common variety of oxygen (oxygen-16) to its rarer isotopes (e.g., oxygen-18), present in compounds, would vary with temperature. Consequently, if one measured the ratio of oxygen-16 to oxygen-18 in an ancient fossil of a sea animal, one could tell the temperature of the ocean water at the time the animal lived. By 1950, Urey and his group had developed the technique to so fine a point that by analyzing the shell layers of a millions-of-years-old fossil (an extinct form of squid), they could determine that the creature was born during a summer, lived four years, and died in the spring.

This "thermometer" has established that 100 million years ago the average world-wide ocean temperature was about 70° F. It cooled slowly to 61° 10 million years later and then rose to 70° again 10 million years after that. Since then, the ocean temperature has declined steadily. Whatever triggered this decline may also be a factor in the extinction of the dinosaurs (who were probably adapted to mild and equable climates) and put a premium on the warm-blooded birds and mammals, who can maintain a constant internal temperature.

Cesare Emiliani, using the Urey technique, studied the shells of foraminifera brought up in cores from the ocean floor. He found that the over-all ocean temperature was about 50° F. 30 million years ago, 43° 20 million years ago, and is now 35°.

What caused these long-term changes in temperature? One possible explanation is the so-called "greenhouse effect" of carbon dioxide. Carbon dioxide absorbs infrared radiation rather strongly. This means that when there are appreciable amounts of it in the atmosphere, it tends to block the escape of heat at night from the sun-warmed earth. The result is that heat accumulates. On the other hand, when the carbon dioxide content of the atmosphere falls, the earth steadily cools.

If the current concentration of carbon dioxide in the air should double (from 0.03 per cent of the air to 0.06 per cent), that small change would suffice to raise the earth's over-all temperature by three degrees and would bring about the complete and quick melting of the continental glaciers. If the carbon dioxide dropped to half the present amount, the temperature would drop sufficiently to bring the glaciers down to New York City again.

Volcanoes discharge large amounts of carbon dioxide into the air; the weathering of rocks absorbs carbon dioxide (thus forming limestone). Here, then, is a possible pair of mechanisms for long-term climatic changes. A period of greater than normal volcanic action might release a large amount of carbon dioxide into the air and initiate a warming of the earth. Contrariwise, an era of mountain-building, exposing large areas of new and unweathered rock to the air, could lower the carbon dioxide concentration in the atmosphere. This is what may have happened at the close of the Mesozoic (the age of reptiles) some 80 million years ago, when the long decline in the earth's temperature began.

But what about the comings and goings of the four ice ages within the last million years? Why was there this rapid alternation of glaciation and melting in comparatively short spells of tens of thousands of years?

In 1920, a Serbian physicist named Milutin Milankovich suggested that slight variations in the earth's relation to the sun might explain the situation. Sometimes the earth's tilt changes a little; sometimes its perihelion (closest approach to the sun in its orbit) is slightly closer than at other times. A combination of these factors, Milankovich argued, could so affect the amount of heat received from the sun by, say, the Northern Hemisphere as to cause a cyclic rise and fall of its average temperature. He thought that such a cycle might last 40,000 years, giving the earth a "Great Spring," "Great Summer," "Great Fall," and "Great Winter," each some 10,000 years in length. Precise dating of coral reefs and deep-sea sediments has shown such temperature shifts.

The difference between Great Summer and Great Winter is small, and the theory implies that only after a long period of over-all temperature decline did the additional small temperature fall of the Great Winter suffice to reduce the Northern Hemisphere's temperature to the point where the ice ages began a million years ago. According to the Milankovich theory, we are now in a Great Summer and, in 10,000 years or so, will begin to enter another Great Winter.

The Milankovich theory has disturbed some geologists, mainly because it implies that the ice ages of the Northern and Southern hemispheres have come at different times, which has not been demonstrated. In recent years, several other theories have been proposed: that the sun has cycles of slight fluctuation in its output of heat; that dust from volcanic eruptions, rather than carbon dioxide, has produced the "greenhouse" warming; and so on. An alternate hypothesis is one advanced by Maurice Ewing of the Lamont Geological Observatory and a colleague, William Donn.

Ewing and Donn ascribe the succession of ice ages in the Northern Hemisphere to the geographical conditions around the North Pole. The Arctic Ocean is nearly surrounded by land. In the mild eons before the recent ice ages began, when this ocean was open water, winds sweeping across it picked up water vapor and dropped snow on Canada and Siberia. As glaciers grew on the land, according to the Ewing-Donn theory, the earth absorbed less heat from the sun, because the cover of ice, as well as clouds resulting from stormier weather, reflected away part of the sunlight. Consequently, the general temperature of the earth dropped. But as it did so, the Arctic Ocean froze over, and, consequently, the winds picked up less moisture from it. Less moisture in the air meant less snow each winter. So the trend was reversed: with less snowy winters, summer melting took the upper hand over winter snowfall. The glaciers retreated until the earth warmed sufficiently to melt the Arctic Ocean to open water again—at which point the cycle started anew with a rebuilding of the glaciers.

It seems a paradox that the melting of the Arctic Ocean, rather than its freezing, should bring on an ice age. Geophysicists, however, find the theory plausible and capable of explaining many things. The main problem about the theory is that it makes the absence of ice ages up to a million years ago more mysterious than ever. But Ewing and Donn have an answer for that. They suggest that during the long period of mildness before the ice ages the North Pole may have been located in the Pacific Ocean. In that case, most of the snow would have fallen in the ocean instead of on land, and no important glaciers could have got started.

The North Pole, of course, has a constant small motion, moving in thirty-foot irregular circles in a period of 435 days or so, as was discovered at the beginning of the twentieth century by the American astronomer Seth Carlo Chandler. It has also drifted thirty feet toward Greenland since 1900. However, such changes—caused perhaps by earthquakes and consequent shifts in the mass-distribution in the globe—are small potatoes.

What is needed for the Ewing-Donn theory are large sweeps, and these might possibly be brought to pass by continental drift. As the crustal plates shift about, the North Pole may at times be enclosed by land or be left in open sea. However, can such changes produced by drift be matched with the occurrence or nonoccurrence of periods of glaciation?

Whatever the cause of the ice ages may have been, it seems now that man himself may be changing the climate in store for the future. The American physicist Gilbert N. Plass has suggested that we may be seeing the last of the ice ages, because the furnaces of civilization are loading the atmosphere with carbon dioxide. A hundred million chimneys are ceaselessly pouring carbon dioxide into the air; the total amount is about 6 billions tons a year—200 times the quantity coming from volcanoes. Plass pointed out that, since 1900, the carbon-dioxide content of our atmosphere has increased about 10 per cent and may increase as much again by the year 2000. This addition to the earth's "greenhouse" shield against the escape of heat, he calculated, should raise the average temperature by about 1.1° C. per century. During the first half of the twentieth century, the average temperature has indeed risen at this rate, according to the available records (mostly in North America and Europe). If the warming continues at the same rate, the continental glaciers may disappear in a century or two.

Investigations during the IGY seemed to show that the glaciers are indeed receding almost everywhere. One of the large glaciers in the Himalayas was reported in 1959 to have receded 700 feet since 1935.

Others had retreated 1,000 or even 2,000 feet. Fish adapted to frigid waters are migrating northward, and warm-climate trees are advancing in the same direction. The sea level is rising slightly each year, as would be expected if the glaciers are melting. The sea level is already so high that, at times of violent storms at high tide, the ocean is not far from threatening to flood the New York subway system.

And yet there seems to be a slight downturn in temperature since the early 1940's, so that half the temperature increase between 1880 and 1940 has been wiped out. This may be due to increasing dust and smog in the air since 1940: particles that cut off sunlight and, in a sense, shade the earth. It would seem that two different types of man-made atmospheric pollution are currently canceling each other's effect, at least in this respect, and at least temporarily.

CHAPTER 4

The Atmosphere

The Shells of Air

Aristotle supposed the world to be made up of four shells, constituting the four elements of matter: earth (the solid ball), water (the ocean), air (the atmosphere), and fire (an invisible outer shell that occasionally became visible in the flashes of lightning). The universe beyond these shells, he said, was composed of an unearthly, perfect fifth element that he called "ether" (from a Latin derivative the name became "quintessence," which means "fifth element").

There was no room in this scheme for nothingness: where earth ended, water began; where both ended, air began; where air ended, fire began; and where fire ended, ether began and continued to the end of the universe. "Nature," said the ancients, "abhors a vacuum" (Latin for "nothingness").

The suction pump, an early invention to lift water out of wells, seemed to illustrate this abhorrence of a vacuum admirably. A piston is fitted tightly within a cylinder. When the pump handle is pushed down, the piston is pulled upward, leaving a vacuum in the lower part of the cylinder. But since nature abhors a vacuum, the surrounding water opens a one-way valve at the bottom of the cylinder and rushes into the vacuum. Repeated pumping lifts the water higher and higher in the cylinder, until it pours out of the pump spout.

According to Aristotelian theory, it should have been possible in this way to raise water to any height. But miners who had to pump water out of the bottoms of mines found that no matter how hard and long they pumped, they could never lift the water higher than thirty-three feet above its natural level.

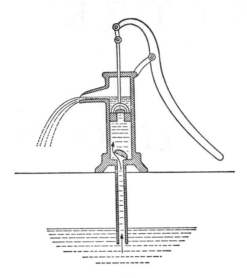

Principle of the water pump. When the handle raises the
piston, a partial vacuum is created in the cylinder, and water
rises into it through a one-way valve. After repeated pumping, the water level is high enough for the water to flow out
of the spout.

Galileo got interested in this puzzle toward the end of his long and
inquisitive life. He could come to no conclusion except that apparently
nature abhorred a vacuum only up to certain limits. He wondered
whether the limit would be lower if he used a liquid denser than water,
but he died before he could try this experiment.

Galileo's students Evangelista Torricelli and Vincenzo Viviani did
perform it in 1644. Selecting mercury (which is thirteen and one half
times as dense as water), they filled a yard-long glass tube with mercury,
stoppered the open end, upended the tube in a dish of mercury, and
removed the stopper. The mercury began to run out of the tube into the
dish, but, when its level had dropped to thirty inches above the level in
the dish, it stopped pouring out of the tube and held at that level.

Thus was constructed the first "barometer." Modern mercury barom-
eters are not essentially different. It did not take long to discover that
the height of the mercury column was not always the same. The English
scientist Robert Hooke pointed out, in the 1660's, that the height of the
mercury column decreased before a storm, thus pointing the way to the
beginning of scientific weather forecasting or "meteorology."

What was holding the mercury up? Viviani suggested that it was the
weight of the atmosphere, pressing down on the liquid in the dish. This
was a revolutionary thought, for the Aristotelian notion had been that
air had no weight, being drawn only to its proper sphere above the earth.
Now it became plain that a thirty-three-foot column of water, or a

thirty-inch column of mercury, measured the weight of the atmosphere—that is, the weight of a column of air of the same cross section from sea level up to as far as the air went.

The experiment also showed that nature did not necessarily abhor a vacuum under all circumstances. The space left in the closed end of the tube after the mercury fell was a vacuum, containing nothing but a very small quantity of mercury vapor. This "Torricellian vacuum" was the first decent vacuum produced by man.

The vacuum was pressed into the service of science almost at once. In 1650, the German scholar Athanasius Kircher demonstrated that sound could not be transmitted through a vacuum, thus upholding an Aristotelian theory (for once). In the next decade, Robert Boyle showed that very light objects will fall as rapidly as heavy ones in a vacuum, thus upholding Galileo's theories of motion against the views of Aristotle.

If air had a finite weight, it must have some finite height. The weight of the atmosphere turned out to be fourteen and seven tenths pounds per square inch; on this basis the atmosphere was just about five miles high—if it was evenly dense all the way up. But, in 1662, Boyle showed that it could not be, because pressure increased air's density. He stood up a tube shaped like the letter "J" and poured some mercury into the mouth of the tube, on the tall side of the J. The mercury trapped a little air in the closed end on the short side. As he poured in more mercury, the air pocket shrank. At the same time its pressure increased, Boyle discovered, for it shrank less and less as the mercury grew weightier. By actual measurement, Boyle showed that reducing the volume of gas to one half doubled its pressure; in other words, the volume varied in inverse ratio to the pressure. This historic discovery, known as "Boyle's law," was the first step in the long series of discoveries about matter that eventually led to the atomic theory.

Since air contracted under pressure, it must be densest at sea level and steadily become thinner as the weight of the overlying air declined toward the top of the atmosphere. This was first demonstrated by the French mathematician Blaise Pascal, who sent his brother-in-law Florin Perier nearly a mile up a mountainside in 1648 and had him carry a barometer and note the manner in which the mercury level dropped as altitude increased.

Theoretical calculations showed that, if the temperature were the same all the way up, the air pressure would decrease tenfold with every twelve miles of rise in altitude. In other words, at 12 miles the column of mercury it could support would have dropped from 30 inches to 3 inches; at 24 miles it would be .3 of an inch; at 36 miles, .03 of an inch and so on. At 108 miles, the air pressure would amount to only 0.000000003 of an inch of mercury. This may not sound like much, but over the whole earth the weight of the air above 108 miles would still total 6 million tons.

Actually all these figures are only approximations, because the air temperature changes with height. Nevertheless, they do clarify the picture, and we can see that the atmosphere has no definite boundary; it simply fades off gradually into the near-emptiness of space. Meteor trails have been detected as high as one hundred miles where the air pressure is only a millionth what it is on the earth's surface, and the air density only a billionth. Yet that is enough to burn these tiny bits of matter to incandescence by friction. And the aurora borealis (Northern Lights), formed of glowing wisps of gas bombarded by particles from outer space, has been located as high as 500 to 600 miles above sea level.

Until the late eighteenth century, it seemed that man would never be able to get any closer to the upper atmosphere than the top of the mountains. The highest mountain close to the centers of scientific research was Mont Blanc in southeastern France, and that was only three miles high. An interesting effort to substitute technology for mountain-climbing came in 1749 when the Scottish astronomer Alexander Wilson attached thermometers to kites, hoping thus to measure atmospheric temperatures at a height. The real breakthrough, however, came in 1782, when the two French brothers Joseph Michel and Jacques Étienne Montgolfier lit a fire under a large bag with an opening underneath and thus filled the bag with hot air. The bag rose slowly; the Montgolfiers had successfully launched man's first balloon! Within a few months balloons were being made with hydrogen, a gas only one fourteenth as dense as air, so that each pound of hydrogen could carry aloft a payload of thirteen pounds. Now gondolas went up carrying animals and, soon, men.

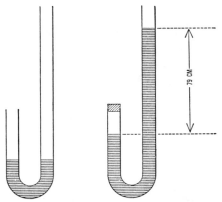

Diagram of Boyle's experiment. When the left arm of the tube is stoppered and more mercury is poured into the right arm, the trapped air is compressed. Boyle showed that the volume of the trapped air varied inversely with the pressure. That is "Boyle's law."

Within a year of the launching of the first balloon, an American named John Jeffries made a balloon flight over London with a barometer and other instruments, plus an arrangement to collect air at various heights. By 1804, the French scientist Joseph Louis Gay-Lussac had ascended nearly four and a half miles and brought down samples of the rarefied air. Such adventures were made a little safer by the French balloonist Jean Pierre Blanchard, who, in 1785, at the very onset of the "balloon age," invented the parachute.

This was nearly the limit for men in an open gondola; three men rose to six miles in 1875, but only one, Gaston Tissandier, survived the lack of oxygen. He was able to describe the symptoms of air deficiency, and that was the birth of "aviation medicine." Unmanned balloons carrying instruments were designed and put into action in 1892, and these could be sent higher and bring back information on temperature and pressure from hitherto unexplored regions.

In the first few miles of altitude rise, the temperature dropped, as was expected. At seven miles or so, it was $-55°$ C. But then came a surprise. Above this level, the temperature did not decrease. In fact, it even rose slightly.

The French meteorologist Leon Phillippe Teisserenc de Bort suggested in 1902 that the atmosphere might have two layers: (1) a turbulent lower layer containing clouds, winds, storms, and all the familiar weather changes (in 1908 he called this layer the "troposphere," from the Greek for "sphere of change") and (2) a quiet upper layer containing sublayers of lighter gases, helium and hydrogen (he named this the "stratosphere," meaning sphere of layers). Teisserenc de Bort called the level at which the temperature ceased to decline the "tropopause"—"end of change," or the boundary between the troposphere and the stratosphere. The tropopause has since been found to vary from an altitude of about ten miles above sea level at the Equator to only five miles above at the poles.

During World War II, high-flying United States bombers discovered a dramatic phenomenon just below the tropopause: the "jet stream," consisting of very strong, steady, west-to-east winds blowing at speeds up to 500 miles per hour. Actually there are two jet streams, one in the Northern Hemisphere at the general latitude of the United States, the Mediterranean, and north China, and one in the Southern at the latitude of New Zealand and Argentina. The streams meander, often debouching into eddies far north or south of their usual course. Airplanes now take advantage of the opportunity to ride on these swift winds. But far more important is the discovery that the jet streams have a powerful influence on the movement of air masses at lower levels. This knowledge at once helped to advance the art of weather forecasting.

But man did not resign his desire for personal exploration to instruments. One could not survive in the thin, cold atmosphere of great heights

—but why expose oneself to that atmosphere? Why not a sealed cabin, within which the pressures and temperatures of earth's surface air could be maintained?

In the 1930's, thanks to sealed cabins, man reached the stratosphere. In 1931, the Piccard brothers (Auguste and Jean Felix), the first of whom later invented the bathyscaphe, rose to 11 miles in a balloon carrying a sealed gondola. Then new balloons of plastic material, lighter and less porous than silk, made it possible to go higher and remain up longer. In 1938, a balloon named Explorer II went to 13 miles, and by the 1960's manned balloons have gone as high as 21½ miles and unmanned balloons almost to 29 miles.

These higher flights showed that the zone of nearly constant temperature did not extend indefinitely upward. The stratosphere came to an end at a height of about twenty miles, for above that the temperature started to rise!

This "upper atmosphere," above the stratosphere, containing only 2 per cent of the earth's total air mass, was in turn penetrated in the 1940's. This time man needed a new type of vehicle altogether—the rocket.

The Chinese, as long ago as the thirteenth century, invented and used small rockets for psychological warfare—to frighten the enemy. Modern Western civilization adapted rockets to a bloodier purpose. In 1801, a British artillery expert, William Congreve, having learned about rockets in the Orient, where Indian troops used them against the British in the 1780's, devised a number of deadly missiles. Some were used against the United States in the War of 1812, notably at the bombardment of Fort McHenry in 1814, which inspired Francis Scott Key to write the "Star-Spangled Banner," singing of "the rockets' red glare." Rocket weapons faded out in the face of improvements in range, accuracy, and power of conventional artillery. However, World War II saw the development of the American bazooka and the Soviet "Katusha," both of which are essentially rocket-propelled packets of explosives. Jet planes, on a much larger scale, also make use of the rocket principle of action and reaction.

Around the beginning of the twentieth century, two men independently conceived a new and finer use of rockets—exploring the upper atmosphere and space. They were a Russian, Konstantin Eduardovich Tsiolkovsky, and an American, Robert Hutchings Goddard. (It is odd indeed, in view of later developments, that a Russian and an American were the first heralds of the age of rocketry, though an imaginative German inventor, Hermann Ganswindt, also advanced even more ambitious, though less systematic and scientific, speculations at this time.)

The Russian was the first in print; he published his speculations and calculations in 1903 to 1913, whereas Goddard did not publish until 1919. But Goddard was the first to put speculation into practice. On

The Montgolfier brothers' hot-air balloon, launched at Versailles, September 19, 1783.

Dimitri Mendeleev, the Russian chemist, ascending in a balloon in 1887 to study the atmosphere.

The Stratoscope II telescope system (*left foreground*) is seen in launch position during inflation of the carrier balloon that will bear it to an altitude of 80,000 feet. At such heights, free of optical interference from the earth's atmosphere, Stratoscope II obtains detailed photos of heavenly bodies.

Astronaut entering a rocket capsule in training for space flight.

March 16, 1926, from a snow-covered farm in Auburn, Massachusetts, he fired a rocket 200 feet into the air. The remarkable thing about his rocket was that it was powered by a liquid fuel, instead of gunpowder. Then, too, whereas ordinary rockets, bazookas, jet planes, and so on make use of the oxygen in the surrounding air, Goddard's rocket, designed to work in outer space, had to carry its own oxidizer in the form of liquid oxygen ("lox," as it is now called in missile-man slang).

Jules Verne, in his nineteenth-century science fiction, had visualized a cannon as a launching device for a trip to the moon, but a cannon expends all its force at once and at the start, when the atmosphere is thickest and offers the greatest resistance. Goddard's rockets moved upward slowly at first, gaining speed and expending final thrust high in the thin atmosphere, where resistance is low. The gradual attainment of speed means that acceleration is kept at bearable levels, an important point for manned vessels.

Unfortunately, Goddard's accomplishment got almost no recognition except from his outraged neighbors, who managed to have him ordered to take his experiments elsewhere. Goddard went off to shoot his rockets in greater privacy, and, between 1930 and 1935, his vehicles attained speeds of as much as 550 miles an hour and heights of a mile and a half. He developed systems for steering a rocket in flight and gyroscopes to keep a rocket headed in the proper direction. Goddard also patented the idea of multistage rockets. Because each successive stage sheds part of the original weight and starts at a high velocity imparted by the preceding stage, a rocket divided into a series of stages can attain much higher speeds and greater heights than could a rocket with the same quantity of fuel all crammed into a single stage.

During World War II, the United States Navy halfheartedly supported further experiments by Goddard. Meanwhile, the German government threw a major effort into rocket research, using as its corps of workers a group of youngsters who had been inspired, primarily, by Hermann Oberth, a Rumanian mathematician who, in 1923, wrote on rockets and spacecraft independently of Tsiolkovsky and Goddard. German research began in 1935 and culminated in the development of the V-2. Under the guidance of the rocket expert Wernher von Braun (who, after World War II, placed his talents at the disposal of the United States), the first true rocket missile was shot off in 1942. The V-2 came into combat use in 1944, too late to win the war for the Nazis, although they fired 4,300 of them altogether, of which 1,230 hit London. Von Braun's missiles killed 2,511 Englishmen and seriously wounded 5,869 others.

On August 10, 1945, almost on the very day of the war's end, Goddard died—just in time to see his spark blaze into flame at last. The United States and the Soviet Union, stimulated by the successes of the

V-2, plunged into rocket research, each carrying off as many German experts in rocketry as could be lured to its side.

By 1949, the United States had fired a captured German V-2 to a height of 128 miles, and, in the same year, its rocket experts sent a WAC-Corporal, the second stage of a two-stage rocket, to 250 miles. The exploration of the upper atmosphere had begun.

Rockets alone would have accomplished little in that exploration had it not been for a companion invention—"telemetering." Telemetering was first applied to atmospheric research, in a balloon, in 1925 by a Russian scientist named Pyotr A. Molchanoff.

Essentially, this technique of "measuring at a distance" entails translating the conditions to be measured (e.g., temperature) into electrical impulses that are transmitted back to earth by radio. The observations take the form of changes in the intensity or spacing of the pulses. For instance, a temperature change affects the electrical resistance of a wire and so changes the nature of the pulse; a change in air pressure similarly is translated into a certain kind of pulse by the fact that air cools the wire, the extent of the cooling depending on the pressure; radiation sets off pulses in a detector, and so on. Nowadays, telemetering has become so elaborate that the rockets seem to do everything but talk, and their intricate messages have to be interpreted by rapid computers.

Rockets and telemetering, then, showed that above the stratosphere, the temperature rose to a maximum of some $-10°$ C. at a height of 30 miles and then dropped again to a low of $-90°$ C. at a height of 50 miles. This region of rise and fall in temperature is called the "mesosphere," a word coined in 1950 by the British geophysicist Sydney Chapman.

Beyond the mesosphere what is left of the thin air amounts to only a few thousandths of 1 per cent of the total mass of the atmosphere. But this scattering of air atoms steadily increases in temperature to an estimated 1,000° at 300 miles and probably to still higher levels above that height. It is therefore called the "thermosphere" ("sphere of heat") —an odd echo of Aristotle's original sphere of fire. Of course, temperature here does not signify heat in the usual sense: it is merely a measure of the speed of the particles.

Above 300 miles we come to the "exosphere," a term first used by Lyman Spitzer in 1949, which may extend to as high as 1,000 miles and gradually merges into interplanetary space.

Increasing knowledge of the atmosphere may enable man to do something about the weather someday and not merely talk about it. Already, a small start has been made. In the early 1940's, the American chemists Vincent Joseph Schaefer and Irving Langmuir noted that very low temperatures could produce nuclei about which raindrops would form. In 1946, an airplane dropped powdered carbon dioxide into a cloud bank in order to form first nuclei and then raindrops ("cloud seeding"). Half an hour later, it was raining. Bernard Vonnegut later improved the technique when he discovered that powdered silver iodide generated on the ground and directed upward worked even better. Rainmakers, of a new scientific variety, are now used to end droughts—or to attempt to end them, for clouds must first be present before they can be seeded. In 1961, Soviet astronomers used cloud seeding to clear a patch of sky through which an eclipse might be glimpsed. It was partially successful.

Those interested in rocketry strove constantly for new and better results. The captured German V-2's were used up by 1952, but by then, larger and more advanced rocket-boosters were being built in both the United States and the Soviet Union, and progress continued.

A new era began when, on October 4, 1957 (within a month of the hundredth anniversary of Tsiolkovsky's birth), the Soviet Union put the first man-made satellite in orbit. Sputnik I traveled around the earth in an elliptical orbit—156 miles above the surface (or 4,100 miles from the earth's center) at perigee and 560 miles away at apogee. An elliptical orbit is something like the course of a roller coaster. In going from apogee (the highest point) to perigee, the satellite slides downhill, so to speak, and loses gravitational potential. This brings an increase in velocity, so that at perigee the satellite starts uphill again at top speed, as a roller coaster does. The satellite loses velocity as it climbs (as does the roller coaster) and is moving at its slowest speed at apogee, before it turns downhill again.

Sputnik I at perigee was in the mesosphere, where the air resistance, though slight, was sufficient to slow the satellite a bit on each trip. On each successive revolution, it failed to attain its previous apogee height. Slowly, it spiraled inward. Eventually, it lost so much energy that it yielded to the earth's pull sufficiently to dive into the denser atmosphere, there to be burned up by friction with the air.

The rate at which a satellite's orbit decays in this way depends partly on the mass of the satellite, partly on its shape, and partly on the density of the air through which it passes. Thus the density of the atmosphere at that level can be calculated. The satellites have given man the first direct measurements of the density of the upper atmosphere. The density proved to be higher than had been thought, but at the altitude of 150 miles, for instance, it is still only one ten-millionth of that at sea level, and, at 225 miles, only one trillionth.

These wisps of air ought not be dismissed too readily, however. Even at a height of 1,000 miles, where the atmospheric density is only one quadrillionth the sea-level figure, that faint breath of air is a billion times as dense as are the gases in outer space itself. The earth's envelope of gases spreads far outward.

The Soviet Union did not remain alone in this field, of course.

Within four months it was joined by the United States, which, on January 30, 1958, launched its first satellite in orbit, "Explorer I." Since then each nation has put hundreds of satellites whirling about the earth, for a variety of purposes. The upper atmosphere, the portion of space in the vicinity of the earth, has been studied by satellite-borne instruments and, in addition, the earth itself has been the target of studies. For one thing, satellites made it possible for the first time in man's history to see our planet (or at least half of it at any one time) as a unit, and to study the air circulation as a whole.

On April 1, 1960, the United States launched the first "weather-eye" satellite, Tiros I ("Tiros" standing for "Television Infrared Observation Satellite"), then Tiros II in November, which, for ten weeks, sent down over 20,000 pictures of vast stretches of the earth's surface and its cloud cover, including pictures of a cyclone in New Zealand and a patch of clouds in Oklahoma that was apparently spawning tornadoes. Tiros III, launched in July 1961, photographed eighteen tropical storms and, in September, showed Hurricane Esther developing in the Caribbean two days before it was located by more orthodox methods. The more sensitive Nimbus I satellite, launched on August 28, 1964, could send back cloud photographs taken at night! By the end of the 1960's, weather forecasting was making routine use of satellite-transmitted data.

Other earthbound uses of satellites have been developed. As early as 1945, the British science-fiction writer Arthur C. Clarke had pointed out that satellites could be used as relays by which radio messages could span continents and oceans, and that as few as three strategically placed satellites could afford world coverage. What then seemed a wild dream began to come true fifteen years later. On August 12, 1960, the United States launched "Echo I," a thin polyester balloon coated with aluminum, which was inflated in space to a diameter of one hundred feet in order to serve as a passive reflector of radio waves. A leader in this successful project was John Robinson Pierce of Bell Telephone Laboratories, who had himself written science-fiction stories under a pseudonym.

On July 10, 1962, "Telstar I" was launched by the United States. It did more than reflect. It received the waves, amplified them, and sent them onward. By use of Telstar, television programs spanned the oceans for the first time (though that did not in itself improve their quality, of course). On July 26, 1963, "Syncom II," a satellite that orbited at a distance of 22,300 miles above the earth's surface, was put in orbit. Its orbital period was just twenty-four hours, so that it hovered indefinitely over the Atlantic Ocean, turning in synchronization with the earth. "Syncom III," placed over the Indian Ocean in similar synchronous fashion, relayed the Olympic Games from Japan to the United States in October 1964.

A still more sophisticated communications satellite, "Early Bird," was launched April 6, 1965, and it made available 240 voice circuits and one TV channel. (In that year, the Soviet Union began to send up communications satellites as well.) Much more is scheduled for the early 1970's, and earth seems on the threshold of becoming "one world," at least as far as communications are concerned.

Satellites have also been launched for the specific purpose of being

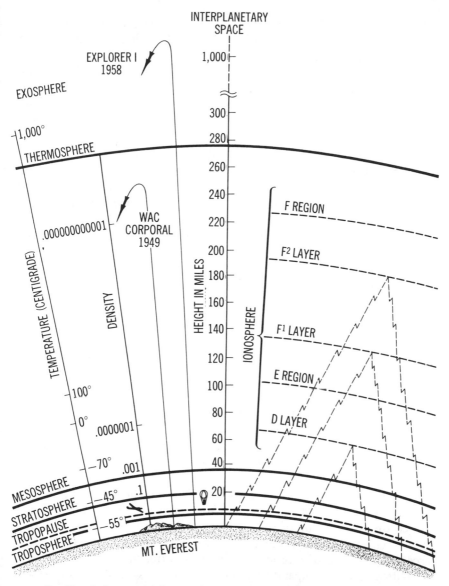

Profile of the atmosphere. The jagged lines indicate the reflection of radio signals from the Kennelly-Heaviside and Appleton layers of the ionosphere. Air density decreases with height and is expressed in percentages of barometric pressure at sea level.

used to determine position on earth. The first such satellite, "Transit 1B," was launched on April 13, 1960.

The Gases in Air

Up to modern times, air was considered a simple, homogeneous substance. In the early seventeenth century, the Flemish chemist Jan Baptista van Helmont began to suspect that there were a number of chemically different gases. He studied the vapor given off by fermenting fruit juice (carbon dioxide)and recognized it as a new substance. Van Helmont was, in fact, the first to use the term "gas"—a word he is supposed to have coined from "chaos," the ancients' word for the original substance out of which the universe was made. In 1756, the Scottish chemist Joseph Black studied carbon dioxide thoroughly and definitely established it as a gas other than air. He even showed that small quantities of it existed in the air. Ten years later, Henry Cavendish studied a flammable gas not found in the atmosphere. It was eventually named hydrogen. The multiplicity of gases was thus clearly demonstrated.

The first to realize that air was a mixture of gases was the French chemist Antoine-Laurent Lavoisier. In experiments conducted in the 1770's, he heated mercury in a closed vessel and found that the mercury combined with part of the air, forming a red powder (mercuric oxide), but four fifths of the air remained a gas. No amount of heating would consume any of this remaining gas. A candle would not burn in it, nor could mice live in it.

Lavoisier decided that air was made up of two gases. The one fifth that combined with mercury in his experiment was the portion of the air that supported life and combustion. This he called "oxygen." The remainder he called "azote," from Greek words meaning "no life." Later it became known as "nitrogen," because the substance was present in sodium nitrate, commonly called "niter." Both gases had been discovered in the previous decade. Nitrogen had been discovered in 1772 by the Scottish physician Daniel Rutherford, and oxygen, in 1774 by the English Unitarian minister Joseph Priestley.

By the mid-nineteenth century, the French chemist Henri Victor Regnault had analyzed air samples from all over the world and discovered the composition of the air to be the same everywhere. The oxygen content was 20.9 per cent, and it was assumed that all the rest (except for a trace of carbon dioxide) was nitrogen.

Nitrogen is a comparatively inert gas; that is, it does not readily combine with other substances. It can, however, be forced into combination, for instance, by heating it with magnesium metal, forming the solid magnesium nitride. Some years after Lavoisier's discovery, Henry Cavendish tried to exhaust the nitrogen by combining it with oxygen under the influence of an electric spark. He failed. No matter what he did, he could not get rid of a small bubble of remaining gas, amounting to less than 1 per cent of the original quantity. Cavendish thought this might be an unknown gas, even more inert than nitrogen. But not all chemists are Cavendishes, and the puzzle was not followed up, so the nature of this residue of air was not discovered for another century.

In 1882, the British physicist Robert John Strutt, Lord Rayleigh, compared the density of nitrogen obtained from air with the density of nitrogen obtained from certain chemicals and found, to his surprise, that the air nitrogen was definitely denser. Could it be that nitrogen obtained from air was not pure but contained small quantities of another, heavier gas? A Scottish chemist, Sir William Ramsay, helped Lord Rayleigh look further into the matter. By this time, they had the aid of spectroscopy. When they heated the small residue of gas left after exhaustion of nitrogen from air and examined its spectrum, they found a new set of bright lines—lines that belonged to no known element. To their newly discovered, very inert element they gave the name "argon" (from a Greek word meaning "inert").

Argon accounted for nearly all of the approximately 1 per cent of unknown gas in air—but not quite all. There were still several "trace constituents" in the atmosphere, each constituting only a few parts per million. During the 1890's Ramsay went on to discover four more inert gases: "neon" (new), "krypton" (hidden), "xenon" (stranger), and helium, which had been discovered more than thirty years before in the sun. In recent decades, the infrared spectroscope has turned up three others: nitrous oxide ("laughing gas"), whose origin is unknown; methane, a product of the decay of organic matter; and carbon monoxide. Methane is released by bogs, and some 45 million tons of the same gas, it has been calculated, are added to the atmosphere each year by the venting of intestinal gases by cattle and other large animals. The carbon monoxide is probably man-made, resulting from the incomplete combustion of wood, coal, gasoline, and so on.

All this, of course, refers to the composition of the lowest reaches of the atmosphere. What about the stratosphere? Teisserenc de Bort believed that helium and hydrogen might exist in some quantity up there, floating on the heavier gases underneath. He was mistaken. In the middle 1930's, Russian balloonists brought down samples of air from the upper stratosphere, and it proved to be made up of oxygen and nitrogen in the same one-to-four mixture as the air of the troposphere.

But there were reasons to believe some unusual gases existed still higher in the upper atmosphere, and one of the reasons was the phenomenon called the "airglow." This is the very feeble general illumination of all parts of the night sky, even in the absence of the moon. The total light of the airglow is considerably greater than that of the stars, but is so diffuse that it is not noticeable except to the delicate light-gathering instruments of the astronomer.

The source of the light had been a mystery for many years. In 1928, the astronomer V. M. Slipher succeeded in detecting in the airglow some mysterious spectral lines that had been found in nebulae in 1864 by William Huggins and were thought to represent an unfamiliar element, named "nebulium." In 1927, through experiments in the laboratory, the American astronomer Ira Sprague Bowen showed that the lines came from "atomic oxygen," that is, oxygen existing as single atoms and not combined in the normal form of the two-atom molecule. Similarly, other strange spectral lines from the aurora turned out to represent atomic nitrogen. Both atomic oxygen and atomic nitrogen in the upper atmosphere are produced by energetic radiation from the sun, which breaks down the molecules into single atoms, something first suggested in 1931 by Sydney Chapman. Fortunately the high-energy radiation is, in this way, absorbed or weakened before it reaches the lower atmosphere.

The airglow, Chapman maintained, comes from the recombination at night of the atoms that are split apart by solar energy during the day. In recombining, the atoms give up some of the energy they absorbed in splitting, so that the airglow is a kind of delayed and very feeble return of sunlight in a new and specialized form. The rocket experiments of the 1950's supplied direct evidence of this. Spectroscopes carried by the rockets recorded the green lines of atomic oxygen most strongly at a height of sixty miles. A smaller proportion of the nitrogen was in the atomic form, because nitrogen molecules hold together more strongly than do oxygen molecules; nevertheless, the red light of atomic nitrogen was strong at a height of ninety-five miles.

Slipher had also found lines in the airglow that were suspiciously like well-known lines emitted by sodium. The presence of sodium seemed so unlikely that the matter was dropped in embarrassment. What would sodium, of all things, be doing in the upper atmosphere? It is not a gas, after all, but a very reactive metal that does not occur alone anywhere on the earth. It is always combined with other elements, most commonly in sodium chloride (table salt). But, in 1938, French scientists established that the lines were indeed identical with the sodium lines. Unlikely or not, sodium had to be in the upper atmosphere. Again rocket experiments clinched the matter: their spectroscopes recorded the yellow light of sodium unmistakably, and most strongly at a height of fifty-five miles. Where the sodium comes from is still a mystery; it may come from ocean salt spray or perhaps from vaporized meteors. Still more puzzling is the fact that lithium—a rarer relative of sodium—was also found, in 1958, to be contributing to the airglow.

In 1956, a team of United States scientists under the leadership of Murray Zelikoff produced an artificial airglow. They fired a rocket that at sixty miles released a cloud of nitric oxide gas. This accelerated the recombination of oxygen atoms in the upper atmosphere. Observers on the ground easily sighted the bright glow that resulted. A similar experiment with sodium vapor also was successful: it created a clearly visible, yellow glow. When Soviet scientists sent "Lunik III" in the direction of the moon in October 1959, they arranged for it to expel a cloud of sodium vapor as a visible signal that it had gone into orbit.

At lower levels in the atmosphere, atomic oxygen disappears, but the solar radiation is still energetic enough to bring about the formation of the three-atom variety of oxygen called "ozone." The ozone concentration is greatest at a height of fifteen miles. Even there, in what is called the "ozonosphere" (first discovered in 1913 by the French physicist Charles Fabry), it makes up only one part in 4 million of the air, but that is enough to absorb ultraviolet light sufficiently to protect life on the earth. The ozone is formed by the action of ultraviolet light on ordinary two-atom oxygen. The ultraviolet radiation is, in this way, consumed: that is why the ozonosphere is earth's shield against it. It is the absorption of ultraviolet by oxygen that raises the temperature of the mesosphere above that of the stratosphere. Near the earth's surface, the concentration of ozone is very low, although it may rise high enough to form an irritating component of "smog."

Further rocket experiments showed that Teisserenc de Bort's speculations concerning layers of helium and hydrogen were not wrong—merely misplaced. From 200 to 600 miles upward, where the atmosphere has thinned out to near-vacuum, there is a layer of helium, now called the "heliosphere." The existence of this layer was first deduced in 1961 by the Belgian physicist Marcel Nicolet from the frictional drag on the Echo I satellite. This was confirmed by actual analysis of the thin-gas surroundings by Explorer XVII, launched on April 2, 1963.

Above the heliosphere is an even thinner layer of hydrogen, the "protonosphere," which may extend upward some 40,000 miles before quite fading off into the general density of interplanetary space.

High temperatures and energetic radiation can do more than force atoms apart or into new combinations. They can chip electrons away from atoms and so "ionize" the atoms. What remains of the atom is called an "ion" and differs from ordinary atoms in carrying an electric charge. The word "ion" comes from a Greek word meaning "traveler." Its origin lies in the fact that when an electric current passes through a solution containing ions, the positively charged ions travel in one direction and the negatively charged ions in the other.

A young Swedish student of chemistry named Svante August Arrhenius was the first to suggest that the ions were charged atoms, as the only means of explaining the behavior of certain solutions that conducted an electric current. His notions, advanced in the thesis he presented for his degree of Doctor of Philosophy in 1884, were so revolutionary that his examiners could scarcely bring themselves to pass him. The charged particles within the atom had not yet been discovered, and the concept of an electrically charged atom seemed ridiculous. Arrhenius got his degree, but with only a minimum passing grade.

When the electron was discovered in the late 1890's, Arrhenius's theory suddenly made startling sense. He was awarded the Nobel Prize in chemistry in 1903 for the same thesis that nineteen years earlier had nearly lost him his doctoral degree. (This sounds like an improbable movie scenario, I admit, but the history of science contains many episodes that make Hollywood seem unimaginative.)

The discovery of ions in the atmosphere did not emerge until after Guglielmo Marconi started his experiments with wireless. When, on December 12, 1901, he sent signals from Cornwall to Newfoundland, across 2,100 miles of the Atlantic Ocean, scientists were startled. Radio waves travel only in a straight line. How had they managed to go around the curvature of the earth and get to Newfoundland?

A British physicist, Oliver Heaviside, and an American electrical engineer, Arthur Edwin Kennelly, soon suggested that the radio signals might have been reflected back from the sky by a layer of charged particles high in the atmosphere. The "Kennelly-Heaviside layer," as it has been called ever since, was finally located in the 1920's. The British physicist Edward Victor Appleton discovered it by paying attention to a curious fading phenomenon in radio transmission. He decided that the fading was the result of interference between two versions of the same signal, one coming directly from the transmitter to his receiver, the other by a roundabout route via reflection from the upper atmosphere. The delayed wave was out of phase with the first, so the two waves partly canceled each other; hence the fading.

It was a simple matter then to find the height of the reflecting layer. All he had to do was to send signals at such a wavelength that the direct signal completely canceled the reflected one—that is, the two signals arrived at directly opposite phases. From the wavelength of the signal used and the known velocity of radio waves, he could calculate the difference in the distances the two trains of waves had traveled. In this way, he determined that the Kennelly-Heaviside layer was some sixty-five miles up.

The fading of radio signals generally occurred at night. Appleton found that shortly before dawn radio waves were not reflected back by the Kennelly-Heaviside layer but were reflected from still higher layers (now sometimes called the "Appleton layers") which begin at a height of 140 miles.

For all these discoveries Appleton received the Nobel Prize in physics in 1947. He had defined the important region of the atmosphere called the "ionosphere," a word introduced in 1930 by the Scottish physicist Robert Alexander Watson-Watt. It includes the later-named mesosphere and thermosphere, and is now divided into a number of layers. From the stratopause up to sixty-five miles or so is the "D region." Above that is the Kennelly-Heaviside layer, called the "D layer." Above the D layer, to a height of 140 miles, is the "E region"—an intermediate area relatively poor in ions. This is followed by the Appleton layers: the "F_1 layer" at 140 miles and "F_2 layer" at 200 miles. The F_1 layer is the richest in ions, the F_2 layer being significantly strong only in the daytime. Above these layers is the "F region."

These layers reflect and absorb only the long radio waves used in ordinary radio broadcasts. The shorter waves, such as those used in television, pass through, for the most part. That is why television broadcasting is limited in range—a limitation which can be remedied by satellite relay stations in the sky, most notably, so far, by the Early Bird satellite, launched in 1965, which allows live television to span oceans and continents. The radio waves from space (e.g., from radio stars) also pass through the ionosphere, fortunately; if they did not, there would be no radio astronomy.

The ionosphere is strongest at the end of the day, after the day-long effect of the sun's radiation, and weakens by dawn because many ions and electrons have recombined. Storms on the sun, intensifying the streams of particles and high-energy radiation sent to the earth, cause the ionized layers to strengthen and thicken. The regions above the ionosphere also flare up into auroral displays. During these electric storms long-distance transmission of radio waves on the earth is disrupted and sometimes blacked out altogether.

It has turned out that the ionosphere is only one of the belts of radiation surrounding the earth. Outside the atmosphere, in what used to be considered "empty" space, man's satellites in 1958 disclosed a startling surprise. To understand it we must make an excursion into the subject of magnetism.

Magnets

Magnets got their name from the ancient Greek town of Magnesia, near which the first "lodestones" were discovered. The lodestone is an iron oxide with natural magnetic properties. Tradition has it that Thales of Miletus, about 550 B.C., was the first philosopher to describe it.

Magnets became something more than a curiosity when it was discovered that a steel needle stroked by a lodestone was magnetized and that, if the needle was allowed to pivot freely in a horizontal plane, it would end up lying along a north–south line. Such a needle was, of course, of tremendous use to mariners; in fact, it became indispensable to ocean navigation, though the Polynesians did manage to cross the Pacific without a compass.

It is not known who first put such a magnetized needle on a pivot and enclosed it in a box to make a compass. The Chinese are supposed to have done it first and passed it on to the Arabs, who, in turn, passed it on to the Europeans. This is all very doubtful and may be only legend. At any rate, in the twelfth century the compass came into use in Europe and was described in detail in 1269 by a French scholar best known by his Latinized name of Peter Peregrinus. Peregrinus named the end of the magnet that pointed north the "north pole" and the other the "south pole."

Naturally, people speculated as to why a magnetized needle should point north. Because magnets were known to attract other magnets, some thought there was a gigantic lodestone mountain in the far north toward which the needle strained. Others were even more romantic and gave magnets a "soul" and a kind of life.

The scientific study of magnets began with William Gilbert, the court physician of Queen Elizabeth I. It was Gilbert who discovered that the earth itself was a giant magnet. He mounted a magnetized needle so that it could pivot freely in a vertical direction (a "dip needle"), and its north pole then dipped toward the ground ("magnetic dip"). Using a spherical lodestone as a model of the earth, he found that the needle behaved in the same way when it was placed over the "northern hemisphere" of his sphere. Gilbert published these findings in 1600 in a classic book entitled *De Magnete*.

In the three and a half centuries that have elapsed since Gilbert's work, no one has ever explained the earth's magnetism to everyone's satisfaction. For a long time scientists speculated that the earth might have a gigantic iron magnet as its core. Although the earth was indeed found to have an iron core, it is now certain that this core cannot be a magnet, because iron, when heated, loses its strong magnetic properties ("ferromagnetism," the prefix coming from the Latin word for iron) at 760° C., and the temperature of the earth's core must be at least 1000° C.

The temperature at which a substance loses its magnetism is called the "Curie temperature," since it was first discovered by Pierre Curie in 1895. Cobalt and nickel, which resemble iron closely in many respects, are also ferromagnetic. The Curie temperature for nickel is 356° C.; for cobalt it is 1075° C. At low temperatures, certain other metals are ferromagnetic. Below – 188° C., dysprosium is ferromagnetic, for instance.

In general, magnetism is a property of the atom itself, but in most materials the tiny atomic magnets are oriented in random directions, so that most of the effect is canceled out. Even so, weak magnetic properties are often evidenced, and the result is "paramagnetism." The strength of magnetism is expressed in terms of "permeability." The permeability of a vacuum is 1.00 and that of paramagnetic substances is between 1.00 and 1.01.

Ferromagnetic substances have much higher permeabilities. Nickel has a permeability of 40, cobalt one of 55, and that of iron is in the thousands. In such substances, the existence of "domains" was postulated in 1907 by the French physicist Pierre Weiss. These are tiny areas about 0.001 to 0.1 centimeters in diameter (which have actually been detected), in which the atomic magnets are so lined up as to reinforce one another, producing strong, over-all fields within the domain. In ordinary nonmagnetized iron, the domains themselves are randomly oriented and cancel one another's effect. When the domains are brought into line by the action of another magnet, the iron is magnetized. The reorientation of domains during magnetism actually produces clicking and hissing noises that can be detected by suitable amplification, this being termed the "Barkhausen effect" after its discoverer, the German physicist Heinrich Barkhausen.

In "antiferromagnetic substances," such as manganese, the domains also line up, but in alternate directions, so that most of the magnetism is canceled. Above a particular temperature, substances lose antiferromagnetism and become paramagnetic.

If the earth's iron core is not itself a permanent magnet because it is above the Curie temperature, then there must be some other way of explaining the earth's ability to affect a compass needle. What that might be grew out of the work of the English scientist Michael Faraday, who discovered the connection between magnetism and electricity.

In the 1820's, Faraday started with an experiment that had been first described by Peter Peregrinus (and which still amuses young students of physics). The experiment consists in sprinkling fine iron filings on a piece of paper above a magnet and gently tapping the paper. The shaken filings tend to line up along arcs from the north pole to the south pole of the magnet. Faraday decided that these marked actual "magnetic lines of force," forming a magnetic "field."

Faraday, who had been attracted to the subject of magnetism by the Danish physicist Hans Christian Oersted's observation in 1820 that an electric current flowing in a wire deflected a nearby compass needle, came to the conclusion that the current must set up magnetic lines of force around the wire.

He felt this to be all the more so since the French physicist André Marie Ampère had gone on to study current-carrying wires immediately after Oersted's discovery. Ampère showed that two parallel wires with the current flowing in the same direction attracted each other; with currents flowing opposite directions they repelled each other. This was very like the fashion in which two magnetic north poles (or two magnetic south poles) repelled each other while a magnetic north pole attracted a magnetic south pole. Better still, Ampère showed that a cylindrical coil of wire with an electric current flowing through it behaved like a bar magnet. In memory of his work, the unit of intensity of electric current was officially named the "ampere" in 1881.

But if all this were so, thought Faraday (who had one of the most efficient intuitions in the history of science) and if electricity can set up a magnetic field so like the real thing that current-carrying wires can act like magnets, should not the reverse be true? Ought not a magnet produce a current of electricity that would be just like the current produced by chemical batteries?

In 1831, Faraday performed the experiment that was to change human history. He wound a coil of wire around one segment of an iron ring and a second coil of wire around another segment of the ring. Then he connected the first coil to a battery. His reasoning was that if he sent a current through the first coil, it would create magnetic lines of force which would be concentrated in the iron ring, and this induced magnetism in turn would produce a current in the second coil. To detect that current, he connected the second coil to a galvanometer—an instrument for measuring electrical currents, which had been devised by the German physicist Johann Salomo Christoph Schweigger in 1820.

The experiment did not work as Faraday had expected. The flow of current in the first coil generated nothing in the second coil. But Faraday noticed that at the moment when he turned on the current, the galvanometer needle kicked over briefly, and it did the same thing, but in the opposite direction, when he turned the current off. He guessed at once that it was the movement of magnetic lines of force across a wire, not the magnetism itself, that set up the current. When a current began to flow in the first coil, it initiated a magnetic field that, as it spread, cut across the second coil, setting up a momentary electric current there. Conversely, when the current from the battery was cut off, the collapsing lines of magnetic force again cut across the wire of the second coil,

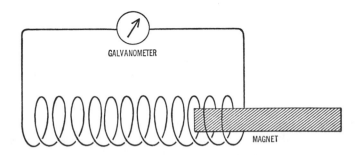

GALVANOMETER

MAGNET

A Faraday experiment on the induction of electricity. When the magnet is moved in or out of the coil of wire, the cutting of its lines of force by the wire produces an electrical current in the coil.

causing a momentary surge of electricity in the direction opposite that of the first flow.

Thus Faraday discovered the principle of electrical induction and created the first "transformer." He proceeded to demonstrate the phenomenon more plainly by using a permanent magnet and moving it in and out of a coil of wire; although no source of electricity was involved, a current flowed in the coil whenever the magnet's lines of force cut across the wire.

Faraday's discoveries not only led directly to the creation of the dynamo for generating electricity but also laid the foundation for James Clerk Maxwell's "electromagnetic" theory, which linked together light and other forms of radiation (such as radio) in a single family of "electromagnetic radiations."

Now the close connection between magnetism and electricity points to a possible explanation of the earth's magnetism. The compass needle

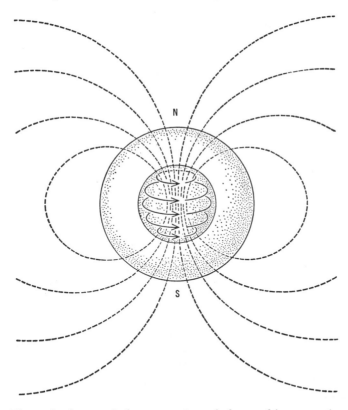

Elsasser's theory of the generation of the earth's magnetic field. Movements of material in the molten nickel-iron core set up electric currents which in turn generate magnetic lines of force. The dotted lines show the earth's magnetic field.

has traced out its magnetic lines of force, which run from the "north magnetic pole," located off northern Canada, to the "south magnetic pole," located at the rim of Antarctica, each being about 15 degrees of latitude from the geographic poles. (The earth's magnetic field has been detected at great heights by rockets carrying "magnetometers.") The new suggestion is that the earth's magnetism may originate in the flow of electric currents deep in its interior.

The physicist Walter Maurice Elsasser has proposed that the rotation of the earth sets up slow eddies in the molten iron core, circling west to east. These eddies have the effect of producing an electric current, likewise circling west to east. Just as Faraday's coil of wire produced magnetic lines of force within the coil, so the circling electric current does in the earth's core. It therefore creates the equivalent of an internal magnet extending north and south. This magnet in turn accounts for the earth's general magnetic field, oriented roughly along the axis of rotation, so that the magnetic poles are near the north and south geographic poles.

The sun also has a general magnetic field, which is two or three times as intense as that of the earth, and local fields apparently associated with the sunspots, which are thousands of times as intense. Studies of these fields (made possible by the fact that intense magnetism affects the wavelength of the light emitted) suggest that there are circular flows of electric charge within the sun.

There are, in fact, many puzzling features concerning sunspots, which may be answered once the causes of magnetic fields on an astronomic scale are worked out. For instance, the number of sunspots on the solar surface wax and wane in an eleven and one half year cycle. This was first established in 1843 by the German astronomer Heinrich Samuel Schwabe, who studied the face of the sun almost daily for seventeen years. Furthermore, the spots appear only at certain latitudes, and these latitudes shift as the cycle progresses. The spots show a certain magnetic orientation that reverses itself in each new cycle. Why all this should be so is still unknown.

Nor must we go to the sun for mysteries in connection with magnetic fields. There are problems here on earth. For instance, why do the magnetic poles not coincide with the geographic poles? The north magnetic pole is off the coast of northern Canada about a thousand miles from the North Pole. Similarly, the south magnetic pole is near the Antarctica shore line west of Ross Sea, about a thousand miles from the South Pole. Furthermore, the magnetic poles are not directly opposite each other on the globe. A line through the earth connecting them (the "magnetic axis") does not pass through the center of the earth.

Again, the deviation of the compass needle from "true north" (i.e., the direction of the North Pole) varies irregularly as one travels east or west. In fact, the compass needle shifted on Columbus' first voyage,

Launching of the first U.S. satellite, Explorer I, on January 31, 1958.

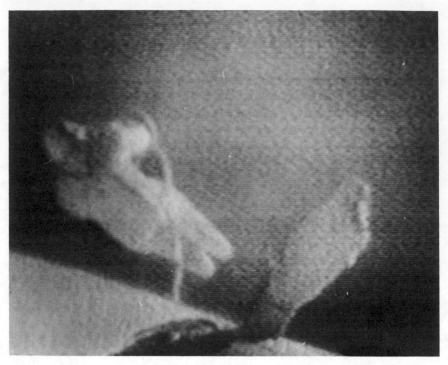

Cosmonaut "walks" in space. Floating weightlessly a short distance from the Soviet spacecraft Voskhod II, Alexei Leonov became the first man to leave capsule in space, March 18, 1965. Projection at right carries a camera. Both Leonov and his ship were traveling some 18,000 miles per hour. In the bottom photograph, Leonov does a somersault. The lifeline allowed him to pull himself back once he pushed away from the ship. Had the line broken, Leonov might have remained in orbit.

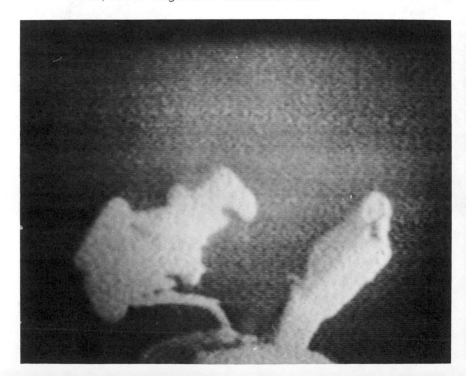

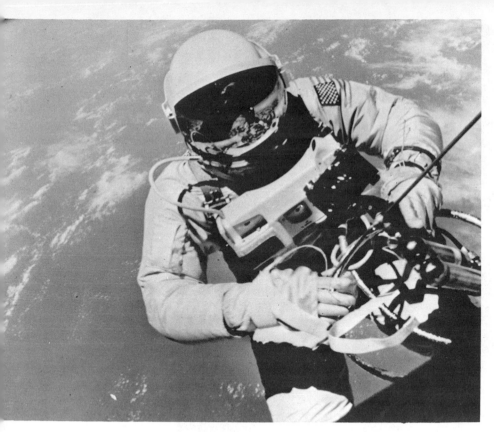

An American "walks" in space. Following the Russian feat, the U.S. Gemini program put astronaut Edward White into free orbit on June 8, 1965. The photographs of astronaut White were taken by Major James McDivitt, who sighted him through a porthole from inside the capsule, and by a movie camera mounted on the craft.

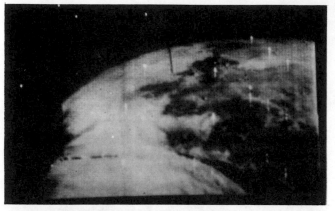

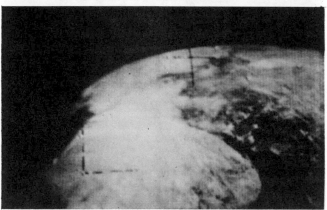

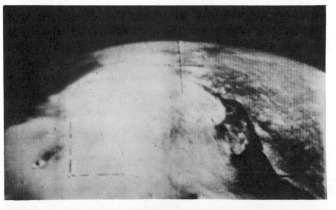

The earth from a rocke
This series of photographs
made by the U.S. meteoro
logical rocket Tiros I, show
the eastern part of th
United States and Canada
the dark area in the lowe
right corner is the St. Law
rence. The pictures wer
made at one-minute interval
from an altitude of abou
450 miles several hundre
miles east of the Atlanti
Coast. The white areas ar
cloud cover.

A Tiros I photograph showing a typhoon (*whirlpool at right*) 1,000 miles east of Australia.

Fully assembled Douglas Delta space vehicle, of the family that orbited four Tiros satellites, the Echo balloon, Telstar, and other scientific payloads, undergoes a vibration test.

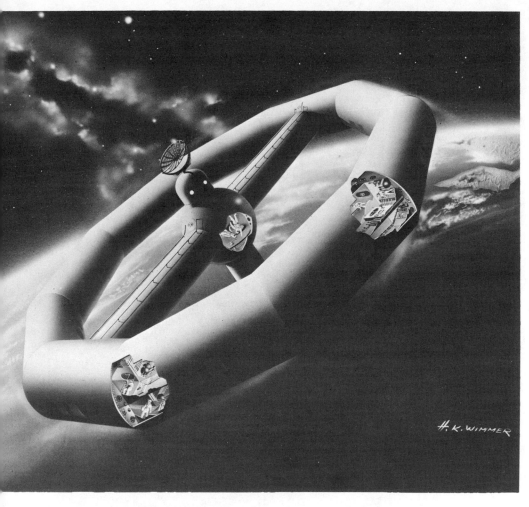

Artist's rendering of space station, such as may one day be placed in permanent orbit. It spins to create centrifugal effect in lieu of nullified gravity; hence for the men aboard, "down" lies toward the circumference of the circular station.

This view of the rising earth greeted the Apollo 8 astronauts as they came from behind the moon after the lunar orbit insertion burn. On the earth 240,000 statute miles away, the sunset terminator bisects Africa.

and Columbus hid this from his crew lest it excite terror that would force him to turn back.

This is one of the reasons why the use of a magnetic compass to determine direction is less than perfect. In 1911, a nonmagnetic method for indicating direction was introduced by the American inventor Elmer Ambrose Sperry. It takes advantage of the tendency of a rapidly turning heavy-rimmed wheel (a "gyroscope," first studied by the same Foucault who had demonstrated the rotation of the earth) to resist changes in its plan of rotation. This can be used to serve as a "gyroscopic compass," which will maintain a fixed direction reference that will serve to guide ships or rockets.

But if the magnetic compass is less than perfect, it has been useful enough to serve mankind for centuries. The deviation of the magnetic needle from the true north can be allowed for. A century after Columbus, in 1581, the Englishman Robert Norman prepared the first map indicating the actual direction marked out by a compass needle ("magnetic declination") in various parts of the world. Lines connecting those points on the planet that show equal declinations ("isogonic lines") run crookedly from north magnetic pole to south magnetic pole.

Unfortunately, such maps must be periodically changed, for even at one spot the magnetic declination changes with time. For instance, the declination at London shifted 32 degrees of arc in two centuries; it was 8 degrees east of north in 1600 and steadily swung around counterclockwise until it was 24 degrees west of north in 1800. Since then it has shifted back and in 1950 was only 8 degrees west of north.

Magnetic dip also changes slowly with time for any given spot on earth, and the map showing lines of equal dip ("isoclinic lines") must also be constantly revised. Moreover, the intensity of earth's magnetic field increases with latitude and is three times as strong near the magnetic poles as in the equatorial regions. This intensity also changes constantly, so that maps showing "isodynamic lines" must also be periodically revised.

Like everything else about the magnetic field, the overall intensity of the field changes. For some time now, the intensity has been diminishing. The field has lost 15 per cent of its total strength since 1670; if this continues, it will reach zero by about the year 4000. What then? Will it continue decreasing, in the sense that it will reverse with the north magnetic pole in Antarctica and the south magnetic pole in the Arctic? In other words, does earth's magnetic field periodically diminish, reverse, intensify, diminish, reverse, and so on?

One way of telling if this can indeed happen is to study volcanic rocks. When lava cools, the crystals form in alignment with the magnetic field. As long ago as 1906, the French physicist Bernard Brunhes noted that some rocks were magnetized in the direction *opposite* to earth's present magnetic field. This finding was largely ignored at the time, but there is no denying it now. The telltale rocks inform us that not only has earth's magnetic field reversed, it has done so many times: nine times in the last 4 million years, at irregular intervals.

The most spectacular finding in this respect is on the ocean floor. If melted rock is indeed pushing up through the Global Rift and spreading out, then as one moves east or west from the Rift one comes across rock that has solidified a progressively longer time ago. By studying the magnetic alignment, one can indeed find reversals occurring in strips, progressively farther from the Rift, at intervals of anywhere from 50,000 to 20 million years. The only rational way of explaining this, so far, is to suppose that there *is* sea-floor spreading and there *are* magnetic-field reversals.

The fact of the reversals is easier to ascertain, however, than the reasons for it.

In addition to long-term drifts of the magnetic field, there are small changes during the course of the day. These suggest some connection with the sun. Furthermore, there are "disturbed days" when the compass needle jumps about with unusual liveliness. The earth is then said to be experiencing a "magnetic storm." Magnetic storms are identical with electric storms and are usually accompanied by an increase in the intensity of auroral displays, an observation reported as long ago as 1759 by the English physicist John Canton.

The aurora borealis (a term introduced in 1621 by the French philosopher Pierre Gassendi, and Latin for "northern dawn") is a beautiful display of moving, colored streamers or folds of light, giving an effect of unearthly splendor. Its counterpart in the Antarctic is called the aurora australis ("southern dawn"). In 1741, the Swedish astronomer Anders Celsius noted its connection with earth's magnetic field. The auroral streamers seem to follow the earth's magnetic lines of force and to concentrate, and become visible, at those points where the lines crowd most closely together—that is, at the magnetic poles. During magnetic storms the northern aurora can be seen as far south as Boston and New York.

Why the aurora should exist was not hard to understand. Once the ionosphere was discovered, it was understood that something (presumably solar radiation of one sort or another) was energizing the atoms in the upper atmosphere and converting them into electrically charged ions. At night, the ions would lose their charge and their energy, the latter making itself visible in the form of auroral light. It was a kind of specialized air-glow, which followed the magnetic lines of force and concentrated near the magnetic poles because that would be expected of electrically charged ions. (The airglow itself involves uncharged atoms and therefore ignores the magnetic field.)

But what about the disturbed days and the magnetic storms? Again the finger of suspicion points to the sun.

Sunspot activity seems to generate magnetic storms. How such a disturbance 93 million miles away could affect the earth was a complete mystery until the spectrohelioscope, invented by the astronomer George Ellery Hale, brought forth a possible answer. This instrument allows the sun to be photographed in light of a particular color—for instance, the red light of hydrogen. Furthermore, it shows the motions or changes taking place on the sun's surface. It gives good pictures of "prominences" and "solar flares"—great bursts of flaming hydrogen. These were first observed during the eclipse of the sun in 1842, which was visible on a line across the width of Europe and was the first solar eclipse to be systematically and scientifically observed.

Before the invention of the spectrohelioscope, only those flares shooting out at right angles to the direction of the earth could be seen. The spectrohelioscope, however, also showed those coming out in our direction from the center of the sun's disk: in hydrogen light these hydrogen-rich bursts appear as light blotches against the darker background of the rest of the disk. It turned out that solar flares were followed by magnetic storms on the earth only when the flare was pointed toward the earth.

Apparently, then, magnetic storms were the result of bursts of charged particles, shot from the flares to the earth across 93 million miles of space. And, as a matter of fact, as long ago as 1896, something like this had been suggested by the Norwegian physicist Olaf Kristian Birkeland.

As a matter of fact, there was plenty of evidence that, wherever the particles might come from, the earth was bathed in an aura of them extending pretty far out in space. Radio waves generated by lightning had been found to travel along the earth's magnetic lines of force at great heights. (These waves, called "whistlers" because they were picked up by receivers as odd whistling noises, had been discovered accidentally by the German physicist Heinrich Barkhausen during World War I.) The radio waves could not follow the lines of force unless charged particles were present.

Yet it did not seem that these charged particles emerged from the sun only in bursts. In 1931, when Sydney Chapman was studying the sun's corona, he was increasingly impressed by its extent. What we can see during a total solar eclipse is only its innermost portion. The measurable concentrations of charged particles in the neighborhood of the earth were, he felt, part of the corona. This meant then, in a sense, that the earth was revolving about the sun within that luminary's extremely attenuated outer atmosphere. Chapman drew the picture of the corona expanding outward into space and being continually renewed at the sun's surface. There would be charged particles continuously streaming out of the sun in all directions, disturbing earth's magnetic field as it passed.

This suggestion became virtually inescapable in the 1950's, thanks to the work of the German astrophysicist Ludwig Franz Biermann. For half a century, it had been thought that the tails of comets, which always pointed generally away from the sun and which increased in length as the comet approached the sun, were formed by the pressure of light from the sun. Such light-pressure does exist, but Biermann showed that it wasn't nearly enough to produce cometary tails. Something stronger and with more of a push was required; this something could scarcely be anything but charged particles. The American physicist Eugene Norman Parker argued further in favor of a steady outflow of particles, with additional bursts at the time of solar flares and, in 1958, named the effect the "solar wind." The existence of this solar wind was finally demonstrated by the Soviet satellites "Lunik I" and "Lunik II," which streaked outward to the neighborhood of the moon in 1959 and 1960, and by the American planetary probe "Mariner II," which in 1962 passed near Venus.

The solar wind is no local phenomenon. There is reason to think it remains dense enough to be detectable at least as far out as the orbit of Saturn. Near the earth the velocity of solar-wind particles varies from 350 to 700 kilometers per second. Its existence represents a loss to the sun of a million tons of matter per second, but though this seems huge in human terms, it is utterly insignificant on the solar scale. In the entire lifetime of the sun less than a hundredth of a per cent of its mass has been lost to the solar wind.

The solar wind may well affect man's everyday life. Beyond its effect on the magnetic field, the charged particles in the upper atmosphere may ultimately have an effect on the details of earth's weather. If so, the ebb and flow of the solar wind may yet become still another weapon in the armory of the weather forecast.

An unforeseen effect of the solar wind was unexpectedly worked out as a result of satellite launchings. One of the prime jobs given to the man-made satellites was to measure the radiation in the upper atmosphere and nearby space, especially the intensity of the cosmic rays (charged particles of particularly high energy). How intense was this radiation up beyond the atmospheric shield? The satellites carried "Geiger counters" (first devised by the German physicist Hans Geiger in 1907 and vastly improved in 1928), which measure particle radiation in the following way. The counter has a box containing gas under a voltage not quite strong enough to send a current through the gas. When a high-energy particle of radiation penetrates into the box, it converts an atom of the gas into an ion. This ion, hurtled forward by the energy of the blow, smashes neighboring atoms to form more ions, which in turn smash their neighbors to form still more. The resulting shower of ions can carry an electric current, and for a fraction of a second a current pulses through the counter. The pulse is telemetered back to earth. Thus the instrument counts the particles, or flux of radiation, at the location where it happens to be.

When the first successful American satellite, "Explorer I," went into orbit on January 31, 1958, its counter detected about the expected concentrations of particles at heights up to several hundred miles. But at higher altitudes (and Explorer I went as high as 1,575 miles) the count fell off; in fact, at times it dropped to zero! This might have been dismissed as due to some peculiar kind of accident to the counter, but Explorer III, launched on March 26, 1958, and reaching an apogee of 2,100 miles, had just the same experience. So did the Soviet Sputnik III, launched on May 15, 1958.

James A. Van Allen of the State University of Iowa, who was in charge of the radiation program, and his aides came up with a possible explanation. The count fell virtually to zero, they decided, not because there was little or no radiation, but because there was too much. The instrument could not keep up with the particles entering it, and it blanked out in consequence. (This would be analogous to the blinding of our eyes by a flash of too-bright light.)

When Explorer IV went up on July 26, 1958, it carried special counters designed to handle heavy loads. One of them, for instance, was shielded with a thin layer of lead (analogous to dark sun-glasses) that would keep out most of the radiation. And this time the counters did tell another story. They showed that the "too-much-radiation" theory was correct. Explorer IV, reaching a height of 1,368 miles, sent down counts which, allowing for the shielding, disclosed that the radiation intensity up there was far higher than scientists had imagined. In fact, it was so intense that it raised a deadly danger to space flight by man.

It became apparent that the Explorer satellites had only penetrated the lower regions of this intense field of radiation. In the fall of 1958 the two satellites shot by the United States in the direction of the moon (so-called "moon probes")—Pioneer I, which went out 70,000 miles, and Pioneer III, which reached 65,000 miles—showed two main bands of radiation encircling the earth. They were named the "Van Allen radiation belts," but were later named the "magnetosphere" in line with the names given other sections of space in the neighborhood of the earth.

It was at first assumed that the magnetosphere was symmetrically placed about the earth, rather like a huge doughnut, and that the magnetic lines of force were themselves symmetrically arranged. This notion was upset when satellite data brought back other news. In 1963, in particular, the satellites "Explorer XIV" and Imp-1" were sent into highly elliptical orbits designed to carry them beyond the magnetosphere if possible.

It turned out that the magnetosphere had a sharp boundary, the "magnetopause," which was driven back upon the earth on the side toward the sun by the solar wind, but which looped back around the earth and extended an enormous distance on the night side. The magneto-pause was some 40,000 miles from the earth in the direction of the sun, but the teardrop tail on the other side may extend outward for a million miles or more. In 1966, the Soviet satellite Luna X, which circled the moon, detected a feeble magnetic field surrounding that world which may actually have been the tail of earth's magnetosphere sweeping past.

The entrapment of charged particles along the magnetic lines of force had been predicted in 1957 by an American-born Greek amateur scientist, Nicholas Christofilos, who made his living as a salesman for an American elevator firm. He had sent his calculations to scientists engaged in such research, but no one had paid much attention to them. (In science, as in other fields, professionals tend to disregard amateurs.) It was only when the professionals independently came up with the same results that Christofilos achieved recognition and was welcomed into the University of California, where he now works. His idea about particle entrapment is now called the "Christofilos effect."

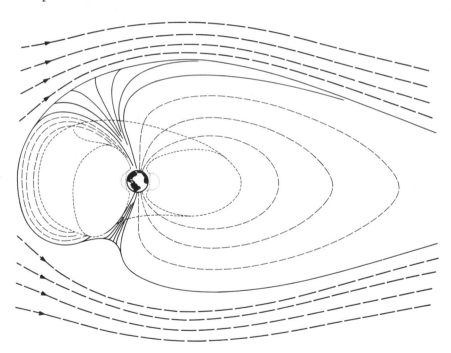

The Van Allen radiation belts, as traced by satellites. They appear to be made up of charged particles trapped in the earth's magnetic field.

To test whether the effect really occurs in space, the United States in August and September of 1958 fired three rockets carrying nuclear bombs 300 miles up and there exploded the bombs—an experiment which was named "Project Argus." The flood of charged particles resulting from the nuclear explosions spread out along the lines of force and were indeed trapped there. The resulting band persisted for a considerable time; Explorer IV detected it during several hundred of its trips around the earth. The cloud of particles also gave rise to feeble auroral displays and disrupted radar for a while.

This was the prelude to other experiments that affected or even altered earth's near-space environment, and some of them met with opposition and vast indignation from sections of the scientific community. A nuclear bomb exploded in space on July 9, 1962, introduced marked changes in the Van Allen belts, changes that showed signs of persisting for a prolonged interval, as some disapproving scientists (such as Fred Hoyle) had predicted. The Soviet Union carried out similar high-altitude tests in 1962. Such tampering with the natural state of affairs may interfere with our understanding of the magnetosphere, and it is unlikely that this experiment will be soon repeated.

Then, too, attempts were made to spread a layer of thin copper needles into orbit about the earth to test their ability to reflect radio signals, in order to establish an unfailing method for long-distance communication. (The ionosphere is disrupted by magnetic storms every once in a while and then radio communication may fail at a crucial moment.)

Despite the objection of radio astronomers who feared interference with the radio signals from outer space, the project ("Project West Ford," after Westford, Massachusetts, where the preliminary work was done) was carried through on May 9, 1963. A satellite containing 400 million copper needles, each three quarters of an inch long and finer than a human hair—fifty pounds worth altogether—was put into orbit. The needles were ejected and then slowly spread into a world-circling band that was found to reflect radio waves just as they had been expected to do. This band remained in orbit for three years. A much thicker band would be required for useful purposes, however, and it is doubtful if the objections of the radio astronomers can be overcome for that.

Naturally, scientists were curious to find out whether there were radiation belts about heavenly bodies other than the earth. One way of determining this was to send satellites upward at velocities great enough to break them loose from the earth's grip altogether (7 miles per second —as compared with a satellite in orbit about earth, which travels a mere 5 miles per second). The first satellite to surpass escape velocity, break away from the earth altogether, and move into orbit about the sun as the first "man-made planet" was the Soviet Union's, Lunik I, launched on January 2, 1959. Their next moon probe, Lunik II, actually hit the moon in September 1959 (the first man-made object to land on a surface of a body other than the earth). They found no signs of radiation belts about the moon.

This was not surprising, for scientists had already surmised that the moon had no magnetic field of any consequence. The over-all density of the moon has long been known to be but 3.3 grams per cubic centimeter (about three fifths that of the earth), and it could not have so low a density unless it were almost entirely silicate, with no iron core to speak of. The lack of a magnetic field would seem to follow, if present theories are correct.

But what of Venus? In size and mass it is almost the earth's twin, and there seems no doubt that it has an iron core. Does it also have a magnetosphere? Both the Soviet Union and the United States attempted to send out "Venus probes" that would, in their orbits, pass close to Venus and send back useful data. The first such probe to be completely successful was Mariner II, launched by the United States on August 27, 1962. It passed within 21,600 miles of Venus on December 14, 1962, and found no signs of a magnetosphere.

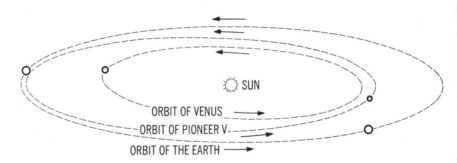

Orbit of the United States' "artificial planet" Pioneer V, launched March 11, 1960, is shown in relation to the sun and the orbits of the earth and Venus. The dot on the rocket's orbit indicates roughly its position on August 9, 1960, when it was closest to the sun.

This does not necessarily deny the presence of an iron core in Venus, since an alternate explanation offers itself at once. The rotation of Venus is very slow, once in about eight months (the fact that the rotation is in the wrong direction has nothing to do with this particular case), and that is not enough to set up the kind of eddies in the core (if it exists) that would account for a magnetic field. Mercury, which rotates once in two months, has been reported to have a weak magnetic field about 1/60 the intensity of earth.

But what of Mars? Being a bit denser than the moon it may have a small iron core, and, since it rotates in twenty four and a half hours, it may have a very weak magnetic field. The United States launched a "Mars Probe"—Mariner IV—on November 28, 1964. In July 1965, it approached Mars closely, and from the data it gathered, it would seem there was no magnetic field to speak of about that planet.

As for the solar system beyond Mars, good evidence turned up quickly to the effect that Jupiter and Saturn, at least, have radiation belts that are both more intense and more extensive than those of the earth. In fact, radio-wave radiation from Jupiter seems to indicate that it possesses a magnetic field at least twelve to sixteen times as strong as that of the earth. In 1965, radio-wave emission was detected from Uranus and Neptune.

One of the more dramatic reasons for intense curiosity in the magnetosphere is, of course, concern for the safety of human pioneers in outer space. In 1959, the United States selected seven men (popularly called "astronauts") to take part in "Project Mercury," which was to place men into orbit about the earth. The Soviet Union also initiated a training program for what they called "cosmonauts."

The honor of reaching this goal first fell to the Soviet Union's cosmonaut Yuri Alexeyevich Gagarin, who was launched into orbit on April 12, 1961 (only three and a half years after the opening of the "Space Age" with Sputnik I), and returned safely after circling the earth in an hour and a half. He was the first "man in space."

The Soviet Union sent other human beings in orbit over the next few years. For a while, the record for endurance was held by Valery F. Bykovsky, who, after being launched on June 14, 1963, circled the earth eighty-two times before coming down. Launched on June 16, 1963, was Valentina V. Tereshkova, the first woman in space. She completed forty-nine orbits.

The first American to be placed in orbit was John Herschel Glenn who, after being launched on February 20, 1962, circled the earth three times. The American record for endurance, so far, is the trip of Leroy Gordon Cooper, who was launched on May 15, 1963, and successfully completed twenty-two orbits.

For years, manned space flights, both in the Soviet Union and the United States, have ended successfully and there were no casualties. There is some question as to whether or not the flights have had some long-term bad effects. Several astronauts, including Glenn, have suffered ailments of the middle ear, although this may not be directly connected with their experiences.

Flights in two-man and three-man capsules were carried through by the United States and the Soviet Union in 1964 and 1965. During the course of one two-man flight on March 18, 1965, the Soviet cosmonaut Aleksei A. Leonov, encased in his spacesuit and holding on to a "life-line" stepped out of his capsule and became the first human being to float freely in space.

Meteors

Even the Greeks knew that "shooting stars" were not really stars, because no matter how many fell, the celestial population of stars remained the same. Aristotle reasoned that a shooting star, being a temporary phenomenon, had to be something within the atmosphere (and this time he was right). These objects were therefore called "meteors," meaning "things in the air." Meteors that actually reach the earth's surface are called "meteorites."

The ancients even witnessed some falls of meteorites to the earth and found them to be lumps of iron. Hipparchus of Nicaea is said to have reported such a fall. The Kaaba, the sacred black stone in Mecca, is supposed to be a meteorite and to have gained its sanctity through its heavenly origin. The *Iliad* mentions a lump of rough iron being awarded as one of the prizes in the funeral games for Patroclus; this must have been meteoric in origin, because the time was the Bronze Age, before the metallurgy of iron ore had been developed. In fact, meteoric iron was probably in use as early as 3000 B.C.

During the eighteenth century, with the Age of Reason in full sway, science made a backward step in this particular respect. The scorners of superstition laughed at stories of "stones from the sky." Farmers who came to the Académie Française with samples of meteorites were politely, but impatiently, shown the door. When two Connecticut scholars in 1807 reported having witnessed a fall, President Thomas Jefferson (in one of his more unfortunate remarks) said that he would sooner believe that two Yankee professors would lie than that stones would fall from heaven.

However, on November 13, 1833, the United States was treated to a meteor shower of the type called "Leonids" because they seem to radiate from a point in the constellation Leo. For some hours it turned the sky into a Roman-candle display more brilliant than any ever seen before or since. No meteorites reached the ground, as far as is known, but the spectacle stimulated the study of meteors, and astronomers turned to it for the first time in all seriousness.

The very next year, the Swedish chemist Jöns Jakob Berzelius began a program for the chemical analyses of meteorites. Eventually such analyses gave astronomers valuable information on the general age of the solar system and even on the overall chemical makeup of the universe.

By noting the times of year when meteors came thickest, and the positions in the sky from which they seemed to come, the meteor watchers were able to work out orbits of various clouds of meteors. In this way they learned that a meteor shower occurred when the earth's orbit intersected the orbit of a meteor cloud.

Could it be that meteor clouds of this sort are actually the debris left over from disintegrated comets? That comets do disintegrate may be accepted since one of them, Biela's Comet, virtually did just that before the eyes of astronomers in the nineteenth century and left a meteor cloud in its orbit.

Comets may be fragile because of their very structure. The American astronomer Fred Lawrence Whipple suggested in 1950 that comets consist of pebbles of rocky material cemented by "ices" of such low-freezing gases as methane or ammonia. Some of the ices evaporate at each close approach to the sun, liberating dust and particles which are swept away from the sun by (as we now know) the solar wind. Eventually, the ices are all gone, and the comet either remains as a rocky core or disintegrates into the meteor cloud formed of its former pebbles. A comet may lose as much as 0.5 per cent of its mass at each approach; even a comet that approaches the sun not-too-closely may last not more than a million years. That comets still exist now that the solar system has lasted nearly 5 billion years, can only be because new comets are constantly entering the inner system from the vast cloud of comets that Oort has postulated to be far out in space.

Most of the meteorites found on the ground (about 1,700 are known altogether, of which 35 weigh over a ton each) were iron, and it seemed that iron meteorites must far outnumber the stony type. This proved to be wrong, however. A lump of iron lying half-buried in a stony field is very noticeable, whereas a stone among other stones is not. When astronomers made counts of meteorites found after they were actually seen to fall, they discovered that the stony meteorites outnumbered iron ones nine to one. (For a time, most stony meteorites were discovered in Kansas, which may seem odd until one realizes that in the stoneless, sedimentary soil of Kansas a stone is as noticeable as a lump of iron would be elsewhere.)

Meteorites seldom do damage. Although about 500 substantial meteorites strike the earth annually (with only some 20 recovered, unfortunately), the earth's surface is large and only small areas are thickly populated. No human being has ever been killed by a meteorite so far as is known, although a woman in Alabama reported being bruised by a glancing blow on November 30, 1955.

Yet meteorites have a devastating potentiality. In 1908, for instance, a strike in northern Siberia gouged out craters up to 150 feet in diameter and knocked down trees for 20 miles around. Fortunately, the meteorite fell in a wilderness; had it fallen from the same part of the sky five hours later in the earth's rotation, it might have hit St. Petersburg, then the capital of Russia. If it had, the city would have been wiped out as thoroughly as by an H-bomb. One estimate is that the total weight of the meteorite was 40,000 tons. The largest strike since then, near Vladivostok (again in Siberia), was in 1947.

There are signs of even heavier strikes in prehistoric times. In Coconino County in Arizona there is a round crater about 4/5 of a mile across and 600 feet deep, surrounded by a lip of earth 100 to 150 feet high. It looks like a miniature crater of the moon. It was long assumed to be an extinct volcano, but a mining engineer named Daniel Moreau Barringer insisted it was the result of a meteoric collision, and the hole now bears the name "Barringer Crater." The crater is surrounded by lumps of meteoric iron—thousands (perhaps millions) of tons of it altogether. Although only a small portion has been recovered so far, more meteoric iron has already been extracted from it and its surroundings than in all the rest of the world. The meteoric origin of the crater was also borne out by the discovery then in 1960 of forms of silica that could only have been produced by the momentary enormous pressures and temperatures accompanying meteoric impact.

Barringer Crater, formed in the desert an estimated 25,000 years ago, has been preserved fairly well. In most parts of the world similar craters would have been obliterated by water and plant overgrowth. Observations from airplanes, for instance, have sighted previously unnoticed circular formations, partly water-filled and partly overgrown, which are almost certainly meteoric. Several have been discovered in Canada, including Brent Crater in Central Ontario and Chubb Crater in northern Quebec, each of which is two miles or more in diameter, and Ashanti Crater in Ghana, which is six miles in diameter. These are perhaps a million years old or more. Fourteen such "fossil craters" are known, and subtle geological signs point to the existence of many more.

The craters of the moon visible to us with telescopes range from holes no larger than Barringer Crater to giants 150 miles across. The moon, lacking air, water, or life, is a nearly perfect museum for craters since they are subject to no wear except from the very slow action of temperature change resulting from the two-week alternation of lunar day and lunar night. Perhaps the earth would be pockmarked like the moon if it were not for the healing action of wind, water, and growing things.

It had been felt, at first, that the craters of the moon were volcanic in origin, but they do not really resemble earthly volcanic craters in structure. By the 1890's, the view that the craters had originated from meteoric strikes came into prominence and has gradually become accepted.

The large "seas" or maria, which are vast, roughly circular stretches that are relatively craterfree, would in this view result from the impact of particularly large meteors. This view was bolstered in 1968 when satellites placed in orbit about the moon showed unexpected deviations in their circumlunar flights. The nature of these deviations forced the conclusion that parts of the lunar surface were denser than average and produced a slight increase in gravitational attraction, to which the satellite flying over those parts responded. These denser-than-average areas, which seemed to coincide with the maria, received the name "mascons" (short for "mass-concentrations"). The most obvious deduction was that the sizable iron meteors that formed the seas were still buried beneath them and were considerably denser than the rocky material that generally made up the moon's crust. At least a dozen mascons were detected within a year of the initial discovery.

The view of the moon as a "dead world" where no volcanic action is possible is, on the other hand, overdrawn. On November 3, 1958, the Russian astronomer N. A. Kozyrev observed a reddish spot in the crater Alphonsus. (William Herschel had reported seeing reddish spots on the moon as early as 1780.) Kozyrev's spectroscopic studies seemed to make it clear that gas and dust had been emitted. Since then, other red spots have been momentarily seen, and it seems quite certain that volcanic activity does occasionally take place on the moon. Dating the total lunar eclipse in December 1964, it was found that as many as 300 craters were hotter than the surrounding landscape, though of course they were not hot enough to glow.

Once the first satellite was put into orbit in 1957, it was only a matter of time before mankind began to investigate the moon at close quarters. The first successful "moon probe"—that is, the first satellite to pass near the moon—was sent up by the Soviet Union on January 2, 1959. It was "Lunik I," the first man-made object to take up an orbit about the Sun. Within two months, the United States had duplicated the feat.

On September 12, 1959, the Soviets sent up Lunik II and aimed it to hit the moon. For the first time in history, a man-made object rested on the surface of another world. Then, a month later, the Soviet satellite Lunik III slipped beyond the moon and pointed a television camera at the side we never see from earth. Forty minutes of pictures of the other side were sent back from a distance of 40,000 miles. They were fuzzy and of poor quality, but they showed something interesting. The other side of the moon had scarcely any maria of the type that are so prominent a feature of our side. Why this asymmetry should exist is not entirely clear. Presumably the maria were formed comparatively late in the moon's history, when one side already faced the earth forever and the large meteors that formed the seas were slanted toward the near face of the moon by earth's gravity.

But lunar exploration was only beginning. In 1964, the United States launched a moon probe, Ranger VII, which was designed to strike the moon's surface, taking photographs as it approached. On July 31, 1964, it completed its mission successfully, taking 4,316 pictures of an area now named "Mare Cognitum" ("Known Sea"). In early 1965, Ranger VIII and Ranger IX had even greater success, if that were possible. These moon probes revealed the moon's surface to be hard (or crunchy, at worst) and not covered by the thick layer of dust some astronomers had suspected might exist. The probes showed those areas that seemed flat when seen through a telescope to be covered by craters too small to be seen from the earth.

The Soviet probe Luna IX succeeded in making a "soft landing" (one not involving the destruction of the object making the landing) on the moon on February 3, 1966, and sent back photographs from ground levels. On April 3, 1966, the Soviets placed "Luna X" in a three-hour orbit about the moon; it measured radioactivity from the lunar surface, and the pattern indicated the rocks of the lunar surface were similar to the basalt that underlies earth's oceans.

American rocketmen followed this lead with even more elaborate rocketry. The first American soft landing on the moon was that of "Surveyor I" on June 1, 1966. By September 1967, Surveyor V was handling and analyzing lunar soil under radio control from earth. It did indeed prove to be basaltlike and to contain iron particles that were probably meteoric in origin.

On August 10, 1966, the first of the American "Lunar Orbiter" probes were sent circling around the moon. (It was these that discovered the mascons.) The Lunar Orbiters took detailed photographs of every part of the moon, so that its surface features everywhere (including the part forever hidden from earth's surface) came to be known in fine detail. In addition, startling photographs were taken of earth as seen from the neighborhood of the moon.

The lunar craters, by the way, have been named for astronomers and other great men of the past. Since most of the names were given by the Italian astronomer Giovanni Battista Riccioli about 1650, it is the older astronomers Copernicus, Tycho, and Kepler, as well as the Greek astronomers Aristotle, Archimedes, and Ptolemy, who are honored by the larger craters.

The other side, first revealed by Lunik III, offered a new chance. The Russians, as was their right, preempted some of the more noticeable features. They named craters not only after Tsiolkovsky, the great prophet of space travel, but also after Lomonosov and Popov, two Russian chemists of the late eighteenth century. They have awarded craters to Western personalities, too, including Maxwell, Hertz, Edison, Pasteur, and the Curies, all of whom are mentioned in this book. One very fitting name placed on the other side of the moon is that of the French pioneer-writer of science fiction, Jules Verne.

In 1970, the other side of the moon was sufficiently well-known to make it possible to name its features systematically. Under the leadership of the American astronomer Donald Howard Menzel, an international body assigned hundreds of names, honoring great men of the past who had contributed to the advance of science in one way or another. Very prominent craters were allotted to such Russians as Mendeleev (who first developed the periodic table that I will discuss in Chapter 5) and Gagarin, who was the first man to be placed in orbit about earth and who had since died in an airplane accident. Other prominent features were used to memorialize the Dutch astronomer Hertzsprung, the French mathematician Galois, the Italian physicist Fermi, the American mathematician Wiener, and the British physicist Cockcroft. In one restricted area, we can find Nernst, Roentgen, Lorentz, Moseley, Einstein, Bohr, and Dalton, all of great importance in the development of the atomic theory and subatomic structure.

Reflecting Menzel's interest in science writing and science fiction, is his just decision to allot a few craters to those who helped rouse the enthusiasm of an entire generation for space flight when orthodox science dismissed it as a chimera. For that reason, there is a crater honoring Hugo Gernsback, who published the first magazines in the United States devoted entirely to science fiction, and another to Willy Ley, who, of all writers, most indefatigably and accurately portrayed the victories and potentialities of rocketry. (Ley died, tragically, six weeks before the first landing on the moon—a landing for which he had waited all his life.)

Yet all lunar exploration by instrument alone had to take a back seat, dramatically, to the greatest of all the rocket feats of the 1960's: manned exploration of space, something we will take up in Chapter 15.

Anyone who is inclined to be complacent about meteors or to think that colossal strikes were just a phenomenon of the solar system's early history might give some thought to the asteroids, or planetoids. Whatever their origin—whether they are surviving planetesimals or remnants of an exploded planet—there are some pretty big ones around and about. Most of them orbit the sun in a belt between Mars and Jupiter. But in 1898 a German astronomer G. Witt discovered one whose orbit, upon calculation, turned out to lie between Mars and the earth. He named it Eros, and ever since planetoids with unusual orbits have been given masculine names. (Those with ordinary orbits, between Mars and Jupiter, are given feminine names even when named after men, e.g., Rockefellia, Carnegia, Hooveria.)

The orbits of Eros and the earth approach to within 13 million miles of each other, which is half the minimum distance between the earth and Venus, our closest neighbor among the full-sized planets. In 1931 Eros reached a point only 17 million miles from the earth and it will make its next close approach in 1975. Several other "earth grazers" have since been found. In 1932 two planetoids named Amor and Apollo were discovered with orbits approaching within 10 million and 7 million miles, respectively, of the earth's orbit. In 1936, there turned up a still-closer planetoid, named Adonis, which could approach to as close as 1.5 million miles from the earth. And in 1937 a planetoid given the name Hermes swam into sight in an orbit which might bring it within 200,000 miles of the earth, or actually closer than the moon. (The calculations of Hermes' orbit may not be entirely reliable, because the object did not stay in sight long and has not been sighted since.)

A particularly unusual earth grazer is Icarus, discovered in 1948 by Walter Baade. It approaches within 4 million miles of the earth, in an elongated cometlike orbit. At aphelion, it recedes as far as the orbit of Mars, and at perihelion it approaches to within 17 million miles of the sun. It is because of this that it has been named after the Greek mythological character who came to grief through flying too close to the sun on wings that were fixed in place with wax. Only certain comets ever approach the sun more closely than does Icarus. One of the large comets of the 1880's approached within less than a million miles of the sun.

Eros, the largest of the earth grazers, is a brick-shaped object perhaps fifteen miles long and five miles broad. Others, such as Hermes, are only about one mile in diameter. Still, even Hermes would gouge out a crater about one hundred miles across if it hit the earth or create tsunamis of unprecedented size if it struck the ocean. Fortunately the odds against an encounter are enormous.

Meteorites, as the only pieces of extraterrestrial matter we can examine, are exciting not only to astronomers, geologists, chemists, and metallurgists, but also to cosmologists, who are concerned with the origins of the universe and the solar system. Among the meteorites are puzzling glassy objects found in several places on earth. The first were found in 1787 in what is now western Czechoslovakia. Australian examples were detected in 1864. They received the name "tektites," from a Greek word for "molten," because they appear to have melted in their passage through the atmosphere.

In 1936, the American astronomer Harvey Harlow Ninninger suggested that tektites are remnants of splashed material forced away from the moon's surface by the impact of large meteors and caught by earth's gravitational field. A particularly widespread strewing of tektites is to be found in Australia and southeast Asia (with many dredged up from the floor of the Indian Ocean). These seem to be the youngest of the tektites, only 700,000 years old. Conceivably, these could have been produced by the great meteoric impact that formed the crater Tycho (the youngest of the spectacular lunar craters) on the moon. The fact that this strike seems to have coincided with the most recent reversal of earth's magnetic field has caused some speculation that the strikingly irregular series of such reversals may mark other such earth-moon catastrophes.

Putting the tektites to one side, meteorites are samples of primitive matter formed in the early history of our system. As such, they give us an independent clock for measuring the age of our system. Their ages can be estimated in various ways, including measurement of products

A schematic drawing of the radiuses of the orbits of most of the sun's planets, indicating their distances from the sun and the positions of Eros and the asteroids. Roughly, each planet is about twice as far from the sun as the one before it.

of radioactive decay. In 1959, John H. Reynolds of the University of California determined the age of a meteorite that had fallen in North Dakota to be 5 billion years, which would therefore be the minimum age of the solar system.

Meteorites make up only a tiny fraction of the matter falling into the earth's atmosphere from space. The small meteors that burn up in the air without ever reaching the ground amount to a far greater aggregate mass. Individually these bits of matter are extremely small; a shooting star as bright as Venus comes into the atmosphere as a speck weighing only one gram (1/28 of an ounce). Some visible meteors are only 1/10,000 as massive as that!

The total number of meteors hitting the earth's atmosphere can be computed, and it turns out to be incredibly large. Each day there are more than 20,000 weighing at least one gram, nearly 200 million others large enough to make a glow visible to the naked eye, and many billions more of still smaller sizes.

We know about these very small "micrometeors" because the air has been found to contain dust particles with unusual shapes and a high nickel content, quite unlike ordinary terrestrial dust. Another evidence of the presence of micrometeors in vast quantities is the faint glow in the heavens called "zodiacal light" (first discovered about 1700 by G. D. Cassini)—so called because it is most noticeable in the neighborhood of the plane of the earth's orbit, where the constellations of the zodiac occur. The zodiacal light is very dim and cannot be seen even on a moonless night unless conditions are favorable. It is brightest near the horizon where the sun has set or is about to rise, and on the opposite side of the sky there is a secondary brightening called the "Gegenschein" (German for "opposite light"). The zodiacal light differs from the airglow: its spectrum has no lines of atomic oxygen or atomic sodium, but is just that of reflected sunlight and nothing more. The reflecting agent presumably is dust concentrated in space in the plane of the planets' orbits—in short, micrometeors. Their number and size can be estimated from the intensity of the zodiacal light.

Micrometeors have now been counted with new precision by means of such satellites as "Explorer XVI," launched December 1962, and "Pegasus I," launched February 16, 1965. To detect them, some of the satellites are covered with patches of a sensitive material that signals each meteoric hit through a change in its electrical resistance. Others record the hits by means of a sensitive microphone behind the skin, picking up the "pings." The satellite counts have indicated that 3,000 tons of meteoric matter enter our atmosphere each day, five sixths of it consisting of micrometeors too small to be detected as shooting stars. These micrometeors may form a thin dust cloud about the earth, one that stretches out, in decreasing density, for 100,000 miles or so before fading out to the usual density of material in interplanetary space.

The Venus probe "Mariner II," launched August 27, 1962, showed the dust concentration in space generally to be only 1/10,000 the concentration near earth—which seems to be the center of a dustball. The American astronomer Fred Lawrence Whipple suggests that the moon may be the source of the cloud, the dust being flung up from the moon's surface by the meteorite beating it has had to withstand. Venus, which has no moon (according to Mariner II), also has no dustball.

The geophysicist Hans Petterson, who has been particularly interested in this meteoric dust, took some samples of air in 1957 on a mountaintop in Hawaii, which is as far from industrial dust-producing areas as one can get on the earth. His findings led him to believe that about 5 million tons of meteoric dust fall on the earth each year. (A similar measurement by James M. Rosen in 1964, making use of instruments borne aloft by balloons, set the figure at 4 million tons, though still others find reason to place the figure at merely 100,000 tons per year.) Hans Petterson tried to get a line on this fall in the past by analyzing cores brought up from the ocean bottom for high-nickel dust. He found that, on the whole, there was more in the upper sediments than in the older ones below, which indicates—though the evidence is still scanty—that the rate of meteoric bombardment may have increased in recent ages. This meteoric dust may possibly be of direct importance to all of us, for, according to a theory advanced by the Australian physicist E. G. Bowen in 1953, this dust serves as nuclei for raindrops. If this is so, then the earth's rainfall pattern reflects the rise and fall of the intensity with which micrometeorites bombard us.

The Origin of Air

Perhaps we should wonder less about how the earth got its atmosphere than about how it managed to hang on to it through all the eons the earth has been whirling and wheeling through space. The answer to the latter question involves something called "escape velocity."

If an object is thrown upward from the earth, the pull of gravity gradually slows it until it comes to a momentary halt and then falls back. If the force of gravity were the same all the way up, the height reached by the object would be proportional to its initial upward velocity; that is, it would reach four times as high when launched with a speed of two miles an hour as it would when it started at one mile an hour (energy increases as the square of the velocity).

But of course the force of gravity does not remain constant: it weakens slowly with height. (To be exact, it weakens as the square of the distance from the earth's center.) Let us say we shoot an object upward with a velocity of one mile per second. It will reach a height of eighty miles before turning and falling (if we ignore air resistance). If we were to fire the same object upward at two miles per second, it would climb higher than four times that distance. At the height of eighty miles, the pull of the earth's gravity is appreciably lower than at ground level, so that the object's further flight would be subject to a smaller gravitational drag. In fact, the projectile would rise to 350 miles, not 320.

Given an initial upward velocity of 6.5 miles per second, an object will climb 25,800 miles. At that point the force of gravity is not more than one fortieth as strong as it is on the earth's surface. If we added just one tenth of a mile per second to the object's initial speed (i.e., launched it at 6.6 miles per second), it would go up to 34,300 miles.

It can be calculated that an object fired up at an initial speed of 6.98 miles per second will never fall back to the earth. Although the earth's gravity will gradually slow the object's velocity, its effect will steadily decline, so that it will never bring the object to a halt (zero velocity) with respect to the earth. (So much for the cliché that "everything that goes up must come down.") Lunik I and Pioneer IV, fired at better than 7 miles per second, will never come down.

The speed of 6.98 miles per second, then, is the earth's "escape velocity." The velocity of escape from any astronomical body can be calculated from its mass and size. From the moon, it is only 1.5 miles per second; from Mars, 3.2 miles per second; from Saturn, 23 miles per second; from Jupiter, the most massive planet in the solar system, it is 38 miles per second.

Now all this has a direct bearing on the earth's retention of its atmosphere. The atoms and molecules of the air are constantly flying about like tiny missiles. Their individual velocities vary a great deal, and the only way they can be described is statistically: for example, giving the fraction of the molecules moving faster than a particular velocity, or giving the average velocity under given conditions. The formula for doing this was first worked out in 1860 by James Clerk Maxwell and the Austrian physicist Ludwig Boltzmann, and it is called the "Maxwell-Boltzmann law."

The mean velocity of oxygen molecules in air at room temperature turns out to be 0.3 mile per second. The hydrogen molecule, being only one sixteenth as heavy, moves on the average four times as fast, or 1.2 miles per second, because, according to the Maxwell-Boltzmann law, the velocity of a particular particle at a particular temperature is inversely proportional to the square root of its molecular weight.

It is important to remember that these are only average velocities. Half the molecules go faster than the average; a certain percentage go more than twice as fast as the average; a smaller percentage more than three times as fast, and so on. In fact, a tiny percentage of the oxygen and hydrogen molecules in the atmosphere go faster than 6.98 miles per second, the escape velocity.

In the lower atmosphere, these speedsters cannot actually escape,

Solar prominences.

Aurora photographed in Alaska during the IGY.

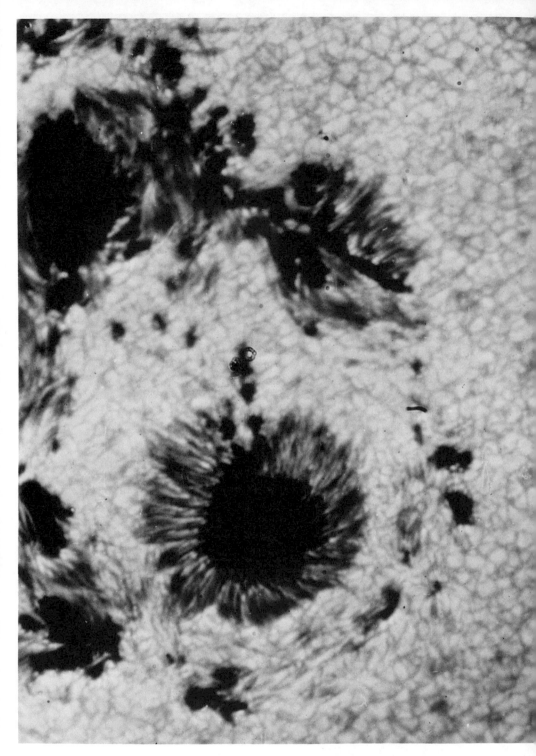

Sun spots photographed with unprecedented sharpness by the U.S. Stratoscope project from a balloon at 80,000 feet. The spots consist of a dark core of relatively cool gases embedded in a strong magnetic field. This group of particularly active spots produced a brilliant aurora and a vigorous magnetic storm on the earth.

August 27, 1954

August 2, 1956

May 25, 1958

July 17, 1961

June 26, 1964

International Quiet Sun Year, a two-year, world-wide scientific enterprise involving 69 nations, began January 1, 1964. This series of photographs depicts the change in the sun's activity from minimum to maximum to minimum. Much important information is obtainable only in periods of infrequent solar disturbances.

WILLAMETTE
Willamette, Oregon
Gift of Mrs. William E. Dodge, 2d.
Found 1902.

A large iron meteorite.

The great meteor crator in Arizona.

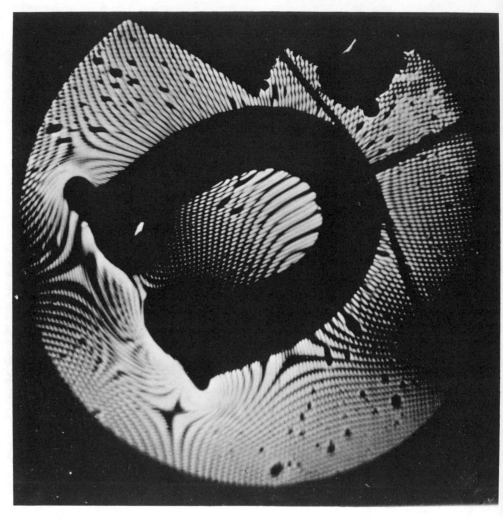

Magnetic field, photographed with an electron microscope by means of a new shadow technique developed by the U.S. National Bureau of Standards. The small horseshoe magnet used here is only about one fourth of an inch wide.

because collisions with their slower neighbors slow them down. But in the upper atmosphere, their chances are much better. First of all, the unimpeded radiation of the sun up there excites a large proportion of them to enormous energy and great speeds. In the second place, the probability of collisions is greatly reduced in the thinner air. Whereas a molecule at the earth's surface travels only four millionths of an inch (on the average) before colliding with a neighbor, at a height of 65 miles its average free path before colliding is 4 inches, and at 140 miles, it is 1,100 yards. There the average number of collisions encountered by an atom or molecule is only 1 per second, against 5 billion per second at sea level. Thus a fast particle at a height of 100 miles or more stands a good chance of escaping from the earth. If it happens to be moving upward, it is moving into regions of lesser and lesser density and experiences an ever smaller chance of collision, so that it may in the end depart into interplanetary space, never to return.

In other words, the earth's atmosphere leaks. But the leakage applies mainly to the lightest molecules. Oxygen and nitrogen are heavy enough so that only a tiny fraction of them achieves the escape velocity, and not much oxygen or nitrogen has been lost from the earth since their original formation. On the other hand, hydrogen and helium are easily raised to escape velocity. Consequently it is not surprising that no hydrogen or helium to speak of remains in the atmosphere of the earth today.

The more massive planets, such as Jupiter and Saturn, can hold even hydrogen and helium, so they may have large and deep atmospheres composed mostly of these elements (which, after all, are the most common substances in the universe). The hydrogen present in vast quantities would react with other elements present, so that carbon, nitrogen, and oxygen would be present only in the form of hydrogen-containing compounds: methane (CH_4), ammonia (NH_3), and water (H_2O) respectively. The ammonia and methane in Jupiter's atmosphere, although present as relatively small-concentration impurities, were first discovered (in 1931, by the German-American astronomer Rupert Wildt) because these compounds produce noticeable absorption bands in the spectra, whereas hydrogen and helium do not. The presence of hydrogen and helium were detected by rather indirect methods in 1952.

On the basis of his findings, Wildt speculated about the structure of Jupiter and the other planets. He suggested that under the thick outermost shell of atmosphere, there was a layer of frozen water, and underneath that a rocky core. Similar structures were suggested for the major planets farther out. Saturn, which was distinctly less dense than Jupiter, would have a thicker atmosphere and a smaller core; Neptune, which was distinctly more dense, a thinner atmosphere and a larger core (for its size). However, all that can actually be seen of Jupiter is its upper atmosphere, and the radio-wave emissions are as yet insufficient to tell us much detail of what goes on below. For instance, it is possible to argue that

Jupiter and the other "gas giants" are hydrogen and helium, for the most part, all the way through to the center, where pressures are so high that hydrogen assumes a metallic form.

Working in the other direction, a small planet like Mars is less able to hold even the comparatively heavy molecules and has an atmosphere only one tenth as dense as our own. The moon, with a smaller escape velocity, cannot hold any atmosphere to speak of and is airless.

Temperature is just as important a factor as gravity. The Maxwell-Boltzmann equation says that the average speed of particles is proportional to the square root of the absolute temperature. If the earth were at the temperature of the sun's surface, all the atoms and molecules in its atmosphere would be speeded up by four to five times, and the earth could no more hold on to its oxygen and nitrogen than it could to hydrogen or helium.

On the other hand, if temperatures were lower, the chance of holding molecules of a particular kind is increased. In 1943, for instance, Kuiper managed to detect an atmosphere of methane on Titan, the largest satellite of Saturn. Titan is not very much larger than the moon, and if it were at the moon's distance from the sun, it would have no atmosphere. At the frigid temperatures of the outer solar system, it manages. It is possible that the other large outer satellites—Neptune's satellite, Triton; and Jupiter's four large satellites, Io, Europa, Ganymede, and Callisto—may have thin atmospheres of some sort, but these have not yet been detected. For the time, Titan remains unique among the satellites of the planetary system.

The earth's possession of an atmosphere is a strong point against the theory that it and the other planets of the solar system originated from some catastrophic accident, such as near-collision between another sun and ours. It argues, rather, in favor of the dust-cloud and planetesimal theory. As the dust and gas of the cloud condensed into planetesimals and these in turn collected to form a planetary body, gas might have been trapped within a spongy mass, like air in a snowbank. The subsequent gravity contraction of the mass might then have squeezed out the gases toward the surface. Whether a particular gas would be held in the earth would depend in part on its chemical reactivity. Helium and neon, though they must have been among the most common gases in the original cloud, are so inert chemically that they form no compounds and would have escaped as gases in short order. Therefore the concentrations of helium and neon on the earth are insignificant fractions of their concentrations in the universe generally. It has been calculated, for instance, that the earth has retained only one out of every 50 billion neon atoms present in the original cloud of gas, and our atmosphere has even fewer, if any, of the original helium atoms. I say "if any" because, while there is a little helium in the atmosphere today, all of it may come from the breakdown of radioactive elements and leakage of helium trapped in cavities underground.

On the other hand, hydrogen, though lighter than helium or neon, has been captured with greater efficiency because it has combined with other substances, notably with oxygen to form water. It is estimated that the earth still has one out of every 5 million hydrogen atoms that were in the original cloud.

Nitrogen and oxygen illustrate the chemical aspect even more neatly. Although the nitrogen molecule and the oxygen molecule are about equal in mass, the earth has held on to 1 out of 6 of the original atoms of highly reactive oxygen but only 1 out of every 800,000 of inert nitrogen.

When we speak of gases of the atmosphere, we have to include water vapor, and here we get into the interesting question of how the oceans originated. In the early stages of the earth's history, even if it was only moderately hot, all the water must have been in the form of vapor. Some geologists believe that the water was then concentrated in the atmosphere as a dense cloud of vapor, and, after the earth cooled, it fell in torrents to form the ocean. On the other hand, some geologists maintain that our oceans have been built up mainly by water seeping up from the earth's interior. Volcanoes show that there still is a great deal of water in the crust, for the gas they discharge is mostly water vapor. If that is so, the oceans may still be growing, albeit slowly.

But was the earth's atmosphere always what it is today, at least since its formation in the first place? It seems very unlikely. For one thing, molecular oxygen, which makes up one-fifth of the volume of the atmosphere, is so active a substance that its presence in free form is extremely unlikely, unless it were continuously being produced. Furthermore, no other planet has an atmosphere anything like our own, so that one is strongly tempted to conclude that earth's atmosphere is the result of unique events (as, for instance, the presence of life on this planet, but not on the others).

Harold Urey has presented detailed arguments in favor of the idea that the original atmosphere was composed of ammonia and methane. Hydrogen, helium, carbon, nitrogen, and oxygen are the predominant elements in the universe, with hydrogen far and away the most common. In the presence of such a preponderance of hydrogen, carbon would be likely to combine with hydrogen to form methane (CH_4), nitrogen with hydrogen to form ammonia (NH_3), and oxygen with hydrogen to form water (H_2O). Helium and excess hydrogen would, of course, escape; the water would form the oceans; the methane and ammonia, as comparatively heavy gases, would be held by the earth's gravity and so constitute the major portion of the atmosphere.

If all the planets with sufficient gravity to hold an atmosphere at all, began with atmospheres of this type, they would nevertheless not all keep such an atmosphere. Ultraviolet radiation from the sun would introduce changes. These changes would be minimal for the outer planets, which in the first place received comparatively little radiation from the distant sun, and in the second place had vast atmospheres capable of absorbing considerable radiation without being perceptibly changed. The outer planets, therefore, would keep the hydrogen/helium/ammonia/methane atmospheres to the present day.

Not so the five inner worlds of Mars, Earth, Moon, Venus, and Mercury. Of these, the Moon and Mercury are too small, too hot, or both to retain any perceptible atmosphere. This leaves Mars, Earth, and Venus, with thin atmospheres of chiefly ammonia, methane, and water to begin with. What would happen?

Ultraviolet radiation striking water molecules in the upper primordial atmosphere of the earth would break them apart to hydrogen and oxygen ("photodissociation"). The hydrogen would escape, leaving oxygen behind. Being reactive, however, the molecules would react with almost any other molecule in the neighborhood. They would react with methane (CH_4) to form carbon dioxide (CO_2) and water (H_2O). They would react with ammonia (NH_3) to form free nitrogen (N_2) and water.

Very slowly, but steadily, the atmosphere would be converted from methane and ammonia to nitrogen and carbon dioxide. The nitrogen would tend to react slowly with the minerals of the crust to form nitrates, leaving carbon dioxide as the major portion of the atmosphere.

Will water continue to photodissociate, however? Will hydrogen continue to escape into space, and will oxygen continue to collect in the atmosphere? And if oxygen does collect and finds nothing to react with (it cannot react further with carbon dioxide), then will it not add a proportion of molecular oxygen to the carbon dioxide present (thus accounting for earth's atmospheric oxygen)? The answer is, No!

Once carbon dioxide becomes the major component of the atmosphere, ultraviolet radiation does not bring about further changes through dissociation of the water molecule. When the oxygen begins to collect in free form, a thin ozone layer is formed in the upper atmosphere. This absorbs the ultraviolet, blocking it from the lower atmosphere and preventing further photodisssociation. A carbon-dioxide atmosphere is stable.

But carbon dioxide introduces the greenhouse effect (see page 157). If the carbon-dioxide atmosphere is thin and if it is relatively far from the sun, the effect is small. This is the case with Mars, for instance. Its atmosphere, chiefly carbon dioxide, is thinner than that of the earth —how much thinner was not realized till the American Mars-probe "Mariner IV" passed close to Mars in July 1965. We now know that the Martian atmosphere is not more than 1/100 as dense as earth's.

Suppose, though, that a planet's atmosphere is more like that of earth, and it is as close to the Sun (or closer). The greenhouse effect will then be enormous: temperatures will rise, vaporizing the oceans to a greater and greater extent. The water vapor will add to the greenhouse effect, accelerating the change, forcing more and more carbon dioxide into the air as well through temperature effects on the crust. In the end, the planet will be enormously hot, will have all its water in the atmosphere in the form of a vapor which will forever hide its surface under eternal clouds, and will have a very thick atmosphere of carbon dioxide.

This is precisely the case with Venus. The American Venus-probe, skimming past Venus in December 1962, corroborated earlier reports, based on radio-wave emission from Venus' atmosphere, that Venus was considerably hotter than would be expected from its position vis-à-vis the sun. The Soviet Union has sent a series of probes actually into Venus' atmosphere beginning in 1967; in December 1970, they managed to place one on the surface itself while the probe's instruments were still working. (The temperature and pressures quickly destroy the instruments.) Venus' surface turns out to be at a temperature of 900° F. or 500° C., which is nearly at the red-hot stage, and its atmosphere, largely carbon dioxide, is about 100 times as dense as that of earth.

Earth did not move in the direction of either Mars or Venus. The nitrogen content of its atmosphere did not soak into the crust, leaving a thin, cold carbon-dioxide wind. Nor did the greenhouse effect turn it into a choking desert world of great heat. Something happened, and that something was the development of life, even while the atmosphere was still in its ammonia/methane stage (see Chapter 12).

Life-induced reactions in earth's oceans broke down nitrogen compounds to liberate molecular nitrogen and thus kept that gas in the atmosphere in large quantities. Furthermore, cells developed the capacity to break down the water molecules to hydrogen and oxygen by using the energy of *visible* light, which is not blocked by ozone. The hydrogen was combined with carbon dioxide to form the complicated molecules that made up the cell, while the oxygen was liberated into the atmosphere. In this way, thanks to life, earth's atmosphere altered from nitrogen-and-carbon-dioxide to nitrogen-and-oxygen. The greenhouse effect was reduced to very little; the earth remained cool, capable of retaining its unique possession of an ocean of liquid water and an atmosphere containing large quantities of free oxygen. .

In fact, our oxygenated atmosphere may be a characteristic only of the last 10 per cent of earth's existence, and even as recently as 600 million years ago, our atmosphere may have had only a tenth as much oxygen as it has now.

But we do have it now, and we may be thankful for the life that made the free atmospheric oxygen possible, and for the life that such oxygen in turn makes possible.

CHAPTER 5

The Elements

The Periodic Table

The early Greek philosophers, whose approach to most problems was theoretical and speculative, decided that the earth was made of a very few "elements," or basic substances. Empedocles of Akragas, about 430 B.C., set the number at four—earth, air, water, and fire. Aristotle, a century later, supposed the heavens to consist of a fifth element, "aether." The successors of the Greeks in the study of matter, the medieval alchemists, got mired in magic and quackery, but they came to shrewder and more reasonable conclusions than the Greeks because they at least handled the materials they speculated about.

Seeking to explain the various properties of substances, the alchemists attached these properties to certain controlling elements that they added to the list. They identified mercury as the element that imparted metallic properties to substances and sulfur as the element that imparted the property of flammability. One of the last and best of the alchemists, the sixteenth-century Swiss physician Theophrastus Bombastus von Hohenheim, better known as Paracelsus, added salt as the element that imparted resistance to heat.

The alchemists reasoned that one substance could be changed into another by merely adding and subtracting elements in the proper proportions. A metal such as lead, for instance, might be changed into gold by adding the right amount of mercury to the lead. The search for the precise technique of converting "base metal" to gold went on for centuries. In the process, the alchemists discovered substances vastly more important than gold—such as the mineral acids and phosphorus.

The mineral acids—nitric acid, hydrochloric acid, and particularly, sulfuric acid—introduced a virtual revolution in alchemical experiments.

These substances were much stronger acids than the strongest previously known (the acetic acid of vinegar), and with them substances could be decomposed without the use of high temperatures and long waits. Even today, the mineral acids, particularly sulfuric acid, are of vital use in industry. It is even said that the extent of the industrialization of a nation can be judged by its annual consumption of sulfuric acid.

Nevertheless, few alchemists allowed themselves to be diverted by these important side issues from what they considered to be the main quest. Unscrupulous members of the craft indulged in outright fakery, producing gold by sleight-of-hand, to win what we would call today "research grants" from rich patrons. This brought the profession into such disrepute that the very word "alchemist" had to be abandoned. By the seventeenth century, "alchemist" had become "chemist" and "alchemy" had graduated to a science called "chemistry."

In the bright birth of science, one of the first of the new chemists was Robert Boyle, the author of Boyle's law of gases (see Chapter 4). In his *The Sceptical Chymist*, published in 1661, Boyle first laid down the specific modern criterion of an element: a basic substance that can be combined with other elements to form "compounds" and that, conversely, cannot be broken down to any simpler substance after it is isolated from a compound.

Boyle retained a medieval view about what the actual elements were, however. For instance, he believed that gold was not an element and could be formed in some way from other metals. So, in fact, did his contemporary, Isaac Newton, who devoted a great deal of time to alchemy. (Indeed, Emperor Francis Joseph of Austria-Hungary subsidized experiments for making gold as late as 1867.)

In the century after Boyle, practical chemical work began to make clear which substances could be broken down into simpler substances and which could not. Henry Cavendish showed that hydrogen would combine with oxygen to form water, so water could not be an element. Later Lavoisier resolved the supposed element air into oxygen and nitrogen. It became plain that none of the Greek "elements" was an element by Boyle's criterion.

As for the elements of the alchemists, mercury and sulfur did indeed turn out to be elements "according to Boyle." But so did iron, tin, lead, copper, silver, gold, and such nonmetals as phosphorus, carbon, and arsenic. And Paracelsus' "element" salt eventually was broken down into two simpler substances.

Of course, the definition of elements depended on the chemistry of the time. As long as a substance could not be broken down by the chemical techniques of the day, it could still be considered an element. For instance, Lavoisier's list of thirty three elements included such items as lime and magnesia. But fourteen years after Lavoisier's death on the guillotine in the French Revolution, the English chemist Humphry

Davy, using an electric current to split the substances, divided lime into oxygen and a new element he called "calcium" and similarly split magnesia into oxygen and another new element he named "magnesium."

On the other hand, Davy was able to show that a green gas that the Swedish chemist Carl Wilhelm Scheele had made from hydrochloric acid was not a compound of hydrochloric acid and oxygen, as had been thought, but a true element, and he named it "chlorine" (from the Greek word for "green").

At the beginning of the nineteenth century, the English chemist John Dalton came up with a radically new way of looking at elements. Oddly enough, this view harked back to some of the Greeks, who had, after all, contributed what has turned out to be perhaps the most important single concept in the understanding of matter.

The Greeks argued about whether matter was continuous or discrete: that is, whether it could be divided and subdivided indefinitely into ever finer dust or would be found in the end to consist of indivisible particles. Leucippus of Miletus and his pupil Democritus of Abdera insisted about 450 B.C. that the latter was the case. Democritus, in fact, gave the particles a name: he called them "atoms" (meaning "nondivisible"). He even suggested that different substances were composed of different atoms or combinations of atoms and that one substance could be converted into another by rearranging the atoms. Considering that all this was only an intelligent guess, one is thunderstruck by the correctness of his intuition. Although the idea may seem obvious today, it was so far from obvious at the time that Plato and Aristotle rejected it out of hand.

It survived, however, in the teachings of Epicurus of Samos, who wrote about 300 B.C., and in the philosophic school ("Epicureanism") to which he gave rise. An important Epicurean was the Roman philosopher Lucretius, who, about 60 B.C., embodied atomic notions in a long poem *On the Nature of Things*. Lucretius' poem survived through the Middle Ages and was one of the earlier works to be printed once that technique had been invented.

The notion of atoms never entirely passed out of the consciousness of Western scholarship. Prominent among the atomists in the dawn of modern science were the Italian philosopher Giordano Bruno and the French philosopher Pierre Gassendi. Bruno had many unorthodox scientific views, such as a belief in an infinite universe with the stars distant suns about which planets revolved, and expressed himself overboldly. He was burned as a heretic in 1600—the outstanding martyr to science of the Scientific Revolution. The Russians have named a crater on the other side of the moon in his honor.

Gassendi's views impressed Boyle. Boyle's own experiments showing that gases could easily be compressed and expanded seemed to show that these gases must be composed of widely spaced particles. Both Boyle and Newton were therefore among the convinced atomists of the seventeenth century.

Dalton showed that the various rules governing the behavior of gases could indeed be explained on the basis of the atomic nature of matter. (He recognized the priority of Democritus by using the word "atoms.") According to Dalton, each element represented a particular kind of atom, and any quantity of the element was made up of identical atoms of this kind. What distinguished one element from another was the nature of its atoms. And the basic physical difference between atoms was in their weight. Thus sulfur atoms were heavier than oxygen atoms, which in turn were heavier than nitrogen atoms, they in turn heavier than carbon atoms, and these in turn heavier than hydrogen atoms.

The Italian chemist Amedeo Avogadro applied the atomic theory to gases in such a way as to show that it made sense to suppose that equal volumes of gas (of whatever nature) were made up of equal numbers of particles. This is "Avogadro's hypothesis." These particles were at first assumed to be atoms, but eventually were shown to be composed, in most cases, of small groups of atoms called "molecules." If a molecule contains atoms of different kinds (as the water molecule, which consists of an oxygen atom and two hydrogen atoms), it is a molecule of a "chemical compound."

Naturally it became important to measure the relative weights of different atoms—to find the "atomic weights" of the elements, so to speak. The tiny atoms themselves were hopelessly beyond the reach of nineteenth-century weighing techniques. But by weighing the quantity of each element separated from a compound, and making deductions from the elements' chemical behavior, it was possible to work out the relative weights of the atoms. The first to go about this systematically was the Swedish chemist Jöns Jacob Berzelius. In 1828, he published a list of atomic weights based on two standards—one giving the atomic weight of oxygen the arbitrary value of 100, the other taking the atomic weight of hydrogen as equal to 1.

Berzelius' system did not catch on at once, but in 1860, at the first International Chemical Congress in Karlsruhe, Germany, the Italian chemist Stanislao Cannizzaro presented new methods for determining atomic weights, making use of Avogadro's hypothesis, which had hitherto been neglected. He described his views so forcefully that the world of chemistry was won over.

The weight of oxygen rather than hydrogen was adopted as the standard, because oxygen could more easily be brought into combination with various elements (and combination with other elements was the key step in the usual method of determining atomic weights). Oxygen's atomic weight was arbitrarily taken by the Belgian chemist Jean Servais Stas, in 1850, as exactly 16, so that the atomic weight of hydrogen, the lightest known element, would be just about one—1.0080, to be exact.

Ever since Cannizzaro's time, chemists have sought to work out atomic weights with greater and greater accuracy. This reached a climax, as far as purely chemical methods were concerned, in the work of the American chemist Theodore William Richards, who, in 1904 and thereafter, determined the atomic weights with an accuracy previously unapproached. For this he received the Nobel Prize in chemistry in 1914. On the basis of later discoveries about the physical constitution of atoms, Richards' figures have since been corrected to still more refined values.

Throughout the nineteenth century, although much work was done on atoms and molecules and scientists generally were convinced of their reality, there existed no direct evidence that they were anything more than convenient abstractions. Some quite prominent scientists, such as the German chemist Wilhelm Ostwald, refused to accept them in any other way. To him, they were useful but not "real."

The reality of molecules was made clear by "Brownian motion." This was first observed in 1827 by the Scottish botanist Robert Brown, who noted that pollen grains suspended in water jiggled erratically. At first it was thought that this was due to the life in the pollen grains, but equally small particles of completely inanimate dyes also showed the motion.

In 1863, it was first suggested that the movement was due to unequal bombardment of the particles by surrounding water molecules. For large objects, a slight inequality in the number of molecules striking from left and from right would not matter. For microscopic objects, bombarded by perhaps only a few hundred molecules per second, a few in excess—this side or that—can induce a perceptible jiggle. The random movement of the tiny particles is almost visible proof of the "graininess" of water, and of matter generally.

Einstein worked out a theoretical analysis of this view of Brownian motion and showed how one could work out the size of the water molecules from the extent of the little jiggling movements of the dye particles. In 1908, the French physicist Jean Perrin studied the manner in which particles settled downward through water under the influence of gravity. The settling was opposed by molecular collisions from below, so that a Brownian movement was opposing gravitational pull. Perrin used this finding to calculate the size of the water molecules by means of the equation Einstein had worked out, and even Ostwald had to give in. For his investigations Perrin received the Nobel Prize for physics in 1926.

So atoms have steadily been translated from semimystical abstractions into almost tangible objects. Indeed, today we can say that man has at last "seen" the atom. This is accomplished with the so-called "field ion microscope," invented in 1955 by Erwin W. Mueller of Pennsylvania

State University. His device strips positively charged ions off an extremely fine needle tip and shoots them to a fluorescent screen in such a way as to produce a 5 million-fold magnified image of the needle tip. This image actually makes the individual atoms composing the tip visible as bright little dots. The technique was improved to the point where images of single atoms could be obtained. The American physicist Albert Victor Crewe reported the detection of individual atoms of uranium and thorium by means of a scanning electron-microscope in 1970.

As the list of elements grew in the nineteenth century, chemists began to feel as if they were becoming entangled in a thickening jungle. Every element had different properties, and they could see no underlying order in the list. Since the essence of science is to try to find order in apparent disorder, scientists hunted for some sort of pattern in the properties of the elements.

In 1862, after Cannizzaro had established atomic weight as one of the important working tools of chemistry, a French geologist, Alexandre Emile Beguyer de Chancourtois, found that he could arrange the elements in the order of increasing atomic weight in a tabular form, such that elements with similar properties fell in the same vertical column. Two years later, a British chemist, John Alexander Reina Newlands, independently arrived at the same arrangement. But both were ignored or ridiculed. Neither could get his suggestions properly published at the time. Many years later, after the importance of the periodic table had become universally recognized, their papers were published at last. Newlands even got a medal.

It was the Russian chemist Dmitri Ivanovich Mendeleev who got the credit for finally bringing order into the jungle of the elements. In 1869, he and the German chemist Julius Lothar Meyer proposed tables of the elements, making essentially the same point that de Chancourtois and Newlands had already made. But Mendeleev received the recognition because he had the courage and confidence to push the idea further than the others.

In the first place, Mendeleev's "periodic table" (so-called because it showed the periodic recurrence of similar chemical properties) was more complicated than that of Newlands and nearer what we now believe to be correct. Second, where the properties of an element placed it out of order according to its atomic weight, he boldly switched the order, on the ground that the properties were more important than the atomic weight. He was eventually proved correct in this. For instance, tellurium, with an atomic weight of 127.61, should, on the weight basis, come after iodine, whose atomic weight is 126.91. But in the columnar table, putting tellurium ahead of iodine places it under selenium, which it closely resembles, and similarly puts iodine under its cousin bromine.

Finally, and most important, where Mendeleev could find no other way to make his arrangement work, he did not hesitate to leave holes in the table and to announce, with what seemed infinite gall, that elements must be discovered that belonged in those holes. He went further. For three of the holes, he described the element that would fit each, utilizing as his guide the properties of the elements above and below the hole in the table. And here Mendeleev had a stroke of luck. Each of his three predicted elements was found in his own lifetime, so that he witnessed the triumph of his system. In 1875, the French chemist Lecoq de Boisbaudran discovered the first of these missing elements and named it "gallium" (after the Latin name for France). In 1879, the

The periodic table of the elements. The shaded areas of the table represent the two rare-earth series: the lanthanides and the actinides, named after their respective first members. The number in the lower right-hand corner of each box indicates the atomic weight of the element. An asterisk marks elements that are radioactive. Each element's atomic number appears at top center of its box.

1								
Hydrogen								
(H) 1.008								
3	**4**							
Lithium	Beryllium							
(Li) 6.939	(Be) 9.012							
11	**12**							
Sodium	Magnesium							
(Na) 22.990	(Mg) 24.312							
19	**20**	**21**	**22**	**23**	**24**	**25**	**26**	**27**
Potassium	Calcium	Scandium	Titanium	Vanadium	Chromium	Manganese	Iron	Cobalt
(K) 39.102	(Ca) 40.08	(Sc) 44.956	(Ti) 47.90	(V) 50.942	(Cr) 51.996	(Mn) 54.938	(Fe) 55.847	(Co) 58.9
37	**38**	**39**	**40**	**41**	**42**	**43***	**44**	**45**
Rubidium	Strontium	Yttrium	Zirconium	Niobium	Molybdenum	Technetium	Ruthenium	Rhodium
(Rb) 85.47	(Sr) 87.62	(Y) 88.905	(Zr) 91.22	(Nb) 92.906	(Mo) 95.94	(Tc) 98.91	(Ru) 101.07	(Rh) 102.9
55	**56**	**57**	**58**	**59**	**60**	**61***	**62**	**63**
Cesium	Barium	Lanthanum	Cerium	Prasodymium	Neodymium	Promethium	Samarium	Europium
(Cs) 132.905	(Ba) 137.34	(La) 138.91	(Ce) 140.12	(Pr) 140.907	(Nd) 144.24	(Pm) 145	(Sm) 150.35	(Eu) 151.
			72	**73**	**74**	**75**	**76**	**77**
			Hafnium	Tantalum	Tungsten	Rhenium	Osmium	Iridium
			(Hf) 178.49	(Ta) 180.948	(W) 183.85	(Re) 186.2	(Os) 190.2	(Ir) 192
87*	**88***	**89***	**90***	**91***	**92***	**93***	**94***	**95***
Francium	Radium	Actinium	Thorium	Protactinium	Uranium	Neptunium	Plutonium	Americiu
(Fr) 223	(Ra) 226.05	(Ac) 227	(Th) 232.038	(Pa) 231	(U) 238.03	(Np) 237	(Pu) 242	(Am) 2
			104*	**105***				
			Rutherfordium	Hahnium				
			(Rf) 259	(Ha) 260				

Swedish chemist Lars Fredrik Nilson found the second and named it "scandium" (after Scandinavia). And in 1886, the German chemist Clemens Alexander Winkler isolated the third and named it "germanium" (after Germany, of course). All three elements had almost precisely the properties predicted by Mendeleev!

With the discovery of X-rays, a new era opened in the history of the periodic table. In 1911, the British physicist Charles Glover Barkla discovered that when X-rays were scattered by a metal, the scattered rays had a sharply defined penetrating power, depending on the metal; in

								2 Helium (He) 4.003
			5 Boron (B) 10.811	6 Carbon (C) 12.011	7 Nitrogen (N) 14.007	8 Oxygen (O) 15.999	9 Fluorine (F) 18.998	10 Neon (Ne) 20.183
			13 Aluminum (Al) 26.982	14 Silicon (Si) 28.086	15 Phosphorus (P) 30.974	16 Sulfur (S) 32.064	17 Chlorine (Cl) 35.453	18 Argon (A) 39.948
28 Nickel (Ni) 58.71	29 Copper (Cu) 63.54	30 Zinc (Zn) 65.37	31 Gallium (Ga) 69.72	32 Germanium (Ge) 72.59	33 Arsenic (As) 74.922	34 Selenium (Se) 78.96	35 Bromine (Br) 79.909	36 Krypton (Kr) 83.80
46 Palladium (Pd) 106.4	47 Silver (Ag)107.870	48 Cadmium (Cd) 112.40	49 Indium (In) 114.82	50 Tin (Sn) 118.69	51 Antimony (Sb) 121.75	52 Tellurium (Te) 127.60	53 Iodine (I) 126.904	54 Xenon (Xe) 131.30
64 Gadolinium (Gd)157.25	65 Terbium (Tb)158.924	66 Dysprosium (Dy) 162.50	67 Holmium (Ho)164.930	68 Erbium (Er) 167.26	69 Thulium (Tm)168.934	70 Ytterbium (Yb) 173.04	71 Lutetium (Lu) 174.97	
78 Platinum (Pt) 195.09	79 Gold (Au)196.967	80 Mercury (Hg) 200.59	81 Thallium (Tl) 204.37	82 Lead (Pb) 207.19	83 Bismuth (Bi) 208.98	84* Polonium (Po) 210	85* Astatine (At) 210	86* Radon (Rn) 222
96* Curium (Cm) 244	97* Berkelium (Bk) 245	98* Californium (Cf) 246	99* Einsteinium (Es) 253	100* Fermium (Fm) 255	101* Mendelevium (Md) 256	102* Nobelium (No) 255	103* Lawrencium (Lw) 257	

other words, each element produced its own "characteristic X-rays." For this discovery Barkla was awarded the Nobel Prize in physics for 1917.

There was some question as to whether X-rays were streams of tiny particles or consisted of wavelike radiations after the manner of light. One way of checking this was to see if X-rays could be diffracted (that is, forced to change direction) by a "diffraction grating" consisting of a series of fine scratches. However, for proper diffraction, the distance between the scratches must be roughly equal to the size of the waves in the radiation. The most finely spaced scratches that could be prepared sufficed for ordinary light, but the penetrating power of X-rays made it likely that, if X-rays were wavelike, the waves would have to be much smaller than those of light. Therefore, no ordinary diffraction gratings would suffice to diffract X-rays.

However, it occurred to the German physicist Max Theodore Felix von Laue that crystals were a natural diffraction grating far finer than any man-made one. A crystal is a solid with a neat geometric shape, with its plane faces meeting at characteristic angles, and with a characteristic symmetry. This visible regularity is the result of an orderly array of atoms making up its structure. There were reasons for thinking that the space between one layer of atoms and the next was about the size of an X-ray wavelength. If so, crystals would diffract X-rays.

Laue experimented and found that X-rays passing through a crystal were indeed diffracted and formed a pattern on a photographic plate that showed that they had the properties of waves. Within the same year, the English physicist William Lawrence Bragg and his equally distinguished father William Henry Bragg developed an accurate method of calculating the wavelength of a particular type of X-ray from its diffraction pattern. Conversely, X-ray diffraction patterns were eventually used to determine the exact orientation of the atom layers that did the diffracting. In this way, X-rays opened the door to a new understanding of the atomic structure of crystals. For their work on X-rays, Laue received the Nobel Prize for physics in 1914, while the Braggs shared the Nobel Prize for physics in 1915.

Then, in 1914, the young English physicist Henry Gwyn-Jeffreys Moseley determined the wavelengths of the characteristic X-rays produced by various metals and made the important discovery that the wavelength decreased in a very regular manner as one went up the periodic table.

This pinned the elements into definite position in the table. If two elements, supposedly adjacent in the table, yielded X-rays that differed in wavelength by twice the expected amount, then there must be a gap between them belonging to an unknown element. If they differed by three times the expected amount, there must be two missing elements. If, on the other hand, the two elements' characteristic X-rays differed by only the expected amount, one could be certain that there was no missing element between the two.

It was now possible to give the elements definite numbers. Until then there had always been the possibility that some new discovery might break into the sequence and throw any adopted numbering system out of kilter. Now there could no longer be unsuspected gaps.

Chemists proceeded to number the elements from 1 (hydrogen) to 92 (uranium). These "atomic numbers" were found to be significant in connection with the internal structures of the atoms (see Chapter 6) and to be more fundamental than the atomic weight. For instance, the X-ray data proved that Mendeleev had been right in placing tellurium (atomic number 52) before iodine (53), in spite of tellurium's higher atomic weight.

Moseley's new system proved its worth almost at once. The French chemist Georges Urbain, after discovering "lutetium" (named after the old Latin name of Paris), had later announced that he had discovered another element which he called "celtium." According to Moseley's system, lutetium was element 71 and "celtium" should be 72. But when Moseley analyzed "celtium's" characteristic X-rays, it turned out to be lutetium all over again. Element 72 was not actually discovered until 1923, when the Danish physicist Dirk Coster and the Hungarian chemist Georg von Hevesy detected it in a Copenhagen laboratory and named it "hafnium," from the Latinized name of Copenhagen.

Moseley was not present for this verification of the accuracy of his method; he had been killed at Gallipoli in 1915 at the age of twenty-eight—certainly one of the most valuable lives lost in World War I. Moseley probably lost a Nobel Prize through his early death. The Swedish physicist Karl Manne George Siegbahn extended Moseley's work, discovering new series of X-rays and accurately determining X-ray spectra for the various elements. He was awarded the Nobel Prize for physics in 1924.

In 1925, Walter Noddack, Ida Tacke, and Otto Berg of Germany filled another hole in the periodic table. After a three-year search through ores containing elements related to the one they were hunting for, they turned up element 75 and named it "rhenium," in honor of the Rhine River. This left only four holes: elements 43, 61, 85, and 87.

It was to take two decades to track those four down. Although chemists did not realize it at the time, they had found the last of the stable elements. The missing ones were unstable species so rare on the earth today that all but one of them would have to be created in the laboratory to be identified. And thereby hangs a tale.

Radioactive Elements

After the discovery of X-rays, many scientists were impelled to investigate these new and dramatically penetrating radiations. One of them was the French physicist Antoine-Henri Becquerel. Henri's father, Alexandre Edmond (the physicist who had first photographed the solar spectrum), had been particularly interested in "fluorescence," which is visible radiation given off by substances after exposure to the ultraviolet rays in sunlight.

The elder Becquerel had, in particular, studied a fluorescent substance called potassium uranyl sulfate (a compound made up of molecules each containing an atom of uranium). Henri wondered whether the fluorescent radiations of the potassium uranyl sulfate contained X-rays. The way to check this was to expose the sulfate to sunlight (whose ultraviolet light would excite the fluorescence) while the compound lay on a photographic plate wrapped in black paper. Since the sunlight could not penetrate the black paper, it would not itself affect the plate, but, if the fluorescence it excited contained X-rays, they *would* penetrate the paper and darken the plate. Becquerel tried the experiment in 1896, and it worked. Apparently there were X-rays in the fluorescence. Becquerel even got the supposed X-rays to pass through thin sheets of aluminum and copper, and that seemed to clinch the matter, for no radiation except X-rays was known to do this.

But then, by a great stroke of good fortune, a seige of cloudy weather intervened. Waiting for the return of sunlight, Becquerel put away his photographic plates, with pinches of sulfate lying on them, in a drawer. After several days, he grew impatient and decided to develop his plates anyway, with the thought that even without direct sunlight some trace of X-rays might have been produced. When he saw the developed pictures, Becquerel experienced one of those moments of deep astonishment and delight that are the dream of all scientists. The photographic plate was deeply darkened by strong radiation! Something other than fluorescence or sunlight was responsible for it. Becquerel decided (and experiments quickly proved) that this something was the uranium in the potassium uranyl sulfate.

This discovery further electrified scientists, already greatly excited by the recent discovery of the X-rays. One of the scientists who at once set out to investigate the strange radiation from uranium was a young Polish-born chemist named Marie Sklodowska, who just the year before had married Pierre Curie, the discoverer of the Curie temperature (see Chapter 4).

Pierre Curie, in collaboration with his brother Jacques, had discovered that certain crystals, when put under pressure, developed a positive electric charge on one side and a negative charge on the other. This phenomenon is called "piezoelectricity" (from a Greek word meaning "to press"). Marie Curie decided to measure the radiation given off from uranium by means of piezoelectricity. She set up an arrangement whereby this radiation would ionize the air between two electrodes, a current would then flow, and the strength of this small current would be measured by the amount of pressure that had to be placed on a crystal to produce a balancing countercurrent. This method worked so well that Pierre Curie dropped his own work at once and, for the rest of his life, joined Marie as an eager second.

It was Marie Curie who suggested the term "radioactivity" to describe the ability of uranium to give off radiations and who went on to demonstrate the phenomenon in a second radioactive substance—thorium. In fast succession, enormously important discoveries were made by other scientists as well. The penetrating radiations from radioactive substances proved to be even more penetrating and more energetic than X-rays; they are now called "gamma rays." Radioactive elements were found to give off other types of radiation also, which led to discoveries about the internal structure of the atom, but this is a story for another chapter (see Chapter 6). What has the greatest bearing on our discussion of the elements is the discovery that the radioactive elements, in giving off the radiation, changed to other elements—a modern version of transmutation.

Marie Curie was the first to come on the implications of this phenomenon, and she did so accidentally. In testing pitchblende for it uranium content, to see if samples of the ore had enough uranium to be worth the refining effort, she and her husband found to their surprise that some of the pieces had more radioactivity than they ought to have even if they were made of pure uranium. This meant, of course, that there had to be other radioactive elements in the pitchblende. These unknown elements could only be present in small quantities, because ordinary chemical analysis did not detect them, so they must be very radioactive indeed.

In great excitement, the Curies obtained tons of pitchblende, set up shop in a small shack, and under primitive conditions and with only their unbeatable enthusiasm to drive them on they proceeded to struggle through the heavy, black ore for the trace quantities of new elements. By July of 1898, they had isolated a trace of black powder 400 times as intensely radioactive as the same quantity of uranium.

This contained a new element with chemical properties like those of tellurium, and it therefore probably belonged beneath it in the periodic table. (It was later given the atomic number 84.) The Curies named it "polonium," after Marie's native land.

But polonium accounted for only part of the radioactivity. More work followed, and, by December of 1898, the Curies had a preparation that was even more intensely radioactive than polonium. It contained still another element, which had properties like those of barium (and was eventually placed beneath barium with the atomic number 88). The Curies called it "radium," because of its intense radioactivity.

They worked on for four more years to collect enough pure radium so that they could see it. Then Marie Curie presented a summary of her work as her Ph.D. dissertation in 1903. It was probably the greatest doctoral dissertation in scientific history. It earned her not one but two Nobel Prizes. Marie and her husband, along with Becquerel, received the Nobel Prize for physics in 1903 for their studies of radioactivity, and, in 1911, Marie alone (her husband having died in a traffic accident in 1906) was awarded the Nobel Prize for chemistry for the discovery of polonium and radium.

Polonium and radium are far more unstable than uranium or thorium, which is another way of saying that they are far more radioactive. More of their atoms break down each second. Their lifetimes are so short that practically all the polonium and radium in the universe should have disappeared within a matter of a million years or so. Why do we still find them in the billions-of-years-old earth? The answer is that radium and polonium are continually being formed in the course of the breakdown of uranium and thorium to lead. Wherever uranium and thorium are found, small traces of polonium and radium are likewise to be found. They are intermediate products on the way to lead as the end product.

Three other unstable elements on the path from uranium and thorium to lead were discovered by means of the careful analysis of pitchblende or by researches into radioactive substances. In 1899, André Louis Debierne, on the advice of the Curies, searched pitchblende for other elements and came up with one he called "actinium" (from the Greek word for "ray"), which eventually received the atomic number 89. The following year, the German physicist Friedrich Ernst Dorn demonstrated that radium, when it broke down, formed a gaseous element. A radioactive gas was something new! Eventually the element was named "radon" (from radium and argon, its chemical cousin) and was given the atomic number 86. Finally, in 1917, two different groups —Otto Hahn and Lise Meitner in Germany and Frederick Soddy and John A. Cranston in England—isolated from pitchblende element 91, named protactinium.

By 1925, then, the score stood at eighty-eight identified elements —eighty-one stable and seven unstable. The search for the missing four —numbers 43, 61, 85, 87—became avid indeed.

Since all the known elements from number 84 to 92 were radioactive, it was confidently expected that 85 and 87 would be radioactive as well. On the other hand, 43 and 61 were surrounded by stable elements, and there seemed no reason to suspect that they were not themselves stable as well. Consequently, they should be found in nature.

Element 43, lying just above rhenium in the periodic table, was expected to have similar properties and to be found in the same ores. In fact, the team of Noddack, Tacke, and Berg, which had discovered rhenium, felt certain that it had also detected X-rays of a wavelength that went along with element 43. So they announced its discovery, too, and named it "masurium," after a region in East Prussia. However, their identification was not confirmed, and in science a discovery is not a discovery unless and until it has been confirmed by at least one independent researcher.

In 1926, two University of Illinois chemists announced that they had found element 61 in ores containing its neighboring elements (60 and 62), and they named their discovery "illinium." The same year, a pair of Italian chemists at the University of Florence thought that they had isolated the same element and named it "florentium." But other chemists could not confirm the work of either group.

A few years later, an Alabama Polytechnic Institute physicist, using a new analytical method of his own devising, reported that he had found small traces of element 87 and of element 85; he called them "virginium" and "alabamine," after his native and adopted states, respectively. But these discoveries could not be confirmed, either.

Events were to show that the "discoveries" of elements 43, 61, 85, and 87 had been mistaken.

The first of the four to be identified beyond doubt was element 43. The American physicist Ernest Orlando Lawrence, who was to receive the Nobel Prize in physics for his invention of the cyclotron (see Chapter 6), made the element in his accelerator by bombarding molybdenum (element 42) with high-speed particles. His bombarded material developed radioactivity, and Lawrence sent it for analysis to the Italian chemist Emilio Gino Segrè, who was interested in the element-43 problem. Segrè and his colleague C. Perrier, after separating the radioactive part from the molybdenum, found that it resembled rhenium in its properties, but was not rhenium. They decided that it could only be element number 43 and that element number 43, unlike its neighbors in the periodic table, was radioactive. Because it is not being produced as a breakdown product of a higher element, virtually none of it is left in the earth's crust, and so Noddack and company were undoubtedly mistaken in thinking they had found it. Segrè and Perrier eventually were given the privilege of naming element 43; they called it "technetium," from a Greek word meaning "artificial," because it was the first man-made element. By 1960, enough technetium had been accumulated to determine its melting point—close to 2200° C. (Segrè was later to receive a Nobel Prize for quite another discovery, having to do with another man-made bit of matter—see Chapter 6.)

In 1939, element number 87 was finally discovered in nature. The French chemist Marguerite Perey isolated it from among the breakdown products of uranium. It was present in extremely small amounts, and only improvements in technique enabled it to be found where earlier it had been missed. She later named the new element "francium," after her native land.

Element 85, like technetium, was produced in the cyclotron, by bombardment of bismuth (element 83). In 1940, Segrè, Dale Raymond Corson, and K. R. MacKenzie isolated element 85 at the University of California, Segrè having by then emigrated from Italy to the United States. World War II interrupted their work on the element, but after the war they returned to it and in 1947 proposed the name "astatine" for the element, from a Greek word meaning "unstable." (By that time, tiny traces of astatine had, like francium, been found in nature among the breakdown products of uranium.)

Meanwhile, the fourth and final missing element, number 61, had been discovered among the products of the fission of uranium, a process that is explained in Chapter 9. (Technetium, too, turned up among these products.) Three chemists at the Oak Ridge National Laboratory —J. A. Marinsky, L. E. Glendenin, and Charles DuBois Coryell— isolated element 61 in 1945. They named it "promethium," after the Greek demigod Prometheus, who had stolen fire for mankind from the sun. Element number 61, after all, had been stolen from sunlike fires of the atomic furnace.

So the list of elements, from 1 to 92, was at last complete. And yet, in a sense, the strangest part of the adventure had only begun. For scientists had broken through the bounds of the periodic table; uranium was not the end.

A search for elements beyond uranium—"transuranium elements" had actually begun as early as 1934. Enrico Fermi in Italy had found that when he bombarded an element with a newly discovered subatomic particle called the "neutron" (see Chapter 6), this often transformed the element into the one of the next higher atomic number. Could uranium be built up to element 93—a totally synthetic element that did not exist in nature? Fermi's group proceeded to attack uranium with neutrons, and they got a product that they thought was indeed element 93. They called it "uranium X."

In 1938, Fermi received the Nobel Prize in physics for his studies in neutron bombardment. At the time, the real nature of his discovery, or its consequences for mankind, was not even suspected. Like that other Italian, Columbus, he had found, not what he was looking for, but something far more important of which he was not aware

Suffice it to say here that, after a series of chases up a number of false trails, it was finally discovered that what Fermi had done was, not to create a new element, but to split the uranium atom into two nearly equal parts. When physicists turned in 1940 to studies of this process, element 93 cropped up as an almost casual result of their experiments. In the mélange of elements that came out of the bombardment of uranium by neutrons, there was one that at first defied identification. Then it dawned on Edwin McMillan of the University of California that perhaps the neutrons released by fission had converted some of the uranium atoms to a higher element, as Fermi had hoped would happen. McMillan and Philip Abelson, a physical chemist, were able to prove that the unidentified element was in fact number 93. The proof of its existence lay in the nature of its radioactivity, as was to be the case in all subsequent discoveries.

McMillan suspected that another transuranium element might be mixed with number 93. The chemist Glenn Theodore Seaborg, together with his co-workers Arthur Charles Wahl and J. W. Kennedy, soon showed that this was indeed so and that the element was number 94.

Since uranium, the supposed end of the periodic table, had been named, at the time of its discovery, for the then newly discovered planet, Uranus, elements 93 and 94 were now named for Neptune and Pluto, planets discovered after Uranus. They were called "neptunium" and "plutonium," respectively. It turned out that they existed in nature, for small traces of neptunium and plutonium were later found in uranium ores. So uranium was not the heaviest natural element after all.

Seaborg and a group at the University of California, in which Albert Ghiorso was prominent, went on to build more transuranium elements, one after the other. By bombarding plutonium with subatomic particles, in 1944 they created elements 95 and 96, named respectively "americium" (after America) and "curium" (after the Curies). When they had manufactured a sufficient quantity of americium and curium to work with, they bombarded those elements and successfully produced number 97 in 1949 and number 98 in 1950. These they named "berkelium" and "californium," after Berkeley and California. In 1951, Seaborg and McMillan shared the Nobel Prize in chemistry for this train of achievements.

The next elements were discovered in more catastrophic fashion. Elements 99 and 100 emerged in the first hydrogen bomb explosion, detonated in the Pacific in November 1952. Although their existence was detected in the explosion debris, the elements were not confirmed and named until after the University of California group made small quantities of both in the laboratory in 1955. The names given them were "einsteinium" and "fermium," for Albert Einstein and Enrico Fermi, both of whom had died some months before. Then the group bombarded a small quantity of einsteinium and formed element 101, which they called "mendelevium," after Mendeleev.

The next step came through a collaboration between California and the Nobel Institute in Sweden. The Institute carried out a particularly complicated type of bombardment that apparently produced a small quantity of element 102. It was named "nobelium," in honor of the Institute, but the experiment has not been confirmed. The element has been formed by methods other than those described by the first group of workers, so that there was a delay before nobelium was officially accepted as the name of the element.

In 1961, a few atoms of element 103 were detected at the University of California, and it was given the name "lawrencium," after E. O. Lawrence, who had recently died. In 1964, a group of Soviet scientists under Georgii Nikolaevich Flerov reported the formation of element 104, and in 1967, the formation of element 105. In both cases, the methods used to form the elements could not be confirmed, and American teams under Albert Ghiorso formed them in other ways. There is a dispute raging over priorities; both groups claim the right to name the elements. The Soviet group has named 104 "kurchatovium," after Igor Vasilievich Kurchatov, who had led the Soviet team that developed their atomic bomb, and who had died in 1960. The American group named 104 "rutherfordium" and 105 "hahnium," after Ernest Rutherford and Otto Hahn, both of whom made key discoveries in subatomic structure.

Each step in this climb up the transuranium scale was harder than the one before. At each successive stage, the element became harder to accumulate and more unstable. When mendelevium was reached, identification had to be made on the basis of seventeen atoms, no more. Fortunately, radiation-detecting techniques were marvelously refined by 1955. The Berkeley scientists actually hooked up their instruments to a firebell, so that every time a mendelevium atom was formed, the characteristic radiation it emitted on breaking down announced the event by a loud and triumphant ring of the bell. (The fire department soon put a stop to this.)

Electrons

When Mendeleev and his contemporaries found that they could arrange the elements in a periodic table composed of families of substances showing similar properties, they had no notion as to why the elements fell into such groups or why the properties were related. Eventually a clear and rather simple answer emerged, but it came only after a long series of discoveries that at first seemed to have nothing to do with chemistry.

It all began with studies of electricity. Faraday performed every experiment with electricity he could think of, and one of the things he tried to do was to send an electric discharge through a vacuum. He was not able to get a vacuum good enough for the purpose. But, by 1854, a German glass blower named Heinrich Geissler had invented an adequate vacuum pump and produced a glass tube enclosing metal electrodes in an unprecedentedly good vacuum. When experimenters succeeded in producing electric discharges in the "Geissler tube," they noticed that a green glow appeared on the tube wall opposite the negative electrode. The German physicist Eugen Goldstein suggested in 1876 that this green glow was caused by the impact on the glass of some sort of radiation originating at the negative electrode, which Faraday had named the "cathode." Goldstein called the radiation "cathode rays."

Were the cathode rays a form of electromagnetic radiation? Goldstein thought so, but the English physicist William Crookes and some others said no: they were a stream of particles of some kind. Crookes designed improved versions of the Geissler tube (called "Crookes tubes"), and with these he was able to show that the rays were deflected by a magnet. This meant that they were probably made up of electrically charged particles.

In 1897, the physicist Joseph John Thomson settled the question beyond doubt by demonstrating that the cathode rays could also be deflected by electric charges. What, then, were these cathode "particles"? The only negatively charged particles known at the time were the negative ions of atoms. Experiments showed that the cathode-ray particles could not possibly be such ions, for they were so strongly deflected by an electromagnetic field that they must have an unthinkably high electric charge or else must be extremely light particles with less than $1/1,000$ the mass of a hydrogen atom. The latter interpretation turned out to fit the evidence best. Physicists had already guessed that the electric current was carried by particles, and so these cathode-ray particles were accepted as the ultimate particles of electricity. They were called "electrons"—a name that had been suggested in 1891 by the Irish physicist George Johnstone Stoney. The electron was finally determined to have $1/1,837$ the mass of a hydrogen atom. (For establishing its existence, Thomson was awarded the Nobel Prize in physics in 1906.)

The discovery of the electron at once suggested that it might be a subparticle of the atom—in other words, that atoms were not the ultimate, indivisible units of matter that Democritus and John Dalton had pictured them to be.

This was a hard pill to swallow, but the lines of evidence converged inexorably. One of the most convincing items was Thomson's showing that negatively charged particles that came out of a metal plate when it was struck by ultraviolet radiation (the "photoelectric effect") were identical with the electrons of the cathode rays. The photoelectric electrons must have been knocked out of the atoms of the metal.

Since electrons could easily be removed from atoms (by other means as well as by the photoelectric effect), it was natural to conclude that they were located in the outer regions of the atom. If this was so, there must be a positively charged region within the atom balancing the electrons' negative charges, because the atom as a whole was normally neutral. It was at this point that investigators began to close in on the solution of the mystery of the periodic table.

To remove an electron from an atom takes a little energy. Conversely, when an electron falls into the vacated place in the atom, it must *give up* an equal amount of energy. (Nature is usually symmetrical, especially when it comes to considerations of energy.) This energy is released in the form of electromagnetic radiation. Now since the energy of radiation is measured in terms of wavelength, the wavelength of the radiation emitted by an electron falling into a particular atom will indicate the force with which the electron is held by that atom. The energy of radiation increases with shortening wavelength: the greater the energy, the shorter the wavelength.

We arrive, then, at Moseley's discovery that metals (i.e., the heavier elements) produced X-rays, each at a characteristic wavelength, which decreased in regular fashion as one went up the periodic table. Each successive element, it seemed, held its electrons more strongly than the one before, which is another way of saying that each had a successively stronger positive charge in its internal region.

Assuming that each unit of positive charge corresponded to the negative charge on an electron, it followed that the atom of each successive element must have one more electron. The simplest way of picturing the periodic table, then, was to suppose that the first element, hydrogen, had 1 unit of positive charge and 1 electron; the second element, helium, 2 positive charges and 2 electrons; the third, lithium, 3 positive charges and 3 electrons; and so on all the way up to uranium, with 92 electrons. So the atomic numbers of the elements turned out to represent the number of electrons in their atoms.

One more major clue and the atomic scientists had the answer to the periodicity of the periodic table. It developed that the electronic radiation of a given element was not necessarily restricted to a single wavelength; it might emit radiations at two, three, four, or even more different wavelengths. These sets of radiations were named the K-series, the L-series, the M-series, and so on. The investigators interpreted this to mean that the electrons were arrayed in "shells" around the positively charged core of the atom. The electrons of the innermost shell were most strongly held, and their removal took the most energy. An electron falling into this shell would emit the most energetic radiation, that is, of the shortest wavelengths, or the K-series. The electrons of the next innermost shell were responsible for the L-series of radiations; the next shell produced the M-series; and so on. Consequently, the shells were called the K-shell, the L-shell, the M-shell, and so on.

By 1925, the Austrian physicist Wolfgang Pauli advanced his "exclusion principle," which explained just how electrons were distributed within each shell, since no two electrons could possess, according to this principle, exactly the same energy and spin. For this, Pauli received the Nobel Prize for physics in 1945.

In 1916, the American chemist Gilbert Newton Lewis worked out the kinships of properties and the chemical behavior of some of the simpler elements on the basis of their shell structure. There was ample evidence, to begin with, that the innermost shell was limited to two electrons. Hydrogen has only one electron; therefore the shell is unfilled. The atom's tendency is to fill this K-shell, and it can do so in a number of ways. For instance, two hydrogen atoms can pool their single electrons and, by sharing the two electrons, mutually fill their K-shells. This is why hydrogen gas almost always exists in the form of a pair of atoms —the hydrogen molecule. To separate the two atoms and free them as "atomic hydrogen" takes a good deal of energy. Irving Langmuir of the General Electric Company, who independently worked out a similar scheme involving electrons and chemical behavior, presented a practical demonstration of the strong tendency of the hydrogen atom to keep its electron shell filled. He made an "atomic hydrogen torch" by blowing hydrogen gas through an electric arc, which split the molecules' atoms apart; when the atoms recombined after passing the arc, they liberated the energy they had absorbed in splitting apart, and this was sufficient to yield temperatures up to 3400° C.!

In helium, element number 2, the K-shell is filled with two electrons. Helium atoms therefore are stable and do not combine with other atoms. When we come to lithium, element 3, we find that two of its electrons fill the K-shell and the third starts the L-shell. The succeeding elements add electrons to this shell one by one: beryllium has 2 electrons in the L-shell, boron has 3, carbon 4, nitrogen 5, oxygen 6, fluorine 7, and neon 8. Eight is the limit for the L-shell, and therefore neon corresponds to helium in having its outermost electron shell filled. And sure enough, it, too, is an inert gas with properties like helium's.

Every atom with an unsatisfied outer shell has a tendency to enter into combination with other atoms in a manner that leaves it with a filled outer shell. For instance, the lithium atom readily surrenders its one L-shell electron so that its outer shell is the filled K, while fluorine tends to seize an electron to add to its seven and complete the L-shell. Therefore lithium and fluorine have an affinity for each other; when they combine, lithium donates its L-electron to fluorine to fill the latter's L-shell. Since the atoms' interior positive charges do not change, lithium, with one electron subtracted, now carries a net positive charge, while fluorine, with one extra electron, carries a net negative charge. The mutual attraction of the opposite charges holds the two ions together. The compound is called lithium fluoride.

L-shell electrons can be shared as well as transferred. For instance, each of two fluorine atoms can share one of its electrons with the other, so that each atom has a total of eight in its L-shell, counting the two shared electrons. Similarly, two oxygen atoms will pool a total of four electrons to complete their L-shells; and two nitrogen atoms will share a total of six. Thus fluorine, oxygen, and nitrogen all form two-atom molecules.

The carbon atom, with only four electrons in its L-shell, will share each of them with a different hydrogen atom, thereby filling the K-shells of the four hydrogen atoms and in turn filling its own L-shell by sharing *their* electrons. This stable arrangement is the methane molecule, CH_4.

In the same way, a nitrogen atom will share electrons with three hydrogen atoms to form ammonia; an oxygen atom will share electrons

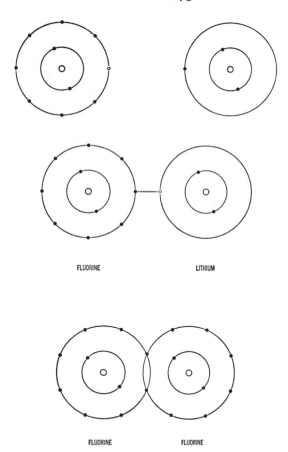

FLUORINE   LITHIUM

FLUORINE   FLUORINE

Transfer and sharing of electrons. Lithium transfers the electron in its outer shell to fluorine in the combination of lithium fluoride; each atom then has a full outer shell. In the fluorine molecule (Fl_2), two electrons are shared, filling both atoms' outer shells.

with two hydrogen atoms to form water; a carbon atom will share electrons with two oxygen atoms to form carbon dioxide; and so on. Almost all the components formed by the elements in the first part of the periodic table can be accounted for on the basis of this tendency to complete the outermost shell by giving up electrons, accepting electrons, or sharing electrons.

The element after neon, sodium, has 11 electrons, and the eleventh must start a third shell. Then follow magnesium, with 2 electrons in the M-shell, aluminum with 3, silicon with 4, phosphorus with 5, sulfur with 6, chlorine with 7, and argon with 8.

Now each element in this group corresponds to one in the preceding series. Argon, with 8 electrons in the M-shell, is like neon (with 8 electrons in the L-shell), and it is an inert gas. Chlorine, having 7 electrons in its outer shell, resembles fluorine very closely in chemical properties. Likewise, silicon resembles carbon, sodium resembles lithium, and so on.

So it goes right through the periodic table. Since the chemical behavior of every element depends on the configuration of electrons in its outermost shell, all those with, say, one electron in the outer shell will react in much the same way chemically. Thus all the elements in the first column of the periodic table—lithium, sodium, potassium, rubidium, cesium, and even the man-made radioactive element francium— are remarkably alike in their chemical properties. Lithium has 1 electron in the L-shell, sodium 1 in the M-shell, potassium 1 in the N-shell, rubidium 1 in the O-shell, cesium 1 in the P-shell, and francium 1 in the Q-shell. Again, all the elements with seven electrons in their respective outer shells—fluorine, chlorine, bromine, iodine, and astatine— resemble one another. The same is true of the last column in the table —the closed-shell group that includes helium, neon, argon, krypton, xenon, and radon.

The Lewis-Langmuir concept works so well that it still serves in its original form to account for the more simple and straightforward varieties of behavior among the elements. However, not all the behavior was quite as simple and straightforward as might be thought.

For instance, each of the inert gases—helium, neon, argon, krypton, xenon, and radon—has eight electrons in the outermost shell (except for helium, which has two electrons in its only shell), and this is the most stable possible situation. Atoms of these elements have a minimum tendency to lose or gain electrons and therefore a minimum tendency to engage in chemical reactions. The gases would be "inert," as their name proclaims.

However, a "minimum tendency" is not really the same as "no tendency," but most chemists forgot this and acted as though it was ultimately impossible for the inert gases to form compounds. This was not true of all of them, of course. As long ago as 1932, the American chemist Linus Pauling considered the ease with which electrons could be removed from different elements and noted that all elements without exception, even the inert gases, can be deprived of electrons. It was just that for this to happen requires more energy in the case of the inert gases than in that of other elements near them in the periodic table.

The amount of energy required to remove electrons among the elements in any particular family decreases with increasing atomic weight, and the heaviest inert gases, xenon and radon, do not have unusually high requirements. It is no more difficult to remove an electron from a xenon atom, for instance, than from an oxygen atom.

Pauling therefore predicted that the heavier inert gases might form chemical compounds with elements that were particularly prone to accept electrons. The element most eager to accept electrons is fluorine, and that seemed to be the natural target.

Now radon, the heaviest inert gas, is radioactive and is unavailable in any but trace quantities. Xenon, however, the next heaviest, is stable and occurs in small quantities in the atmosphere. The best chance, therefore, would be to attempt to form a compound between xenon and fluorine. However, for thirty years nothing was done in this respect, chiefly because xenon was expensive and fluorine very hard to handle, and chemists felt they had better things to do than chase this particular will-o'-the-wisp.

In 1962, however, the British-Canadian chemist Neil Bartlett, working with a new compound, platinum hexafluoride (PtF_6), found that it was remarkably avid for electrons, almost as much so as fluorine itself. This compound would take electrons away from oxygen, an element that is normally avid to gain electrons rather than lose them. If PtF_6 could take electrons from oxygen, it ought to be able to take them from xenon too. The experiment was tried, and xenon fluoroplatinate ($XePtF_6$), the first compound of an inert gas, was reported.

Other chemists at once sprang into the fray, and a number of xenon compounds with fluorine, with oxygen, or with both were formed, the most stable being xenon difluoride (XeF_2). A compound of krypton and fluorine, krypton tetrafluoride (KrF_4), has also been formed, as well as a radon fluoride. Compounds with oxygen were also formed. There were, for instance, xenon oxytetrafluoride ($XeOF_4$), xenic acid (H_2XeO_4), and sodium perxenate (Na_4XeO_6). Most interesting, perhaps, was xenon trioxide (Xe_2O_3), which explodes easily and is dangerous. The smaller inert gases—argon, neon, and helium—are more resistant to sharing their electrons than the larger ones, and they remain inert [for all anything chemists can do even yet.]

Chemists quickly recovered from the initial shock of finding that the inert gases could form compounds. Such compounds fit into the general picture after all. Consequently, there is now a general reluctance

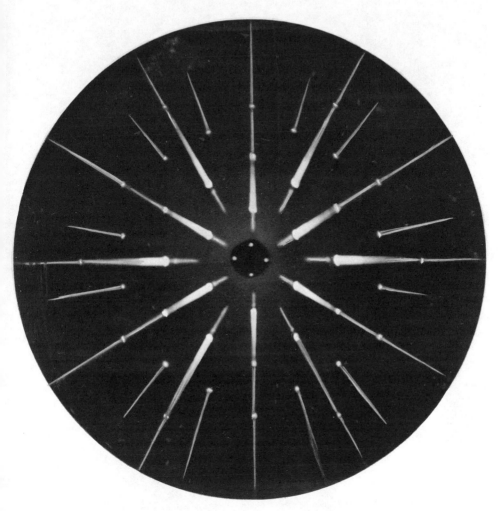

A single ice crystal photographed by X-ray diffraction, showing the symmetry and balance of the physical forces holding the structure together.

Uranium ore. The black portion is pitchblende (uranium oxide); the inset in the lower left corner is an autoradiograph produced by pitchblende's radioactivity.

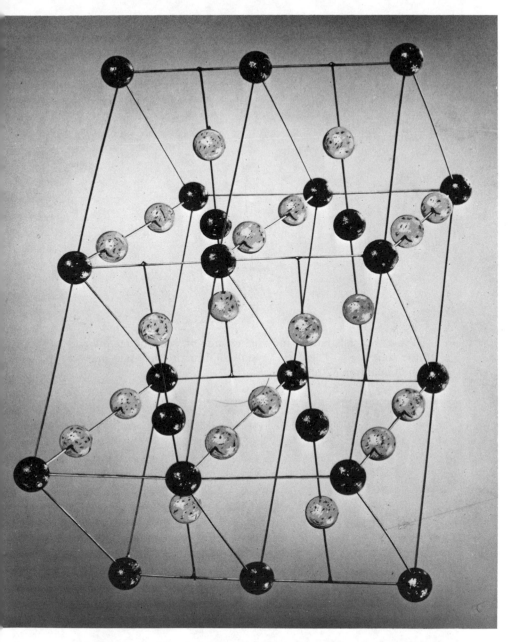

Molecular model of titanium oxide in crystalline form, which can serve as a transistor. Removal of one of the oxygen atoms (light balls) will make the material semiconducting.

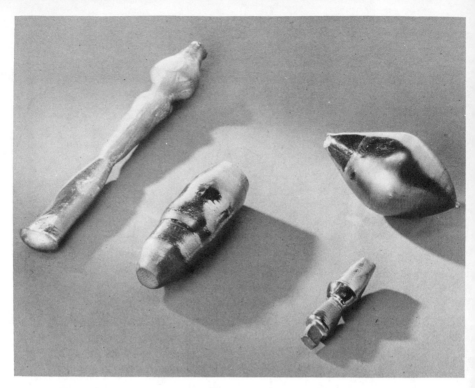

Typical crystals under study in research on semiconductors and the solid state. These are crystals of bismuth telluride and an indium-antimony alloy.

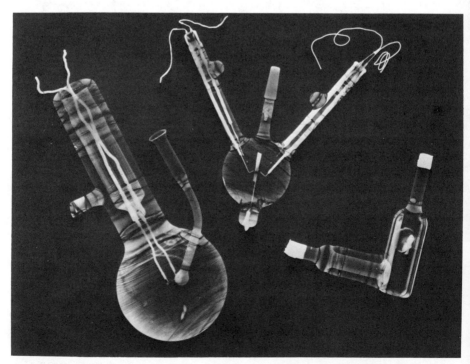

Ultrahigh-vacuum system components used to study chemical reactions between simple gases and atomically clean metals or semiconductor surfaces.

to speak of the gases as "inert gases." The alternate name of "noble gases" is now preferred, and one speaks of "noble gas compounds" and "noble gas chemistry." (I think this is a change for the worse. After all, the gases are still inert, even if not completely inert. The concept "noble," in this context, implies "standoffish" or "disinclined to mix with the common herd," and this is just as inappropriate as "inert" and, moreover, does not suit a democratic society.)

In addition to the fact that the Lewis–Langmuir scheme was applied too rigidly to the inert gases, it can scarcely be applied at all to many of the elements with atomic numbers higher than 20. In particular, refinements had to be added to deal with a very puzzling aspect of the periodic table having to do with the so-called "rare earths"—elements 57 to 71, inclusive.

To go back a bit, the early chemists considered any substance that was insoluble in water and unchanged by heat to be an "earth" (a hangover of the Greek view of "earth" as an element). Such substances included what we would today call calcium oxide, magnesium oxide, silicon dioxide, ferric oxide, aluminum oxide, and so on—compounds which actually constitute about 90 per cent of the earth's crust. Calcium oxide and magnesium oxide are slightly soluble and in solution display "alkaline" properties (that is, opposite to those of acids), and so they were called the "alkaline earths"; when Humphry Davy isolated the metals calcium and magnesium from these earths, they were named alkaline earth metals. The same name was eventually applied to all the elements that fall into the column of the periodic table containing magnesium and calcium: that is, to beryllium, strontium, barium, and radium.

The puzzle to which I have referred began in 1794, when a Finnish chemist, Johan Gadolin, examined an odd rock that had been found near the Swedish hamlet Ytterby and decided that it was a new "earth." Gadolin gave this "rare earth" the name "yttria," after Ytterby. Later the German chemist Martin Heinrich Klaproth found that yttria could be divided into two "earths," for one of which he kept the name yttria, while he named the other "ceria" (after the newly discovered planetoid Ceres). But the Swedish chemist Carl Gustav Mosander subsequently broke these down further into a series of different earths. All eventually proved to be oxides of new elements named the "rare-earth metals." By 1907, fourteen such elements had been identified. In order of increasing atomic weight they are:

lanthanum (from a Greek word meaning "hidden")
cerium (from Ceres)
praseodymium (from the Greek for "green twin," after a green line in its spectrum)
neodymium ("new twin")

samarium (from "samarskite," the mineral in which it was found)
europium (from Europe)
gadolinium (in honor of Johan Gadolin)
terbium (from Ytterby)
dysprosium (from a Greek word meaning "hard to get at")
holmium (from Stockholm)
erbium (from Ytterby)
thulium (from Thule, an old name for Scandinavia)
ytterbium (from Ytterby)
lutetium (from Lutetia, an old name for Paris).

On the basis of their X-ray properties, these elements were assigned the atomic numbers from 57 (lanthanum) to 71 (lutetium). As I related earlier, there was a gap at 61 until the missing element, promethium, emerged from the fission of uranium. It made the fifteenth in the list.

Now the trouble with the rare-earth elements is that they apparently cannot be made to fit into the periodic table. It is fortunate that only four of them were definitely known when Mendeleev proposed the table; if they had all been on hand, the table might have been altogether too confusing to be accepted. There are times, even in science, when ignorance is bliss.

The first of the rare-earth metals, lanthanum, matches up all right with yttrium, number 39, the element above it in the table. (Yttrium, though found in the same ores as the rare earths and similar to them in properties, is not a rare-earth metal. It is, however, named after Ytterby. Four elements honor that hamlet—which is overdoing it.) The confusion begins with the rare earth after lanthanum, namely, cerium, which ought to resemble the element following yttrium, that is, zirconium. But it does nothing of the sort; instead, it resembles yttrium again. And the same is true of all fifteen of the rare-earth elements: they strongly resemble yttrium and one another (in fact, they are so alike chemically that at first they could not be separated except by the most tedious procedures), but they are not related to any other elements preceding them in the table. We have to skip the whole rare-earth group and go on to hafnium, element 72, to find the element related to zirconium, the one after yttrium.

Baffled by this state of affairs, chemists could do no better than to group all the rare-earth elements into one box beneath yttrium and list them individually in a kind of footnote to the table.

The answer to the puzzle finally came as a result of details added to the Lewis-Langmuir picture of the electron-shell structure of the elements.

In 1921, C. R. Bury suggested that the shells were not necessarily limited to eight electrons apiece. Eight always sufficed to satisfy the outer shell. But a shell might have a greater capacity when it was not on the outside. As one shell built on another, the inner shells might absorb more electrons, and each succeeding shell might hold more than the one before. Thus the K-shell's total capacity would be 2 electrons, the L-shell's 8, the M-shell's 18, the N-shell's 32, and so on—the step-ups going according to a pattern of successive squares multiplied by two (i.e., $2 \times 1, 2 \times 4, 2 \times 9, 2 \times 16$, etc.).

This view was backed up by a detailed study of the spectra of the elements. The Danish physicist Niels Henrik David Bohr showed that each electron shell was made up of subshells at slightly different energy levels. In each succeeding shell, the spread of the subshells was greater, so that soon the shells overlapped. As a result, the outermost subshell of an interior shell (say the M-shell) might actually be farther from the center, so to speak, than the innermost subshell of the next shell beyond it (i.e., the N-shell). This being so, the N-shell's inner subshell might fill with electrons while the M-shells' outer subshell was still empty.

An example will make this clearer. The M-shell, according to the theory, is divided into three subshells, whose capacities are 2, 6, and 10 electrons respectively, making a total of 18. Now argon, with eight electrons in its M-shell, has filled only two inner subshells. And, in fact, the M-shell's third, or outermost, subshell will not get the next electron in the element-building process, because it lies beyond the innermost subshell of the N-shell. That is, in potassium, the element after argon, the nineteenth electron goes, not into the outermost subshell of M, but into the innermost subshell of N. Potassium, with one electron in its N-shell, resembles sodium, which has one electron in its M-shell. Calcium, the next element (20), has two electrons in the N-shell and resembles magnesium, which has two in the M-shell. But now the innermost subshell of the N-shell, having room for only two electrons, is full. The next electrons to be added can start filling the outermost subshell of the M-shell, which so far has not been touched. Scandium (21) begins the process, and zinc (30) completes it. In zinc, the outermost subshell of the M-shell has at last acquired its complement of 10 electrons. The 30 electrons of zinc are distributed as follows: 2 in the K-shell, 8 in the L-shell, 18 in the M-shell, and 2 in the N-shell. At this point, electrons can resume the filling of the N-shell. The next electron gives the N-shell three electrons and forms gallium (31), which resembles aluminum, with three in the M-shell.

The point of all this is that elements 21 to 30, formed on the road to filling a subshell which had been skipped temporarily, are "transitional" elements. Note that calcium resembles magnesium and gallium resembles aluminum. Now magnesium and aluminum are adjacent members of the periodic table (numbers 12 and 13). But calcium (20) and gallium (31) are not. Between them lie the transitional elements, and these introduce a complication in the periodic table.

The N-shell is larger than the M-shell and is divided into four sub-shells instead of three: they can hold 2, 6, 10, and 14 electrons, respectively. Krypton, element 36, fills the two innermost subshells of the N-shell, but here the innermost subshell of the overlapping O-shell intervenes, and, before electrons can go on to N's two outer subshells, they must fill that one. The element after krypton, rubidium (37), has its thirty-seventh electron in the O-shell. Strontium (38) completes the filling of the two-electron O subshell. Thereupon a new series of transitional elements proceeds to fill the skipped third subshell of the N-shell. With cadmium (48) this is completed; now N's fourth and outermost subshell is skipped while electrons fill O's second innermost subshell, ending with xenon (54).

But even now N's fourth subshell must bide its turn, for by this stage the overlapping has become so extreme that even the P-shell inter-poses a subshell that must be filled before N's last. After xenon come cesium (55) and barium (56), with one and two electrons, respectively, in the P-shell. It is still not N's turn: the fifty-seventh electron, surpris-ingly, goes into the third subshell of the O-shell, creating the element lanthanum. Then, and only then, an electron at long last enters the

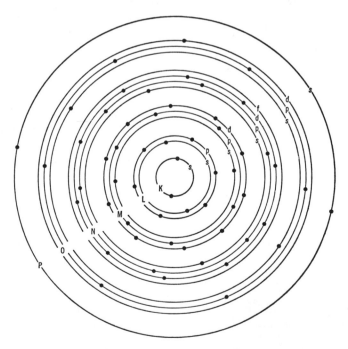

The electron shells of lanthanum. Note that the fourth sub-shell of the N-shell has been skipped and is empty.

outermost subshell of the N-shell. One by one the rare-earth elements add electrons to the N-shell until element 71, lutetium, finally fills it. Lutetium's electrons are arranged thus: 2 in the K-shell, 8 in the L-shell, 18 in the M-shell, 32 in the N-shell, 9 in the O-shell (two subshells full plus one electron in the next subshell), and 2 in the P-shell (innermost subshell full).

Now at last we begin to see why the rare-earth elements, and some other groups of transitional elements, are so alike. The decisive thing that differentiates elements, as far as their chemical properties are concerned, is the configuration of electrons in their outermost shell. For instance, carbon, with four electrons in its outermost shell, and nitrogen, with five, are completely different in their properties. On the other hand, in sequences where electrons are busy filling inner subshells while the outermost shell remains unchanged, the properties vary less. Thus iron, cobalt, and nickel (elements 26, 27, and 28), all of which have the same outer-shell electronic configuration—an N subshell filled with two electrons—are a good deal alike in chemical behavior. Their internal electronic differences (in an M subshell) are largely masked by their surface electronic similarity. And this goes double for the rare-earth elements. Their differences (in the N-shell) are buried under, not one, but

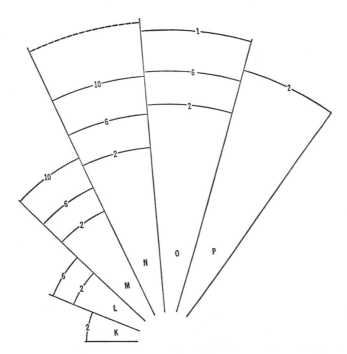

Schematic representation of the overlapping of electron shells and subshells in lanthanum. The outermost subshell of the N-shell has yet to be filled.

two outer electronic configurations (in the O-shell and the P-shell), which in all these elements are identical. Small wonder that the elements are chemically as alike as peas in a pod.

Because the rare-earth metals have so few uses, and are so difficult to separate, chemists made little effort to do so—until the uranium atom was fissioned. Then it became an urgent matter indeed, because radioactive varieties of some of these elements were among the main products of fission, and in the atomic bomb project it was necessary to separate and identify them quickly and cleanly.

The problem was solved in short order by use of a chemical technique first devised in 1906 by a Russian botanist named Mikhail Semenovich Tswett. He named it "chromatography" ("writing in color"). Tswett had found that he could separate plant pigments, chemically very much alike, by washing them down a column of powdered limestone with a solvent. He dissolved his mixture of plant pigments in petroleum ether and poured this on the limestone. Then he proceeded to pour in clear solvent. As the pigments were slowly washed down through the limestone powder, each pigment moved down at a different rate, because each differed in strength of adhesion to the powder. The result was that they separated into a series of bands, each of a different color. With continued washing, the separated substances trickled out separately at the bottom of the column, one after the other.

The world of science for many years ignored Tswett's discovery, possibly because he was only a botanist and only a Russian, while the leaders of research on separating difficult-to-separate substances at the time were German biochemists. But, in 1931, a German biochemist, Richard Willstätter, rediscovered the process, whereupon it came into general use. (Willstätter had received the 1915 Nobel Prize in chemistry for his excellent work on plant pigments. Tswett, so far as I know, has gone unhonored.)

Chromatography through columns of powder was found to work on almost all sorts of mixtures—colorless as well as colored. Aluminum oxide and starch proved to be better than limestone for separating ordinary molecules. Where ions are separated, the process is called ion exchange, and compounds known as zeolites were the first efficient agents applied for this purpose. Calcium and magnesium ions could be removed from "hard" water, for instance, by pouring the water through a zeolite column. The calcium and magnesium ions adhere to the zeolite and are replaced in solution by the sodium ions originally present on the zeolite, so "soft" water drips out of the bottom of the column. The sodium ions of zeolite have to be replenished from time to time by pouring in a concentrated solution of salt (sodium chloride). In 1935, a refinement came with the development of "ion-exchange resins." These synthetic substances can be designed for the job to be done. For instance, certain resins will substitute hydrogen ions for positive ions, while others substitute hydroxyl ions for negative ions; a combination of both types will remove most of the salts from sea water. Kits containing such resins were part of the survival equipment on life rafts during World War II.

It was the American chemist Frank Harold Spedding who adapted ion-exchange chromatography to the separation of the rare earths. He found that these elements came out of an ion-exchange column in the reverse order of their atomic number, so that they were not only quickly separated but also identified. In fact, the discovery of promethium, the missing element 61, was confirmed in this way from the tiny quantities found among the fission products.

Thanks to chromatography, purified rare-earth elements can now be prepared by the pound or even by the ton. It turns out that the rare earths are not particularly rare: the rarest of them (excepting promethium) are more common than gold or silver, and the most abundant —lanthanum, cerium, and neodymium—are more plentiful than lead. Together the rare-earth metals make up a larger percentage of the earth's crust than copper and tin combined. So scientists have pretty well dropped the term "rare earths" and now call this series of elements the "lanthanides," after its lead-off member. To be sure, the individual rare earths have not had many uses in the past, but in 1965 certain europium-yttrium compounds turned out to be particularly useful as red-sensitive "phosphors" in color television. Obviously, big things may come of this.

As if to reward the chemists and physicists for their decipherment of the rare-earth mystery, the new knowledge provided a key to the chemistry of the elements at the end of the periodic table, including the man-made ones.

The series of heavy elements in question begins with actinium, number 89. In the table it falls under lanthanum. Actinium has two electrons in the Q-shell, just as lanthanum has two electrons in the P-shell. Actinium's eighty-ninth and last electron entered the P-shell, just as lanthanum's fifty-seventh and last entered the O-shell. Now the question is: Do the elements after actinium continue to add electrons to the P-shell and remain ordinary transition elements? Or do they, perchance, follow the pattern of the elements after lanthanum, where the electrons dive down to fill the skipped subshell below? If the latter is true, then actinium may start a new series of "rare-earth metals."

The natural elements in this series are actinium, thorium, protactinium, and uranium. They were not much studied before 1940. What little was known about their chemistry suggested that they were ordinary transition elements. But when the man-made elements neptunium and plutonium were added to the list and studied intensively, these two showed a strong chemical resemblance to uranium. This prompted Glenn Seaborg to propose that the heavy elements were in fact following the rare-earth pattern and filling the buried, unfilled subshell. As more transuranium elements were added to the list, studies of their chemistry bore out this view, and it is now generally accepted.

The shell being filled is the fourth subshell of the O-shell. With lawrencium, element number 103, that subshell is filled. All the elements from actinium to lawrencium share much the same chemical properties and resemble lanthanum and the lanthanides. With element 104, the 104th electron will have to be added to the P-shell, and its properties should be like those of hafnium. It will be the final touch that clinches the existence of the second rare-earth series, and that is why chemists look forward so eagerly to obtaining and studying element 104.

They already have one independent confirmation. Ion-exchange chromatography separates the transuranium elements beautifully and in an exactly analogous manner to the separation of the lanthanides.

In token of the parallelism, the heavier "rare-earth metals" are now called "actinides," just as the members of the first series are called lanthanides.

Gases

From the dawn of chemistry, it was recognized that many substances could exist in the form of a gas, liquid, or solid, depending on the temperature. Water is the most common example: sufficiently cooled, it becomes solid ice, and sufficiently heated, it becomes gaseous steam. Van Helmont, who first used the word "gas," differentiated between substances that were gases at ordinary temperatures, such as carbon dioxide, and those that, like steam, were gases only at elevated temperatures. He called the latter "vapors," and we still speak of "water vapor" rather than "water gas."

The study of gases, or vapors, continued to fascinate chemists, partly because they lent themselves to quantitative studies. The rules governing their behavior were simpler and more easily worked out than those governing the behavior of liquids and solids.

In 1787, the French physicist Jacques Alexandre César Charles discovered that, when a gas was cooled, each degree of cooling caused its volume to contract by about 1/273 of the volume it had at 0° C., and, conversely, each degree of warming caused it to expand by the same 1/273. The expansion with warmth raised no logical difficulties, but, if shrinkage with cold were to continue according to Charles' law (as it is called to this day), at −273° C. a gas should have shrunk to nothing! This paradox did not particularly bother chemists, for they realized that Charles' law could not hold all the way down, and they had no way of getting to very low temperatures to see what happened.

The development of the atomic theory, picturing gases as collections of molecules, presented the situation in new terms. The volume was now seen to depend on the velocity of the molecules. The higher the temperature, the faster they moved, the more "elbow room" they required, and the greater the volume. Conversely, the lower the temperature, the more slowly they moved, the less room they required, and the smaller the volume. In the 1860's, the British physicist William Thomson, who had just been raised to the peerage as Lord Kelvin, suggested that it was the molecules' average energy content that declined by 1/273 for every degree of cooling. Whereas volume could not be expected to disappear completely, energy could. Thomson maintained that at $-273°$ C. the energy of molecules would sink to zero. Therefore $-273°$ C. must represent the lowest possible temperature. So this temperature (now put at $-273.16°$ C. according to refined modern measurements) would be "absolute zero," or, as it is often stated, "zero Kelvin." On this absolute scale the melting point of ice is 273° K.

Naturally, among physicists there would be great interest in trying to reach absolute zero. There is something about any distant horizon that calls for conquest. Men had been exploring extremes of coldness even before Thomson defined the ultimate goal. This exploration involved attempts to liquefy gases. Michael Faraday had found that even at ordinary temperatures some gases could be liquefied by putting them under pressure; he had liquefied chlorine, sulfur dioxide, and ammonia in this way in the 1820's. Now, once liquefied, a gas could act as a cooling agent. When the pressure above the liquid was slowly reduced, the gas evaporated, and the evaporation absorbed heat from the remaining liquid. (When you blow on a moistened finger, the coolness you feel is the effect of the water evaporation drawing heat from the finger.) The general principle is well known today as the basis of modern refrigeration.

As early as 1755, the Scottish chemist William Cullen had produced ice mechanically by forming a vacuum over quantities of water, enforcing rapid evaporation of the water, and, of course, cooling to the freezing point. Nowadays, an appropriate gas is liquefied by a compressor and then circulated in coils or pipe where, as the liquid evaporates, it withdraws heat from the surrounding space.

Water itself is inappropriate for the purpose, as the ice that forms would clog the pipes. In 1834, an American inventor, Jacob Perkins, patented (in Great Britain) the use of ether as a refrigerant. Other gases such as ammonia and sulfur dioxide also came into use. All these refrigerants had the disadvantage of being poisonous or flammable. In 1930, however, the American chemist Thomas Midgley discovered dichlorodifluoromethane (CF_2Cl_2), better known under the trade-name of "Freon." This is nontoxic and nonflammable and suits the purpose perfectly. With Freon, home refrigeration became widespread and commonplace.

Refrigeration applied, in moderation, to large volumes is "air conditioning," so called because the air is also conditioned, i.e., filtered and dehumidified. The first practical air-conditioning unit was designed in 1902 by the American inventor Willis H. Carrier; since World War II air conditioning has become so common in the major American cities as to be nearly universal.

But the refrigeration principle can be carried to extremes, too. If a liquefied gas is enclosed in a well-insulated container, so that its evaporation draws heat only from the liquid itself, very low temperatures can be attained. By 1835, physicists had reached temperatures as low as $-110°$ C. ($163°$ K.).

Hydrogen, oxygen, nitrogen, carbon monoxide, and some other common gases, however, defied liquefaction at this temperature even with the use of high pressures. For a time, their liquefaction was despaired of, and they were called "the permanent gases."

In 1869, however, the Irish physicist Thomas Andrews deduced from his experiments that every gas had a "critical temperature" above which it could not be liquefied even under pressure. This was later put on a firm theoretical basis by the Dutch physicist, Johannes Diderik Van der Waals, who, in this fashion, earned the 1910 Nobel Prize for physics.

To liquefy any gas one had to be certain, therefore, that one was working at a temperature below the critical value, or it was labor thrown out. Efforts were made to reach still lower temperatures to conquer the stubborn gases. A "cascade" method, lowering temperatures by steps, turned the trick. First, liquefied sulfur dioxide, cooling through exaporation, was used to liquefy carbon dioxide, then the liquid carbon dioxide to liquefy a more resistant gas, and so on. In 1877, the Swiss physicist Raoul Pictet finally managed to liquefy oxygen, at a temperature of $-140°$ C. ($133°$ K.) and under a pressure of 500 atmospheres (7,500 pounds per square inch). The French physicist Louis Paul Cailletet at about the same time liquefied, not only oxygen, but also nitrogen and carbon monoxide. Naturally these liquids made it possible to go on at once to still lower temperatures. The liquefaction point of oxygen at ordinary air pressure was eventually found to be $-183°$ C. ($90°$ K.), that of carbon monoxide $-190°$ C. ($83°$ K.), and that of nitrogen $-195°$ C. ($78°$ K.).

Hydrogen resisted all efforts at liquefaction until 1900. The Scottish chemist James Dewar then accomplished the feat by bringing a new stratagem into play. Lord Kelvin (William Thomson) and the English physicist James Prescott Joule had shown that even in the gaseous state a gas could be cooled simply by letting it expand and preventing heat from leaking into the gas from outside, provided the temperature was low enough to begin with. Dewar therefore cooled compressed hydrogen to a temperature of $-200°$ C. in a vessel surrounded by liquid nitrogen, let this superfrigid hydrogen expand and cool further, and repeated the cycle again and again by conducting the ever-cooling hydrogen back through pipes. The compressed hydrogen, subjected to this "Joule-Thomson effect," finally became liquid at a temperature of about $-240°$ C. ($33°$ K.). At still lower temperatures he managed to obtain solid hydrogen.

To preserve his superfrigid liquids, Dewar devised special silver-coated glass flasks. These were double-walled with a vacuum between. Heat could be lost (or gained) through a vacuum only by the comparatively slow process of radiation, and the silver coating reflected the incoming (or, for that matter, outgoing) radiation. Such "Dewar flasks" are the direct ancestor of the household Thermos bottle.

By 1895, the British inventor William Hampson and the German physicist Carl Lindé had developed methods of liquefying air on a commercial scale. Pure liquid oxygen, separated from the nitrogen, became a highly useful article. Its main use, in terms of quantity, was in blowtorches, principally for welding. But more dramatic were its services in medicine (e.g., oxygen tents), in aviation, in submarines, and so on.

With the coming of rocketry, liquefied gases suddenly rose to new heights of glamor. Rockets require an extremely rapid chemical reaction, yielding large quantities of energy. The most convenient type of fuel is a combination of a liquid combustible, such as alcohol or kerosene, and liquid oxygen. Oxygen, or some alternate oxidizing agent, must be carried by the rocket in any case, because it runs out of any natural supply of oxygen when it leaves the atmosphere. And the oxygen must be in liquid form, since liquids are denser than gases and more oxygen can be squeezed into the fuel tanks in liquid form than in gaseous. Consequently, liquid oxygen has come into high demand in rocketry.

The efficiency of a mixture of fuel and oxidizer is measured by a quantity known as the "specific impulse." This represents the number of pounds of thrust produced by the combustion of one pound of the fuel-oxidizer mixture in one second. For a mixture of kerosene and oxygen, the specific impulse is equal to 242. Since the payload a rocket can carry depends on the specific impulse, there has been an avid search for more efficient combinations. The best chemical fuel, from this point of view, is liquid hydrogen. Combined with liquid oxygen, it can yield a specific impulse equal to 350 or so. If liquid ozone or liquid fluorine could be used in place of oxygen, the specific impulse could be raised to something like 370.

Research to find even better fuels for rockets is being pursued in several directions. Certain light metals, such as lithium, boron, magnesium, aluminum, and, particularly, beryllium, deliver more energy on combining with oxygen than even hydrogen does. Some of these are rare, however, and all involve technical difficulties in the burning, difficulties arising from smokiness, oxide deposits, and so on.

Attempts are also being made to work out new solid fuels that serve as their own oxidizers (like gunpowder, which was the first rocket propellant, but much more efficient). Such fuels are called "monopropellants," since they need no separate supply of oxidizer and make up the one propellant required. Fuels that also require oxidizers are "bipropellants" (two propellants). Monopropellants, it is hoped, would be easy to store and handle and would burn in a rapid but controlled fashion. The principal difficulty is probably that of developing a monopropellant with a specific impulse approaching those of the bipropellants.

Another possibility is atomic hydrogen, which Langmuir put to use in his blowtorch. It had been calculated that a rocket engine operating on the recombination of hydrogen atoms into molecules could develop a specific impulse of more than 1,300. The main problem is how to store the atomic hydrogen. So far the best hope seems to be to cool the free atoms very quickly and very drastically immediately after they are formed. Researches at the National Bureau of Standards seem to show that free hydrogen atoms are best preserved if trapped in a solid material at extremely low temperatures—say frozen oxygen or argon. If it could be arranged to push a button, so to speak, to let the frozen gases start warming up and evaporating, the hydrogen atoms would be freed and allowed to recombine. If such a solid could hold even as much as 10 per cent of its weight in free hydrogen atoms, the result would be a better fuel than any we now possess. But, of course, the temperature would have to be very low indeed—considerably below that of liquefied hydrogen. These solids would have to be kept at about $-272°$ C., or just one degree above absolute zero.

In another direction altogether lies the possibility of driving ions backward (rather than the exhaust gases of burnt fuel). The individual ions, of tiny mass, would produce tiny impulses, but this could be continued over long periods. A ship placed in orbit by the high but short-lived force of chemical fuel could then, in the virtually frictionless medium of space, slowly accelerate under the long-lived lash of ions to nearly light velocity. The material best suited to such an ionic drive is cesium, the substance that can most easily be made to lose electrons and form cesium ion. An electric field can then be made to accelerate the cesium ion and shoot it out the rocket opening.

But to return to the world of low temperature. Even the liquefaction and solidification of hydrogen did not represent the final victory. By the time hydrogen yielded, the inert gases had been discovered; of these the lightest, helium, remained a stubborn holdout against liquefaction at the lowest temperatures attainable. Then, in 1908, the Dutch physicist Heike Kammerlingh Onnes finally subdued helium. He carried the Dewar system one step further. Using liquid hydrogen, he cooled helium gas under pressure to about $-255°$ C. ($18°$ K.) and then let the gas expand to cool itself further. By this method he liquefied the gas. Thereafter, by letting the liquid helium evaporate, he got down to the temperature at which helium could be liquefied under normal atmospheric pressure (4.2° K.) and even to temperatures as low as 0.7° K. For his low-temperature work, Onnes received the Nobel Prize in physics in 1913. (Nowadays the liquefaction of helium is a simpler matter. In 1947, the American chemist Samuel Cornette Collins invented the "cryostat," which, by alternate compressions and expansions, can produce as much as two gallons of liquid helium an hour.) Onnes, however, did more than reach new depths of temperature. He was the first to show that unique properties of matter existed at those depths.

One of these properties is the strange phenomenon called "superconductivity." In 1911, Onnes was testing the electrical resistance of mercury at low temperatures. It was expected that resistance to an electric current would steadily decrease as the removal of heat reduced the normal vibration of the atoms in the metal. But at 4.12° K. the mercury's electrical resistance suddenly disappeared altogether! An electric current coursed through it without any loss of strength whatever. It was soon found that other metals also could be made superconductive. Lead, for instance, became superconductive at 7.22°K. An electric current of several hundred amperes set up in a lead ring kept at that temperature by liquid helium went on circling through the ring for two and a half years with absolutely no detectable decrease in quantity.

As temperatures were pushed lower and lower, more metals were added to the list of superconductive materials. Tin became superconductive at 3.73° K., aluminum at 1.20° K., uranium at 0.8° K., titanium at 0.53° K., hafnium at 0.35° K. (Some 1,400 different elements and alloys are now known to display superconductivity.) But iron, nickel, copper, gold, sodium, and potassium must have still lower transition points—if they can be made superconductive at all—because they have not been reduced to this state at the lowest temperatures reached. The highest transition point found for a metal is that of technetium, which becomes superconductive at temperatures under 11.2° K.

A low-boiling liquid can easily maintain substances immersed in it at the temperature of its boiling point. To attain lower temperatures, the aid of a still-lower-boiling liquid must be called upon. Liquid hydrogen boils at 20.4° K., and it would be most useful to find a superconducting substance with a transition temperature at least this high. Only then can superconductivity be studied in systems cooled by liquid hydrogen. Failing that, only the one lower-boiling liquid, liquid helium—much rarer, more expensive, and harder to handle—must be used. A few alloys, particularly those involving the metal niobium, have transition temperatures higher than those of any pure metal. Finally, in 1968, an alloy of niobium, aluminum, and germanium was found that remained superconductive at 21° K. Superconductivity at liquid-hydrogen temperatures became feasible—but just barely.

A useful application of superconductivity suggests itself at once in connection with magnetism. A current of electricity through a coil of wire around an iron core can produce a strong magnetic field—the greater the current, the stronger the field. Unfortunately, the greater the current, the greater the heat produced under ordinary cicumstances, and this puts a limit to what can be done. In superconductive wires, however, electricity flows without producing heat and, it would seem, more and more electric current could be squeezed into the wires to produce unprecedentedly strong "electromagnets" at only a fraction of the power that must be expended under ordinary conditions. There is, however, a catch.

Along with superconductivity goes another property involving magnetism. At the moment that a substance becomes superconductive, it also becomes perfectly "diamagnetic"; that is, it excludes the lines of force of a magnetic field. This was discovered by W. Meissner in 1933 and is therefore called the "Meissner effect." By making the magnetic field strong enough, however, one can destory the substance's superconductivity and the hope for supermagnetism, even at temperatures well below its transition point. It is as if, once enough lines of force have been concentrated in the surroundings, some at last manage to penetrate the substance, and when that happens, gone is the superconductivity as well.

Attempts have been made to find superconductive substances that will tolerate high magnetic fields. There is, for instance, a tin-niobium alloy with the high transition temperature of 18° K. It can support a magnetic field of some 250,000 gauss, which is high indeed. This fact was discovered in 1954, but it was only in 1960 that techniques were developed for forming wires of this ordinarily brittle alloy. A compound of vanadium and gallium may do even better, and superconductive electromagnets reaching field intensities of 500,000 gauss have been constructed.

Another startling phenomenon at low temperatures was discovered in helium itself. It is called "superfluidity."

Helium is the only known substance that cannot be frozen solid, even at absolute zero. There is a small irreducible energy content, even at absolute zero, which cannot possibly be removed (so that the energy content is "zero" in a practical sense), but which is enough to keep the extremely "nonsticky" atoms of helium free of each other, and therefore liquid. Actually, the German physicist Hermann Walther Nernst showed in 1905 that it is not the energy of a substance that becomes zero at absolute zero, but a closely related property: the "entropy." For this he received the 1920 Nobel Prize in chemistry. This does not mean, however, that solid helium doesn't exist under any conditions. It can be produced at temperatures below 1° K., by a pressure of about twenty-five atmospheres.

In 1935, Willem Hendrik Keesom and his sister, A. P. Keesom, working at the Onnes laboratory in Leyden, found that liquid helium at a temperature below 2.2° K. conducted heat almost perfectly. It conducted heat so quickly, at the speed of sound, in fact, that all parts of the helium were always at the same temperature. It would not boil—as any ordinary liquid will by reason of localized hot spots forming bubbles of vapor—because there were no localized hot spots in the liquid helium (if you can speak of hot spots in connection with a liquid below 2° K.). When it evaporated, the top of the liquid simply slipped off quietly—peeling off, so to speak, in sheets.

The Russian physicist Peter Leonidovich Kapitza went on to investigate this property and found that the reason helium conducted heat so well was that it flowed with remarkable ease, carrying the heat from one part of itself to another almost instantaneously, at least 200 times as rapidly as copper, the next best heat conductor. It flowed even more easily than a gas, having a viscosity only 1/1,000 that of gaseous hydrogen, and it would leak through apertures so tiny that they stopped a gas. Furthermore, the superfluid liquid would form a film on glass and flow along it as quickly as it would pour through a hole. If an open container of the liquid was placed in a larger container filled to a lower level, the fluid would creep up the side of the glass and over the rim into the outer container, until the levels in both were equalized.

Helium is the only substance that exhibits this phenomenon of superfluidity. In fact, the superfluid behaves so differently from the way helium itself does above 2.2° K. that it has been given a separate name, helium II, to distinguish it from liquid helium above that temperature, called helium I.

Only helium permits investigation of temperatures close to absolute zero, and, consequently, it has become a very important element in both pure and applied science. The atmospheric supply is negligible, and the most important sources are natural gas wells into which helium, formed from uranium and thorium breakdown in the earth's crust, sometimes seeps. The gas produced by the richest known well (in New Mexico) is 7.5 per cent helium.

Spurred by the odd phenomena discovered in the neighborhood of absolute zero, physicists have naturally made every effort to get down as close to absolute zero as possible and expand their knowledge of what is now known as "cryogenics." The evaporation of liquid helium can, under special conditions, produce temperatures as low as 0.5° K. (Temperatures at such a level, by the way, are measured by special methods involving electricity—e.g., by the size of the current generated in a thermocouple, by the resistance of a wire made of some nonsuperconductive metal, by changes in magnetic properties, or even by the speed of sound in helium. The measurement of extremely low temperatures is scarcely easier than their attainment.) Temperatures substantially lower than 0.5° have been reached by a technique first suggested in 1925 by the Dutch physicist Peter Joseph Wilhelm Debye. A "paramagnetic" substance (i.e., a substance that concentrates lines of magnetic force) is placed almost in contact with liquid helium, separated from it by helium gas, and the temperature of the whole system is reduced to about 1° K. The system is then placed within a magnetic field. The molecules of the paramagnetic substance line up parallel to the field's lines of force and in doing so give off heat. This heat is removed by further slight evaporation of the surrounding helium. Now the magnetic field is removed. The paramagnetic molecules immediately fall into a random orientation. In going from an ordered to a random orientation, the molecules must absorb heat, and the only thing they can absorb it from is the liquid helium. The temperature of the liquid helium therefore drops.

This can be repeated and repeated, each time lowering the temperature of the liquid helium, the technique being perfected by the American chemist William Francis Giauque, who received the Nobel Prize for chemistry in 1949 in consequence. In this way, a temperature of 0.00002° K. was reached in 1957.

In 1962, the German-British physicist Heinz London and his co-workers, suggested the possibility of using a new device to attain still lower temperatures. Helium occurs in two varieties, helium 4 and helium 3. Ordinarily they mix perfectly, but at temperatures below about 0.8° K., they separate, with the helium 3 in a top layer. Some of the helium 3 is in the bottom layer with the helium 4, and it is possible to cause helium 3 to shift back and forth across the boundary, lowering the temperature each time in a fashion analogous to the shift between liquid and vapor in the case of an ordinary refrigerant such as Freon. Cooling devices making use of this principle were first constructed in the Soviet Union in 1965.

The Russian physicist Isaak Yakovievich Pomeranchuk suggested in 1950 a method of deep cooling using other properties of helium 3, while as long ago as 1934, the Hungarian-British physicist Nicholas Kurti suggested the use of magnetic properties similar to those taken advantage of by Giauque, but involving the atomic nucleus—the innermost structure of the atom—rather than entire atoms and molecules.

As a result of the use of these new techniques, temperatures as low as 0.000001° K. have been attained. And as long as physicists find themselves within a millionth of a degree of absolute zero, might they not just get rid of what little entropy is left and finally reach the mark itself?

No! Absolute zero is unattainable, something Nernst demonstrated in his Nobel-Prize-winning treatment of the subject (sometimes referred to as "the Third Law of Thermodynamics"). In any lowering of temperature, only part of the entropy can be removed. In general, removing half of the entropy of a system is equally difficult regardless of what the total is. Thus it is just as hard to go from 300° K. (about room temperature) to 150° K. (colder than any temperature Antarctica attains) as to go from 20° to 10° K. It is then just as hard to go from 10° to 5° K. and from 5° to 2.5° K. and so on. Having attained a millionth of a degree above absolute zero, the task of going from that to half-a-millionth of a degree is as hard as going from 300° to 150° K., and if that is attained it is an equally difficult task to go from half-a-millionth to a quarter-of-a-millionth, and so on forever. Absolute zero lies at an infinite distance no matter how closely it seems to be approached.

One of the new scientific horizons opened up by the work on liquefaction of gases was the development of an interest in producing high pressures. It seemed that putting various kinds of matter (not only gases) under great pressure might bring out some fundamental information about the nature of matter and also about the interior of the earth. At a depth of 7 miles, for instance, the pressure is 1,000 atmospheres; at 400 miles, 200,000 atmospheres; at 2,000 miles, 1,400,000 atmospheres; and at the center of the earth, 4,000 miles down, it reaches 3,000,000 atmospheres. (Of course, Earth is a rather small planet. The central pressures within Saturn are estimated to be over 50 million atmospheres; within the even larger Jupiter, 100 million.)

The best that nineteenth-century laboratories could do was about 3,000 atmospheres, attained by E. H. Amagat in the 1880's. But, in 1905, the American physicist Percy Williams Bridgman began to devise new methods that soon reached pressures of 20,000 atmospheres and burst the tiny metal chambers he used for his experiments. He went to stronger materials and eventually succeeded in producing pressures of over a million atmospheres. For his work on high pressure he received the Nobel Prize in physics in 1946.

Under his ultrahigh pressures, Bridgman was able to force the atoms and molecules of a substance into more compact arrangements, which were sometimes retained after the pressure was released. For instance, he converted ordinary yellow phosphorus, a nonconductor of electricity, into a black, conducting form of phosphorus. He brought about startling changes even in water. Ordinary ice is less dense than liquid water. Using high pressure, Bridgman produced a series of ices ("ice-II," "ice-III," etc.) that were, not only denser than the liquid, but were ice only at temperatures well above the normal freezing point of water. Ice-VII is a solid at temperatures higher than the boiling point of water.

The word "diamond" brings up the most glamorous of all the high-pressure feats. Diamond, of course, is crystallized carbon, as is also graphite. (When an element appears in two different forms, these forms are "allotropes." Diamond and graphite are the most dramatic example of the phenomenon. Ozone and ordinary oxygen are another example.) The chemical nature of diamond was first proved in 1772 by

Lavoisier and some fellow French chemists. They pooled their funds to buy a diamond and proceeded to heat it to a temperature high enough to burn it up. The gas that resulted was found to be carbon dioxide. Later the British chemist Smithson Tennant showed that the amount of carbon dioxide measured could be produced only if diamond was pure carbon, and, in 1799, the French chemist Guyton de Morveau clinched the case by converting a diamond into a lump of graphite.

That was an unprofitable maneuver, but now why could not matters be reversed? Diamond is 55 per cent denser than graphite. Why not put graphite under pressure and force the atoms composing it into the tight packing characteristic of diamond?

Many efforts were made and, like the alchemists, a number of experimenters reported successes. The most famous was the claim of the French chemist Ferdinand Frédéric Henri Moissan. In 1893, he dissolved graphite in molten cast iron and reported that he found small

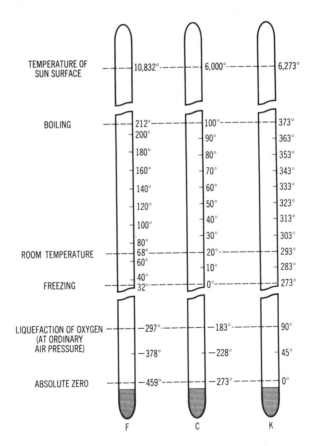

A comparison of the Fahrenheit, Centigrade, and Kelvin
thermometric scales.

diamonds in the mass after it cooled. Most of the objects were black, impure, and tiny, but one was colorless and almost a millimeter long. These results were widely accepted, and, for a long time, Moissan was considered to have manufactured synthetic diamonds. However, his results were never successfully repeated.

The search for synthetic diamonds was not without its side victories, however. In 1891, the American inventor Edward Goodrich Acheson stumbled upon silicon carbide, to which he gave the trade name Carborundum. This proved harder than any substance then known but diamond, and it has been a much-used abrasive, that is, a substance used for grinding and polishing, ever since.

The efficiency of an abrasive depends on its hardness. An abrasive can polish or grind substances less hard than itself, and diamond, as the hardest substance, is the most useful in this respect. The hardness of various substances is commonly measured on the "Mohs scale," introduced by the German mineralogist Friedrich Mohs in 1818. This assigns minerals numbers from 1, for talc, to 10, for diamond. A mineral of a particular number is able to scratch all those with lower numbers. On the Mohs scale, Carborundum is given the number 9. The divisions are not equal, however. On an absolute scale, the difference in hardness between 10 (diamond) and 9 (Carborundum) is four times greater than the difference between 9 (Carborundum) and 1 (talc).

In the 1930's, chemists finally worked out the pressure requirements for converting graphite to diamond. It turned out that the conversion called for a pressure of at least 10,000 atmospheres, and even then it would be impracticably slow. Raising the temperature would speed the conversion, but would also raise the pressure requirements. At 1500° C., a pressure of at least 30,000 atmospheres would be necessary. All this proved that Moissan and his contemporaries, under the conditions they used, could no more have produced diamonds than the alchemists could have produced gold. (There is some evidence that Moissan was actually a victim of one of his assistants, who, tiring of the tedious experiments, decided to end them by planting a real diamond in the cast-iron mixture.)

Aided by Bridgman's pioneering work in attaining the necessary high temperatures and pressures, scientists at the General Electric Company finally accomplished the feat in 1955. Pressures of 100,000 atmospheres or more were produced, along with temperatures of up to 2500° C. In addition, a small quantity of metal, such as chromium, was used to form a liquid film across the graphite. It was on this film that the graphite turned to diamond. In 1962, a pressure of 200,000 atmospheres and a temperature of 5000° C. could be attained. Graphite was then turned to diamond directly, without the use of a catalyst.

The synthetic diamonds are too small and impure to be used as gems, but they are now produced commercially as abrasives and cutting tools and, indeed, are a major source of such products. By the end of the decade, an occasional small diamond of gem quality could be produced.

A newer product made by the same sort of treatment can supplement the use of diamond. A compound of boron and nitrogen (boron nitride) is very similar in properties to graphite (except that boron nitride is white instead of black). Subjected to the high temperatures and pressures that convert graphite to diamond, the boron nitride undergoes a similar conversion. From a crystal arrangement like that of graphite, the atoms of boron nitride are converted to one like that of diamond. In its new form it is called "borazon." Borazon is about four times as hard as Carborundum. In addition it has the great advantage of being more resistant to heat. At a temperature of 900° C. diamond burns up but borazon comes through unchanged. Over twenty new materials, in addition to diamond and borazon, had been formed by pressure work in the 1960's.

Metals

Most of the elements in the periodic table are metals. As a matter of fact, only about twenty of the 102 elements can be considered definitely nonmetallic. Yet the use of metals came relatively late in the history of the human species. One reason is that, with rare exceptions, the metallic elements are combined in nature with other elements and are not easy to recognize or extract. Primitive man at first used only materials that could be manipulated by simple treatments such as carving, chipping, hacking, and grinding. This restricted his materials to bones, stones, and wood.

His introduction to metals may have come in the form of discoveries of meteorites, or of small nuggets of gold, or metallic copper in the ashes of fires built on rocks containing a copper ore. In any case, people who were curious enough (and lucky enough) to find these strange new substances and look into ways of handling them would discover many advantages in them. Metal differed from rock in that it had an attractive luster when polished. It could be beaten into sheets and drawn into wire. It could be melted and poured into a mold to solidify. It was much more beautiful and adaptable than rock and ideal for ornaments. Metals probably were fashioned into ornaments long before they were put to any other use.

Because they were rare, attractive, and did not alter with time, these metals were valued and bartered until they became a recognized medium of exchange. Originally, pieces of metal (gold, silver, or copper) had to be weighed separately in trading transactions, but, by 700 B.C., standardized weights of metal stamped in some official government fashion were issued in the Asia Minor kingdom of Lydia and the Aegean island of Aegina. Coins are still with us today.

What really brought metals into their own was the discovery that some of them would take a sharper cutting edge than stone could, and they would maintain that edge under conditions that would ruin a stone ax. Moreover, metal was tough. A blow that would splinter a wooden club or shatter a stone ax would only slightly deform a metal object of similar size. These advantages more than compensated for the fact that metal was heavier than stone and harder to obtain.

The first metal obtained in reasonable quantity was copper, which was in use by 4000 B.C. Copper itself is too soft to make useful weapons or armor (though it will make pretty ornaments), but it was often found alloyed with a little arsenic or antimony, which resulted in a substance that was harder than the pure metal. Then samples of copper ore must have been found that contained tin. The copper-tin alloy (bronze) was hard enough for purposes of weaponry. Men soon learned to add the tin deliberately. The Bronze Age replaced the Stone Age in Egypt and western Asia about 3000 B.C. and in southeastern Europe by 2000 B.C. Homer's *Iliad* and *Odyssey* commemorate that period of culture.

Iron was known as early as bronze, but for a long time meteorites were the only source. It remained no more than a precious metal, limited to occasional use, until methods were discovered for smelting iron ore and thus obtaining iron in unlimited quantities. The difficulty lay in working with fires hot enough and methods suitable enough to add carbon to the iron and harden it into the form we now call "steel." Iron smelting began somewhere in Asia Minor about 1400 B.C. and developed and spread slowly.

An iron-weaponed army could rout a bronze-armed one, for iron swords would cut through bronze. The Hittites of Asia Minor were the first to use iron weapons to any extent, and they had a period of power in western Asia. Then the Assyrians succeeded the Hittites. By 800 B.C., they had a completely ironized army which was to dominate western Asia and Egypt for two and a half centuries. At about the same time, the Dorians brought the Iron Age to Europe by invading Greece and defeating the Achaeans, who committed the error of clinging to the Bronze Age.

Iron is obtained essentially by heating iron ore (usually a ferric oxide) with carbon. The carbon atoms carry off the oxygen of the ferric oxide, leaving a lump of pure iron behind. In ancient times, the temperatures used did not melt the iron, and the product was a tough metal that could be worked into the desired shape by hammering—that is, "wrought iron." Iron metallurgy on a larger scale came into being in the Middle Ages. Special furnaces were used, and higher temperatures that melted the iron. The molten iron could be poured into molds to form castings, so it was called "cast iron." This was much less expensive than wrought iron and much harder, too, but it was brittle and could not be hammered. Increasing demand for iron of either form helped to deforest England, for instance, which consumed its wood in the iron-smelting furnaces. But then, in 1780, the English iron-worker Abraham Darby showed that coke (carbonized coal) would work as well as, or better than, charcoal (carbonized wood). The pressures on the forests eased in this direction, and the more-than-century-long domination of coal as an energy source began.

It was not until late in the eighteenth century that chemists, thanks to the French physicist René Antoine Ferchault de Réaumur, finally realized that it was the carbon content that dictated the toughness and hardness of iron. To maximize those properties the carbon content ought to be between 0.2 and 1.5 per cent; the steel that then results is harder and tougher and generally stronger than either cast iron or wrought iron. But until the mid-nineteenth century, high-quality steel could be made only by the complicated procedure of carefully adding the appropriate quantity of carbon to wrought iron (itself comparatively expensive). Steel remained therefore a luxury metal, used only where no substitute could be found—as in swords and springs.

The Age of Steel was ushered in by a British engineer named Henry Bessemer. Originally interested primarily in cannon and projectiles, Bessemer invented a system of rifling intended to enable cannon to shoot farther and more accurately. Napoleon III of France was interested and offered to finance further experiments. But a French artillerist killed the idea by pointing out that the propulsive explosion Bessemer had in mind would shatter the cast-iron cannons used in those days. Bessemer, chagrined, turned to the problem of creating stronger iron. He knew nothing of metallurgy, so he could approach the problem with a fresh mind. Cast iron was brittle because of its carbon content. Therefore the problem was to reduce the carbon.

Why not burn the carbon away by melting the iron and sending a blast of air through it? This seemed at first thought a ridiculous idea. Would not the air blast cool the molten metal and cause it to solidify? Bessemer tried it anyway, and he found that quite the reverse was true. As the air burned the carbon, the combustion gave off heat and the temperature of the iron rose rather than fell. The carbon burned off nicely. By proper controls, steel could be produced in quantity and comparatively cheaply.

In 1856, Bessemer announced his "blast furnace." Ironmakers adopted the method with enthusiasm, then dropped it in anger when they found that inferior steel was being formed. Bessemer discovered that the iron ore used by the industry contained phosphorus (which had been absent from his own ore samples). Although Bessemer explained to the ironmakers that phosphorus had betrayed them, they refused to be twice-bitten. Bessemer therefore had to borrow money and set up his own steel works in Sheffield. Importing phosphorus-free iron ore from Sweden, he speedily produced steel at a price that undersold the other ironmakers.

In 1875, the British metallurgist Sidney Gilchrist Thomas discovered that by lining the interior of the furnace with limestone and magnesia, he could easily remove the phosphorus from the molten iron. After this, almost any iron ore could be used in the manufacture of steel. Meanwhile, the German-British inventor Karl Wilhelm Siemens developed the "open-hearth method" in 1868, in which pig iron was heated with iron ore; this process also could take care of the phosphorus content.

The Age of Steel then got under way. The name is no mere phrase. Without steel, skyscrapers, suspension bridges, great ships, railroads, and many other modern constructions would be almost unthinkable, and, despite the rise of other metals, steel still remains the preferred metal in a host of everyday uses, from automobile bodies to knives.

(It is a mistake, of course, to think that any single advance can bring about a major change in the way of life of humanity. This is always the result of a whole complex of interrelated advances. For instance, all the steel in the world could not make skyscrapers practical without the existence of that too-often-taken-for-granted device, the elevator. In 1861, the American inventor Elisha Graves Otis patented a hydraulic elevator, and in 1889, the company he founded installed the first electrically run elevators in a New York commercial building.)

With steel cheap and commonplace, it became possible to experiment with the addition of other metals ("alloy steel") to see if steel could be still further improved. The British metallurgist Robert Abbott Hadfield pioneered in this direction. In 1882, he found that adding manganese to steel to the extent of 13 per cent produced a harder alloy, which could be used in machinery for particularly brutal jobs, such as rock-crushing. In 1900, a steel alloy containing tungsten and chromium was found to retain its hardness well at high temperatures, even red heat, and this alloy proved a boon for high-speed tools. Today there are innumerable other alloy steels for particular jobs, employing such metals as molybdenum, nickel, cobalt, and vanadium.

The great difficulty with steel is its vulnerability to corrosion—a process that returns iron to the crude state of the ore whence it came. One way of combating this is to shield the metal by painting it or by plating it with a metal less likely to corrode, such as nickel, chromium, cadmium, or tin. A more effective method is to form an alloy that does not corrode. In 1913, the British metallurgist Harry Brearley discovered such an alloy by accident. He was looking for steel alloys that would be particularly suitable for gun barrels. Among the samples he discarded as unsuitable was a nickel-chromium alloy. Months later, he happened to notice that these particular pieces in his scrap heap were as bright as ever, although the rest were rusted. That was the birth of "stainless steel." It is too soft and too expensive for use in large-scale construction, but it serves admirably in cutlery and small appliances where nonrusting is more important than hardness.

Since something like a billion dollars a year is spent over the world in the not too successful effort to keep iron and steel from corroding, the search for a general rust inhibitor goes on unabated. One interesting recent discovery is that pertechnetates, compounds containing technetium, protect iron against rusting. Of course, this rare, man-made element may never be common enough to be used on any substantial scale, but it offers an invaluable research tool. Its radioactivity allows chemists to follow its fate and to observe what happens to it on the iron surface. If this use of technetium leads to a new understanding which will help solve the corrosion problem, that achievement alone will pay back in a matter of months all the money invested in research on the synthetic elements over the last quarter century.

One of iron's most useful properties is its strong ferromagnetism. Iron itself is an example of a "soft magnet." It is easily magnetized under the influence of an electric or magnetic field; that is, its magnetic domains (see Chapter 4) are easily lined up. It is also easily demagnetized when the field is removed, and the domains fall into random orientation again. This ready loss of magnetism can be useful, as in electromagnets, where the iron core is magnetized easily with the current on, but *should* be as easily demagnetized when the current goes off.

Since World War II, a new class of soft magnets has been developed. These are the "ferrites," an example being nickel ferrite ($NiFe_2O_4$) and manganese ferrite ($MnFe_2O_4$), which are used in computers as elements which must gain or lose magnetism with the utmost ease and rapidity.

"Hard magnets," with domains which are difficult to orient or, once oriented, to disorient, will, once magnetized, retain the property over long periods. Various steel alloys are the commonest examples though particularly strong, hard magnets have been found among alloys that contain little or no iron. The best known example of this is "alnico," discovered in 1931, one variety of which is made of aluminum, nickel, and cobalt (the name of the alloy being derived from the first two letters of each of the substances), plus a bit of copper.

In the 1950's, techniques were developed to use powdered iron as a magnet, the particles being so small as to consist of individual domains. These could be oriented in molten plastic, which would then be allowed to solidify, holding the domains fixed in their orientation. Such "plastic magnets" are very easy to shape and mold, but can be made adequately strong as well.

We have seen in recent decades the emergence of enormously useful new metals—metals that were almost useless and even unknown up to a century or so ago and in some cases up to our own generation. The most striking example is aluminum. Aluminum is the most common of all metals—60 per cent more common than iron. But it is also exceedingly difficult to extract from its ores. In 1825, Hans Christian Oersted (who had discovered the connection between electricity and magnetism) separated a little aluminum in impure form. Thereafter, many chemists tried unsuccessfully to purify the metal, until the French chemist Henri Etienne Sainte-Clair Deville in 1854 finally devised a method of obtaining aluminum in reasonable quantities. Aluminum is so active chemically that he had to use metallic sodium (even more active) to break aluminum's grip on its neighboring atoms. For a while aluminum sold for a hundred dollars a pound, making it practically a precious metal. Napoleon III indulged himself in aluminum cutlery and had an aluminum rattle fashioned for his infant son; and in the United States, as a mark of the nation's great esteem for George Washington, the Washington Monument was capped with a slab of solid aluminum.

In 1886, Charles Martin Hall, a young student of chemistry at Oberlin College, was so impressed by his professor's statement that anyone who could discover a cheap method of making aluminum would make a fortune that he decided to try his hand at it. In a home laboratory in his woodshed, Hall set out to apply Humphry Davy's early discovery that an electric current sent through a molten metal could separate the metal ions by depositing them on the cathode plate. Looking for a material that could dissolve aluminum, he stumbled across cryolite, a mineral found in reasonable quantity only in Greenland. (Nowadays synthetic cryolite is available.) Hall dissolved aluminum oxide in cryolite, melted the mixture, and passed an electric current through it. Sure enough, pure aluminum collected on the cathode. Hall rushed to his professor with his first few ingots of the metal. (To this day they are treasured by the Aluminum Company of America.)

As it happened, a young French chemist named Paul Louis Toussaint Héroult, who was just Hall's age (twenty-two), discovered the same process in the same year. (To complete the coincidence, Hall and Héroult both died in 1914.)

The Hall-Héroult process made aluminum an inexpensive metal, though it was never to be as cheap as steel, because useful aluminum ore is less common than useful iron ore, and electricity (the key to aluminum) is more expensive than coal (the key to steel). Nevertheless, aluminum has two great advantages over steel. First, it is light—only one third the weight of steel. Second, in aluminum's case corrosion merely takes the form of a thin, transparent film over its surface, which protects deeper layers from corrosion without affecting the metal's appearance.

Pure aluminum is rather soft, but alloying can take care of that. In 1906, the German metallurgist Alfred Wilm made a tough alloy by adding a bit of copper and a smaller bit of magnesium to the aluminum.

He sold his patent rights to the Durener Metal Works in Germany, and they gave the alloy the name Duralumin.

Engineers quickly realized how valuable a light but strong metal could be in aircraft. After the Germans introduced Duralumin in zeppelins during World War I and the British learned its composition by analyzing the alloy in a crashed zeppelin, use of this new metal spread over the world. Because Duralumin was not quite as corrosion-resistant as aluminum itself, metallurgists covered it with thin sheets of pure aluminum, forming the product called Alclad.

Today there are aluminum alloys which, weight for weight, are stronger than some steels. Aluminum has tended to replace steel wherever lightness and corrosion resistance are more important than brute strength. It has become, as everyone knows, almost a universal metal, used in airplanes, rockets, railway trains, automobiles, doors, screens, house siding, paint, kitchen utensils, foil wrapping and what not.

And now we have magnesium, a metal even lighter than aluminum. Its main use is in airplanes, as you might expect; as early as 1910, Germany was making use of magnesium-zinc alloys for that purpose. After World War I, magnesium-aluminum alloys came into increasing use.

Only about one fourth as abundant as aluminum and more active chemically, magnesium is harder to obtain from ores. But fortunately there is a rich source in the ocean. Magnesium, unlike aluminum or iron, is present in sea water in quantity. The ocean carries dissolved matter to the amount of 3.5 per cent of its mass. Of this dissolved material, 3.7 per cent is magnesium ion. The ocean as a whole, therefore, contains about 2 quadrillion (2,000,000,000,000,000) tons of magnesium, or all we could use for the indefinite future.

The problem was to get it out. The method chosen was to pump sea water into large tanks and add calcium oxide (also obtained from the sea, i.e., from oyster shells). The calcium oxide reacts with the water and the magnesium ion to form magnesium hydroxide, which is insoluble and therefore precipitates out of solution. The magnesium hydroxide is converted to magnesium chloride by treatment with hydrochloric acid, and the magnesium metal is then separated from the chlorine by means of an electric current.

In January of 1941, the Dow Chemical Company produced the first ingots of magnesium from sea water, and the stage was laid for a tenfold increase in magnesium production during the war years.

As a matter of fact, any element that can be extracted profitably from sea water may be considered in virtually limitless supply since, after use, it eventually returns to the sea. It has been estimated that if 100 million tons of magnesium were extracted from sea water each year for a million years, the magnesium content of the ocean would drop from its present figure of 0.13 to 0.12 per cent.

If steel was the "wonder metal" of the mid-nineteenth century, aluminum of the early twentieth century, and magnesium of the mid-twentieth century, what will the next new wonder metal be? The possibilities are limited. There are only seven really common metals in the earth's crust. Besides iron, aluminum, and magnesium, they are sodium, potassium, calcium, and titanium. Sodium, potassium, and calcium are far too active chemically to be used as construction metals. (For instance, they react violently with water.) That leaves titanium, which is about one eighth as abundant as iron.

Titanium has an extraordinary combination of good qualities. It is only a little more than half as heavy as steel, stronger, weight for weight, than aluminum or steel, resistant to corrosion, and able to withstand high temperatures. For all these reasons, titanium is now being used in aircraft, ships, and guided missiles wherever these properties can be put to good use.

Why was mankind so slow to discover the value of titanium? The reason is much the same as for aluminum and magnesium. It reacts too readily with other substances, and in its impure forms—combined with oxygen or nitrogen—it is an unprepossessing metal, brittle and seemingly useless. Its strength and other fine qualities emerge only when it is isolated in really pure form (in a vacuum or under an inert gas). The effort of metallurgists has succeeded to the point where a pound of titanium which would have cost $3,000 in 1947, cost $2 in 1969.

The search need not, however, be for new wonder-metals. The older metals (and some nonmetals, too) can be made far more "wonderful" than they are now.

In Oliver Wendell Holmes' poem "The Deacon's Masterpiece," the story is told of a "one-hoss shay" (one-horse buggy) which was carefully made in such a way as to have no weakest point. In the end, the shay went all at once—decomposing into a powder. But it had lasted a hundred years.

The atomic structure of crystalline solids, both metal and nonmetal, is rather like the one-hoss shay situation. A metal's crystals are riddled with submicroscopic clefts and scratches. Under pressure, a fracture will start at one of these weak points and spread through the crystal. If, like the deacon's wonderful one-hoss shay, a crystal could be built with no weak points, it would have great strength.

Such no-weak-point crystals do form as tiny fibers called "whiskers" on the surface of crystals. Tensile strengths of carbon whiskers have been found to run as high as 1,400 tons per square inch, which is from fifteen to seventy times the tensile strength of steel. If methods could be designed for manufacturing defect-free metal in quantity, we would find ourselves with materials of astonishing strength. In 1968, for instance, Soviet scientists produced a tiny defect-free crystal of tungsten that would sustain a load of 1,635 tons per square inch, as compared to 213 tons per square inch for the best steel. And even if defect-free substances were not available in bulk, the addition of defect-free fibers to ordinary metals would reinforce and strengthen them.

Then, too, as late as 1968, an interesting new method was found for combining metals. The two methods of historic interest were alloying, where two or more metals are melted together and form a more-or-less-homogeneous mixture, and plating, where one metal is bound firmly to another (a thin layer of expensive metal is usually bound to the surface of a bulky volume of cheaper metal, so that the surface is, for instance, as beautiful and corrosion-resistant as gold but the whole nearly as cheap as copper).

The American metallurgist Newell C. Cook and his associates were attempting to plate a silicon layer on a platinum surface, using molten alkali fluoride as the liquid in which the platinum was immersed. The expected plating did not occur. What happened, apparently, was that the molten fluoride removed the very thin film of bound oxygen ordinarily present on even the most resistant metals and presented the platinum surface "naked" to the silicon atoms. Instead of binding themselves to the surface on the other side of the oxygen atoms, they worked their way *into* the surface. The result was that a thin outer layer of the platinum became an alloy.

Cook followed this new direction and found that many substances could be combined in this way to form a "plating" of alloy on pure metal (or on another alloy). Cook called the process "metalliding" and quickly showed its usefulness. Thus, copper to which 2 to 4 per cent of beryllium is added in the form of an ordinary alloy, become extraordinarily strong. The same result can be achieved if copper is "beryllided" at the cost of much less of the relatively rare beryllium. Again, steel metallided with boron ("boriding") is hardened. The addition of silicon, cobalt, and titanium, also produces useful properties.

Wonder metals, in other words, if not found in nature can be created by human ingenuity.

CHAPTER 6

The Particles

The Nuclear Atom

As I pointed out in the preceding chapter, it was known by 1900 that the atom was not a simple, indivisible particle, but contained at least one subatomic particle—the electron, identified by J. J. Thomson. Thomson suggested that electrons were stuck like raisins in the positively charged main body of the atom.

But very shortly it developed that there were also other subparticles within the atom. When Becquerel discovered radioactivity, he identified some of the radiation emitted by radioactive substances as consisting of electrons, but other emissions were discovered as well. The Curies in France and Ernest Rutherford in England found one that was less penetrating than the electron stream. Rutherford called this radiation "alpha rays" and gave the electron emission the name "beta rays." The flying electrons making up the latter radiation are, individually, "beta particles." The alpha rays were also found to be made up of particles and these were called "alpha particles." "Alpha" and "beta" are, of course, the first two letters of the Greek alphabet.

Meanwhile the French chemist Paul Ulrich Villard discovered a third form of radioactive emission, which was named "gamma rays" after the third letter of the Greek alphabet. The gamma rays were quickly identified as radiation resembling X-rays, but with shorter wavelengths.

Rutherford learned by experiment that a magnetic field deflected alpha particles much less than it did beta particles. Furthermore, they were deflected in the opposite direction, which meant that the alpha particle had a positive charge, as opposed to the electron's negative one. From the amount of deflection, it could be calculated that the alpha particle must have at least twice the mass of the hydrogen ion, which

281

possessed the smallest known positive charge. The amount of deflection would be affected both by the particle's mass and by its charge. If the alpha particle's positive charge was equal to that of the hydrogen ion, its mass would be two times that of the hydrogen ion; if its charge was double that, it would be four times as massive as the hydrogen ion, and so on.

Rutherford settled the matter in 1909 by isolating alpha particles. He put some radioactive material in a thin-walled glass tube surrounded by a thick-walled glass tube, with a vacuum between. The alpha particles could penetrate the thin inner wall but not the thick outer one. They bounced back from the outer wall, so to speak, and in doing so lost energy and therefore were no longer able to penetrate the thin walls either. Thus they were trapped between. Now Rutherford excited the alpha particles by means of an electric discharge so that they glowed. They then showed the spectral lines of helium. (It has become evident that alpha particles produced by radioactive substances in the soil are the source of the helium in natural gas wells.) If the alpha particle is helium, its mass must be four times that of hydrogen. This, in turn, means that its positive charge amounts to two units, taking the hydrogen ion's charge as the unit.

Rutherford later identified another positive particle in the atom. This one had actually been detected, but not recognized, many years before. In 1886, the German physicist Eugen Goldstein, using a cathode-ray tube with a perforated cathode, had discovered a new radiation that

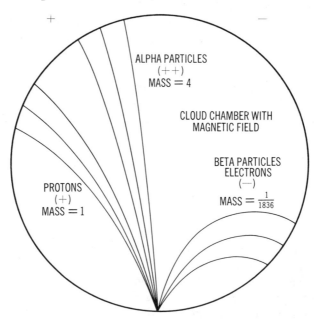

Deflection of particles by a magnetic field.

streamed through the holes of the cathode in the direction opposite to the cathode rays themselves. He called it "Kanalstrahlen" ("channel rays"). In 1902, this radiation served as the first occasion when the Doppler-Fizeau effect (see Chapter 1) was detected in any earthly source of light. The German physicist Johannes Stark placed a spectroscope in such a fashion that the rays raced toward it and demonstrated the violet shift. For this research, he was awarded the Nobel Prize for physics in 1919.

Since channel rays move in a direction opposite to the negatively charged cathode rays, Thomson suggested that this radiation be called "positive rays." It turned out that the particles of the "positive rays" could easily pass through matter. They were therefore judged to be much smaller in volume than ordinary ions or atoms. The amount of their deflection by a magnetic field indicated that the smallest of these particles had the same charge and mass as a hydrogen ion, assuming that this ion carried the smallest possible unit of positive charge. The positive-ray particle was therefore deduced to be the fundamental positive particle—the opposite number of the electron. Rutherford named it the "proton" (from the Greek word for "first").

The proton and the electron do indeed carry equal, though opposite, electric charges, although the proton is 1,836 times as massive as the electron. It seemed likely, then, that an atom was composed of protons and electrons, mutually balancing their charges. It also appeared that the protons were in the interior of the atom, for whereas electrons could easily be peeled off, protons could not. But now the big question was: what sort of structure did these particles of the atom form?

Rutherford himself came upon the beginning of the answer. Between 1906 and 1908, he kept firing alpha particles at a thin foil of metal (such as gold or platinum) to probe its atoms. Most of the projectiles passed right through undeflected (as bullets might pass through the leaves of a tree). But not all: Rutherford found that on the photographic plate that served as his target behind the metal, there was an unexpected scattering of hits around the central spot, and some particles bounced back! It was as if some of the bullets had not passed through leaves alone but had ricocheted off something more substantial.

Rutherford decided that what they had hit was some sort of dense core, which occupied only a very small part of the volume of the atom. Most of an atom's volume, it seemed, must be occupied by electrons. As alpha particles charged through the foil of metal, they usually encountered only electrons, and they brushed aside this froth of light particles, so to speak, without being deflected. But once in a while an alpha particle might happen to hit an atom's denser core, and then it was deflected. That this happened only very occasionally showed that the atomic cores must be very small indeed, because a projectile passing through the metal foil must encounter many thousands of atoms.

It was logical to suppose that the hard core was made up of protons. Rutherford pictured the protons of an atom as crowded into a tiny "atomic nucleus" at the center. (It has since been demonstrated that this nucleus has a diameter of little more than 1/100,000 that of the whole atom.)

This, then, is the basic model of the atom: a positively charged nucleus taking up very little room, but containing almost all the mass of the atom, surrounded by a froth of electrons taking up nearly all the volume of the atom, but containing practically none of its mass. For his extraordinary pioneering work on the ultimate nature of matter, Rutherford received the Nobel Prize in chemistry in 1908.

It now became possible to describe specific atoms and their behavior in more definite terms. For instance, the hydrogen atom possesses but a single electron. If this is removed, the proton that remains immediately attaches itself to some neighboring molecule. But when the bare hydrogen nucleus does not find an electron to share in this fashion, it acts as a proton—that is to say, a subatomic particle—and in that form it can penetrate matter and react with other nuclei if it has enough energy.

Helium, with two electrons, does not give one up so easily. As I mentioned in the preceding chapter, its two electrons form a closed shell, and the atom is therefore inert. If helium is stripped of both electrons, however, it becomes an alpha particle—that is, a subatomic particle carrying two units of positive charge.

The third element, lithium, has three electrons in its atom. Stripped of one or two, it is an ion. If all three of its electrons are removed, it, too, becomes a bare nucleus, carrying a three-unit positive charge.

The number of units of positive charge in the nucleus of an atom has to be exactly equal to the number of electrons it normally contains, for the atom as a whole is ordinarily neutral. And, in fact, the atomic numbers of the elements are based on their units of positive charge rather than of negative charge, because the number of an atom's electrons may easily be made to vary in ion formation, whereas the number of its protons can be altered only with great difficulty.

This scheme of the construction of atoms had hardly been worked out when a new conundrum arose. The number of units of positive charge on a nucleus did not balance at all with the nucleus' weight, or mass, except in the case of the hydrogen atom. The helium nucleus, for instance, had a positive charge of two but was known to have four times the mass of the hydrogen nucleus. And the situation got worse and worse as one went down the table of elements, until, by the time uranium was reached, one had a nucleus with a mass equal to 238 protons but a charge equal to only 92.

How could a nucleus containing four protons (as the helium nucleus was supposed to) have only two units of positive charge? The first, and simplest, guess was that two units of its charge were neutralized by the presence in the nucleus of negatively charged particles of negligible weight. Naturally the electron sprang to mind. The puzzle might be straightened out if one assumed that the helium nucleus consisted of four protons and two neutralizing electrons, leaving a net positive charge of two—and so on all the way to uranium, whose nucleus would have 238 protons and 146 electrons, netting 92 units of positive charge. The whole idea was given encouragement by the fact that radioactive nuclei were actually known to emit electrons—i.e., beta particles.

This view of matter prevailed for more than a decade, until a better answer came in a roundabout way from other investigations. But, in the meantime, some serious objections to the hypothesis arose. For one thing, if the nucleus was built essentially of protons, with the light electrons contributing practically nothing to the mass, how was it that the relative masses of the various nuclei did not come to whole numbers? According to the measured atomic weights, the nucleus of the chlorine atom, for instance, had a mass of 35½ times that of the hydrogen nucleus. Did that mean it contained 35½ protons? No scientist (then or now) could accept the idea of half a proton.

Actually, this particular question had an answer that was discovered even before the main issue was solved. It makes an interesting story in itself.

Isotopes

As early as 1816, an English physician named William Prout had suggested that all atoms were built up from the hydrogen atom. As time went on and the atomic weights were worked out, Prout's theory fell by the wayside, because it developed that many elements had fractional weights (taking oxygen as the standard at 16). Chlorine, as I have mentioned, has an atomic weight of about 35.5—35.453, to be more exact. Other examples are antimony, 121.75; barium, 137.34; boron, 10.811; cadmium, 112.40.

Around the turn of the century there came a series of puzzling observations that was to lead to the explanation. The Englishman William Crookes (he of the Crookes tube) separated from uranium a small quantity of a substance that proved much more radioactive than uranium itself. He suggested that uranium was not radioactive at all—only this impurity, which he called "uranium X." Henri Becquerel, on the other hand, discovered that the purified, feebly radioactive uranium somehow increased in radioactivity with time. After it was left standing for a while, the active uranium X could be extracted from it, again and again. In other words, uranium was converted by its own radioactivity to the still more active uranium X.

Then Rutherford similarly separated a strongly radioactive "thorium X" from thorium and found that thorium, too, went on producing more thorium X. It was already known that the most famous radioactive element of all, radium, broke down to the radioactive gas radon. So Rutherford and his assistant, the chemist Frederick Soddy, concluded that radioactive atoms, in the process of emitting their particles, generally transformed themselves into other varieties of radioactive atoms.

Chemists began searching for such transformations and came up with quite an assortment of new substances, giving them such names as radium A, radium B, mesothorium I, mesotherium II, and actinium C. All of them were grouped into three series, depending on their atomic ancestry. One series arose from the breakdown of uranium, another from that of thorium, and a third from that of actinium (later it turned out that actinium itself had a predecessor, named "protactinium"). Altogether, some forty members of these series were identified, each distinguished by its own peculiar pattern of radiation. But the end product of all three series was the same: each chain of substances eventually broke down to the same stable element—lead.

Now obviously these forty substances could not all be separate elements; between uranium (92) and lead (82) there were only ten places in the periodic table, and all but two of these belonged to known elements. The chemists found, in fact, that though the substances differed in radioactivity, some of them were identical with one another in chemical properties. For instance, as early as 1907 the American chemists Herbert Newby McCoy and W. H. Ross showed that "radiothorium," one of the disintegration products of thorium, showed precisely the same chemical behavior as thorium. "Radium D" behaved chemically exactly like lead; in fact, it was often called "radiolead." All this suggested that the substances in question were actually varieties of the same element: radiothorium a form of thorium, radiolead a member of a family of leads, and so on.

In 1913, Soddy gave clear expression of this idea and developed it further. He showed that when an atom emitted an alpha particle it changed into an element two places lower in the list of elements; when it emitted a beta particle it changed into an element one place higher. On this basis, "radiothorium" would indeed fall in thorium's place in the table, and so would the substances called "uranium X_1" and "uranium Y": all three would be varieties of element 90. Likewise, "radium D," "radium B," "thorium B," and "actinium B" would all share lead's place as varieties of element 82.

To the members of a family of substances sharing the same place in the periodic table Soddy gave the name "isotope" (from Greek words meaning "same position"). Soddy received the Nobel Prize in chemistry in 1921.

The proton-electron model of the nucleus fitted in beautifully with Soddy's isotope theory. Removal of an alpha particle from a nucleus would reduce the positive charge of that nucleus by two—exactly what was needed to move it two places down in the periodic table. On the other hand, the ejection of an electron (beta particle) from a nucleus would leave an additional proton unneutralized and thus increase the nucleus' positive charge by one unit. That amounted to raising the atomic number by one, so the element would move to the next higher position in the periodic table.

How is it that when thorium breaks down to "radiothorium," after going through not one but three disintegrations, the product is still thorium? Well, in the process the thorium atom loses an alpha particle, then a beta particle, then a second beta particle. If we accept the proton building-block idea, this means it has lost four electrons (two supposedly contained in the alpha particle) and four protons. (The actual situation differs from this picture, but in a way that does not affect the result.) The thorium nucleus started with 232 protons and 142 electrons (supposedly). Having lost four protons and four electrons, it is reduced to 228 protons and 138 electrons. This still leaves the atomic number 90, the same as before. So "radiothorium," like thorium, has ninety planetary electrons circling around the nucleus. Since the chemical properties of an atom are controlled by the number of its planetary electrons, thorium and "radiothorium" behave the same chemically, regardless of their difference in atomic weight (232 against 228).

The isotopes of an element are identified by their atomic weight, or "mass number." Thus ordinary thorium is called thorium 232, while "radiothorium" is thorium 228. Similarly, the radioactive isotopes of lead are known as lead 210 ("radium D"), lead 214 ("radium B"), lead 212 ("thorium B"), and lead 211 ("actinium B").

The notion of isotopes was found to apply to stable elements as well as to radioactive ones. For instance, it turned out that the three radioactive series I have mentioned ended in three different forms of lead. The uranium series ended in lead 206, the thorium series in lead 208, and the actinium series in lead 207. Each of these was an "ordinary," stable isotope of lead, but the three leads differed in atomic weight.

Proof of the existence of stable isotopes came from a device invented by an assistant of J. J. Thomson named Francis William Aston. It was an arrangement that separated isotopes very sensitively by virtue of the difference in deflection of their ions by a magnetic field; Aston called it a "mass spectrograph." In 1919, using an early version of this instrument, Thomson showed that neon was made up of two varieties of atom, one with a mass number of 20, the other with a mass number of 22. Neon 20 was the common isotope; neon 22 came with it in the ratio of one atom in ten. (Later a third isotope, neon 21, was discovered, amounting to only one atom in 400 in the neon of the atmosphere.)

Now the reason for the fractional atomic weights of the elements at last became clear. Neon's atomic weight of 20.183 represented the composite weight of the three differently weighted isotopes making up the element as it was found in nature. Each individual atom had an integral mass number, but the average mass number—the atomic weight —was fractional.

Aston proceeded to show that several common stable elements were indeed mixtures of isotopes. He found that chlorine, with a fractional atomic weight of 35.453, was made up of chlorine 35 and chlorine 37, in the "abundance ratio" of four to one. Aston was awarded the Nobel Prize in chemistry in 1922.

In his address accepting the Prize, Aston clearly forecast the possibility of making use of the energy bound in the atomic nucleus, foreseeing both nuclear power plants and nuclear bombs (see Chapter 9). In 1935, the Canadian-American physicist Arthur Jeffrey Dempster used Aston's instrument to take a long step in that direction. He showed that, although 993 of every 1,000 uranium atoms were uranium 238, the remaining seven were uranium 235. This was a discovery fraught with a significance soon to be realized.

Thus, after a century of false trails, Prout's idea was finally vindicated. The elements *were* built of uniform building blocks—if not of hydrogen atoms, at least of units with hydrogen's mass. The reason the elements did not bear this out in their weights was that they were mixtures of isotopes containing different numbers of building blocks. In fact, even oxygen, whose atomic weight of sixteen was used as the standard for measuring the relative weights of the elements, was not a completely pure case. For every 10,000 atoms of common oxygen 16, there were twenty atoms of an isotope with a weight equal to 18 units and four with the mass number 17.

Actually there are a few elements consisting of a "single isotope." (This is a misnomer: to speak of an element as having only one isotope is like saying a woman has given birth to a "single twin.") The elements of this kind include beryllium, all of whose atoms have the mass number 9; fluorine, made up solely of fluorine 19; aluminum, solely aluminum 27; and a number of others. A nucleus with a particular structure is now called a "nuclide," following the suggestion made in 1947 by the American chemist Truman Paul Kohman. One can properly say that an element such as aluminum is made up of a single nuclide.

Ever since Rutherford identified the first nuclear particle (the alpha particle), physicists have busied themselves poking around in the nucleus, trying either to change one atom into another or to break it up to see what it is made of. At first they had only the alpha particle to work with. Rutherford made excellent use of it.

One of the fruitful experiments Rutherford and his assistants carried out involved firing alpha particles at a screen coated with zinc sulfide. Each hit produced a tiny scintillation (an effect first discovered by Crookes in 1903), so that the arrival of single particles could be witnessed and counted with the naked eye. Pursuing this technique, the experimenters put up a metal disk that would block the alpha particles from reaching the zinc sulfide screen so that the scintillations stopped. When hydrogen was introduced into the apparatus, scintillations appeared on the screen despite the blocking metal disk. Moreover, these new scintillations differed in appearance from those produced by alpha particles. Since the metal disk stopped alpha particles, some other radiation must be penetrating it to reach the screen. The radiation, it was decided, must consist of fast protons. In other words, the alpha particles would now and then make a square hit on the nucleus of a hydrogen atom and send it careening forward, as one billiard ball might send another forward on striking it. The struck protons, being relatively light, would shoot forward at great velocity and so could penetrate the metal disk and strike the zinc sulfide screen.

This detection of single particles by scintillation is an example of a "scintillation counter." To make such counts, Rutherford and his assistants first had to sit in the dark for fifteen minutes in order to sensitize their eyes and then make their painstaking counts. Modern scintillation counters do not depend on the human eye and mind. Instead, the scintillations are converted to electric pulses that are then counted electronically. The final result need merely be read off from appropriate dials. The counting may be made more practical where scintillations are numerous, by using electric circuits that allow only one in two or in four (or even more) scintillations to be recorded. Such "scalers" (which scaled down the counting, so to speak) were first devised by the English physicist C. E. Wynn-Williams in 1931. Since World War II, organic substances have substituted for zinc sulfide and have proved preferable.

In Rutherford's original scintillation experiments, there came an unexpected development. When his experiment was performed with nitrogen instead of hydrogen as the target for the alpha-particle bombardment, the zinc sulfide screen still showed scintillations exactly like those produced by protons. Rutherford could only conclude that the bombardment had knocked protons out of the nitrogen nucleus.

To try to find out just what had happened, Rutherford turned to the "Wilson cloud chamber." This device had been invented in 1895 by the Scottish physicist Charles Thomson Rees Wilson. A glass container fitted with a piston is filled with moisture-saturated air. When the piston is pulled outward, the air abruptly expands and therefore cools. At the reduced temperature, it is supersaturated with the moisture. Now any charged particle will cause the water vapor to condense on it. If a particle dashes through the chamber, ionizing atoms in it, a foggy line of droplets will mark its wake.

The nature of this track can tell a great deal about the particle. The light beta particle leaves a faint, wavering path; the particle is knocked about even in passing near electrons. The much more massive alpha particle makes a straight, thick track. If it strikes a nucleus and rebounds, the path has a sharp bend in it. If it picks up two electrons and becomes a neutral helium atom, its track ends. Aside from the size and character of its track, there are other ways of identifying a particle in the cloud chamber. Its response to an applied magnetic field tells whether it is positively or negatively charged, and the amount of curve indicates its mass and energy. By now physicists are so familiar with photographs of all sorts of tracks that they can read them off as if they were primer print. For the development of his cloud chamber, Wilson shared the Nobel Prize in physics in 1927.

The cloud chamber has been modified in several ways since its invention and cousin instruments have been devised. The original cloud chamber was not usable after expansion until the chamber had been reset. In 1939, A. Langsdorf, in the United States, devised a "diffusion cloud chamber," in which warm alcohol vapor diffused into a cooler region in such a way that there was always a supersaturated region and tracks could be observed continuously.

Then came the "bubble chamber," a device similar in principle. In it, superheated liquids under pressure are used rather than supersaturated gas. The path of the charged particle is marked by a line of vapor bubbles in the liquid rather than liquid droplets in vapor. The inventor, the American physicist Donald Arthur Glaser, is supposed to have gotten the idea by studying a glass of beer in 1953. If so, it was a most fortunate glass of beer for the world of physics and for him, for Glaser received the Nobel Prize for physics in 1960 for the invention of the bubble chamber.

The first bubble chamber had only been a few inches in diameter. Within the decade, bubble chambers six feet long were being used. Bubble chambers, like diffusion cloud chambers, are constantly set for action. In addition, since many more atoms are present in a given volume of liquid than of gas, more ions are produced in a bubble chamber, which is thus particularly well adapted to the study of fast and short-lived particles. Within a decade of its invention, bubble chambers were producing hundreds of thousands of photographs per week. Ultra-short-lived particles were discovered in the 1960's that would have gone undetected without the bubble chamber.

Liquid hydrogen is an excellent liquid with which to fill bubble chambers, because the hydrogen nuclei are so simple (consisting of single protons) as to introduce a minimum of added complication. Bubble chambers 12 feet across and 7 feet high, using as much as 6,400 gallons of liquid hydrogen, now exist, and a 200-liter liquid-helium bubble-chamber is in operation in Great Britain.

Although the bubble chamber is more sensitive to short-lived particles than the cloud chamber, it has its shortcomings. Unlike the cloud chamber, the bubble chamber cannot be triggered by desired events. It must record everything wholesale, and uncounted numbers of tracks must be searched through for those of significance. The search was on, then, for some method of detecting tracks that combined the selectivity of the cloud chamber with the sensitivity of the bubble chamber.

This need was met eventually by the "spark chamber," in which incoming particles ionize gas and set off electric currents through neon gas that is crossed by many metal plates. The currents show up as a visible line of sparks, marking the passage of the particles, and the device can be adjusted to react only to those particles under study. The first practical spark-chamber was constructed in 1959 by the Japanese physicists S. Fukui and S. Miyamoto. In 1963, Soviet physicists improved it further, heightening its sensitivity and flexibility. Short streamers of light are produced that, seen on end, make a virtually continuous line (rather than the separate sparks of the spark chamber). The modified device is therefore a "streamer chamber." It can detect events that take place within the chamber, and particles that streak off in any direction, where the original spark chamber fell short in both respects.

But, leaving modern sophistication in studying the flight of sub-atomic particles, we must turn back half a century to see what happened when Rutherford bombarded nitrogen nuclei with alpha particles within one of the original Wilson cloud chambers. The alpha particle would leave a track that would end suddenly in a fork. Plainly this represented a collision with a nitrogen nucleus. One branch of the fork would be comparatively thin, representing a proton shooting off. The other branch, a short, heavy track, represented what was left of the nitrogen nucleus, rebounding from the collision. But there was no sign of the alpha particle itself. It seemed that it must have been absorbed by the nitrogen nucleus, and this supposition was later verified by the British physicist Patrick Maynard Stuart Blackett, who is supposed to have taken more than 20,000 photographs in the process of collecting eight such collisions (surely an example of superhuman patience, faith, and persistence). For this and other work in the field of nuclear physics, Blackett received the Nobel Prize in physics in 1948.

The fate of the nitrogen nucleus could now be deduced. When it absorbed the alpha particle, its mass number of 14 and positive charge of 7 were raised to 18 and 9, respectively. But since the combination immediately lost a proton, the mass number dropped to 17 and the positive charge to 8. Now the element with a positive charge of 8 is oxygen, and the mass number 17 belongs to the isotope oxygen 17. In other words, Rutherford had, in 1919, transmuted nitrogen into oxygen. This was the first man-made transmutation in history. The dream of the alchemists had been fulfilled, though in a manner they could not possibly have foreseen.

Alpha particles from radioactive sources had limits as projectiles: they were not nearly energetic enough to break into nuclei of the heavier elements, whose high positive charges exercise a strong repulsion against positively charged particles. But the nuclear fortress had been breached, and more energetic attacks were to come.

New Particles

The matter of attacks on the nucleus brings us back to the question of the make-up of the nucleus. In 1930, two German physicists, Walther Bothe and H. Becker, reported that they had released from the nucleus a mysterious new radiation of unusual penetrating power. They had produced it by bombarding beryllium atoms with alpha particles. The year before Bothe had devised methods for using two or more counters in conjunction—"coincidence counters." These could be used to identify nuclear events taking place in a millionth of a second. For this and other work he shared in the Nobel Prize for physics in 1954.

Two years later the Bothe-Becker discovery was followed up by the French physicists Frédéric and Irène Joliot-Curie. (Irène was the daughter of Pierre and Marie Curie, and Joliot had added her name to his on marrying her.) They used the new-found radiation from beryllium to bombard paraffin, a waxy substance composed of hydrogen and carbon. The radiation knocked protons out of the paraffin.

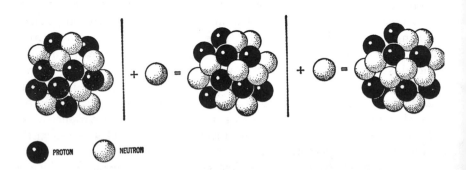

Nuclear makeup of oxygen 16, oxygen 17, and oxygen 18.
They contain eight protons each and, in addition, eight,
nine, and ten neutrons, respectively.

The English physicist James Chadwick quickly suggested that the radiation consisted of particles. To determine their size, he bombarded boron atoms with them, and from the increase in mass of the new nucleus he calculated that the particle added to the boron had a mass about equal to the proton. Yet the particle itself could not be detected in a Wilson cloud chamber. Chadwick decided that the explanation must be that the particle had no electric charge (an uncharged particle produces no ionization and therefore condenses no water droplets).

So Chadwick concluded that a completely new particle had turned up—a particle with just about the same mass as a proton but without any charge, or, in other words, electrically neutral. The possibility of such a particle had already been suggested, and a name had even been proposed —the "neutron." Chadwick accepted that name. For his discovery of the neutron, he was awarded the Nobel Prize in physics in 1935.

The new particle at once solved certain doubts that theoretical physicists had had about the proton-electron model of the nucleus. The German theoretical physicist Werner Heisenberg announced that the concept of a nucleus consisting of protons and neutrons, rather than protons and electrons, gave a much more satisfactory picture—more in accord with what the mathematics of the case said the nucleus should be like.

Furthermore, the new model fitted the facts of the periodic table of elements just as neatly as the old one had. The helium nucleus, for instance, would consist of two protons and two neutrons, which explained its mass of four and nuclear charge of two units. And the concept accounted for isotopes in very simple fashion. For example, the chlorine-35 nucleus would have 17 protons and 18 neutrons; the chlorine-37 nucleus, 17 protons and 20 neutrons. This would give both the same nuclear charge, and the extra weight of the heavier isotope would lie in its two extra neutrons. Likewise, the three isotopes of oxygen would differ only in their numbers of neutrons: oxygen 16 would have eight protons and eight neutrons; oxygen 17, eight protons and nine neutrons; oxygen 18, eight protons and ten neutrons.

In short, every element could be defined simply by the number of protons in its nucleus, which is equivalent to the atomic number. All the elements except hydrogen, however, also had neutrons in the nucleus, and the mass number of a nuclide was the sum of its protons and neutrons. Thus the neutron joined the proton as a basic building block of matter. For convenience, both are now lumped together under the general term "nucleons," a term first used in 1941 by the Danish physicist Christian Moller. From this came "nucleonics," suggested in 1944 by the American engineer Zay Jeffries to represent the study of nuclear science and technology.

This new understanding of nuclear structure has resulted in additional classifications of nuclides. Nuclides with equal numbers of protons are, as has just been explained, isotopes. Similarly, nuclides with equal numbers of neutrons (as, for instance, hydrogen 2 and helium 3, each containing one neutron in the nucleus) are "isotones." Nuclides with equal total number of nucleons, and therefore of equal mass numbers, as calcium 40 and argon 40, are "isobars."

The proton-neutron model of the nucleus is not likely to be seriously upset in the future. At first it left unexplained the fact that radioactive nuclei emitted electrons, but that question was soon cleared up, as I shall explain shortly.

Nevertheless, in a very important respect the discovery of the neutron disappointed physicists. They had been able to think of the universe as being built of just two fundamental particles—the proton and the electron. Now a third had to be added. To scientists, every retreat from simplicity is regrettable.

The worst of it was that, as things turned out, this was only the beginning. Simplicity's backward step quickly became a headlong rout. There were more particles to come.

For many years physicists had been studying the mysterious "cosmic rays" from space, first discovered in 1911 by the Austrian physicist Victor Francis Hess on balloon flights high in the atmosphere.

The presence of such radiation was detected by an instrument so simple as to hearten those who sometimes feel that modern science can progress only by use of unbelievably complex devices. The instrument was an "electroscope," consisting of two pieces of thin gold foil attached to a metal rod within a metal housing fitted with windows. (The ancestor of this device was constructed as long ago as 1706 by the English physicist Francis Hauksbee.)

If the metal rod is charged with static electricity, the pieces of gold foil separate. Ideally, they would remain separated forever, but ions in the surrounding atmosphere slowly conduct away the charge so that the leaves gradually collapse toward each other. Energetic radiation, such as X-rays, gamma rays, or streams of charged particles, produces the ions necessary for such charge leakage. Even if the electroscope is well shielded, there is still a slow leakage, indicating the presence of a very penetrating radiation not directly related to radioactivity. It was this penetrating radiation, which increased in intensity, that Hess noticed as he rose high in the atmosphere. Hess shared the Nobel Prize for physics in 1936 for this discovery.

The American physicist Robert Andrews Millikan, who collected a great deal of information on this radiation (and gave it the name "cosmic rays"), decided that it must be a form of electromagnetic radiation. Its penetrating power was such that some of it could even pass through several feet of lead. To Millikan this suggested that the radiation was like the penetrating gamma rays, but with an even shorter wavelength.

Others, notably the American physicist Arthur Holly Compton, contended that the cosmic rays were particles. There was a way to investigate the question. If they were charged particles, they should be deflected by the earth's magnetic field as they approached the earth from outer space. Compton studied the measurements of cosmic radiation at various latitudes and found that it did indeed curve with the magnetic field: it was weakest near the magnetic equator and strongest near the poles, where the magnetic lines of force dipped down to the earth.

The "primary" cosmic particles, as they enter our atmosphere, carry fantastically high energies. Most of them are protons, but some are nuclei of heavier elements. In general, the heavier the nucleus, the rarer it is among the cosmic particles. Nuclei as complex as those making up iron atoms were detected quickly enough, and in 1968, nuclei as complex as those of uranium were detected. The uranium nuclei make up only one particle in 10 million. A few very high-energy electrons are also included.

When the primary particles hit atoms and molecules of the air, they smash these nuclei and produce all sorts of "secondary" particles. It is this secondary radiation (still very energetic) that we detect near the earth, but balloons sent to the upper atmosphere have recorded the primary radiation.

Now it was as a result of cosmic-ray research that the next new particle—after the neutron—was discovered. This discovery had actually been predicted by a theoretical physicist. Paul Adrien Maurice Dirac had reasoned, from a mathematical analysis of the properties of sub-atomic particles, that each particle should have an "antiparticle." (Scientists like nature to be not only simple but also symmetrical.) Thus there ought to be an "antielectron," exactly like the electron except that it had a positive instead of a negative charge, and an "antiproton" with a negative instead of a positive charge.

Dirac's theory did not make much of a splash in the scientific world when he proposed it in 1930. But sure enough, two years later the "antielectron" actually turned up. The American physicist Carl David Anderson was working with Millikan on the problem of whether cosmic rays were electromagnetic radiation or particles. By then most people were ready to accept Compton's evidence that they were charged particles, but Millikan was an extraordinarily hard loser, and he was not satisfied that the issue was settled. Anderson undertook to find out whether cosmic rays entering a Wilson cloud chamber would be bent by a strong magnetic field. To slow down the rays sufficiently so that the curvature, if any, could be detected, Anderson placed in the chamber a lead barrier about a quarter of an inch thick. He found that the cosmic radiation crossing the chamber after it came through the lead did make a curved track. But he also found something else. In their passage through the lead, the energetic cosmic rays knocked particles out of the lead atoms. One of these particles made a track just like that of an electron. But it curved in the wrong direction! Same mass but opposite charge. There it was—Dirac's "antielectron." Anderson called his discovery the "positron." It is an example of the secondary radiation produced by cosmic rays, but in 1963 it was found that positrons were included among the primary radiations as well.

Left to itself, the positron is as stable as the electron (why not, since it is identical with the electron except for electric charge?) and could exist indefinitely. It is not, however, left to itself, for it comes into existence in a universe filled with electrons. As it streaks along, it almost immediately (say, within a millionth of a second) finds itself in the neighborhood of one.

For a moment, there may be an electron-positron association—a situation in which the two particles circle each other about a mutual center of force. In 1945, the American physicist Arthur Edward Ruark suggested that this two-particle system be called "positronium," and in 1951, the Austrian-American physicist Martin Deutsch was able to detect positronium through the characteristic gamma-radiation it gave up.

However, even if a positronium system forms, it remains in existence for only a 10-millionth of a second, at most. The dance ends in the combination of the electron and positron. When the two opposite bits of matter combine they cancel each other, leaving no matter at all ("mutual annihilation"); only energy, in the form of gamma rays, is left behind. This confirmed Albert Einstein's suggestion that matter could be converted into energy and vice versa. Indeed, Anderson soon succeeded in detecting the reverse phenomenon: gamma rays suddenly disappearing and giving rise to an electron-positron pair. This is called "pair production."(Anderson, along with Hess, received the Nobel Prize in physics in 1936.)

The Joliot-Curies shortly afterward came across the positron in another connection, and in so doing made an important discovery. Bombarding aluminum atoms with alpha particles, they found that the procedure produced not only protons but also positrons. This in itself was interesting but not fabulous. When they stopped the bombardment, however, the aluminum kept right on emitting positrons! The emission faded off with time. Apparently they had created a new radioactive substance in the target.

The Joliot-Curies interpreted what had happened in this way. When an aluminum nucleus absorbed an alpha particle, the addition of two protons changed aluminum (atomic number 13) to phosphorus (atomic number 15). Since the alpha particle contained four nucleons altogether, the mass number would go up by four—from aluminum 27 to phosphorus 31. Now if the reaction knocked a proton out of this nucleus, the reduction of its atomic number and mass number by one would change it to another element—namely, silicon 30.

Since an alpha particle is the nucleus of helium and a proton the nucleus of hydrogen, we can write the following equation of this "nuclear reaction":

aluminum 27 + helium 4 → silicon 30 + hydrogen 1

Notice that the mass numbers balance: 27 plus 4 equals 30 plus 1. So do the atomic numbers, for aluminum's is 13 and helium's 2, making 15 together, while silicon's atomic number of 14 and hydrogen's 1 also add up to 15. This balancing of both mass numbers and atomic numbers is a general rule of nuclear reactions.

The Joliot-Curies assumed that neutrons as well as protons had been formed in the reaction. If phosphorus 31 emitted a neutron instead of a proton, the atomic number would not change, though the mass number would go down one. In that case the element would remain phosphorus but become phosphorus 30. This equation would read:

aluminum 27 + helium 4 → phosphorus 30 + neutron 1

Since the atomic number of phosphorus is 15 and that of the neutron is 0, again the atomic numbers on both sides of the equation also balance.

Both processes—alpha absorption followed by proton emission, and alpha absorption followed by neutron emission—take place when aluminum is bombarded by alpha particles. But there is one important distinction between the two results. Silicon 30 is a perfectly well-known isotope of silicon, making up a little more than 3 per cent of the silicon in nature. But phosphorus 30 does not exist in nature. The only known natural form of phosphorus is phosphorus 31. Phosphorus 30, in short, is a radioactive isotope with a brief lifetime that exists today only when it is produced artificially; in fact, it was the first such isotope made by man. The Joliot-Curies received the Nobel Prize in chemistry in 1935 for their discovery of artificial radioactivity.

The unstable phosphorus 30 that the Joliot-Curies had produced by bombarding aluminum quickly broke down by emitting positrons. Since the positron, like the electron, has practically no mass, this emission did not change the mass number of the nucleus. However, the loss of one positive charge did reduce its atomic number by one, so that it was converted from phosphorus to silicon.

Where does the positron come from? Are positrons among the components of the nucleus? The answer is no. What happens is that a proton within the nucleus changes to a neutron by shedding its positive charge, which is released in the form of a speeding positron.

Now the emission of beta particles—the puzzle we encountered earlier in the chapter—can be explained. This comes about as the result of a process just the reverse of the decay of a proton into a neutron. That is, a neutron changes into a proton. The proton-to-neutron change releases a positron and, to maintain the symmetry, the neutron-to-proton change releases an electron (the beta particle). The release of a negative charge is equivalent to the gain of a positive charge and accounts for the formation of a positively charged proton from an uncharged neutron. But how does the uncharged neutron manage to dig up a negative charge and send it flying outward?

Actually, if it were just a negative charge, the neutron could not do so. Two centuries of experience have taught physicists that neither a negative electric charge nor a positive electric charge can be created out of nothing. Neither can either type of charge be destroyed. This is the law of "conservation of electric charge."

However, a neutron does not create only an electron in the process of producing a beta particle; it creates a proton as well. The uncharged neutron disappears, leaving in its place a positively charged proton and a negatively charged electron. The two new particles, *taken together*, have an over-all electric charge of zero. No *net* charge has been created. Similarly, when a positron and electron meet and engage in mutual annihilation, the charge of the positron and electron, *taken together*, is zero to begin with.

When a proton emits a positron and changes into a neutron, the original particle (the proton) is positively charged, and the final particles (the neutron and positron), taken together, have a positive charge.

It is also possible for a nucleus to absorb an electron. When this happens, a proton within the nucleus changes to a neutron. An electron plus a proton (which, taken together, have a charge of zero) form a neutron, which has a zero charge. The electron captured is from the innermost electron shell of the atom, since the electrons of that shell are closest to the nucleus and most easily gathered in. The innermost shell is the K-shell (see page 245) and the process is therefore called "K-capture." An electron from the L-shell then drops into the vacant spot, and an X-ray is emitted. It is by these X-rays that K-capture can be detected. This was first accomplished in 1938 by the American physicist Luis W. Alvarez. Ordinary nuclear reactions involving the nucleus alone are usually not affected by chemical change, which affects electrons only. Since K-capture affects electrons as well as nuclei, the chance of its occurring can be somewhat altered as a result of chemical change.

All of these particle interactions satisfy the law of conservation of electric charge and must also satisfy quite a number of other conservation laws. Any particle interaction that violates none of the conservation laws will eventually occur, physicists suspect, and an observer with the proper tools and proper patience will detect it. Those events that violate a conservation law are "forbidden" and will not take place. Nevertheless, physicists are occasionally surprised to find that what had seemed a conservation law is not as rigorous or as universal as had been thought. We shall come across examples of that.

Once the Joliot-Curies had created the first artificial radioactive isotope, physicists proceeded merrily to produce whole tribes of them. In fact, radioactive varieties of every single element in the periodic table have now been formed in the laboratory. In the modern periodic table, each element is really a family, with stable and unstable members, some found in nature, some only in the laboratory.

For instance, hydrogen comes in three varieties. First there is ordinary hydrogen, containing a single proton. In 1932, the chemist Harold Urey succeeded in isolating a second. He did it by slowly evaporating a large quantity of water, working on the theory that he would be left in the end with a concentration of the heavier form of hydrogen that was suspected to exist. Sure enough, when he examined the last few drops of unevaporated water spectroscopically, he found a faint line in the spectrum in exactly the position predicted for "heavy hydrogen."

Heavy hydrogen's nucleus is made up of one proton and one neutron. Having a mass number of two, the isotope is hydrogen 2. Urey named the atom "deuterium," from a Greek word meaning "second," and the nucleus a "deuteron." A water molecule containing deuterium is called "heavy water." Because deuterium has twice the mass of ordinary hydrogen, heavy water has higher boiling and freezing points than ordinary water. Whereas ordinary water boils at 100° C., and freezes at 0° C., heavy water boils at 101.42° C. and freezes at 3.79° C. Deuterium itself has a boiling point of 23.7° K. as compared with 20.4° K. for ordinary hydrogen. Deuterium occurs in nature in the ratio of one part to 6,000 parts of ordinary hydrogen. For his discovery of deuterium, Urey received the Nobel Prize in chemistry in 1934.

The deuteron turned out to be a valuable particle for bombarding nuclei. In 1934, the Australian physicist Marcus Lawrence Elwin Oliphant and the Austrian chemist Paul Harteck, attacking deuterium itself with deuterons, produced a third form of hydrogen, made up of one proton and two neutrons. The reaction went:

hydrogen 2 + hydrogen 2 → hydrogen 3 + hydrogen 1

The new "superheavy" hydrogen was named "tritium," from the Greek word for "third," and its nucleus is a "triton." Its boiling point is 25.0° K., and its melting point 20.5° K. Pure tritium oxide ("superheavy water") has been prepared, and its melting point is 4.5° C. Tritium is radioactive and breaks down comparatively rapidly. It exists in nature, being formed as one of the products of the bombardment of the atmosphere by cosmic rays. In breaking down, it emits an electron and changes to helium 3, a stable but rare isotope of helium.

Helium 3 differs from ordinary helium 4 in some interesting ways, particularly in the fact that it does not display the same properties of superconductivity and superfluidity in the liquid state, discussed in the preceding chapter. Atmospheric helium contains only 0.00013 per cent of helium 3, all originating, no doubt, from the breakdown of tritium. (Tritium, because it is unstable, is even rarer. It is estimated that only three and one-half pounds exist all told in the atmosphere and oceans.) The helium-3 content of helium obtained in natural gas wells, where cosmic rays have had less opportunity to form tritium, is even smaller in percentage.

These two isotopes, helium 3 and helium 4, are not the only heliums.

Nuclei of ordinary hydrogen, deuterium, and tritium.

Physicists have created two radioactive forms: helium 5, one of the most unstable nuclei known, and helium 6, also very unstable.

And so it goes. By now the list of known isotopes has grown to about 1,400 altogether. Over 1100 of these are radioactive and many of them have been created by new forms of atomic artillery far more potent than the alpha particles from radioactive sources which were the only projectiles at the disposal of Rutherford and the Joliot-Curies.

The sort of experiment performed by the Joliot-Curies in the early 1930's seemed a matter of the scientific ivory-tower at the time, but it has come to have a highly practical application. Suppose a set of atoms of one kind, or of many, are bombarded with neutrons. A certain percentage of each kind of atom will absorb a neutron, and a radioactive atom will generally result. This radioactive element will decay, giving off subatomic radiation in the form of particles or gamma rays.

Every different type of atom will absorb neutrons to form a different type of radioactive atom, giving off different and characteristic radiation. The radiation can be detected with great delicacy. From its type and from the rate at which its production declines, the radioactive atom giving it off can be identified and, therefore, so can the original atom before it absorbed a neutron. Substances can be analyzed in this fashion ("neutron-activation analysis") with unprecedented precision: amounts as small as a trillionth of a gram of a particular nuclide are detectable.

Neutron-activation analysis can be used to determine delicate differences in impurities-content in samples of particular pigments from different centuries and in this way can determine the authenticity of a supposedly old painting, using only the barest fragment of its pigment.

Other delicate decisions of this sort can be made: even hair from Napoleon's century-and-a-half-old corpse was studied and found to contain suspicious quantities of arsenic (which perhaps he took medicinally).

Particle Accelerators

Dirac had predicted not only an antielectron (the positron) but also an antiproton. But to produce an antiproton would take vastly more energy. The energy needed was proportional to the mass of the particle. Since the proton was 1,836 times as massive as the electron, the formation of an antiproton called for at least 1,836 times as much energy as the formation of a positron. The feat had to wait for the development of a device for accelerating subatomic particles to sufficiently high energies.

At the time of Dirac's prediction, the first steps in this direction had just been taken. In 1928, the English physicists John D. Cockcroft and Ernest Walton, working in Rutherford's laboratory, developed a "voltage multiplier," a device for building up electric potential, which could drive the charged proton up to an energy of nearly 400,000 electron volts. (One electron volt is equal to the energy developed by an electron accelerated across an electric field with a potential of one volt.) With protons accelerated in this machine they were able to break up the lithium nucleus, and for this work, they were awarded the Nobel Prize for physics in 1951.

Meanwhile the American physicist Robert Jemison Van de Graaff was creating another type of accelerating machine. Essentially, it operated by separating electrons from protons and depositing them at opposite ends of the apparatus by means of a moving belt. In this way the "Van de Graaff electrostatic generator" developed a very high electric potential between the opposite ends; Van de Graaff got it up to 8 million volts. Electrostatic generators can easily accelerate protons to a speed amounting to 24 million electron volts (physicists now invariably abbreviate million electron volts to "Mev").

The dramatic pictures of the Van de Graaff electrostatic generator producing huge sparks caught the popular imagination and introduced the public to "atom smashers." It was popularly viewed as a device to produce "man-made lightning," although, of course, it was much more than that. (A generator designed to produce artificial lightning and nothing more had actually been built in 1922 by the German-American electrical engineer Charles Proteus Steinmetz.)

The energy that can be reached in such a machine is restricted by practical limits on the attainable potential. However, another scheme for accelerating particles shortly made its appearance. Suppose that, instead of firing particles with one big shot, you accelerated them with a series of small pushes. If each successive push was timed just right, it would increase the speed each time, just as pushes on a child's swing will send it higher and higher if they are applied "in phase" with the swing's oscillations.

This idea gave birth, in 1931, to the "linear accelerator." The particles are driven down a tube divided into sections. The driving force is an alternating electric field, so managed that as the particles enter each successive section, they get another push. Since the particles speed up as they go along, each section must be longer than the one before, so that the particles will take the same time to get through it and will be in phase with the timing of the pushes.

It is not easy to keep the timing just right, and anyway there is a limit to how long a tube you can make practicably, so the linear accelerator did not catch on in the 1930's. One of the things that pushed it into the background was that Ernest Orlando Lawrence of the University of California conceived a better idea.

Instead of driving the particles down a straight tube, why not whirl them around in a circular path? A magnet could bend them in such a path. Each time they completed a half circle, they would be given a kick by the alternating field, and in this setup the timing would not be so difficult to control. As the particles speeded up, their path would be bent less sharply by the magnet, so they would move in ever wider circles and perhaps take the same time for each round trip. At the end of their spiraling flight, the particles would emerge from the circular chamber

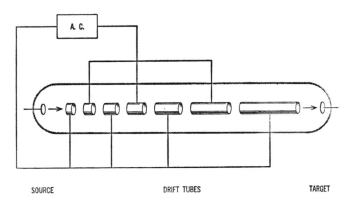

SOURCE   DRIFT TUBES   TARGET

Principle of the linear accelerator. A high-frequency alternating charge alternately pushes and pulls the charged particles in the successive drive tubes, accelerating them in one direction.

(actually divided into semicircular halves, called "dees") and attack their target.

Lawrence's compact new device was named the "cyclotron." His first model, less than a foot in diameter, could accelerate protons to energies of nearly 1.25 Mev. By 1939 the University of California had a cyclotron, with magnets five feet across, capable of raising particles to some 20 Mev, twice the speed of the most energetic alpha particles emitted by radioactive sources. In that year Lawrence received the Nobel Prize in physics for his invention.

The cyclotron itself had to stop at about 20 Mev, because at that energy the particles were traveling so fast that the mass increase with velocity—an effect predicted by Einstein's Theory of Relativity—became appreciable. This increase in mass caused the particles to start lagging and falling out of phase with the electrical kicks. But there was a cure for this, and it was worked out in 1945 independently by the Soviet physicist Vladimir Iosifovich Veksler and the California physicist Edwin

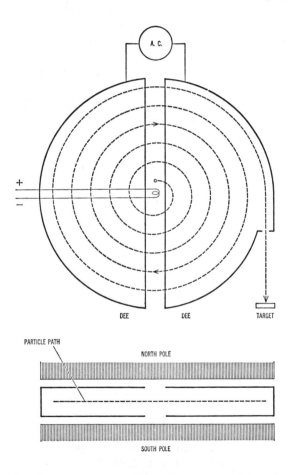

Principle of the cyclotron, shown in top view (*above*) and side view (*below*). Particles injected from the source are given a kick in each dee by the alternating charge and are bent in their spiral path by a magnet.

Mattison McMillan. The cure was simply to synchronize the alternations of the electric field with the increase in mass of the particles. This modification of the cyclotron was called the "synchrocyclotron." By 1946 the University of California had built one which accelerated particles to energies of 200 to 400 Mev. Later larger synchrocyclotrons in the United States and in the Soviet Union raised the energies to 700 to 800 Mev.

Meanwhile the acceleration of electrons had been getting separate attention. To be useful in smashing atoms, the light electrons had to be raised to much higher speeds than protons (just as a ping-pong ball has to be moving much faster than a golf ball to do as much damage). The cyclotron would not work for electrons, because at the high velocities needed to make the electrons effective, their increase in mass was too great. In 1940 the American physicist Donald William Kerst designed an electron-accelerating device which balanced the increasing mass with an electric field of increasing strength. The electrons were kept in the same circular path instead of spiraling outward. This instrument was named the "betatron," after beta particles. Betatrons now generate electron velocities up to 340 Mev.

They have been joined by another instrument of slightly different design called the "electron synchrotron." The first of these was built in England in 1946 by F. K. Goward and D. E. Barnes. These raise electron energies to the 1,000 Mev mark, but cannot go higher because electrons moving in a circle radiate energy at increasing rates as velocity is increased. This radiation produced by an accelerating particle is called "Bremsstrahlung," a German word meaning "braking radiation."

Taking a leaf from the betatron and electron synchrotron, physicists working with protons began about 1947 to build "proton synchrotrons," which likewise kept their particles in a single circular path. This helped save on weight. Where particles move in outwardly spiraling paths, a magnet must extend the entire width of the spiral to keep the magnetic force uniform throughout. With the path held in a circle, the magnet need be only large enough to cover a narrow area.

Because the more massive proton does not lose energy with motion in a circular path as rapidly as does the electron, physicists set out to surpass the 1,000-Mev mark with a proton synchroton. This value of 1,000 Mev is equal to a billion electron volts—abbreviated to Bev. (In Great Britain a billion is a million million, so Bev does not mean the same thing as in the United States; for 1,000 Mev the British use the shorthand Gev, the G coming from "giga," Greek for "giant.")

In 1952, the Brookhaven National Laboratory on Long Island completed a proton synchroton that reached 2 to 3 Bev. They called it the "cosmotron," because it had arrived at the main energy range of particles in the cosmic rays. Two years later the University of California brought in its "Bevatron," capable of producing particles of between 5 and 6 Bev. Then, in 1957, the Soviet Union announced its "phasotron" had got to 10 Bev.

But by now these machines seem puny in comparison with accelerators of a newer type, called the "strong-focusing synchrotron." The limitation on the bevatron type is that particles in the stream fly off into the walls of the channel in which they travel. The new type counteracts this tendency by means of alternating magnetic fields of different shape which keep focusing the particles in a narrow stream. The idea was first suggested by Christofilos, whose "amateur" abilities outshone the professionals here as well as in the case of the Christofilos effect. This, incidentally, further decreased the size of the magnet required for the energy levels attained. Where particle energy was increased fiftyfold, the weight of the magnet involved was less than doubled.

In November 1959, the European Committee for Nuclear Research (CERN), a cooperative agency of twelve nations, completed in Geneva a strong-focusing synchrotron which reached 24 Bev and produced large pulses of particles (containing 10 billion protons) every three seconds. This synchrotron is nearly three city blocks in diameter, and one round trip through it is two fifths of a mile. In the three-second period during which the pulse builds up, the protons travel half a million times around that track. The instrument has a magnet weighing 3,500 tons and costs 30 million dollars.

It is not, however, the last word. Brookhaven has completed an even larger machine that has moved well above the 30-Bev mark, and the Soviet Union now has one that is over a mile in diameter and that attained over 70 Bev when put into operation in 1967. American physicists are supervising the construction of one that will be three miles in diameter and will attain 300 Bev, while others of 1,000 Bev are dreamed of.

The linear accelerator, or "linac," has also undergone a revival. Improvements in technique have removed the difficulties that plagued the early models. For extremely high energies, a linear accelerator has some advantages over the cyclic type. Since electrons do not lose energy when traveling in a straight line, a linac can accelerate electrons more powerfully and focus beams on targets more sharply. Stanford University has built a linear accelerator two miles long which can reach energies of perhaps 45 Bev.

Nor is sheer size the only answer to greater power. The notion has been repeatedly broached of having two accelerators in tandem so that a stream of energetic particles collides head-on with another stream moving in the opposite direction. This will quadruple the energies involved over the collision of one such stream with a stationary object. Such a tandem-accelerator may be the next step.

With merely the Bevatron, man at last came within reach of creating the antiproton. The California physicists set out deliberately to produce and detect it. In 1955, Owen Chamberlain and Emilio G. Segrè, after bombarding copper with protons of 6.2 Bev hour after hour, definitely caught the antiproton—in fact, sixty of them. It was far from easy to identify them. For every antiproton produced, 40,000 particles of other types came into existence. But by an elaborate system of detectors, so designed and arranged that only an antiproton could touch all the bases, they recognized the particle beyond question. For their achievement, Chamberlain and Segrè received the Nobel Prize in physics in 1959.

The antiproton is as evanescent as the positron—at least in our universe. Within a tiny fraction of a second after it is created, the particle is snatched up by some normal, positively charged nucleus. There the antiproton and one of the protons of the nucleus annihilate each other, turning into energy and minor particles. In 1965, enough energy was concentrated to reverse the process and produce a proton-antiproton pair.

Once in a while a proton and an antiproton have only a near collision instead of a direct one. When that happens, they mutually neutralize their respective charges. The proton is converted to a neutron, which is fair enough. But the antiproton becomes an "antineutron"! What can an "antineutron" be? The positron is the opposite of the electron by virtue of its opposite charge, and the antiproton is likewise "anti" by virtue of its charge. But what gives the uncharged antineutron the quality of oppositeness?

Here we have to digress a little into the subject of the spin of particles. A particle may usually be viewed as spinning on its axis, like a top or the earth or the sun or our Galaxy or, for all we know, the universe itself. This particle spin was first suggested in 1925 by the Dutch physicists George Eugene Uhlenbeck and Samuel Abraham Goudsmit. In spinning, the particle generates a tiny magnetic field; such fields have been measured and thoroughly explored, notably by the German physicist Otto Stern and the American physicist Isidor Isaac Rabi who received the Nobel Prizes in physics in 1943 and 1944, respectively, for their work on this phenomenon.

Spin is measured in such a way that the spin of an electron or a proton is said to be equal to one-half. When this is doubled, it becomes an odd integer (1). The energies of particles whose spin, on being doubled, becomes an odd integer in this fashion can be dealt with according to a system of rules worked out independently, in 1926, by Fermi and Dirac. These rules make up the "Fermi-Dirac statistics," and particles (such as the electron and proton) that obey these are called "fermions." The neutron is also a fermion.

There also exist particles whose spin, when doubled, is an even number. Their energies can be dealt with by another set of rules devised by Einstein and by the Indian physicist S. N. Bose. Particles that

Linear accelerator at the University of California Radiation Laboratory in Berkeley.

The Soviet Union's phasotron, which accelerates particles to 10 billion electron volts.

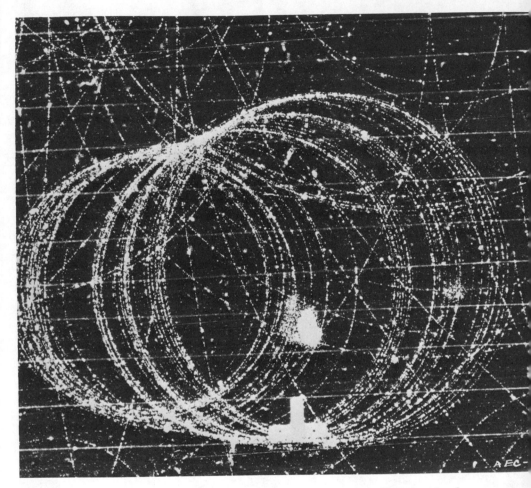

Tracks of electrons and positrons formed in a bubble chamber by high-energy gamma rays. The circular pattern was made by an electron revolving in the magnetic field.

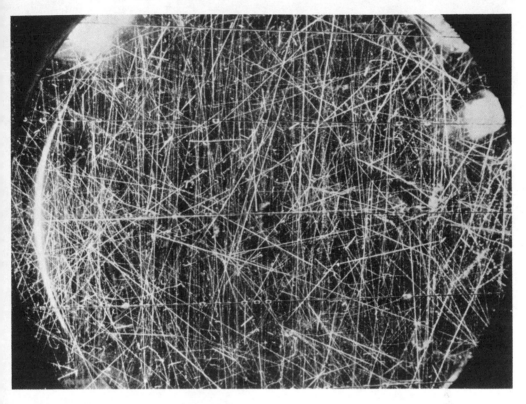

Tracks of nuclear particles produced by 1,2 Bev protons from the Cosmotron smashing atoms in a brass target.

Brookhaven's synchrotron, which went into operation in the early 1960's and produced a proton beam with an energy of more than 30 Bev. Using the "strong-focusing" principle, it is called the Alternating Gradient Synchrotron (AGS). The pipe running diagonally from the upper left to the lower right of the picture is the channel through which the protons are injected from a linear accelerator into the circular synchrotron.

follow the "Bose-Einstein statistics" are "bosons." The alpha particle, for instance, is a boson.

These classes of particles have different properties. For instance, the Pauli exclusion principle (see Chapter 5) applies, not only to electrons, but to all fermions. It does not, however, apply to bosons.

It is easy to understand how a charged particle sets up a magnetic field, but not so easy to see why the uncharged neutron should. Yet it unquestionably does. The most direct evidence of this is that when a neutron beam strikes magnetized iron, it behaves differently from the way it does when the iron is not magnetized. The neutron's magnetism remains a mystery; physicists suspect that the neutron contains positive and negative charges which add up to zero, but which somehow manage to set up a magnetic field when the particle spins.

In any case, the spin of the neutron gives us the answer to the question as to what the antineutron is. It is simply a neutron with its spin direction reversed; its south magnetic pole, say, is up instead of down. Actually the proton and antiproton and the electron and positron show exactly the same pole-reversed phenomenon.

Antiparticles can undoubtedly combine to form "antimatter," as ordinary particles form ordinary matter. The first actual example of antimatter was produced at Brookhaven in 1965. There the bombardment of a beryllium target with 7 Bev protons produced combinations of antiprotons and antineutrons, something that was an "antideuteron." "Antihelium-3" has since been produced and undoubtedly, if enough pains are taken, still more complicated antinuclei can be formed. The principle is clear, however, and no physicist doubts it. Antimatter can exist.

But does it exist in actuality? Are there masses of antimatter in the universe? If there were, they would not betray themselves from a distance. Their gravitational effects and the light they produce would be exactly like that of ordinary matter. If, however, they encountered ordinary matter, the massive annihilation reactions that result ought to be most noticeable. Astronomers have therefore taken to looking speculatively at distant galaxies to see if any unusual activity might betray matter-antimatter interactions. What about the exploding galaxies? What about Messier 87 with a bright jet of luminosity sticking out of its globular main body? What about the enormous energies pouring out of quasars? For that matter, what about the meteoric strike in Siberia in 1908 and its destructive consequences? Was it just a meteor or was its destructiveness the result of the fact that it was a piece (and perhaps a not-very-large piece) of antimatter?

There are some questions that are even more fundamental. Why should there be some chunks of antimatter in a universe composed mainly of matter? Since matter and antimatter are equivalent in all respects but that of electromagnetic oppositeness, any force which would create one would have to create the other, and the universe should be made up of equal quantities of each. If they were intimately mixed, particles of matter and antimatter would annihilate each other, but what if some effect served to separate them after their creation?

In this connection, there is a suggestion made by the Swedish physicist Oskar Klein which has been popularized by his compatriot Hannes Alfven. Suppose the universe came into being in the form of a very rarefied collection of particles spread out over a sphere a trillion light-years or more in diameter. The particles might indeed be half ordinary ones and half antiparticles, but under such conditions of rarefaction, they would move freely and hardly ever collide and annihilate one another.

If there were magnetic fields in this thin-universe, particles in a particular region would tend to curve in one direction and antiparticles in the other, so that they would tend to separate. Eventually, after separation, they would collect into galaxies and antigalaxies. If sizable chunks of matter and antimatter met and began to interact—either while the galaxies were forming or after they had formed—the radiation liberated at their volumes of junction would, by its pressure, drive them apart. Finally, then, there would be evenly spread out galaxies and antigalaxies.

Mutual gravitational attraction would now draw them together in a "contracting universe." The further the universe contracted, the greater the chance of galaxy-antigalaxy collisions, and the greater the energetic radiation produced. Finally, when the universe shrank to a diameter of a billion light-years or so, the radiation produced would have sufficient pressure to explode it and drive the galaxies and anti-galaxies forcibly apart. It is this expansion stage in which we now find ourselves, according to Klein and Alfven, and what we interpret as evidence in favor of the "big bang" is evidence in favor of that crucial moment when the universe-antiuniverse produces enough energy to blow itself up, so to speak.

If this is so, then half the galaxies we can see are antigalaxies. But which half? And how can we tell? There are no answers to that—so far.

A consideration of antimatter on a somewhat less majestic scale, brings up the question of cosmic rays again. Most of the cosmic-ray particles have energies between 1 and 10 Bev. This might be accounted

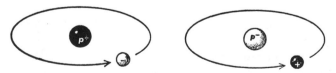

An atom of hydrogen and an atom of its antimatter counter-part, consisting of an antiproton and a positron.

for by matter-antimatter interaction, but a few cosmic particles run much higher: 20 Bev, 30 Bev, 40 Bev. Physicists at the Massachusetts Institute of Technology have even detected some with the colossal energy of 20 billion Bev! Numbers such as this are more than the mind can grasp, but we may get some idea of what that energy means when we calculate that the amount of energy represented by 10 trillion Bev would be enough to enable a single submicroscopic particle to raise a one-ton weight two inches.

Ever since cosmic rays were discovered, people have wondered where they came from and how they arise. The simplest concept is that somewhere in the Galaxy, perhaps in our sun, perhaps farther away, there are nuclear reactions going on which shoot forth particles with the huge energies we find them possessing. Indeed, bursts of mild cosmic rays occur every other year or so (as was first discovered in 1942) in connection with flares from the sun. What then of such sources as supernovae, pulsars, and quasars? But there is no known nuclear reaction that could produce anything like billions and trillions of Bev. The most energetic one we can conceive of would be the mutual annihilation of heavy nuclei of matter and antimatter, and this would liberate at most 250 Bev.

The alternative is to suppose, as Fermi did, that some force in space accelerates the cosmic particles. They may come originally with moderate energies from explosions such as supernovae and gradually be accelerated as they travel through space. The most popular theory at present is that they are accelerated by cosmic magnetic fields, acting like gigantic synchrotrons. Magnetic fields do exist in space, and our Galaxy as a whole is thought to possess one, although this can at best be but 1/20,000 as intense as the magnetic field associated with the earth.

Traveling through this field, the cosmic particles would be slowly accelerated in a curved path. As they gained energy, their paths would swing out wider and wider until the most energetic ones would whip right out of the Galaxy. Although most of the particles would never reach this escape trajectory, because they would lose energy by collisions with other particles or with large bodies, some would. Indeed, the most energetic cosmic particles that reach us may be passing through our Galaxy after having been hurled out of other galaxies in this fashion.

More New Particles

The discovery of the antiparticles did not really disturb physicists; on the contrary, it was a pleasing confirmation of the symmetry of the universe. What did disturb them was a quick succession of discoveries showing that the proton, the electron, and the neutron were not the only "elementary particles" they had to worry about.

The first of these complications had arisen even before the neutron was discovered. It had to do with the emission of beta particles by radioactive nuclei. The particle emitted by a radioactive nucleus generally carries a considerable amount of energy. Where does the energy come from? It is created by conversion of a little of the nucleus' mass into energy; in other words, the nucleus always loses a little mass in the act of expelling the particle. Now physicists had long been troubled by the fact that often the beta particle emitted in a nucleus' decay did not carry enough energy to account for the amount of mass lost by the nucleus. In fact, the electrons weren't all equally deficient. They emerged with a wide spectrum of energies, the maximum (attained by very few electrons) being almost right, but all the others falling short to a smaller or greater degree. Nor was this a necessary concomitant of subatomic particle-emission. Alpha particles emitted by a particular nuclide possessed equal energies in expected quantities. What, then, was wrong with beta-particle emission? What had happened to the missing energy?

Lise Meitner, in 1922, was the first to ask this question with suitable urgency and by 1930, Niels Bohr, for one, was ready to abandon the great principle of conservation of energy, at least as far as it applied to subatomic particles. In 1931, however, Wolfgang Pauli, in order to save conservation of energy, suggested a solution to the riddle of the missing energy. His solution was very simple: another particle carrying the missing energy came out of the nucleus along with the beta particle. This mysterious second particle had rather strange properties. It has no charge and no mass; all it had, as it sped along at the velocity of light, was a certain amount of energy. It looked, in fact, like a fictional item created just to balance the energy books.

And yet, no sooner had it been proposed than physicists were sure that the particle existed. When the neutron was discovered and found to break down into a proton, releasing an electron which, as in beta decay, also carried a deficiency of energy, they were still surer. Enrico Fermi in Italy gave the putative particle a name—"neutrino," Italian for "little neutral one."

The neutron furnished physicists with another piece of evidence for the existence of the neutrino. As I have mentioned, almost every particle has a spin. The amount of spin is expressed in multiples of one half, plus or minus, depending on the direction of the spin. Now the proton, the neutron, and the electron have each one a spin of one half. If, then, the neutron, with spin one half, gives rise to a proton and an electron, each with spin one half, what happens to the law of conservation of angular momentum? There is something wrong here. The proton and the electron may total their spins to one (if both spin in the same direction) or to zero (if their spins are opposite), but any way you slice it their spins cannot add up to one half. Again, however, the neutrino comes to the rescue. Let the spin of the neutron be $+\frac{1}{2}$. Let the proton's spin be $+ \frac{1}{2}$ and the electron's $- \frac{1}{2}$, for a net of 0. Now give the neutrino the spin $+ \frac{1}{2}$, and the books are neatly balanced.

$$+ \tfrac{1}{2} \text{ (n)} = + \tfrac{1}{2} \text{ (p)} - \tfrac{1}{2} \text{ (e)} + \tfrac{1}{2} \text{ (neutrino)}.$$

There is still some more balancing to do. A single particle (the neutron) has formed two particles (the proton and the electron), and if we include the neutrino, actually three particles. It seems more reasonable to suppose that the neutron is converted into two particles and an antiparticle, or a net of one particle. In other words, what we really need to balance is not a neutrino but an antineutrino.

The neutrino itself would arise from the conversion of a proton into a neutron. There the products would be a neutron (particle), a positron (antiparticle), and a neutrino (particle). This, too, balances the books.

The most important proton-to-neutron conversions are those involved in the nuclear reactions that go on in the sun and other stars. Stars therefore emit fast floods of neutrinos, and it is estimated that perhaps 6 to 8 per cent of their energy is carried off in this way. This, however, is only true for such stars as our sun. In 1961, the American physicist Hong Yee Chiu suggested that, as the central temperatures of a star rise, additional neutrino-producing reactions become important. As a star progresses in its evolutionary course toward a hotter and hotter core (see Chapter 1), a larger and larger proportion of its energy is carried off by neutrinos.

There is crucial importance in this. The ordinary method of transmitting energy, by photons, is slow. Photons interact with matter, and they make their way out from the sun's core to its surface only after uncounted myriads of absorptions and re-emissions. Consequently, although the sun's central temperature is 15,000,000° C., its surface is only 6,000° C. The substance of the sun is a good heat insulator.

Neutrinos, however, virtually do not interact with matter. It has been calculated that the average neutrino could pass through 100 light-years of solid lead with only a 50 per cent chance of being absorbed. This means that any neutrinos formed in the sun's core leave at once and at the speed of light, reaching the sun's surface, without interference, in less than three seconds and speeding off. (Any that move in our direction pass through us without affecting us in any way. This is true, day or night, for at night, when the bulk of the earth is between ourselves and the sun, the neutrinos can pass through the earth and ourselves as easily as through ourselves alone.)

By the time a central temperature of 6,000,000,000° K. is reached, Chiu calculates, most of the star's energy is being pumped into neutrinos. The neutrinos leave at once, carrying the energy with them, and the sun's center cools drastically. It is this, perhaps, which leads to the catastrophic contraction that then makes itself evident as a supernova.

Antineutrinos are produced in any neutron-to-proton conversion, but these do not go on (as far as is known) on the vast scale that leads to such floods of neutrinos from every star. The most important sources of antineutrinos are from natural radioactivity and uranium fission (which I shall discuss in more detail in Chapter 9).

Naturally physicists could not rest content until they had actually tracked down the neutrino; scientists are never happy to accept phenomena or laws of nature entirely on faith (as concepts such as the "soul" must be). But how detect an entity as nebulous as the neutrino—an object with no mass, no charge, and practically no propensity to interact with ordinary matter?

Still, there was some slight hope. Although the probability of a neutrino reacting with any particle is exceedingly small, it is not quite zero. To be unaffected in passing through one-hundred light-years of lead is just a measure of the average, but there will be some neutrinos that react with a particle before they go that far, and a few—an almost unimaginably small proportion of the total number—that will be stopped within the equivalent of 1/10 inch of lead.

In 1953, a group of physicists led by Clyde L. Cowan and Frederick Reines of the Los Alamos Scientific Laboratory set out to try the next-to-impossible. They erected their apparatus for detecting neutrinos next to a large fission reactor of the Atomic Energy Commission on the Savannah River in Georgia. The reactor would furnish streams of neutrons, which, hopefully, would release floods of antineutrinos. To catch them, the experimenter used large tanks of water. The plan was to let the antineutrinos bombard the protons (hydrogen nuclei) in the water and detect the results of the capture of an antineutrino by a proton.

What would happen? When a neutron breaks down, it yields a proton, an electron, and an antineutrino. Now a proton's absorption of an antineutrino should produce essentially the reverse. That is to say, the proton should be converted to a neutron, emitting a positron in the process. So there were two things to be looked for: (1) the creation of neutrons, and (2) the creation of positrons. The neutrons could be detected by dissolving a cadmium compound in the water, for when cadmium absorbs neutrons, it emits gamma rays of a certain characteristic energy. And the positrons could be identified by their annihilating interaction with electrons, which would yield certain other gamma rays. If the experimenters' instruments detected gamma rays of exactly these two telltale energies and separated by the proper time interval, they could be certain that they had caught antineutrinos.

The experimenters arranged their ingenious detection devices, waited patiently, and, in 1956, exactly a quarter century after Pauli's invention of the particle, they finally trapped the antineutrino. The newspapers and even some learned journals called it simply the "neutrino."

To get the real neutrino, we need some source that is rich in neutrinos. The obvious one is the sun. What system can be used to detect the neutrino as opposed to the antineutrino? One possibility (following a suggestion of the Italian physicist Bruno Pontecorvo) begins with chlorine-37, which makes up about ¼ of all chlorine atoms. Its nucleus contains 17 protons and 20 neutrons. If one of those neutrons absorbs a neutrino, it becomes a proton (and emits an electron). The nucleus will then have 18 protons and 19 neutrons and will be argon-37.

To form a sizable target of chlorine-neutrons, one might use liquid chlorine, but that is a very corrosive and toxic substance, and to keep it liquid would present a problem in refrigeration. Instead, chlorine-containing organic compounds can be used; one called tetrachloroethylene is a good one for the purpose.

The American physicist Raymond R. Davis made use of such a neutrino trap in 1956 to show that there really was a difference between the neutrino and the antineutrino. Assuming the two particles were different, the trap would detect only neutrinos and not antineutrinos. When it was set up near a fission reactor in 1956 under conditions where it would certainly detect antineutrinos (if antineutrinos were identical to neutrinos), it did *not* detect them.

The next step was to try to detect neutrinos from the sun. A huge tank containing 100,000 gallons of tetrachloroethylene was used for the purpose. It was set up in a deep mine in South Dakota. There was enough earth above it to absorb any particles emerging from the sun except neutrinos. (Consequently, we have the odd situation that in order to study the sun we must burrow deep, deep into the bowels of the earth.) The tank is then exposed to the solar neutrinos for several months to allow enough argon-37 to accumulate to be detectable. The tank is then flushed with helium for 22 hours and the tiny quantity of argon-37 in the helium gas determined. By 1968, solar neutrinos were indeed detected, but in less than half the amounts expected from current theories as to what is going on inside the sun. However, the experimental techniques involved here are fantastically difficult and it is, as yet, early days.

Our list of particles has grown, then, to eight: proton, neutron, electron, neutrino, and their respective antiparticles. This, however, did not exhaust the list. Further particles began to seem necessary to physicists if they were to explain how the particles in the nucleus hung together.

Ordinary attractions between protons and electrons, between one atom and another, between one molecule and another, could be explained by electromagnetic forces—the mutual attraction of opposite electric charges. This would not suffice for the nucleus, where the only charged particles present were protons. Indeed, by electromagnetic reasoning, we would suppose that the protons, all positively charged, should repel one another violently and that any atomic nucleus should explode with shattering force the instant it was formed (if it ever could be formed in the first place).

Clearly, some other force must be involved, something much stronger than the electromagnetic force and capable of overpowering it. The superior strength of this "nuclear force" can be easily demonstrated by the following consideration. The atoms of a strongly bound molecule, such as that of carbon monoxide, can be pried apart by the application of only eleven electron volts of energy. That quantity of energy suffices to handle a strong manifestation of electromagnetic force.

On the other hand, the proton and neutron making up a deuteron, one of the most weakly bound of all nuclei, require 2 million electron volts for disruption. Making allowance for the fact that particles within the nucleus are much closer to one another than atoms within a molecule, it is still fair to conclude that the nuclear force is 130 times as strong as the electromagnetic force.

But what is the nature of this nuclear force? The first fruitful lead came in 1932 when Werner Heisenberg suggested that the protons were held together by "exchange forces." He pictured the protons and neutrons in the nucleus as continually interchanging identity, so that any given particle was first a proton, then a neutron, then a proton, and so on. This might keep the nucleus stable in the same way that you might be able to hold a hot potato by tossing it quickly from hand to hand. Before a proton could "realize" (so to speak) that it was a proton and try to flee its neighbor protons, it had become a neutron and could stay where it was. Naturally it could get away with this only if the changes took place exceedingly quickly, say within a trillionth of a trillionth of a second.

Another way of looking at it is to imagine two particles, exchanging a third. Each time particle A emits the exchange-particle it moves backward to conserve momentum. Each time particle B accepts the exchange-particle it is pushed backward for the same reason. As the exchange-particle bounces back and forth, particles A and B move farther and farther apart so that they seem to experience a repulsion. If, on the other hand, the exchange-particle moves around boomerang-fashion, from the rear of particle A to the rear of particle B, then the two particles would be pushed closer together and seem to experience an attraction.

It would seem by Heisenberg's theory that all forces of attraction and repulsion would be the result of exchange particles. In the case of electromagnetic attraction and repulsion the exchange particle is the photon, which, as we shall see in the next chapter, is a massless particle associated with light and electromagnetic radiation generally. It can be argued that it is because the photon is massless that electromagnetic attraction and repulsion is long-range, lessening in intensity only as the square of the distance, and therefore important over interstellar and even intergalactic distances.

By this reasoning, the gravitational force, which also is long-range and also lessens in intensity as the square of the distance, should involve the continual exchange of massless particles. Physicists have named such a particle the "graviton."

The gravitational force is much, much weaker than the electromagnetic force. A proton and an electron attract each other gravitationally with only about $1/10^{39}$ as much force as they attract each other electromagnetically. The graviton must be correspondingly less energetic than the photon and must therefore be unimaginably difficult to detect.

Nevertheless, the American physicist Joseph Weber has been trying to detect the graviton since 1957. His most recent attempts have made use of a pair of aluminum cylinders 153 centimeters long and 66 centimeters wide, suspended by a wire in a vacuum chamber. The gravitons (which would be detected in wave-form) would displace those cylinders slightly, and a measuring system for detecting a displacement of a hundred-trillionth of a centimeter is used. The feeble waves of the gravitons, coming from deep in space, ought to wash over the entire planet, and cylinders separated by great distances ought to be affected simultaneously. In 1969, Weber announced he had detected the effects of gravitational waves. If so, the question is what—even out in space— could represent fluctuations in gravitational force sufficient to produce detectable waves? Are they events involving neutron stars, black holes, or what? We don't know.

But back to the nuclear force. Unlike the electromagnetic and the gravitational fields, it was short-range. Although extremely strong within the nucleus, it vanished almost completely outside the nucleus. For that reason, no massless exchange-particle would do.

In 1935, the Japanese physicist Hideki Yukawa mathematically analyzed the problem. An exchange particle possessing mass would produce a short-range force-field. The mass would be in inverse ratio to the range: the greater the mass, the shorter the range. It turned out that the mass of the appropriate particle lay somewhere between that of the proton and the electron; Yukawa estimated it to be between 200 and 300 times the mass of an electron.

Barely a year later, this very kind of particle was discovered. At the California Institute of Technology, Carl Anderson (the discoverer of the positron), investigating the tracks left by secondary cosmic rays, came across a short track that was more curved than a proton's and less curved than an electron's. In other words, the particle had an intermediate mass. Soon more such tracks were detected, and the particles were named "mesotrons," or "mesons" for short.

Eventually other particles in this intermediate mass range were discovered, and this first one was distinguished as the "mu meson" or the "muon." ("Mu" is one of the letters of the Greek alphabet; almost all of them have now been used in naming subatomic particles.) As in the case of the particles mentioned earlier, the muon comes in two varieties, a negative and a positive.

The negative muon, 206.77 times as massive as the electron (and therefore about one-ninth as massive as a proton) is the particle; the positive muon is the antiparticle. The negative muon and positive muon correspond to the electron and positron, respectively. Indeed, by 1960 it had become evident that the negative muon was identical with the electron in almost every way except mass. It was a "heavy electron." Similarly, the positive muon was a "heavy positron."

There is no explanation, so far, for the identity, but it carries through to the point where, as was discovered in 1953, negative muons can replace electrons in atoms to form "muonic atoms." Similarly, positive muons could replace positrons in antimatter.

Positive and negative muons will undergo mutual annihilation and may briefly circle about a mutual center of force before doing so— just as is true of positive and negative electrons. A more interesting situation, however, was discovered in 1960 by the American physicist Vernon Willard Hughes. He detected a system in which the electron circled a positive muon, a system he called "muonium." (A positron circling a negative muon would be "antimuonium.")

The muonium atom (if it may be called that) is quite analogous to hydrogen 1, in which an electron circles a positive proton, and the two are similar in many of their properties. Although muons and electrons seem to be identical except for mass, that mass-difference is enough to keep the electron and the positive muon from being true opposites, so that one will not annihilate the other. Muonium, therefore, doesn't have the kind of instability that positronium has. Muonium endures longer and would endure forever (if undisturbed from without) were it not for the fact that the muon itself does not. After two millionths of a second or so, the muon decays and the muonium atom ceases to exist.

Another similarity between muons and electrons is this: just as heavy particles may produce electrons plus antineutrinos (as when a neutron is converted to a proton) or positrons plus neutrinos (as when a proton is converted to a neutron) so heavy particles can interact to form negative muons plus antineutrinos or positive muons plus neutrinos. For years, physicists took it for granted that the neutrinos that accompanied electrons and positrons and those that accompanied negative and positive muons were identical. In 1962, however, it was learned that the neutrinos never crossed over, so to speak; the electron's neutrino was never involved in any interaction that would form a muon, and the muon's neutrino was never involved in any interaction that would form an electron or positron.

In short, physicists found themselves with two pairs of chargeless, massless particles, the electron's antineutrino and the positron's neutrino plus the negative muon's antineutrino and the positive muon's neutrino. What the difference between the two neutrinos and between the two antineutrinos might be is more than anyone can tell at the moment, but they are different.

The muons differ from the electron and positron in another respect, that of stability. The electron or positron, left to itself, will remain unchanged indefinitely. The muon is unstable, however, and breaks down after an average lifetime of a couple of millionths of a second. The negative muon breaks down to an electron (plus an antineutrino of the electron variety and a neutrino of the muon variety), while the positive muon does the same in reverse, producing a positron, an electron-neutrino, and a muon-antineutrino.

Since the muon is a kind of heavy electron, it cannot very well be the nuclear cement Yukawa was looking for. Electrons are not found within the nucleus and therefore neither should the muon. This was discovered to be true on a purely experimental basis, long before the near identity of muon and electron was suspected; muons simply showed no tendency to interact with nuclei. For a while, Yukawa's theory seemed to be tottering.

In 1947, however, the British physicist Cecil Frank Powell discovered another type of meson in cosmic-ray photographs. It was a little more massive than the muon and proved to possess about 273 times the mass of an electron. The new meson was named a "pi meson" or a "pion."

The pion was found to react strongly with nuclei and to be just the particle predicted by Yukawa. (Yukawa was awarded the Nobel Prize in physics in 1949, and Powell received it in 1950.) Indeed, there was a positive pion that acted as the exchange force between protons and neutrons, and there was a corresponding antiparticle, the negative pion, which performed a similar service for antiprotons and antineutrons. Both are even shorter lived than muons; after an average lifetime of about one fortieth of a microsecond, they break up into muons plus neutrinos of the muon variety. (And, of course, the muon breaks down further to electrons and additional neutrinos.) There is also a neutral pion, which is its own antiparticle. (There is, in other words, only one variety of that particle.) It is extremely unstable, breaking down in less than a quintillionth of a second to form a pair of gamma rays.

Despite the fact that a pion "belongs" within the nucleus, it will fleetingly circle a nucleus before interacting with it, sometimes, to form a "pionic atom." This was detected in 1952. Indeed, any pair of negative and positive particles or particle-systems can be made to circle each other, and in the 1960's, physicists have studied a number of evanescent "exotic atoms" in order to gain some notion as to the details of particle structure.

Since its discovery, the pion has grown quite important to the physicists' view of the subatomic world. Even free protons and neutrons may be surrounded by tiny clouds of pions and may even be composed of them. The American physicist Robert Hofstadter investigated nuclei with extremely energetic electrons produced by a linear accelerator. He has suggested that both proton and neutron consist of cores made up of mesons. As a result of his work in this field, he shared in the Nobel Prize in physics in 1961.

As the number of known particles grew, it became necessary for physicists to divide them into groups. The lightest were called "leptons" from a Greek word meaning "small." These include the electrons, the muons, their antiparticles, and their neutrinos. The photon is usually included also, as a particle with zero mass and zero charge but with a spin equal to 1. (In this way, the photon differs from the various neutrinos, which also have zero mass and charge, but which have a spin of one-half.) The photon is considered to be its own antiparticle. The graviton is also included, differing from the other massless, chargeless particles by having a spin of 2.

The fact that photons and gravitons are their own antiparticles helps explain why it is so difficult to detect whether a distant galaxy is matter or antimatter. Much of what we receive from a distant galaxy is photons and gravitons; a galaxy of antimatter emits exactly the same photons and gravitons that a galaxy of matter would. There are no antiphotons and antigravitons that might act as distinctive fingerprints of antimatter. We ought to receive neutrinos, however,—or antineutrinos. A preponderance of neutrinos would mark matter; one of antineutrinos, antimatter. With the development and improvement of techniques for detecting neutrinos or antineutrinos from outer space, it may become possible someday to pin down this matter of the existence and location of antigalaxies.

Among the leptons, those that do not carry an electric charge also do not possess mass. These chargeless, massless particles are all stable. Left to themselves, each will endure unchanged (as far as we know) forever. For some reason, charge can only exist when mass exists, but particles with mass tend to break down to particles with lesser mass. Thus a muon tends to break down to an electron. An electron (or positron) is, as far as we know, the least massive particle that can exist. For it to break down further would mean the loss of mass altogether, and this would also require the loss of electric charge. Since the law of conservation of electric charge makes it impossible to lose charge, the electron cannot break down. An electron and a positron can undergo mutual annihilation, for the opposite charges will cancel each other, but either one left to itself would, as far as we know, exist eternally.

More massive than the leptons are the "mesons," which are a family that no longer includes the muon even though that particle was the original meson. Among the mesons now are the pions and a newer variety, the "K-mesons" or "kayons." These were first detected in 1952 by two Polish physicists, Marian Danysz and Jerzy Pniewski. These are about 970 times as massive as an electron and, therefore, about half the mass of a proton or neutron. The kayon comes in two varieties, a positive kayon and a neutral kayon, and each has an antiparticle associated with it. They are unstable, of course, breaking down in about a microsecond to pions.

Above the mesons are the "baryons" (from a Greek word meaning "heavy"). Until the 1950's, the proton and the neutron were the only specimens known. Beginning in 1954, however, a series of still more massive particles (sometimes called "hyperons") were discovered. It is the baryon particles that have particularly proliferated in recent years, in fact, and the proton and neutron are but the lightest of a large variety.

There is a "law of conservation of baryon number," physicists have discovered, for in all particle breakdowns, the net number of baryons (that is, baryons minus antibaryons) remains the same. The breakdown is always from a more massive to a less massive particle and that explains why the proton is stable and is the *only* baryon to be stable. It happens to be the lightest baryon. If it broke down it would have to cease being a baryon and that would break the law of conservation of baryon number. For the same reason, an antiproton is stable, because it is the lightest antibaryon. Of course, a proton and an antiproton can engage in mutual annihilation since, taken together, they make up one baryon plus one antibaryon for a net baryon number of zero.

The first baryons beyond the proton and neutron to be discovered were given Greek names. There was the "lambda particle," the "sigma particle," and the "xi particle." The first came in one variety, a neutral particle; the second in three varieties, positive, negative, and neutral; the third in two varieties, negative and neutral. Every one of these had an associated antiparticle, making a dozen particles altogether. All were exceedingly unstable; none could live for more than a hundredth of a microsecond or so; and some, such as the neutral sigma particle, broke down after a hundred trillionth of a microsecond.

The lambda particle, which is neutral, can replace a neutron in a nucleus to form a "hypernucleus"—an entity which endures less than a billionth of a second. The first to be discovered was a hyper-tritium nucleus made up of a proton, a neutron, and a lambda particle. This was located among the products of cosmic radiation by Danysz and Pniewski in 1952. In 1963, Danysz reported hypernuclei containing two lambda particles. What's more, negative hyperons can be made to replace electrons in atomic structure, as was first reported in 1968. Such massive electron-replacements circle the nucleus at such close quarters as to spend their time actually within the nuclear outer regions.

But all these are the comparatively stable particles; they live long enough to be directly detected and to be easily awarded a lifetime and personality of their own. In the 1960's, the first of a whole series of particles was detected by Alvarez (who received the Nobel Prize in physics in 1968 as a result). These were so short-lived that their existence could only be deduced from the necessity of accounting for their breakdown products. Their half-lives are something of the order of a few trillionths of a trillionth of a second, and one might wonder whether they are really individual particles or merely a combination of two or more particles, pausing to nod at each other before flashing by.

These ultra-short-lived entities are called "resonance particles" and, counting them, over 150 different subatomic particles are now known; physicists are uncertain as to how many remain to be found. The situation among the particles is about what it was a century ago among the elements, before Mendeleev had proposed the periodic table.

Some physicists believe that the various mesons and baryons are not truly independent particles—that baryons can absorb and emit mesons to attain various levels of excitement, and that each excited baryon might easily be mistaken for a different particle. (We have a similar situation among atoms, where a particular atom may absorb or emit photons to reach various levels of electronic excitement—except that an excited hydrogen-atom is still recognized as a hydrogen atom.)

What is needed, then, is some sort of periodic table for subatomic particles—something that would group them into families consisting of a basic member or members with other particles that are excited states of that basic member or members.

Something of the sort was proposed in 1961 by the American physicist Murray Gell-Mann and the Israeli physicist Yuval Ne'emen, who were working independently. Groups of particles were put together in a pattern that depended on their various properties into a beautifully symmetric fashion, which Gell-Mann called the "eightfold way" but which is more formally referred to as "SU#3." In particular, one such grouping needed one more particle for completion. That particle, if it was to fit into the group, had to have a particular mass and a particular set of other properties. The combination was not a likely one for a particle to have; yet, in 1964, a particle (the "omega-minus") was detected with just the predicted set of properties, and in succeeding years it was detected dozens of times. In 1971 its antiparticle, the "antiomega-minus," was detected.

Even if baryons are divided into groups and a subatomic periodic table is set up, there would still be enough different particles to give physicists the urge to find something still simpler and more fundamental. In 1964, Gell-Mann—having endeavored to work out the simplest way of accounting for all the baryons with a minimum number of more fundamental "sub-baryonic particles"—came up with the notion of "quarks." He got this name because he found that only three different quarks were necessary and that different combinations of the three quarks were needed to make up all the known baryons. This reminded him of a line from *Finnegan's Wake* by James Joyce, which goes "Three quarks for Musther Mark."

In order to account for the known properties of baryons, the three different quarks had to have specific properties of their own. The most astonishing property was a fractional electric charge. All known particles had either no electric charge, an electric charge exactly equal to that of the electron (or positron), or an electric charge equal to some exact multiple of the electron (or positron). The known charges, in other words, were 0, $+1$, -1, $+2$, -2, and so on. One quark, however (the "p-quark"), had a charge of $+2/3$, while the other two (the "n-quark" and the "lambda-quark") had a charge of $-1/3$ apiece. The n-quark and the lambda-quark were distinguished from each other by something called the "strangeness number." Whereas the n-quark (and the p-quark, too) had a strangeness number of 0, the lambda-quark had a strangeness number of -1.

Each quark had its "antiquark." There was the anti-p-quark, with a charge of $-2/3$ and a strangeness number of 0; the anti-n-quark, with a charge of $+1/3$ and a strangeness number of 0; and the anti-lambda-quark, with a charge of $+1/3$ and a strangeness number of $+1$.

Now then, one can imagine a proton built up of two p-quarks and an n-quark, while a neutron is built up of two n-quarks and a p-quark (which is why we have the "p" and "n" prefixes to the quarks). A lambda particle is made up of a p-quark, an n-quark, and a lambda-quark (which accounts for the last named prefix), an omega-minus particle is made up of three lambda-quarks, and so on. One can even combine quarks by twos to form the different mesons.

The question is, though, however convenient the quarks may be mathematically, do they really exist? That is, we may agree that there are four quarters to a dollar, but does that mean that somewhere in a paper dollar bill there are four metallic quarters? One way of answering the question would be to strike a proton, neutron, or other particle, with such energy as to cause it to fly apart into its constituent quarks. Unfortunately, the binding forces holding quarks together are far higher than those holding baryons together (just as those are far higher than the forces holding atoms together), and man does not yet dispose of enough energy to break up the baryon. There are cosmic-ray particles with enough energy to do so (if it can be done), but though some reports of quarklike particles among cosmic-ray products have been heard, these are not widely accepted.

At the moment of writing, the quark hypothesis must be regarded as interesting—but speculative.

The K-mesons and the hyperons introduced physicists to a fourth field of force different from the three already known: gravitational, electromagnetic, and nuclear.

Of these three, the nuclear force is by far the strongest, but it acts

THE LONGER-LIVED SUBATOMIC PARTICLES

FAMILY NAME	PARTICLE NAME	SYMBOL	MASS	SPIN	ELECTRIC CHARGE
	PHOTON	γ (GAMMA RAY)	0	1	NEUTRAL
	GRAVITON	————	0	2	NEUTRAL
ELECTRON FAMILY	ELECTRON'S NEUTRINO	ν_e	0	½	NEUTRAL
	ELECTRON	e^-	1	½	NEGATIVE
MUON FAMILY	MUON'S NEUTRINO	ν_μ	0 (?)	½	NEUTRAL
	MUON	μ^-	206.77	½	NEGATIVE
MESONS	PION	π^+	273.2	0	POSITIVE
		π^-	273.2	0	NEGATIVE
		π^0	264.2		
	KAON	K^+	966.6	0	POSITIVE
		K^0	974	0	NEUTRAL
BARYONS	NUCLEON	p (PROTON)	1836.12	½	POSITIVE
		n (NEUTRON)	1838.65	½	NEUTRAL
	LAMBDA	Λ^0	2128.8	½	NEUTRAL
	SIGMA	Σ^+	2327.7	½	POSITIVE
		Σ^-	2340.5	½	NEGATIVE
		Σ^0	2332	½	NEUTRAL
	XI	Ξ^-	2580	½	NEGATIVE
		Ξ^0	2570	½	NEUTRAL

* The K^0 meson has two different lifetimes. All other particles have only one.

After a table in K. W. Ford. *The World of Elementary Particles*. Reprinted through the courtesy of Blaisdell Publishing Company, a division of Ginn and Company.

ANTIPARTICLE	NO. OF DISTINCT PARTICLES	AVERAGE LIFETIME (SECONDS)	TYPICAL MODE OF DECAY
SAME PARTICLE	1	INFINITE	——
SAME PARTICLE	1	INFINITE	——
$\overline{\nu_e}$	2	INFINITE	——
e^+ (POSITRON)	2	INFINITE	——
$\overline{\nu_\mu}$	2	INFINITE	——
μ^+	2	2.212×10^{-6}	$\mu^- \to e^- + \nu_e + \nu_\mu$
π^- SAME AS π^+ THE $\pi^°$ PARTICLES	3	2.55×10^{-8} 2.55×10^{-8} 1.9×10^{-16}	$\pi^+ \to \mu^+ + \nu_\mu$ $\pi^- \to \mu^- + \nu_\mu$ $\pi^° \to \gamma + \gamma$
$\overline{K^-}$ (NEGATIVE) $\overline{K^°}$	4	1.22×10^{-8} 1.00×10^{-10} and* 6×10^{-8}	$K^+ \to \pi^+ + \pi^°$ $K^° \to \pi^+ + \pi^-$
$\overline{p}$ (NEGATIVE) $\overline{n}$	4	INFINITE 1013	$n \to p + e^- + \nu_e$
$\overline{\Lambda^°}$	2	2.51×10^{-10}	$\Lambda^° \to p + \pi^-$
$\overline{\Sigma^+}$ (NEGATIVE) $\overline{\Sigma^-}$ (POSITIVE) $\overline{\Sigma^°}$	6	8.1×10^{-11} 1.6×10^{-10} ABOUT 10^{-20}	$\Sigma^+ \to n + \pi^+$ $\Sigma^- \to n + \pi^-$ $\Sigma^° \to \Lambda^° + \gamma$
$\overline{\Xi^-}$ (POSITIVE) $\overline{\Xi^°}$	4	1.3×10^{-10} ABOUT 10^{-10}	$\Xi^- \to \Lambda^° + \pi^-$ $\Xi^° \to \Lambda^° + \pi^°$
	33		

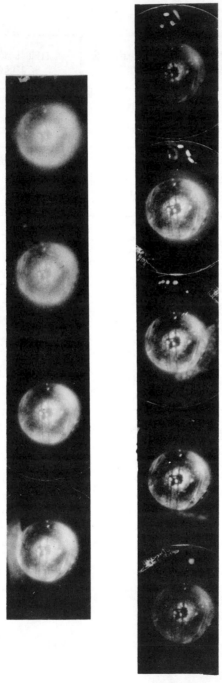

Radioactive tracers in a sample of material here make their own picture by means of a technique called "tracer micrography," developed by the U. S. National Bureau of Standards. Electrons emitted by the radioactive atoms are focused by a magnetic lens upon a photographic film, which thus shows the distribution of the radioactive material in the sample.

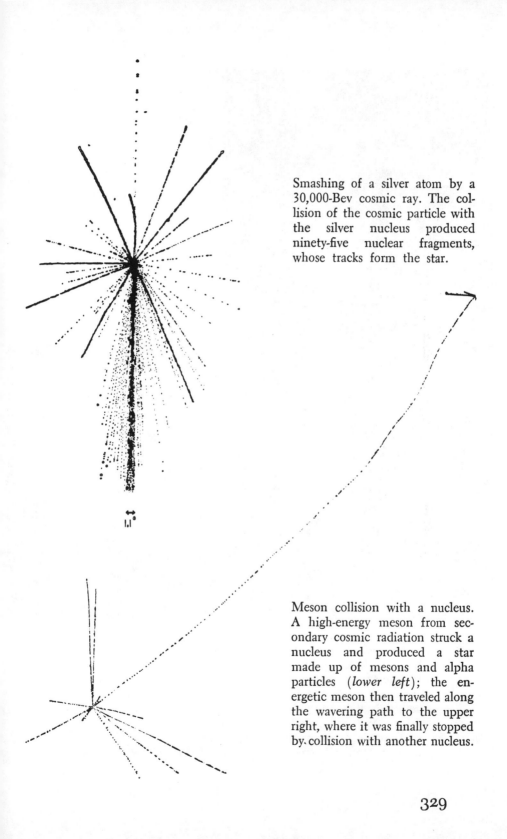

Smashing of a silver atom by a 30,000-Bev cosmic ray. The collision of the cosmic particle with the silver nucleus produced ninety-five nuclear fragments, whose tracks form the star.

Meson collision with a nucleus. A high-energy meson from secondary cosmic radiation struck a nucleus and produced a star made up of mesons and alpha particles (*lower left*); the energetic meson then traveled along the wavering path to the upper right, where it was finally stopped by collision with another nucleus.

329

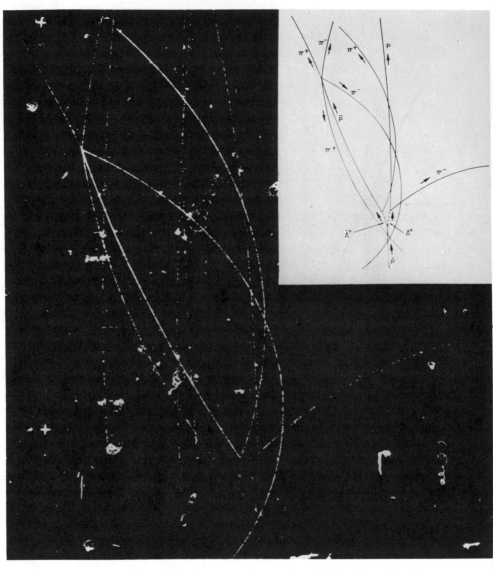

Antiprotons and lambda particles produced this picture in the large bubble chamber at the University of California's Radiation Laboratory. As the diagram shows, an antiproton (p̄) from the Bevatron entered at the bottom; where its track ends there is a gap (*dashed lines*) that represents the travel of the undetectable neutral lambda and antilambda particles. The antilambda then decayed into a positive pi meson and an antiproton, which went on to produce four pi mesons (*upper left*); the lambda decayed into a proton and a negative pi meson (*right side of the fork*).

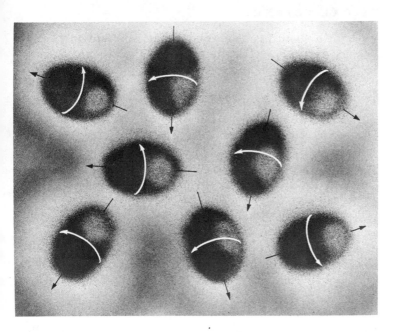

Spinning protons in this schematic drawing are oriented in random directions. The white arrow shows the direction of the spin.

Protons lined up by a steady magnetic field. Those oriented in
the opposite-to-normal direction (*arrows pointing downward*)
are in the excited spin state.

only over an extremely short distance. Whereas the electromagnetic and gravitational forces decrease only as the square of the distance, nuclear forces drop off so rapidly with distance that the force between two nucleons falls almost to zero if they are separated by a distance greater than their own diameter. (This is as if the earth's gravity were to become practically zero 4,000 miles above the surface.) Consequently interactions between particles under the influence of nuclear forces must take place very quickly.

For instance, imagine a pi meson and a proton approaching each other. If the nuclear force is to cause them to interact, it must do so while they are within a proton's width of each other. A proton's width is about 0.0000000000001 centimeter. Flying mesons are traveling at almost the speed of light, which is 30 billion centimeters a second. Thus the pi meson will be within the influence of the nuclear force for only about 0.000000000000000000000001 second (a hundred billionth of a trillionth of a second). And yet, even in this short time, the nuclear force brings about an interaction. The pi meson and the proton can react to produce a lambda hyperon and a K-meson.

This is an example of what physicists call a "strong interaction." Baryons and those mesons subject to strong interactions are lumped together as "hadrons," from a Greek word meaning "bulky." A "weak interaction" is one that requires a considerably longer time. The theory of such interactions was first worked out in 1934 by Fermi. An example of such an interaction is the breakdown of a K-meson or a hyperon. This takes one ten-billionth of a second or so. That may seem a breathlessly short time, but compared to the time it takes for a pi meson and proton to interact, it is very long. It is, in fact, about a trillion times as long as the physicists had expected, considering the speed of most nuclear interactions.

They concluded that the "weak interactions" were governed by forces much weaker than the nuclear forces, and they took to calling some of the particles that broke down as a result of weak interactions "strange particles." This name applied primarily to K-mesons and hyperons.

Gell-Mann gave different particles something called "strangeness number" (which we met earlier in connection with quarks), awarding the numbers in such a way that strangeness is conserved in all particle interactions. That is, the net strangeness number is the same afterward as before.

This fourth type of interaction should be mediated by way of an exchange particle, too. The gravitational, electromagnetic, and strong interactions, have the graviton, the photon, and the pion as theirs; the weak interaction should have something called the "W-particle." It is also called an "intermediate boson," because it ought to obey Bose-Einstein statistics and because it ought to have an intermediate rate of decay. So far the W-particle has not actually been detected.

Nuclear physicists now deal with about a dozen conservation laws. Some are the familiar conservation laws of nineteenth-century physics: the conservation of energy, the conservation of momentum, the conservation of angular momentum, and the conservation of electric charge. Then there are conservation laws that are less familiar: the conservation of strangeness, the conservation of baryon number, the conservation of isotopic spin, and so on.

The strong interactions seem to obey all these conservation laws; in the early 1950's, physicists took it for granted that the laws were universal and irrevocable. But they were not. In the case of weak interactions, some of the conservation laws are not obeyed.

The particular conservation law that was shattered was the "conservation of parity." Parity is a strictly mathematical property that cannot be described in concrete terms; suffice it to say that the property refers to a mathematical function that has to do with the wave characteristics of a particle and its position in space. Parity has two possible values—"odd" and "even." The key point we must bear in mind is that parity has been considered a basic property that, like energy or momentum, is subject to the law of conservation: in any reaction or change, parity must be conserved. That is to say, when particles interact to form new particles, the parity on both sides of the equation (so it was thought) must balance, just as mass numbers must, or atomic numbers, or angular momentum.

Let me illustrate. If an odd-parity particle and an even-parity particle interact to form two other particles, one of the new particles must be odd parity and the other even parity. If two odd-parity particles form two new particles, both of the new ones must be odd or both even. Conversely, if an even-parity particle breaks down to form two particles, both must be even parity or both must be odd parity. If it forms three particles, either all three have even parity or one has even parity and the other two have odd parity. (You may be able to see this more clearly if you consider the odd and even numbers, which follow similar rules. For instance, an even number can only be the sum of two even numbers or of two odd numbers, but never the sum of an even number and an odd one.) This is what is meant by the "conservation of parity."

The beginning of the trouble came when it was found that K-mesons sometimes broke down to two pi mesons (which, since the pi meson has odd parity, added up to even parity) and sometimes gave rise to three pi mesons (adding up to odd parity). Physicists concluded that there were two types of K-meson, one of even parity and one of odd parity; they named the two the "theta meson" and the "tau meson," respectively.

Now in every respect except the parity result, the two mesons were identical: the same mass, the same charge, the same stability, the same everything. It was hard to believe that there could be two particles with exactly the same properties. Was it possible that the two were actually the same and that there was something wrong with the idea of the conservation of parity? In 1956, two young Chinese physicists working in the United States, Tsung Dao Lee and Chen Ning Yang, made precisely that suggestion. They proposed that, although the conservation of parity held in strong interactions, it might break down in weak interactions, such as the decay of K-mesons.

As they worked out this possibility mathematically, it seemed to them that if the conservation of parity broke down, the particles involved in weak interactions should show a "handedness," something first pointed out in 1927 by the Hungarian physicist Eugene Wigner. Let me explain.

Your right hand and left hand are opposites. One can be considered the mirror image of the other: in a mirror the right hand looks like a left hand. If all hands were symmetrical in every respect, the mirror image would be no different from the direct image, and there would be no such distinction as "right" and "left" hand. Very well, then, let us apply this to a group of particles emitting electrons. If electrons come out in equal numbers in all directions, the particle in question has no "handedness." But if most of them tend to go in a preferred direction— say up rather than down—then the particle is not symmetrical. It shows a "handedness": if we look at the emissions in a mirror, the preferred direction will be reversed.

The thing to do, therefore, was to observe a collection of particles that emit electrons in a weak interaction (say some particle that decays

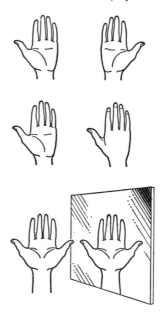

Mirror-image asymmetry and symmetry illustrated by hands.

by beta emission) and see if the electrons came out in a preferred direction. Lee and Yang asked an experimental physicist at Columbia University, Chien-Shiung Wu, to perform the experiment.

She set up the necessary conditions. All the electron-emitting atoms had to be lined up in the same direction if a uniform direction of emission was to be detected; this was done by means of a magnetic field and the material was kept at a temperature near absolute zero.

Within forty-eight hours the experiment yielded the answer. The electrons were indeed emitted asymmetrically. The conservation of parity did break down in weak interactions. The "theta meson" and the "tau meson" were one and the same particle, breaking down with odd parity in some cases, with even parity in others. Other experimenters soon confirmed the overthrow of parity, and for their bold conjecture the theoretical physicists Lee and Yang received the Nobel Prize in physics in 1957.

If symmetry breaks down with respect to weak interactions, perhaps it will break down elsewhere. The universe as a whole may be left handed (or right-handed) after all. Alternatively, there may be two universes, one left-handed, the other right-handed; one composed of matter, the other of antimatter.

Physicists are now viewing the conservation laws in general with a new cynicism. Any one of them might, like conservation of parity, apply under some conditions and not under others.

Parity, after its fall, was combined with "charge conjugation," another mathematical property assigned to subatomic particles, which governed its status as a particle or antiparticle, and the two together were thought to be conserved. Added also was still another symmetry, which implied that the law governing subatomic events was the same whether time runs forward or backward. The whole may be called "CPT-conservation." In 1964, however, nuclear reactions were discovered that violated CPT-conservation. This affects "time reversal." It means you can distinguish between time running forward (on the subatomic scale) and time running backward—something physicists had thought to be impossible. To avoid this dilemma, a possible fifth force, weaker even than the gravitational force, was postulated. Such a fifth force, however, ought to produce certain detectable effects. In 1965, those effects were sought and not found. Physicists were left with their time dilemma. But there is no reason to despair. Such problems seem always to lead, eventually, to new and deeper understanding of the universe.

Inside the Nucleus

Now that so much has been learned about the general make-up and nature of the nucleus, there is great curiosity as to its structure, particularly the fine structure inside. First of all, what is its shape? Because it is so small and so tightly packed with neutrons and protons, physicists naturally assume that it is spherical. The fine details of the spectra of atoms suggest that many nuclei have a spherical distribution of charge. Some do not: they behave as if they have two pairs of magnetic poles, and these nuclei are said to have "quadrupole moments." But their deviation from the spherical is not very large. The most extreme case is that of the nuclei of the lanthanides, in which the charge distribution seems to make up a prolate spheroid (football-shaped, in other words). Even here the long axis is not more than 20 per cent greater than the short axis.

As for the internal structure of the nucleus, the simplest model pictures it as a tightly packed collection of particles much like a drop of liquid, where the particles (molecules) are packed closely with little space between, where the density is virtually even throughout, and where there is a sharp surface boundary.

This "liquid-drop model" was first worked out in detail in 1936 by Niels Bohr. It suggests a possible explanation of the absorption and emission of particles by some nuclei. When a particle enters the nucleus, one can suppose, it distributes its energy of motion among all the closely packed particles, so that no one particle receives enough energy immediately to break away. After perhaps a quadrillionth of a second, when there has been time for billions of random collisions, some particle accumulates sufficient energy to fly out of the nucleus.

The model could also account for the emission of alpha particles by the heavy nuclei—that is, the unstable elements with atomic numbers above 83. In these large nuclei, the short-range nuclear forces may not reach all the way across the nucleus; hence the force of repulsion between positive particles can take effect. As a result, portions of the nucleus in the form of the two-proton, two-neutron alpha particle (a very stable combination) may break off spontaneously from the surface of the nucleus. After the nucleus has decayed to a size such that the nuclear force overwhelms the force of repulsion, the nucleus becomes stable.

The liquid-drop model suggests another form of nuclear instability. When a large drop of liquid suspended in another liquid is set wobbling by currents in the surrounding fluid, it tends to break up into smaller spheres, often into roughly equal halves. We can think of the fission of uranium as closely analogous to this process. The fissionable nucleus, when struck by a neutron, begins to wobble, in a manner of speaking. It may stretch out into the shape of a dumbbell (as a liquid drop would), and in that case the nuclear attractive forces would not reach from one end of the dumbbell to the other, with the result that the repulsive force would drive the two portions apart. Bohr offered this explanation when nuclear fission was discovered.

Other nuclei besides uranium 235 ought to be (and proved to be) subject to fission if they receive enough input of energy. In fact, if a nucleus is large enough for the repulsive forces to become important, it ought occasionally to fission even without the input of energy. (This is like saying that the droplike nucleus is always vibrating and wobbling, and every once in a while the vibration is strong enough to produce the dumbbell and bring about the break.)

In 1940, two Soviet physicists, G. N. Flerov and K. A. Petrjak, discovered indeed that the heavier isotope of uranium, U-238, sometimes fissions spontaneously, without the addition of any particle. Uranium exhibits instability mainly by emitting alpha particles, but in a pound of uranium there are four spontaneous fissions per second while about 8 million nuclei are emitting alpha particles.

Spontaneous fission also takes place in uranium 235, in protactinium, in thorium, and, more frequently, in the transuranium elements. As nuclei get larger and larger, the probability of spontaneous fission increases. In the heaviest elements of all—einsteinium, fermium, and mendelevium—it becomes the most important method of breakdown, far outweighing alpha-particle emission.

Another popular model of the nucleus likens it to the atom as a whole, picturing the nucleons within the nucleus, like the electrons around the nucleus, as occupying shells and subshells, each affecting the others only slightly. This is called the "shell model."

How can there be room for independent shells of nucleons in the tiny, tightly packed nucleus? Well, however it is managed, the evidence suggests that there is some "empty space" there. For instance, in a mesonic atom the meson may actually circle in an orbit within the nucleus for a short time. And Robert Hofstadter has found that the nucleus consists of a high-density core surrounded by a "skin" of gradually decreasing density. The thickness of the skin is about half the radius of the nucleus, so that it actually makes up seven eighths of the volume.

By analogy with the situation in the atom's electronic shells, one may suppose that the nuclei with filled outer nucleonic shells should be more stable than those whose outer shells are not filled. The simplest theory would indicate that nuclei with 2, 8, 20, 40, 70, or 112 protons or neutrons, would be particularly stable. This, however, does not fit observation. The German-American physicist Maria Goeppert Mayer took account of the spin of the protons and neutrons and showed how this would affect the situation. It turned out that nuclei containing 2, 8, 20, 50, 82, or 126 protons or neutrons, would then be particularly stable, and that fit the observations. Nuclei with 28 or 40 protons or neutrons, would be fairly stable. All others would be less stable, if stable at all. These "shell numbers" are sometimes called "magic numbers" (with 28 or 40 occasionally referred to as "semi-magic numbers.")

Among the magic-number nuclei are helium 4 (2 protons and 2

neutrons), oxygen 16 (8 protons and 8 neutrons), and calcium 40 (20 protons and 20 neutrons), all especially stable and more abundant in the universe than other nuclei of similar size.

As for the higher magic numbers, tin has ten stable isotopes, each with 50 protons, and lead has four, each with 82 protons. There are five stable isotopes (each of a different element) with 50 neutrons each, and seven stable isotopes with 82 neutrons each. In general, the detailed predictions of the nuclear-shell theory work best near the magic numbers. Midway between (as in the case of the lanthanides and actinides), the fit is poor. But just in the midway regions, nuclei are farthest removed from the spherical (and shell theory assumes spherical shape) and are most markedly ellipsoidal. The 1963 Nobel Prize for physics was awarded to Goeppert Mayer and to two others: Wigner, and the German physicist J. Hans Daniel Jensen, who also contributed to the theory.

In general, as nuclei grow more complex, they become rarer in the universe, or less stable, or both. The most complex stable isotopes are lead 208 and bismuth 209, each with the magic number of 126 neutrons, and lead, with the magic number of 82 protons in addition. Beyond that all nuclides are unstable and with (in general) shortening half-lives as the number of protons, neutrons, or both are increased. But the shortening is not invariable. Thorium 232 has a half-life of 14 billion years and is all but stable. Even californium 251 has a half-life in the centuries.

Could very complex nuclides—those more complex than any that have been observed or synthesized—be sufficiently stable to permit formation in relatively large quantities? There are, after all, magic numbers beyond 126, and these might result in certain relatively stable supercomplex atoms. Calculations have shown that, in particular, an element with 114 protons and 184 neutrons might have a surprisingly long half-life; physicists who have now reached element number 105 are seeking out ways of attaining this isotope of element 114 which, chemically, ought to resemble lead.

CHAPTER 7

The Waves

Light

Of all the helpful attributes of nature, the one that man probably appreciates most is light. According to the Bible, the first words of God were, "Let there be light," and the sun and the moon were created primarily to serve as sources of light: "And let them be for lights in the firmament of the heaven to give light upon the earth."

The scholars of ancient and medieval times were completely in the dark as to the nature of light. They speculated that it consisted of particles emitted by the glowing object or perhaps by the eye itself. The only facts about it that they were able to establish were that light traveled in a straight path, that it was reflected from a mirror at an angle equal to that at which the beam struck the mirror, and that a light beam was bent ("refracted") when it passed from air into glass, water, or some other transparent substance.

The first important experiments on the nature of light were conducted by Isaac Newton in 1666. He let a beam of sunlight, entering a dark room through a chink in a blind, fall obliquely on one face of a triangular glass prism. The beam was bent when it entered the glass and then bent still farther in the same direction when it emerged from a second face of the prism. Newton caught the emerging beam on a white screen to see the effect of the reinforced refraction. He found that instead of forming a spot of white light, the beam was now spread out in a band of colors—red, orange, yellow, green, blue, and violet, in that order.

Newton deduced that ordinary white light was a mixture of different kinds of light which, separately, affect our eyes so as to produce the sensation of different colors. The spread-out band of its components was called a "spectrum," from a Latin word meaning "ghost."

Newton decided that light consisted of tiny particles ("corpuscles") traveling at enormous speed. This would explain why light traveled in straight lines and cast sharp shadows. It was reflected by a mirror because the particles bounced off the surface, and it was bent on entering a refracting medium (such as water or glass) because the particles traveled faster in such a medium than in air.

Still, there were some awkward questions. Why should the particles of green light, say, be refracted more than those of yellow light? Why was it that two beams of light could cross without affecting each other— that is, without the particles colliding?

In 1678, the Dutch physicist Christian Huyghens (a versatile scientist who had built the first pendulum clock and done important work in astronomy) suggested an opposing theory, namely, that light consisted of tiny waves. If it was made up of waves, there was no difficulty about explaining the different amount of refraction of different kinds of light through a refracting medium, provided it was assumed that light traveled more slowly through the refracting medium than through air. The amount of refraction would vary with the length of the waves: the shorter the wavelength, the greater the refraction. This meant that violet light (the most refracted) had a shorter wavelength than blue light, blue shorter than green, and so on. It was this difference in wavelength that distinguished the colors to the eye. And, of course, if light consisted of waves, two beams could cross without trouble. (After all, sound waves and water waves crossed without losing their identity.)

But Huyghens' wave theory was not very satisfactory either. It did not explain why light rays traveled in straight lines and cast sharp shadows, nor why light waves could not go around obstacles, as water waves and sound waves could. Furthermore, if light consisted of waves, how could it travel through a vacuum, as it certainly did in coming to us through space from the sun and stars? What medium was it waving?

For about a century, the two theories contended with each other. Newton's "corpuscular theory" was by far the more popular, partly because it seemed on the whole more logical, and partly because it had the support of Newton's great name. But in 1801 an English physician and physicist, Thomas Young, performed an experiment that swung opinion the other way. He projected a narrow beam of light through two closely spaced holes toward a screen behind. If light consisted of particles, presumably the two beams emerging through the holes would simply produce a brighter region on the screen where they overlapped and less bright regions where they did not. But this was not what Young found. The screen showed a series of bands of light, each separated from the next by a dark band. It seemed that in these dark intervals, the light of the two beams together added up to darkness!

The wave theory would easily explain this. The bright band represented the reinforcement of waves of one beam by waves of the other; in other words, there the two sets of waves were "in phase," both peaks together and strengthening each other. The dark bands, on the other hand, represented places where the waves were "out of phase," the trough of one canceling the peak of the other. Instead of reinforcing each other, the waves at these places interfered with each other, leaving the net light energy at those points zero.

From the width of the bands and the distance between the two holes through which the beams issued, it was possible to calculate the length of light waves, say of red light or violet or colors between. The wavelengths turned out to be very small indeed. The wavelength of red light, for example, came to about 0.000075 centimeter. (Nowadays the wavelengths of light are expressed in a convenient unit suggested by Ångstrom. The unit, called the angstrom—abbreviated Å—is one hundred-millionth of a centimeter. Thus the wavelength of red light is about 7500 angstrom units, the wavelength of violet light is about 3900 angstrom units, and the color wavelengths of the visible spectrum lie between these numbers.)

The shortness of the wavelengths is very important. The reason light waves travel in straight lines and cast sharp shadows is that they are incomparably smaller than ordinary objects; waves can curve around an obstruction only when that obstruction is not much larger than the wavelength. Even bacteria, for instance, are vastly wider than a wavelength of light, so light can define them sharply under a microscope. Only objects somewhere near a wavelength of light in size (for example, viruses and other submicroscopic particles) are small enough for light waves to pass around them.

It was the French physicist Augustin Jean Fresnel who showed (in 1818) that if an interfering object is small enough, a light wave will indeed travel around it. In that case, the light produces what is called a "diffraction" pattern. For instance, the very fine parallel lines of a "diffraction grating" act as a series of tiny obstacles that reinforce one another. Since the amount of diffraction depends on the wavelength, a spectrum is produced. From the amount by which any color or portion of the spectrum is diffracted and from the known separation of the scratches on the glass, the wavelength can again be calculated.

Fraunhofer pioneered in the use of such diffraction gratings, an advance generally forgotten in the light of his more famous discovery of spectral lines. The American physicist Henry Augustus Rowland invented concave gratings and developed techniques for ruling them with as many as 20,000 lines to the inch. It was this that made it possible for the prism to be supplanted in spectroscopy.

Between such experimental findings and the fact that Fresnel worked out the mathematics of wave motion systematically, the wave theory of light seemed established and the corpuscular theory smashed —seemingly for good.

Not only were light waves accepted as existing, their length was measured with increasingly good precision. By 1827, the French physicist Jacques Babinet was suggesting that the wavelength of light—an unalterable physical quantity—be used as the standard for measurement of length, instead of the various man-made standards that were in fact used. This suggestion did not become practicable, however, until the 1880's, when the German-American physicist Albert Abraham Michelson invented an instrument called the "interferometer," which could measure the wavelengths of light with unprecedented accuracy. In 1893, Michelson measured the wavelength of the red line in the cadmium spectrum and found it to be 1/1,553,164 meters long.

A measure of uncertainty still existed when it was discovered that elements consisted of different isotopes, each contributing a line of slightly different wavelength. As the twentieth century progressed, however, the spectral lines of individual isotopes were measured. In the 1930's, the lines of krypton 86 were measured. This isotope, being that of gas, could be dealt with at low temperatures where atomic motion was slowed, with less consequent thickening to the line.

In 1960, the krypton-86 line was adopted by the General Conference of Weights and Measures as the fundamental standard of length. The meter has been redefined as equal to 1,650,763.73 wavelengths of this spectral line. This has increased the precision of measurement of length a thousandfold. The old standard meter bar could be measured, at best, to within one part in a million, whereas the light wave can be measured to within one part in a billion.

Light obviously travels at tremendous speeds. If you put out a light, it gets dark everywhere at once, as nearly as can be made out. This is not quite as true for sound, for instance. If you watch a man in the distance chopping wood, you do not hear the stroke until some moments after the ax strikes. Sound has clearly taken a certain amount of time to travel to the ear. In fact, its speed of travel is easy to measure: it amounts to 1,090 feet per second, or about 750 miles per hour, in the air at sea level.

Galileo was the first to try to measure the speed of light. Standing on one hill while an assistant stood on another, he would uncover a lantern; as soon as the assistant saw the flash, he would signal by uncovering a light of his own. Galileo did this at greater and greater distances, assuming that the time it took the assistant to make his response would remain uniform and therefore any increase in the interval between his uncovering his own lantern and seeing the responding flash would represent the time taken by the light to cover the extra distance. The idea was sound, but of course light travels much too fast for Galileo to have detected any difference by this crude method.

In 1676, the Danish astronomer Olaus Roemer did succeed in timing the speed of light—on an astronomical distance scale. Studying Jupiter's eclipses of its four large satellites, Roemer noticed that the interval between successive eclipses became longer when the earth was moving away from Jupiter and became shorter when it was moving toward Jupiter in its orbit. Presumably the difference in eclipse times reflected the difference in distance between the earth and Jupiter; that is, it would be a measure of the distance in the time that light took to travel between Jupiter and the earth. From a rough estimate of the size of the earth's orbit, and from the maximum discrepancy in the eclipse timing, which Roemer took to represent the time it took light to cross the full width of the earth's orbit, he calculated the speed of light. His estimate came to 140,000 miles per second, remarkably good for what might be considered a first try and high enough to evoke the disbelief of his contemporaries.

Roemer's results were, however, confirmed a half century later from a completely different direction. In 1728, the British Astronomer Royal James Bradley found that stars seemed to shift position because of the earth's motion; not through parallax, but because the velocity of the earth's motion about the sun was a measurable (though small) fraction of the speed of light. The analogy usually used is that of a man under an umbrella striding through a rainstorm. Even though the drops are falling vertically, the man must tip the umbrella forward, for he is stepping into the drops. The faster he walks, the farther he must tip the umbrella. Similarly, the earth moves into the light rays falling from the

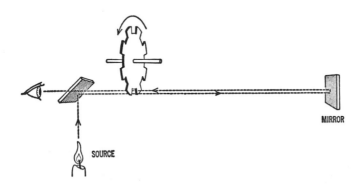

Fizeau's arrangement for measuring the speed of light. Light reflected by the semimirror near the source passes through a gap in the rapidly spinning toothed wheel to a distant mirror (*right*) and is reflected back to the next tooth or the next gap.

343

stars, and the astronomer must tip his telescope a little bit, and in different directions, as the earth changes its direction of motion. From the amount of tip (the "aberration of light"), Bradley could estimate the value of the speed of light and got a higher, and better, value than Roemer had.

Eventually, scientists obtained still more accurate measurements by applying refinements of Galileo's original idea. In 1849, the French physicist Armand Hippolyte Louis Fizeau set up an arrangement whereby a light was flashed to a mirror 5 miles away and reflected back to the observer. The elapsed time for the 10-mile round trip of the flash was not much more than 1/20,000 of a second, but Fizeau was able to measure it by placing a rapidly rotating toothed wheel in the path of the light beam. When the wheel turned at a certain speed, the flash going out between the two teeth would hit the next tooth when it came back from the mirror, and so Fizeau, behind the wheel, would not see it. When the wheel was speeded up, the returning flash would not be blocked but would come through the next gap between teeth. Thus, by controlling and measuring the speed of the turning wheel, Fizeau was able to calculate the elapsed time, and therefore the speed of travel, of of the flash of light.

A year later, Jean Foucault (who was soon to perform his pendulum experiment; see Chapter 3) refined the measurement by using a rotating mirror instead of a toothed wheel. Now the elapsed time was measured by a slight shift in the angle of reflection by the rapidly turning mirror. Foucault got a value of 187,000 miles per second for the speed of light in air. In addition, Foucault used his method to determine the speed of light through various liquids. He found the speed to be markedly less than the speed of light in air. This fitted Huyghen's wave theory, too.

Still greater precision in the measurement of light's velocity came

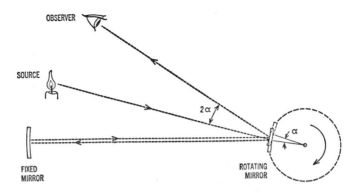

Foucault's method. The amount of rotation of the mirror, instead of Fizeau's toothed wheel, gave the speed of the light's travel.

with the work of Michelson, who over a period of more than forty years, starting in 1879, applied the Fizeau-Foucault approach with greater and greater refinement. He eventually sent light through a vacuum rather than through air (even air slows it up slightly), using evacuated steel pipes up to a mile long for the purpose. He measured the speed of light in a vacuum to be 186,284 miles per second. He was also to show that all wavelengths of light travel at the same speed in a vacuum.

In 1963, still more precise measurements have placed the speed of light at 186,281.7 miles per second. With the speed of light known with such amazing precision, it became possible to use light, or at least forms of light, to measure distance.

Imagine a short pulse of light moving outward, striking some obstacle, being reflected backward, and being received at the point where it had issued forth an instant before. What is needed is a wave form of low enough frequency to penetrate fog, mist, and cloud, but of high enough frequency to be reflected efficiently. The ideal range was found to be in the microwave (very short radio wave) region, with wavelengths of from 0.5 to 100 centimeters. From the time lapse between emission of the pulse and return of the echo, the distance of the reflecting object can be estimated.

A number of physicists worked on devices making use of this principle, but the Scottish physicist Robert Alexander Watson-Watt was the first to make it thoroughly practicable. By 1935, he had made it possible to follow an airplane by the microwave reflections it sent back. The system was called "*ra*dio *d*etection *a*nd *r*anging," the word "range" meaning "to determine the distance of." The phrase was abbreviated to "ra. d. a. r.," or "radar." (A word, such as radar, that is constructed out of the initials of a phrase is called an "acronym." Acronyms are becoming more and more common in the modern world, particularly in science and technology.)

The world first became conscious of radar when it was learned that it was by use of that device that the British had been able to detect oncoming Nazi planes during the Battle of Britain, despite night and fog. To radar therefore belongs at least part of the credit of the British victory.

Since World War II, radar has had numerous peacetime uses. It has been used to detect rainstorms and has helped the weatherman in this respect. It has turned up mysterious reflections called "angels," which turned out to be, not heavenly messengers, but flocks of birds, so that now radar is used in the study of bird migrations.

And, as described in Chapter 2, it was radar reflections from Venus and Mercury that gave astronomers new knowledge concerning the rotations of those planets and, with regard to Venus, information about the nature of the surface.

Through all the mounting evidence of the wave nature of light, a nagging question kept bothering the physicists. How was light transmitted through a vacuum? Other kinds of wave—sound, for instance—required a material medium. (From our observation platform here on earth, we could never hear an explosion on the moon or anywhere else in space, however loud, because sound waves cannot travel across space.) Yet here was light traveling through a vacuum more easily than through matter, and reaching us from galaxies billions of light-years away.

The classical scientists were always uncomfortable about the notion. of "action at a distance." Newton, for instance, worried about how the force of gravity could operate through space. As a possible explanation, he revived the Greeks' idea of an "ether" filling the heavens and speculated that perhaps the force of gravity might somehow be conducted by the ether.

Trying to account for the travel of light waves through space, physicists decided that light, too, must be conducted by the supposed ether. They began to speak of the "luminiferous ether." But this idea at once ran into a serious difficulty. Light waves are transverse waves: that is, they undulate at right angles to the direction of travel, like the ripples on the surface of water, in contrast to the "longitudinal" motion of sound waves. Now physical theory said that only a *solid* medium could convey transverse waves. (Transverse water waves travel on the water surface, a special case, but cannot penetrate the body of the liquid.) Therefore the ether had to be solid, not gaseous or liquid. Not only must it be extremely rigid; to transmit waves at the tremendous speed of light, it had to be far more rigid than steel. What is more, this rigid ether had to permeate ordinary matter—not merely the vacuum of space but gases, water, glass, and all the other transparent substances through which light could travel.

To cap it all, this solid, super-rigid material had to be so frictionless, so yielding, that it did not interfere in the slightest with the motion of the smallest planetoid or the flicker of an eyelid!

Yet despite the difficulties introduced by the notion of the ether, it seemed so useful. Faraday, who had no mathematical background at all but who had marvelous insight, worked out the concept of lines of force—lines along which a magnetic field had equal strength—and, visualizing these as elastic distortions of the ether, used them to explain magnetic phenomena.

In the 1860's, Clerk Maxwell, a great admirer of Faraday, set about supplying the mathematical analysis that would account for the lines of force. In doing so, he evolved a set of four simple equations that among them described almost all phenomena involving electricity and magnetism. These equations, advanced in 1864, not only described the interrelationship of the phenomena of electricity and magnetism, but showed the two could not be separated. Where an electric field existed, there had to be a magnetic field, too, at right angles, and vice versa. There was, in fact, only a single "electromagnetic field." Furthermore, in considering the implications of his equations, Maxwell found that a changing electric field had to induce a changing magnetic field, which in turn had to induce a changing electric field, and so on; the two leapfrogged, so to speak, and the field progressed outward in all directions. The result was a radiation possessing the properties of a wave-form. In short, Maxwell predicted the existence of "electromagnetic radiation" with frequencies equal to that in which the electromagnetic field waxed and waned.

It was even possible for Maxwell to calculate the velocity at which such an electromagnetic wave would have to move. He did this by taking into consideration the ratio of certain corresponding values in the equations describing the force between electric charges and the force between magnetic poles. This ratio turned out to be precisely equal to the velocity of light, and Maxwell could not accept that as a mere coincidence. Light was an electromagnetic radiation, and along with it were other radiations with wavelengths far longer, or far shorter, than that of ordinary light—and all these radiations involved the ether.

Maxwell's equations, by the way, introduced a problem that is still with us. They seemed to emphasize a complete symmetry between the phenomena of electricity and magnetism: what was true of one was true of the other. Yet in one fundamental way, the two seemed different. Particles existed which carried one or the other of the two opposed charges—positive or negative, but not both. Thus the electron carried a negative electric charge only, while the positron carried a positive electric charge only. Analogously, ought not there be particles with a north magnetic pole only, and others with a south magnetic pole only? These "magnetic monopoles" have, however, never been found. Every particle involving a magnetic field has always possessed both a north and a south magnetic pole. Theory seems to indicate that the separation of the monopoles would require enormous energies of which only cosmic rays can dispose, but cosmic-ray research has as yet revealed no sign of them.

But back to the ether which, at the height of its power, met its Waterloo as a result of an experiment undertaken to test another classical question as knotty as "action at a distance," namely, the question of "absolute motion."

By the nineteenth century, it had become perfectly plain that the earth, the sun, the stars, and in fact all objects in the universe were in motion. Where, then, could you find a fixed reference point, one that was at "absolute rest," to determine "absolute motion"—the foundation on which Newton's laws of motion were based? There was one possibility.

Newton had suggested that the fabric of space itself (the ether, presumably) was at rest, so that one could speak of "absolute space." If the ether was motionless, perhaps one could find the "absolute motion" of an object by determining its motion in relation to the ether.

In the 1880's, Albert Michelson conceived an ingenious scheme to do just that. If the earth was moving through a motionless ether, he reasoned, then a beam of light sent in the direction of its motion and reflected back should travel a shorter distance than one sent out at right angles and reflected back. To make the test, Michelson invented the "interferometer," a device with a semimirror that lets half of a light beam through in the forward direction and reflects the other half at right angles. Both beams are then reflected back by mirrors to an eyepiece at the source. If one beam has traveled a slightly longer distance than the other, they arrive out of phase and form interference bands. This instrument is an extremely sensitive measurer of differences in length—so sensitive, in fact, that it can measure the growth of a plant from second to second and the diameter of some stars that seem to be dimensionless points of light in even the largest telescope.

Michelson's plan was to point the interferometer in various directions with respect to the earth's motion and detect the effect of the ether by the amount by which the split beams were out of phase on their return.

In 1887, with the help of the American chemist Edward Williams Morley, Michelson set up a particularly delicate version of the experiment. Stationing the instrument on a stone floating on mercury, so that it could

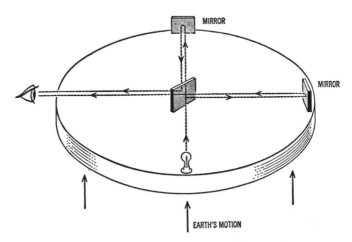

Michelson's interferometer. The semimirror (*center*) splits the light beam, reflecting one half and letting the other half go straight ahead. If the two reflecting mirrors (*at right and straight ahead*) are at different distances, the returning beams of light will arrive at the observer out of phase.

348

be turned in any direction easily and smoothly, they projected their beam in various directions with respect to the earth's motion. They discovered practically no difference! The interference bands were virtually the same no matter in what direction they pointed the instrument or however many times they performed the experiment. (It should be said here that more recent experiments along the same line with still more delicate instruments have shown the same negative results.)

The foundations of physics tottered. Either the ether was moving with the earth, which made no sense at all, or there was no such thing as the ether. In either case there was no "absolute motion" or "absolute space." "Classical" physics—the physics of Newton—had had the rug pulled out from under it. Newtonian physics still held in the ordinary world: planets still moved in accordance with his law of gravitation, and objects on earth still obeyed his law of inertia and of action-and-reaction. It was just that the classical explanations were incomplete, and physicists must be prepared to find phenomena that did not obey the classical "laws." The observed phenomena, both old and new, would remain, but the theories accounting for them would have to be broadened and refined.

The "Michelson-Morley experiment" is probably the most important experiment-that-did-not-work in the whole history of science. Michelson was awarded the Nobel Prize in physics in 1907—the first American scientist to receive a Nobel Prize.

Relativity

In 1893, the Irish physicist George Francis FitzGerald came up with a novel explanation to account for the negative results of the Michelson-Morley experiment. He suggested that all matter contracted in the direction of its motion and that the amount of contraction increased with the rate of motion. According to this interpretation, the interferometer was always shortened in the direction of the earth's "true" motion by an amount which exactly compensated for the difference in distance that the light beam would have to travel. Moreover, all possible measuring devices, including human sense organs, would be "foreshortened" in just the same way. FitzGerald's explanation almost made it look as if nature conspired to keep man from measuring absolute motion by introducing an effect that just canceled out any differences he might try to use to detect that motion.

This frustrating phenomenon became known as the "FitzGerald contraction." FitzGerald worked out an equation for it. An object moving at 7 miles per second (about the speed of our fastest present rockets) would contract by only about two parts per billion in the direction of flight. But at really high speeds the contraction would be substantial.

At 93,000 miles per second (half the speed of light) it would be 15 per cent; at 163,000 miles per second (⅞ the speed of light), 50 per cent. That is, a one-foot ruler moving past us at 163,000 miles per second would seem only six inches long to us—provided we knew a method of measuring its length as it flew by. And at the speed of light, 186,282 miles per second, its length in the direction of motion would be zero. Since presumably there can be no length shorter than zero, it would follow that the speed of light in a vacuum is the greatest possible velocity in the universe.

The Dutch physicist Hendrik Antoon Lorentz soon carried Fitz-Gerald's idea one step further. Thinking about cathode rays, on which he was working at the time, he reasoned that if the charge of a charged particle were compressed into a smaller volume, the mass of the particle should increase. Therefore a flying particle foreshortened in the direction of its travel by the FitzGerald contraction would have to increase in mass.

Lorentz presented an equation for the mass increase that turned out to be very similar to FitzGerald's equation for shortening. At 93,000 miles per second, an electron's mass would be increased by 15 per cent; at 163,000 miles per second, by 100 per cent (that is, its mass would be doubled); and at the speed of light, its mass would be infinite. Again it seemed that no speed greater than that of light could be possible, for how could mass be more than infinite?

The FitzGerald length effect and the Lorentz mass effect are so closely connected that the equations are often lumped together as the "Lorentz-FitzGerald equations."

If the FitzGerald contraction could not be measured, the Lorentz mass effect could be—indirectly. The ratio of an electron's mass to its charge can be determined from its deflection by a magnetic field. As an electron's velocity increased, the mass would increase, but there was no reason to think that the charge would; therefore, its mass–charge ratio should increase. By 1900, the German physicist W. Kauffman discovered that this ratio increased with velocity in such a way as to indicate that the electron's mass increased just as predicted by the Lorentz-FitzGerald equations. Later and better measurements showed the agreement to be just about perfect.

In discussing the speed of light as a maximum velocity, we must remember that it is the speed of light in a vacuum (186,282 miles per second) that is important here. In transparent material media, light moves more slowly. Its velocity in such a medium is equal to its velocity in a vacuum divided by the index of refraction of the medium. (The "index of refraction" is a measure of the extent by which a light beam, entering the material obliquely from a vacuum, is bent.)

In water, with an index of refraction of about 1.3, the speed of light is 186,282 divided by 1.3, or about 143,000 miles per second. In glass (index of refraction about 1.5), the speed of light is 124,000 miles per second; while in diamond (index of refraction, 2.4) the speed of light is a mere 78,000 miles per second.

It is possible for subatomic particles to travel through a particular transparent medium at a velocity greater than that of light in that medium (though *not* greater than that of light in a vacuum). When particles travel through a medium in this fashion, they throw back a wake of bluish light much as an airplane traveling at supersonic velocities throws back a wake of sound.

The existence of such radiation was observed by the Russian physicist Paul Alekseyevich Cherenkov (his name is also spelled Cerenkov) in 1934; in 1937, the theoretical explanation was offered by the Russian physicists Ilya Mikhailovich Frank and Igor Yevgenevich Tamm. All three shared the Nobel Prize for physics in 1958 as a result.

Particle detectors have been devised that detect the "Cerenkov radiation," and these "Cerenkov counters" are particularly well adapted to study particularly fast particles such as those making up the cosmic rays.

While the foundations of physics were still rocking, a second explosion took place.

This time the innocent question that started all the trouble had to do with the radiation emitted by matter when it is heated. (Although the radiation in question is usually in the form of light, physicists speak of the problem as "black-body radiation." All this means is that they are thinking of an ideal body that absorbs light perfectly—without reflecting any of it away, as a perfectly black body would do—and also radiates perfectly.) The Austrian physicist Josef Stefan showed in 1879 that the total radiation emitted by a body depended only on its temperature (not at all on the nature of its substance) and that in ideal circumstances the radiation was proportional to the fourth power of the absolute temperature: i.e., doubling the absolute temperature would increase its total radiation sixteen-fold ("Stefan's law"). It was also known that as the temperature rose, the predominant radiation moved toward shorter wavelengths. As a lump of steel is heated, for instance, it starts by radiating chiefly in the invisible infrared, then glows dim red, then bright red, then orange, then yellow-white, and finally, if it could somehow be kept from vaporizing at that point, it would be blue-white.

In 1893, the German physicist Wilhelm Wien worked out a theory of the energy distribution of black-body radiation, that is, of the amount of energy radiated at each particular wavelength range. It provided a formula that accurately described the distribution of energy at the violet end of the spectrum, but not at the red end. (For his work on heat he received the Nobel Prize in physics in 1911.) On the other hand, the English physicists Lord Rayleigh and James Jeans worked up an equation that described the distribution at the red end of the spectrum, but failed completely at the violet end. In short, the best theories available could explain one-half of the radiation or the other, but not both at once.

The German physicist Max Karl Ernst Ludwig Planck tackled the problem. He found that, in order to make the equations fit the facts, he had to introduce a completely new notion. He suggested that radiation consisted of small units or packets, just as matter was made up of atoms. He called the unit of radiation the "quantum" (after the Latin word for "how much?"). Planck argued that radiation could be absorbed only in whole numbers of quanta. Furthermore, he suggested that the amount of energy in a quantum depended on the wavelength of the radiation. The shorter the wavelength, the more energetic the quantum; or, to put it another way, the energy content of the quantum is inversely proportional to the wavelength.

Now the quantum could be related directly to the frequency of a given radiation. Like the quantum's energy content, the frequency is inversely proportional to the radiation's wavelength. If both the frequency and the quantum's energy content were inversely proportional to the wavelength, then the two were directly proportional to each other. Planck expressed this by means of his now-famous equation:

$$e = h\nu$$

The symbol "e" stands for the quantum energy, "ν" (the Greek letter "nu") for the frequency, and "h" for "Planck's constant," which gives the proportional relation between quantum energy and frequency.

The value of h is extremely small, and so is the quantum. The units of radiation are so small, in fact, that light looks continuous to us, just as ordinary matter seems continuous. But at the beginning of the twentieth century the same fate befell radiation as had befallen matter at the beginning of the nineteenth: both now had to be accepted as discontinuous.

Planck's quanta cleared up the connection between temperature and the wavelengths of emitted radiation. A quantum of violet light was twice as energetic as a quantum of red light, and naturally it would take more heat energy to produce violet quanta than red quanta. Equations worked out on the basis of the quantum explained the radiation of a black body very neatly at both ends of the spectrum.

Eventually Planck's quantum theory was to do a great deal more: it was to explain the behavior of atoms, of the electrons in atoms, and of nucleons in the atoms' nuclei. Planck was awarded the Nobel Prize in physics in 1918.

Planck's theory made little impression on physicists when it was first announced in 1900. It was too revolutionary to be accepted at once. Planck himself seemed appalled at what he had done. But five years

later a young German-born Swiss physicist named Albert Einstein verified the existence of his quanta.

The German physicist Philipp Lenard had found that when light
struck certain metals, it caused the metal surface to emit electrons, as if
the force of the light kicked electrons out of the atoms. The phenomenon acquired the name "photoelectric effect," and for its discovery
Lenard received the Nobel Prize for physics in 1905. When physicists began to experiment with it, they found, to their surprise, that increasing
the intensity of the light did not give the kicked-out electrons any more
energy. But changing the wavelength of light did affect them: blue light,
for instance, caused the electrons to fly out at greater speed than yellow
light did. A very dim blue light would kick out fewer electrons than a
bright yellow light would, but those few "blue-light" electrons would
travel with greater speed than any of the "yellow-light" electrons. On
the other hand, red light, no matter how bright, failed to knock out any
electrons at all from some metals.

None of this could be explained by the old theories of light. Why
should blue light do something red light could not do?

Einstein found the answer in Planck's quantum theory. To absorb
enough energy to leave the metal surface, an electron had to be hit by a
quantum of a certain minimum size. In the case of an electron held only
weakly by its atom (e.g., in cesium) even a quantum of red light would
do. Where atoms held electrons more strongly, yellow light was required,
or blue light, or even ultraviolet. And in any case, the more energetic the
quantum, the more speed it would give to the electron it kicked out.

Here was a case where the quantum theory explained a physical
phenomenon with perfect simplicity, whereas the prequantum view of
light had remained helpless. Other applications of quantum mechanics
followed thick and fast. For his explanation of the photoelectric effect
(not for his theory of relativity) Einstein was awarded the Nobel Prize
in physics in 1921.

In his "Special Theory of Relativity," presented in 1905 and evolved
in his spare time while he worked as examiner at the Swiss patent office,
Einstein proposed a new fundamental view of the universe based on an
extension of the quantum theory. He suggested that light traveled
through space in quantum form (the "photon"), and thus he resurrected
the concept of light consisting of particles. But this was a new kind of
particle. It had properties of a wave as well as of a particle, and sometimes it showed one set of properties and sometimes the other.

This has been made to seem a paradox, or even a kind of mysticism,
as if the true nature of light passes all possible understanding. That is
not so. To illustrate with an analogy, a man may have many aspects: husband, father, friend, businessman. Depending on circumstances and on
his surroundings, he behaves like a husband, father, friend, or businessman. You would not expect him to exhibit his husbandly behavior toward

a customer or his businesslike behavior toward his wife, and yet that makes him neither a paradox nor more than one man.

In the same way, radiation has both corpuscular and wave properties. In some capacities, the corpuscular properties are particularly pronounced; in others, the wave properties. This dual character gives a more satisfactory account of radiation than either set of properties alone can.

The discovery of the wave nature of light had led to all the triumphs of nineteenth-century optics, including spectroscopy. But it had also required physicists to imagine the existence of the ether. Now Einstein's particle-wave view kept all the nineteenth-century victories (including Maxwell's equations), but made it unnecessary to assume that the ether existed. Radiation could travel through a vacuum by virtue of its particle attributes, and the ether idea, killed by the Michelson-Morley experiment, could now be buried.

Einstein introduced a second important idea in his Special Theory of Relativity: that the speed of light in a vacuum never varied, regardless of the motion of its source. In Newton's view of the universe, a light beam from a source moving toward an observer should seem to travel more quickly than one from a source moving in any other direction. In Einstein's view, this did not happen, and from that assumption he was able to derive the Lorentz-FitzGerald equations. He showed that the increase of mass with velocity, which Lorentz had applied only to charged particles, could be applied to all objects of any sort. He reasoned further that increases in velocity would not only foreshorten length and increase mass but also slow the pace of time: in other words, clocks would slow down along with the shortening of yardsticks.

The most fundamental aspect of Einstein's theory was its denial of the existence of "absolute space" and "absolute time." This may sound like nonsense: How can the human mind learn anything at all about the universe if it has no point of departure? Einstein answered that all we needed to do was to pick a "frame of reference" to which the events of the universe could be related. Any frame of reference (the earth motionless, or the sun motionless, or we ourselves motionless, for that matter) would be equally valid, and we could simply choose the frame that was most convenient. It is more convenient to calculate planetary motions in a frame of reference in which the sun is motionless than in one in which the earth is motionless—but no more "true."

Thus measurements of space and time are "relative" to some arbitrarily chosen frame of reference—and that is the reason for naming Einstein's idea the "theory of relativity."

To illustrate. Suppose we on the earth were to observe a strange planet ("Planet X"), exactly like our own in size and mass, go whizzing past us at 163,000 miles per second relative to ourselves. If we could measure its dimensions as it shot past, we would find that it was foreshortened by 50 per cent in the direction of its motion. It would be an ellipsoid rather than a sphere and would, on further measurement, seem to have twice the mass of the earth.

Yet to an inhabitant of Planet X, it would seem that he himself and his own planet were motionless. The earth would seem to be moving past *him* at 163,000 miles per second, and it would appear to have an ellipsoidal shape and twice the mass of *his* planet.

One is tempted to ask which planet would *really* be foreshortened and doubled in mass, but the only possible answer is: that depends on the frame of reference. If you find that frustrating, consider that a man is small compared to a whale and large compared to a beetle. Is there any point in asking what a man is *really*, large or small?

For all its unusual consequences, relativity explains all the known phenomena of the universe at least as well as prerelativity theories do. But it goes further: it explains easily some phenomena that the Newtonian outlook explained poorly or not at all. Consequently, Einstein has been accepted over Newton, not as a replacement so much as a refinement. The Newtonian view of the universe can still be used as a simplified approximation that works well enough in ordinary life and even in ordinary astronomy, as in placing satellites in orbit. But when it comes to accelerating particles in a synchrotron, for example, we find that we must take account of the Einsteinian increase of mass with velocity to make the machine work.

Einstein's view of the universe so mingled space and time that either concept by itself became meaningless. The universe was four-dimensional, with time one of the dimensions (but behaving not quite like the ordinary spatial dimensions of length, breadth, and height). The four-dimensional fusion is often referred to as "space-time." This notion was first developed by one of Einstein's teachers, the Russian-German mathematician Hermann Minkowski, in 1907.

With time, as well as space, up to odd tricks in relativity, one aspect of relativity that still provokes arguments among physicists is Einstein's notion of the slowing of clocks. A clock in motion, he said, keeps time more slowly than a stationary one. In fact, all phenomena that change with time change more slowly when moving than when at rest, which is the same as saying that time itself is slowed. At ordinary speeds the effect is negligible, but at 163,000 miles per second a clock would seem (to an observer watching it fly past) to take two seconds to tick off one second. And at the speed of light, time would stand still.

The time-effect is more disturbing than those involving length and weight. If an object shrinks to half its length and then returns to normal, or if it doubles its weight and then returns to normal, no trace is left behind to indicate the temporary change, and opposing viewpoints need not quarrel.

Time, however, is a cumulative thing. If a clock on Planet X seems to be running at half-time for an hour because of its great speed, and if it is then brought to rest, it will resume its ordinary time-rate, but it will bear the mark of being half an hour slow! Well then, if two ships pass each other, and each considers the other to be moving at 163,000 miles per second and to be moving at half-time, when the two ships come together again, observers on each ship will expect the clock on the other ship to be half an hour slower than their own. But it isn't possible for each clock to be slower than the other? What, then, would happen? This problem is called the "clock paradox."

Actually, it isn't a paradox at all. If one ship just flashed by the other and both crews swore the other ship's clock was slow, it wouldn't matter which clock was "really" slow, because the two ships would separate forever. The two clocks would never be brought to the same place at the same time in order to be matched, and the clock paradox would never arise. Indeed, Einstein's Special Theory of Relativity only applies to uniform motion, so it is only the steady separation we are talking about.

Suppose, though, the two ships *did* come together after the flash-past, so that the clocks *could* be compared. In order for that to happen, something new must be added. At least one ship must accelerate. Suppose ship B did so: slowing down, traveling in a huge curve to point itself in the direction of A, then speeding up until it catches up with A. Of course, B might choose to consider itself at rest; by its chosen frame of reference, it is A that does all the changing, speeding up backwards to come to A. If the two ships were all there were to the universe, then indeed the symmetry would keep the clock paradox in being.

However, A and B are *not* all there is to the universe, and that upsets the symmetry. When B accelerates, it is doing so with reference not only to A, but to all the rest of the universe besides. If B chooses to consider itself at rest, it must consider not only A, but all the galaxies without exception, to be accelerating with respect to itself. It is B against the universe, in short. Under those circumstances, it is B's clock that ends up half an hour slow, not A's.

This affects notions of space travel. If astronauts leaving earth speed up to near the speed of light, their time would be much slower than ours. They might reach a distant destination and return in what seemed to them weeks, though on the earth many centuries would have passed. If time really slows in motion, a person might journey even to a distant star in his own lifetime. But of course he would have to say good-by to his own generation and the world he knew. He would return to a world of the future.

In the Special Theory of Relativity Einstein did not deal with accelerated motion or gravitation. These were treated in his "General Theory of Relativity," published in 1915. The General Theory presented a completely altered view of gravitation. It was viewed as a property of space, rather than as a force between bodies. As the result of the presence of matter, space became curved, and bodies followed the line of least resistance among the curves, so to speak. Strange as Einstein's idea seemed, it was able to explain something that the Newtonian law of gravity had not been able to explain.

The greatest triumph of Newton's law of gravity had come in 1846. The planet Uranus, discovered in 1781, had a slightly erratic orbit around the sun. A half century of observation made that unmistakable. Astronomers decided that some still undiscovered planet beyond it must be exerting a gravitational pull on it. The British astronomer John Couch Adams and the French astronomer Urbain Jean Joseph Leverrier calculated the position of this hypothetical planet, using Newton's theories as a basis. In 1846, the German astronomer Johann Gottfried Galle pointed his telescope at the spot indicated by Leverrier and, sure enough, there was a new planet—since named Neptune.

After that, nothing seemed capable of shaking Newton's law of gravity. And yet one planetary motion remained unexplained. The planet Mercury's point of nearest approach to the sun ("perihelion") changed from one trip to the next; it was never in the same place twice in the planet's "yearly" revolutions around the sun. Astronomers were able to account for most of this irregularity as due to "perturbations" of its orbit by the pull of the neighboring planets.

Indeed, there had been some feeling in the early days of work with the theory of gravitation that perturbations arising from the shifting pull of one planet on another might eventually act to break up the delicate mechanism of the solar system. In the earliest decades of the nineteenth century, however, the French astronomer Pierre Simon Laplace showed that the solar system was not as delicate as all that. The perturbations were all cyclic, and orbital irregularities never increased to more than a certain amount in any direction. In the long run, the solar system is stable, and astronomers were more certain than ever that all particular irregularities could be worked out by taking perturbations into account.

This, however, did not work for Mercury. After all perturbations were allowed for, there was still an unexplained one-way shift of Mercury's perihelion by an amount equal to 43 seconds of arc per century. This motion, discovered by Leverrier in 1845, is not much: in 4,000 years it adds up only to the width of the moon. It was enough, however, to upset astronomers.

Leverrier suggested that this deviation might be caused by a small, undiscovered planet closer to the sun than Mercury. For decades astronomers searched for the supposed planet (called "Vulcan"), and many were the reports of its discovery. All the reports turned out to be mistaken. Finally it was agreed that Vulcan did not exist.

Then Einstein's General Theory of Relativity supplied the answer.

It showed that the perihelion of any revolving body should have a motion beyond that predicted by Newton's law. When this new calculation was applied to Mercury, the planet's shift of perihelion fit it exactly. Planets farther from the sun than Mercury should show a progressively smaller shift of perihelion. In 1960, the perihelion of Venus' orbit had been found to be advancing about 8 seconds of arc per century; this shift fits Einstein's theory almost exactly.

More impressive were two unexpected new phenomena that only Einstein's theory predicted. First, Einstein maintained that an intense gravitational field should slow down the vibrations of atoms. The slow-down would be evidenced by a shift of spectral lines toward the red (the "Einstein shift"). Casting about for a gravitational field strong enough to produce this effect, astronomers thought of the dense white-dwarf stars. They looked at the spectra of white dwarfs and did indeed find the predicted shift of lines.

The verification of Einstein's second prediction was even more dramatic. His theory said a gravitational field would bend light rays. Einstein calculated that a ray of light just skimming the sun's surface would be bent out of a straight line by 1.75 seconds of arc. How could that be checked? Well, if stars beyond the sun and just off its edge could be observed during an eclipse of the sun and their positions compared with what they were against the background when the sun did not interfere, any shift resulting from bending of their light should show up. Since Einstein had published his paper on general relativity in 1915, the test had to wait until after the end of World War I. In 1919, the British Royal Astronomical Society organized an expedition to make the test by witnessing a total eclipse visible from the island of Principe, a small Portuguese-owned island off West Africa. The stars did shift position. Einstein had been verified again.

APPARENT

TRUE

The gravitational bending of light waves, postulated by Einstein in the General Theory of Relativity.

By this same principle, if one star were directly behind another, the light of the farther star would bend about the nearer in such a way that the farther star would appear larger than it really is. The nearer star would act as a "gravitational lens." Unfortunately, the apparent size of stars is so minute that an eclipse of a distant star by a much closer one (as seen from Earth) is extremely rare, although some astronomers have speculated that the puzzling properties of quasars may be due to gravitational lens-effects. An eclipse of this sort should take place in 1988. No doubt astronomers will be watching.

The three great victories of Einstein's General Theory were all astronomic in nature. Scientists longed to discover a way to check it in the laboratory under conditions they could vary at will. The key to such a laboratory demonstration arose in 1958, when the German physicist, Rudolf Ludwig Mössbauer, showed that, under certain conditions, a crystal could be made to produce a beam of gamma rays of sharply defined wavelength. Ordinarily, the atom emitting the gamma ray recoils, and this recoil broadens the band of wavelengths produced. In crystals under certain conditions, the crystal acts as a single atom: the recoil is distributed among all the atoms and sinks to virtually nothing, so that the gamma ray emitted is exceedingly sharp. Such a sharp-wavelength beam can be absorbed with extraordinary efficiency by a crystal similar to that which produced it. If the gamma rays are of even slightly different wavelength from that which the crystal would naturally produce, it would not be absorbed. This is called the "Mössbauer effect."

If such a beam of gamma rays is emitted downward so as to fall with gravity, the General Theory of Relativity requires it to gain energy so that its wavelength becomes shorter. In falling just a few hundred feet, it should gain enough energy for the decrease in wavelength of the gamma rays, though very minute, to become sufficiently large so that the absorbing crystal will no longer absorb the beam.

Furthermore, if the crystal emitting the gamma ray is moved upward while the emission is proceeding, the wavelength of the gamma ray is increased through the Doppler-Fizeau effect. The velocity at which the crystal is moved upward can be adjusted so as to just neutralize the effect of gravitation on the falling gamma ray, which will then be absorbed by the crystal on which it impinges.

Experiments conducted in 1960 and later made use of the Mössbauer effect to confirm the General Theory with great exactness. They were the most impressive demonstration of its validity that has yet been seen; as a result, Mössbauer was awarded the 1961 Nobel Prize for physics.

Despite all this, the claim to validity of Einstein's General Theory remains tenuous. The confirmations remain borderline. In 1961, the American physicist Robert Henry Dicke evolved a more complex concept he calls the "scalar-tensor theory," which treats gravitation not as a geometric effect, as Einstein's theory does, but as a combination of two fields of different properties. The two theories predict phenomena so nearly alike as to be virtually indistinguishable. In the summer of 1966, Dicke measured the sphericity of the sun and, by very delicate measurement, claimed to have detected a slight equatorial bulge. This bulge would account for 8 per cent of the observed advance of Mercury's perihelion and would destroy the excellent fit of the General Theory. This would weaken Einstein's theory but leave Dicke's unaffected.

On the other hand, although both theories predict that light (or radio) waves would be slowed when passing a massive object, they differ somewhat in the degree of slowing they predict. In 1970, radio signals were reflected from planetary probes at a time when they were about to pass behind the sun (as viewed from earth) at a known distance. The time it took for the radio waves to arrive back would measure the degree to which they were slowed as they skimmed by the sun in each direction. The results, as reported, were considerably closer to Einstein's prediction than to Dicke's, but the matter was still not conclusive.

Heat

So far in this chapter I have been neglecting a phenomenon that usually accompanies light in our everyday experience. Almost all luminous objects from a star to a candle give off heat as well as light.

Heat was not studied, other than qualitatively, before modern times. It was enough for a person to say, "It is hot," or "It is cold," or "This is warmer than that." To subject temperature to quantitative measure, it was first necessary to find some measurable change that seemed to take place uniformly with change in temperature. One such change was found in the fact that substances expand when warmed and contract when cooled.

Galileo was the first to try to make use of this fact to detect changes in temperature. In 1603, he inverted a glass tube of heated air into a bowl of water. As the air in the tube cooled to room temperature, it contracted and drew water up the tube, and there Galileo had his "thermometer" (from Greek words meaning "heat measure"). When the temperature of the room changed, the water level in the tube changed. If the room warmed, the air in the tube expanded and pushed the water level down; if it grew cooler, the air contracted and the water level moved up. The only trouble was that the basin of water into which the tube had been inserted was open to the air and the air pressure kept changing. That also shoved the water level up and down, independently of temperature, confusing the results. The thermometer was the first important scientific instrument to be made of glass.

By 1654, the Grand Duke of Tuscany, Ferdinand II, had evolved a thermometer which was independent of air pressure. It contained a liquid sealed into a bulb to which a straight tube was attached. The contraction and expansion of the liquid itself was used as the indication of temperature change. Liquids change their volume with temperature much less than gases do, but by using a sizable reservoir of liquid and a filled bulb, so that the liquid could expand only up a very narrow tube, the rise and fall within that tube, for even tiny volume changes, could be made considerable.

The English physicist Robert Boyle did much the same thing about the same time, and he was the first to show that the human body had a constant temperature, markedly higher than the usual room temperature. Others demonstrated that certain physical phenomena always took place at some fixed temperature. Before the end of the seventeenth century this was found to be so in the case of the melting of ice and the boiling of water.

The first liquids used in thermometry were water and alcohol. Since water froze too soon and alcohol boiled away too easily, the French physicist Guillaume Amontons resorted to mercury. In his device, as in Galileo's, the expansion and contraction of air caused the mercury level to rise or fall.

Then in 1714 the German physicist Gabriel Daniel Fahrenheit combined the advances of the Grand Duke and of Amontons by enclosing mercury in a tube and using its own expansion and contraction with temperature as the indicator. Furthermore, Fahrenheit put a graded scale on the tube to allow the temperature to be read quantitatively.

There is some argument as to exactly how Fahrenheit arrived at the particular scale he used. He set zero, according to one account, at the lowest temperature he could get in his laboratory, attained by mixing salt and melting ice. He then set the freezing point of pure water at 32 and its boiling point at 212. This had two advantages. First, the range of temperature over which water was liquid came to 180°, which seemed a natural number to use in connection with "degrees." (It is the number of degrees in a semicircle.) Second, body temperature came near a round 100°; normally it is 98.6° Fahrenheit, to be exact.

So constant is body temperature normally, that if it is more than a degree or so above the average value, the body is said to run a fever and there is a clear feeling of illness. In 1858, the German physician Karl August Wunderlich introduced the procedure of frequent checks on body temperature as an indication of the course of disease. In the next decade, the British physician Thomas Clifford Allbutt invented the "clinical thermometer" in which there is a constriction in the narrow tube containing the mercury. The mercury thread rises to a maximum when placed in the mouth, but does not fall when the thermometer is removed. The mercury thread simply divides at the constriction, leaving the portion above with its reading held constant. In Great Britain and the United States, the Fahrenheit scale is still used. We are familiar with it in everyday affairs such as weather-reporting and in the use of clinical thermometers.

In 1742, however, the Swedish astronomer Anders Celsius adopted a different scale. In its final form, this set the freezing point of water at 0 and its boiling point at 100. Because of the hundredfold division of the temperature range in which water was liquid, this is called the "centigrade scale," from Latin words meaning "hundred steps." Most people still speak of measurements on this scale as "degrees centigrade," but scientists, at an international conference in 1948, renamed the scale after the inventor, following the Fahrenheit precedent. Officially, then, one should speak of the "Celsius scale" and of "degrees Celsius." The symbol C. still holds. It was Celsius' scale that won out in most of the civilized world. Scientists, in particular, found the Celsius scale convenient.

Temperature measures the intensity of heat but not its quantity. Heat will always flow from a place of higher temperature to a place of lower temperature until the temperatures are equal, just as water will flow from a higher level to a lower one until the levels are equal. This is true regardless of the relative amounts of heat contained in the bodies involved. Although a bathtub of lukewarm water contains far more heat than a burning match, when the match is placed near the water, heat goes from the match to the water, not vice versa.

Joseph Black, who had done important work on gases (see Chapter 4), was the first to make clear the distinction between temperature and heat. In 1760, he announced that various substances were raised in temperature by different amounts when a given amount of heat was poured into them. To raise the temperature of a gram of iron by one degree Celsius takes three times as much heat as to warm a gram of lead by one degree. And beryllium requires three times as much heat as iron.

Furthermore, Black showed it was possible to pour heat into a substance without raising its temperature at all. When ice is heated, it begins to melt, but it does not rise in temperature. Heat will eventually melt all the ice, but the temperature of the ice itself never goes above 0° C. The same thing happens in the case of boiling water at 100° C. As heat is poured into the water, more and more of it boils away as vapor, but the temperature of the liquid does not change.

The development of the steam engine (see Chapter 8), which came at about the same time as Black's experiments, intensified the interest of scientists in heat and temperature. They began to speculate about the nature of heat, as earlier they had speculated about the nature of light.

In the case of heat, as of light, there were two theories. One held that heat was a material substance which could be poured or shifted from one substance to another. It was named "caloric," from the Latin for "heat." According to this view, when wood was burned the caloric in the wood passed into the flame and from that into a kettle above the flame and from that into the water in the kettle. As water filled with caloric, it was converted to steam.

In the late eighteenth century, two famous observations gave rise to the theory that heat was a form of vibration. One was published by the American physicist and adventurer Benjamin Thompson, a Tory who fled the country during the Revolution, was given the title Count Rumford and then proceeded to knock around Europe. While supervising the boring of cannon in Bavaria in 1798, he noticed that quantities of heat were being produced. He found that enough heat was being generated to bring eighteen pounds of water to the boiling point in less than three hours. Where was all the caloric coming from? Thompson decided that heat must be a vibration set up and intensified by the mechanical friction of the borer against the cannon.

The next year the chemist Humphry Davy performed an even more significant experiment. Keeping two pieces of ice below the freezing point, he rubbed them together, not by hand but by a mechanical contrivance, so that no caloric could flow into the ice. By friction alone, he melted some of the ice. He, too, concluded that heat must be a vibration and not a material. Actually, this experiment should have been conclusive, but the caloric theory, though obviously wrong, persisted to the middle of the nineteenth century.

Nevertheless, although the nature of heat was misunderstood, scientists learned some important things about it, just as the investigators of light turned up interesting facts about the reflection and refraction of light beams before they knew its nature. Jean Baptiste Joseph Fourier and Nicholas Léonard Sadi Carnot in France studied the flow of heat and made important advances. In fact, Carnot is usually considered the founder of the science of "thermodynamics" (from Greek words meaning "movement of heat"). He placed the working of steam engines on a firm theoretical foundation.

Carnot did his work in the 1820's. By the 1840's, physicists were concerned with the manner in which the heat that was put into steam could be converted into the mechanical work of moving a piston. Was there a limit to the amount of work that could be obtained from a given amount of heat? And what about the reverse process: How was work converted to heat?

Joule spent thirty-five years converting various kinds of work into heat, doing very carefully what Rumford had earlier done clumsily. He measured the amount of heat produced by an electric current. He heated water and mercury by stirring them with paddle wheels, or by forcing water through narrow tubes. He heated air by compressing it, and so on. In every case he calculated how much mechanical work had been done on the system and how much heat was obtained as a result. He found that a given amount of work, of any kind, always produced a given amount of heat, which was called the "mechanical equivalent of heat."

Since heat could be converted into work, it must be considered a form of "energy" (from Greek words meaning "containing work"). Electricity, magnetism, light, and motion could all be used to do work, so they, too, were forms of energy. And work itself, being convertible into heat, was a form of energy.

This emphasized something that had been more or less suspected since Newton's time: that energy was "conserved" and could neither be created nor destroyed. Thus, a moving body has "kinetic energy" ("energy of motion"), a term introduced by Lord Kelvin in 1856. Since a body moving upward is slowed by gravity, its kinetic energy slowly disappears. However, as the body loses kinetic energy, it gains energy of position, for, by virtue of its location high above the surface of the earth, it can eventually fall and regain kinetic energy. In 1853, the Scottish physicist William John Macquorn Rankine named this energy of position "potential energy." It seemed that a body's kinetic energy plus its potential energy (its "mechanical energy") remained nearly the same during the course of its movement, and this was called "conservation of mechanical energy." However, mechanical energy was not *perfectly* conserved. Some was lost to friction, to air resistance, and so on.

What Joule's experiments showed above all was that such conservation could be made exact when heat was taken into account, for, when mechanical energy was lost to friction or air resistance, it appeared as heat. Take that heat into account, and one can show, without qualification, that no new energy is created and no old energy destroyed. The first person actually to put this notion into words was Heinrich von Helmholtz. In 1847, von Helmholtz enunciated the "law of conservation of energy," which states that energy can be converted from one form to another but cannot be created or destroyed. Whenever a certain amount of energy seems to disappear in one place, an equivalent amount must appear in another. This is also called "the first law of thermodynamics." It remains a foundation block of modern physics, undisturbed by either quantum theory or relativity.

Now, although any form of work can be converted entirely into heat, the reverse is not true. When heat is turned to work, some of it is unusable and is unavoidably wasted. In running a steam engine, the heat of the steam is converted into work only until the temperature of the steam is reduced to the temperature of the environment; after that, although there is much remaining heat in the cold water formed from the steam, no more of it can be converted to work. Even in the temperature range at which work can be extracted, some of the heat does not go into work but is used up in heating the engine and the air around it, in overcoming friction between the piston and the cylinder, and so on.

In any energy conversion—e.g., electric energy into light energy, or magnetic energy into energy of motion—some of the energy is wasted. It is not lost; that would be contrary to the first law. But it is converted to heat that is dissipated in the environment.

The capacity of any system to perform work is its "free energy." The portion of the energy that is unavoidably lost as nonuseful heat is reflected in the measurement of the "entropy"—a term first used in 1850 by the German physicist Rudolf Julius Emmanuel Clausius.

Clausius pointed out that in any process involving a flow of energy there is always some loss, so that the entropy of the universe is continually increasing. This continual increase of entropy is called the "second law of thermodynamics." It is sometimes referred to as the "running-down of the universe" or the "heat-death of the universe." Fortunately, the quantity of usable energy (supplied almost entirely by the stars, which are, of course, "running down" at a tremendous rate) is so vast that there is enough for all purposes for many billions of years.

A clear understanding of the nature of heat finally came with the understanding of the atomic nature of matter. It developed from the realization that the molecules composing a gas were in continual motion, bouncing off one another and off the walls of their container. The first investigator who attempted to explain the properties of gases from this standpoint was the Swiss mathematician Daniel Bernoulli, in 1738, but he was ahead of his times. In the mid-nineteenth century, Maxwell and Boltzmann (see page 222) worked out the mathematics adequately and established the "kinetic theory of gases" ("kinetic" comes from a Greek word meaning "motion"). The theory showed heat to be equivalent to the motion of molecules. Thus the caloric theory of heat received its deathblow. Heat was seen to be a vibrational phenomenon: the movement of molecules in gases and liquids or the jittery to-and-fro trembling of molecules in solids.

When a solid is heated to a point where the to-and-fro trembling is strong enough to break the bonds that hold neighboring molecules together, the solid melts and becomes a liquid. The stronger the bond between neighboring molecules in a solid, the more heat is needed to make it vibrate violently enough to break the bond. This means that the substance has a higher melting point.

In the liquid state, the molecules can move freely past one another. When the liquid is heated further, the movements of the molecules finally become sufficiently energetic to set them free of the body of the liquid altogether, and then the liquid boils. Again the boiling point is higher where the intermolecular forces are stronger.

In converting a solid to a liquid, all of the energy of heat goes into breaking the intermolecular bonds. This is why the heat absorbed by melting ice does not raise the ice's temperature. The same is true of a liquid being boiled.

Now we can distinguish between heat and temperature easily. Heat is the total energy contained in the molecular motions of a given quantity of matter. Temperature represents the average energy of motion per molecule in that matter. Thus a quart of water at 60° C. contains twice as much heat as a pint of water at 60° C. (twice as many molecules are vibrating), but the quart and pint have the same temperature, for the average energy of molecular motion is the same in each case.

There is energy in the very structure of a chemical compound—that is, in the bonding forces that hold an atom or iron or molecule to its neighbor. If these bonds are broken and rearranged into new bonds involving less energy, the excess of energy will make its appearance as heat or light or both. Sometimes the energy is released so quickly that an explosion is the result.

It is possible to calculate the chemical energy contained in any substance and show what the amount of heat released in any reaction must be. For instance, the burning of coal involves breaking the bonds between carbon atoms in the coal and the bonds between the oxygen molecules' atoms, with which the carbon recombines. Now the energy of the bonds in the new compound (carbon dioxide) is less than that of the bonds in the original substances that formed it. This difference, which can be measured, is released as heat and light.

In the 1870's, the American physicist Josiah Willard Gibbs worked out the theory of "chemical thermodynamics" in such detail that this branch of science was brought from virtual nonexistence to complete maturity at one stroke.

The long paper in which Gibbs described his reasoning was far above the heads of others in America, and it was published in the *Transactions of the Connecticut Academy of Arts and Sciences* only after considerable hesitation. Even afterward, its close-knit mathematical argument and the retiring nature of Gibbs himself combined to keep the subject under a bushel basket until the Russian-German physical chemist Wilhelm Ostwald discovered the work in 1883, translated the paper into German, and proclaimed the importance of Gibbs to the world.

As an example of the importance of Gibbs' work, his equations demonstrated the simple, but rigorous, rules governing the equilibrium between different substances existing simultaneously in more than one phase (i.e., in both solid form and in solution, in two immiscible liquids and a vapor, and so on). This "phase rule" is the breath of life to metallurgy and to many other branches of chemistry.

Mass to Energy

With the discovery of radioactivity in 1896 (see Chapter 5), a totally new question about energy arose at once. The radioactive substances uranium and thorium were giving off particles with astonishing energies. Moreover, Marie Curie found that radium was continually emitting heat in substantial quantities: an ounce of radium gave off 4,000 calories per hour, and this would go on hour after hour, week after week, decade after decade. The most energetic chemical reaction known could not produce a millionth of the energy liberated by radium. And, what was no less surprising, this production of energy, unlike chemical reactions, did not depend on temperature: it went on just as well at the very low temperature of liquid hydrogen as it did at ordinary temperatures!

Quite plainly an altogether new kind of energy, very different from chemical, was involved here. Fortunately physicists did not have to wait long for the answer. Once again, it was supplied by Einstein, in his Special Theory of Relativity.

Einstein's mathematical treatment of energy showed that mass could be considered a form of energy—a very concentrated form, for a very small quantity of mass would be converted into an immense quantity of energy.

Einstein's equation relating mass and energy is now one of the most famous equations in the world. It is:

$$e = mc^2$$

Here "e" represents energy (in ergs), "m" represents mass (in grams) and "c" represents the speed of light (in centimeters per second).

Since light travels at 30 billion centimeters per second, the value of c^2 is 900 billion billion. This means that the conversion of one gram of mass energy will produce 900 billion billion ergs. The erg is a small unit of energy not translatable into any common terms, but we can get an idea of what this number means when we are told that the energy in one gram of mass is sufficient to keep a 1,000-watt electric-light bulb running for 2,850 years. Or, to put it another way, the complete conversion of a gram of mass into energy would yield as much as the burning of 2,000 tons of gasoline.

Einstein's equation destroyed one of the sacred conservation laws of science. Lavoisier's "law of conservation of mass" had stated that matter could neither be created nor destroyed. Actually, every energy-releasing chemical reaction changes a small amount of mass into energy: the products, if they could be weighed with utter precision, would not quite equal the original matter. But the mass lost in ordinary chemical reactions is so small that no technique available to the chemists of the nineteenth century could conceivably have detected it. Physicists, however, were now dealing with a completely different phenomenon, the nuclear reaction of radioactivity rather than the chemical reaction of burning coal. Nuclear reactions released so much energy that the loss of mass was large enough to be measured.

By postulating the interchange of mass and energy, Einstein merged the laws of conservation of energy and of mass into one law—the conservation of mass-energy. The first law of thermodynamics not only still stood: it was more unassailable than ever.

The conversion of mass to energy was confirmed experimentally by Francis W. Aston through his mass spectrograph. This could measure the mass of atomic nuclei very precisely by the amount of their deflection by a magnetic field. What Aston did with an improved instrument in 1925, was to show that the various nuclei were not exact multiples of the masses of the neutrons and protons that composed them.

Let us consider the masses of these neutrons and protons for a moment. For a century, the masses of atoms and subatomic particles generally have been measured on the basis of allowing the atomic weight of oxygen to be exactly 16.00000 (see Chapter 5). In 1929, however, William Giauque had showed that oxygen consisted of three isotopes, oxygen 16, oxygen 17, and oxygen 18 and that the atomic weight of oxygen was the weighted average of the mass numbers of these three isotopes.

To be sure, oxygen 16 was by far the most common of the three, making up 99.759 per cent of all oxygen atoms. This meant that if oxygen had the over-all atomic weight of 16.00000, the oxygen-16 isotope had a mass number of *almost* 16. (The masses of the small quantities of oxygen 17 and oxygen 18 brought the value up to 16.) Chemists, for a generation after the discovery, did not let this disturb them, but kept the old basis for what came to be called "chemical atomic weights."

Physicists, however, reacted otherwise. They preferred to set the mass of the oxygen-16 isotope at exactly 16.0000 and determine all other masses on that basis. On this basis, the "physical atomic weights" could be set up. On the oxygen-16 equals 16 standard, the atomic weight of oxygen itself, with its traces of heavier isotopes, was 16.0044. In general the physical atomic weights of all elements would be 0.027 per cent higher than their chemical atomic weight counterparts.

In 1961, physicists and chemists reached a compromise. It was agreed to determine atomic weights on the basis of allowing the carbon-12 isotope to have a mass of 12.0000. This based the atomic weights on a characteristic mass number and made them as fundamental as possible. In addition, this base made the atomic weights almost exactly what they were under the old system. Thus on the carbon-12 equals 12 standard, the atomic weight of oxygen is 15.9994.

Well, then, let us start with a carbon-12 atom, with its mass equal to 12.0000. Its nucleus contains six protons and six neutrons. From mass-spectrographic measurements it becomes evident that, on the carbon-12 equals 12 standard, the mass of a proton is 1.007825 and that of a neutron is 1.008665. Six protons, then, should have a mass of 6.046950 and six neutrons 6.051990. Together the twelve nucleons should have a mass of 12.104940. But the mass of the carbon-12 is 12.00000. What has happened to the missing 0.104940?

This disappearing mass is the "mass defect." The mass defect divided by the mass number gives the mass defect per nucleon, or the "packing fraction." The mass has not really disappeared, of course. It has been converted into energy in accordance with Einstein's equation so that the mass defect is also the "binding energy" of the nucleus. To break a nucleus down into individual protons and neutrons would require the input of an amount of energy equal to the binding energy, since an amount of mass equivalent to that energy would have to be formed.

Aston determined the packing fraction of many nuclei, and he found that it increased rather quickly from hydrogen up to elements in the neighborhood of iron and then decreased, rather slowly, for the rest of the periodic table. In other words, the binding energy per nucleon was highest in the middle of the periodic table. This meant that conversion of an element at either end of the table into one nearer the middle should release energy.

Take uranium 238 as an example. This nucleus breaks down by a series of decays to lead 206. In the process, it emits eight alpha particles. (It also gives off beta particles, but these are so light they can be ignored.) Now the mass of lead 206 is 205.9745 and that of eight alpha particles totals 32.0208. Altogether these products add up to a mass of 237.9953. But the mass of uranium 238, from which they came, is 238.-0506. The difference, or loss of mass, is 0.0553. That loss of mass is just enough to account for the energy released when uranium breaks down.

When uranium breaks down to still smaller atoms, as it does in fission, a great deal more energy is released. And when hydrogen is converted to helium, as it is in stars, there is an even larger fractional loss of mass and a correspondingly richer development of energy.

Physicists began to look upon the mass-energy equivalence as a very reliable bookkeeping. For instance, when the positron was discovered in 1934, its mutual annihilation with an electron produced a pair of gamma rays whose energy was just equal to the mass of the two particles. Furthermore, as Blackett was first to point out, mass could be created out of appropriate amounts of energy. A gamma ray of the proper energy, under certain circumstances, would disappear and give rise to an "electron-positron pair," created out of pure energy. Larger amounts of energy, supplied by cosmic particles or by particles fired out of proton synchrotons (see Chapter 6), would bring about the creation of more massive particles, such as mesons and antiprotons.

It is no wonder that when the bookkeeping did not balance, as in the emission of beta particles of less than the expected energy, physicists invented the neutrino to balance the energy account rather than tamper with Einstein's equation (see Chapter 6).

If any further proof of the conversion of mass to energy was needed, the atomic bomb provided the final clincher.

Particles and Waves

In the 1920's, dualism reigned supreme in physics. Planck had shown that radiation was particlelike as well as wavelike. Einstein had shown that mass and energy were two sides of the same coin, and that space and time were inseparable. Physicists began to look for other dualisms.

In 1923, the French physicist Louis Victor de Broglie was able to show that, just as radiation had the characteristics of particles, so the particles of matter, such as electrons, should display the characteristics of waves. The waves associated with these particles, he predicted, would have a wavelength inversely related to the mass times the velocity (that is, the momentum) of the particle. The wavelength associated with electrons of moderate speed, de Broglie calculated, ought to be in the X-ray region.

In 1927, even this surprising prediction was borne out. Clinton Joseph Davisson and Lester Halbert Germer of the Bell Telephone Laboratories were bombarding metallic nickel with electrons. As the result of a laboratory accident, which had made it necessary to heat the nickel for a long time, the metal was in the form of large crystals, which were ideal for diffraction purposes because the spacing between atoms in a crystal is comparable to the very short wavelengths of electrons. Sure enough, the electrons passing through those crystals behaved not as particles but as waves. The film behind the nickel showed interference patterns, alternate bands of fogging and clarity, just as it would have shown if X-rays rather than electrons had gone through the nickel.

Interference patterns were the very thing that Young had used more than a century earlier to prove the wave nature of light. Now they proved the wave nature of electrons. From the measurements of the interference bands, the wavelength associated with the electron could be calculated, and it turned out to be 1.65 angstrom units, almost exactly what de Broglie had calculated it ought to be.

In the same year the British physicist George Paget Thomson, working independently and using different methods, also showed that electrons had wave properties.

De Broglie received the Nobel Prize in physics in 1929, and Davisson and Thomson shared the Nobel Prize in physics in 1937.

This entirely unlooked-for discovery of a new kind of dualism was put to use almost at once in microscopy. Ordinary optical microscopes, as I have mentioned, cease to be useful at a certain point because there is a limit to the size of objects that light waves can define sharply. As objects get smaller, they also get fuzzier, because the light waves begin to pass around them—something first pointed out by the German physicist Ernst Karl Abbe in 1878. (For the same reason, the long radio-waves give a fuzzy picture even of large objects in the sky.) The cure, of course, is to try to find shorter wavelengths with which to resolve the smaller objects. Ordinary-light microscopes can distinguish two dots 1/5,000 millimeter apart, but ultraviolet microscopes can distinguish dots 1/10,000 millimeter apart. X-rays would be better still, but there are no lenses for X-rays. This problem can be solved, however, by using the waves associated with electrons, which have about the same wavelength as X-rays but are easier to manipulate. For one thing, a magnetic field can bend the "electron rays," because the waves are associated with a charged particle.

Just as the eye can see an expanded image of an object if the light rays involved are appropriately manipulated by lenses, so a photograph can register an expanded image of an object if electron waves are appropriately manipulated by magnetic fields. And, since the wavelengths associated with electrons are far smaller than those of ordinary light, the resolution obtainable with an "electron microscope" at high magnification is much greater than that available to an ordinary microscope.

A crude electron microscope capable of magnifying 400 times was made in Germany in 1932 by Ernst Ruska and Max Knoll, but the first really usable one was built in 1937 at the University of Toronto by James Hillier and Albert F. Prebus. Their instrument could magnify an object 7,000 times, whereas the best optical microscopes reach their limit with a magnification of about 2,000. By 1939, electron microscopes were commercially available and, eventually, Hillier and others developed electron microscopes capable of magnifying up to 2,000,000 times.

A "proton microscope," if one were developed, would magnify to a far greater extent than does an electron microscope, because the waves associated with protons are shorter. In a sense, the proton synchrotron is a kind of proton microscope, probing the interior of the nucleus with its speeding protons. The greater the speed of a proton, the greater its momentum and the shorter the wavelength associated with it. Protons with an energy of one Mev can "see" the nucleus, while at 20 Mev they can begin to "see" detail within the nucleus. This is another reason why physicists are eager to pile more and more electron volts into their atom smashers—so that they may "see" the ultrasmall more clearly.

It ought not be too surprising that this particle-wave dualism works in reverse and that phenomena ordinarily considered wavelike in nature would have particle characteristics as well. Planck and Einstein had already shown radiation to consist of quanta, which, in a fashion, are particles. In 1923, Compton, the physicist who was to demonstrate the particle nature of cosmic rays (see Chapter 6), showed that such quanta possessed some down-to-earth particle qualities. He found that X-rays, on being scattered by matter, lost energy and became longer in wavelength. This was just what might be expected of a radiation "par-

371

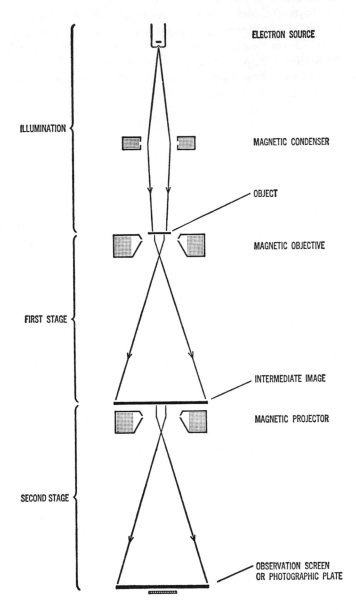

ELECTRON SOURCE

ILLUMINATION

MAGNETIC CONDENSER

OBJECT

MAGNETIC OBJECTIVE

FIRST STAGE

INTERMEDIATE IMAGE

MAGNETIC PROJECTOR

SECOND STAGE

OBSERVATION SCREEN
OR PHOTOGRAPHIC PLATE

Diagram of electron microscope. The magnetic condenser directs the electrons in a parallel beam. The magnetic objective functions like a convex lens, producing an enlarged image, which is then further magnified by a magnetic projector. The image is projected on a fluorescent observation screen or a photographic plate.

ticle" bouncing off a matter particle; the matter particle is pushed forward, gaining energy, and the X-ray veers off, losing energy. This "Compton effect" helped establish the wave-particle dualism.

The matter waves had important consequences for theory, too. For one thing, they cleared up some puzzles about the structure of the atom.

In 1913, Niels Bohr had pictured the hydrogen atom as consisting of a central nucleus surrounded by an electron that could circle that nucleus in any one of a number of orbits. These orbits were in fixed position; if a hydrogen electron dropped from an outer orbit to an inner one, it lost energy, emitting that energy in the form of a quantum possessing a fixed wavelength. If the electron was to move from an inner electron to an outer one, it would have to absorb a quantum of energy, but only one of a fixed size and wavelength that was just enough to move it by the proper amount. That was why hydrogen could absorb or emit only certain wavelengths of radiation, producing characteristic lines in its spectrum. Bohr's scheme, which was made gradually more complex over the next decade, was highly successful in explaining many facts about the spectra of various elements, and he was awarded the Nobel Prize in physics in 1922 for his theory. The German physicists James Franck and Gustav Hertz (the latter a nephew of Heinrich Hertz), whose studies on collisions between atoms and electrons lent an experimental foundation to Bohr's theories, shared the Nobel Prize in physics in 1925.

Bohr had no explanation of why the orbits were fixed in the positions they held. He simply chose the orbits that would give the correct results, so far as absorption and emission of the actually observed wavelengths of light were concerned.

In 1926, the German physicist Erwin Schrödinger decided to take another look at the atom in the light of the de Broglie theory of the wave nature of particles. Considering the electron as a wave, he decided that the electron did not circle around the nucleus as a planet circles around the sun but constituted a wave that curved all around the nucleus, so that it was in all parts of its orbit at once, so to speak. It turned out that, on the basis of the wavelength predicted by de Broglie for an electron, a whole number of electron waves would exactly fit the orbits outlined by Bohr. Between the orbits, the waves would not fit in a whole number but would join up "out of phase"; and such orbits could not be stable.

Schrödinger worked out a mathematical description of the atom called "wave mechanics" or "quantum mechanics," and this became a more satisfactory method of looking at the atom than the Bohr system had been. Schrödinger shared the Nobel Prize in 1933 with Dirac, the author of the theory of antiparticles (see Chapter 6), who also contributed to the development of this new picture of the atom. The German physicist Max Born, who contributed further to the mathemat-

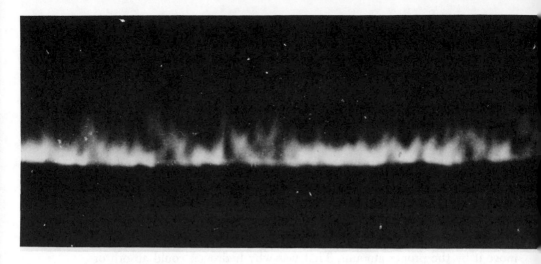

Magnetic domains photographed with an electron microscope by a special technique. The line of light shows the fine edge of a thin, magnetized piece of steel; the faint areas represent the deflection of electrons by the magnetic domains.

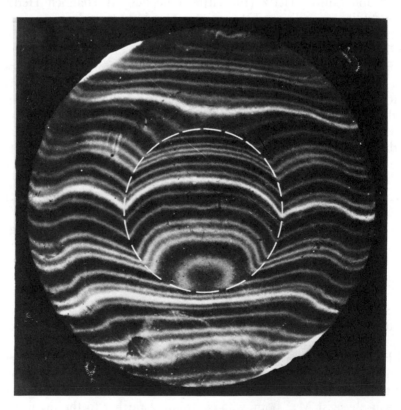

Contour map of a polished piece of metal, made by means of the interference effect of light waves. The contour lines here show differences on only one millionth of an inch.

Surface of a human hair, photographed with an electron microscope.

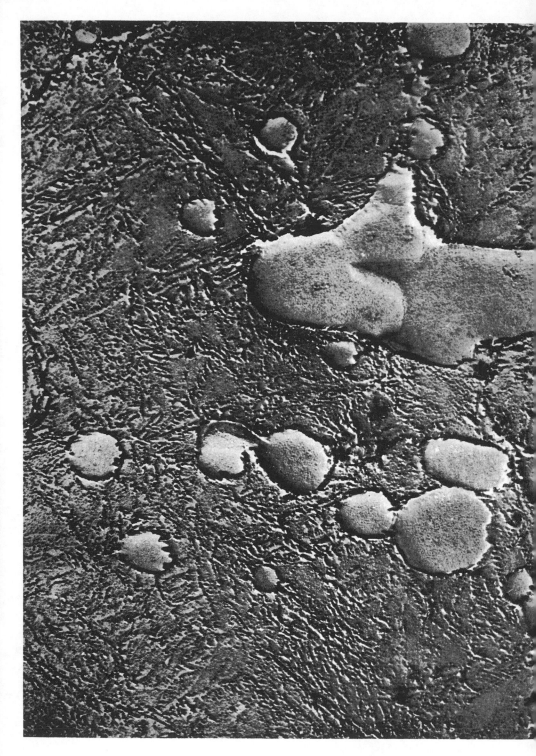

Electron micrograph of the surface of a piece of etched steel.

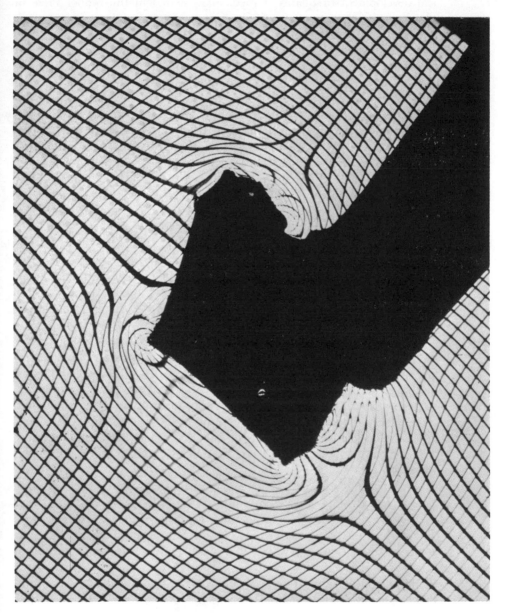

Electric field around a charged crystal is photographed with the electron microscope by means of a shadow technique. The method uses a fine wire mesh; the distortion of the net, caused by deflection of electrons, shows the shape and strength of the electric field.

ical development of quantum mechanics, shared in the Nobel Prize in physics in 1954.

By this time the electron had become a pretty vague "particle." And this vagueness was soon to grow worse. Werner Heisenberg of Germany proceeded to raise a profound question that projected particles, and physics itself, almost into a realm of the unknowable.

Heisenberg had presented his own model of the atom. He had abandoned all attempts to picture the atom as composed either of particles or of waves. He decided that any attempt to draw an analogy between atomic structure and the structure of the world about us was doomed to failure. Instead, he described the energy levels or orbits of electrons purely in terms of numbers, without a trace of picture. Since he used a mathematical device called a "matrix" to manipulate his numbers, his system was called "matrix mechanics."

Heisenberg received the Nobel Prize in physics in 1932 for his contributions to quantum mechanics, but his "matrix" system was less popular with physicists than Schrödinger's wave mechanics, since the latter seemed just as useful as Heisenberg's abstractions, and it is difficult for even a physicist to force himself to abandon all attempts to picture what he is talking about.

By 1944, physicists seemed to have done the correct thing, for the Hungarian-American mathematician John von Neumann presented a line of argument that seemed to show that matrix mechanics and wave mechanics were mathematically equivalent. Everything that was demonstrated by one could be equally well demonstrated by the other. Why not choose the less abstract version therefore? (Nevertheless, in 1964, Dirac raised the question as to whether the two are really equivalent. He thinks not and favors Heisenberg over Schrödinger; the matrices over the waves.)

After having introduced matrix mechanics (to jump back in time again), Heisenberg went on to consider the problem of describing the position of a particle. How could one determine where a particle was? The obvious answer is: Look at it. Well, let us imagine a microscope that could make an electron visible. We must shine a light or some appropriate kind of radiation on it to see it. But an electron is so small that a single photon of light striking it would move it and change its position. In the very act of measuring its position, we would have changed that position.

This is a phenomenon that occurs in ordinary life. When we measure the air pressure in a tire with a gauge, we let a little air out of the tire and change the pressure slightly in the act of measuring it. Likewise when we put a thermometer in a bathtub of water to measure the temperature, the thermometer's absorption of heat changes the temperature slightly. A meter measuring electric current takes away a little current for moving the pointer on the dial. And so it goes in every measurement of any kind that we make.

However, in all ordinary measurements the change in the subject we are measuring is so small that we can ignore it. The situation is quite different when we come to look at the electron. Our measuring device now is at least as large as the thing we are measuring; there is no usable measuring agent smaller than the electron. Consequently our measurement must inevitably have, not a negligible, but a decisive, effect on the object measured. We could stop the electron and so determine its position at a given instant. But in that case we could not know what its motion or velocity was. On the other hand, we might record its velocity, but then we could not fix its position at any given moment.

Heisenberg showed that there is no way of devising a method of pinpointing the position of a subatomic particle unless you are willing to be quite uncertain as to its exact motion. And, in reverse, there is no way of pinpointing a particle's exact motion unless you are willing to be quite uncertain as to its exact position. To calculate both exactly, at the same instant of time, is impossible.

If this is so, then even at absolute zero, there cannot be complete lack of energy. If energy reached zero and particles became completely motionless, then only position need be determined since velocity could be taken as zero. It would be expected, therefore, that some residual "zero-point energy" must remain, even at absolute zero, to keep particles in motion and, so to speak, uncertain. It is this zero-point energy, which cannot be removed, that is sufficient to keep helium liquid even at absolute zero (see Chapter 5).

In 1930, Einstein showed that the uncertainty principle, which stated it was impossible to reduce the error in position without increasing the error in momentum, implied it was also impossible to reduce the error in measurement of energy without increasing the uncertainty of time during which the measurement could take place. He thought he could use this as a springboard for the disproof of the uncertainty principle, but Bohr proceeded to show that Einstein's attempted disproof was wrong.

Indeed, Einstein's version of uncertainty proved very useful, since it meant that in subatomic processes, the law of conservation of energy could be violated for very brief periods of time, provided all was brought back to the conservational state by the end of those periods—the greater the deviation from conservation, the briefer the time-interval allowed. Yukawa used this notion in working out his theory of pions (see Chapter 6). It was even possible to explain certain subatomic phenomena by assuming that particles were produced out of nothing in defiance of energy conservation, but ceased to exist before the time allotted for their detection, so that they were only "virtual particles." The theory of virtual particles was worked out in the late 1940's by three men: the American physicists Julian Schwinger and Richard Phillips Feynman, and the Japanese physicist Sin-itiro Tomonaga. The three were jointly awarded the 1965 Nobel Prize in physics in consequence.

The uncertainty principle has profoundly affected the thinking of physicists and philosophers. It had a direct bearing on the philosophical question of "causality" (that is, the relationship of cause and effect). But its implications for science are not those that are commonly supposed. One often reads that the principle of indeterminacy removes all certainty from nature and shows that science after all does not and never can know what is really going on, that scientific knowledge is at the mercy of the unpredictable whims of a universe in which effect does not necessarily follow cause. Whether or not this interpretation is valid from the standpoint of philosophy, the principle of uncertainty has in no way shaken the attitude of scientists toward scientific investigation. If, for instance, the behavior of the individual molecules in a gas cannot be predicted with certainty, nevertheless on the average the molecules do obey certain laws, and their behavior can be predicted on a statistical basis, just as insurance companies can calculate reliable mortality tables even though it is impossible to predict when any particular individual will die.

In most scientific observations, indeed, the indeterminacy is so small compared with the scale of the measurements involved that it can be neglected for all practical purposes. One can determine simultaneously both the position and the motion of a star, of a planet, of a billiard ball, or even of a grain of sand, with complete satisfactory accuracy.

As for the uncertainty among the subatomic particles themselves this does not hinder but actually helps physicists. It has been used to explain facts about radioactivity and about the absorption of subatomic particles by nuclei, as well as many other subatomic events, more reasonably than would have been possible without the uncertainty principle.

The uncertainty principle means that the universe is more complex than was thought, but not that it is irrational.

CHAPTER 8

The Machine

Fire and Steam

The first law of thermodynamics states that energy cannot be created out of nothing. But there is no law against turning one form of energy into another. The whole civilization of mankind has been built upon finding new sources of energy and harnessing it in ever more efficient and sophisticated ways. In fact, the greatest single discovery in man's history involved methods for converting the chemical energy of a fuel such as wood into heat and light.

It was perhaps half a million years ago that our manlike ancestors "discovered" fire. No doubt they had encountered—and been put to flight by—lightning-ignited brush fires and forest fires before that. But the discovery of its virtues did not come until curiosity overcame fear. Some pre-man may have been attracted to the quietly burning remnants of such a fire and found amusement in playing with it, in feeding it sticks, and in watching the dancing flames. At night he would have appreciated the light and warmth of the fire, and the fact that it kept other animals away. Eventually he would learn to make a fire himself by rubbing dry sticks together, the more easily and surely to use it, to warm his camp or cave with it, to roast his game in order to make it easier to chew and better-tasting, and so on.

Fire provided man with a practically limitless supply of energy, which is why it is considered the greatest single human discovery—the one that hastened man's rise above the state of an animal. Yet curiously enough, for many thousands of years—in fact, up to the Industrial Revolution—man realized only a small part of its possibilities. He used it to light and warm his home, to cook his food, to work metals and make pottery and glass—and that was about all.

Meanwhile he was discovering other sources of energy. And some of the most important of them were developed during the so-called "Dark Ages." It was in medieval times that man began to burn the black rock called coal in his metallurgical furnaces, to harness the wind with windmills, to use water mills for grinding grain, to employ magnetic energy in the compass, and to use explosives in warfare.

About 670 A.D., a Syrian alchemist, Callinicus, is believed to have invented "Greek fire," a primitive incendiary bomb composed of sulfur and naphtha, which was credited with saving Constantinople from its first siege by the Moslems. Gunpowder arrived in Europe in the thirteenth century. Roger Bacon described it about 1280 A.D., but it had been known in Asia for centuries before that and may have been introduced to Europe by the Mongol invasions beginning in 1240 A.D. In any case, artillery powered by gunpowder came into use in Europe in the fourteenth century, and cannons are supposed to have appeared first at the battle of Crécy in 1346 A.D.

The most important of all the medieval inventions is the one credited to Johann Gutenberg of Germany. About 1450 A.D., he cast the first movable type and thereby introduced printing as a powerful force in human affairs. He also devised printer's ink, in which carbon black was suspended in linseed oil, rather than, as hitherto, in water. Together

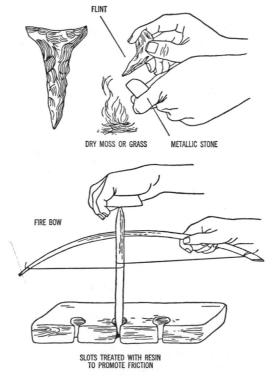

FLINT

DRY MOSS OR GRASS   METALLIC STONE

FIRE BOW

SLOTS TREATED WITH RESIN
TO PROMOTE FRICTION

Early firemaking methods.

with the replacement of parchment by paper (which had been invented by a Chinese eunuch, Ts'ai Lun—according to tradition—about 50 A.D. and which reached modern Europe, by way of the Arabs, in the thirteenth century), this made possible the large-scale production and distribution of books and other written material. No invention prior to modern times was adopted so rapidly. Within a generation of the discovery, 40,000 books were in print.

The recorded knowledge of mankind was no longer buried in royal collections of manuscripts but was made accessible in libraries available to all who could read. Pamphlets began to create and give expression to public opinion. (Printing was largely responsible for the success of Martin Luther's revolt against the Papacy, which might otherwise have been nothing more than a private quarrel.) And it was printing that created one of the prime instruments that gave rise to science as we know it. That indispensable instrument is the wide communication of ideas. Science had been a matter of personal communications among a few devotees; now it became a major field of activity, which enlisted more and more workers, elicited the prompt and critical testing of theories, and ceaselessly opened new frontiers.

The great turning point in man's harnessing of energy came at the end of the seventeenth century, although there had been a dim foreshadowing in ancient times. The Greek inventor, Hero of Alexandria, sometime during the first centuries A.D. (his life cannot be pinned down even to a particular century), built a number of devices that ran on steam power. He used the expanding push of steam to open temple doors, whirl spheres, and so on. The ancient world, then deep in decline, could not follow up this premature advance.

Then, over fifteen centuries later, a new and vigorously expanding society had a second chance. It arose out of the increasingly acute necessity of pumping water out of mines that were being driven ever deeper, The old hand pump (see Chapter 4) made use of a vacuum to lift water; and as the seventeenth century proceeded, men came to appreciate, ever more keenly, just how great the power of a vacuum was (or, rather, the power of air pressure that was called into play by the existence of a vacuum).

In 1650, for instance, the German physicist (and mayor of the city of Magdeburg) Otto von Guericke invented an air pump worked by muscle power. He proceeded to put two flanged metal hemispheres together and to pump the air out from between them by means of a nozzle that one hemisphere possessed. As the air pressure within dropped lower and lower, the air pressure from without, no longer completely counterbalanced, pushed the hemispheres together more and more powerfully. At the end, two teams of horses straining in opposite directions could not pull the hemispheres apart, but when air was allowed to re-enter, they fell apart of themselves. This experiment was conducted before important people including on one occasion, the German emperor himself, and it made a big splash.

Now it occurred to several inventors: Why not use steam instead of muscle power to create the vacuum? Suppose one filled a cylinder (or similar vessel) with water and heated the water to a boil. Steam, as it formed, would push out the water. If the vessel was cooled (e.g., by means of cold water played on the outside surface), the steam in the vessel would condense to a few drops of water and leave a virtual vacuum. The water that one wanted to raise (e.g., out of a flooded mine) could then rise through a valve into this evacuated vessel.

The first to translate this idea into a practical working device was an English military engineer named Thomas Savery. His "steam engine" (the word "engine" originally meant any ingenious device, and came from the same Greek root as "ingenious") could be used to pump water out of a mine or a well or to drive a water wheel, so he called it "The Miner's Friend." But it was dangerous (because the high pressure of the steam might burst the vessels and pipes) and very inefficient (because the heat of the steam was lost each time the container was cooled). Seven years after Savery patented his engine in 1698, an English blacksmith named Thomas Newcomen built an improved engine that operated at low steam pressure; it had a piston in a cylinder and employed air pressure to push down the piston.

Newcomen's engine was not very efficient either, and the steam engine remained a minor gadget for more than sixty years until a Scottish instrument maker named James Watt found the way to make it effective. Hired by the University of Glasgow to fix a model of a Newcomen engine that was not working properly, Watt fell to thinking about the device's wasteful use of fuel. Why, after all, should the steam vessel have to be cooled off each time? Why not keep the steam chamber steam-hot at all times and lead the steam into a separate condensing chamber that could be kept cold? Watt went on to add a number of other improvements: employing steam pressure to help push the piston, devising a set of mechanical linkages that kept the piston moving in a straight line, hitching the back-and-forth motion of the piston to a shaft that turned a wheel, and so on. By 1782 his steam engine, which got at least three times as much work out of a ton of coal as Newcomen's, was ready to be put to work as a universal work horse.

In the times after Watt, steam-engine efficiency was continually increased, chiefly through the use of hotter and hotter steam at higher and higher pressure. Carnot's founding of thermodynamics (see Chapter 7) arose mainly out of the realization that the maximum efficiency with which any heat engine could be run was proportional to the difference in temperature between the hot reservoir (steam, in the usual case) and the cold.

The first application of the steam engine to some labor more dramatic than that of pumping water out of mines was the steamship. In 1787 the American inventor John Fitch built a steamboat that worked, but it failed as a financial venture and Fitch died unknown and unappreciated. Robert Fulton, a more able promoter than Fitch, launched his steamship, the "Clermont," in 1807 with so much more fanfare and support that he came to be considered the inventor of the steamship, though actually he was no more the builder of the first such ship than Watt was the builder of the first steam engine.

Fulton should, perhaps, better be remembered for his strenuous attempts to build underwater craft. His submarines were not practical, but they anticipated a number of modern developments. He built one called the "Nautilus," which probably served as inspiration for Jules Verne's fictional submarine of the same name in *Twenty Thousand Leagues under the Sea*, published in 1870. That, in turn, was the inspiration for the naming of the first nuclear-powered submarine (see Chapter 9).

By the 1830's, steamships were crossing the Atlantic and were being driven by the screw propellor, a considerable improvement over the side paddle wheels. And by the 1850's, the speedy and beautiful Yankee Clippers had begun to furl their sails and to be replaced by steamers in the merchant fleets and navies of the world.

Meanwhile the steam engine had also begun to dominate land transportation. In 1814, the English inventor George Stephenson (owing a good deal to the prior work of an English engineer, Richard Trevithick) built the first practical steam locomotive. The in-and-out working of steam-driven pistons could turn metal wheels along steel rails as they could turn paddle wheels in the water. And in 1830, the American manufacturer Peter Cooper built the first steam locomotive in the Western Hemisphere. For the first time in history land travel became as convenient as sea travel, and overland commerce could compete with seaborne trade. By 1840, the railroad had reached the Mississippi River, and, by 1869, the full width of the United States was spanned by rail.

British inventors also led in introducing the steam engine into factories to run machinery. With this Industrial Revolution (a term introduced in 1837 by the French economist Jérôme Adolphe Blanqui), man completed his graduation from muscle power to mechanical power.

Electricity

In the nature of things, the steam engine is suitable only for large-scale, steady production of power. It cannot efficiently deliver energy in small packages or intermittently at the push of a button: a "little" steam engine, in which the fires were damped down or started up on demand, would be an absurdity. But the same generation that saw the development of steam power also saw the discovery of a means of transforming energy into precisely the form I have mentioned—a ready store of energy which could be delivered anywhere, in small amounts or large, at the push of a button. This form, of course, is electricity.

The Greek philosopher Thales, about 600 B.C., noted that a fossil resin found on the Baltic shores, which we call amber and they called "elektron," gained the ability to attract feathers, threads, or bits of fluff when it was rubbed with a piece of fur. It was William Gilbert of England, the investigator of magnetism (see Chapter 3), who first suggested that this attractive force be called "electricity," from the Greek word "elektron." Gilbert found that, in addition to amber, some other materials, such as glass, gained electric properties on being rubbed.

In 1733, the French chemist Charles Francis de Cisternay Du Fay discovered that if two amber rods, or two glass rods, were electrified by rubbing, they repelled each other. However, an electrified glass rod attracted an electrified amber rod. If the two were allowed to touch, both lost their electricity. He felt this showed there were two kinds of electricity, "vitreous" and "resinous."

The American scholar Benjamin Franklin, who became intensely interested in electricity, suggested that it was a single fluid. When glass was rubbed, electricity flowed into it, making it "positively charged"; on the other hand, when amber was rubbed, electricity flowed out of it, and it therefore became "negatively charged." And when a negative rod made

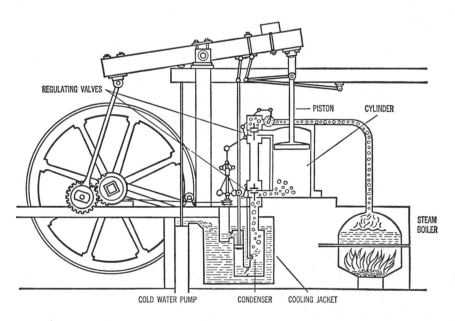

REGULATING VALVES

PISTON   CYLINDER

STEAM BOILER

COLD WATER PUMP   CONDENSER   COOLING JACKET

Watt's steam engine.

contact with a positive one, the electric fluid would flow from the positive to the negative until a neutral balance was achieved.

This was a remarkably shrewd speculation. If we substitute the word electrons for Franklin's "fluid" and reverse the direction of flow (actually electrons flow from the amber to the glass), his guess was essentially correct.

A French inventor named John Théophile Desaguliers suggested in 1740 that substances through which the electric fluid traveled freely (e.g., metals) be termed "conductors" and those through which it did not move freely (e.g., glass and amber) be called "insulators."

Experimenters found that a large electric charge could gradually be accumulated in a conductor if it was insulated from loss of electricity by glass or a layer of air. The most spectacular device of this kind was the "Leyden jar." It was first devised in 1745 by the German scholar E. Georg von Kleist, but it was first put to real use at the University of Leyden in Holland, where it was independently constructed a few months later by the Dutch scholar Peter van Musschenbroek. The Leyden jar is an example of what is today called a "condenser," or "capacitor," that is, two conducting surfaces, separated by a small thickness of insulator, within which one can store a quantity of electric charge.

In the case of the Leyden jar, the charge is built up on tinfoil coating a glass jar, via a brass chain stuck into the jar through a stopper. When you touch the charged jar, you get a startling electric shock. The Leyden jar can also produce a spark. Naturally, the greater the charge on a body, the greater its tendency to escape. The force driving the electrons away from the region of highest excess (the "negative pole") toward the region of greatest deficiency (the "positive pole") is the "electromotive force" (EMF) or "electric potential." If the electric potential

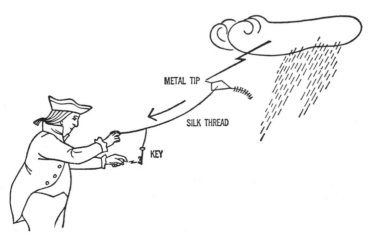

Franklin's experiment.

becomes high enough, the electrons will even jump an insulating gap between the negative and positive poles. Thus they will leap across an air gap, producing a bright spark and a crackling noise. The light of the spark is caused by the radiation resulting from the collisions of innumerable electrons with air molecules, and the noise arises from the expansion of the quickly heated air, followed by the clap of cooler air rushing into the partial vacuum momentarily produced.

Naturally one wondered whether lightning and thunder were the same phenomenon, on a vast scale, as the little trick performed by a Leyden jar. A British scholar, William Wall, had suggested just this in 1708. This thought was sufficient to prompt Benjamin Franklin's famous experiment in 1752. The kite he flew in a thunderstorm had a pointed wire, to which he attached a silk thread which could conduct electricity down from the thunderclouds. When Franklin put his hand near a metal key tied to the silk thread, the key sparked. Franklin charged it again from the clouds, then used it to charge a Leyden jar, obtaining the same kind of charged Leyden jar in this fashion as in any other. Thus Franklin demonstrated that the thunderclouds were charged with electricity and that thunder and lightning were indeed the effect of a Leyden-jar-in-the-sky in which the clouds formed one pole and the earth another.

The luckiest thing about the experiment, from Franklin's personal standpoint, was that he survived. Some others who tried it were killed, because the induced charge on the kite's pointed wire accumulated to the point of producing a fatally intense discharge to the body of the man holding the kite.

Franklin at once followed up this advance in theory with a practical application. He devised the "lightning rod," which was simply an iron rod attached to the highest point of a structure and connected to wires leading to the ground. The sharp point bled off electric charges from the clouds above, as Franklin showed by experiment, and, if lightning did strike, the charge was carried safely to the ground.

Lightning damage diminished drastically as the rods rose over structures all over Europe and the American colonies, no small accomplishment. Yet even today, two billion lightning flashes strike each year, killing (it is estimated) twenty people a day and hurting eighty more.

Franklin's experiment had two electrifying (please pardon the pun) effects. In the first place, the world at large suddenly became interested in electricity. Second, it put the American colonies on the map, culturally speaking. For the first time an American had actually displayed sufficient ability as a scientist to impress the cultivated Europeans of the Age of Reason. When, a quarter century later, Franklin represented the infant United States at Versailles and sought assistance, he won respect, not only as the simple envoy of a new republic, but also as a mental giant who had tamed the lightning and brought it humbly to earth. That flying kite contributed more than a little to the cause of American independence.

Following Franklin's work, electrical research advanced by leaps. Quantitative measurements of electrical attraction and repulsion were carried out in 1785 by the French physicist Charles Augustin de Coulomb. He showed that this attraction (or repulsion) between given charges varied inversely as the square of the distance. In this, electrical attraction resembled gravitational attraction. In honor of this finding, the "coulomb" has been adopted as a name for a common unit of quantity of electricity.

Shortly thereafter, the study of electricity took a new, startling, and very fruitful turning. What we have been looking at above is, of course, "static electricity." This refers to an electric charge that is placed on an object and then stays there. The discovery of an electric charge that moved, of electric currents or "dynamic electricity," began with the Italian anatomist Luigi Galvani. In 1791, he accidentally discovered that thigh muscles from dissected frogs would contract if simultaneously touched by two different metals (thus adding the verb "to galvanize" to the English language).

The muscles behaved as though they had been stimulated by an electric spark from a Leyden jar, and so Galvani assumed that muscles contained something he called "animal electricity." Others, however, suspected that the origin of the electric charge might lie in the junction of the two metals rather than in muscle. In 1800, the Italian physicist Alessandro Volta studied combinations of dissimilar metals, connected not by muscle tissue but by simple solutions.

He began by using chains of dissimilar metals connected by bowls half-full of salt water. To avoid too much liquid too easily spilled, he prepared small disks of copper and of zinc, piling them alternately. He also made use of cardboard disks moistened with salt water so that his "voltaic pile" consisted of silver, cardboard, zinc, silver, cardboard, zinc, silver, and so on. From such a setup, electric current could be drawn off continuously.

Any series of similar items indefinitely repeated may be called a battery. Volta's instrument was the first "electric battery." It may also be called an "electric cell." It was to take a century before scientists would understand how chemical reactions involved electron transfers and how to interpret electric currents in terms of shifts and flows of electrons. Meanwhile, however, they made use of the current without understanding all its details.

Humphry Davy used an electric current to pull apart the atoms of tightly bound molecules and was able for the first time, in 1807 and 1808, to prepare such metals as sodium, potassium, magnesium, calcium, strontium, and barium. Faraday (Davy's assistant and protégé) went on to work out the general rules of such molecule-breaking "electrolysis" and, this, a half century later, was to guide Arrhenius in his working out of the hypothesis of ionic dissociation (see Chapter 4).

The manifold uses of dynamic electricity in the century and a half since Volta's battery, seems to have placed static electricity in the shade and to have reduced it to a mere historical curiosity. Not so, for knowledge and ingenuity need never be static. By 1960, the American inventor Chester Carlson had perfected a practical device for copying material by attracting carbon-black to paper through localized electrostatic action. Such copying, involving no solutions or wet media, is called "xerography" (from Greek words meaning "dry writing") and has revolutionized office procedures.

The names of the early workers in electricity have been immortalized in the names of the units used for various types of measurement involving electricity. I have already mentioned "coulomb" as a unit of quantity of electricity. Another unit is the "faraday," for 96,500 coulombs is equal to one faraday. Faraday's name is used a second time, for a "farad" is a unit of electrical capacity. Then, too, the unit of electrical intensity (the quantity of electric current passing through a circuit in a given time) is called the "ampere," after the French physicist Ampère (see Chapter 4). One ampere is equal to one coulomb per second. The unit of electromotive force (the force that drives the current) is the "volt," after Volta.

A given EMF did not always succeed in driving the same quantity of electricity through different circuits. It would drive a great deal of current through good conductors, little current through poor conductors, and virtually no current through nonconductors. In 1827, the German mathematician George Simon Ohm studied this "resistance" to electrical flow and showed that it could be precisely related to the amperes of current flowing through a circuit under the push of a known EMF. The resistance could be determined by taking the ratio of volts to amperes. This is "Ohm's law," and the unit of electrical resistance is the "ohm," so that one ohm is equal to one volt divided by one ampere.

The conversion of chemical energy to electricity, as in Volta's battery and the numerous varieties of its descendants, has always been relatively expensive because the chemicals involved are not common or cheap. For this reason, although electricity could be used in the laboratory with great profit in the early nineteenth century, it could not be applied to large-scale uses in industry.

There have been sporadic attempts to make use of the chemical reactions involved in the burning of ordinary fuels as a source of electricity. Fuels such as hydrogen (or, better still, coal) are much cheaper than metals such as copper and zinc. As long ago as 1839, the English scientist William Grove devised an electric cell running on the combination of hydrogen and oxygen. It was interesting, but not practical. In recent years, physicists have been working hard to prepare practical varieties of such "fuel cells." The theory is all set; it is only the practical problems that must be ironed out, and these are proving most refractory.

When the large-scale use of electricity came into being in the latter half of the nineteenth century, it is not surprising, then, that it did not arrive by way of the electric cell. As early as the 1830's Faraday had produced electricity by means of the mechanical motion of a conductor across the lines of force of a magnet (see Chapter 4). In such an "electric generator" or "dynamo" (from a Greek word for "power"), the kinetic energy of motion could be turned into electricity. Such motion could be kept in being by steam power, which in turn could be generated by burning fuel. Thus, much more indirectly than in a fuel cell, the energy of burning coal or oil (or even wood) could be converted into electricity. By 1844, large, clumsy versions of such generators were being used to power machinery.

What was needed were ever stronger magnets, so that motion across the intensified lines of force could produce larger floods of electricity. These stronger magnets were obtained, in turn, by the use of electric currents. In 1823, the English electrical experimenter, William Sturgeon, wrapped eighteen turns of bare copper wire about a U-shaped iron bar and produced an "electromagnet." When the current was on, the magnetic field it produced was concentrated in the iron bar which could then lift twenty times its own weight of iron. With the current off, it was no longer a magnet and would lift nothing.

In 1829, the American physicist Joseph Henry improved this gadget vastly by using insulated wire. Once the wire was insulated it was possible to wind it in close loops over and over without fear of short circuits. Each loop increased the intensity of the magnetic field and the power of the electromagnet. By 1831, Henry had produced an electromagnet, of no great size, that could lift over a ton of iron.

The electromagnet was clearly the answer to better electrical gen-

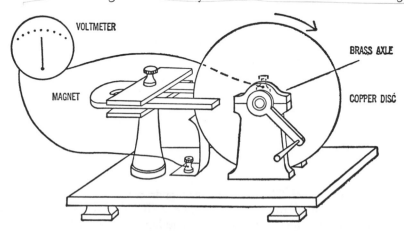

Faraday's "dynamo." The rotating copper disk cuts the magnet's lines of force, inducing a current on the voltmeter.

erators. In 1845, the English physicist Charles Wheatstone made use of such an electromagnet for this purpose. Better understanding of the theory behind lines of force came about with Maxwell's mathematical interpretation of Faraday's work (see pages 163–165) in the 1860's, and, in 1872, the German electrical engineer Friedrich von Hefner-Alteneck designed the first really efficient generator. At last electricity could be produced cheaply and in floods, and it could be done, not only from burning fuel, but from falling water.

For the work that led to the early application of electricity to technology, the lion's share of the credit must fall to Joseph Henry. Henry's first application of electricity was the invention of telegraphy. He devised a system of relays which made it possible to transmit an electric current over miles of wire. The strength of a current declines fairly rapidly as it travels at constant voltage across longer and longer stretches of resisting wire; what Henry's relays did was to use the dying signal to activate a small electromagnet that operated a switch that turned on a boost in power from stations placed at appropriate intervals. Thus a message consisting of coded pulses of electricity could be sent for a considerable distance. Henry actually built a telegraph that worked.

Because he was an unwordly man, who believed that knowledge should be shared with the world and therefore did not patent his discoveries, Henry got no credit for this invention. The credit fell to the

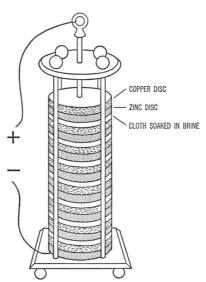

COPPER DISC
ZINC DISC
CLOTH SOAKED IN BRINE

Volta's battery. The two different metals in contact give rise to a flow of electrons, which are conducted from one "cell" to the next by the salt-soaked cloth. The familiar "dry battery" or "flashlight battery" of today, involving carbon and zinc, was first devised by Bunsen (of spectroscopy fame) in 1841.

artist (and eccentric religious bigot) Samuel Finley Breese Morse. With Henry's help, freely given (but later only grudgingly acknowledged), Morse built the first practical telegraph in 1844. Morse's main original contribution to telegraphy was the system of dots and dashes known as the "Morse code."

Henry's most important development in the field of electricity was the electric motor. He showed that electric current could be used to turn a wheel, just as the turning of a wheel can generate current in the first place. And an electrically driven wheel (or motor) could be used

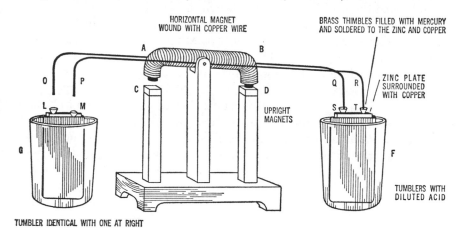

Henry's motor. The upright bar magnet *D* attracts the wire-wound magnet *B*, pulling the long metal probes *Q* and *R* into the brass thimbles *S* and *T*, which act as terminals for the wet cell *F*. Current flows into the horizontal magnet, producing an electromagnetic field that pulls *A* and *C* together. The whole process is then repeated on the opposite side. Thus the horizontal bars oscillate up and down.

to run machinery. The motor could be carried anywhere; it could be turned on or off at will (without waiting to build up a head of steam); and it could be made as small as one wished.

The catch was that electricity had to be transported from the generating station to the place where the motor was to be used. Some way had to be found to cut down the loss of electrical energy (taking the form of dissipated heat) as it traveled over wires.

One answer was the "transformer." The experimenters with currents found that electricity suffered far less loss if it was transmitted at a low rate of flow. So the output from the generator was stepped up to a high voltage by means of a transformer that, while multiplying the voltage, say, three times, reduces the current (rate of flow) to one third. At the receiving station, the voltage can be stepped down again so that the current is correspondingly increased for use in motors.

The transformer works by using the "primary" current to induce a current at high voltage in a secondary coil. This induction requires varying the magnetic field through the second coil. Since a steady current will not do this, the current used is a continually changing one that builds up to a maximum and then drops to zero and starts building in the opposite direction—in other words, an "alternating current."

Alternating current (ac) did not win out over direct current (dc) without a struggle. Thomas Alva Edison, the greatest name in electricity in the final decades of the nineteenth century, championed dc and established the first dc generating station in New York in 1882 to supply current for the electric light he had invented. He fought ac on the ground that it was more dangerous (pointing out, for instance, that it was used in electric chairs). He was bitterly opposed by Nikola Tesla, an engineer who had worked for Edison and been shabbily treated. Tesla developed a successful system of ac in 1888. In 1893, George Westinghouse, also a believer in ac, won a crucial victory over Edison by obtaining for his electric company the contract to develop the Niagara Falls power plants on an ac basis. In the following decades, Steinmetz established the theory of alternating currents on a firm mathematical basis. Today alternating current is all but universal in systems of power distribution. (In 1966, to be sure, engineers at General Electric devised a direct-current transformer—something long held to be impossible—but it involves liquid-helium temperatures and a low efficiency. It is fascinating, theoretically, but of no likely commercial use right now.)

Electrical Gadgets

The steam engine is a "prime mover." It takes energy already existing in nature (the chemical energy of wood, oil, or coal) and turns it into work. The electric motor is not; it converts electricity into work, but the electricity must itself be formed from the energy of burning fuel or falling water. For this reason, electricity is more expensive than steam for heavy jobs. Nevertheless, it can be used for the purpose. At the Berlin Exhibition of 1879, an electric-powered locomotive (using a third rail as its source of current) successfully pulled a train of coaches. Electrified trains are common now, especially for rapid transit within cities, for the added expense is more than made up for by increased cleanliness and smoothness of operation.

Where electricity really comes into its own, however, is where it performs tasks that steam cannot. There is, for instance, the telephone, patented by the Scottish-born inventor Alexander Graham Bell in 1876. In the telephone mouthpiece, the speaker's sound waves strike a thin steel diaphragm and make it vibrate in accordance with the pattern of the waves. The vibrations of the diaphragm in turn set up an analogous pattern in an electric current, by way of carbon powder. When the diaphragm presses on the carbon powder, the powder conducts more current; when the diaphragm moves away, it conducts less. Thus the electric current strengthens and weakens in exact mimicry of the sound waves. At the telephone receiver the fluctuations in the strength of the current actuate an electromagnet that makes a diaphragm vibrate and reproduce the sound waves.

In 1877, a year after the invention of the telephone, Edison patented his "phonograph." The first records had the grooves scored on tinfoil wrapped around a rotating cylinder. The American inventor Charles Sumner Tainter substituted wax cylinders in 1885 and then Emile Berliner introduced wax-coated disks in 1887. In 1925, recordings began to be made by means of electricity through the use of a "microphone," which translated sound into a mimicking electric current via a piezoelectric crystal instead of a metal diaphragm—the crystal allowing a better quality of reproduction of the sound. In the 1930's, the use of radio tubes for amplification was introduced. Then, in the post-World War II era, came the long-playing record, a "hi-fi," and "stereophonic" sound, which have had the effect, so far as the sound itself is concerned, of practically removing all mechanical barriers between the orchestra or singer and the listener!

"Tape-recording" of sound was invented in 1898 by a Danish electrical engineer named Valdemar Poulsen, but had to await certain technical advances to become practical. An electromagnet, responding to an electric current carrying the sound pattern, magnetizes a powder coating on a tape or wire moving past it, and the playback is accomplished through an electromagnet which picks up this pattern of magnetism and translates it again into a current that will reproduce the sound.

Of all the tricks performed by electricity, certainly the most popular was its turning night into day. Mankind had fought off the daily crippling darkness-after-sundown with the campfire, the torch, and the candle; for a hundred thousand years or so, the level of artificial light remained dim and flickering. Then, in the nineteenth century came whale oil, kerosene, and gas, and man-made light became somewhat stronger. Now electricity brought to pass a far better kind of lighting—safer, more convenient, and as brilliant as one could wish.

The problem was to heat a filament by electricity to an incandescent glow. It seemed simple, but many tried and failed to produce a durable lamp. Naturally, the filament had to glow in the absence of oxygen or be oxidized to destruction almost at once. The first attempts to remove oxygen involved the straightforward route of removing air. By 1875, Crookes (in connection with his work on cathode rays; see Chapter 5) had devised methods for producing a good enough vacuum for this purpose, and with sufficient speed and economy. Nevertheless, the filaments used remained unsatisfactory. They broke too easily. In 1878, Thomas Edison, fresh from his triumph in creating the phonograph, announced that he would tackle the problem. He was only thirty-one, but such was his reputation as an inventor that his announcement caused the stocks of gas companies to tumble on the New York and London stock exchanges.

After hundreds of experiments and fabulous frustrations, Edison finally found a material that would serve as the filament—a scorched cotton thread. On October 21, 1879, he lit his bulb. It burned for forty continuous hours. On the following New Year's Eve, Edison put his lamps on triumphant public display by lighting up the main street of Menlo Park, N. J., where his laboratory was located. He quickly patented his lamp and began to produce it in quantity.

Yet Edison was not the sole inventor of the incandescent lamp. At least one other inventor had about an equal claim—Joseph Swan of England, who exhibited a carbon-filament lamp at a meeting of the Newcastle-on-Tyne Chemical Society on December 18, 1878, but did not get his lamp into production until 1881.

Edison proceeded to work on the problem of providing homes with a steady and sufficient supply of electricity for his lamps—a task which took as much ingenuity as the invention of the lamp itself. Two major improvements were later made in the lamp. In 1910, William David Coolidge of the General Electric Company adopted the heat-resisting metal tungsten as the material for the filament, and, in 1913, Irving Langmuir introduced the inert gas nitrogen in the lamp to prevent the evaporation and breaking of the filament that occurred in a vacuum.

Argon (the use of which was introduced in 1920) serves the purpose even better than nitrogen, for argon is completely inert. Krypton, another inert gas, is still more efficient, allowing a lamp filament to reach higher temperatures and burn more brightly without loss of life expectancy.

Other kinds of electric lamp have, of course, been developed. The so-called "neon lights" (introduced by the French chemist Georges

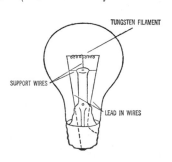

Incandescent lamp.

Claude in 1910), are tubes in which an electric discharge excites atoms of neon gas to emit a bright, red glow. The "sun lamp" contains mercury vapor, which when excited by a discharge yields radiation rich in ultraviolet light; this can be used not only to produce a tan but also to kill bacteria or generate fluorescence. And the latter in turn leads to fluorescent lighting, introduced in its contemporary form in 1939 at the New York World's Fair. Here the ultraviolet light from mercury vapor excites fluorescence in a "phosphor" coating the inside of the tube. Since this cool light wastes little energy in heat, it consumes less electric power.

A 40-watt fluorescent tube supplies as much light, and far less heat, than a 150-watt incandescent light. Since World War II, therefore, there has been a massive swing toward the fluorescent. The first fluorescent tubes made use of beryllium salts as phosphors. This resulted in cases of serious poisoning ("berylliosis") induced by breathing dusts containing these salts or by introducing the substance through cuts caused by broken tubes. After 1949, other far less dangerous phosphors were used.

The latest promising development is a method that converts electricity directly into light without the prior formation of ultraviolet light. In 1936, the French physicist Georges Destriau discovered that an intense alternating current could make a phosphor, such as zinc sulfide, glow. Electrical engineers are now distributing the phosphor through plastic or glass and are using this phenomenon, called "electroluminescence," to develop glowing panels. Thus a luminescent wall or ceiling could light a room, bathing it in a soft, colored glow.

Probably no invention involving light has given mankind more enjoyment than photography. This had its earliest beginnings in the observation that light, passing through a pinhole into a small dark chamber ("camera obscura" in Latin), will form a dim, inverted image of the scene outside the chamber. Such a device was constructed about 1550 by an Italian alchemist, Giambattista della Porta. This is the "pinhole camera."

In a pinhole camera, the amount of light entering is very small. If,

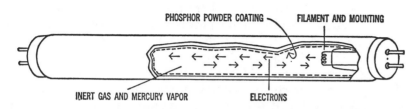

PHOSPHOR POWDER COATING   FILAMENT AND MOUNTING

INERT GAS AND MERCURY VAPOR   ELECTRONS

Fluorescent lamp. A discharge of electrons from the filament excites the mercury vapor in the tube, producing ultraviolet radiation. The ultraviolet makes the phosphor glow.

however, a lens is substituted for the pinhole, a considerable quantity of light can be brought to a focus, and the image is then much brighter. With that accomplished it is necessary to find some chemical reaction that will respond to light. A number of men labored in this cause, including, most notably, the Frenchmen Joseph Nicephore Niepce and Louis Jacques Mande Daguerre, and the Englishman William Henry Fox Talbot. By the mid-nineteenth century, permanent images painted in chemicals could be produced.

The image is focused on an emulsion of a silver compound smeared (at first) on a glass plate. The light produces a chemical change in the compound, the amount of change being proportional to the intensity of the light at any given point. In the developing process, the chemical developer converts those parts changed by the light into metallic silver, again to an extent proportional to the intensity of light. The unaffected silver compound is then dissolved away, leaving a "negative" on which the image appears as a pattern of blackening in various degrees. Light projected through the negative reverses the light and dark spots and forms the positive image. Photography went on to prove its value in human documentation almost at once when, in the 1850's, the British photographed Crimean war scenes and when, in the next decade, the American photographer Matthew Brady, with what we would now consider impossibly primitive equipment, took classic photographs of the American Civil War in action.

Throughout the nineteenth century, the process was gradually made faster and simpler. The American inventor George Eastman developed dry plates (in place of the original moist emulsion) and then adopted plastic film as the backing for the emulsion. More sensitive emulsions were created, so that faster shots could be made and the subject did not need to "pose."

Since World War II picture-taking has been further simplified by means of the "Land camera," invented by Edwin Herbert Land of the Polaroid Corporation. It uses two films on which the negative and positive are developed automatically by chemicals incorporated in the film.

In the early twentieth century, a process of color photography was developed by the Luxembourg-born French physicist Gabriel Lippmann, which won him the Nobel Prize for physics in 1908. That proved a false alarm, however, and practical color photography was not developed until 1936. This second, and successful, try was based on the observation, in 1855, by Maxwell and von Helmholtz that any color in the spectrum could be produced by combining red, green, and blue light. On this principle, the color film is composed of emulsions in three layers, one sensitive to the red, one to the green, and one to the blue components of the image. Three separate but superimposed pictures are formed, each reproducing the intensity of light in its part of the spectrum as a pattern of black-and-white shading. The film is then developed in three successive stages, using red, blue, and green dyes to deposit the appropriate colors on the negative. Each spot in the picture is a specific combination of red, green, and blue, and the brain interprets these combinations to reconstitute the full range of color.

In 1959, Land presented a new theory of color vision. The brain, he maintained, does not require a combination of three colors to create the impression of full color. All it needs is two different wavelengths, or sets of wavelengths, one longer than the other by a certain minimum amount. For instance, one of the sets of wavelengths may be an entire spectrum, or white light. Because the average wavelength of white light is in the yellow-green region, it can serve as the "short" wavelength. Now a picture reproduced through a combination of white light and red light (serving as the long wavelength) comes out in full color. Land has also made pictures in full color with filtered green light and red light and with other appropriate dual combinations.

The invention of motion pictures came from an observation first made by the English physician Peter Mark Roget in 1824. He noted that the eye forms a persistent image, which lasts for an appreciable fraction of a second. After the inauguration of photography, many experimenters, particularly in France, made use of this fact to create the illusion of motion by showing a series of pictures in rapid succession. Everyone is familiar with the parlor gadget consisting of a series of picture cards which, when riffled rapidly, make a figure seem to move and perform acrobatics. If a series of pictures, each slightly different from the one before, is flashed on a screen at intervals of about one sixteenth of a second apart, the persistence of the successive images in the eye will cause them to blend together and so give the impression of continuous motion.

It was Edison who produced the first "movie." He photographed a series of pictures on a strip of film and then ran the film through a projector, which showed each in succession with a burst of light. The first motion picture was put on display for public amusement in 1894, and, in 1914, theaters showed the full-length motion picture, *The Birth of a Nation.*

To the silent movies, a sound track was added in 1927. The "sound track" also takes the form of light: the wave pattern of music and the actor's speech is converted into a varying current of electricity by a microphone, and this current lights a lamp which is photographed along with the action of the motion picture. When the film, with this track of light at one side, is projected on the screen, the brightening and dimming of the lamp in the pattern of the sound waves is converted back to an electric current by means of a "phototube," using the photoelectric effect, and the current in turn is reconverted to sound.

Within two years after the first "talking picture," *The Jazz Singer,* silent movies were a thing of the past, and so, almost, was vaudeville. By the late 1930's, the "talkies" had added color. In addition, the 1950's saw the development of wide-screen techniques and even a short-lived fad for three-dimensional (3D) effects, involving two pictures thrown on the same screen. By wearing polarized spectacles, an observer saw a separate picture with each eye, thus producing a stereoscopic effect.

Internal-Combustion Engines

While petroleum gave way to electricity in the field of artificial illumination, it became indispensable for another technical development that revolutionized modern life as deeply, in its way, as did the introduction of electrical gadgetry. This development was the internal-combustion engine, so-called because in such an engine, fuel is burned within the cylinder so that the gases formed push the piston directly. Ordinary steam engines are "external-combustion engines," the fuel being burned outside and the steam being then led, ready-formed, into the cylinder.

This compact device, with small explosions set off within the cylinder, made it possible to apply motive power to small vehicles in ways for which the bulky steam engine was not well-suited. To be sure, steam-driven "horseless carriages" were devised as long ago as 1786, when William Murdock, a partner of James Watt, built one. A century later, the American inventor Francis Edgar Stanley invented the famous "Stanley Steamer," which for a while competed with the early cars equipped with internal combustion machines. The future, however, lay with the latter.

Actually, some internal-combustion engines were built at the beginning of the nineteenth century, before petroleum came into common use. They burned turpentine vapors or hydrogen as fuel. But it was only with gasoline, the one vapor-producing liquid that is both combustible and obtainable in large quantities, that such an engine could become more than a curiosity.

The first practical internal-combustion engine was built in 1860 by a French inventor Etienne Lenoir; in 1876, the German technician Nikolaus August Otto built a "four-cycle" engine. First a piston fitting tightly in a cylinder is pushed outward, so that a mixture of gasoline and air is sucked into the vacated cylinder. Then the piston is pushed in again to compress the vapor. At the point of maximum compression the vapor is ignited and explodes. The explosion drives the piston outward, and it is this powered motion that drives the engine. It turns a wheel which pushes the piston in again to expel the burned residue or "exhaust"—the fourth and final step in the cycle. Now the wheel moves the piston outward to start the cycle over again.

A Scottish engineer named Dugald Clerk almost immediately added

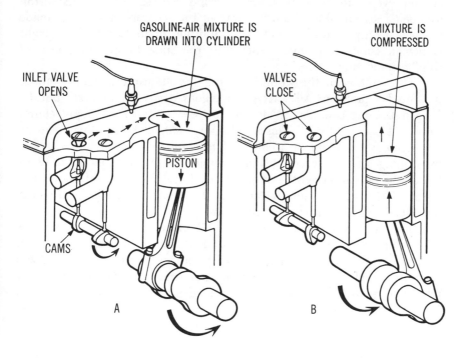

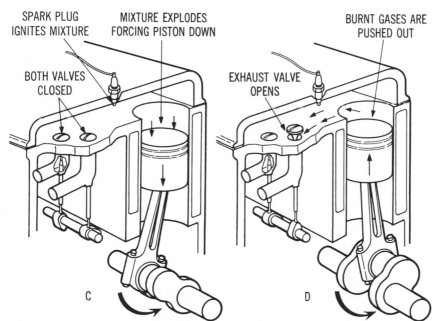

Nikolaus Otto's "four-cycle" engine, built in 1876.

an improvement. He hooked up a second cylinder, so that its piston was being driven while the other was in the recovery stage: this made the power output steadier. Later the addition of more cylinders (eight is now a common number) increased the smoothness and power of this "reciprocating engine."

The ignition of the gasoline-air mixture at just the right moment presented a problem. All sorts of ingenious devices were used, but by 1923 it became common to depend on electricity. The supply comes from a "storage battery." This is a battery that, like any other, delivers electricity as the result of a chemical reaction. But it can be recharged by sending an electric current through it in the direction opposite to the discharge; this current reverses the chemical reaction and allows the chemicals to produce more electricity. The reverse current is provided by a small generator driven by the engine.

The most common type of storage battery has plates of lead and lead oxide in alternation, with cells of fairly concentrated sulfuric acid. It was invented by the French physicist Gaston Planté in 1859 and was put into its modern form in 1881 by the American electrical engineer Charles Francis Brush. More rugged and more compact storage batteries have been invented since, as, for instance, a nickel-iron battery developed by Edison about 1905, but none can compete with the lead battery in economy.

The electric voltage supplied by the storage battery is stored in the magnetic field of a transformer called an "induction coil," and the collapse of this field provides the stepped-up voltage that produces the ignition spark across the gap in the familiar spark plugs.

Once an internal-combustion engine starts firing, inertia will keep it moving between power strokes. But outside energy must be supplied to start the engine. At first it was done by hand (e.g., the automobile crank), and outboard motors and power lawn mowers are still started by yanking a cord. The "self-starter" in modern automobiles is powered by the storage battery, which supplies the energy for the first few turns of the engine.

The first practical automobiles were built, independently, in 1885 by the German engineers Gottlieb Daimler and Karl Benz. But what really made the automobile, as a common conveyance, was the invention of "mass production."

The prime originator of this technique was Eli Whitney, who merits more credit for it than for his more famous invention of the cotton gin. In 1789, Whitney received a contract from the Federal Government to make guns for the army. Up to that time guns had been manufactured individually, each from its own fitted parts. Whitney conceived the notion of making the parts uniform, so that a given part would fit any gun. This single, simple innovation—manufacturing standard, interchangeable parts for a given type of article—was perhaps as responsible as any other factor for the creation of modern mass-production industry. When power tools came in, they made it possible to stamp out standard parts in practically unlimited numbers.

It was the American engineer Henry Ford who first exploited the concept to the full. He had built his first automobile (a two-cylinder job) in 1892, then had gone to work for the Detroit Automobile Company in 1899 as chief engineer. The company wanted to produce custom-made cars, but Ford had another notion. He resigned in 1902 to produce cars on his own—in quantity. In 1909, he began to turn out the Model T Ford and by 1913, he began to manufacture it on the Whitney plan —car after car, each just like the one before and all made with the same parts.

Ford saw that he could speed up production by using human workers as one used machines, performing the same small job over and over with uninterrupted regularity. The American inventor Samuel Colt (who had invented the revolver or "six-shooter") had taken the first steps in this direction in 1847, and the automobile manufacturer Ransom E. Olds had applied the system to the motor car in 1900. Olds lost his financial backing, however, and it fell to Ford to carry this movement to its fruition. Ford set up the "assembly line," with workers adding parts to the construction as it passed them on moving belts until the finished car rolled off at the end of the line. Two economic advances were achieved by this system: high wages for the workers and cars that could be sold at amazingly low prices.

By 1913, Ford was manufacturing 1,000 Model T's a day. Before the line was discontinued in 1927, 15 million had been turned out, and the price had dropped to 290 dollars. The passion for yearly change then won out, and Ford was forced to join the parade of variety and superficial novelty that has raised the price of automobiles tenfold and lost Americans much of the advantage of mass production.

In 1892, the German mechanical engineer Rudolf Diesel introduced a modification of the internal-combustion engine that was simpler and more economical of fuel. He put the fuel-air mixture under high pressure, so that the heat of compression alone was enough to ignite it. The "diesel engine" made it possible to use higher-boiling fractions of petroleum, which do not knock. Because of the higher compression used, the engine must be more solidly constructed and is therefore considerably heavier than the gasoline engine. Once an adequate fuel-injection system was developed in the 1920's it began to gain favor for trucks, tractors, buses, ships, and locomotives, and is new undisputed king of heavy transportation.

Improvements in gasoline itself further enhanced the efficiency of the internal-combustion engine. Gasoline is a complex mixture of molecules made up of carbon and hydrogen atoms ("hydrocarbons"), some of which burn more quickly than others. Too quick a burning rate is undesirable, for then the gasoline-air mixture explodes in too many places at once, producing "engine knock." A slower rate of burning produces an even expansion of vapor that pushes the piston smoothly and effectively.

The amount of knock produced by a given gasoline is measured as its "octane rating," by comparing it to the knock produced by a hydrocarbon called "iso-octane," which is particularly low in knock production, mixed with "normal heptane," which is particularly high in knock production. One of the prime functions of gasoline refining is, among many other things, to produce a hydrocarbon mixture with a high octane rating.

Automobile engines have been designed through the years with a higher and higher "compression ratio"; that is, the gasoline-air mixture is compressed to greater and greater densities before ignition. This milks the gasoline of more power, but it also encourages knock, so that gasoline of continually higher octane rating has had to be developed.

The task has been made easier by the use of chemicals that, when added in small quantities to the gasoline, reduce knock. The most efficient of these "anti-knock compounds" is "tetraethyl lead," a lead compound first introduced for the purpose in 1925. Gasoline containing it is "leaded gasoline" or "ethyl gas." If tetraethyl lead were present alone, the lead oxides formed during gasoline combustion would foul and ruin the engine. For this reason, ethylene bromide is also added. The lead atom of tetraethyl lead combines with the bromide atom of ethylene bromide to form lead bromide, which, at the temperature of the burning gasoline, is vaporized and expelled with the exhaust.

Diesel fuels are tested for ignition delay after compression (too great a delay is undesirable) by comparison with a hydrocarbon called "cetane," which contains sixteen carbon atoms in its molecule as compared with eight for "iso-octane." For diesel fuels, therefore, one speaks of a "cetane number."

The greatest triumph of the internal-combustion engine came, of course, in the air. By the 1890's, man had achieved the age-old dream—older than Daedalus and Icarus—of flying on wings. Gliding had become an avid sport of the aficionados. The first man-carrying glider was built in 1853 by the English inventor George Cayley. The "man" it carried, however, was only a boy. The first important practitioner of the sport, the German engineer Otto Lilienthal, was killed in 1896 during a glider flight. Meanwhile, a violent urge to take off in powered flight had begun.

The American physicist and astronomer Samuel Pierpont Langley tried in 1902 and 1903 to fly a glider powered by an internal-combustion engine and came within an ace of succeeding. Had his money not given out, he might have got into the air on the next try. As it was, the honor was reserved for the brothers Orville and Wilbur Wright, bicycle manufacturers who had taken up gliders as a hobby.

On December 17, 1903, at Kitty Hawk, N. C., the Wright brothers got off the ground in a propeller-driven glider and stayed in the air for fifty-nine seconds, flying 852 feet. It was the first airplane flight in history, and it went almost completely unnoticed by the world at large.

There was considerably more public excitement after the Wrights had achieved flights of twenty-five miles and more and when, in 1909, the French engineer Louis Blériot crossed the English Channel in an airplane. The air battles and exploits of World War I further stimulated the imagination and the biplanes of that day, with their two wings held precariously together by struts and wires, were familiar to a generation of post-World War I movie-goers. The German engineer Hugo Junkers designed a successful monoplane just after the war and the thick single wing, without struts, took over completely. (In 1939, the Russian-American engineer Igor Ivan Sikorsky built a multiengined plane and designed the first helicopter, a plane with upper vanes that made vertical takeoffs and landings and even hovering practical.)

But, through the early 1920's, the airplane remained more or less a curiosity—merely a new and more horrible instrument of war and a plaything of stunt flyers and thrill-seekers. Aviation did not come into its own until Charles Augustus Lindbergh in 1927 flew nonstop from New York to Paris. The world went over the feat, and the development of bigger and safer airplanes began.

Two major innovations have been effected in the airplane engine since it was established as a means of transportation. The first was the adoption of the gas-turbine engine. In this engine the hot, expanding gases of the fuel drive a wheel by their pressure against its blades, instead of driving pistons in cylinders. The engine is simple, cheaper to run, and less vulnerable to trouble, and it needed only the development of alloys that could withstand the high temperatures of the gases to become a practicable affair. Such alloys were devised by 1939. Since then "turboprop" planes, using a turbine engine to drive the propellers, have become increasingly popular.

But they are now being superseded, at least for long flights, by the second major development—the jet plane. In principle the driving force here is the same as the one that makes a toy balloon dart forward when its mouth is opened and the air escapes. This is action-and-reaction: the motion of the expanding, escaping air in one direction results in equal motion, or thrust, in the opposite direction, just as the forward movement of the bullet in a gun barrel makes the gun kick backward in recoil. In the jet engine, the burning of the fuel produces hot, high-pressure gases that drive the plane forward with great force as they stream backward through the exhaust. A rocket is driven by exactly the same means, except that it carries its own supply of oxygen to burn the fuel.

Patents for "jet propulsion" were taken out by a French engineer, René Lorin, as early as 1913, but at the time it was a completely impractical scheme for airplanes. Jet propulsion is economical only at speeds of more than 400 miles an hour. In 1939, an Englishman, Frank Whittle, flew a reasonably practical jet plane, and, in January 1944, jet planes were put into war use by Great Britain and the United States aganist the "buzz-bombs," Germany's V-1 weapon, a pilotless robot plane carrying explosives in its nose.

After World War II, military jets were developed that approached the speed of sound. The speed of sound depends on the natural elasticity of air molecules, their ability to snap back and forth. When the plane approaches that speed, the air molecules cannot get out of the way, so to speak, and are compressed ahead of the plane, which then undergoes a variety of stresses and strains. There was talk of the "sound barrier" as though it were something physical that could not be approached without destruction. However, tests in wind tunnels led the way to more efficient streamlining, and, on October 14, 1947, an American X-1 rocket plane, piloted by Charles E. Yeager, "broke the sound barrier"; for the first time in history, man surpassed the speed of sound. The air battles of the Korean War in the early 1950's were fought by jet planes moving at such velocities that comparatively few planes were shot down.

The ratio of the velocity of an object to the velocity of sound (which is 740 miles per hour at 0° C.) in the medium through which the object is moving is the "Mach number" after the Austrian physicist Ernst Mach, who first investigated, theoretically, the consequences of motion at such velocities in the mid-nineteenth century. By the 1960's, airplane velocities surpassed Mach 5. This was done by the experimental rocket plane X-15, the rockets of which pushed it high enough, for short periods of time, to allow its pilots to qualify as astronauts. Military planes travel at lower velocities and commercial planes at lower velocities still.

A plane traveling at "supersonic velocities" (over Mach 1) carries its sound waves ahead of it since it travels more quickly than the sound waves alone would. If close enough to the ground to begin with, the cone of compressed sound-waves may intersect the ground with a loud

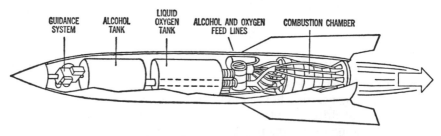

A simple liquid-fueled rocket.

James Watt's steam engine.

John Fitch's steamboat.

George Stephenson's steam locomotive.

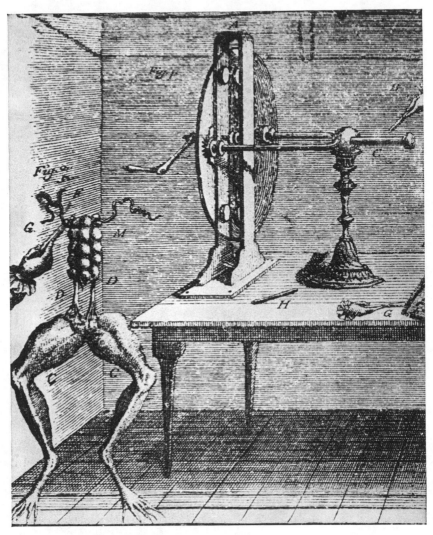

Galvani's experiment, which led to the discovery of electric currents. Electricity from his static-electricity machine made the frog's leg twitch; he found that touching the nerve with two different metals also caused the leg to twitch.

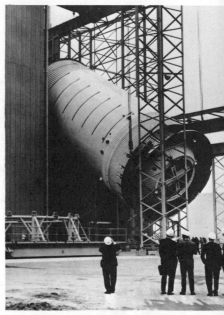

S-11 Hydrogen-fueled booster vehicle, the 81.5-foot tall second stage of NASA's Saturn V Apollo Moon Rocket, was assembled vertically in tower (*top left*) at Huntsville, Alabama; it is seen in turnover station (*top right*) and aboard truck that took it to Cape Kennedy where it was installed atop the giant Saturn V. It is thought to be the most powerful hydrogen-fuel booster in any nation's space program.

"sonic boom." (The crack of a bullwhip is a miniature sonic boom, since, properly manipulated, the tip of such a whip can be made to travel at supersonic velocities.)

Radio

In 1888, Heinrich Hertz conducted the famous experiments that detected radio waves, predicted twenty years earlier by James Clerk Maxwell (see Chapter 7). What he did was to set up a high-voltage alternating current that surged into first one, then another of two metal balls separated by a small air gap. Each time the potential reached a peak in one direction or the other, it sent a spark across the gap. Under these circumstances, Maxwell's equations predicted, electromagnetic radiation should be generated. Hertz used a receiver consisting of a simple loop of wire with a small air gap at one point to detect that energy. Just as the current gave rise to radiation in the first coil, so the radiation ought to give rise to a current in the second coil. Sure enough, Hertz was able to detect small sparks jumping across the gap to his detector coil, placed across the room from the radiating coil. Energy was being transmitted across space.

By moving his detector coil to various points in the room, Hertz was able to tell the shape of the waves. Where sparks came through brightly, the waves were at peak or trough. Where sparks did not come through at all, they were midway. Thus he could calculate the wavelength of the radiation. He found that the waves were tremendously longer than those of light.

In the decade following, it occurred to a number of people that the "Hertzian waves" might be used to transmit messages from one place to another, for the waves were long enough to go around obstacles. In 1890, the French physicist Édouard Branly made an improved receiver by replacing the wire loop with a glass tube filled with metal filings to

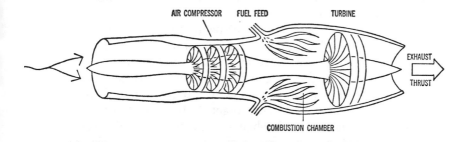

A turbojet engine. Air is drawn in, compressed, and mixed with fuel, which is ignited in the combustion chamber. The expanding gases power a turbine and produce thrust.

411

which wires and a battery were attached. The filings would not carry the battery's current unless a high-voltage alternating current was induced in the filings, as Hertzian waves would do. With this receiver he was able to detect Hertzian waves at a distance of 150 yards. Then the English physicist Oliver Joseph Lodge (who later gained a dubious kind of fame as a champion of spiritualism), modified this device and succeeded in detecting signals at a distance of half a mile and in sending messages in Morse code.

The Italian inventor Guglielmo Marconi discovered that he could improve matters by connecting one side of the generator and receiver to the ground and the other to a wire, later called an "antenna" (because it resembled, I suppose, an insect's feeler). By using powerful generators, Marconi was able to send signals over a distance of nine miles in 1896, across the English Channel in 1898, and across the Atlantic in 1901. Thus was born what the British still call "wireless telegraphy" and the Americans named "radiotelegraphy," or "radio" for short.

Marconi worked out a system for excluding "static" from other sources and tuning in only on the wavelength generated by the transmitter. For his inventions, Marconi shared the Nobel Prize in physics in 1909 with the German physicist Karl Ferdinand Braun, who also contributed to the development of radio.

The American physicist Reginald Aubrey Fessenden proceeded to develop a special generator of high-frequency alternating currents (doing away with the spark-gap device) and to devise a system of "modulating" the radio wave so that it carried a pattern mimicking sound waves. What was modulated was the amplitude (or height) of the waves; consequently this was called "amplitude modulation," now known as AM radio. On Christmas Eve, 1906, music and speech came out of a radio receiver for the first time.

The early radio enthusiasts had to sit over their sets wearing earphones. Some means of strengthening, or "amplifying," the signal was needed, and the answer was found in a discovery that Edison had made —his only discovery in "pure" science.

In one of his experiments, looking toward improving the electric lamp, Edison, in 1883, sealed a metal wire into a light bulb near the hot filament. To his surprise, electricity flowed from the hot filament to the metal wire across the air gap between them. Because this phenomenon had no utility for his purposes, Edison, a practical man, merely wrote it up in his notebooks and forgot it. But the "Edison effect" became very important indeed when the electron was discovered and it became clear that current across a gap meant a flow of electrons. The British physicist Owen Willans Richardson showed, in experiments conducted between 1900 and 1903, that electrons "boiled" out of metal filaments heated in vacuum. For this, he eventually received the Nobel Prize for physics in 1928.

In 1904, the English electrical engineer John Ambrose Fleming put the Edison effect to brilliant use. He surrounded the filament in a bulb with a cylindrical piece of metal (called a "plate"). Now this plate could act in either of two ways. If it was positively charged, it would attract the electrons boiling off the heated filament and so would create a circuit that carried electric current. But if the plate was negatively charged, it would repel the electrons and thus prevent the flow of current. Suppose, then, that the plate was hooked up to a source of alternating current. When the current flowed in one direction, the plate would get a positive charge and pass current in the tube; when the alternating current changed direction, the plate would acquire a negative charge and no current would flow in the tube. Thus the plate would pass current in only one direction; in effect, it would convert alternating to direct current. Because such a tube acts as a valve for the flow of current, the British logically call it a "valve." In the United States, it

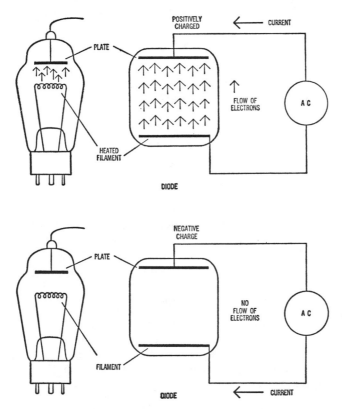

Principle of the vacuum-tube diode.

is vaguely called a "tube." Scientists took to calling it a "diode," because it has two electrodes—the filament and the plate.

The diode serves in a radio set as a "rectifier," changing alternating current to direct where necessary. In 1907, the American inventor Lee De Forest went a step further. He inserted a third electrode in the tube, making a "triode" out of it. The third electrode is a perforated plate ("grid") between the filament and the plate. The grid attracts electrons and speeds up the flow from the filament to the plate (through the holes in the grid). A small increase in the positive charge on the grid will result in a large increase in the flow of electrons from the filament to the plate. Consequently, even the small charge added by weak radio signals will increase the current flow greatly, and this current will mirror all the variations imposed by the radio waves. In other words, the triode acts as an "amplifier." Triodes and even more complicated modifications of the tube have become essential equipment, not only for radio sets, but for all sorts of electronic equipment.

One more step was needed to make radio sets completely popular. During World War I, the American electrical engineer Edwin Howard Armstrong developed a device for lowering the frequency of a radio wave. This was intended, at the time, for detecting aircraft, but after the war it was put to use in radio receivers. Armstrong's "superheterodyne receiver" made it possible to tune in clearly on an adjusted frequency by the turn of one dial, where previously it had been a complicated task to adjust reception over a wide range of possible frequencies. In 1921, regular radio programs were begun by a station in Pittsburgh. Other stations were set up in rapid succession, and with the control of sound level and station tuning reduced to the turn of a dial, radio sets became hugely popular. By 1927, telephone conversations could be carried on across oceans, with the help of radio, and "wireless telephony" was a fact.

There remained the problem of static. The systems of tuning introduced by Marconi and his successors minimized "noise" from thunderstorms and other electrical sources, but did not eliminate it. Again it was Armstrong who found an answer. In place of amplitude modulation, which was subject to interference from the random amplitude modulations of the noise sources, he substituted frequency modulation. That is, he kept the amplitude of the radio carrier wave constant and superimposed a variation in frequency on it. Where the sound wave was large in amplitude, the carrier wave was made low in frequency, and vice versa. Frequency modulation (FM) virtually eliminated static, and FM radio came into popularity after World War II for programs of serious music.

Television was an inevitable sequel to radio, just as talking movies were to the silents. The technical forerunner of television was the transmission of pictures by wire. This entailed translating a picture into an electric current. A narrow beam of light passed through the picture on a photographic film to a phototube behind. Where the film was comparatively opaque, a weak current was generated in the phototube; where it was clearer, a large current was formed. The beam of light swiftly "scanned" the picture from left to right, line by line, and produced a varying current representing the entire picture. The current was sent over wires and at the destination reproduced the picture on film by

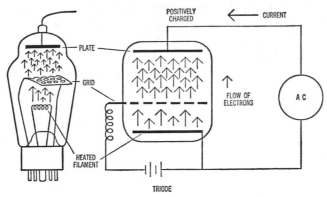

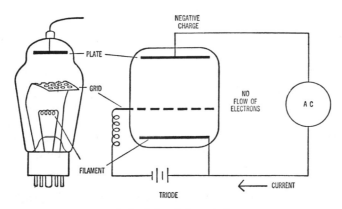

Principle of the triode.

a reverse process. Such "wire-photos" were transmitted between London and Paris as early as 1907.

Television is the transmission of a "movie" instead of still photographs—either "live" or from a film. The transmission must be extremely fast, which means that the action must be scanned very rapidly. The light–dark pattern of the image is converted into a pattern of electrical impulses by means of a camera using, in place of film, a coating of metal that emits electrons when light strikes it.

415

A form of television was first demonstrated in 1926 by the Scottish inventor John Logie Baird. However, the first practical television camera was the "iconoscope," patented in 1938 by the Russian-born American inventor Vladimir Kosma Zwòrykin. In the iconoscope, the rear of the camera is coated with a large number of tiny cesium-silver droplets. Each emits electrons as the light beam scans across it, in proportion to the brightness of the light. The iconoscope was later replaced by the "image orthicon"—a refinement in which the cesium-silver screen is thin enough so that the emitted electrons can be sent forward to strike a thin glass plate that emits more electrons. This "amplification" increases the sensitivity of the camera to light, so that strong lighting is not necessary.

The television receiver is a variety of cathode-ray tube. A stream of electrons shot from a filament ("electron-gun") strikes a screen coated with a fluorescent substance, which glows in proportion to the intensity of the electron stream. Pairs of electrodes controlling the direction of the stream cause it to sweep across the screen from left to right in a series of hundreds of horizontal lines, each slightly below the one before, and the entire "painting" of a picture on the screen in this fashion is completed in a thirtieth of a second. The beam goes on painting successive pictures at the rate of thirty per second. At no instant of time is there more than one dot on the screen (bright or dark, as the case may be); yet, thanks to the persistence of vision, we see not only complete pictures but an uninterrupted sequence of movement and action.

Experimental television was broadcast in the 1920's, but television did not become practical in the commercial sense until 1947. Since then it has virtually taken over the field of entertainment.

In the mid-1950's two refinements were added. By the use of three types of fluorescent material on the television screen, designed to react to the beam in red, blue, and green colors, color television was introduced. And "video tape,'" a type of recording with certain similarities to the sound track on a movie film, made it possible to reproduce recorded programs or events with better quality than could be obtained from motion-picture film.

The vacuum tube, the heart of all the electronic devices, eventually became a limiting factor. Usually the components of a device are steadily improved in efficiency as time goes on—which means that they are stepped up in power and flexibility and reduced in size and mass. (This is sometimes called "miniaturization.") But the vacuum tube became a bottleneck in the road to miniaturization. And then, quite by accident, an unexpected solution turned up.

In the 1940's, several scientists at the Bell Telephone Laboratories grew interested in the substances known as "semiconductors." These substances, such as silicon and germanium, conduct electricity only moderately well, and the problem was to find out why that should be.

The Bell Lab investigators discovered that such conductivity as they possessed was enhanced by traces of impurities mixed with the element in question.

Let us consider a crystal of pure germanium. Each atom has four electrons in its outermost shell, and in the regular array of atoms in the crystal each of the four electrons pairs up with an electron of a neighboring germanium atom, so that all the electrons are paired in stable bonds. Because this arrangement is similar to that in diamond, germanium, silicon, and other such substances are called "adamantine substances," from an old word for "diamond."

If, now, a little bit of arsenic is introduced into this contented adamantine arrangement, the picture grows complicated. Arsenic has five electrons in its outermost shell. An arsenic atom taking the place of a germanium atom in the crystal will be able to pair four of its five electrons with the neighboring atoms, but the fifth can find no electron to pair with. It is left "on the loose." Now if an electric voltage is applied to this crystal, the loose electron will wander in the direction of the positive electrode. It will not move as freely as would electrons in a conducting metal, but the crystal will conduct electricity better than a nonconductor such as sulfur or glass.

This is not very startling, but now we come to a case which is somewhat more odd. Let us add a bit of boron, instead of arsenic, to the germanium. The boron atom has only three electrons in its outermost shell. These three can pair up with the electrons of three neighboring germanium atoms. But what happens to the electron of the boron atom's fourth germanium neighbor? That electron is paired with a "hole"! The word "hole" is used advisedly, because this site, where the electron would find a partner in a pure germanium crystal, does in fact behave like a vacancy. If a voltage is applied to the boron-contaminated crystal, the next neighboring electron, attracted **to**ward the positive electrode, will move into the hole. In doing so, it leaves a hole where it was, and the electron next farther away from the positive electrode moves into *that* hole. And so the hole, in effect, travels steadily toward the negative electrode, moving exactly like an electron, but in the opposite direction. In short, it has become a conveyor of electric current.

To work well, the crystal must be almost perfectly pure with just the right amount of the specified impurity (i.e., arsenic or boron). The germanium-arsenic semiconductor, with a wandering electron, is said to be "n-type"—n for "negative." The germanium-boron semiconductor, with a wandering hole that acts as if it were positively charged, is "p-type"—p for "positive."

Unlike ordinary conductors, the electrical resistance of semiconductors drops as the temperature rises. This is because higher temperatures weaken the hold of atoms on electrons and allow them to drift more freely. (In metallic conductors, the electrons are already free enough at ordinary temperatures. Raising the temperature introduces more random movement and impedes their flow in response to the electric field.) By determining the resistance of a semiconductor, temperatures can be measured that are too high to be conveniently measured in other fashions. Such temperature-measuring semiconductors are called "thermistors."

But semiconductors in combination can do much more. Suppose we make a germanium crystal with one half p-type and the other half n-type. If we connect the n-type side to a negative electrode and the p-type side to a positive electrode, the electrons on the n-type side will move across the crystal toward the positive electrode, while the holes on the p-type side will travel in the opposite direction toward the negative electrode. Thus a current flows through the crystal. Now let us reverse the situation—that is, connect the n-type side to the positive electrode and the p-type to the negative electrode. This time the electrons of the n-side travel toward the positive electrode—which is to say, away from the p-side—and the holes of the p-side similarly move in the direction away from the n-side. As a result, the border regions at the junction between the n- and p-sides lose their free electrons and holes. This amounts to a break in the circuit, and no current flows.

In short, we now have a setup that can act as a rectifier. If we hook up alternating current to this dual crystal, the crystal will pass the current in one direction, but not in the other. Therefore alternating current will be converted to direct current. The crystal serves as a diode, just as a vacuum tube (or "valve") does.

With this device, electronics returned full circle to the first type of rectifier used for radio—namely, the crystal with a "cat's whisker." But the new type of crystal was far more effective and versatile. And it had impressive advantages over the vacuum tube. It was lighter, much less bulky, stronger, invulnerable to shocks, and it did not heat up—all of which gave it a much longer life than the tube. The new device was named, at the suggestion of John Robinson Pierce of the Bell Lab, the "transistor," because it *trans*ferred a signal across a re*sistor*.

In 1948, William Shockley, Walter H. Brattain, and John Bardeen at the Bell Lab went on to produce a transistor which could act as an amplifier. This was a germanium crystal with a thin p-type section sandwiched between two n-type ends. It was in effect a triode with the equivalent of a grid between the filament and the plate. By controlling the positive charge in the p-type center, holes could be sent across the junctions in such a manner as to control the electron flow. Furthermore, a small variation in the current of the p-type center would cause a large variation in the current across the semiconductor system. The semiconductor triode could thus serve as an amplifier, just as a vacuum tube triode did. Shockley and his co-workers Brattain and Bardeen received the Nobel Prize in physics in 1956.

However well transistors might work in theory, their use in practice required certain comcomitant advances in technology. (This is invariably true in applied science.) Efficiency in transistors depended very strongly on the use of materials of extremely high purity, so that the nature and concentration of deliberately added impurities could be carefully controlled.

Fortunately, William G. Pfann introduced the technique of zone-refining in 1952. A rod of, let us say, germanium, is placed in the hollow of a circular heating element, which softens and begins to melt a section of the rod. The rod is drawn through the hollow so that the molten zone moves along it. The impurities in the rod tend to remain in the molten zone and are therefore literally washed to the ends of the rod. After a few passes of this sort, the main body of the germanium rod is unprecedentedly pure.

By 1953, tiny transistors were being used in hearing aids, making them so small that they could be fitted inside the ear. In short order, the transistor—steadily developed so that it could handle higher frequencies, withstand higher temperatures, and be reduced to almost microscopic size—took over many functions of the vacuum tube. Perhaps the most notable example is its use in electronic computers, which have been greatly reduced in size and improved in reliability. In the process, new substances have been developed with useful semiconductor properties.

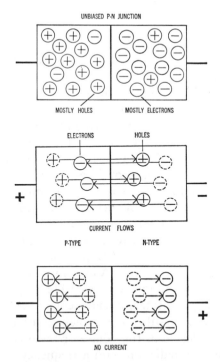

Principle of the junction transistor.

Indium phosphide and gallium arsenide, for instance, have been developed for use in transistors designed to work at high temperatures.

Nor do transistors represent the ultimate in miniaturization. In 1953, a simple two-wire mechanism, operating at liquid helium temperatures, was developed. It could act as a switch by the setting-up and breaking-down of superconductivity in one wire by changes in the magnetic field of the other wire. Such switches are called "cryotrons."

In addition, there are tiny devices in which two thin films of metals such as aluminum and lead are separated by a thin film of insulator such as aluminum oxide. At temperatures in the superconductive range, current will flow, tunneling through the insulator if the voltage is high enough. The amount of current can be delicately controlled by voltage, temperature, and magnetic field intensity. Such "tunnel sandwiches" offer another route to miniaturization.

As an indication of the interconnectedness of science, the new development of rocketry, which demands tremendous structures, also demands miniaturization to the full, for the payload that is eventually placed in orbit is small and must be crammed as full as can be with instrumentation.

Masers and Lasers

Perhaps the most fascinatingly novel of recent devices begins with investigations involving the ammonia molecule (NH_3). The three hydrogen atoms of the ammonia molecule can be viewed as occupying the three apexes of an equilateral triangle, whereas the single nitrogen atom is some distance above the center of the triangle.

It is possible for the ammonia molecule to vibrate. That is, the nitrogen atom can move through the plane of the triangle to an equivalent position on the other side, then back to the first side, and so on, over and over. The ammonia molecule can, in fact, be made to vibrate back and forth with a natural frequency of 24 billion times a second.

This vibration period is extremely constant, much more so than the period of any man-made vibrating device; much more constant, even, than the movement of astronomical bodies. Such vibrating molecules can be made to control electric currents, which will in turn control time-measuring devices with unprecedented precisions, something that was first demonstrated in 1949 by the American physicist Harold Lyons. By the mid-1950's such "atomic clocks" were surpassing all ordinary chronometers. Accuracies in time-measurement of one second in 100,000 years were reached in 1964 with a maser making use of hydrogen atoms.

The ammonia molecule in the course of these vibrations liberates a beam of electromagnetic radiation with a frequency of 24 billion cycles per second. This radiation has a wavelength of 1.25 centimeters and is in the microwave region. Another way of looking at this fact is to imagine the ammonia molecule is capable of occupying one of two energy levels, with the energy difference equal to that of a photon representing a 1.25-centimeter radiation. If the ammonia molecule drops from the higher energy level to the lower, it emits a photon of this size. If a molecule in the lower energy level absorbs a photon of this size, it rises to the higher energy level.

But what if an ammonia molecule is already in the higher energy level and is exposed to such photons? As early as 1917, Einstein had pointed out that if a photon of just the right size struck such an upper-level molecule, the molecule would be nudged back down to the lower level and would emit a photon of exactly the size and moving in exactly the direction of the entering photon. There would be two identical photons where only one had existed before. This was confirmed experimentally in 1924.

Ammonia exposed to microwave radiation could, therefore, undergo two possible changes: molecules could be pumped up from lower level to higher or be nudged down from higher level to lower. Under ordinary conditions, the former process would predominate, for only a very small percentage of the ammonia molecules would, at any one instant, be at the higher energy level.

Suppose, though, that some method were found to place all or almost all the molecules in the upper energy level. Then it would be the movement from higher level to lower that would predominate. Indeed, something quite interesting would happen. The incoming beam of microwave radiation would supply a photon that would nudge one molecule downward. A second photon would be released, and the two would speed on, striking two molecules, so that two more were released. All four would bring about the release of four more, and so on. The initial photon would let loose a whole avalanche of photons, all of exactly the same size and moving in exactly the same direction.

In 1953, the American physicist Charles Hard Townes devised a method for isolating ammonia molecules in the high-energy level and subjected them to stimulation by microwave photons of the correct size. A few photons entered, and a flood of such photons left. The incoming radiation was thus greatly amplified.

The process was described as "microwave amplification by stimulated emission of radiation," and, from the initials of this phrase, the instrument came to be called a "maser."

Solid masers were soon developed; solids in which electrons could be made to take up one of two energy levels. The first masers, both gaseous and solid, were intermittent. That is, they had to be pumped up to the higher energy level first, then stimulated. After a quick burst of radiation, nothing more could be obtained until the pumping process had been repeated.

To circumvent this, it occurred to the Dutch-American physicist Nicolaas Bloembergen to make use of a three-level system. If the material chosen for the core of the maser can have electrons in any of three energy levels—a lower, a middle, and an upper—then pumping and emission can go on simultaneously. Electrons are pumped up from the lowest energy level to the highest. Once at the highest, proper stimulation will cause them to drop down: first to the middle level, then to the lower. Photons of different size are required for pumping and for stimulated emission, and the two processes will not interfere with each other. Thus, we end with a continuous maser.

As microwave amplifiers, masers can be used as very sensitive detectors in radio astronomy, where exceedingly feeble microwave beams received from outer space will be greatly intensified with great fidelity to the original radiation characteristics. (Reproduction without loss of original characteristics is to reproduce with little "noise." Masers are extraordinarily "noiseless" in this sense of the word.) They have carried their usefulness into outer space, too. A maser was carried on board the Soviet satellite "Cosmos 97," launched November 30, 1965, and did its work well.

For his work, Townes received the 1964 Nobel Prize for physics, sharing it with two Soviet physicists, Nicolai Gennediyevich Basov and Aleksandr Mikhailovich Prochorov, who had worked independently on maser theory.

In principle, the maser technique could be applied to electromagnetic waves of any wavelength, notably to those of visible light. Townes pointed out the possible route of such applications to light wavelengths in 1958. Such a light-producing maser might be called an "optical maser." Or else, this particular process might be called "*light amplification by stimulated emission of radiation*," and the new set of initials "laser" might be used. It is the latter word that has grown popular.

The first successful laser was constructed in 1960 by the American physicist Theodore Harold Maiman. He used a bar of synthetic ruby for the purpose, this being, essentially, aluminum oxide with a bit of chromium oxide added. If the ruby bar is exposed to light, the electrons of the chromium atoms are pumped to higher levels and, after a short while, begin to fall back. The first few photons of light emitted (with a wavelength of 694.3 millimicrons) stimulate the production of other such photons, and the bar suddenly emits a beam of deep red light four times as intense as light at the sun's surface. Before 1960 was over, continuous lasers were prepared by an Iranian physicist, Ali Javan, working at Bell Laboratories. He used a gas mixture (neon and helium) as the light source.

The laser made possible light in a completely new form. The light was the most intense that had ever been produced, and the most narrowly monochromatic (single wavelength), but it was even more than that.

Ordinary light, produced in any other fashion, from a wood fire to the sun or to a firefly, consists of relatively short wave packets. They can be pictured as short bits of waves pointing in various directions. Ordinary light is made up of countless numbers of these.

The light produced by a stimulated laser, however, consists of photons of the same size and moving in the same direction. This means that the wave packets are all of the same frequency, and, since they are lined up precisely end to end, so to speak, they melt together. The light appears to be made up of long stretches of waves of even amplitude (height) and frequency (width). This is "coherent light," because the wave packets seem to stick together. Physicists had learned to prepare coherent radiation for long wavelengths. It had never been done for light, though, until 1960.

The laser was so designed, moreover, that the natural tendency of the photons to move in the same direction was accentuated. The two ends of the ruby tube were accurately machined and silvered so as to serve as plane mirrors. The emitted photons flashed back and forth along the rod, knocking out more photons at each pass, until they had built up sufficient intensity to burst through the end that was more lightly silvered. Those that did come through were precisely those that had been emitted in a direction exactly parallel to the long axis of the rod, for those would move back and forth, striking the mirrored ends over and over. If any photon of proper energy happened to enter the rod in a different direction (even a very slightly different direction) and started a train of stimulated photons in that different direction, these would quickly pass out the sides of the rod after only a few reflections at most.

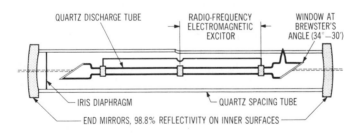

QUARTZ DISCHARGE TUBE
RADIO-FREQUENCY ELECTROMAGNETIC EXCITOR
WINDOW AT BREWSTER'S ANGLE (34°–30')
IRIS DIAPHRAGM
QUARTZ SPACING TUBE
END MIRRORS, 98.8% REFLECTIVITY ON INNER SURFACES

Continuous-wave laser with concave mirrors and Brewster angle windows on discharge tube. The tube is filled with a gas whose atoms are raised to high-energy states by electromagnetic excitation. These atoms are then stimulated to emit energy of a certain wavelength by the introduction of a light beam. Acting like a pipe organ, the resonant cavity builds up a train of coherent waves between the end mirrors. The thin beam that escapes is the laser ray. After a drawing in *Science*, October 9, 1964.

A beam of laser light is made up of coherent waves so parallel that it can travel through long distances without diverging to uselessness. It could be focused finely enough to heat a pot of coffee a thousand miles away. Laser beams even reached to the moon, in 1962, spreading out to a diameter of only two miles after having crossed nearly a quarter of a million miles of space!

Once the laser was devised, interest in its further development was nothing short of explosive. Within a few years individual lasers capable of producing coherent light in hundreds of different wavelengths, from the near ultraviolet to the far infrared, were developed.

Laser action was obtained from a wide variety of solids, from metallic oxides, fluorides, tungstates, from semiconductors, from liquids, from columns of gas. Each variety had its advantages and disadvantages.

In 1964, the first "chemical laser" was developed by the American physicist Jerome V. V. Kasper. In such a laser, the source of energy is a chemical reaction (in the case of the first, the dissociation of CF_3I by a pulse of light). The advantage of the chemical laser over the ordinary variety is that the energy-yielding chemical reaction can be incorporated with the laser itself, and no outside energy-source is needed. This is analogous to a battery-powered device as compared with one that must be plugged into a wall socket. There is an obvious gain in portability here, to say nothing of the fact that chemical lasers seem to be considerably more efficient than the ordinary variety (12 per cent or more, as compared with 2 per cent or less).

Organic lasers—those in which a complex organic dye is used as the source of coherent light—were first developed in 1966 by John R. Lankard and Peter Sorokin. The complexity of the molecule makes it possible to produce light by a variety of electronic reactions and therefore in a variety of wavelengths. A single organic laser can be "tuned" to deliver any wavelength within a range, rather than find itself confined to a single wavelength as is true of the others.

The narrowness of the beam of laser light means that a great deal of energy can be focused into an exceedingly small area; in that area, the temperature reaches extreme levels. The laser can vaporize metal for quick spectral investigation and analysis and can weld, cut, or punch holes of any desired shape through high-melting substances. By shining laser beams into the eye, surgeons have succeeded in welding loosened retinas so rapidly that surrounding tissues have no time to be affected by heat and, in similar fashion, to destroy tumors.

To show the vast range of laser applications, Arthur L. Shawlow has developed the trivial (but impressive) laser-eraser, which in an intensely brief flash evaporates the typewriter ink of the formed letters without so much as scorching the paper beneath; at the other extreme, laser interferometers can make unprecedentedly refined measurements. When earth-strains intensify, this can be detected by separated lasers, where shifts in the interference fringes of their light will detect tiny earth-movements with a delicacy of one part in a trillion. Then, too, the first men on the moon left a reflector system designed to bounce back laser beams to earth. By such a method, the distance to the moon may be determined with greater accuracy than the distance, in general, from point to point on earth's surface.

One possible application that created excitement from the beginning has been the use of laser beams as carrier beams in communications. The high frequency of coherent light, as compared with that of the coherent radio-waves used in radio and television today, holds forth the promise of being able to crowd many thousands of channels into the space that now holds one channel. The prospect arises that every human being on earth may have his own personal wavelength. Naturally, the laser light must be modulated. Varying electric currents produced by sound must be translated into varying laser light (either through changes in its amplitude on its frequency, or perhaps just by turning it on and off), which can in turn be used to produce varying electric current elsewhere. Such systems are being developed.

It may be that since light is so much more subject than radio waves to interference by clouds, mist, fog, and dust, that it will be necessary to conduct laser light through pipes containing lenses (to reconcentrate the beam at intervals) and mirrors (to reflect it around corners). However, a carbon-dioxide laser has been developed that produces continuous laser-beams of unprecedented power that are far enough in the infrared to be little affected by the atmosphere. Atmospheric communication may also be possible then.

A still more fascinating application of laser beams that is very much here-and-now involves a new kind of photography. In ordinary photography a beam of ordinary light reflected from an object falls on a photographic film. What is recorded is the cross-section of the light, which is by no means all the information it can potentially contain.

Suppose instead that a beam of light is split in two. One part strikes an object and is reflected with all the irregularities that this object would impose on it. The second part is reflected from a mirror with no irregularities. The two parts meet at the photographic film, and the interference of the various wavelengths is recorded. In theory, the recording of this interference would include all the data concerning each light beam. The photograph that records this interference pattern seems to be blank when developed, but if light is shone upon the film and passes through and takes on the interference characteristics, it produces an image containing the complete information. The image is as three-dimensional as was the surface from which light was reflected, and an ordinary photograph can be taken of the image from various angles that show the change in perspective.

This notion was first worked out by the Hungarian-British physicist

Dennis Gabor in 1947, when he was trying to work out methods for the sharpening of images produced by electron microscopes. He called it "holography," from a Latin word meaning "the whole writing."

While Gabor's idea was theoretically sound, it could not be implemented, because ordinary light would not do. With wavelengths of all sizes moving in all directions, the interference-fringes produced by the two beams of light, would be so chaotic as to yield no information at all. It would be like producing a million dim images all superimposed in slightly different positions.

The introduction of laser light changed everything. In 1965, Emmet N. Leith and Juris Upatnieks, at the University of Michigan, were able to produce the first holograms. Since then, the technique has been sharpened to the point where holography in color has become possible, and where the photographed interference-fringes can successfully be viewed with ordinary light. Microholography promises to add a new dimension (literally) to biological investigations, and where it will end, none can predict.

CHAPTER 9

The Reactor

Nuclear Fission

The rapid advances in technology in the twentieth century have been bought at the expense of a stupendous increase in our consumption of the earth's energy resources. As the underdeveloped nations, with their billions of people, join the already industrialized countries in high living, the rate of consumption of fuel will jump even more spectacularly. Where will mankind find the energy supplies needed to support his civilization?

We have already seen a large part of the earth's timber disappear. Wood was man's first fuel. By the beginning of the Christian era, much of Greece, northern Africa, and the Near East had been ruthlessly deforested, partly for fuel, partly to clear the land for animal herding and agriculture. The uncontrolled felling of the forests was a double-barreled disaster. Not only did it destroy the wood supply, but the drastic uncovering of the land meant a more or less permanent destruction of fertility. Most of these ancient regions, which once supported man's most advanced cultures, are sterile and unproductive now, populated by a ground-down and backward people.

The Middle Ages saw the gradual deforestation of western Europe, and modern times have seen the much more rapid deforestation of the North American continent. Almost no great stands of virgin timber remain in the world's temperate zones except in Canada and Siberia.

It seems unlikely that man will ever be able to get along without wood. Building lumber and paper will always be necessities.

As for fuel, coal and oil have taken wood's place. Coal was mentioned by the Greek botanist Theophrastus as long ago as 200 B.C., but the first records of actual coal mining in Europe do not date back before the twelfth century. By the the seventeenth century, England, deforested and desperately short of wood for its navy, began to shift to the large-scale use of coal for fuel, a switch that lay the groundwork for the Industrial Revolution.

The shift was slow elsewhere. Even in 1800, wood supplied 94 per cent of the fuel needs in the young, forest-rich United States. In 1885, wood supplied only 50 per cent of the fuel needs, and by the 1960's, only 3 per cent. The balance, moreover, has shifted beyond coal to oil and natural gas. In 1900, the energy supplied by coal in the United States was ten times that supplied by oil and gas together. Half a century later, coal supplied only one third the energy supplied by oil and gas. Coal, oil, and gas are "fossil fuels," relics of plant life eons old, and cannot be replaced once they are used up. With respect to coal and oil, man is living on his capital at an extravagant rate.

The oil, particularly, is going fast. The world is now burning a million barrels of oil each hour, and the rate of consumption is rising rapidly. Although well over a trillion barrels remain in the earth, it is estimated that by 1980 oil production will reach its peak and begin to decline. Of course, additional oil can be formed by the combination of the more common coal with hydrogen under pressure. This process was first developed by the German chemist Friedrich Bergius in the 1920's, and he shared in the Nobel Prize for chemistry in 1931 as a result. The coal reserve is large indeed, perhaps as large as 7 trillion tons, but not all of it is easy to mine. By the twenty-fifth century or sooner, coal may become an expensive commodity.

We can expect new finds. Perhaps surprises in the way of coal and oil await us in Australia, in the Sahara, even in Antarctica. Moreover, improvements in technology may make it economical to exploit thinner and deeper coal seams, to plunge more and more deeply for oil, and to extract oil from oil shale and from subsea reserves.

No doubt we shall also find ways to use our fuel more efficiently. The process of burning fuel to produce heat to convert water to steam to drive a generator to create electricity wastes a good deal of energy along the way. Most of these losses could be side-stepped if heat could be converted directly into electricity. The possibility of doing this appeared as long ago as 1823, when a German physicist, Thomas Johann Seebeck, observed that, if two different metals were joined in a closed circuit and if the junction of the two elements were heated, a compass needle in the vicinity would be deflected. This meant that the heat was producing an electric current in the circuit ("thermoelectricity"), but Seebeck misinterpreted his own work, and his discovery was not followed up.

With the coming of semiconductor techniques, however, the old "Seebeck effect" underwent a renaissance. Current thermoelectric devices make use of semiconductors. Heating one end of a semiconductor creates an electric potential in the material: in a p-type semiconductor the cold end becomes negative; in an n-type it becomes positive. Now if these two types of semiconductor are joined in a U-shaped structure, with the n-p junction at the bottom of the U, heating the bottom will cause the upper end of the p branch to gain a negative charge and the upper end of the n branch to acquire a positive charge. As a result, current will flow from one end to the other, and it will be generated so long as the temperature difference is maintained. (In reverse, the use of a current can bring about a temperature drop, so that a thermoelectric device can also be used as a refrigerator.)

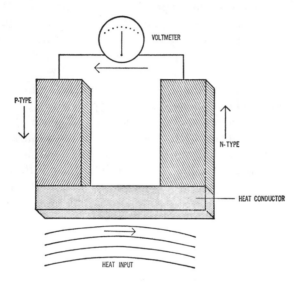

The thermoelectric cell. Heating the conductor causes electrons to flow toward the cold end of the n-type semi-conductor and from the cold to the warm region of the p-type. If a circuit is formed, current flows in the direction shown by the arrows. Thus heat is converted to electrical energy.

The thermoelectric cell, requiring no expensive generator or bulky steam engine, is portable and could be set up in isolated areas as a small-scale supplier of electricity. All it needs as an energy source is a kerosene heater. Such devices are reported to be used routinely in rural areas of the Soviet Union.

Notwithstanding all possible increases in the efficiency of using fuel and the likelihood of new finds of coal and oil, these sources of energy are definitely limited. The day will come, and not far in the future, when neither coal nor oil can serve as an important large-scale energy source.

And yet man's energy needs will continue and even be far larger than those of today. What can be done?

One possibility is to make increasing use of renewable energy sources: to live on the earth's energy income rather than its capital. Wood can be such a resource if forests are grown and harvested as a crop, though wood alone could not come anywhere near meeting all our energy needs. We could also make much more use of wind power and water power, though these again could never be more than subsidiary sources of energy. The same must be said about certain other potential sources of energy in the earth, such as tapping the heat of the interior (e.g., in hot springs) or harnessing the ocean tides.

Far more important, for the long run, is the possibility of directly tapping some of the vast energy pouring on the earth from the sun. This "insolation" produces energy at a rate that is some 50,000 times as great as man's current rate of energy consumption. In this respect, one particularly promising device is the "solar battery," which also makes use of semiconductors.

As developed by the Bell Telephone Laboratories in 1954, the solar battery is a flat sandwich of n-type and p-type semiconductors. Sunlight striking the plate knocks some electrons out of place. The transfer is connected, as an ordinary battery would be, in an electric circuit. The freed electrons move toward the positive pole and holes move toward the negative pole, thus constituting a current. The solar battery can develop electric potentials of up to half a volt, and up to 9 watts of power, from each square foot exposed to the sun. This is not much, but the beauty

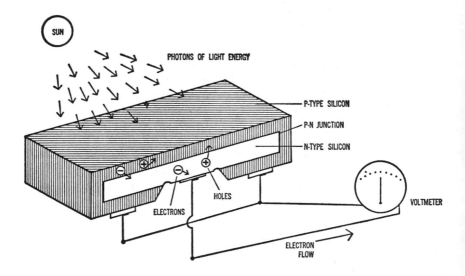

A solar battery cell. Sunlight striking the thin wafer frees electrons, thus forming electron-hole pairs. The p-n junction acts as a barrier, or electric field, separating electrons from holes. A potential difference therefore develops across the junction, and current then flows through the wire circuit.

of the solar battery is that it has no liquids, no corrosive chemicals, no moving parts—it just keeps on generating electricity indefinitely merely by lying in the sun.

The artificial satellite Vanguard I, launched by the United States on March 17, 1958, was the first to be equipped with a solar battery to power its radio signals.

The amount of energy falling upon one acre of a generally sunny area of the earth is 9.4 million kilowatt-hours per year. If substantial areas in the earth's desert regions, such as Death Valley and the Sahara, were covered with solar batteries and electricity-storing devices, they could provide the world with its electricity needs for an indefinite time —for as long, in fact, as the human race is likely to endure, if it does not commit suicide.

But the tapping of solar energy, it seems, is not likely to be achieved on any great scale in this generation or the next. Fortunately we have an immense source of energy, here on the earth, which can tide us over for hundreds of years after we run out of inexpensive coal and oil. It is the energy in the atomic nucleus.

Nuclear energy is commonly called "atomic energy," but that is a misnomer. Strictly speaking, atomic energy is the energy yielded by chemical reactions, such as the burning of coal and oil, because they involve the behavior of the atom as a whole. The energy released by changes in the nucleus is of a totally different kind and vastly greater in magnitude.

Soon after the discovery of the neutron by Chadwick in 1932, physicists realized that they had a wonderful key for unlocking the atomic nucleus. Since it had no electric charge, the neutron could easily penetrate the charged nucleus. Physicists immediately began to bombard various nuclei with neutrons to see what nuclear reactions they could bring about; among the most ardent investigators with this new tool was Enrico Fermi of Italy. In the space of a few months, he had prepared new radioactive isotopes of thirty-seven different elements.

Fermi and his associates discovered that they got better results if they slowed down the neutrons by passing them through water or paraffin first. Bouncing off protons in the water or paraffin, the neutrons are slowed just as a billiard ball is by hitting other billiard balls. When a neutron is reduced to "thermal" speed (the normal speed of motion of atoms), it has a greater chance of being absorbed by a nucleus, because it remains in the vicinity of the nucleus longer. Another way of looking at it is to consider that the wavelength of the wave associated with the neutron is longer, for the wavelength is inversely proportional to the momentum of the particle. As the neutron slows down, its wavelength increases. To put it metaphorically, the neutron grows fuzzier and takes up more volume. It therefore hits a nucleus more easily, just as a bowling ball has more chance of hitting a tenpin than a golf ball would have.

The probability that a given species of nucleus will capture a neutron is called its "cross section." This term, metaphorically, pictures the nucleus as a target of a particular size. It is easier to hit the side of a barn with a baseball than it is to hit a foot-wide board at the same distance. The cross sections of nuclei under neutron bombardment are reckoned in trillion-trillionths of a square centimeter (10^{-24} square centimeter). That unit, in fact, was named a "barn" by the American physicists M. G. Holloway and C. P. Baker in 1942. The name served to hide what was really going on in those hectic wartime days.

When a nucleus absorbs a neutron, its atomic number is unchanged (because the charge of the nucleus remains the same), but its mass number goes up by one unit. Hydrogen 1 becomes hydrogen 2, oxygen 17 becomes oxygen 18, and so on. The energy delivered to the nucleus by the neutron as it enters may "excite" the nucleus—that is, increase its energy content. This surplus energy is then emitted as a gamma ray.

The new nucleus often is unstable. For example, when aluminum 27 takes in a neutron and becomes aluminum 28, one of the neutrons in the new nucleus soon changes to a proton (by emitting an electron). This increase in the positive charge of the nucleus transforms the aluminum (atomic number 13) to silicon (atomic number 14).

Because neutron bombardment is an easy way of converting an element to the next higher one, Fermi decided to bombard uranium to see if he could form an artificial element—number 93. In the products of the bombardment of uranium, he and his co-workers did find signs of new radioactive substances. They thought they had made element 93, and called it "uranium X." But how could the new element be identified positively? What sort of chemical properties should it have?

Well, element 93, it was thought, should fall under rhenium in the periodic table, so it ought to be chemically similar to rhenium. (Actually, though no one realized it at the time, element 93 belonged in a new rare-earth series, which meant that it would resemble uranium, not rhenium—see Chapter 5. Thus, the search for its identification got off on the wrong foot entirely.) If it were like rhenium, perhaps the tiny amount of "element 93" created might be identified by mixing the products of the neutron bombardment with rhenium and then separating out the rhenium by chemical methods. The rhenium would act as a "carrier," bringing out the chemically similar "element 93" with it. If the rhenium proved to have radioactivity attached to it, this would indicate the presence of element 93.

The German physicist Otto Hahn and the Austrian physicist Lise Meitner, working together in Berlin, pursued this line of experiment. Element 93 failed to show up with rhenium. Hahn and Meitner then went on to try to find out whether the neutron bombardment had transformed uranium into other elements near it in the periodic table. At this point, in 1938, Germany occupied Austria, and Miss Meitner, who, until then, as an Austrian national, had been safe despite the fact that she was Jewish, was forced to flee from Hitler's Germany to the safety of Stockholm. Hahn continued his work with the German physicist Fritz Strassman.

Several months later Hahn and Strassman found that barium, when added to the bombarded uranium, carried off some radioactivity. They decided that this radioactivity must belong to radium, the element below barium in the periodic table. The conclusion was, then, that the neutron bombardment of uranium changed some of it to radium.

But this radium turned out to be peculiar stuff. Try as they would, Hahn and Strassman could not separate it from the barium. In France, Irène Joliot-Curie and her co-worker P. Savitch undertook a similar task and also failed.

And then Meitner, the refugee in Scandinavia, boldly cut through the riddle and broadcast a thought that Hahn was voicing in private but was hesitating to publish. In a letter published in the British journal *Nature* in January of 1939, she suggested that the "radium" could not be separated from the barium because no radium was there. The supposed radium was actually radioactive barium: it was barium that had been formed in the neutron bombardment of uranium. This radioactive barium decayed by emitting a beta particle and formed lanthanum. (Hahn and Strassman had found that ordinary lanthanum added to the products brought out some radioactivity, which they assigned to actinium; actually it was radioactive lanthanum.)

But how could barium be formed from uranium? Barium was only a middleweight atom. No known process of radioactive decay could transform a heavy element into one only about half its weight. Meitner made so bold as to suggest that the uranium nucleus had split in two. The absorption of a neutron had caused it to undergo what she termed "fission." The two elements into which it had split, she said, were barium and element 43, the element above rhenium in the periodic table. A nucleus of barium and one of element 43 (later named technetium) would make up a nucleus of uranium. What made it a particularly daring suggestion was that neutron bombardment only supplied 6 million electron-volts, and the main thought of the day concerning nuclear structure made it seem that hundreds of millions would be required.

Meitner's nephew, Otto Robert Frisch, hastened to Denmark to place the new theory before Bohr, even in advance of publication. Bohr had to face the surprising ease with which this would require the nucleus to split, but fortunately he was evolving the liquid-drop theory of nuclear structure, and it seemed to him that this would explain it. (In later years the liquid-drop theory, taking into account the matter of nuclear shells, was to explain even the fine details of nuclear fission and why the nucleus broke into unequal halves.)

In any case, theory or not, Bohr grasped the implications at once. He was just leaving to attend a conference on theoretical physics in Washington, and there he told physicists what he had heard in Denmark of the fission suggestion. In high excitement, the physicists went back to their laboratories to test the hypothesis, and within a month half a dozen experimental confirmations were announced. The Nobel Prize for chemistry went to Hahn in 1944 as a result.

And so began the work that led to the most terrible weapon of destruction ever devised.

The Nuclear Bomb

The fission reaction released an unusual amount of energy, vastly more than did ordinary radioactivity. But it was not solely the additional energy that made fission so portentous a phenomenon. More important was the fact that it released two or three neutrons. Within two months after the Meitner letter, the awesome possibility of a "nuclear chain reaction" had occurred to a number of physicists.

The expression "chain reaction" has acquired an exotic meaning, but actually it is a very common phenomenon. The burning of a piece of paper is a chain reaction. A match supplies the heat required to start it; once the burning has begun, this supplies the very agent, heat, needed to maintain and spread the flame. Burning brings about more burning on an ever-expanding scale.

That is exactly what happens in a nuclear chain reaction. One neutron fissions a uranium nucleus; this releases two neutrons that can produce two fissions that release four neutrons which can produce four fissions, and so on. The first atom to fission yields 200 Mev of energy; the next step yields 400 Mev, the next 800 Mev, the next 1,600 Mev, and so on. Since the successive stages take place at intervals of about a 50-trillionth of a second, you see that within a tiny fraction of a second a staggering amount of energy will be released. (The actual average number of neutrons produced per fission is 2.47, so matters go even more quickly than this simplified calculation indicates.) The fission of one ounce of uranium produces as much energy as the burning of 90 tons of coal or of 2,000 gallons of fuel oil. Peacefully used, uranium fission could relieve all our immediate worries about vanishing fossil fuels and man's mounting consumption of energy.

But the discovery of fission came just before the world was plunged into an all-out war. The fissioning of an ounce of uranium, physicists estimated, would yield as much explosive power as 600 tons of TNT.

The thought of the consequences of a war fought with such weapons was horrible, but the thought of a world in which Nazi Germany laid its hands on such an explosive before the Allies did was even more horrible.

The Hungarian-American physicist Leo Szilard, who had been thinking of nuclear chain reactions for years, foresaw the possible future with complete clarity. He and two other Hungarian-American physicists, Eugene Wigner and Edward Teller, prevailed on the gentle and pacific Einstein in the summer of 1939 to write a letter to President Franklin Delano Roosevelt, pointing out the potentialities of uranium fission and suggesting that every effort be made to develop such a weapon before the Nazis managed to do so.

The letter was written on August 2, 1939, and was delivered to the President on October 11, 1939. Between those dates, World War II had erupted in Europe. Physicists at Columbia University, under the supervision of Fermi, who had left Italy for America the previous year, worked to produce sustained fission in a large quantity of uranium.

Eventually, the government of the United States itself took action in the light of Einstein's letter. On December 6, 1941, President Roosevelt (taking a huge political risk in case of failure) authorized the organization of a giant project, under the deliberately noncommittal name of "Manhattan Engineer District" for the purpose of devising an atom bomb. The next day, the Japanese attacked Pearl Harbor, and the United States was at war.

As was to be expected, practice did not by any means follow easily from theory. It took a bit of doing to arrange a uranium chain reaction. In the first place, you had to have a substantial amount of uranium, refined to extreme purity so that neutrons would not be wasted in absorption by impurities. Uranium is a rather common element in the earth's crust, averaging about 2 grams per ton of rock, which makes it 400 times as common as gold. But it is well spread out, and there are few places in the world where it occurs in rich ores or even in reasonable concentration. Furthermore, before 1939 uranium had had almost no uses, and no methods for its purification had been worked out. Less than an ounce of uranium metal had been produced in the United States.

The laboratories at Iowa State College, under the leadership of Spedding, went to work on the problem of purification by ion-exchange resins (see Chapter 5), and in 1942 began to produce reasonably pure uranium metal.

That, however, was only a first step. Now the uranium itself had to be broken down to separate out its more fissionable fraction. The isotope uranium 238 (U-238) has an even number of protons (92) and an even number of neutrons (146). Nuclei with even numbers of nucleons are more stable than those with odd numbers. The other isotope in natural uranium—uranium 235—has an odd number of neutrons (143). Bohr had therefore predicted that it would fission more readily than uranium

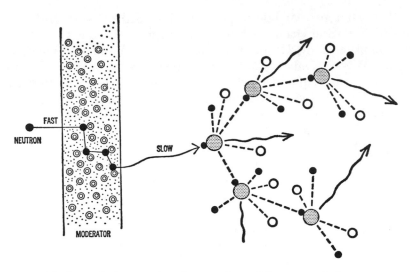

Nuclear chain reaction in uranium. The gray circles are uranium nuclei, the black dots neutrons, the wavy arrows gamma rays, and the small circles fission fragments.

238. In 1940, a research team under the leadership of the American physicist John Ray Dunning isolated a small quantity of uranium 235 and showed that Bohr's conjecture was true. U-238 fissions only when struck by fast neutrons of more than a certain energy, but U-235 would undergo fission upon absorbing neutrons of any energy, all the way down to simple thermal neutrons.

The trouble was that in purified natural uranium only one atom in 140 is U-235, the rest being U-238. This meant that most of the neutrons released by fission of U-235 would be captured by U-238 atoms without producing fission. Even if the uranium were bombarded with neutrons fast enough to split U-238, the neutrons released by the fissioning U-238 would not be energetic enough to carry on a chain reaction in the remaining atoms of this more common isotope. In other words, the presence of U-238 would cause the chain reaction to damp and die. It would be like trying to burn wet leaves.

There was nothing for it, then, but to try for a large-scale separation of U-235 from the U-238, or at least the removal of enough U-238 to effect a substantial enrichment of the U-235 content in the mixture. The physicists attacked this problem by several methods, each of them offering only thin prospects of success. The one that eventually worked out best was "gaseous diffusion." This remained the method of choice, though fearfully expensive, until 1960. A West German scientist then developed a much cheaper technique of U-235 isolation by centrifugation, the heavier molecules being thrown outward and the lighter ones, containing U-235, lagging behind. This process may make nuclear bombs cheap enough for minor powers to manufacture, a consummation not entirely to be desired.

The uranium-235 atom is 1.3 per cent less massive than the uranium-238 atom. Consequently if the atoms were in the form of a gas, the U-235 atoms would move about slightly faster than the U-238 atoms. This meant they might be separated by reason of their faster diffusion through a series of filtering barriers. But first uranium had to be converted to a gas. About the only way to get it in this form was to combine it with fluorine and make uranium hexafluoride, a volatile liquid composed of one uranium atom and six fluorine atoms. In this compound a molecule containing U-235 would be less than 1 per cent lighter than one containing U-238; but that difference proved to be sufficient to make the method work.

The uranium hexafluoride gas was forced through porous barriers under pressure. At each barrier, the molecules containing U-235 got through a bit faster, on the average, and so with every passage through the successive barriers the advantage in favor of U-235 grew. To obtain sizable amounts of almost pure uranium-235 hexafluoride required thousands of barriers, but well-enriched concentrations of U-235 could be achieved with a much smaller number of barriers.

By 1942, it was reasonably certain that the gaseous diffusion method (and one or two others) could produce "enriched uranium" in quantity, and separation plants (costing a billion dollars each, and consuming as much electricity as all of New York City) were built at the secret city of Oak Ridge, Tennessee, sometimes called Dogpatch by irreverent scientists, after the mythical town of Al Capp's Li'l Abner.

Meanwhile, the physicists were calculating the "critical size" that would be needed to maintain a chain reaction in a lump of enriched uranium. If the lump was small, too many neutrons would escape from its surface before being absorbed by U-235 atoms. To minimize this loss by leakage, the volume of the lump had to be large in proportion to its surface. At a certain "critical size," enough neutrons would be intercepted by U-235 atoms to keep a chain reaction going.

The physicists also found a way to make efficient use of the available neutrons. "Thermal" (ie., slow) neutrons, as I have mentioned, are more readily absorbed by uranium 235 than are fast ones. The experimenters therefore used a "moderator" to slow the neutrons from the rather high speeds they had on emerging from the fission reaction. Ordinary water would have been an excellent slowing agent, but unfortunately the nuclei of ordinary hydrogen hungrily snap up neutrons. Deuterium (hydrogen 2) fills the bill much better; it has practically no tendency to absorb neutrons. Consequently the fission experimenters became very interested in preparing supplies of heavy water.

Up to 1943, it was prepared by electrolysis for the most part. Ordinary water split into hydrogen and oxygen more readily than did heavy water, so that, if a large supply of water were electrolyzed, the final bit of water was rich in heavy water and could be preserved. After 1943, careful distillation was the favored method. Ordinary water had the lower boiling point, so that the last bit of unboiled water was rich in heavy water.

Heavy water was indeed valuable in the early 1940's. There is a thrilling story of how Joliot-Curie managed to smuggle France's supply of that liquid out of the country ahead of the invading Nazis in 1940. A hundred gallons of it, which had been prepared in Norway, did fall into the hands of the German Nazis. It was destroyed by a British commando raid in 1942.

Still, heavy water had drawbacks: it might boil away when the chain reaction got hot, and it would corrode the uranium. The scientists seeking to create a chain-reacting system in the Manhattan Project decided to use carbon, in the form of very pure graphite, as the moderator.

Another possible moderator, beryllium, had the disadvantage of toxicity. Indeed, the disease, berylliosis, was first recognized in the early 1940's in one of the physicists working on the atom bomb.

Now let us imagine a chain reaction. We start things off by sending a triggering stream of neutrons into the assembly of moderator and enriched uranium. A number of uranium-235 atoms undergo fission, releasing neutrons which go on to hit other uranium-235 atoms. They in turn fission and turn loose more neutrons. Some neutrons will be absorbed by atoms other than uranium 235; some will escape from the pile altogether. But if from each fission one neutron, and exactly one, takes effect in producing another fission, then the chain reaction will be self-sustaining. If the "multiplication factor" is more than one, even very slightly more (e.g., 1.001), the chain reaction will rapidly build up to an explosion. This is good for bomb purposes but not for experimental purposes. Some device had to be worked out to control the rate of fissions. That could be done by sliding in rods of a substance such as cadmium, which has a high cross section for neutron capture. The chain reaction develops so rapidly that the damping cadmium rods could not be slid in fast enough, were it not for the fortunate fact that the fissioning uranium atoms do not emit all their neutrons instantly. About one neutron in 150 is a "delayed neutron" emitted a few minutes after fission, since it emerges, not directly from the fissioning atoms, but from the smaller atoms formed in fission. When the multiplication factor is only slightly above one, this delay is sufficient to give time for applying the controls.

In 1941, experiments were conducted with uranium-graphite mixtures, and enough information was gathered to lead physicists to decide that, even without enriched uranium, a chain reaction might be set up if only the lump of uranium were made large enough.

Physicists set out to build a uranium chain reactor of critical size at the University of Chicago. By that time some six tons of pure uranium were available; this amount was eked out with uranium oxide. Alternate layers of uranium and graphite were laid down one on the other, fifty-seven layers in all, with holes through them for insertion of the cadmium control rods. The structure was called a "pile"—a noncommittal code name that did not give away its function. (During World War I, the newly designed armored vehicles on caterpillar treads were referred to as "tanks" for the same purpose of secrecy. The name "tank" stuck, but "atomic pile" fortunately gave way eventually to the more descriptive name "nuclear reactor.")

The Chicago pile, built under the football stadium, measured 30 feet wide, 32 feet long, and 21½ feet high. It weighed 1,400 tons and contained 52 tons of uranium, as metal and oxide. (Using pure uranium 235, the critical size would have been, it is reported, no more than 9 ounces.) On December 2, 1942, the cadmium control rods were slowly pulled out. At 3:45 P.M. the multiplication factor reached one—a self-sustaining fission reaction was underway. At that moment mankind (without knowing it) entered the "Atomic Age."

The physicist in charge was Enrico Fermi, and Eugene Wigner presented him with a bottle of Chianti in celebration. Arthur Compton, who was at the site, made a long-distance telephone-call to James Bryant Conant at Harvard, announcing the success. "The Italian navigator," he said, "has entered the new world." Conant asked, "How were the natives?" The answer came at once: "Very friendly!"

It is curious that the first Italian navigator discovered one new world in 1492, and the second discovered another in 1942; those who value the mystic interplay of numbers make much of this coincidence.

Meanwhile another fissionable fuel had turned up. Uranium 238, upon absorbing a thermal neutron, forms uranium 239, which breaks down quickly to neptunium 239, which in turn breaks down almost as quickly to plutonium 239.

Now the plutonium 239 nucleus has an odd number of neutrons (145) and is more complex than uranium 235, so it should be highly unstable. It seemed a reasonable guess that plutonium, like uranium 235, might undergo fission with thermal neutrons. In 1941, this was confirmed experimentally. Still uncertain whether the preparation of uranium 235 would prove practical, the physicists decided to hedge their bets by trying to make plutonium in quantity.

Special reactors were built in 1943 at Oak Ridge and at Hanford, in the State of Washington, for the purpose of manufacturing plutonium.

These reactors represented a great advance over the first pile in Chicago. For one thing, the new reactors were designed so that the uranium could be removed from the pile periodically. The plutonium produced could be separated from the uranium by chemical methods, and the fission products, some of them strong neutron absorbers, could be removed. In addition, the new reactors were water-cooled to prevent overheating. (The Chicago pile could operate only for short periods, because it was cooled merely by air.)

By 1945, enough purified uranium 235 and plutonium 239 were available for the construction of bombs. This portion of the task was undertaken at a third secret city, Los Alamos, New Mexico, under the leadership of the American physicist J. Robert Oppenheimer.

For bomb purposes it was desirable to make the nuclear chain reaction mount as rapidly as possible. This called for making the reaction go with fast neutrons, to shorten the intervals between fissions, so the moderator was omitted. The bomb was also enclosed in a massive casing to hold the uranium together long enough for a large proportion of it to fission.

Since a critical mass of fissionable material would explode spontaneously (sparked by stray neutrons from the air), the bomb fuel was divided into two or more sections. The triggering mechanism was an ordinary explosive (TNT?) which drove these sections together when the bomb was to be detonated. One arrangement was called "the thin man"—a tube with two pieces of uranium 235 at its opposite ends. Another, "the fat man," had the form of a ball in which a shell composed of fissionable material was "imploded" toward the center, making a dense critical mass held together momentarily by the force of the implosion and by a heavy outer casing called the "tamper." The tamper also served to reflect back neutrons into the fissioning mass and, therefore, to reduce the critical size.

To test such a device on a minor scale was impossible. The bomb had to be above critical size or nothing. Consequently, the first test was the explosion of a full-scale nuclear-fission bomb, usually called, incorrectly, an "atom bomb" or "A-bomb." At 5:30 A.M. on July 16, 1945, at Alamogordo, New Mexico, a bomb was exploded with truly horrifying effect; it had the explosive force of 20,000 tons of TNT. The physicist I. I. Rabi, on being asked later what he had witnessed, is reported to have said mournfully, "I can't tell you, but don't expect to die a natural death." (It is only fair to add that the gentleman so addressed by Rabi did die a natural death some years later.)

Two more fission bombs were prepared. One, a uranium bomb called "Little Boy," 10 feet long by 2 feet wide and weighing 4½ tons, was dropped on Hiroshima on August 6, 1945; it was set off by radar echo. Days later, the second, a plutonium bomb, 11 by 5 feet, weighing 5 tons, and named "Fat Man," was dropped on Nagasaki. Together,

Fission of a uranium atom. The white streak in the middle of this photographic plate represents the tracks of the two atoms flying apart from the central point where the uranium atom split in two. The plate was soaked in a uranium compound and bombarded with neutrons, which produced the fission caught in this picture. The other white dots are randomly developed silver grains. The picture was made in the Eastman Kodak Research Laboratories.

Radioactivity made visible. On the tray is some tantalum made radioactive in the Brookhaven reactor; the glowing material is shielded here under several feet of water. The radioactive tantalum will be placed in the pipe shown and then transferred to a large lead container for use as a 1,000-curie source of radioactivity for industrial purposes.

Drawing of the first chain reactor, built under the Chicago football stadium.

The Chicago reactor under construction. This was the only photograph made during the building of the reactor. The rods in the holes are uranium, and the reactor's nineteenth layer, consisting of solid graphite blocks, is in process of being laid on.

The reactor core of one of the first nuclear power plants, built in Shippingport, Pennsylvania.

The first hydrogen bomb, exploded at Bikini on March 1, 1954.

Nuclear explosions can move great quantities of earth and should prove useful in such projects as the digging of the new canal in the Central American Isthmus. Above, a crater created by the explosion of a buried nuclear charge of 100-kiloton force. Below, an aerial view of an experiment with a row of thirteen chemical charges. Digging the canal may require as many as 300 nuclear explosions.

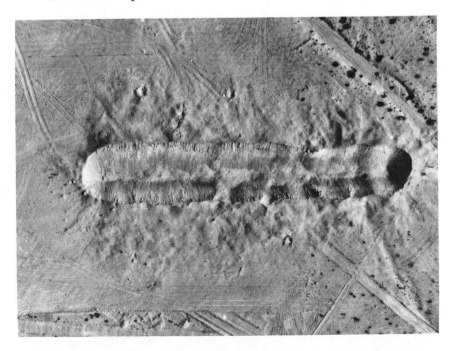

Another view of 100-kiloton nuclear crater, part of the U. S. program dubbed "Project Plowshare." Buried charge that created this crater was about five times more powerful than bomb dropped on Hiroshima. Moving 12 million cubic yards of earth, it produced hole with depth of 320 feet and diameter of 1,200 feet.

the two bombs had the explosive force of 35,000 tons of TNT. With the bombing of Hiroshima, the Atomic Age, already nearly three years old, broke on the consciousness of the world.

For four years after that, Americans lived under the delusion that there was an "atom-bomb secret" which could be kept from other nations forever if only security measures were made tight enough. Actually, the facts and theories of nuclear fission had been matters of public record since 1939, and the Soviet Union was fully engaged in research on the subject in 1940. If World War II had not occupied her lesser resources to a far greater extent than it occupied the greater resources of the un-invaded United States, the U.S.S.R. might have made an atomic bomb by 1945, as we did. As it was, the Soviet Union exploded her first atomic bomb on September 22, 1949, to the dismay and unnecessary amazement of most Americans. It had six times the power of the Hiroshima bomb and had an explosive effect equal to 210,000 tons of TNT.

On October 3, 1952, Great Britain became the third atomic power by exploding a test bomb of its own; on February 13, 1960, France joined the "atomic club" as the fourth member, setting off a plutonium bomb in the Sahara. On October 16, 1964, the People's Republic of China (Communist China) announced the explosion of an atomic bomb, and thus became the fifth member.

The bomb became more versatile, too. In 1953, the United States, for the first time, fired a fission bomb from a cannon, rather than dropping it from a plane. Thus "atomic artillery" (or "tactical atomic weapons") was developed.

Meanwhile the fission bomb had been reduced to triviality. Man had succeeded in setting off another energetic nuclear reaction which made superbombs possible.

In the fission of uranium, 0.1 per cent of the mass of the uranium atom is converted to energy. But in the fusion of hydrogen atoms to form helium, fully 0.5 per cent of their mass is converted to energy, as had first been pointed out in 1915 by the American chemist William Draper Harkins. At temperatures in the millions of degrees, the energy of protons is high enough to allow them to fuse. Thus two protons may unite and, after emitting a positron and a neutrino (a process which converts one of the protons to a neutron), become a deuterium nucleus. A deuterium nucleus may then fuse with a proton to form a tritium nucleus, which can fuse with still another proton to form helium 4. Or deuterium and tritium nuclei will combine in various ways to form helium 4.

Because such nuclear reactions take place only under the stimulus of high temperatures, they are referred to as "thermonuclear reactions." In the 1930's, the one place where the necessary temperatures were believed to exist was at the center of stars. In 1938, the German-born physicist Hans Albrecht Bethe (who had left Hitler's Germany for the United

States in 1935) proposed that fusion reactions were responsible for the energy that the stars radiated. It was the first completely satisfactory explanation of stellar energy since Helmholtz had raised the question nearly a century earlier.

Now the uranium fission bomb provided the necessary temperatures on the earth. It could serve as a match hot enough to ignite a fusion chain reaction in hydrogen. For a while it looked very doubtful that the reaction could actually be made to work in the form of a bomb. For one thing, the hydrogen fuel, in the form of a mixture of deuterium and tritium, had to be condensed to a dense mass, which meant that it had to be liquefied and kept at a temperature only a few degrees above absolute zero. In other words, what would be exploded would be a massive refrigerator. Furthermore, even assuming a hydrogen bomb could be made, what purpose would it serve? The fission bomb was already devastating enough to knock out cities; a hydrogen bomb would merely pile on destruction and wipe out whole civilian populations.

Nevertheless, despite the unappetizing prospects, the United States and the Soviet Union felt compelled to go on with it. The United States Atomic Energy Commission proceeded to produce some tritium fuel, set up a 65-ton fission-fusion contraption on a coral atoll in the Pacific, and on November 1, 1952, produced the first thermonuclear explosion (a "hydrogen bomb" or "H-bomb") on our planet. It fulfilled all the ominous predictions: the explosion yielded the equivalent of 10 million tons of TNT (10 "megatons")—500 times the puny 20-"kiloton" energy of the Hiroshima bomb. The blast wiped out the atoll.

The Russians were not far behind; on August 12, 1953, they also produced a successful thermonuclear explosion, and it was light enough to be carried in a plane. We did not produce a portable one until early 1954. Where we developed the fusion bomb 7½ years after the fission bomb, the Soviets took only 5 years.

Meanwhile a scheme for generating a thermonuclear chain reaction in a simpler way and packing it into a portable bomb had been conceived. The key to this reaction was the element lithium. When the isotope lithium 6 absorbs a neutron, it splits into nuclei of helium and tritium, giving forth 4.8 Mev of energy in the process. Suppose, then, that a compound of lithium and hydrogen (in the form of the heavy isotope deuterium) is used as the fuel. This compound is a solid, so there is no need for refrigeration to condense the fuel. A fission trigger would provide neutrons to split the lithium. And the heat of the explosion would cause the fusion of the deuterium present in the compound and of the tritium produced by the splitting of lithium. In other words, several energy-yielding reactions would take place: the splitting of lithium, the fusion of deuterium with deuterium, and the fusion of deuterium with tritium.

Now besides releasing tremendous energy, these reactions would also yield a great number of surplus neutrons. It occurred to the bomb builders: Why not use the neutrons to fission a mass of uranium? Even common uranium 238 could be fissioned with fast neutrons (though less readily than U-235). The heavy blast of fast neutrons from the fusion reactions might fission a considerable number of U-238 atoms. Suppose one built a bomb with a U-235 core (the igniting match), a surrounding explosive charge of lithium deuteride, and around all this a blanket of uranium 238 which would also serve as explosive. That would make a really big bomb. The U-238 blanket could be made almost as thick as one wished, because there is no critical size at which uranium 238 will undergo a chain reaction spontaneously. The result is sometimes called a "U-bomb."

The bomb was built; it was exploded at Bikini in the Marshall Islands on March 1, 1954; and it shook the world. The energy yield was around 15 megatons. Even more dramatic was a rain of radioactive particles that fell on twenty-three Japanese fishermen in a fishing boat named "The Lucky Dragon." The radioactivity destroyed the cargo of fish, made the fishermen ill, eventually killed one, and did not exactly improve the health of the rest of the world.

Since 1954, fission-fusion-fission bombs have become items in the armaments of the United States, the Soviet Union, and Great Britain. In 1967, China became the fourth member of the fusion club, having made the transition from fission in only three years. The Soviet Union has exploded hydrogen bombs in the 50- to 100-megaton range and the United States is perfectly capable of building such bombs, or even larger ones, at short notice.

There are hints, too, that it may be possible to design a hydrogen bomb that would deliver a highly concentrated stream of neutrons, rather than heat. This would destroy life without doing much damage to property. Such a "neutron bomb" or "N-bomb" seems desirable to those who worry about property and hold life cheap.

Nuclear Power

The dramatic use of nuclear power in the form of unbelievably destructive bombs has done more to present the scientist in the role of an ogre than anything else that has occurred since the beginnings of science.

In a way this portrayal has its justifications, for no arguments or rationalizations can change the fact that scientists did indeed construct the atomic bomb, knowing from the beginning its destructive powers and that it would probably be put to use.

It is only fair to add that this was done under the stress of a great war against ruthless enemies and with an eye to the frightful possibility that a man as maniacal as Adolf Hitler might get such a bomb first. It must also be added that, on the whole, the scientists working on the bomb were deeply disturbed about it and that many opposed its use, while some even left the field of nuclear physics afterward in what can only be described as remorse. Far fewer pangs of conscience were felt by most of the political and military leaders who made the actual decision to use the bombs.

Furthermore, we cannot and should not subordinate the fact that in releasing the energy of the atomic nucleus scientists put at man's disposal a power that can be used constructively as well as destructively. It is important to emphasize this in a world and at a time in which the threat of nuclear destruction has put science and scientists on the shamefaced defensive, and in a country like the United States, which has a rather strong Rousseauan tradition against book learning as a corrupter of the simple integrity of man in a state of nature.

Even the explosion of an atomic bomb need not be purely destructive. Like the lesser chemical explosives long used in mining and in the construction of dams and highways, nuclear explosives could be vastly helpful in construction projects. All kinds of dreams of this sort have been advanced: excavating harbors, digging canals, breaking up underground rock-formations, preparing heat reservoirs for power—even the long-distance propulsion of spaceships. In the sixties, however, the furor for such far-out hopes died down. The prospects of the danger of radioactive contamination, or of unlooked-for expense, or both, served as dampers.

Yet one constructive use of nuclear power that was realized lay in the kind of chain reaction that was born under the football stadium at the University of Chicago. A controlled nuclear reactor can develop huge quantities of heat, which, of course, can be drawn off by a "coolant," such as water or even molten metal, to produce electricity or heat a building.

Experimental nuclear reactors that produced electricity were built in Great Britain and the United States within a few years after the war. The United States now has a fleet of nuclear-powered submarines, the first of which, the U.S.S. "Nautilus" (having cost 50 million dollars), was launched in January 1954. This vessel, as important for its day as Fulton's "Clermont" was in its time, introduced engines with a virtually unlimited source of power that permits submarines to remain underwater for indefinitely long periods, whereas ordinary submarines must surface frequently to recharge their batteries by means of diesel generators that require air for their working. Furthermore, where ordinary submarines travel at a speed of eight knots, a nuclear submarine travels at twenty knots or more.

The first "Nautilus" reactor core lasted for 62,500 miles; included among those miles was a dramatic demonstration. The "Nautilus" made an underwater crossing of the Arctic Ocean in 1958. This trip demonstrated that the ocean depth at the North Pole was 13,410 feet (2½ miles), far deeper than had been thought previously. A second, larger nuclear submarine, the U.S.S. "Triton," circumnavigated the globe underwater along Magellan's route in eighty-four days, between February and May of 1960.

The Soviet Union also possesses nuclear submarines, and, in December 1957, it launched the first nuclear-powered surface vessel, the "Lenin," an ice-breaker. Shortly before that, the United States had laid the keel for a nuclear-powered surface vessel, and, in July 1959, the U.S.S. "Long Beach" (a cruiser) and the "Savannah" (a merchant ship) were launched. The "Long Beach" is powered by two nuclear reactors.

Less than ten years after the launching of the first nuclear vessels, the United States had sixty-one nuclear submarines and four nuclear surface ships operating, being built, or authorized for future building. And yet, except for submarines, enthusiasm for nuclear propulsion also waned. In 1967, the *Savannah* was retired after two years of life. It took $3 million a year to run her, and this was considered too expensive.

But it is not the military alone who must be served. The first nuclear reactor built for the production of electric power for civilian use was put into action in the Soviet Union in June of 1954. It was a small one, with a capacity of not more than 5,000 kilowatts. By October 1956, Great Britain had its "Calder Hall" plant in operation, with a capacity of more than 50,000 kilowatts. The United States was third in the field. On May

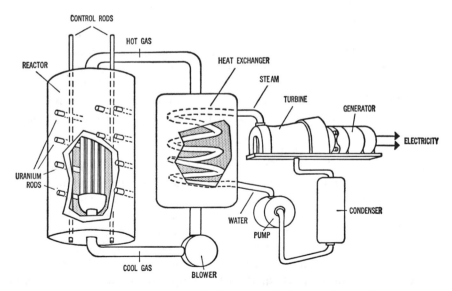

A nuclear power plant of the gas-cooled type, shown in a schematic design. The reactor's heat here is transferred to a gas, which may be a vaporized metal circulating through it, and the heat is then used to convert water to steam.

26, 1958, Westinghouse completed a small nuclear reactor for the production of civilian electric power at Shippingport, Pennsylvania, with a capacity of 60,000 kilowatts. Other reactors quickly followed both in the United States and elsewhere.

Within little more than a decade, there were nuclear reactors in a dozen countries, and nearly half the supply of civilian electricity in the United States was being supplied by fissioning nuclei. Even outer space was invaded, for a satellite powered by a small reactor was launched on April 3, 1965. And yet the problem of radioactive contamination is a serious one. When the 1970's opened, public opposition to the continued proliferation of nuclear power plants was becoming louder.

If fission eventually replaces coal and oil as the world's chief source of energy, how long will the new fuel last? Not very long, if we have to depend entirely on the scarce fissionable material uranium 235. But, fortunately, man can create other fissionable fuels with uranium 235 as a starter.

We have seen that plutonium is one of these man-made fuels. Suppose we build a small reactor with enriched uranium fuel and omit the moderator, so that fast neutrons will stream into a surrounding jacket of natural uranium. These neutrons will convert uranium 238 in the jacket into plutonium. If we arrange things so that few neutrons are wasted, from each fission of a uranium-235 atom in the core we may get more than one plutonium atom manufactured in the jacket. In other words, we will breed more fuel than we consume.

The first such "breeder reactor" was built under the guidance of the Canadian-American physicist Walter H. Zinn at Arco, Idaho, in 1951. It was called EBR-1 (Experimental Breeder Reactor-1). Besides proving the workability of the breeding principle, it produced electricity. It was retired as obsolescent (so fast is progress in this field) in 1964.

Breeding could multiply the fuel supply from uranium many times, because all of the common isotope of uranium, uranium 238, would become potential fuel.

The element thorium, made up entirely of thorium 232, is another potential fissionable fuel. Upon absorbing fast neutrons, it is changed to the artificial isotope thorium 233, which soon decays to uranium 233. Now uranium 233 is fissionable by slow neutrons, and will maintain a self-sustaining chain reaction. Thus thorium can be added to the fuel supply, and thorium appears to be about five times as abundant as uranium in the earth. In fact, it has been estimated that the top hundred yards of the earth's crust contains an average of 12,000 tons of uranium and thorium per square mile. Naturally, not all of this material is easily available.

All in all, the total amount of power conceivably available from the uranium and thorium supplies of the earth is about twenty times that available from the coal and oil we have left.

Radioactivity

The arrival of the Atomic Age brought to man a hazard almost entirely new to his experience. The unlocking of the nucleus released floods of nuclear radiations. To be sure, life on the earth had always been exposed to natural radioactivity and cosmic rays. But man's concentration of naturally radioactive substances, such as radium, which ordinarily exist as greatly diluted traces in the earth's crust, vastly compounded the danger. Some early workers with X-rays and radium even received lethal doses: both Marie Curie and her daughter Irène Joliot-Curie died of leukemia from their exposures, and there is the famous case of the watch-dial painters in the 1920's who died as the result of pointing their radium-tipped brushes with their lips.

The fact that the general incidence of leukemia has doubled in the last two decades may be due, partly, to the increasing use of X-rays for numerous purposes. The incidence of leukemia in doctors, who are likely to be so exposed, is twice that of the general public. In radiologists, who are medical specialists in the use of X-rays, the incidence is ten times greater. It is no wonder that attempts are being made to substitute other techniques, such as those making use of ultrasonic sound, for X-rays. The coming of fission added new force to the danger. Whether in bombs or in power reactors, it unleashes radioactivity on a scale that could make the entire atmosphere, the oceans, and everything we eat, drink, or breathe dangerous to human life. Fission has introduced a form of pollution that will tax man's ingenuity to control.

When the uranium or plutonium atom splits, its "fission products" take various forms. The fragments may include isotopes of barium, or technetium, or any of a number of other possibilities. All told, some 200 different radioactive fission products have been identified. These are troublesome in nuclear technology, for some strongly absorb neutrons and place a damper on the fission reaction. For this reason, the fuel in a reactor must be removed and purified every once in a while.

In addition, these fission fragments are all dangerous to life in varying degrees, depending on the energy and nature of the radiation. Alpha particles taken into the body, for instance, are more dangerous than beta particles. The rate of decay also is important: a nuclide that breaks down rapidly will bombard the receiver with more radiation per second or per hour than one that breaks down slowly.

The rate of breakdown of a radioactive nuclide is something that can be spoken of only when large numbers of the nuclide are involved. An individual nucleus may break down at any time—the next instant or a billion years hence or any time in between—and there is no way of predicting when it will. Each radioactive species, however, has an average rate of breakdown, so if a large number of atoms is involved, it is possible to predict with great accuracy what proportion of them will break down in any unit of time. For instance, let us say that experiment shows that, in a given sample of an atom we shall call X, the atoms are breaking down at the rate of one out of two per year. At the end of a year, 500 of every 1,000 original X atoms in the sample would be left as X atoms; at the end of two years, 250; at the end of three years, 125; and so on. The time it takes for half of the original atoms to break down is called that particular atom's "half-life" (an expression introduced by Rutherford in 1904); consequently, the half-life of atom X is one year. Every radioactive nuclide has its own characteristic half-life, which never changes under ordinary conditions. (The only kind of outside influence that can change it is bombardment of the nucleus with a particle or the extremely high temperature in the interior of a star—in other words, a violent event capable of attacking the nucleus per se.)

The half-life of uranium 238 is 4.5 billion years. It is not surprising, therefore, that there is still uranium 238 left in the universe, despite the decay of uranium atoms. A simple calculation will show that it will take a period more than six times as long as the half-life to reduce a particular quantity of a radioactive nuclide to 1 per cent of its original quantity. Even 30 billion years from now there will still be two pounds of uranium left from each ton of it now in the earth's crust.

Although the isotopes of an element are practically identical chemically, they may differ greatly in their nuclear properties. Uranium 235, for instance, breaks down six times as fast as uranium 238; its half-life is only 710 million years. It can be reasoned, therefore, that in eons gone by, uranium was much richer in uranium 235 than it is today. Six billion years ago, for instance, uranium 235 would have made up about 70 per cent of natural uranium. Mankind is not, however, just catching the tail end of the uranium 235. Even if he had been delayed another million years in discovering fission, the earth would still have 99.9 per cent as much uranium 235 then as it has now.

Clearly any nuclide with a half-life of less than 100 million years would have declined to the vanishing point in the long lifetime of the universe. This explains why we cannot find more than traces of plutonium today. The longest-lived plutonium isotope, plutonium 244, has a half-life of only 70 million years.

The uranium, thorium, and other long-lived radioactive elements thinly spread through the rocks and soil produce small quantities of radiation, which is always present in the air about us. Man is even slightly radioactive himself, for all living tissue contains traces of a comparatively rare, unstable isotope of potassium (potassium 40), which has a half-life of 1.3 billion years. (Potassium 40, as it breaks down, produces some argon 40, and this probably accounts for the fact that argon

40 is by far the most common inert-gas nuclide existing on earth. Potassium–argon ratios have been used to test the age of meteorites.)

The various naturally occurring radioactive nuclides make up what is called "background radiation" (to which cosmic rays also contribute). The constant exposure to natural radiation probably has played a part in evolution by producing mutations and may be partly responsible for the affliction of cancer. But living organisms have lived with it for millions of years. Nuclear radiation has become a serious hazard only in our own time, first as man began to experiment with radium, and then with the coming of fission and nuclear reactors.

By the time the atomic-energy project began, physicists had learned from painful experience how dangerous nuclear radiation was. The workers in the project were therefore surrounded with elaborate safety precautions. The "hot" fission products and other radioactive materials were placed behind thick shielding walls, and looked at only through lead glass. Instruments were devised to handle the materials by remote control. Each person was required to wear strips of photographic film or other detecting devices to "monitor" his accumulated exposure. Extensive animal experiments were carried out to estimate the "maximum permissible exposure." (Mammals are more sensitive to radiation than are other forms of life, but man is averagely resistant, for a mammal.)

Despite everything, accidents happened, and a few nuclear physicists died of "radiation sickness" from massive doses. Yet there are risks in every occupation, even the safest; the nuclear-energy workers have actually fared better than most, thanks to increasing knowledge of what the hazards are and care in avoiding them.

But a world full of nuclear power reactors, spawning fission products by the ton and the thousands of tons, will be a different story. How will all that deadly material be disposed of?

A great deal of it is short-lived radioactivity which fades away to harmlessness within a matter of weeks or months; it can be stored for that time and then dumped. Most dangerous are the nuclides with half-lives of one to thirty years. They are short-lived enough to produce intense radiation, yet long-lived enough to be hazardous for generations. A nuclide with a thirty-year half-life will take two centuries to lose 99 per cent of its activity.

Fission products can be put to good use. As sources of energy, they can power small devices or instruments. The particles emitted by the radioactive isotope are absorbed and their energy converted to heat. This in turn produces electricity in thermocouples. Batteries that produce electricity in this fashion are radioisotope power generators, usually referred to as SNAP ("Systems for Nuclear Auxiliary Power") or, more dramatically, as "atomic batteries." They can be as light as four pounds, generate up to sixty watts, and last for years. SNAP batteries have been used in satellites; in Transit 4A and Transit 4B, for instance, which were put in orbit by the United States in 1961 to serve, ultimately, as navigational aids.

The isotope most commonly used in SNAP batteries is strontium 90, which will soon be mentioned in another connection. Isotopes of plutonium and curium are also used in some varieties.

Radionuclides also have large potential uses in medicine (e.g., for treatment of cancer), in killing bacteria and preserving food, and in many fields of industry, including chemical manufacturing. For instance, the Hercules Powder Company has designed a reactor to use radiation in the production of the antifreeze ethylene glycol.

Yet when all is said and done, no conceivable uses could employ more than a small part of the vast quantities of fission products that power reactors will discharge. This represents an important difficulty in connection with nuclear power generally. The more obvious danger of explosions due to a sudden uncontrolled fission reaction (a "nuclear excursion," as it is called) has always been in the minds of planners. It is to their credit that this has almost never happened, although one such case did indeed kill three men in Idaho in 1961 and spread radioactive contamination over the station. The matter of fission products, however, is far more difficult to handle. It is estimated that every 200,000 kilowatts of nuclear-produced electricity will involve the production of a pound and a half of fission products per day. What to do with it? Already the United States has stored millions of gallons of radioactive liquid underground and it is estimated that by 2000 A.D. as much as half a million gallons of radioactive liquid will require disposal each day! Both the United States and Great Britain have dumped concrete containers of fission products at sea. There have been proposals to drop the radioactive wastes in oceanic abysses, to store them in old salt mines, to incarcerate them in molten glass, and bury the solidified material. But there is always the nervous thought that in one way or another the radioactivity will escape in time and contaminate the soil or the seas. One particularly haunting nightmare is the possibility that a nuclear-powered ship might be wrecked and spill its accumulated fission products into the ocean. The sinking of the American nuclear submarine U.S.S. "Thresher" in the North Atlantic on April 10, 1963, lent new substance to this fear, although in this case such contamination apparently did not take place.

If radioactive pollution by peaceful nuclear energy is a potential danger, at least it will be kept under control, and probably successfully, by every possible means. But there is a pollution which has already spread over the world and which, indeed, in a nuclear war might be broadcast deliberately. This is the fallout from atomic bombs.

Fallout is produced by all nuclear bombs, even those not fired in anger. Because fallout is carried around the world by the winds and brought to earth by rainfall, it is virtually impossible for any nation to explode a nuclear bomb in the atmosphere without detection. In the event of a nuclear war, fallout in the long run might produce more casualties and do more damage to living things in the world at large than the fire and blast of the bombs themselves would wreak on the countries attacked.

Fallout is divided into three types: "local," "tropospheric," and "stratospheric." Local fallout results from ground explosions in which radioactive isotopes are adsorbed on particles of soil and settle out quickly within a hundred miles of the blast. Air blasts of fission bombs in the kiloton range send fission products into the troposphere. These settle out in about a month, being carried some thousands of miles eastward by the winds in that interval of time.

The huge output of fission products from the thermonuclear super-bombs is carried into the stratosphere. Such stratospheric fallout takes a year or more to settle and is distributed over a whole hemisphere, falling eventually on the attacker as well as the attacked.

The intensity of the fallout from the first superbomb, exploded in the Pacific on March 1, 1954, caught scientists by surprise. They had not expected the fallout from a fusion bomb to be so "dirty." Seven thousand square miles were seriously contaminated, an area nearly the size of Massachusetts. But the reason became clear when they learned that the fusion core was supplemented with a blanket of uranium 238 that was fissioned by the neutrons. Not only did this multiply the force of the explosion, but it gave rise to a vastly greater cloud of fission products than a simple fission bomb of the Hiroshima type.

The fallout from the bomb tests to date has added only a small amount of radioactivity to the earth's background radiation. But even a small rise above the natural level may increase the incidence of cancer, cause genetic damage, and shorten the average life expectancy slightly. The most conservative estimators of the hazards agree that, by increasing the mutation rate (see Chapter 12 for a discussion of mutations), fallout is storing up a certain amount of trouble for future generations.

One of the fission products is particularly dangerous for human life. This is strontium 90 (half-life, twenty-eight years), the isotope so useful in SNAP generators. Strontium 90 falling on the soil and water is taken up by plants and thereafter incorporated into the bodies of those animals (including man) that feed directly or indirectly on the plants. Its peculiar danger lies in the fact that strontium, because of its chemical similarity to calcium, goes to the bones and lodges there for a long time. The minerals in bone have a slow "turnover"; that is, they are not replaced nearly as rapidly as are the substances in the soft tissues. For that reason strontium 90, once absorbed, may remain in the body for a major part of a person's lifetime.

Strontium 90 is a brand-new substance in our environment; it did not exist on the earth in any detectable quantity until man fissioned the uranium atom. But today, within less than a generation, some strontium 90 has become incorporated in the bones of every human being on earth and, indeed, in all vertebrates. Considerable quantities of it are still floating in the stratosphere, sooner or later to add to the concentration in our bones.

The strontium-90 concentration is measured in "strontium units" (S.U.). One S.U. is one micromicrocurie of strontium 90 per gram of calcium in the body. A "curie" is a unit of radiation (named in honor of the Curies, of course) originally meant to be equivalent to that produced by a gram of radium in equilibrium with its breakdown product, radon. It is now more generally accepted as meaning 37 billion disintegrations per second. A micromicrocurie is a trillionth of a curie, or 2.12 disintegrations per minute. A strontium unit would therefore mean 2.12 disintegrations per minute per gram of calcium present in the body.

The concentration of strontium 90 in the human skeleton varies greatly from place to place and among individuals. Some persons have

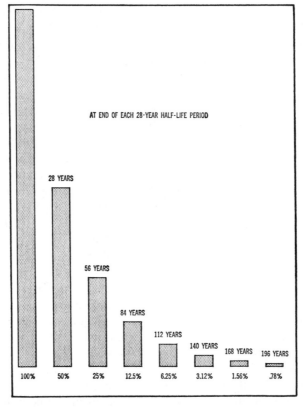

Decay of strontium 90 over approximately 200 years.

been found to have as much as seventy-five times the average amount. Children average at least four times as high a concentration as adults, because of the higher turnover of material in their growing bones. Estimates of the averages themselves vary, because they are based mainly on estimates of the amounts of strontium 90 found in the diet. (Incidentally, milk is not a particularly hazardous food, from this point of view, because calcium obtained from vegetables has more strontium 90 associated with it. The cow's "filtration system" eliminates some of the strontium it gets in its plant fodder.) The estimates of the average strontium-90 concentration in the bones of people in the United States in 1959 ranged from less than one strontium unit to well over five strontium units. (The "maximum permissible" was established by the International Commission on Radiation Protection at 67 S.U.) But the averages mean little, particularly since strontium 90 may collect in "hot spots" in the bones and reach a high enough level there to initiate leukemia or cancer.

The importance of radiation effects has, among other things, resulted in the adoption of a number of types of unit designed to measure these effects. One such, the "roentgen," named in honor of the discoverer of X-rays, is based on the number of ions produced by the X-rays or gamma rays being studied. More recently, the "rad" (short for "radiation") has been introduced. It represents the absorption of 100 ergs per gram of any type of radiation.

The nature of the radiation is of importance. A rad of massive particles is much more effective in inducing chemical change in tissues than a rad of light particles; hence, energy in the form of alpha particles is more dangerous than the same energy in the form of electrons.

Chemically, the damage done by radiation is caused chiefly by the breakdown of water molecules (which make up most of the mass of living tissue) into highly active fragments ("free radicals") that, in turn, react with the complicated molecules in tissue. Damage to bone marrow, interfering with blood cell production, is a particularly serious manifestation of "radiation sickness," which, if far enough advanced, is irreversible and leads to death.

Many eminent scientists firmly believe that the fallout from the bomb tests represents an important peril to the human race. The American chemist Linus Pauling has argued that the fallout from a single superbomb may lead to 100,000 deaths from leukemia and other diseases in the world, and he has pointed out that radioactive carbon 14, produced by the neutrons from a nuclear explosion, constitutes a serious genetic danger. He has, for this reason, been extremely active in pushing for cessation of testing of nuclear bombs; he endorses all movements designed to lessen the danger of war and to encourage disarmament. On the other hand, some scientists, including the Hungarian-American physicist Edward Teller, minimize the seriousness of the fallout hazard.

The sympathy of the world generally lies with Pauling, as might be indicated by the fact that he was awarded the Nobel Peace Prize in 1963. (Nine years earlier, Pauling had won a Nobel Prize in chemistry; he thus joins Marie Curie as the only members of the select group who have been awarded two Nobel Prizes.)

In the fall of 1958, the United States, the U.S.S.R., and Great Britain suspended bomb testing by a gentleman's agreement (which, however, did not prevent France from exploding her first atomic bomb in the spring of 1960). For three years, things looked rosy; the concentration of strontium 90 reached a peak and leveled off about 1960 at a point well below what is estimated to be the maximum consistent with safety. Even so, some 25 million curies of strontium 90 and cesium 137 (another dangerous fission product) had been delivered into the atmosphere during the thirteen years of nuclear testing when some 150 bombs of all varieties were exploded. Only two of these were exploded in anger, but the results were dire indeed.

In 1961, without warning, the Soviet Union ended the moratorium and began testing again. Since the U.S.S.R. exploded thermonuclear bombs of unprecedented power, the United States felt that it was forced to begin testing again. World public opinion, sharpened and concentrated by the relief of the moratorium, reacted with great indignation.

On October 10, 1963, therefore, the three chief nuclear powers signed a partial test-ban treaty (*not* a mere gentleman's agreement) in which nuclear bomb explosions in the atmosphere, in space, or underwater were banned. Only underground explosions were permitted since these did not produce fallout. This has been the most hopeful move in the direction of human survival since the opening of the Atomic Age. The chief danger now, assuming the test-ban treaty is observed by all signatories, is that France and the People's Republic of China (the newest members of the atomic club) have refused to sign the treaty so far.

Nuclear Fusion

For more than twenty years, nuclear physicists have had in the back of their minds a dream even more attractive than turning fission to constructive uses. It is the dream of harnessing fusion energy. Fusion, after all, is the engine that makes our world go round: the fusion reactions in the sun are the ultimate source of all our forms of energy and of life itself. If somehow we could reproduce and control such reactions on the earth, all our energy problems would be solved. Our fuel supply would be as big as the ocean, for the fuel would be hydrogen.

Oddly enough, this would not be the first time hydrogen will have been used as a fuel. Not long after hydrogen was discovered and its

Soviet experiments in fusion. Russian workers are shown installing ALFA, an apparatus with a doughnut-shaped chamber in which they conducted experiments with plasmas, preparatory to producing thermonuclear reactions.

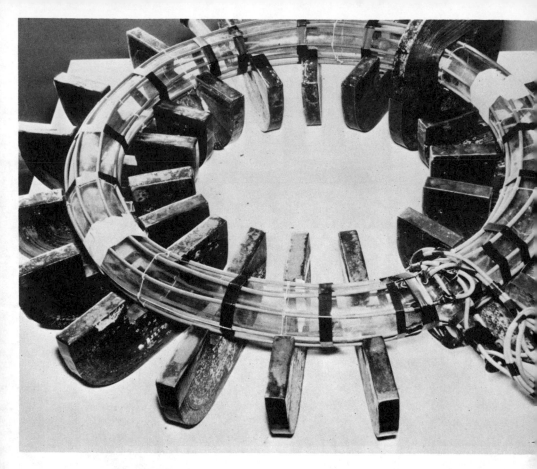

Los Alamos experiments in fusion. This is a tube of the "Perhapsatron," one of the American devices for studying pinch effects on plasmas. The pieces of iron are the halves of the magnetic cores (one is complete) that will ring the tube.

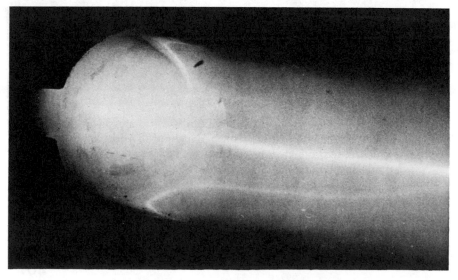

Pinched plasma in the Perhapsatron tube.

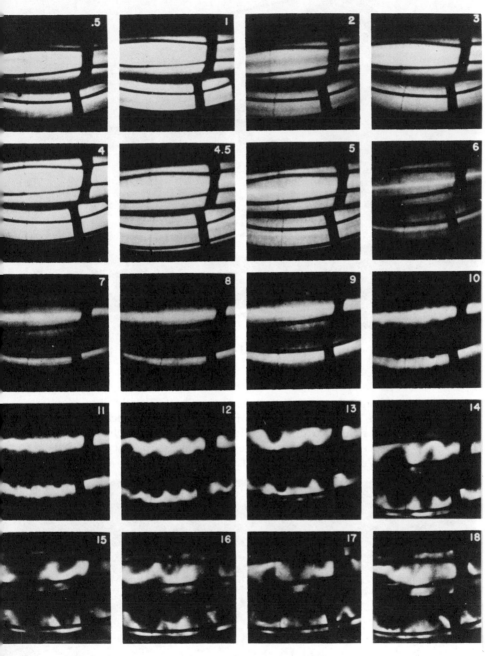

The life and death of a pinch. This series of pictures shows the brief history of a wisp of plasma in the magnetic field of the Perhapsatron. Each photograph gives two views of the plasma, one from the side and one from below through a mirror. The pinch broke down in millionths of a second; the number on each picture is the time in microseconds.

The Bell solar battery, here used to furnish power to a telephone line.

A thermoelectric cell for converting heat directly into electricity. The heat of sunlight is focused by the concave mirror at the right on a disk of semiconducting material where it produces a flow of electrons. These experiments were performed at the Westinghouse Research Laboratories.

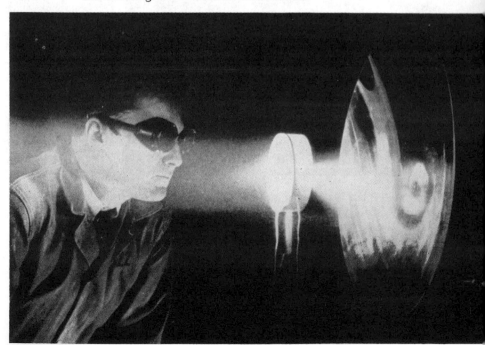

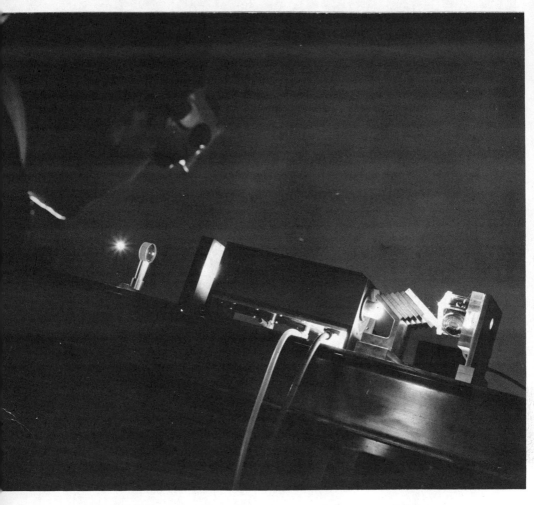

Laser beam, invisible because of the nature of the waves of light of which it consists, is an extremely powerful and highly concentrated form of radiant energy. The laser's uses currently include measurements of the orbital position and launching trajectory of space vehicles; the instrument holds spectacular promise in communications and other widely diversified areas. This particular instrument, the K-2Q Laser, was developed by the Korad Corporation. It can accurately measure distance to within a foot or less over ranges exceeding 300 miles. The sparkle comes from particles exposed to the intense energy of the laser beam.

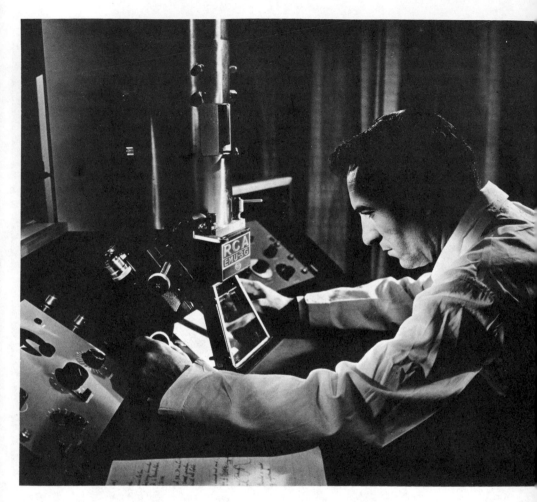

An electron microscope, in capable hands.

properties studied, it gained a place as a chemical fuel. The American scientist Robert Hare devised an oxyhydrogen torch in 1801, and the hot flame of hydrogen burning in oxygen has served industry ever since. Now, however, as a nuclear fuel, a much more glittering possibility lay before it.

Fusion power would be immensely more convenient than fission power. Pound for pound, a fusion reactor would deliver about five to ten times as much power as a fission reactor. A pound of hydrogen, on fusion, could produce 35 million kilowatt-hours of energy. Furthermore, fusion produces no radioactive ashes. Finally, a fusion reaction, in the event of any conceivable malfunction, could only collapse and go out, whereas a fission reaction might conceivably (though not very probably) go out of control and into a full explosion.

Of the three isotopes of hydrogen, hydrogen-1 is the most common, but it is also the one most difficult to force into fusion. It is the particular fuel of the sun, but the sun has it by the trillions of cubic miles, together with an enormous gravity field to hold it together and central temperatures in the many millions of degrees. Only a tiny percentage of the hydrogen within the sun is fusing at any given moment, but given the vast mass present, even a tiny percentage is enough.

Hydrogen-3 is the easiest to bring to fusion, but it exists in such tiny quantities and can be made only at so fearful an expenditure of energy that it is hopeless to think of it as a practical fuel all by itself.

That leaves hydrogen-2, which is easier to handle than hydrogen-1 and much more common that hydrogen-3. In all the hydrogen of the world, only one atom out of 6,000 is deuterium, but that is enough. There is 35 trillion tons of deuterium in the ocean, enough to supply man with ample energy for all the foreseeable future.

Yet there are problems. That might seem surprising, since fusion bombs exist. If we can make hydrogen fuse, why can't we make a reactor as well as a bomb? Ah, but to make a fusion bomb, we need to use a fission-bomb igniter and then let it go. To make a fusion reactor, we need a gentler igniter, obviously, and we must then keep the reaction going at a constant, controlled—and nonexplosive—rate.

The first problem is the less difficult. Heavy currents of electricity, high-energy sound-waves, laser beams, and so on, can all produce temperatures in the millions of degrees very briefly. There is no doubt that the required temperature will be reached.

Maintaining the temperature while keeping the (it is to be hoped) fusing hydrogen in place is something else. Obviously no material container can hold a gas at anything like a temperature of over 100 million degrees. Either the container would vaporize or the gas would cool. The first step toward a solution is to reduce the density of the gas to far below normal pressure; this cuts down the heat content, though the energy of the particles remains high. The second step is a concept of great ingenuity. A gas at very high temperature has all the electrons stripped off its atoms; it is a "plasma" (a term introduced by Irving Langmuir in the early 1930's) made up of electrons and bare nuclei. Since it consists entirely of charged particles, why not use a strong magnetic field, taking the place of a material container, to hold it? The fact that magnetic fields could restrain charged particles and "pinch" a stream of them together

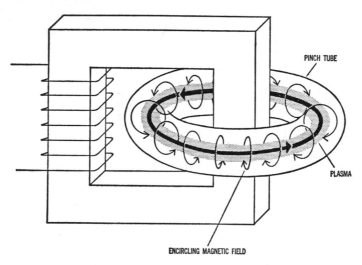

Magnetic bottle designed to hold a hot gas of hydrogen nuclei (a plasma). The ring is called a "torus."

had been known since 1907, when it was named the "pinch effect." The "magnetic bottle" idea was tried and it worked—but only for the briefest instant. The wisps of plasma pinched in the bottle immediately writhed like a snake, broke up, and died out.

Another approach is to have a magnetic field stronger at the ends of the tube so that plasma is pushed back and kept from leaking. This is also found wanting. It doesn't seem as though much is wanting. If only plasma at 100 million degrees can be held in place for about a second, the fusion reaction would start and energy would pour out of the system. That energy could be used to make the magnetic field firmer and more powerful and to keep the temperature at the proper level. The fusion reaction would then be self-sustaining, with the very energy it produced serving to keep it going. But to keep the plasma from leaking for just a second is more than can be done as yet.

Since the plasma leakage takes place with particular ease at the end of the tube, why not remove the ends by giving the tube a doughnut shape? A particularly useful design is a doughnut-shaped tube ("torus") twisted into a figure eight. This figure-eight device was first designed in 1951 by Spitzer and is called a "stellarator." An even more hopeful device was designed by the Soviet physicist Lev Andreevich Artsimovich. It is called "Toroidal Kamera Magnetic," a name which is abbreviated to "Tokamak."

The Tokamak only works with very rarefied gases, but the Soviets, using hydrogen at one-millionth atmospheric density, have achieved a temperature of 100 million degrees for one-hundredth of a second. Hydrogen so rare must be held in place for longer than a second, but if the Soviets can make hydrogen-2 just ten times denser and then hold on for a second, they might be able to make it.

American physicists are also working with Tokamaks and in addition with a device called "Scyllac," which is designed to hold denser gas and therefore require a smaller containment period.

For nearly twenty years physicists have been inching toward fusion power. Progress has been slow, but as yet no signs of a definite dead-end have appeared.

Meanwhile, practical applications of fusion research are to be found. Plasma torches emitting jets at temperatures up to 50,000° C. in absolute silence can far outdo ordinary chemical torches. And it is suggested that the plasma torch is the ultimate waste-disposal unit. In its flame everything, *everything*, would be broken down to its constituent elements, and all the elements would be available for recycling and for conversion into useful materials again.

Part Two

THE
Biological
Sciences

CHAPTER 10

The Molecule

Organic Matter

The term molecule (from a Latin word meaning "small mass") originally meant the ultimate, indivisible unit of a substance, and in a sense it *is* an ultimate particle, because it cannot be broken down without losing its identity. To be sure, a molecule of sugar or of water can be divided into single atoms or groups, but then it is no longer sugar or water. Even a hydrogen molecule loses its characteristic chemical properties when it is broken down into its two component hydrogen atoms.

Just as the atom has furnished chief excitement in twentieth-century physics, so the molecule has been the subject of equally exciting discoveries in chemistry. Chemists have been able to work out detailed pictures of the structure of even very complex molecules, to identify the roles of specific molecules in living systems, to create elaborate new molecules, and to predict the behavior of a molecule of a given structure with amazing accuracy.

By the mid-twentieth century, the complex molecules that form the key units of living tissue, the proteins and nucleic acids, were being studied with all the techniques made possible by an advanced chemistry and physics. The two sciences, "biochemistry" (the study of the chemical reactions going on in living tissue) and "biophysics" (the study of the physical forces and phenomena involved in living processes), merged to form a brand new discipline—"molecular biology." Through the findings of molecular biology, modern science has in a single generation of effort all but wiped out the borderline between life and nonlife.

Yet less than a century and a half ago, the structure of not even the simplest molecule was understood. About all the chemists of the early nineteenth century could do was to separate all matter into two great

473

categories. They had long been aware (even in the days of the alchemists) that substances fell into two sharply distinct classes with respect to their response to heat. One group—for example, salt, lead, water—remained basically unchanged after being heated. Salt might glow red-hot when heated, lead might melt, water might vaporize—but when they were cooled back to the original temperature, they were restored to their original form, none the worse, apparently, for their experience. On the other hand, the second group of substances—for example, sugar, olive oil—were changed permanently by heat. Sugar became charred when heated and remained charred after it was cooled again; olive oil was vaporized and the vapor did not condense on cooling. Eventually the scientists noted that the heat-resisting substances generally came from the inanimate world of the air, ocean, and soil, while the combustible substances usually came from the world of life, either from living matter directly or from dead remains. In 1807, the Swedish chemist Jöns Jakob Berzelius named the combustible substances "organic" (because they were derived, directly or indirectly, from the living organisms) and all the rest "inorganic."

Early chemistry focused mainly on the inorganic substances. It was the study of the behavior of inorganic gases that led to the development of the atomic theory. Once that theory was established, it soon clarified the nature of inorganic molecules. Analysis showed that inorganic molecules generally consisted of a small number of different atoms in definite proportions. The water molecule contained two atoms of hydrogen and one of oxygen; the salt molecule contained one atom of sodium and one of chlorine; sulfuric acid contained two atoms of hydrogen, one of sulfur, and four of oxygen, and so on.

When the chemists began to analyze organic substances, the picture seemed quite different. Two substances might have exactly the same composition and yet show distinctly different properties. (For instance, ethyl alcohol is composed of two carbon atoms, one oxygen atom, and six hydrogen atoms; so is dimethyl ether—yet one is a liquid at room temperature while the other is a gas.) The organic molecules contained many more atoms than the simple inorganic ones, and there seemed to be no rhyme or reason in the way they were combined. Organic compounds simply could not be explained by the straightforward laws of chemistry that applied so beautifully to inorganic substances.

Berzelius decided that the chemistry of life was something apart which obeyed its own set of subtle rules. Only living tissue, he said, could make an organic compound. His point of view is an example of "vitalism."

Then in 1828 the German chemist Friedrich Wöhler, a student of Berzelius, produced an organic substance in the laboratory! He was heating a compound called ammonium cyanate, which was then generally considered inorganic. Wöhler was thunderstruck to discover that, on being heated, this material turned into a white substance identical in properties with "urea," a component of urine. According to Berzelius' views, only the living kidney could form urea, and yet Wöhler had formed it from inorganic material merely by applying a little heat.

Wöhler repeated the experiment many times before he dared publish his discovery. When he finally did, Berzelius and others at first refused to believe it. But other chemists confirmed the results. Furthermore, they proceeded to synthesize many other organic compounds from inorganic precursors. The first to bring about the production of an organic compound from its elements was the German chemist Adolph Wilhelm Hermann Kolbe, who produced acetic acid in this fashion in 1845. It was this that really killed Berzelius' version of vitalism. More and more it became clear that the same chemical laws applied to inorganic and organic molecules alike. Eventually the distinction between organic and inorganic substances was given a simple definition: all substances containing carbon (with the possible exceptions of a few simple compounds such as carbon dioxide) are called organic; the rest are inorganic.

To deal with the complex new chemistry, chemists needed a simple shorthand for representing compounds, and fortunately Berzelius had already suggested a convenient, rational system of symbols. The elements were designated by abbreviations of their Latin names. Thus C would stand for carbon, O for oxygen, H for hydrogen, N for nitrogen, S for sulfur, P for phosphorus, and so on. Where two elements began with the same letter, a second letter was used to distinguish them: e.g., Ca for calcium, Cl for chlorine, Cd for cadmium, Co for cobalt, Cr for chromium, and so on. In only a comparatively few cases are the Latin or Latinized names (and initials) different from the English, thus: iron ("ferrum") is Fe; silver ("argentum") Ag; gold ("aurum") Au; copper ("cuprum") Cu; tin ("stannum") Sn; mercury ("hydragyrum") Hg; antimony ("stibium") Sb; sodium ("natrium") Na; and potassium ("kalium") K.

With this system it is easy to symbolize the composition of a molecule. Water is written H_2O (thus indicating the molecule to consist of two hydrogen atoms and an oxygen atom); salt, $NaCl$; sulfuric acid, H_2SO_4, and so on. This is called the "empirical formula" of a compound; it tells what the compound is made of but says nothing about its structure, that is, the manner in which the atoms of the molecule are interconnected.

Baron Justus von Liebig, a coworker of Wöhler's, went on to work out the composition of a number of organic chemicals, thus applying "chemical analysis" to the field of organic chemistry. He would carefully burn a small quantity of an organic substance and trap the gases formed (chiefly CO_2 and water vapor, H_2O) with appropriate chemicals. Then he would weigh the chemicals used to trap the combustion products to see how much weight had been added by the trapped products. From that weight he could determine the amount of carbon, hydrogen, and oxygen in the original substance. It was then an easy matter to calculate, from the atomic weights, the numbers of each type of atom in the molecule. In this way, for instance, he established that the molecule of ethyl alcohol had the formula C_2H_6O.

Liebig's method could not measure the nitrogen present in organic compounds, but the French chemist Jean Baptiste André Dumas devised a combustion method which did collect the gaseous nitrogen released from substances. He made use of his methods to analyze the gases of the atmosphere with unprecedented accuracy in 1841.

The methods of "organic analysis" were made more and more delicate until veritable prodigies of refinement were reached in the "microanalytical" methods of the Austrian chemist Fritz Pregl. He devised techniques, in the early 1900's, for the accurate analysis of quantities of organic compounds barely visible to the naked eye and received the Nobel Prize for chemistry in 1923 in consequence.

Unfortunately, determining only the empirical formulas of organic compounds was not very helpful in elucidating their chemistry. In contrast to inorganic compounds, which usually consisted of two or three atoms or at most a dozen, the organic molecules were often huge. Liebig found that the formula of morphine was $C_{17}H_{19}O_3N$, and of strychnine, $C_{21}H_{22}O_2N_2$.

Chemists were pretty much at a loss to deal with such large molecules or make head or tail of their formulas. Wöhler and Liebig tried to group atoms into smaller collections called "radicals" and to work out theories to show that various compounds were made up of specific radicals in different numbers and combinations. Some of the systems were most ingenious, but none really explained enough. It was particularly difficult to explain why two compounds with the same empirical formula, such as ethyl alcohol and dimethyl ether, should have different properties.

This phenomenon was first dragged into the light of day in the 1820's by Liebig and Wöhler. The former was studying a group of compounds called "fulminates," the latter a group called "isocyanates," and the two turned out to have identical empirical formulas—the elements were present in equal parts, so to speak. Berzelius, the chemical dictator of the day, was told of this and was reluctant to believe it until, in 1830, he discovered some examples for himself. He named such compounds, with different properties but with elements present in equal parts, "isomers" (from Greek words meaning "equal parts"). The structure of organic molecules was indeed a puzzle in those days.

The chemists, lost in the jungle of organic chemistry, began to see daylight in the 1850's when they noted that each atom could combine with only a certain number of other atoms. For instance, the hydrogen atom apparently could attach itself to only one atom: it could form hydrogen chloride, HCl, but never HCl$_2$. Likewise chlorine and sodium could each take only a single partner, so they formed NaCl. An oxygen atom, on the other hand, could take two atoms as partners—for instance, H$_2$O. Nitrogen could take on three: e.g., NH$_3$ (ammonia). And carbon could combine with as many as four: e.g., CCl$_4$ (carbon tetrachloride).

In short, it looked as if each type of atom had a certain number of hooks by which it could hang on to other atoms. The English chemist Edward Frankland called these hooks "valence" bonds, from a Latin word meaning "power," to signify the combining powers of the elements.

The German chemist Friedrich August Kekulé von Stradonitz saw that if carbon were given a valence of 4 and if it were assumed that carbon atoms could use those valences, in part at least, to join up in chains, then a map could be drawn through the organic jungle. His technique was made more visual by the suggestion of a Scottish chemist, Archibald Scott Couper, that these combining forces between atoms ("bonds," as they are usually called) be pictured in the form of small dashes. In this way, organic molecules could be built up like so many "Tinker-toy" structures.

In 1861, Kekulé published a textbook with many examples of this system, which proved its convenience and value. The "structural formula" became the hallmark of the organic chemist.

For instance, the methane (CH$_4$), ammonia (NH$_3$), and water (H$_2$O) molecules, respectively, could be pictured this way:

$$H - \underset{\underset{\displaystyle H}{|}}{\overset{\overset{\displaystyle H}{|}}{C}} - H \qquad H - \underset{\overset{\displaystyle H}{|}}{N} - H \qquad H - O - H$$

Organic molecules could be represented as chains of carbon atoms with hydrogen atoms attached along the sides. Thus butane (C$_4$H$_{10}$) would have the structure:

$$H - \underset{\overset{|}{H}}{\overset{\overset{H}{|}}{C}} - \underset{\overset{|}{H}}{\overset{\overset{H}{|}}{C}} - \underset{\overset{|}{H}}{\overset{\overset{H}{|}}{C}} - \underset{\overset{|}{H}}{\overset{\overset{H}{|}}{C}} - H$$

Oxygen or nitrogen might enter the chain in the following manner, picturing the compounds methyl alcohol (CH$_4$O) and methylamine (CH$_5$N), respectively:

$$H - \underset{\overset{|}{H}}{\overset{\overset{H}{|}}{C}} - O - H \qquad\qquad H - \underset{\overset{|}{H}}{\overset{\overset{H}{|}}{C}} - \underset{\overset{H}{|}}{N} - H$$

477

An atom possessing more than one hook, such as carbon with its four, need not use each of them for a different atom: it might form a double or triple bond with one of its neighbors, as in ethylene (C_2H_4) or acetylene (C_2H_2):

$$H - \overset{\overset{\displaystyle H}{|}}{C} = \overset{\overset{\displaystyle H}{|}}{C} - H \qquad\qquad\qquad H - C \equiv C - H$$

Now it became easy to see how two molecules could have the same number of atoms of each element and still differ in properties. The two isomers must differ in the arrangement of those atoms. For instance, the structural formulas of ethyl alcohol and dimethyl ether, respectively, could be written:

$$H - \overset{\overset{\displaystyle H}{|}}{\underset{\underset{\displaystyle H}{|}}{C}} - \overset{\overset{\displaystyle H}{|}}{\underset{\underset{\displaystyle H}{|}}{C}} - O - H \qquad\qquad H - \overset{\overset{\displaystyle H}{|}}{\underset{\underset{\displaystyle H}{|}}{C}} - O - \overset{\overset{\displaystyle H}{|}}{\underset{\underset{\displaystyle H}{|}}{C}} - H$$

The greater the number of atoms in a molecule, the greater the number of possible arrangements and the greater the number of isomers. For instance, heptane, a molecule made up of seven carbon atoms and sixteen hydrogen atoms, can be arranged in nine different ways; in other words, there can be nine different heptanes, each with its own properties. These nine isomers resemble one another fairly closely, but it is only a family resemblance. Chemists have prepared all nine of these isomers but have never found a tenth—good evidence in favor of the Kekulé system.

A compound containing forty carbon atoms and eighty-two hydrogen atoms could exist in some 62.5 trillion arrangements, or isomers. And organic molecules of this size are by no means uncommon.

Only carbon atoms can hook to one another to form long chains. Other atoms do well if they can form a chain as long as half a dozen or so. That is why inorganic molecules are usually simple, and why they rarely have isomers. The greater complexity of the organic molecule introduces so many possibilities of isomerism that nearly 2 million organic compounds are known, new ones are being formed daily, and a virtually limitless number await discovery.

Structural formulas are now universally used as indispensable guides to the nature of organic molecules. As a short cut, chemists often write the formula of a molecule in terms of the groups of atoms ("radicals") that make it up, such as the methyl (CH_3) and methylene (CH_2) radicals. Thus the formula for butane can be written as $CH_3CH_2CH_2CH_3$.

478

The Details of Structure

In the latter half of the nineteenth century, chemists discovered a particularly subtle kind of isomerism which was to prove very important in the chemistry of life. The discovery emerged from the oddly asymmetrical effect that certain organic compounds had on rays of light passing through them.

A cross section of a ray of ordinary light would show that the waves of which it consists undulate in all planes—up and down, from side to side, and obliquely. Such light is called "unpolarized." But when light passes through a crystal of the transparent substance called Iceland spar, for instance, it is refracted in such a way that the light emerges "polarized." It is as if the array of atoms in the crystal allows only certain planes of undulation to pass through (just as the palings of a fence might allow a person moving sideways to squeeze through but not one coming up to them broadside on). There are devices, such as the "Nicol prism," invented by the Scottish physicist William Nicol in 1829, that lets light through in only one plane. This has now been replaced, for most purposes, by materials such as Polaroid (crystals of a complex of quinine sulfate and iodine, lined up with axes parallel and embedded in nitrocellulose) first produced in the 1930's by Edwin Land.

Reflected light is often partly plane-polarized, as was first discovered in 1808 by the French physicist, Étienne Louis Malus. (He invented the term "polarization" through the application of a remark of Newton's about the poles of light particles, one occasion where Newton was wrong —but the name remains anyway.) The glare of reflected light from win-

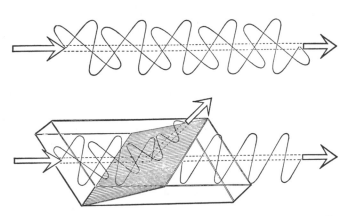

The polarization of light. The waves of light normally oscillate in all planes (*top*). The Nicol prism (*bottom*) lets through the oscillations in only one plane, reflecting away the others. The transmitted light is plane-polarized.

dows of buildings and cars, and even from paved highways, can therefore be cut to bearable levels by the use of Polaroid sunglasses.

Early in the nineteenth century the French physicist Jean Baptiste Biot had discovered that when plane-polarized light passed through quartz crystals, the plane of polarization was twisted. That is, the light went in undulating in one plane and came out undulating in a different plane. A substance that does this is said to display "optical activity." Some quartz crystals twisted the plane clockwise ("dextrorotation") and some counterclockwise ("levorotation"). Biot found that certain organic compounds, such as camphor and tartaric acid, did the same thing. He thought it likely that some kind of asymmetry in the arrangement of the atoms in the molecules was responsible for the twisting of light. But for several decades this suggestion remained purely speculative.

In 1844, Louis Pasteur (only twenty-two at the time) took up this interesting question. He studied two substances: tartaric acid and racemic acid. Both had the same chemical composition, but tartaric acid rotated the plane of polarized light while racemic acid did not. Pasteur suspected that the crystals of salts of tartaric acid would prove to be asymmetric and those of racemic acid would be symmetric. Examining both sets of crystals under the microscope, he found to his surprise that both were asymmetric. But the racemate crystals had two versions of the asymmetry. Half of them were the same shape as those of the tartrate and the other half were mirror images. Half of the racemate crystals were left-handed and half right-handed, so to speak.

Pasteur painstakingly separated the left-handed racemate crystals from the right-handed and then dissolved each kind separately and sent light through each solution. Sure enough, the solution of the crystals possessing the same asymmetry as the tartrate crystals twisted the plane of polarized light just as the tartrate did, with the same specific rotation. Those crystals *were* tartrate. The other set twisted the plane of polarized light in the opposite direction, with the same amount of rotation. The reason the original racemate had shown no rotation of light, then, was that the two opposing tendencies canceled each other.

Pasteur next reconverted the two separated types of racemate salt to acid again by adding hydrogen ions to the respective solutions. (A salt, by the way, is a compound in which some hydrogen ions of the acid molecule are replaced by other positively charged ions, such as those of sodium or potassium.) He found that each of these racemic acids was now optically active—one rotating polarized light in the same direction as tartaric acid did (for it *was* tartaric acid) and the other in the opposite direction.

Other pairs of such mirror-image compounds ("enantiomorphs," from Greek words meaning "opposite shapes") were found. In 1863, the German chemist Johannes Wislicenus found that lactic acid (the acid of sour milk) formed such a pair. Furthermore, he showed the properties of the two forms to be identical *except* for the action on polarized light. This has turned out to be generally true of enantiomorphs.

So far, so good, but where did the asymmetry lie? What was there about the two molecules that made them mirror images of each other? Pasteur could not say. And although Biot, who had suggested the existence of molecular asymmetry, lived to be eighty-eight, he did not live long enough to see his intuition vindicated.

It was in 1874, twelve years after Biot's death, that the answer was finally presented. Two young chemists, a twenty-two-year-old Dutchman named Jacobus Hendricus Van't Hoff and a twenty-seven-year-old Frenchman named Joseph Achille Le Bel, independently advanced a new theory of the carbon valence bonds that explained how mirror-image molecules could be constructed. (Later in his career, Van't Hoff studied the behavior of substances in solution and showed how the laws governing their behavior resembled the laws governing the behavior of gases. For this achievement he was the first man, in 1901, to be awarded the Nobel Prize in chemistry.)

Kekulé had drawn the four bonds of the carbon atom all in the same plane, not necessarily because this was the way they were actually arranged but because it was the convenient way of drawing them on a flat piece of paper. Van't Hoff and Le Bel now suggested a three-dimensional model in which the bonds were directed in two mutually perpendicular planes, two in one plane and two in the other. A good way to picture this is to imagine the carbon atom as standing on any three of its bonds as legs, in which case the fourth bond points vertically upward (see the drawing on the next page). If you suppose the carbon atom to be at the center of a tetrahedron (a four-sided geometrical figure with triangular sides), then the four bonds point to the four vertexes of the figure. The model is therefore called the "tetrahedral carbon atom."

Now let us attach to these four bonds two hydrogen atoms, a chlorine atom, and a bromine atom. Regardless of which atom we attach to which bond, we will always come out with the same arrangement. Try it and see. With four toothpicks stuck into a marshmallow (the carbon atom) at the proper angles, you could represent the four bonds. Now suppose you stick two black olives (the hydrogen atoms), a green olive (chlorine), and a cherry (bromine) on the ends of the toothpicks in any order. Let us say that when you stand this on three legs with a black olive on the fourth pointing upward, the order on the three standing legs in the clockwise direction is black olive, green olive, cherry. You might now switch the green olive and cherry so that the order runs black olive, cherry, green olive. But all you need to do to see the same order as before is to turn the structure over so that the black olive serving as one of the supporting legs sticks up in the air and the one that was in the air rests on the table. Now the order of the standing legs again is black olive, green olive, cherry.

In other words, when at least two of the four atoms (or groups of atoms) attached to carbon's four bonds are identical, only one structural arrangement is possible. (Obviously this is true when three or all four of the attachments are identical.)

But when all four of the attached atoms (or groups of atoms) are different, the situation changes. Now two different structural arrangements are possible—one the mirror image of the other. For instance, suppose you stick a cherry on the upward leg and a black olive, a green olive, and a cocktail onion on the three standing legs. If you then switch the black olive and green olive so that the clockwise order runs green olive, black olive, onion, there is no way you can turn the structure to make the order come out black olive, green olive, onion, as it was before you made the switch. Thus with four different attachments you can always form two different structures, mirror images of each other. Try it and see.

Van't Hoff and Le Bel thus solved the mystery of the asymmetry of optically active substances. The mirror-image substances that rotated light in opposite directions were substances containing carbon atoms with four different atoms or groups of atoms attached to the bonds. One of the two possible arrangements of these four attachments rotated polarized light to the right; the other rotated it to the left.

More and more evidence beautifully supported Van't Hoff's and Le Bel's tetrahedral model of the carbon atom, and, by 1885, their theory (thanks, in part, to the enthusiastic support of the respected Wislicenus) was universally accepted.

The notion of three-dimensional structure also was applied to atoms other than carbon. The German chemist Viktor Meyer applied it successfully to nitrogen, while the English chemist William Jackson Pope applied it to sulfur, selenium, and tin. The German-Swiss chemist Alfred Werner added other elements and, indeed, beginning in the 1890's, worked out a "coordination theory" in which the structure of complex inorganic substances was explained by careful consideration of the distribution of atoms and atom groupings about some central atom. For this work, Werner was awarded the Nobel Prize in chemistry for 1913.

The two racemic acids that Pasteur had isolated were named d-tartaric acid (for "dextrorotatory") and l-tartaric acid (for "levorotatory"), and mirror-image structural formulas were written for them. But which

The tetrahedral carbon atom.

was which? Which was actually the right-handed and which the left-handed compound? There was no way of telling at the time.

To provide chemists with a reference, or standard of comparison, for distinguishing right-handed and left-handed substances, the German chemist Emil Fischer chose a simple compound called "glyceraldehyde," a relative of the sugars, which were among the most thoroughly studied of the optically active compounds. He arbitrarily assigned left-handedness to one form which he named L-glyceraldehyde, and right-handedness to its mirror image, named D-glyceraldehyde. His structural formulas for them were:

$$\begin{array}{ccc} \text{CHO} & & \text{CHO} \\ | & & | \\ \text{H} - \text{C} - \text{OH} & & \text{HO} - \text{C} - \text{H} \\ | & & | \\ \text{CH}_2\text{OH} & & \text{CH}_2\text{OH} \end{array}$$

D-Glyceraldehyde   L-Glyceraldehyde

Any compound that could be shown by appropriate chemical methods (rather careful ones) to have a structure related to L-glyceraldehyde would be considered in the "L-series" and would have the prefix "L" attached to its name, regardless of whether it was levorotatory or dextrorotatory as far as polarized light was concerned. As it turned out, the levorotatory form of tartaric acid was found to belong to the D-series instead of the L-series. (Nowadays, a compound that falls in the D-series structurally but rotates light to the left has its name prefixed by "D(−)." Similarly, we have "D(+)," "L(−)," and "L(+).")

This preoccupation with the minutiae of optical activity has turned out to be more than a matter of idle curiosity. As it happens, almost all the compounds occurring in living organisms contain asymmetric carbon atoms. And in every such case the organism makes use of only one of the two mirror-image forms of the compound. Furthermore, similar compounds generally fall in the same series. For instance, virtually all the simple sugars found in living tissue belong to the D-series, while virtually all the amino acids (the building blocks of proteins) belong to the L-series.

In 1955, a chemist named J. M. Bijvoet finally determined what structure tended to rotate polarized light to the left, and vice versa. It turned out that Fischer had, by chance, guessed right in naming the levorotatory and dextrorotatory forms.

For some years after the secure establishment of the Kekulé system of structural formulas, one compound with a rather simple molecule resisted formulation. That compound was benzene (discovered in 1825 by Faraday). Chemical evidence showed that it consisted of six carbon atoms and six hydrogen atoms. What happened to all the extra carbon bonds? (Six carbon atoms linked to one another by single bonds could hold 14 hydrogen atoms, and they do in the well-known compound called hexane, C_6H_{14}.) Evidently the carbon atoms in benzene were linked together by double or triple bonds. Thus benzene might have a structure such as $CH \equiv C - CH = CH - CH = CH_2$. But the trouble was that the known compounds with that sort of structure had properties quite different from those of benzene. Besides, all the chemical evidence seemed to indicate that the benzene molecule was very symmetrical, and six carbons and six hydrogens could not be arranged in a chain in any reasonably symmetrical fashion.

In 1865, Kekulé himself came up with the answer. He related some years later that the vision of the benzene molecule came to him while he was riding on a bus and sunk in a reverie, half-asleep. In his dream, chains of carbon atoms seemed to come alive and dance before his eyes, and then suddenly one coiled on itself like a snake. Kekulé awoke from his reverie with a start and could have cried "Eureka!" He had the solution: the benzene molecule was a ring.

Kekulé suggested that the six carbon atoms of the molecule were arranged as follows:

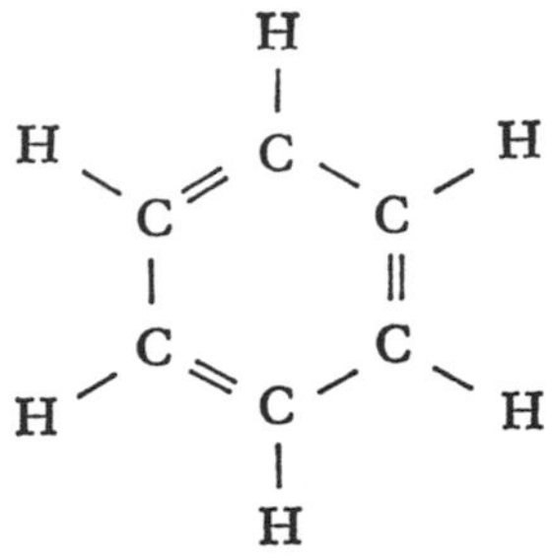

Here at last was the required symmetry. It explained, among other things, why the substitution of another atom for one of benzene's hydrogen atoms always yielded just one unvarying product. Since all the carbons in the ring were indistinguishable from one another in structural terms, no matter where you made the substitution for a hydrogen atom on the ring you would get the same product. Second, the ring structure showed that there were just three ways in which you could replace two hydrogen atoms on the ring: you could make the substitutions on two adjacent carbon atoms in the ring, on two separated by a single skip, or on two separated by a double skip. Sure enough, it was found that just three doubly substituted benzene isomers could be made.

Kekulé's blueprint of the benzene molecule, however, presented an awkward question. Generally, compounds with double bonds are more reactive, which is to say more unstable, than those with only single bonds. It is as if the extra bond were ready and more than willing to desert the attachment to the carbon atom and form a new attachment.

Double-bonded compounds readily add on hydrogen or other atoms and can even be broken down without much difficulty. But the benzene ring is extraordinarily stable—more stable than carbon chains with only single bonds. (In fact, it is so stable and common in organic matter that molecules containing benzene rings make up an entire class of organic compounds, called "aromatic," all the rest being lumped together as the "aliphatic" compounds.) The benzene molecule resists taking on more hydrogen atoms and is hard to break down.

The nineteenth-century organic chemists could find no explanation for this queer stability of the double bonds in the benzene molecule, and it disturbed them. The point may seem a small one, but the whole Kekulé system of structural formulas was endangered by the recalcitrance of the benzene molecule. The failure to explain this one conspicuous paradox made all the rest uncertain.

The closest approach to a solution prior to the twentieth century was that of the German chemist Johannes Thiele. In 1899, he suggested that when double bonds and single bonds alternated, the nearer ends of a pair of double bonds somehow neutralized each other and canceled each other's reactive nature. Consider, as an example, the compound "butadiene," which contains, in simplest form, the case of two double bonds separated by a single bond ("conjugated double bonds"). Now if two atoms are added to the compound, they add onto the end carbons, as shown in the formula below. Such a view explained the nonreactivity of benzene, since the three double bonds of the benzene rings, being arranged in a ring, neutralize each other completely.

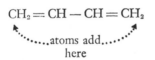

$$CH_2 = CH - CH = CH_2$$

......atoms add......
here

Some forty years later, a better answer was found by way of the new theory of chemical bonds that pictured atoms as linked together by sharing electrons.

The chemical bond, which Kekulé had drawn as a dash between atoms, came to be looked upon as representing a shared pair of electrons (see Chapter 5). Each atom that formed a combination with a partner shared one of its electrons with the partner, and the partner reciprocated by donating one of *its* electrons to the bond. Carbon, with four electrons in its outer shell, could form four attachments; hydrogen could donate its one electron to a bond with one other atom, and so on.

Now the question arose: How were the electrons shared? Obviously, two carbon atoms share the pair of electrons between them equally, because each atom has an equal hold on electrons. On the other hand, in a combination such as H_2O, the oxygen atom, which has a stronger hold on electrons than a hydrogen atom, takes possession of the greater share of the pair of electrons it has in common with each hydrogen atom. This means that the oxygen atom, by virtue of its excessive portion of electrons, has a slight excess of negative charge. By the same token, the hydrogen atom, suffering from an electron deficiency, has a slight excess of positive charge. A molecule containing an oxygen-hydrogen pair, such as water or ethyl alcohol, possesses a small concentration of negative charge in one part of the molecule and a small concentration of positive charge in another. It possesses two poles of charge, so to speak, and is called a "polar molecule."

This view of molecular structure was first proposed in 1912 by the Dutch chemist Peter Joseph Wilhelm Debye, who later pursued his research in the United States. He used an electric field to measure the amount of separation of poles of electric charge in a molecule. In such a field, polar molecules line themselves up with the negative ends pointing toward the positive pole and the positive ends toward the negative pole, and the ease with which this is done is the measure of the "dipole moment" of the molecule. By the early 1930's, measurements of dipole moments had become routine, and in 1936, for this and other work, Debye was awarded the Nobel Prize in chemistry.

The new picture explained a number of things that earlier views of molecular structure could not. For instance, it explained some anomalies of the boiling points of substances. In general, the greater the molecular weight, the higher the boiling point. But this rule is commonly broken. Water, with a molecular weight of only 18, boils at 100° C., whereas propane, with more than twice this molecular weight (44), boils at the much lower temperature of –42° C. Why should that be? The answer is that water is a polar molecule with a high dipole moment, while propane is "nonpolar"—it has no poles of charge. Polar molecules tend to orient themselves with the negative pole of one molecule adjacent to the positive pole of its neighbor. The resulting electrostatic attraction between neighboring molecules makes it harder to tear the molecules apart, and so such substances have relatively high boiling points. This accounts for the fact that ethyl alcohol has a much higher boiling point (78° C.) than its isomer dimethyl ether, which boils at –24° C., although both substances have the same molecular weight (46). Ethyl alcohol has a large dipole moment and dimethyl ether only a small one. Water has a dipole moment even larger than that of ethyl alcohol.

When de Broglie and Schrödinger formulated the new view of electrons not as sharply defined particles but as packets of waves (see Chapter 7), the idea of the chemical bond underwent a further change. In 1939, the American chemist Linus Pauling presented a quantum-mechanical concept of molecular bonds in a book entitled *The Nature of the Chemical Bond*. His theory finally explained, among other things, the paradox of the stability of the benzene molecule.

Pauling pictured the electrons that form a bond as "resonating" between the atoms they join. He showed that under certain conditions it was necessary to view an electron as occupying any one of a number of positions (with varying probability). The electron, with its wavelike properties, might then best be presented as being spread out into a kind of blur, representing the weighted average of the individual probabilities of position. The more evenly the electron was spread out, the more stable was the compound. Such "resonance stabilization" was most likely to occur when the molecule possessed conjugated bonds in one plane and when the existence of symmetry allowed a number of alternative positions for the electron (viewed as a particle). The benzene ring is planar and symmetrical, and Pauling showed that the bonds of the ring were not really double and single in alternation, but that the electrons were smeared out, so to speak, into an equal distribution which resulted in all the bonds being alike and all being stronger and less reactive than ordinary single bonds.

The resonance structures, though they explain chemical behavior satisfactorily, are difficult to present in simple symbolism on paper. Therefore the old Kekulé structures, although now understood to represent only approximations of the actual electronic situation, are still universally used and will undoubtedly continue to be used through the foreseeable future.

Organic Synthesis

After Kolbe had produced acetic acid, there came in the 1850's a chemist who went systematically and methodically about the business of synthesizing organic substances in the laboratory. He was the Frenchman Pierre Eugène Marcelin Berthelot. He prepared a number of simple organic compounds from still simpler inorganic compounds such as carbon monoxide. Berthelot built his simple organic compounds up through increasing complexity until he finally had ethyl alcohol, among other things. It was "synthetic ethyl alcohol," to be sure, but absolutely indistinguishable from the "real thing," because it *was* the real thing.

Ethyl alcohol is an organic compound familiar to all and highly valued by most. No doubt the thought that the chemist could make ethyl alcohol from coal, air, and water (coal to supply the carbon, air the oxygen, and water the hydrogen), without the necessity of fruits or grain as a starting point, must have created enticing visions and endowed the chemist with a new kind of reputation as a miracle worker. At any rate, it put organic synthesis on the map.

For chemists, however, Berthelot did something even more significant. He began to form products that did not exist in nature. He took "glycerol," a compound discovered by Scheele in 1778 and obtained from the breakdown of the fats of living organisms, and combined it with acids not known to occur naturally in fats (although they occurred naturally elsewhere). In this way he obtained fatty substances which were not quite like those that occurred in organisms.

Thus Berthelot laid the groundwork for a new kind of organic chemistry—the synthesis of molecules that nature could not supply. This meant the possible formation of a kind of "synthetic" which might be a substitute—perhaps an inferior substitute—for some natural compound that was hard or impossible to get in the needed quantity. But it also meant the possibility of "synthetics" which were improvements on anything in nature.

This notion of improving on nature in one fashion or another, rather than merely supplementing it, has grown to colossal proportions since Berthelot showed the way. The first fruits of the new outlook were in the field of dyes.

The beginnings of organic chemistry were in Germany. Wöhler and Liebig were both German, and other men of great ability followed them. Before the middle of the nineteenth century, there were no organic chemists in England even remotely comparable to those in Germany. In fact, English schools had so low an opinion of chemistry that they taught the subject only during the lunch recess, not expecting (or even perhaps desiring) many students to be interested. It is odd, therefore, that the first feat of synthesis with world-wide repercussions was actually carried through in England.

It came about in this way. In 1845, when the Royal College of Science in London finally decided to give a good course in chemistry, it imported a young German to do the teaching. He was August Wilhelm von Hofmann, only twenty-seven at the time, and he was hired at the suggestion of Queen Victoria's husband, the Prince Consort Albert (who was himself of German birth).

Hofmann was interested in a number of things, among them coal tar, which he had worked with on the occasion of his first research project under Liebig. Coal tar is a black, gummy material given off by coal when it is heated strongly in the absence of air. The tar is not an attractive material, but it is a valuable source of organic chemicals. In the 1840's, for instance, it served as a source of large quantities of reasonably pure benzene and of a nitrogen-containing compound called "aniline," related to benzene, which Hofmann had been the first to obtain from coal tar.

About ten years after he arrived in England, Hofmann came across a seventeen-year-old boy studying chemistry at the college. His name was William Henry Perkin. Hofmann had a keen eye for talent and knew enthusiasm when he saw it. He took on the youngster as an assistant and set him to work on coal-tar compounds. Perkin's enthusi-

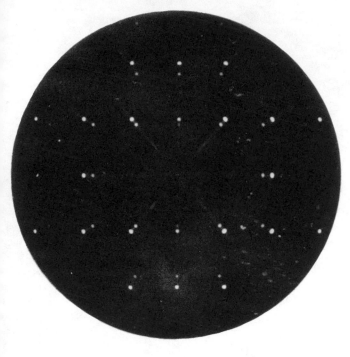

Structure of urea is indicated in this X-ray diffraction picture. It shows the positions of the atoms in a single layer of a urea crystal, looking down at the layer.

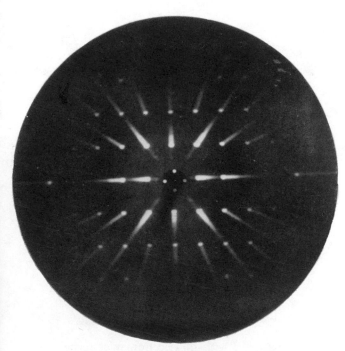

Eight-sided ring of an organic compound called cylo-octa-tetra-ene, shown by X-ray diffraction. The ring is similar to the benzene ring, with alternating single and double bonds, but it is eight-sided instead of six-sided.

Silk fibers as photographed with the electron microscope. The magnification here is about 6,000 times.

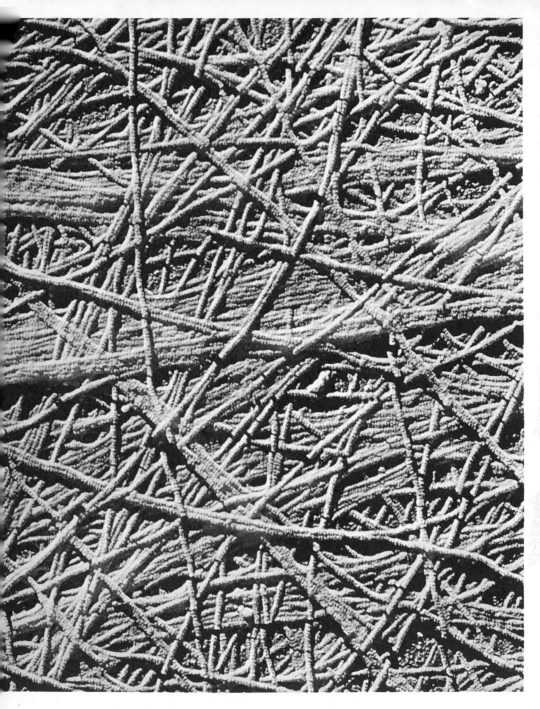

Collagen fibers shown in an electron micrograph. Note how the fibers are collected in bundles.

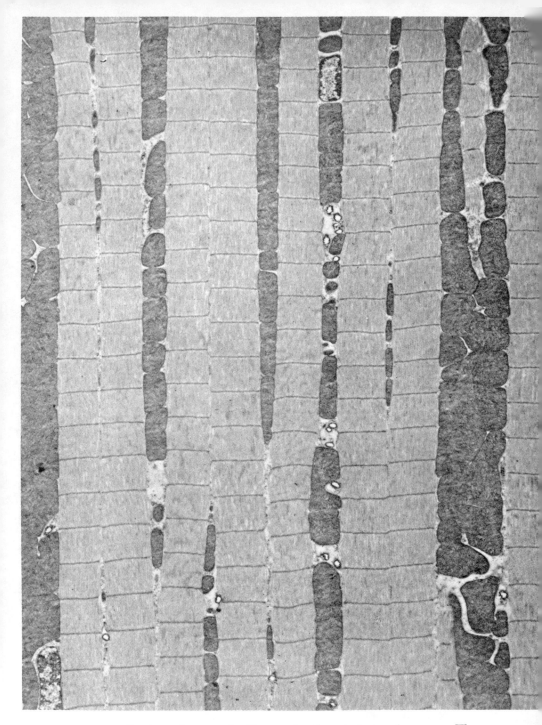

Section of a muscle fiber under the electron microscope. The gray, striated structures are the fibrils, and the dark bodies are mitochondria, which contain enzymes that carry out energy-yielding reactions. This tissue is from the flight muscles of a beetle.

asm was tireless. He set up a laboratory in his home and worked there as well as at school.

Hofmann, who was also interested in medical applications of chemistry, mused aloud one day in 1856 on the possibility of synthesizing quinine, a natural substance used in the treatment of malaria. Now those were the days before structural formulas had come into their own. The only thing known about quinine was its composition, and no one at the time had any idea of just how complicated its structure was. (It was not till 1908 that the structure was correctly deduced.)

Blissfully ignorant of its complexity, Perkin, at the age of eighteen, tackled the problem of synthesizing quinine. He began with allyltoluidine, one of his coal-tar compounds. This molecule seemed to have about half the numbers of the various types of atoms that quinine had in its molecule. If he put two of these molecules together and added some missing oxygen atoms (say by mixing in some potassium dichromate, known to add oxygen atoms to chemicals with which it was mixed), Perkin thought he might get a molecule of quinine.

Naturally this approach got Perkin nowhere. He ended with a dirty, red-brown goo. Then he tried aniline in place of allyltoluidine and got a blackish goo. This time, though, it seemed to him that he caught a purplish glint in it. He added alcohol to the mess, and the colorless liquid turned a beautiful purple. At once Perkin thought of the possibility that he had discovered something that might be useful as a dye.

Dyes had always been greatly admired, and expensive, substances. There were only a handful of good dyes—dyes that stained fabric permanently and brilliantly and did not fade or wash out. There was dark blue indigo, from the indigo plant and the closely related "woad" for which Britain was famous in early Roman times; there was "Tyrian purple," from a snail (so-called because ancient Tyre grew rich on its manufacture—in the later Roman Empire the royal children were born in a room with hangings dyed with Tyrian purple, whence the phrase "born to the purple"); and there was reddish alizarin, from the madder plant ("alizarin" came from Arabic words meaning "the juice"). To these inheritances from ancient and medieval times later dyers had added a few tropical dyes and inorganic pigments (today used chiefly in paints).

This explains Perkin's excitement about the possibility that his purple substance might be a dye. At the suggestion of a friend, he sent a sample to a firm in Scotland which was interested in dyes, and quickly the answer came back that the purple compound had good properties. Could it be supplied cheaply? Perkin proceeded to patent the dye (there was considerable argument as to whether an eighteen-year-old could obtain a patent, but eventually he obtained it), to quit school, and to go into business.

His project was not easy. Perkin had to start from scratch, preparing design. Within six months, however, he was producing what he named "aniline purple"—a compound not found in nature and superior to any natural dye in its color range.

French dyers, who took to the new dye more quickly than did the more conservative English, named the color "mauve," from the mallow (Latin name "malva"), and the dye itself came to be known as "mauveine." Quickly it became the rage (the period being sometimes referred to as the "Mauve Decade"), and Perkin grew rich. At the age of twenty-three he was the world authority on dyes.

The dam had broken. A number of organic chemists, inspired by Perkin's astonishing success, went to work synthesizing dyes, and many succeeded. Hofmann himself turned to this new field, and, in 1858, he synthesized a red-purple dye which was later given the name "magenta" by the French dyers (then, as now, arbiters of the world's fashions). The dye was named for the Italian city where the French defeated the Austrians in a battle in 1859.

Hofmann returned to Germany in 1865, carrying his new interest in dyes with him. He discovered a group of violet dyes still known as "Hofmann's violets." By the mid-twentieth century, no less than 3,500 synthetic dyes were in commercial use.

Chemists also synthesized the natural dyestuffs in the laboratory. Karl Graebe of Germany and Perkin both synthesized alizarin in 1869 (Graebe applying for the patent one day sooner than Perkin), and in 1880 the German chemist Adolf von Baeyer worked out a method of synthesizing indigo. (For his work on dyes von Baeyer received the Nobel Prize in chemistry in 1905.)

Perkin retired from business in 1874, at the age of thirty-five, and returned to his first love, research. By 1875, he had managed to synthesize coumarin (a naturally occurring substance which has the pleasant odor of new-mown hay); this served as the beginning of the synthetic perfume industry.

Perkin alone could not maintain British supremacy against the great development of German organic chemistry, and, by the turn of the century, "synthetics" became almost a German monopoly. It was a German chemist, Otto Wallach, who carried on the work on synthetic perfumes that Perkin had started. In 1910, Wallach was awarded the Nobel Prize in chemistry for his investigations. The Croatian chemist Leopold Ruzicka, teaching in Switzerland, first synthesized musk, an important component of perfumes. He shared the Nobel Prize in chemistry in 1938. However, during World War I, Great Britain and the United States, shut off from the products of the German chemical laboratories, were forced to develop chemical industries of their own.

Achievements in synthetic organic chemistry could not have pro-his own starting materials from coal tar with equipment of his own ceeded at anything better than a stumbling pace if chemists had had to depend upon fortunate accidents such as the one that had been seized upon by Perkin. Fortunately the structural formulas of Kekulé, presented three years after Perkin's discovery, made it possible to prepare blueprints, so to speak, of the organic molecule. No longer did chemists have to prepare quinine by sheer guesswork and hope; they had methods for attempting to scale the structural heights of the molecule step by step, with advance knowledge of where they were headed and what they might expect.

Chemists learned how to alter one group of atoms to another; to open up rings of atoms and to form rings from open chains; to split groups of atoms in two; and to add carbon atoms one by one to a chain. The specific method of doing a particular architectural task within the organic molecule is still often referred to by the name of the chemist who first described the details. For instance, Perkin discovered a method of adding a two-carbon atom group by heating certain substances with chemicals named acetic anhydride and sodium acetate. This is still called the "Perkin reaction." Perkin's teacher, Hofmann, discovered that a ring of atoms which included a nitrogen could be treated with a substance called methyl iodide in the presence of silver compound in such a way that the ring was eventually broken and the nitrogen atom removed. This is the "Hofmann degradation." In 1877, the French chemist Charles Friedel, working with the American chemist James Mason Crafts, discovered a way of attaching a short carbon chain to a benzene ring by the use of heat and aluminum chloride. This is now known as the "Friedel-Crafts reaction."

In 1900, the French chemist Victor Grignard discovered that magnesium metal, properly used, could bring about a rather large variety of different joinings of carbon chains; he presented the discovery in his doctoral dissertation. For the development of these "Grignard reactions" he shared in the Nobel Prize in chemistry in 1912. The French chemist Paul Sabatier, who shared it with him, had discovered (with J. B. Senderens) a method of using finely divided nickel to bring about the addition of hydrogen atoms in those places where a carbon chain possessed a double bond. This is the "Sabatier-Senderens reduction."

In 1928, the German chemists Otto Diels and Kurt Alder discovered a method of adding the two ends of a carbon chain to opposite ends of a double bond in another carbon chain, thus forming a ring of atoms. For the discovery of this "Diels-Alder reaction," they shared the Nobel Prize for chemistry in 1950.

In other words, by noting the changes in the structural formulas of substances subjected to a variety of chemicals and conditions, organic chemists worked out a slowly growing set of ground rules on how to change one compound into another at will. It was not easy. Every compound and every change had its own peculiarities and difficulties. But the main paths were blazed, and the skilled organic chemist found them clear signs toward progress in what had formerly seemed a jungle.

Knowledge of the manner in which particular groups of atoms behaved could also be used to work out the structure of unknown compounds. For instance, when simple alcohols react with metallic sodium and liberate hydrogen, only the hydrogen linked to an oxygen atom is released, not the hydrogens linked to carbon atoms. On the other hand, some organic compounds will take on hydrogen atoms under appropriate conditions while others will not. It turns out that compounds that add hydrogen generally possess double or triple bonds and add the hydrogen at those bonds. From such information a whole new type of chemical analysis of organic compounds arose; the nature of the atom groupings was determined, rather than just the numbers and kinds of various atoms present. The liberation of hydrogen by the addition of sodium signified the presence of an oxygen-bound hydrogen atom in the compound; the acceptance of hydrogen meant the presence of double or triple bonds. If the molecule was too complicated for analysis as a whole, it could be broken down into simpler portions by well-defined methods; the structures of the simpler portions could be worked out and the original molecule deduced from those.

Using the structural formula as a tool and guide, chemists could work out the structure of some useful naturally occurring organic compound (analysis) and then set about duplicating it or something like it in the laboratory (synthesis). One result was that something which was rare, expensive or difficult to obtain in nature might become cheaply available in quantity in the laboratory. Or, as in the case of the coal-tar dyes, the laboratory might create something that fulfilled a need better than did similar substances found in nature.

One startling case of a deliberate improvement on nature involves cocaine, found in the leaves of the coca plant, which is native to Bolivia and Peru, but is now grown chiefly in Java. Like the compounds strychnine, morphine, and quinine, all mentioned earlier, cocaine is an example of an "alkaloid," a nitrogen-containing plant product that, in small concentration, has profound physiological effects on man. Depending on the dose, alkaloids can cure or kill. The most famous alkaloid death of all times was that of Socrates, who was killed by "coniine," an alkaloid in hemlock.

The molecular structure of the alkaloids is, in some cases, extraordinarily complicated, but that just sharpened chemical curiosity. The English chemist Robert Robinson tackled the alkaloids systematically. He worked out the structure of morphine (for all but one dubious atom) in 1925, and the structure of strychnine in 1946. He received the Nobel Prize for chemistry in 1947 as recognition of the value of his work.

Robinson had merely worked out the structure of alkaloids without using that structure as a guide to their synthesis. The American chemist Robert Burns Woodward took care of that. With his American colleague William von Eggers Doering, he synthesized quinine in 1944. It was the wild goose chase after this particular compound by Perkin that had had such tremendous results. And, if you are curious, here is the structural formula of quinine:

No wonder it stumped Perkin.

That Woodward and von Doering solved the problem is not merely a tribute to their brilliance. They had at their disposal the new electronic theories of molecular structure and behavior worked out by men such as Pauling. Woodward went on to synthesize a variety of complicated molecules which had, before his time, represented hopeless challenges. In 1954, for instance, he synthesized strychnine.

Long before the structure of the alkaloids had been worked out, however, some of them, notably cocaine, were of intense interest to medical men. The South American Indians, it had been discovered, would chew coca leaves, finding it an antidote to fatigue and a source of happiness-sensation. The Scottish physician Robert Christison introduced the plant to Europe. (This is not the only gift to medicine on the part of the witch doctors and herb-women of prescientific societies. There are also quinine and strychnine, already mentioned, as well as opium, digitalis, curare, atropine, strophanthidin, and reserpine. In addition, the smoking of tobacco, the chewing of betel nuts, the drinking of alcohol, and the taking of such drugs as marijuana and peyote are all inherited from primitive societies.)

Cocaine was not merely a general happiness-producer. Doctors discovered that it deadened the body, temporarily and locally, to sensations of pain. In 1884, the American physician Carl Koller discovered that cocaine could be used as a pain-deadener when added to the mucuous membranes around the eye. Eye operations could then be performed without pain. Cocaine could also be used in dentistry, allowing teeth to be extracted without pain.

This fascinated doctors, for one of the great medical victories of the nineteenth century had been that over pain. In 1799, Humphry Davy had prepared the gas "nitrous oxide" (N_2O) and studied its effects. He found that when it was inhaled it released inhibitions so that men breathing it would laugh, cry, or otherwise act foolishly. Its common name is "laughing gas," for that reason.

In the early 1840's, an American scientist, Gardner Quincy Cotton, discovered that nitrous oxide deadened the sensation of pain, and, in 1844, an American dentist, Horace Wells, used it in dentistry. By that time, something better had entered the field.

The American surgeon Crawford Williamson Long in 1842 had used ether to put a patient to sleep during tooth extractions. In 1846, the American dentist William Thomas Green Morton conducted a surgical operation under ether at the Massachusetts General Hospital. Morton usually gets the credit for the discovery, because Long did not describe his feat in the medical journals until after Morton's public demonstration, and Well's earliest public demonstrations with nitrous oxide had been only indifferent successes.

The American poet and physician Oliver Wendell Holmes suggested that pain-deadening compounds be called "anesthetics" (from Greek words meaning "no feeling"). Some people at the time felt that anesthetics were a sacrilegious attempt to avoid pain inflicted on mankind by God, but if anything was needed to make anesthesia respectable, it was its use by the Scottish physician James Young Simpson for Queen Victoria of England during childbirth.

Anesthesia had finally converted surgery from torture-chamber butchery to something that was at least humane and, with the addition of antiseptic conditions, even life-saving. For that reason, any further advance in anesthesia was seized on with great interest. Cocaine's special interest was that it was a "local anesthetic," deadening pain in a specific area without inducing general unconsciousness and lack of sensation, as in the case of such "general anesthetics" as ether.

There are several drawbacks to cocaine, however. In the first place, it can induce troublesome side-effects and can even kill patients sensitive to it. Second, it can bring about addiction and has to be used skimpily and with caution. (Cocaine is one of the dangerous "dopes" or "narcotics" that deaden not only pain but other unpleasant sensations and give the user the illusion of euphoria. The user may become so accustomed to this that he may require increasing doses and, despite the actual bad effect upon his body, become so dependent on the illusions it carries with it that he cannot stop using it without developing painful "withdrawal symptoms." Such "drug addiction" for cocaine and other drugs of this sort is an important social problem. Up to twenty tons of cocaine are produced illegally each year and sold with tremendous profits to a few and tremendous misery to many, despite world-wide efforts to stop the traffic.) Third, the molecule is fragile, and heating cocaine to sterilize it of any bacteria leads to changes in the molecule that interfere with its anesthetic effects.

The structure of the cocaine molecule is rather complicated:

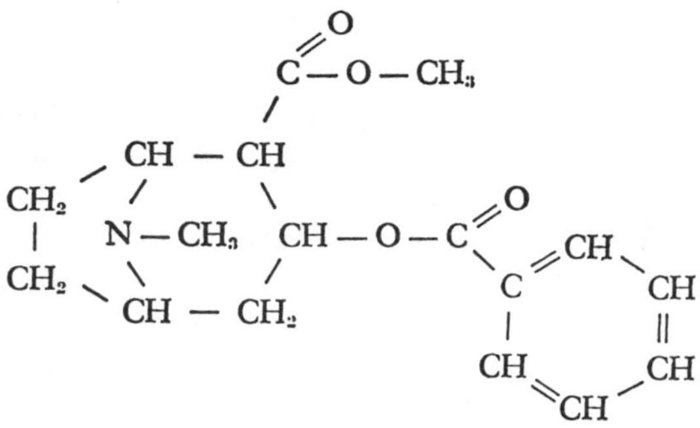

$$
\begin{array}{c}
\text{O} \\
\parallel \\
\text{C}-\text{O}-\text{CH}_3 \\
\diagup \\
\text{CH}-\text{CH}
\end{array}
$$

The double ring on the left is the fragile portion, and that is the difficult one to synthesize. (The synthesis of cocaine was not achieved until 1923, when the German chemist Richard Willstätter managed it.) However, it occurred to chemists that they might synthesize similar compounds in which the double ring was not closed. This would make the compound both easier to form and more stable. The synthetic substance might possess the anesthetic properties of cocaine, perhaps without the undesirable side effects.

For some twenty years, German chemists tackled the problem, turning out dozens of compounds, some of which were pretty good. The most successful modification was obtained in 1909, when a compound with the following formula was prepared:

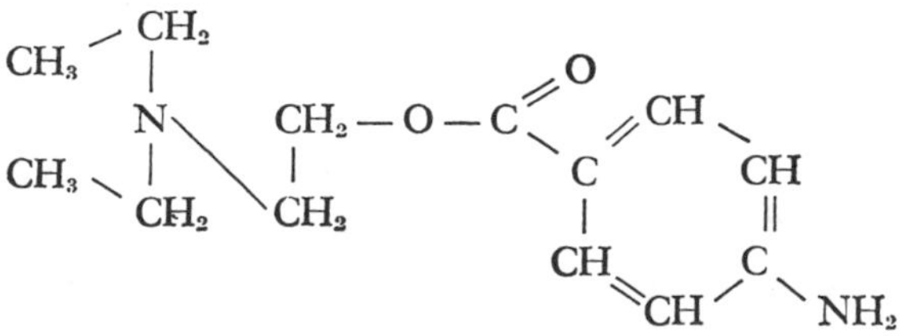

Compare this with the formula for cocaine and you will see the similarity, and also the important fact that the double ring no longer exists. This simpler molecule—stable, easy to synthesize, with good anesthetic properties and very little in the way of side effects—does not exist in nature. It is a "synthetic substitute" far better than the real thing. It is called "procaine," but is better known to the public by the trade-name Novocaine.

Perhaps the most effective and best-known of the general pain-deadeners is morphine. Its very name is from the Greek word for "sleep." It is a purified derivative of the opium juice or "laudanum" used for centuries by peoples, both civilized and primitive, to combat the pains and tensions of the workaday world. As a gift to the pain-wracked, it is heavenly, but it, too, carries the deadly danger of addiction. An attempt to find a substitute backfired. In 1898, a synthetic derivative, "diacetylmorphine," better known as "heroin" was introduced in the

499

belief that it would be safer. Instead, it turned out to be the most dangerous dope of all.

Less dangerous "sedatives" (sleep-inducers) are chloral hydrate and, particularly, the barbiturates. The first example of this latter group was introduced in 1902, and they are now the most common constituents of "sleeping pills." Harmless enough when used properly, they can nevertheless induce addiction, and an overdose can cause death. In fact, because death comes quietly as the end product of a gradually deepening sleep, barbiturate overdosage is a rather popular method of suicide, or attempted suicide.

The most common sedative, and the longest in use, is, of course, alcohol. Methods of fermenting fruit juice and grain were known in prehistoric times, as was distillation to produce stronger liquors than could be produced naturally. The value of light wines in areas where the water supply is nothing but a short cut to typhoid fever and cholera, and the social acceptance of drinking in moderation, make it difficult to treat alcohol as the drug it is, although it induces addiction as surely as morphine and through sheer quantity of use does much more harm. Legal prohibition of sale of liquor seems to be unhelpful; certainly the American experiment of "Prohibition" (1920–33) was a disastrous failure. Nevertheless, alcoholism is more and more being treated as the disease it is rather than as a moral disgrace. The acute symptoms of alcoholism ("delirium tremens") are probably not so much due to the alcohol itself as to the vitamin deficiencies induced in those who eat little while drinking much.

Man now has at his disposal all sorts of synthetics of great potential use and misuse: explosives, poison gases, insecticides, weed-killers, antiseptics, disinfectants, detergents, drugs—almost no end of them, really. But synthesis is not merely the handmaiden of consumer needs. It can also be placed at the service of pure chemical research.

It often happens that a complex compound, produced either by living tissue or by the apparatus of the organic chemist, can only be assigned a tentative molecular structure, after all possible deductions have been drawn from the nature of the reactions it undergoes. In that case, a way out is to synthesize a compound by means of reactions designed to yield a molecular structure like the one that has been deduced. If the properties of the resulting compound are identical with the compound being investigated in the first place, the assigned structure becomes something in which a chemist can place his confidence.

An impressive case in point involves hemoglobin, the main component of the red blood cells and the pigment that gives the blood its red color. In 1831, the French chemist L. R. LeCanu split hemoglobin into two parts, of which the smaller portion, called "heme," made up 4 per cent of the mass of hemoglobin. Heme was found to have the empirical formula $C_{34}H_{32}O_4N_4Fe$. Such compounds as heme were known to occur in other vitally important substances, both in the plant and animal kingdoms, and so the structure of the molecule was a matter of great moment to biochemists. For nearly a century after LeCanu's isolation of heme, however, all that could be done was to break it down into smaller molecules. The iron atom (Fe) was easily removed, and what was left then broke up into pieces roughly a quarter the size of the original molecule. These fragments were found to be "pyrroles"—molecules built on rings of five atoms, of which four are carbon and one nitrogen. Pyrrole itself has the following structure:

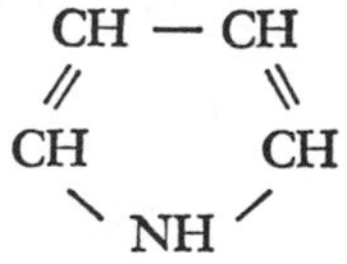

$$CH - CH$$
$$CH \qquad CH$$
$$NH$$

The pyrroles actually obtained from heme possessed small groups of atoms containing one or two carbon atoms attached to the ring in place of one or more of the hydrogen atoms.

In the 1920's, the German chemist Hans Fischer tackled the problem further. Since the pyrroles were one quarter the size of the original heme, he decided to try to combine four pyrroles and see what he got. What he finally succeeded in getting was a four-ring compound which he called "porphin" (from a Greek word meaning "purple," because of its purple color). Porphin would look like this:

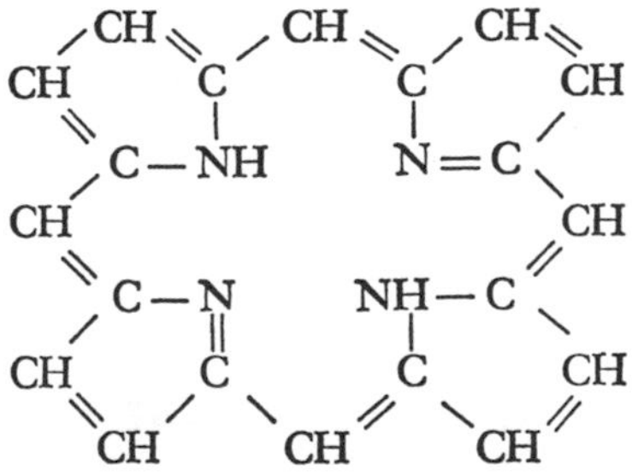

$$
\begin{array}{c}
CH \quad CH \quad CH \\
CH \quad C \quad C \quad CH \\
C-NH \quad N=C \\
CH \quad CH \\
C-N \quad NH-C \\
CH \quad C \quad C \quad CH \\
CH \quad CH \quad CH
\end{array}
$$

However, the pyrroles obtained from heme in the first place contained small "side chains" attached to the ring. These remained in place when the pyrroles were joined to form porphin. The porphin with various side chains attached make up a family of compounds called the "porphyrins." It was obvious to Fischer upon comparing the properties of heme with those of the porphyrins he had synthesized that heme (minus its iron atom) was a porphyrin. But which one? No fewer than fifteen compounds could be formed from the various pyrroles obtained from heme, according to Fischer's reasoning, and any one of those fifteen might be heme itself.

A straightforward answer could be obtained by synthesizing all fifteen and testing the properties of each one. Fischer put his students to work preparing, by painstaking chemical reactions that allowed only a particular structure to be built up, each of the fifteen possibilities. As each different porphyrin was formed, he compared its properties with those of the natural porphyrin of heme.

In 1928, he discovered that the porphyrin numbered nine in his series was the one he was after. The natural variety of porphyrin is therefore called "porphyrin IX" to this day. It was a simple procedure to convert porphyrin IX to heme by adding iron. Chemists at last felt confident that they knew the structure of that important compound. Here is the structure of heme, as worked out by Fischer:

$$\text{Heme structure (porphyrin IX with Fe)}$$

For his achievement Fischer was awarded the Nobel Prize in chemistry in 1930.

All the triumphs of synthetic organic chemistry through the nineteenth century and the first half of the twentieth century, great as they were, were won by means of the same processes used by the alchemists of ancient times—mixing and heating substances. Heat was the one sure way of adding energy to molecules and making them interact, but the interactions were usually random in nature and took place by way of briefly existent, unstable intermediates, whose nature could only be guessed at.

What chemists needed was a more refined, more direct method for producing energetic molecules—a method that would produce a group of molecules all moving at about the same speed in about the same direction. This would remove the random nature of interactions, for whatever one molecule would do, all would do. One way would be to accelerate ions in an electric field, much as subatomic particles are accelerated in cyclotrons.

In 1964, the German-American chemist Richard Leopold Wolfgang accelerated ions and molecules to high energies and, by means of what might be called a "chemical accelerator," produced ion speeds that heat would produce only at temperatures of from 10,000° to 100,000° C. Furthermore, the ions were all traveling in the same direction.

If the ions so accelerated are provided with a supply of electrons they can snatch up, they will be converted to neutral molecules which will still be traveling at great speeds. Such neutral beams were produced by the American chemist Leonard Wharton in 1969.

As to the brief intermediate stages of a chemical reaction, computers could help. It was necessary to work out the quantum-mechanical equations governing the state of the electrons in different atom-combinations and to work out the events that would take place on collision. In 1968, for instance, a computer guided by the Italian-American chemist Enrico Clementi "collided" ammonia and hydrochloric acid on closed-circuit television to make ammonium chloride, with the computer working out the events that must take place. The computer indicated that the ammonium chloride which was formed could exist as a high-pressure gas at 700° C. This was not previously known, but was proved experimentally a few months later.

In the last decade, chemists have developed brand-new tools, both theoretically and experimentally. Intimate details of reactions not hitherto available will be known, and new products—unattainable before or at least attainable only in small lots—will be formed. We may be at the threshold of unexpected wonders.

Polymers and Plastics

When we consider molecules like those of heme and quinine, we are approaching a complexity with which even the modern chemist can cope only with great difficulty. The synthesis of such a compound requires so many steps and such a variety of procedures that we can hardly expect to produce it in quantity without the help of some living organism (other than the chemist). This is nothing about which to get an inferiority complex, however. Living tissue itself approaches the limit of its capacity at this level of complexity. Few molecules in nature are more complex than heme and quinine.

To be sure, there are natural substances composed of hundreds of thousands, even millions, of atoms, but these are not really individual molecules, constructed in one piece, so to speak. Rather, these large molecules are built up of units strung together like beads in a necklace. Living tissue usually synthesizes some small, fairly simple compound and then merely hooks the units together in chains. And that, as we shall see, the chemist also is capable of doing.

In living tissue this union of small molecules ("condensation") is usually accompanied by the over-all elimination of two hydrogen atoms and an oxygen atom (which combine to form a water molecule) at each point of junction. Invariably, the process can be reversed (both in the body and in the test tube): by the addition of water, the units of the chain can be loosened and separated. This reverse of condensation is called "hydrolysis," from Greek words meaning "loosening through water." In the test tube the hydrolysis of these long chains can be hastened by a variety of methods, the most common being the addition of a certain amount of acid to the mixture.

The first investigation of the chemical structure of a large molecule dates back to 1812, when a Russian chemist, Gottlieb Sigismund Kirchhoff, found that boiling starch with acid produced a sugar identical in properties with glucose, the sugar obtained from grapes. In 1819 the French chemist Henri Braconnot also obtained glucose by boiling various plant products such as sawdust, linen, and bark, all of which contain a compound called "cellulose." It was easy to guess that both starch and cellulose were built of glucose units, but the details of the molecular structure of starch and cellulose had to await knowledge of the molecular structure of glucose. At first, before the days of structural formulas, all that was known of glucose was its empirical formula, $C_6H_{12}O_6$. This proportion suggested that there was one water molecule, H_2O, attached to each of the six carbon atoms. Hence glucose, and compounds similar to it in structure, were called "carbohydrates" ("watered carbon").

The structural formula of glucose was worked out in 1886 by the German chemist Heinrich Kiliani. He showed that its molecule consisted of a chain of six carbon atoms, to which hydrogen atoms and oxygen-hydrogen groups were separately attached. There were no intact water combinations anywhere in the molecule.

Over the next decade or so the German chemist Emil Fischer studied glucose in detail and worked out the exact arrangement of the oxygen-hydrogen groups around the carbon atoms, four of which were asymmetric. There are sixteen possible arrangements of these groups, and therefore sixteen possible optical isomers, each with its own properties. Chemists have, indeed, made all sixteen, only a few of which actually occur in nature. It was as a result of his work on the optical activity of these sugars that Fischer suggested the establishment of the L-series and D-series of compounds. For putting carbohydrate chemistry on a firm structural foundation, Fischer received the Nobel Prize in chemistry in 1902.

Here are the structural formulas of glucose and of two other common sugars, fructose and galactose:

$$
\begin{array}{ccc}
\text{CH}=\text{O} & \text{CH}_2-\text{OH} & \text{CH}=\text{O} \\
| & | & | \\
\text{H}-\text{C}-\text{OH} & \text{C}=\text{O} & \text{H}-\text{C}-\text{OH} \\
| & | & | \\
\text{HO}-\text{C}-\text{H} & \text{HO}-\text{C}-\text{H} & \text{HO}-\text{C}-\text{H} \\
| & | & | \\
\text{H}-\text{C}-\text{OH} & \text{H}-\text{C}-\text{OH} & \text{HO}-\text{C}-\text{H} \\
| & | & | \\
\text{H}-\text{C}-\text{OH} & \text{H}-\text{C}-\text{OH} & \text{H}-\text{C}-\text{OH} \\
| & | & | \\
\text{CH}_2-\text{OH} & \text{CH}_2-\text{OH} & \text{CH}_2-\text{OH} \\
\\
\text{glucose} & \text{fructose} & \text{galactose}
\end{array}
$$

Once chemists knew the structure of the simple sugars, it was relatively easy to work out the manner in which they were built up into more complex compounds. For instance, a glucose molecule and a fructose can be condensed to the "double-sugar" sucrose—the sugar we use at the table. Glucose and galactose combine to form lactose, which occurs in nature only in milk.

There is no reason why such condensations cannot continue indefinitely, and in starch and cellulose they do. Each consists of long chains of glucose units, condensed in a particular pattern.

The details of the pattern are important, because although both compounds are built up of the same unit, they are profoundly different. Starch in one form or another forms the major portion of humanity's diet, while cellulose is completely inedible. The difference in the pattern of condensation, as painstakingly worked out by chemists, is analogous to the following: Suppose a glucose molecule is viewed as either right-side-up (when it may be symbolized as "u") or upside-down (symbolized as "n"). The starch molecule can then be viewed as consisting of a string of glucose molecules after this fashion ". . . . uuuuuuuuu . . . .", while cellulose consists of ". . . . unununun . . . ." The body's digestive juices possess the ability to hydrolyze the "uu" linkage of starch, breaking it up to glucose, which we can then absorb to obtain energy. Those same juices are helpless to touch the "un" linkage of cellulose, and any cellulose we ingest travels through the alimentary canal and out.

There are certain microorganisms that can digest cellulose, though none of the higher animals can. Some of these microorganisms live in the intestinal tracts of ruminants and termites, for instance. It is thanks to these small helpers that cows can live on grass, and termites live on wood. The microorganisms form glucose from cellulose in quantity, use what they need, and the host uses the overflow. The microorganisms supply the processed food, while the host supplies the raw material and the living quarters. This form of cooperation between two forms of life for mutual benefit is called "symbiosis," from Greek words meaning "life together."

Christopher Columbus discovered South American natives playing with balls of a hardened plant juice. Columbus and the other explorers who visited South America over the next two centuries were fascinated by these bouncy balls (obtained from the sap of trees in Brazil). Samples were brought back to Europe eventually as a curiosity. About 1770, Joseph Priestley (soon to discover oxygen) found that a lump of this bouncy material would rub out pencil marks, so he invented the uninspired name of "rubber," still the English word for the substance. The British call it "India rubber, because it came from the "Indies" (the original name of Columbus' new world).

People eventually found other uses for rubber. In 1823, a Scotsman named Charles Macintosh patented garments made of a layer of rubber between two layers of cloth for use in rainy weather, and raincoats are still sometimes called "mackintoshes" (with an added "k").

The trouble with rubber as used in this way, however, was that in warm weather it became gummy and sticky, while in cold weather it was leathery and hard. A number of individuals tried to discover ways of treating rubber so as to remove these undesirable characteristics. Among them was an American named Charles Goodyear, who was innocent of chemistry but worked stubbornly along by trial and error. One day in 1839, he accidentally spilled a mixture of rubber and sulfur on a hot stove. He scraped it off as quickly as he could and found, to his amazement, that the heated rubber-sulfur mixture was dry even while it was still warm. He heated it and cooled it and found that he had a sample of rubber that did not turn gummy with heat or leathery with cold but remained soft and springy throughout.

This process of adding sulfur to rubber is now called "vulcanization" (after Vulcan, the Roman god of fire). Goodyear's discovery founded the rubber industry. It is sad to have to report that Goodyear himself never reaped a reward despite this multimillion dollar discovery. He spent his life fighting for patent rights and died deeply in debt.

Knowledge of the molecular structure of rubber dates back to 1879, when a French chemist, Gustave Bouchardat, heated rubber in the absence of air and obtained a liquid called "isoprene." Its molecule is composed of five carbon atoms and eight hydrogen atoms, arranged as follows:

$$CH_2 = \overset{\displaystyle CH_3}{\overset{\displaystyle |}{C}} - CH = CH_2$$

A second type of plant juice ("latex"), obtained from certain trees in southeast Asia, yields a substance called "gutta percha." This lacks the elasticity of rubber, but when it is heated in the absence of air it, too, yields isoprene.

Both rubber and gutta percha are made up of thousands of isoprene units. As in the case of starch and cellulose, the difference between them lies in the pattern of linkage. In rubber, the isoprene units are joined in the ". . . uuuuu . . ." fashion and in such a way that they form coils, which can straighten out when pulled, thus allowing stretching. In gutta percha, the units join in the ". . . unununununun . . ." fashion, and these form chains that are straighter to begin with and therefore much less stretchable.

A simple sugar molecule, such as glucose, is a "monosaccharide" (Greek for "one sugar"); sucrose and lactose are "disaccharides" ("two sugars"); and starch and cellulose are "polysaccharides" ("many sugars"). Because two isoprene molecules join to form a well-known type of compound called "terpene" (obtained from turpentine), rubber and gutta percha are called "polyterpenes."

The general term for such compounds was invented by Berzelius (a great inventor of names and symbols) as far back as 1830. He called the basic unit a "monomer" ("one part") and the large molecule a "polymer" ("many parts"). Polymers consisting of many units (say more than a

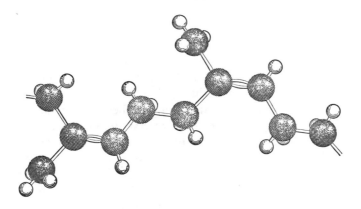

The gutta percha molecule, a portion of which is shown here, is made up of thousands of isoprene units. The first five carbon atoms at the left (*black balls*) and the eight hydrogen atoms bonded to them make up an isoprene unit.

hundred) are now called "high polymers." Starch, cellulose, rubber, and gutta percha are all examples of high polymers.

Polymers are not clear-cut compounds but are complex mixtures of molecules of different sizes. The average molecular weight can be determined by several methods. One involves measurement of "viscosity" (the ease or difficulty with which a liquid flows under a given pressure). The larger the molecule and the more elongated it is, the more it contributes to the "internal friction" of a liquid and the more it makes it pour like molasses, rather than like water. The German chemist Hermann Staudinger worked out this method in 1930 as part of his general work on polymers, and in 1953, he was awarded the Nobel Prize for chemistry for his contribution toward the understanding of these giant molecules.

In 1913, two Japanese chemists discovered that natural fibers such as those of cellulose diffracted X-rays, just as a crystal does. The fibers are not crystals in the ordinary sense, but they are "microcrystalline" in character. That is, the long chains of units making up their molecules tend to run in parallel bundles for longer or shorter distances, here and there. Over the course of those parallel bundles, atoms are arranged in a repetitive order as they are in crystals, and X-rays striking those sections of the fiber are diffracted.

So polymers have come to be divided into two broad classes—crystalline and amorphous.

In a crystalline polymer, such as cellulose, the strength of the individual chains is increased by the fact that parallel neighbors are joined together by chemical bonds. The resulting fibers have considerable tensile strength. Starch is crystalline, too, but far less so than is cellulose. It therefore lacks the strength of cellulose or its capacity for fiber formation.

Rubber is an amorphous polymer. Since the individual chains do not line up, cross-links do not occur. If heated, the various chains can vibrate independently and slide freely over and around one another. Consequently, rubber or a rubberlike polymer will grow soft and sticky and eventually melt with heat. (Stretching rubber straightens the chains and introduces a certain amount of microcrystalline character. Stretched rubber has considerable tensile strength, therefore.) Cellulose and starch, in which the individual molecules are bound together here and there, cannot undergo the same independence of vibration, so there is no softening with heat. They remain stiff until the temperature is high enough to induce vibrations that shake the molecule apart so that charring and smoke emission take place.

At temperatures below the gummy, sticky stage, amorphous polymers are often soft and springy. At still lower temperatures, however, they become hard and leathery, even glassy. Raw rubber is dry and elastic only over a rather narrow temperature range. The addition of sulfur to the extent of 5 to 8 per cent provides flexible sulfur links from chain to chain, which reduce the independence of the chains and thus prevent gumminess at moderate heat. They also increase the free play between the chains at moderately low temperatures; therefore the rubber

The rubber molecule has a structure which is indicated here by a model of a portion of the molecule containing four isoprene units.

Vulcanization of rubber in one of the early plants of the mid-
nineteenth century. At the left is a pile of solid rubber tires.

A modern plastics plant. This one produces vinyl resins.

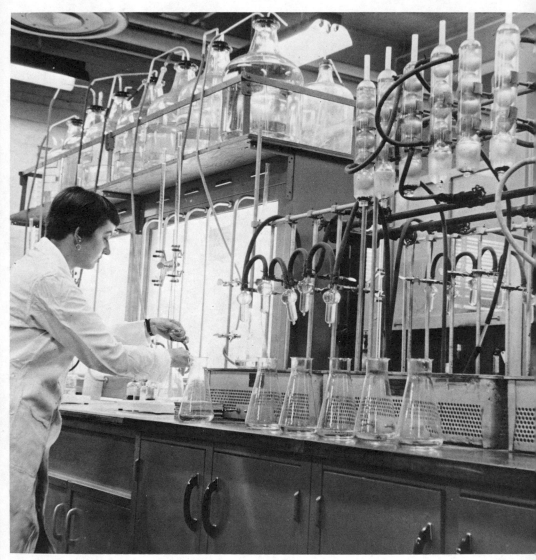

Glassware in an industrial chemical laboratory.

does not harden. The addition of greater amounts of sulfur, up to 30 to 50 per cent, will bind the chains so tightly that the rubber grows hard. It is then known as "hard rubber" or "ebonite."

(Even vulcanized rubber will turn glassy if the temperature is lowered sufficiently. An ordinary rubber ball, dipped in liquid air for a few moments, will shatter if thrown against a wall. This is a favorite demonstration in introductory chemistry courses.)

Various amorphous polymers show different physical properties at a given temperature. At room temperature natural rubber is elastic, various resins are glassy and solid, and chicle (from the sapodilla tree of South America) is soft and gummy (it is the chief ingredient of chewing gum).

Aside from our food, which is mainly made up of high polymers (meat, starch, and so on), probably the one polymer that man has depended on longest is cellulose. It is the major component of wood, which has been indispensable as a fuel and a construction material. Wood's cellulose is also used to make paper. In the pure fibrous forms of cotton and linen, cellulose has been man's most important textile material. And the organic chemists of the mid-nineteenth century naturally turned to cellulose as a raw material for making other giant molecules.

One way of modifying cellulose is by attaching the "nitrate group" of atoms (a nitrogen atom and three oxygen atoms) to the oxygen-hydrogen combinations ("hydroxyl groups") in the glucose units. When this was done, by treating cellulose with a mixture of nitric acid and sulfuric acid, an explosive of until-then unparalleled ferocity was created. The explosive was discovered by accident in 1846 by a German-born Swiss chemist named Christian Friedrich Schönbein (who, in 1839, had discovered ozone). He had spilled an acid mixture in the kitchen (where he was forbidden to experiment but where he had taken advantage of his wife's absence to do just that), and he snatched up his wife's cotton apron, so the story goes, to wipe up the mess. When he hung the apron over the fire to dry, it went poof, leaving nothing behind.

Schönbein recognized the potentialities at once, as can be told from the name he gave the compound, which in English translation is "guncotton." (It is also called "nitrocellulose.") Shönbein peddled the recipe to several governments. Ordinary gunpowder was so smoky that it blackened the gunners, fouled the cannon, which then had to be swabbed between shots, and raised such a pall of smoke that after the first volleys battles had to be fought by dead reckoning. War offices therefore leaped at the chance to use an explosive which was not only more powerful but also smokeless. Factories for the manufacture of guncotton began to spring up. And almost as fast as they sprang up, they blew up. Guncotton was too eager an explosive; it would not wait for the cannon. By the early 1860's, the abortive guncotton boom was over, figuratively as well as literally.

Later, however, methods were discovered for removing the small quantities of impurities that encouraged guncotton to explode. It then became reasonably safe to handle. The English chemist Dewar (of liquefied gas fame) and a co-worker, Frederick Augustus Abel, introduced the technique, in 1889, of mixing it with nitroglycerine, and adding Vaseline to the mixture to make it moldable into cords (the mixture was called "cordite"). That, finally, was a useful smokeless powder. The Spanish-American War of 1898 was the last of any consequence fought with ordinary gunpowder.

(The machine age added its bit to the horrors of gunnery, also. In the 1860's, the American inventor Richard Gatling produced the first "machine gun" for the rapid firing of bullets, and this was improved by another American inventor, Hiram Stevens Maxim, in the 1880's. The "Gatling gun" gave rise to the slang term "gat" for gun. It and its descendant the "Maxim gun" gave the unabashed imperialists of the late nineteenth century an unprecedented advantage over the "lesser breeds," to use Rudyard Kipling's offensive phrase, of Africa and Asia. "Whatever happens, we have got / The Maxim gun and they have not!" went a popular jingle.)

"Progress" of this sort continued in the twentieth century. The most important explosive in World War I was "trinitrotoluene" familiarly abbreviated as TNT. In World War II, an even more powerful explosive, "cyclonite," came into use. Both contained the nitro group (NO_2) rather than the nitrate group (ONO_2). As lords of war, however, all chemical explosives gave way to nuclear bombs in 1945 (see Chapter 9).

Nitroglycerine, by the way, was discovered in the same year as was guncotton. An Italian chemist named Ascanio Sobrero treated glycerol with a mixture of nitric acid and sulfuric acid and knew he had something when he nearly killed himself in the explosion that followed. Sobrero, lacking Schönbein's promotional impulses, felt nitroglycerine to be too dangerous a substance to deal with and virtually suppressed information about it. But within ten years a Swedish family, the Nobels, took to manufacturing it as a "blasting oil" for use in mining and construction work. After a series of accidents, including one which took the life of a member of the family, Alfred Bernhard Nobel, the brother of the victim, discovered a method of mixing nitroglycerine with an absorbent earth called "kieselguhr" or "diatomaceous earth" (kieselguhr consists largely of the tiny skeletons of one-celled organisms called diatoms). The mixture consisted of three parts of nitroglycerine to one of kieselguhr, but such was the absorptive power of the latter that the mixture was virtually a dry powder. A stick of this impregnated earth (dynamite) could be dropped, hammered, even burned, without explosion. When set off by a percussion cap (electrically, and from a distance), it displayed all the shattering force of pure nitroglycerine.

Percussion caps contain sensitive explosives that detonate by heat or by mechanical shock and are therefore called "detonators." The strong shock of the detonation sets off the less sensitive dynamite. It might seem as though the danger were merely shifted from nitroglycerine to detonators, but it is not so bad as it sounds, since the detonator is only needed in tiny quantities. The detonators most used are mercury fulminate ($HgC_2N_2O_2$) and lead azide (PbN_6).

Sticks of dynamite eventually made it possible to carve the American West into railroads, mines, highways, and dams at a rate unprecedented in history. Dynamite, and other explosives he discovered, made a millionaire of the lonely and unpopular Nobel (who found himself, against his humanitarian will, regarded as a "merchant of death"). When he died in 1896, he left behind a fund out of which the famous Nobel Prizes, amounting to over 40,000 dollars each, were to be granted each year in five fields: chemistry, physics, medicine and physiology, literature, and peace. The first prizes were awarded on December 10, 1901, the fifth anniversary of his death, and these have now become the greatest honor any scientist can receive. (It is a pity that Nobel did not think to set up the category "astronomy and earth sciences" so that such men as Shapley, Hubble, and others might have been properly rewarded for their work.)

Considering the nature of human society, explosives continued to take up a sizable fraction of the endeavor of great scientists. Since almost all explosives contain nitrogen, the chemistry of that element and its compounds was of key importance. (It is also, it must be admitted, of key importance to life as well.)

The German chemist Wilhelm Ostwald, who was interested in chemical theory rather than in explosives, studied the rates at which chemical reactions proceeded. He applied the mathematical principles associated with physics to chemistry, thus being one of the founders of "physical chemistry." Toward the turn of the century, he worked out new methods for converting ammonia (NH_3) to nitrogen oxides, which could then be used to manufacture explosives. For his theoretical work, particularly on catalysis, Ostwald received the Nobel Prize for chemistry in 1909.

The ultimate source of usable nitrogen was, in the early decades of the twentieth century, the nitrate deposits in the desert of northern Chile. During World War I, these fields were placed out of reach of Germany by the British Navy. However, the German chemist Fritz Haber had devised a method by which the molecular nitrogen of the air could be combined with hydrogen under pressure, to form the ammonia needed for the Ostwald process. This "Haber process" was improved by the German chemist Karl Bosch, who supervised the building of plants during World War I for the manufacture of ammonia. Haber received the Nobel Prize for chemistry in 1918 and Bosch shared one in 1931. By the late 1960's, the United States alone was manufacturing 12 million tons of ammonia per year by the Haber process.

But let us return to modified cellulose. Clearly, it was the addition of the nitrate group that made for explosiveness. In guncotton all of the available hydroxyl groups were nitrated. What if only some of them were nitrated? Would they not be less explosive? Actually, such partly nitrated cellulose proved not to be explosive at all. However, it did burn very readily; the material was eventually named "pyroxylin" (from Greek words meaning "firewood").

Pyroxylin could be dissolved in mixtures of alcohol and ether. (This was discovered independently by the French scholar Louis Nicolas Ménard and an American medical student named J. Parkers Maynard— and an odd similarity in names that is.) When the the alcohol and ether evaporated, the pyroxylin was left behind as a tough, transparent film, which was named "collodion." Its first use was as a coating over minor cuts and abrasions; it was called "new skin." However, the adventures of pyroxylin were only beginning. Much more lay ahead.

Pyroxylin itself is brittle in bulk. But the English chemist Alexander Parkes found that if it was dissolved in alcohol and ether and mixed with a substance such as camphor, the evaporation of the solvent left behind a hard solid that became soft and malleable when heated. It could then be modeled into some desired shape which it would retain when cooled and hardened. So nitrocellulose was transformed into the first artificial "plastic," and the year in which this was done was 1865. Camphor, which introduced the plastic properties into an otherwise brittle substance, was the first "plasticizer."

What brought plastics to the attention of the public and made it more than a chemical curiosity was its dramatic introduction into the billiard parlor. Billiard balls were then made from ivory, a commodity which could be obtained only over an elephant's dead body—a point that naturally produced problems. In the early 1860's, a prize of 10,000 dollars was offered for the best substitute for ivory that would fulfill the billiard ball's manifold requirements of hardness, elasticity, resistance to heat and moisture, lack of grain, and so on. The American inventor John Wesley Hyatt was one of those who went out for the prize. He made no progress until he heard of Parkes' trick of plasticizing pyroxylin to a moldable material that would set as a hard solid. Hyatt set about working out improved methods of manufacturing the material, using less of the expensive alcohol and ether and more in the way of heat and pressure. By 1869, Hyatt was turning out cheap billiard balls of this material, which he called "celluloid." It won him the prize.

Celluloid turned out to have significance away from the pool table. It was versatile indeed. It could be molded at the temperature of boiling water; it could be cut, drilled, and sawed at lower temperatures; it was strong and hard in bulk but could also be produced in the form of thin flexible films that served for shirt collars, baby rattles, and so on. In the form of still thinner and more flexible films it could be used as a base for silver compounds in gelatin, and thus it became the first practical photographic film.

The one fault of celluloid was that, thanks to its nitrate groups, it had a tendency to burn with appalling quickness, particularly when in the form of thin film. It was the cause of a number of fire tragedies.

The substitution of acetate groups (CH_3COO^-) for nitrate groups led to the formation of another kind of modified cellulose called "cellulose acetate." Properly plasticized, this has properties as good or almost as good as those of celluloid, plus the saving grace of being much less apt to burn. Cellulose acetate came into use just before World War I, and after the war it completely replaced celluloid in the manufacture of photographic film and many other items.

Within half a century after the development of celluloid, chemists emancipated themselves from dependence on cellulose as the base for plastics. As early as 1872, Baeyer (who was later to synthesize indigo) had noticed that when phenols and aldehydes were heated together, a gooey, resinous mass resulted. Since he was interested only in the small molecules he could isolate from the reaction, he ignored this mess at the bottom of the flask (as nineteenth-century organic chemists typically tended to do when goo fouled up their glassware). Thirty-seven years later, the Belgian-born American chemist Leo Hendrik Baekeland, experimenting with formaldehyde, found that under certain conditions the reaction would yield a resin that on continued heating under pressure became first a soft solid, then a hard, insoluble substance. This resin could be molded while soft and then be allowed to set into a hard, permanent shape. Or, once hard, it could be powdered, poured into a mold and set into one piece by heat and pressure. Very complex forms could be cast easily and quickly. Furthermore, the product was inert and impervious to most environmental vicissitudes.

Baekeland named his product "Bakelite," after his own name. Bakelite belongs to the class of "thermosetting plastics," which, once they set on cooling, cannot be softened again by heating (though, of course, they can be destroyed by intense heat). Materials such as the cellulose derivatives, which can be softened again and again, are called "thermoplastics." Bakelite has numerous uses—as an insulator, an adhesive, a laminating agent, and so on. Although the oldest of the thermosetting plastics, it is still the most used.

Bakelite was the first production, in the laboratory, of a useful high polymer from small molecules. For the first time the chemist had taken over this particular task completely. It does not, of course, represent synthesis in the sense of the synthesis of heme or quinine, where chemists must place every last atom into just the proper position, almost one at a time. Instead, the production of high polymers requires merely that the small units of which they are composed be mixed under the proper conditions. A reaction is then set up in which the units form a chain automatically, without the specific point-to-point intervention of the chemist. The chemist can, however, alter the nature of the chain indirectly by varying the starting materials or the proportions among them, or by the addition of small quantities of acids, alkalies, or various substances that act as "catalysts" and tend to guide the precise nature of the reaction.

With the success of Bakelite, chemists naturally turned to other possible starting materials in search of more synthetic high polymers that might be useful plastics. And, as time went on, they succeeded many times over.

British chemists discovered in the 1930's, for instance, that the gas ethylene ($CH_2 = CH_2$), under heat and pressure, would form very long chains. One of the two bonds in the double bond between the carbon atoms opens up and attaches itself to a neighboring molecule. With this happening over and over again, the result is a long-chain molecule called "polythene" in England and "polyethylene" in the United States.

The paraffin-wax molecule is a long chain made up of the same units, but the molecule of polyethylene is even longer. Polyethylene is therefore like wax, but more so. It has the cloudy whiteness of wax, the slippery feel, the electrical insulating properties, the waterproofness, and the lightness (it is about the only plastic that will float on water). It is, however, at its best, much tougher than paraffin and much more flexible.

As it was first manufactured, polyethylene required dangerous pressures, and the product had a rather low melting point—just above the boiling point of water. It softened to uselessness at temperatures below the melting point. Apparently this was due to the fact that the carbon chain had branches which prevented the molecules from forming close-packed, crystalline arrays. In 1953, a German chemist named Karl Ziegler found a way to produce unbranched polyethylene chains, and without the need for high pressures. The result was a new variety of polyethylene, tougher and stronger than the old, and capable of withstanding boiling-water temperatures without softening too much. Ziegler accomplished this by using a new type of catalyst—a resin with ions of metals such as aluminum or titanium attached to negatively charged groups along the chain.

On hearing of Ziegler's development of metal-organic catalysts for polymer formation, the Italian chemist Giulio Natta began applying the technique to propylene (ethylene to which a small one-carbon methyl group, CH_3-, was attached). Within ten weeks, he had found that in the resultant polymer all the methyl groups face in the same direction, rather than (as was usual in polymer formation before that time) facing, in random fashion, in either direction. Such "isotactic polymers" (the name was proposed by Mrs. Natta) proved to have useful properties, and these can now be manufactured virtually at will. Chemists can design polymers, in other words, with greater precision than ever before. For their work in this field, Ziegler and Natta shared the 1963 Nobel Prize for chemistry.

The atomic-bomb project contributed another useful high polymer in the form of an odd relative of polyethylene. In the separation of uranium 235 from natural uranium, the nuclear physicists had to combine the uranium with fluorine in the gaseous compound uranium hexafluoride. Fluorine is the most active of all substances and will attack almost anything. Looking for lubricants and seals for their vessels that would be impervious to attack by fluorine, the physicists resorted to "fluorocarbons"—substances in which the carbon was already combined with fluorine (replacing hydrogen).

Until then, fluorocarbons had been only laboratory curiosities. The first (and simplest) of this type of molecule, "carbon tetrafluoride" (CF_4), had only been obtained in pure form in 1926. The chemistry of these interesting substances was now pursued intensively. Among the fluorocarbons studied was "tetrafluoroethylene" ($CF_2 = CF_2$), which had first been synthesized in 1933 and is, as you see, ethylene with its four hydrogens replaced by four fluorines. It was bound to occur to someone that tetrafluoroethylene might polymerize as ethylene itself did. After the war, Du Pont chemists produced a long-chain polymer which was as monotonously $CF_2CF_2CF_2$ . . . as polyethylene was $CH_2CH_2CH_2$. . . . Its trade name is Teflon, the "tefl" being an abbreviation of "tetrafluoro-."

Teflon is like polyethylene, only more so. The carbon-fluorine bonds are stronger than the carbon–hydrogen bonds and offer even less opportunity for the interference of the environment. Teflon is insoluble in everything, unwettable by anything, an extremely good electrical insulator, and considerably more resistant to heat than is even the new and improved polyethylene. Teflon's best-known application, so far as the housewife is concerned, is as a coating upon frying pans, thus enabling food to be fried without fat, since fat will not stick to the standoffish fluorocarbon polymer.

An interesting compound that is not quite a fluorocarbon is Freon (CF_2Cl_2), introduced in 1932 as refrigerant. It is more expensive than the ammonia or sulfur dioxide used in large-scale freezers, but, on the other hand, Freon is nonodorous, nontoxic, and nonflammable, so that accidental leakage introduces a minimum of danger. It is through Freon that room air conditioners have become so characteristic a part of the American scene since World War II.

Plastic properties do not, of course, belong solely to the organic world. One of the most ancient of all plastic substances is glass. The large molecules of glass are essentially chains of silicon and oxygen atoms; that is, –Si–O–Si–O–Si–O–Si–, and so on indefinitely. Each silicon atom in the chain has two unoccupied bonds to which other groups can be added. The silicon atom, like the carbon atom, has four valence bonds. The silicon–silicon bond, however, is weaker than the carbon–carbon bond, so that only short silicon chains can be formed, and those (in compounds called "silanes") are unstable. The silicon–oxygen bond is a strong one, however, and such chains are even more stable than those of carbon. In fact, since the earth's crust is half oxygen and a quarter silicon, the solid ground we stand upon may be viewed as essentially a silicon–oxygen chain.

Although the beauties and usefulness of glass (a kind of sand, made transparent) are infinite, it possesses the great disadvantage of being breakable. And in the process of breaking, it produces hard, sharp pieces which can be dangerous, even deadly. With untreated glass in the windshield of a car, a crash may convert the auto into a shrapnel bomb.

Glass can be prepared, however, as a double sheet between which is placed a thin layer of a transparent polymer, which hardens and acts as an adhesive. This is "safety glass," for when it is shattered, even into powder, each piece is held firmly in place by the polymer. None goes flying out on death-dealing missions. Originally, as far back as 1905, collodion was used as the binder, but nowadays that has been replaced for the most part by polymers built of small molecules such as vinyl chloride. (Vinyl chloride is like ethylene, except that one of the hydrogen atoms is replaced by a chlorine atom.) The "vinyl resin" is not discolored by light, so safety glass can be trusted not to develop a yellowish cast with time.

Then there are the transparent plastics that can completely replace glass, at least in some applications. In the middle 1930's, Du Pont polymerized a small molecule called methyl methacrylate and cast the polymer that resulted (a "polyacrylic plastic") into clear, transparent sheets. The trade names of these products are Plexiglas and Lucite. Such "organic glass" is lighter than ordinary glass, more easily molded, less brittle, and simply snaps instead of shattering when it does break. During World War II, molded transparent plastic sheets came into important use as windows and transparent domes in airplanes, where lightness and nonbrittleness are particularly useful. To be sure, the polyacrylic plastics have their disadvantages. They are affected by organic solvents, are more easily softened by heat than glass is, and are easily scratched. Polyacrylic plastics used in the windshields of cars, for instance, would quickly scratch under the impact of dust particles and become dangerously hazy. Consequently, glass is not likely ever to be replaced entirely. In fact, it is actually developing new versatility. Glass fibers have been spun into textile material that has all the flexibility of organic fibers and the inestimable further advantage of being absolutely fireproof.

In addition to glass substitutes, there is also what might be called a glass compromise. As I said, each silicon atom in a silicon–oxygen chain has two spare bonds for attachment to other atoms. In glass those other atoms are oxygen atoms, but they need not be. What if carbon-containing groups are attached instead of oxygen? You will then have an inorganic chain with organic offshoots, so to speak—a compromise between an organic and an inorganic material. As long ago as 1908, the English chemist Frederic Stanley Kipping formed such compounds, and they have come to be known as "silicones."

During World War II, long-chain "silicone resins" came into prominence. Such silicones are essentially more resistant to heat than purely organic polymers. By varying the length of the chain and the nature of the side chains, a list of desirable properties not possessed by glass itself can be obtained. For instance, some silicones are liquid at room temperature and change very little in viscosity over large ranges of temperature. (That is, they do not thin out with heat or thicken with cold.) This is a particularly useful property for a hydraulic fluid—the type of fluid used to lower landing gear on airplanes, for instance. Other silicones form soft, puttylike sealers that do not harden or crack at the low temperatures of the stratosphere and are remarkably water-repellent. Still other silicones serve as acid-resistant lubricants, and so on.

By the late 1960's, plastics of all sorts were being used at the rate of over 7 million tons a year, creating a serious problem as far as waste-disposal is concerned.

A possible polymer, utterly unexpected and of potentially fascinating theoretical implications, was announced in 1962. In that year the Soviet physicist Boris Vladimirovich Deryagin reported that water in very thin tubes seemed to have most peculiar properties. Chemists generally were skeptical, but eventually investigators in the United States confirmed Deryagin's findings. What seemed to happen was that under constricted conditions, water molecules lined up in orderly fashion, with the atoms approaching each other more closely than under ordinary conditions. The structure resembles a polymer composed of H_2O units, and the expression "polywater" came to be used for it.

Polywater was 1.4 times as dense as ordinary water, could be heated to 500° C. before being made to boil, and froze to a glassy ice only at —40° C. What gave it particular interest to biologists was the speculation that polywater might exist in the constricted confines of the cell interior and that its properties might be a key to some life processes.

However, reports soon began to filter out of chemistry laboratories that polywater might be ordinary water that had dissolved sodium silicate out of glass, or might be contaminated with perspiration. In short, polywater may only be impure water. The weight of the evidence seems to be shifting in the direction of the negative, so that polywater, after a brief and exciting life, may be dismissed—but the controversy, at the time of writing, is not yet quite over.

Fibers

In the story of organic synthesis, a particularly interesting chapter is that of the synthetic fibers. The first artificial fibers (like the first bulk plastics) were made from cellulose as the starting material. Naturally, the chemists began with cellulose nitrate, since it was available in reasonable quantity. In 1884, Hilaire Bernigaud de Chardonnet, a French chemist, dissolved cellulose nitrate in a mixture of alcohol and ether and forced the resulting thick solution through small holes. As the solution sprayed out, the alcohol and ether evaporated, leaving behind the cellulose nitrate as a thin thread of collodion. (This is essentially the manner in which spiders and silkworms spin their threads. They eject a liquid through tiny orifices and this becomes a solid fiber on exposure to air.) The cellulose-nitrate fibres were too flammable for use, but the nitrate groups could be removed by appropriate chemical treatment, and the result was a glossy cellulose thread that resembled silk.

De Chardonnet's process was expensive, of course, what with nitrate groups being first put on and then taken off, to say nothing of the dangerous interlude while they were in place and of the fact that the alcohol-ether mixture used as solvent was also dangerously flammable. In 1892 methods were discovered for dissolving cellulose itself. The English chemist Charles Frederick Cross, for instance, dissolved it in carbon disulfide and formed a thread from the resulting viscous solution (named "viscose"). The trouble was that carbon disulfide is flammable, toxic, and evil-smelling. In 1903, a competing process employing acetic acid as part of the solvent, and forming a substance called cellulose acetate, came into use.

These artificial fibers were first called "artificial silk," but were later named "rayon" because their glossiness reflected rays of light. The two chief varieties of rayon are usually distinguished as "viscose rayon" and "acetate rayon."

Viscose, by the way, can be squirted through a slit to form a thin, flexible, waterproof, transparent sheet—"cellophane"—a process invented in 1908 by a French chemist, Jacques Edwin Brandenberger. Some synthetic polymers also can be extruded through a slit for the same purpose. Vinyl resins, for instance, yielded the covering material known as Saran.

It was in the 1930's that the first completely synthetic fiber was born.

Let me begin by saying a little about silk. Silk is an animal product made by certain caterpillars which are exacting in their requirements for food and care. The fiber must be tediously unraveled from their cocoons. For these reasons, silk is expensive and cannot be turned out on a mass-production basis. It was first produced in China more than 2,000 years ago, and the secret of its preparation was jealously guarded by the Chinese, so that it could be kept a lucrative monopoly for export. However, secrets cannot be kept forever, despite all security measures. The secret spread to Korea, Japan, and India. Ancient Rome received silk by the long overland route across Asia, with middlemen levying tolls every step of the way; thus the fiber was beyond the reach of anyone except the most wealthy. In 550 A.D. silkworm eggs were smuggled into Constantinople, and silk production in Europe got its start. Nevertheless, silk has always remained more or less a luxury item. Moreover, until recently there was no good substitute for it. Rayon could imitate its glossiness but not its sheerness or strength.

After World War I, when silk stockings became an indispensable item of the feminine wardrobe, the pressure for greater supplies of silk or of some adequate substitute became very strong. This was particularly true in the United States, where silk was used in greatest quantity and where relations with the chief supplier, Japan, were steadily deteriorating. Chemists dreamed of somehow making a fiber that could compare with it.

Silk is a protein, Its molecule is built up of monomers called "amino acids," which in turn contain "amino" ($- NH_2$) and "carboxyl" ($- COOH$) groups. The two groups are joined by a carbon atom between them; labeling the amino group a and the carboxyl group c, and symbolizing the intervening carbon by a hyphen, we can write an amino acid like this: a - c. These amino acids polymerize in head-to-tail fashion; that is, the amino group of one condenses with the carboxyl group of the next. Thus the structure of the silk molecule runs like this: . . . a - c . a - c . a - c . a - c . . . .

In the 1930's, a Du Pont chemist named Wallace Hume Carothers was investigating molecules containing amine groups and carboxyl groups in the hope of discovering a good method of making them condense in such a way as to form molecules with large rings. (Such molecules are of importance in perfumery.) Instead, he found them condensing to form long-chain molecules.

Carothers had already suspected that long chains might be possible and he was not caught napping. He lost little time in following up this development. He eventually formed fibers from adipic acid and hexamethylenediamine. The adipic acid molecule contains two carboxyl groups separated by four carbon atoms, so it can be symbolized as:

c - - - - c. Hexamethylenediamene consists of two amine groups separated by six carbon atoms, thus: a - - - - - - a. When Carothers mixed the two substances together, they condensed to form a polymer like this:

. . . a - - - - - - a . c - - - - c . a - - - - - - a . c - - - - c . a - - - - - - a . . . .

The points at which condensation took place had the c . a configuration found in silk, you will notice.

At first the fibers produced were not much good. They were too weak. Carothers decided that the trouble lay in the presence of the water produced in the condensation process. The water set up a counteracting hydrolysis reaction which prevented polymerization from going very far. Carothers found a cure for this: he arranged to carry on the polymerization under low pressure, so that the water vaporized (and was easily removed by letting it condense on a cooled glass surface held close to the reacting liquid and so slanted as to carry the water away (a "molecular still"). Now the polymerization could continue indefinitely. It formed nice long, straight chains, and in 1935 Carothers finally had the basis for a dream fiber.

The polymer formed from adipic acid and hexamethylenediamine was melted and extruded through holes. It was then stretched so that the fibers would lie side by side in crystalline bundles. The result was a glossy, silklike thread that could be used to weave a fabric as sheer and beautiful as silk, and even stronger. This first of the completely synthetic fibers was named "nylon." Carothers did not live to see his discovery come to fruition, however. He died in 1937.

Du Pont announced the existence of the synthetic fiber in 1938 and began producing it commercially in 1939. During World War II, the United States Armed Forces took all the production of nylon for parachutes and for a hundred other purposes. But after the war nylon completely replaced silk for hosiery; indeed, women's stockings are now called "nylons."

Nylon opened the way to the production of many other synthetic fibers. Acrylonitrile, or vinyl cyanide ($CH_2 = CHCN$), can be made to polymerize into a long chain like that of polyethylene but with cyanide groups (completely nonpoisonous in this case) attached to every other carbon. The result, introduced in 1950, is "Orlon." If vinyl chloride ($CH_2 = CHCl$) is added, so that the eventual chain contains chlorine atoms as well as cyanide groups, "Dynel" results. Or the addition of acetate groups, through the use of vinyl acetate ($CH_2 = CHOOCCH_3$), produces "Acrilan."

The British in 1941 made a "polyester" fiber, in which the carboxyl group of one monomer condenses with the hydroxyl group of another. The result is the usual long chain of carbon atoms, broken in this case by the periodic insertion of an oxygen in the chain. The British call it "Terylene," but in the United States it has appeared under the name of "Dacron."

524

These new synthetic fibers are more water-repellent than most of the natural fibers; thus they resist dampness and are not easily stained. They are not subject to destruction by moths or beetles. Some are crease-resistant and can be used to prepare "wash-and-wear" fabrics.

Rubbers

It is a bit startling to realize that man has been riding on rubber wheels for only about a hundred years. For thousands of years he had ridden on wooden or metal rims. When Goodyear's discovery made vulcanized rubber available, it occurred to a number of people that rubber rather than metal might be wrapped around wheels. In 1845, a British engineer, Robert William Thomson, went this idea one better: he patented a device consisting of an inflated rubber tube that would fit over a wheel. By 1890, "tires" were routinely used for bicycles, and, in 1895, they were placed on horseless carriages.

Amazingly enough, rubber, though a soft, relatively weak substance, proved to be much more resistant to abrasion than wood or metal. This durability, coupled with its shock-absorbing qualities and the air-cushioning idea, introduced man to unprecedented riding comfort.

As the automobile increased in importance, the demand for rubber for tires grew astronomical. In half a century, the world production of rubber increased 42-fold. You can judge the quantity of rubber in use for tires today when I tell you that, in the United States, they leave no less than 200,000 tons of abraded rubber on the highways each year, in spite of the relatively small amount abraded from the tires of an individual car.

The increasing demand for rubber introduced a certain insecurity in the war resources of many nations. As war was mechanized, armies and supplies began to move on rubber, and rubber could be obtained in significant quantity only from the Malayan peninsula, far removed from the "civilized" nations most apt to engage in "civilized" warfare. (The Malayan peninsula is not the natural habitat of the rubber tree. The tree was transplanted there, with great success, from Brazil, where the original rubber supply steadily diminished.) The supply of the United States was cut off at the beginning of its entry into World War II when the Japanese overran Malaya. American apprehensions in this respect were responsible for the fact that the very first object rationed during the war emergency, even before the attack on Pearl Harbor, was rubber tires.

Even in World War I, when mechanization was just beginning, Germany was hampered by being cut off from rubber supplies by Allied sea power.

By the time of World War I, then, there was reason to consider the possibility of constructing a synthetic rubber. The natural starting material for such a synthetic rubber was isoprene, the building block of natural rubber. As far back as 1880, chemists had noted that isoprene, on standing, tended to become gummy, and if acidified, would set into a rubberlike material. Kaiser Wilhelm II eventually had the tires of his official automobile made of such material, as a kind of advertisement of Germany's chemical virtuosity.

However, there were two catches to the use of isoprene as the starting material for synthesizing rubber. First, the only major source of isoprene was rubber itself. Second, when isoprene polymerizes, it is most likely to do so in a completely random manner. The rubber chain possesses all the isoprene units oriented in the same fashion: - - - uuuuuuuuu - - -. The gutta percha chain has them oriented in strict alternation: - - - - unununununun - - - -. When isoprene is polymerized in the laboratory under ordinary conditions, however, the u's and n's are mixed randomly, forming a material which is neither rubber nor gutta percha. Lacking the flexibility and resilience of rubber, it is useless for automobile tires (except possibly for imperial automobiles used on state occasions).

Eventually, catalysts like those that Ziegler introduced in 1953 for manufacturing polyethylene made it possible to polymerize isoprene to a product almost identical with natural rubber, but by that time many useful synthetic rubbers, very different chemically from natural rubber, had been developed.

The first efforts, naturally, concentrated on attempts to form polymers from readily available compounds resembling isoprene. For instance, during World War I, under the pinch of the rubber famine, Germany made use of dimethylbutadiene:

$$CH_2 = C - C = CH_2$$
$$| \quad |$$
$$CH_3 \quad CH_3$$

Dimethylbutadiene differs from isoprene (see page 484) only in containing a methyl group (CH_3) on both middle carbons of the four-carbon chain instead of on only one of them. The polymer built of dimethylbutadiene, called "methyl rubber," could be formed cheaply and in quantity. Germany produced about 2,500 tons of it during World War I. While it did not stand up well under stress, it was nonetheless the first of the usable synthetic rubbers.

About 1930, both Germany and the Soviet Union tried a new tack. They used as the monomer, butadiene, which has no methyl group at all:

$$CH_2 = CH - CH = CH_2$$

With sodium metal as a catalyst, they formed a polymer called "Buna" (from "*bu*tadiene" and N*a* for sodium).

Buna rubber was a synthetic rubber which could be considered satisfactory in a pinch. It was improved by the addition of other monomers, alternating with butadiene at intervals in the chain. The most successful addition was "styrene," a compound resembling ethylene but with a benzene ring attached to one of the carbon atoms. This product was called Buna S. Its properties were very similar to those of natural rubber, and, in fact, thanks to it, Germany's armed forces suffered no serious rubber shortage in World War II. The Soviet Union also supplied itself with rubber in the same way. The raw materials could be obtained from coal or petroleum.

The United States was later in developing synthetic rubber in commercial quantities, perhaps because it was in no danger of a rubber famine before 1941. But after Pearl Harbor it took up synthetic rubber with a vengeance. It began to produce buna rubber and another type of synthetic rubber called "neoprene," built up of "chloroprene":

$$CH_2 = C - CH = CH_2$$
$$|$$
$$Cl$$

This molecule, as you see, resembles isoprene except for the substitution of a chlorine atom for the methyl group.

The chlorine atoms, attached at intervals to the polymer chain, confer upon neoprene certain resistances that natural rubber does not have. For instance, it is more resistant to organic solvents such as gasoline: it does not soften and swell nearly as much as would natural rubber. Thus neoprene is actually preferable to rubber for such uses as gasoline hoses. Neoprene first clearly demonstrated that in the field of synthetic rubbers, as in many other fields, the product of the test tube need not be a mere substitute for nature, but could be an improvement.

Amorphous polymers with no chemical resemblance to natural rubber but with rubbery qualities have now been produced, and they offer a whole constellation of desirable properties. Since they are not actually rubbers, they are called "elastomers" (an abbreviation of "elastic polymer").

The first rubber-unlike elastomer had been discovered in 1918. This was a "polysulfide rubber"; its molecule was a chain composed of pairs of carbon atoms alternating with groups of four sulfur atoms. The substance was given the name "Thiokol," the prefix coming from the Greek word for sulfur. The odor involved in its preparation held it in abeyance for a long time, but eventually it was put into commercial production.

Elastomers have also been formed from acrylic monomers, fluorocarbons, and silicones. Here, as in almost every field he touches, the organic chemist works as an artist, using materials to create new forms and improve upon nature.

CHAPTER 11

The Proteins

Key Molecules of Life

Early in their study of living matter, chemists noticed that there was a large group of substances that behaved in a peculiar manner. Heating changed these substances from the liquid to the solid state, instead of the other way round. The white of eggs, a substance in milk (casein), and a component of the blood (globulin) were among the things that showed this property. In 1777, the French chemist Pierre Joseph Macquer put all the substances that coagulated on heating into a special class that he called "albuminous," after "albumen," the name the Roman encyclopedist Pliny had given to egg white.

When the nineteenth-century organic chemists undertook to analyze the albuminous substances, they found these compounds considerably more complicated than other organic molecules. In 1839, the Dutch chemist Gerardus Johannes Mulder worked out a basic formula, $C_{40}H_{62}O_{12}N_{10}$, which he thought the albuminous substances had in common. He believed that the various albuminous compounds were formed by the addition of small sulfur-containing groups or phosphorus-containing groups to this central formula. Mulder named his root formula "protein" (a word suggested to him by the inveterate word-coiner Berzelius), from a Greek word meaning "of first importance." Presumably the term was merely meant to signify that this core formula was of first importance in determining the structure of the albuminous substances, but as things turned out, it proved to be a very apt word for the substances themselves. The "proteins," as they came to be known, were soon found to be of key importance to life.

Within a decade after Mulder's work, the great German organic chemist Justus von Liebig had established that proteins were even more essential for life than carbohydrates or fats; they supplied not only carbon, hydrogen, and oxygen, but also nitrogen, sulfur, and often phosphorus, which were absent from fats and carbohydrates.

The attempts of Mulder and others to work out complete empirical formulas for proteins were doomed to failure at the time they were made. The protein molecule was far too complicated to be analyzed by the methods available. However, a start had already been made on another line of attack that was eventually to reveal, not only the composition, but also the structure of proteins. Chemists had begun to learn something about the building blocks of which they were made.

In 1820, Henri Braconnot, having succeeded in breaking down cellulose into its glucose units by heating the cellulose in acid (see Chapter 10), decided to try the same treatment with gelatin, an albuminous substance. The treatment yielded a sweet, crystalline substance. Despite Braconnot's first suspicions, this turned out to be not a sugar, but a nitrogen-containing compound, for ammonia (NH_3) could be obtained from it. Nitrogen-containing substances are conventionally given names ending in "-ine," and the compound isolated by Braconnot is now called "glycine," from the Greek word for "sweet."

Shortly afterward Braconnot obtained a white, crystalline substance by heating muscle tissue with acid. He named this one "leucine," from the Greek word for "white."

Eventually, when the structural formulas of glycine and leucine were worked out, they were found to have a basic resemblance:

$$
NH_2 - CH_2 - C \overset{\displaystyle O}{\underset{\displaystyle OH}{\diagup}}
$$

glycine

$$
\begin{array}{c}
CH_3 \diagdown \quad \diagup CH_3 \\
CH \\
| \\
CH_2 \\
| \\
NH_2 - CH - C \overset{\displaystyle O}{\underset{\displaystyle OH}{\diagup}}
\end{array}
$$

leucine

Each compound, as you see, has at its ends an amine group (NH_2) and a carboxyl group (COOH). Because the carboxyl group gives acid properties to any molecule that contains it, molecules of this kind were named "amino acids." Those that have the amine group and carboxyl group linked together by a single carbon atom between them, as both these molecules have, are called "alpha-amino acids."

As time went on, chemists isolated other amino acids from proteins. For instance, Liebig obtained one from the protein of milk (casein), which he called "tyrosine" (from the Greek word for "cheese"; casein itself comes from the Latin word for "cheese").

The differences among the various alpha-amino acids lie entirely in the nature of the atom grouping attached to that single carbon atom between the amine and the carboxyl groups. Glycine, the simplest of all the amino acids, has only a pair of hydrogen atoms attached there. The others all possess a carbon-containing "side chain" attached to that carbon atom.

$$
\begin{array}{c}
\text{OH} \\
| \\
\text{C} \\
\diagup \quad \diagdown \\
\text{CH} \qquad \text{CH} \\
| \qquad \quad \| \\
\text{CH} \qquad \text{CH} \\
\diagdown \quad \diagup \\
\text{C} \\
| \\
\text{CH}_2 \\
| \\
\text{NH}_2 - \text{CH} - \text{C} \underset{\text{OH}}{\overset{\text{O}}{\diagup}}
\end{array}
$$

I shall give the formula of just one more amino acid, which will be useful in connection with matters to be discussed later in the chapter. It is "cystine," discovered in 1899 by the German chemist K. A. H. Mörner. This is a double-headed molecule containing two atoms of sulfur:

$$
\begin{array}{c}
\text{NH}_2 - \text{CH} - \text{C} \underset{\text{OH}}{\overset{\text{O}}{\diagup}} \\
| \\
\text{CH}_2 \\
| \\
\text{S} \\
| \\
\text{S} \\
| \\
\text{CH}_2 \\
| \\
\text{NH}_2 - \text{CH} - \text{C} \underset{\text{OH}}{\overset{\text{O}}{\diagup}}
\end{array}
$$

Actually, cystine had first been isolated in 1810 by the English chemist William Hyde Wollaston from a bladder stone, and it had been named cystine from the Greek word for "bladder" in consequence. What Mörner did was to show that this century-old compound was a component of protein as well as the substance in bladder stones.

Cystine is easily "reduced" (a term that, chemically, is the opposite of "oxidized"). This means that it will easily add on two hydrogen atoms, which fall into place at the S–S bond. The molecule then divides into two halves, each containing an –SH ("mercaptan" or "thiol") group. This reduced half is "cysteine," and it is easily oxidized back to cystine.

The general fragility of the thiol group is such that it is important to the functioning of a number of protein molecules. (A delicate balance and a capability of moving this way or that under slight impulse is the hallmark of the chemicals most important to life; the members of the thiol group are among the atomic combinations that contribute to this ability.) The thiol group is particularly sensitive to damage by radiation and the administration of cysteine either immediately before or immediately after exposure to radiation protects somewhat against radiation sickness. The injected cysteine undergoes the chemical changes to which, otherwise, important cellular components might be exposed. This is a very slight ray of hope in connection with an extremely dark cloud of worry.

Altogether, nineteen important amino acids (that is, occurring in most proteins) have now been identified. The last of these was discovered in 1935 by the American chemist William Cumming Rose. It is unlikely that any other common ones remain to be found.

By the end of the nineteenth century, biochemists had become certain that proteins were giant molecules built up of amino acids, just as cellulose was constructed of glucose and rubber of isoprene units. But there was this important difference: whereas cellulose and rubber were made with just one kind of building block, a protein was built from a number of different amino acids. That meant that working out its structure would pose special and subtle problems.

The first problem was to find out just how the amino acids were joined together in the protein chain molecule. Emil Fischer made a start on the problem by linking amino acids together in chains, in such a way that the carboxyl group of one amino acid was always joined to the amino group of the next. In 1901, he achieved his first such condensation, linking one glycine molecule to another with the elimination of a molecule of water:

$$NH_2 - CH_2 - \overset{\displaystyle O}{\overset{\|}{C}} - OH + NH_2 - CH_2 - \overset{\displaystyle O}{\overset{\|}{C}} \diagdown OH$$

$$\downarrow$$

$$NH_2 - CH_2 - \overset{\displaystyle O}{\overset{\|}{C}} - NH - CH_2 - \overset{\displaystyle O}{\overset{\|}{C}} \diagdown OH + H_2O$$

This is the simplest condensation possible. By 1907, Fischer had synthesized a chain made up of 18 amino acids, 15 of them glycine and the remaining 3 leucine. This molecule did not show any of the obvious properties of proteins, but Fischer felt that was only because the chain was not long enough. He called his synthetic chains "peptides," from a

Greek word meaning "digest," because he believed that proteins broke down into such groups when they were digested. Fischer named the combination of the carboxyl's carbon with the amine group a "peptide link."

In 1932, the German biochemist Max Bergmann (a pupil of Fischer's) devised a method of building up peptides from various amino acids. Using Bergmann's method, the Polish-American biochemist Joseph Stewart Fruton prepared peptides that could be broken down into smaller fragments by digestive juices. Since there was good reason to believe that digestive juices would hydrolyze (split by the addition of water) only one kind of molecular bond, this meant that the bond between the amino acids in the synthetic peptides must be of the same kind as the one joining amino acids in true proteins. The demonstration laid to rest any lingering doubts as to the validity of Fischer's peptide theory of protein structure.

Still, the synthetic peptides of the early decades of the twentieth century were very small and nothing like proteins in their properties. Fischer had made one consisting of eighteen amino acids, as I have said; in 1916, the Swiss chemist Emil Abderhalden went him one better by preparing a peptide with nineteen amino acids, but that held the record for thirty years. And chemists knew that such a peptide must be a tiny fragment indeed compared with the size of a protein molecule, because the molecule weights of proteins were enormous.

Consider, for instance, hemoglobin, a protein of the blood. Hemoglobin contains iron, making up just 0.34 per cent of the weight of the molecule. Chemical evidence indicates that the hemoglobin molecule has four atoms of iron, so the total molecular weight must be about 67,000; four atoms of iron, with a total weight of 4×55.85, would come to 0.34 per cent of such a molecular weight. Consequently, hemoglobin must contain about 550 amino acids (the average molecular weight of the amino acids being about 120). Compare that with Abderhalden's puny nineteen. And hemoglobin is only an average-sized protein.

The best measurement of the molecular weights of proteins has been obtained by whirling them in a centrifuge, a spinning device that pushes particles outward from the center by centrifugal force. When the centrifugal force is more intense than the earth's gravitational force, particles suspended in a liquid will settle outward away from the center at a faster rate than they would settle downward under gravity. For instance, red blood corpuscles will settle out quickly in such a centrifuge, and fresh milk will separate into two fractions, the fatty cream and the denser skim milk. These particular separations will take place slowly under ordinary gravitational forces, but centrifugation speeds them up.

Protein molecules, though very large for molecules, are not heavy enough to settle out of solution under gravity; nor will they settle out rapidly in an ordinary centrifuge. But in the 1920's, the Swedish chemist

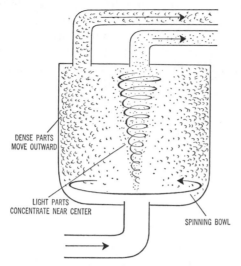

Principle of the centrifuge.

Theodor Svedberg developed an "ultracentrifuge" capable of separating molecules according to their weight. This high-speed device whirls at more than 10,000 revolutions per second and produces centrifugal forces up to 900,000 times as intense as the gravitational force at the earth's surface. For his contributions to the study of suspensions, Svedberg received the Nobel Prize in chemistry in 1926.

With the ultracentrifuge, chemists were able to determine the molecular weights of a number of proteins on the basis of their rate of sedimentation (measured in "svedbergs" in honor of the chemist). Small proteins turned out to have molecular weights of only a few thousand and to contain perhaps not more than fifty amino acids (still decidedly more than nineteen). Other proteins have molecular weights in the hundreds of thousands and even in the millions, which means that they must consist of thousands or tens of thousands of amino acids. The possession of such large molecules put proteins into a class of substances that have only been studied systematically from the mid-nineteenth century onward.

The Scottish chemist Thomas Graham was the pioneer in this field through his interest in "diffusion," that is, in the manner in which the molecules of two substances, brought into contact, will intermingle. He began by studying the rate of diffusion of gases through tiny holes or fine tubes. By 1831, he was able to show that the rate of diffusion of a gas was inversely proportional to the square root of its molecular weight ("Graham's law"). (It was through the operation of Graham's law that uranium 235 was separated from uranium 238, by the way.)

In following decades, Graham passed to the study of the diffusion of dissolved substances. He found that solutions of such compounds as salt, sugar, or copper sulfate would find their way through a blocking sheet of parchment (presumably containing submicroscopic holes). On the other hand, solutions of such materials as gum arabic, glue, and gelatin would not. Clearly, the giant molecules of the latter group of substances would not fit through the holes in the parchment.

Graham called materials that could pass through parchment (and that happened to be easily obtained in crystalline form) "crystalloids." Those that did not, such as glue (in Greek, "kolla"), he called "colloids." The study of giant molecules (or giant aggregates of atoms, even where these did not form distinct molecules) thus came to be known as "colloid chemistry." Because proteins and other key molecules in living tissue are of giant size, colloid chemistry is of particular importance to "biochemistry" (the study of the chemical reactions proceeding in living tissue).

Advantage can be taken of the giant size of protein molecules in a number of ways. Suppose that pure water is on one side of a sheet of parchment and a colloidal solution of protein on the other. The protein molecules cannot pass through the parchment; moreover, they block the passage of some of the water molecules, which might otherwise move through. For this reason, water moves more readily into the colloidal portion of the system than out of it. Fluid builds up on the side of the protein solution and sets up an "osmotic pressure."

In 1877, the German botanist Wilhelm Pfeffer showed how one could measure this osmotic pressure and from that determine the molecular weight of a giant molecule. It was the first reasonably good method for estimating the size of such molecules.

Again, protein solutions could be placed in bags made of "semipermeable membranes" (membranes with pores large enough to permit the passage of small, but not of large molecules). If these were placed in running water, small molecules and ions would pass through the membrane and be washed away, while the large protein molecule would remain behind. This process of "dialysis" is the simplest method of purifying protein solutions.

Molecules of colloidal size are large enough to scatter light; small molecules cannot. Furthermore, light of short wavelength is more efficiently scattered than that of long wavelength. The first to note this effect, in 1869, was the Irish physicist John Tyndall; in consequence, it is called the "Tyndall effect." The blue of the sky is explained now by the scattering effect of dust particles in the atmosphere upon the short-wave sunlight. At sunset, when light passes through a greater thickness of atmosphere rendered particularly dusty by the activity of the day, enough light is scattered to leave only the red and orange, thus accounting for the beautiful ruddy color of sunsets.

Light passing through a colloidal solution is scattered so that it can be seen as a visible cone of illumination when viewed from the side.

Solutions of crystalloidal substances do not show such a visible cone of light when illuminated and are "optically clear." In 1902, the Austro-German chemist Richard Adolf Zsigmondy took advantage of this observation to devise an "ultramicroscope," which viewed a colloidal solution at right angles, with individual particles (too small to be seen in an ordinary microscope) showing up as bright dots of light. For his endeavor, he received the Nobel Prize for chemistry in 1925.

The protein chemists naturally were eager to synthesize long, "polypeptide" chains, with the hope of producing proteins. But the methods of Fischer and Bergmann allowed only one amino acid to be added at a time—a procedure that seemed at the time to be completely impractical. What was needed was a procedure that would cause amino acids to join up in a kind of chain reaction, such as Baekeland had used in forming his high-polymer plastics. In 1947, both the Israeli chemist E. Katchalski and the Harvard chemist Robert Woodward (who had synthesized quinine) reported success in producing polypeptides through chain-reaction polymerization. Their starting material was a slightly modified amino acid. (The modification eliminated itself neatly during the reaction.) From this beginning, they built up synthetic polypeptides consisting of as many as a hundred or even a thousand amino acids.

These chains are usually composed of only one kind of amino acid, such as glycine or tyrosine, and are therefore called "polyglycine" or "polytyrosine." It is also possible, by beginning with a mixture of two modified amino acids, to form a polypeptide containing two different amino acids in the chain. But these synthetic constructions resemble only the very simplest kind of protein: for example, "fibroin," the protein in silk.

Some proteins are as fibrous and crystalline as cellulose or nylon. Examples are fibroin; keratin, the protein in hair and skin; and collagen, the protein in tendons and in connective tissue. The German physicist R. O. Herzog proved the crystallinity of these substances by showing that they diffracted X-rays. Another German physicist, R. Brill, analyzed the pattern of the diffraction and determined the spacing of the atoms in the polypeptide chain. The British biochemist William Thomas Astbury and others in the 1930's obtained further information about the structure of the chain by means of X-ray diffraction. They were able to calculate with reasonable precision the distances between adjacent atoms and the angles at which adjacent bonds were set. And they learned that the chain of fibroin was fully extended: that is, the atoms were in as nearly a straight line as the angles of the bonds between them would permit.

This full extension of the polypeptide chain is the simplest possible arrangement. It is called the "beta configuration." When hair is stretched, its keratin molecule, like that of fibroin, takes up this configuration. (If hair is moistened, it can be stretched up to three times its original length.) But in its ordinary, unstretched state, keratin shows a more complicated arrangement, called the "alpha configuration."

In 1951, Linus Pauling and Robert Brainard Corey of Cal Tech suggested that, in the alpha configuration, polypeptide chains took a helical shape (a shape like that of a spiral staircase). After building various models to see how the structure would arrange itself if all the bonds between atoms lay in their natural directions without strain, they decided that each turn of the helix would have the length of 3.6 amino acids, or 5.4 angstrom units.

What enables a helix to hold its structure? Pauling suggested that the agent is the so-called "hydrogen bond." As we have seen, when a hydrogen atom is attached to an oxygen or a nitrogen atom, the latter holds the major share of the bonding electrons, so that the hydrogen atom has a slight positive charge and the oxygen or nitrogen a slight negative charge. In the helix, it appears, a hydrogen atom periodically occurs close to an oxygen or nitrogen atom on the turn of the helix immediately above or below it. The slightly positive hydrogen atom is attracted to its slightly negative neighbor. This attraction has only one-twentieth of the force of an ordinary chemical bond, but it is strong enough to hold the helix in place. However, a pull on the fiber easily uncoils the helix and thereby stretches the fiber.

We have considered so far only the "backbone" of the protein molecule—the chain that runs . . . CCNCCNCCNCCN . . . . But the various side chains of the amino acids also play an important part in protein structure.

All the amino acids except glycine have at least one asymmetric carbon atom—the one between the carboxyl group and the amine group. Thus each could exist in two optically active isomers. The general formulas of the two isomers are:

$$
\begin{array}{cc}
\overset{\displaystyle O}{\underset{\displaystyle \|}{}} & \overset{\displaystyle O}{\underset{\displaystyle \|}{}} \\
C - OH & C - OH \\
| & | \\
H - C - NH_2 & NH_2 - C - H \\
| & | \\
\text{side chain} & \text{side chain}
\end{array}
$$

However, it seems quite certain from both chemical and X-ray analysis that polypeptide chains are made up only of L-amino acids. In this situation, the side chains stick out alternately on one side of the backbone and then the other. A chain composed of a mixture of both isomers would not be stable, because, whenever an L-amino and a D-amino acid were next to each other, there would be two side chains sticking out on the same side, which would crowd them and strain the bonds.

The side chains are important factors in holding neighboring peptide chains together. Wherever a negatively charged side chain on one chain is near a positively charged side chain on its neighbor, they will form an electrostatic link. The side chains also provide hydrogen bonds that can serve as links. And the double-headed amino acid cystine (see page 489) can insert one of its amine-carboxyl sequences in one chain and the other in the next. The two chains are then tied together by the two sulfur atoms in the side chain (the "disulfide link"). The binding together of polypeptide chains accounts for the strength of protein fibers. It explains the remarkable toughness of the apparently fragile spider web and the fact that keratin can form structures as hard as fingernails, tiger claws, alligator scales, and rhinoceros horns.

All this nicely describes the structure of protein fibers. What about proteins in solution? What sort of structure do they have?

They certainly possess a definite structure, but it is extremely delicate, for gentle heating or stirring of the solution or the addition of a bit of acid or alkali or any of a number of other environmental stresses will "denature" a dissolved protein. That is, the protein loses its ability to perform its natural functions, and many of its properties change. Furthermore, denaturation usually is irreversible: for instance, a hard-boiled egg can never be un-hard-boiled again.

It seems certain that denaturation involves the loss of some specific configuration of the polypeptide backbone. Just what feature of the structure is destroyed? X-ray diffraction will not help us when proteins are in solution, but other techniques are available.

In 1928, for instance, the Indian physicist Chandrasekhara Venkata Raman found that light scattered by molecules in solution was, to some extent, altered in wavelength. From the nature of the alteration, deductions could be made as to the structure of the molecule. For this discovery of the "Raman effect," Raman received the 1930 Nobel Prize for physics. (The altered wavelengths of light are usually referred to as the "Raman spectrum" of the molecule doing the scattering.)

Another delicate technique was developed twenty years later, one that was based on the fact that atomic nuclei possess magnetic properties. Molecules exposed to a high intensity magnetic field will absorb certain frequencies of radio waves. From such absorption, referred to as "nuclear magnetic resonance" and frequently abbreviated as NMR, information concerning the bonds between atoms can be deduced. In particular, NMR techniques can locate the position of the small hydrogen atoms within molecules, something X-ray diffraction cannot do. NMR techniques were worked out in 1946 by two teams, working independently, one under E. M. Purcell (later to be the first to detect the radio waves emitted by the neutral hydrogen atom in space; see Chapter 2) and the other under the Swiss-American physicist Felix Bloch. Purcell and Bloch shared the Nobel Prize for physics in 1952 for this feat.

To return, then, to the question of the denaturation of proteins in solution. The American chemists Paul Mead Doty and Elkan Rogers Blout used light-scattering techniques on solutions of synthetic polypeptides and found that they had a helical structure. By changing the acidity of the solution, they could break down the helices into randomly curved coils; by readjusting the acidity, they could restore the helices. And they showed that the conversion of the helices to random coils reduced the amount of the solution's optical activity. It was even possible to show which way a protein helix is twisted: it runs in the direction of a right-handed screw thread.

All this suggests that the denaturation of a protein involves the destruction of its helical structure.

Amino Acids in the Chain

What I have described so far represents an over-all look at the structure of the protein molecule—the general shape of the chain. What about the details of its construction? For instance, how many amino acids of each kind are there in a given protein molecule?

We might break down a protein molecule into its amino acids (by heating it with acid) and then determine how much of each amino acid is present in the mixture. Unfortunately, some of the amino acids resemble each other chemically so closely that it is almost impossible to get clear-cut separations by ordinary chemical methods. The amino acids can, however, be separated neatly by chromatography (see Chapter 5). In 1941, the British biochemists Archer John Porter Martin and Richard Laurence Millington Synge pioneered the application of chromatography to this purpose. They introduced the use of starch as the packing material in the column. In 1948, the American biochemists Stanford Moore and William Howard Stein brought the starch chromatography of amino acids to a high pitch of efficiency.

After the mixture of amino acids has been poured into the starch column and all the amino acids have attached themselves to the starch particles, they are slowly washed down the column with fresh solvent. Each amino acid moves down the column at its own characteristic rate. As each emerges at the bottom separately, the drops of solution of that amino acid are caught in a container. The solution in each container is then treated with a chemical that turns the amino acid into a colored product. The intensity of the color is a measure of the amount of the particular amino acid present. This color intensity is measured by an instrument called a "spectrophotometer," which indicates the intensity by means of the amount of light of that particular wavelength that is absorbed.

(Spectrophotometers can be used for other kinds of chemical

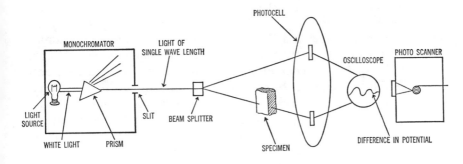

A spectrophotometer. The beam of light is split into two
so that one beam passes through the specimen being analyzed and the other goes directly to the photocell. Since the
weakened beam that has passed through the specimen liberates fewer electrons in the photocell than the unabsorbed
beam does, the two beams create a difference in potential
that measures the amount of absorption of the light by
the specimen.

analysis, by the way. If light of successively increased wavelength is sent
through a solution, the amount of absorption changes smoothly, rising
to maxima at some wavelengths and falling to minima at others. The
result is an "absorption spectrum." A given atomic group has its own
characteristic absorption peak or peaks. This is especially true in the region
of the infrared, as was first shown by the American physicist William
Weber Coblentz shortly after 1900. His instruments were too crude to
make the technique practical then, but since World War II, the "infrared
spectrophotometer," designed to scan, automatically, the spectrum from
two to forty microns and to record the results has come into increasing
use for analysis of the structure of complex compounds. "Optical
methods" of chemical analysis, involving radio-wave absorption, light
absorption, light scattering, and so on, are extremely delicate and non-
destructive—the sample survives the inspection, in other words—and
are completely replacing the classical analytical methods of Liebig,
Dumas, and Pregl that were mentioned in the previous chapter.)

The measurement of amino acids with starch chromatography is
quite satisfactory, but by the time this procedure was developed, Martin
and Synge had worked out a simpler method of chromatography. It is
called "paper chromatography." The amino acids are separated on a sheet
of filter paper (an absorbent paper made of particularly pure cellulose).
A drop or two of a mixture of amino acids is deposited near a corner of
the sheet, and this edge of the sheet is then dipped into a solvent, such
as butyl alcohol. The solvent slowly creeps up the paper through capillary
action. (Dip the corner of a blotter into water and see it happen your-

539

self.) The solvent picks up the molecules in the deposited drop and sweeps them along the paper. As in column chromatography, each amino acid moves up the paper at a characteristic rate. After a while the amino acids in the mixture become separated in a series of spots on the sheet. Some of the spots may contain two or three amino acids. To separate these, the filter paper, after being dried, is turned around ninety degrees from its first position, and the new edge is now dipped into a second solvent which will deposit the components in separate spots. Finally, the whole sheet, after once again being dried, is washed with chemicals that cause the patches of amino acids to show up as colored or darkened spots. It is a dramatic thing to see: all the amino acids, originally mixed in a single solution, are now spread out over the length and breadth of the paper in a mosaic of colorful spots. Experienced biochemists can identify each amino acid by the spot it occupies, and thus they can read the composition of the original protein almost at a glance. By dissolving a spot they can even measure how much of that particular amino acid was present in the protein.

(Martin, along with A. T. James, applied the principles of this technique to the separation of gases in 1952. Mixtures of gases or vapors may be passed through a liquid solvent or over an adsorbing solid by means of a current of inert "carrier gas," such as nitrogen or helium. The mixture is pushed through and emerges at the other end separated. Such "gas chromatography" is particularly useful because of the speed

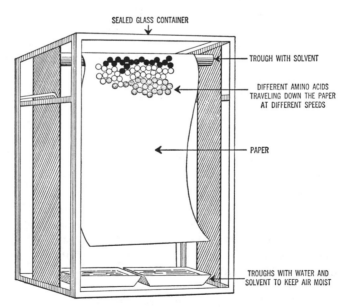

Paper chromatography.

of its separations and the great delicacy with which it can detect trace impurities.)

Chromatographic analysis yielded accurate estimates of the amino-acid contents of various proteins. For instance, the molecule of a blood protein called "serum albumin" was found to contain 15 glycines, 45 valines, 58 leucines, 9 isoleucines, 31 prolines, 33 phenylalanines, 18 tyrosines, 1 tryptophan, 22 serines, 27 threonines, 16 cystines, 4 cysteines, 6 methionines, 25 arginines, 16 histidines, 58 lysines, 46 aspartic acids, and 80 glutamic acids—a total of 526 amino acids of 18 different types built into a protein with a molecular weight of about 69,000. (In addition to these eighteen, there is one other common amino acid—alanine.)

The German-American biochemist Erwin Brand suggested a system of symbols for the the amino acids which is now in general use. To avoid confusion with the symbols of the elements, he designated each amino acid by the first three letters of its name, instead of just the initial. There are a few special variations: cystine is symbolized CyS, to show that its two halves are usually incorporated in two different chains; cysteine is CySH, to distinguish it from cystine; and isoleucine is Ileu rather than Iso, for "iso" is the prefix of many chemical names.

In this shorthand, the formula of serum albumin can be written: $Gly_{15} Val_{45} Leu_{58} Ileu_9 Pro_{31} Phe_{33} Tyr_{18} Try_1 Ser_{22} Thr_{27} CyS_{32} CySH_4 Met_6 Arg_{25} His_{16} Lys_{58} Asp_{46} Glu_{80}$. This is, you will admit, more concise, though certainly nothing to be rattled off.

Discovering the empirical formula of a protein was only half the battle—in fact, much less than half. Now came the far more difficult task of deciphering the structure of a protein molecule. There was every reason to believe that the properties of every protein depended on exactly how—in what order—all those amino acids are arranged in the molecular chain. This presents the biochemist with a staggering problem. The number of possible arrangements in which nineteen amino acids can be placed in a chain (even assuming that only one of each is used) comes to nearly 120 million billion. If you find this hard to believe, try multiplying out 19 times 18 times 17 times 16, and so on, which is the way the number of possible arrangements is calculated. And if you do not trust the arithmetic, get nineteen checkers, number them 1 to 19, and see in how many different orders you can arrange them. I guarantee you will not continue the game long.

When you have a protein of the size of serum albumin, composed of more than 500 amino acids, the number of possible arrangements comes out to something like 10^{600}—that is, 1 followed by 600 zeros. This is a completely fantastic number—far more than the number of sub-atomic particles in the entire known universe, or, for that matter, far more than the universe could hold if it were packed solid with such particles.

Nevertheless, although the task of finding out which one of all those possible arrangements a serum albumin molecule actually possesses may seem hopeless, this sort of problem has actually been tackled and solved.

In 1945, the British biochemist Frederick Sanger set out to determine the order of amino acids in a peptide chain. He started by trying to identify the amino acid at one end of the chain—the amine end.

Obviously, the amine group of this end amino acid (called the "N-terminal amino acid") is free: that is, not attached to another amino acid. Sanger made use of a chemical that combines with a free amine group but not with an amine group that is bound to a carboxyl group. This produces a DNP (dinitrophenyl) derivative of the peptide chain. With DNP he could label the N-terminal amino acid, and since the bond holding this combination together is stronger than those linking the amino acids in the chain, he could break up the chain into its individual amino acids and isolate the one with the DNP label. As it happens, the DNP group has a yellow color, so this particular amino acid, with its DNP label, shows up as a yellow spot on a paper chromatogram.

Thus, Sanger was able to separate and identify the amino acid at the amine end of a peptide chain. In a similar way, he identified the amino acid at the other end of the chain—the one with a free carboxyl group, called the "C-terminal amino acid." He was also able to peel off a few other amino acids one by one and identify the "end sequence" of a peptide chain in several cases.

Now Sanger proceeded to attack the peptide chain all along its length. He worked with insulin, a protein that has the merit of being very important to the functioning of the body and which has the added virtue of being rather small for a protein, having a molecular weight of only 6,000 in its simplest form. DNP treatment showed this molecule to consist of two peptide chains, for it contained two different N-terminal amino acids. The two chains were joined together by cystine molecules. By a chemical treatment that broke the bond between the two sulfur atoms in the cystine, Sanger split the insulin molecule into its two peptide chains, each intact. One of the chains had glycine as the N-terminal amino acid (call it the G-chain), and the other had phenylalanine as the N-terminal amino acid (the P-chain.) The two could now be worked on separately.

Sanger and a co-worker, Hans Tuppy, first broke up the chains into individual amino acids and identified the twenty-one amino acids that made up the G-chain and the thirty that composed the P-chain. Next, to learn some of the sequences, they broke the chains, not into individual amino acids, but into fragments consisting of two or three. This could be done by partial hydrolysis, breaking only the weaker bonds in the chain, or by attacking the insulin with certain digestive substances which broke only certain links between amino acids and left the others intact.

By these devices Sanger and Tuppy broke each of the chains into a large number of different pieces. For instance, the P-chain yielded 48 different fragments, 22 of which were made up of two amino acids (dipeptides), 14 of three, and 12 of more than three.

The various small peptides, after being separated, could then be broken down into their individual amino acids by paper chromatography. Now the investigators were ready to determine the order of the amino acids in these fragments. Suppose they had a dipeptide consisting of valine and isoleucine. The question would be: Was the order Val-Ileu or Ileu-Val? In other words, was valine or isoleucine the N-terminal amino acid? (The amine group, and consequently the N-terminal unit, is conventionally considered to be at the left end of a chain.) Here the DNP label could provide the answer. If it was present on the valine, that would be the N-terminal amino acid, and the arrangement in the dipeptide would then be established to be Val-Ileu. If it was present on the isoleucine, it would be Ileu-Val.

The arrangement in a fragment consisting of three amino acids also could be worked out. Say its components were leucine, valine, and glutamic acid. The DNP test could first identify the N-terminal amino acid. If it was, say, leucine, the order had to be either Leu-Val-Glu or Leu-Glu-Val. Each of these combinations was then synthesized and deposited as a spot on a chromatogram to see which would occupy the same place on the paper as did the fragment being studied.

As for peptides of more than three amino acids, these could be broken down to smaller fragments for analysis.

After thus determining the structures of all the fragments into which the insulin molecule had been divided, the next step was to put the pieces together in the right order in the chain—in the fashion of a jigsaw puzzle. There were a number of clues to work with. For instance, the G-chain was known to contain only one unit of the amino acid alanine. In the mixture of peptides obtained from the breakdown of G-chains, alanine was found in two combinations: alanine-serine and cystine-alanine. This meant that in the intact G-chain the order must be CyS-Ala-Ser.

By means of such clues, Sanger and Tuppy gradually put the pieces together. It took a couple of years to identify all the fragments definitely and arrange them in a completely satisfactory sequence, but by 1952 they had worked out the exact arrangement of all the amino acids in the G-chain and the P-chain. They then went on to establish how the two chains were joined together. In 1953, their final triumph in deciphering the structure of insulin was announced. The complete structure of an important protein molecule had been worked out for the first time. For this achievement, Sanger was awarded the Nobel Prize in chemistry in 1958.

Biochemists immediately adopted Sanger's methods to determine the structure of other protein molecules. Ribonuclease, a protein molecule consisting of a single peptide chain with 124 amino acids, was conquered in 1959, and the protein unit of tobacco mosaic virus, with 158 amino acids, in 1960. In 1964, trypsin, a protein with 223 amino acids, was deciphered. By 1967, the technique was actually automated. The Swedish-Australian biochemist P. Edman devised a "sequenator" which could work on 5 milligrams of pure protein, peeling off and identifying the amino acids one by one. Sixty amino acids of the myoglobin chain were identified in this fashion in four days.

Once the amino-acid order in a polypeptide chain was worked out, it became possible to attempt to put together amino acids in just that right order. Naturally, the beginning was a small one. The first protein to be synthesized in the laboratory was "oxytocin," a hormone with important functions in the body. Oxytocin is extremely small for a protein molecule: it consists only of eight amino acids. In 1953, the American biochemist Vincent du Vigneaud succeeded in synthesizing a peptide chain exactly like that thought to represent the oxytocin molecule. And, indeed, the synthetic peptide showed all the properties of the natural hormone. Du Vigneaud was awarded the Nobel Prize in chemistry in 1955.

More complicated protein-molecules were synthesized as the years passed, but in order to synthesize a specific molecule with particular amino acids arranged in a particular order, the string had to be "threaded," so to speak, one at a time. That was as difficult in the 1950's as it had been a half-century earlier in Fischer's time. Each time a particular amino-acid was coupled to a chain, the new compound had to be separated from all the rest by tedious procedures, and then a new start had to be made to add one more particular amino acid. At each step a good part of the material was lost in side-reactions, and only small quantities of even simple chains were formed.

Beginning in 1959, however, a team under the leadership of the American biochemist Robert Bruce Merrifield, struck out in a new direction. An amino acid, the beginning of the desired chain, as bound to beads of polystyrene resin. These beads were insoluble in the liquid being used and could be separated from everything else by simple filtration. A new solution would be added containing the next amino-acid, which would bind to the first. Again a filtration, then another. The steps in between additions were so simple and quick that they could be automated with almost nothing lost. In 1965, the molecule of insulin was synthesized in this fashion; in 1969, it was the turn of the still longer chain of ribonuclease with all its 124 amino acids. Then, in 1970, the Chinese-American biochemist Cho Hao Li synthesized the 188-amino-acid chain of human growth hormone.

With the protein molecule understood, so to speak, as a string of amino acids, it became desirable to take a still more sophisticated view.

544

What was the exact manner in which that amino acid chain bent and curved? What was the exact shape of the protein molecule?

Tackling this problem were the Austrian-English chemist Max Ferdinand Perutz and his English colleague John Cowdery Kendrew. Perutz took hemoglobin, the oxygen-carrying protein of blood, containing something like 12,000 atoms, as his province. Kendrew took on myoglobin, a muscle protein similar in function to hemoglobin but only about a quarter the size. As their tool, they used X-ray diffraction studies.

Perutz used the device of combining the protein molecules with a massive atom, such as that of gold or mercury, which was particularly efficient in diffracting X-rays. This gave him clues from which he could more accurately deduce the structure of the molecule without the massive atom. By 1959, myoglobin, and then hemoglobin, the year after, fell into place. It became possible to prepare three-dimensional models in which every single atom could be located in what seemed very likely to be the correct place. In both cases, the protein structure was clearly based upon the helix. As a result, Perutz and Kendrew shared the Nobel Prize in chemistry in 1962.

There is reason to think that the three-dimensional structures worked out by the Perutz-Kendrew techniques are after all determined by the nature of the string of amino acids. The amino-acid string has, so to speak, natural "crease-points," and when they bend, certain interconnections inevitably take place and keep it properly folded. What these folds and interconnections are, can be determined if all the interatomic distances are worked out and the angles at which the connecting bonds are placed are determined. This can be done, but it is a tedious job indeed. Here, too, computers have been called in to help, and these have not only made the calculation but thrown the results on a screen.

What with one thing or another, the list of protein molecules whose shapes are known in three-dimensional detail, is growing rapidly. Insulin, which started the new forays into molecular biology, had its three-dimensional shape worked out by the English biochemist Dorothy Crowfoot Hodgkin in 1969.

Enzymes

There are useful consequences following from the complexity and almost infinite variety of protein molecules. Proteins have a multitude of different functions to perform in living organisms.

One major function is to provide the structural framework of the body. Just as cellulose serves as the framework of plants, so fibrous proteins act in the same capacity for the complex animals. Spiders spin gossamer threads and insect larvae spin cocoon threads of protein fibers. The scales of fish and reptiles are made up mainly of the protein keratin. Hair, feathers, horns, hoofs, claws, and finger nails—all merely modified scales —also contain keratin. Skin owes its strength and toughness to its high content of keratin. The internal supporting tissues—cartilage, ligaments, tendons, even the organic framework of bones—are made up largely of protein molecules, such as collagen and elastin. Muscle is made of a complex fibrous protein called actomyosin.

In all these cases, the protein fibers are more than a cellulose substitute. They are an improvement; they are stronger and more flexible. Cellulose will do to support a plant, which is not called on for any motion more complex than swaying with the wind. But protein fibers must be designed for the bending and flexing of the appendages of the body, for rapid motions and vibrations, and so on.

The fibers, however, are among the simplest of the proteins, in form as well as function. Most of the other proteins have more subtle and more complicated jobs to do.

To maintain life in all its aspects, numerous chemical reactions must proceed in the body. These must go on at high speed and in great variety, each reaction meshing with all the others, for it is not upon any one reaction, but upon all together, that life's smooth workings must depend. Moreover, all the reactions must proceed under the mildest of environments: without high temperatures, strong chemicals, or great pressures. The reactions must be under strict yet flexible control, and they must be constantly adjusted to the changing characteristics of the environment and the changing needs of the body. The undue slowing down, or speeding up, of even one reaction out of the many thousands would more or less seriously disorganize the body.

All this is accomplished by protein molecules.

Toward the end of the eighteenth century, chemists, following the leadership of Lavoisier, began to study reactions in a quantitative way— in particular, to measure the rates at which chemical reactions proceeded. They quickly noted that reaction rates could be changed drastically by comparatively minor changes in the environment. For instance, when Kirchhoff found that starch could be converted to sugar in the presence of acid, he noticed that while the acid greatly speeded up this reaction, it was not itself consumed in the process. Other examples of this were soon discovered. The German chemist Johann Wolfgang Döbereiner found that finely divided platinum (called "platinum black") encouraged the combination of hydrogen and oxygen to form water—a reaction which without this help could take place only at a high temperature. Döbereiner even designed a self-igniting lamp in which a jet of hydrogen, played upon a surface coated with platinum black, caught fire.

Because the "hastened reactions" were usually in the direction of breaking down a complex substance to a simpler one, Berzelius named the phenomenon "catalysis" (from Greek words essentially meaning "break down"). Thus, platinum black came to be called a catalyst for the combination of hydrogen and oxygen, and acid a catalyst for the hydrolysis of starch to glucose.

Catalysis has proved of the greatest importance in industry. For instance, the best way of making sulfuric acid (the most important single inorganic chemical next to air, water, and, perhaps, salt) involves the burning of sulfur, first to sulfur dioxide (SO_2), then to sulfur trioxide (SO_3). The step from the dioxide to the trioxide would not proceed at more than a snail's pace without the help of a catalyst such as platinum black. Finely divided nickel (which has replaced platinum black in most cases, because it is cheaper) and such compounds as copper chromite, vanadium pentoxide, ferric oxide, and manganese dioxide also are important catalysts. In fact, a great deal of the success of an industrial chemical process depends on finding just the right catalyst for the reaction involved. It was the discovery of a new type of catalyst by Ziegler that revolutionized the production of polymers.

How is it possible for a substance, sometimes present only in very small concentrations, to bring about large quantities of reaction without itself being changed?

Well, one kind of catalyst does in fact take part in the reaction, but in a cyclic fashion, so that it is continually restored to its original form. An example is vanadium pentoxide (V_2O_5), which can catalyze the change of sulfur dioxide to sulfur trioxide. Vanadium pentoxide passes on one of its oxygen to SO_2, forming SO_3 and changing itself to vanadyl oxide (V_2O_4). But the vanadyl oxide rapidly reacts with oxygen in the air and is restored to V_2O_5. The vanadium pentoxide thus acts as a middleman, handing an oxygen atom to sulfur dioxide, taking another from the air, handing that to sulfur dioxide, and so on. The process is so rapid that a small quantity of vanadium pentoxide will suffice to bring about the conversion of large quantities of sulfur dioxide, and in the end we will appear still to have the vanadium pentoxide unchanged.

In 1902, the German chemist George Lunge suggested that this sort of thing was the explanation of catalysis in general. In 1916, Irving Langmuir went a step farther and advanced an explanation for the catalytic action of substances, such as platinum, that are so nonreactive that they cannot be expected to engage in ordinary chemical reactions. Langmuir suggested that excess valence bonds at the surface of platinum metal would seize hydrogen and oxygen molecules. While held imprisoned in close proximity on the platinum surface, the hydrogen and oxygen molecules would be much more likely to combine to form water molecules than in their ordinary free condition as gaseous molecules. Once a water molecule was formed, it would be displaced from the platinum surface by hydrogen and oxygen molecules. Thus the process of seizure of hydrogen and oxygen, their combination into water, release of the water, seizure of more hydrogen and oxygen, and formation of more water could continue indefinitely.

This is called "surface catalysis." Naturally, the more finely divided the metal, the more surface a given mass will provide and the more effectively catalysis can proceed. Of course, if any extraneous substance attaches itself firmly to the surface bonds of the platinum, it will "poison" the catalyst.

All surface catalysts are more or less selective, or "specific." Some easily absorb hydrogen molecules and will catalyze reactions involving hydrogen; others easily absorb water molecules and catalyze condensa-tions or hydrolyses, and so on.

The ability of surfaces to add on layers of molecules ("adsorption") is widespread and can be put to uses other than catalysis. Silicon dioxide prepared in spongy form ("silica gel") will adsorb large quantities of water. Packed in with electronic equipment, the performance of which would suffer under conditions of high humidity, it acts as a "dessicant," keeping humidity low.

Again, finely divided charcoal ("activated carbon") will adsorb organic molecules readily; the larger the organic molecule, the more readily. Activated carbon can be used to decolorize solutions, for it would adsorb the colored impurities (usually of high molecular weight) leaving behind the desired substance (usually colorless and of com-paratively low molecular weight).

Activated carbon is also used in gas masks, a use foreshadowed by an English physician, John Stenhouse, who first prepared a charcoal air filter in 1853. The oxygen and nitrogen of air pass through such a mass unaffected, but the relatively large molecules of poison gases are ad-sorbed.

The organic world, too, has its catalysts. Indeed, some of them have been known for thousands of years, though not by that name. They are as old as bread-making and wine-making.

Bread dough, left to itself and kept from contamination by outside influences, will not rise. Add a lump of "leaven" (from a Latin word meaning "rise"), and bubbles begin to appear, lifting and lightening the dough. The common English word for leaven is "yeast," possibly descended from a Sanskrit word meaning "to boil."

Yeast also hastens the conversion of fruit juices and grain to alcohol. Here again, the conversion involves the formation of bubbles, so the process is called "fermentation," from a Latin word meaning "boil." The yeast preparation is often referred to as "ferment."

It was not until the seventeenth century that the nature of leaven was discovered. In 1680, for the first time, van Leeuwenhoek saw yeast cells. For the purpose, he made use of an instrument that was to revolu-tionize biology—the "microscope." It was based on the bending and

548

focusing of light by lenses. Instruments using combinations of lenses ("compound microscopes") were devised as early as 1590 by a Dutch spectacle maker, Zacharias Janssen. The early microscopes were useful in principle, but the lenses were so imperfectly ground that the objects magnified were almost useless, fuzzy blobs. Van Leeuwenhoek ground tiny but perfect lenses that magnified quite sharply up to 200 times. He used single lenses ("simple microscope").

With time, the practice of using good lenses in combinations (for a compound microscope is, potentially at least, much stronger than a simple one) spread, and the world of the very little opened up further. A century and a half after Leeuwenhoek, a French physicist, Charles Cagniard de la Tour, using a good compound microscope, studied the tiny bits of yeast intently enough to catch them in the process of reproducing themselves. The little blobs were alive. Then, in the 1850's, yeast became a dramatic subject of study.

France's wine industry was in trouble. Aging wine was going sour and becoming undrinkable, and millions of francs were being lost. The problem was placed before the young dean of the Faculty of Sciences at the University of Lille, in the heart of the vineyard area. The young dean was Louis Pasteur, who had already made his mark by being the first to separate optical isomers in the laboratory.

Pasteur studied the yeast cells in the wine under the microscope. It was obvious to him that the cells were of varying types. All the wine contained yeast that brought about fermentation, but those wines that went sour contained another type of yeast in addition. It seemed to Pasteur that the souring action did not get under way until the fermentation was completed. Since there was no need for yeast after the necessary fermentation, why not get rid of all the yeast at that point and avoid letting the wrong kind make trouble?

He therefore suggested to a horrified wine industry that the wine be heated gently after fermentation, in order to kill all the yeast in it. Aging, he predicted, would then proceed without souring. The industry reluctantly tried his outrageous proposal, and found to its delight that souring ceased, while the flavor of the wine was not in the least damaged by the heating. The wine industry was saved. Furthermore, the process of gentle heating ("pasteurization") was later applied to milk, also, to kill any disease germs present.

Other organisms besides yeast hasten breakdown processes. In fact, a process analogous to fermentation takes place in the intestinal tract. The first man to study digestion scientifically was the French physicist René Antoine Ferchault de Réaumur. He used a hawk as his experimental subject, and in 1752 he made it swallow small metal tubes containing meat; the tubes protected the meat from any mechanical grinding action, but they had openings, covered by gratings, so that chemical processes in the stomach could act on the meat. Réaumur found that when the hawk regurgitated these tubes, the meat was partly dissolved, and a yellowish fluid was present in the tubes.

In 1777, the Scottish physician Edward Stevens isolated fluid from the stomach ("gastric juice") and showed that the dissolving process could be made to take place outside the body, thus divorcing it from the direct influence of life.

Clearly the stomach juices contained something that hastened the breakdown of meat. In 1834, the German naturalist Theodor Schwann added mercuric chloride to the stomach juice and precipitated a white powder. After freeing the powder of the mercury compound, and dissolving what was left, he found he had a very concentrated digestive juice. He called the powder he had discovered "pepsin," from the Greek word meaning "digest."

Meanwhile, two French chemists, Anselme Payen and Jean François Persoz, had found in malt extract a substance that could bring about the conversion of starch to sugar more rapidly than could acid. They called this "diastase," from a Greek word meaning "to separate," because they had separated it from malt.

For a long time, chemists made a sharp distinction between living ferments such as yeast cells and nonliving, or "unorganized," ferments such as pepsin. In 1878, the German physiologist Wilhelm Kühne suggested that the latter be called "enzymes," from Greek words meaning "in yeast," because their activity was similar to that brought about by the catalyzing substances in yeast. Kühne did not realize how important, indeed universal, that term "enzyme" was to become.

In 1897, the German chemist Eduard Buchner ground yeast cells with sand to break up all the cells and succeeded in extracting a juice that he found could perform the same fermentative tasks that the original yeast cells could. Suddenly the distinction between the ferments inside and outside of cells vanished. It was one more breakdown of the vitalists' semimystical separation of life from nonlife. The term "enzyme" was now applied to all ferments.

For this discovery Buchner received the Nobel Prize in chemistry in 1907.

Now it was possible to define an enzyme simply as an organic catalyst. Chemists began to try to isolate enzymes and find out what sort of substances they were. The trouble was that the amount of enzyme in cells and natural juices was very small, and the extracts obtained were invariably mixtures in which it was hard to tell what was an enzyme and what was not.

Many biochemists suspected that enzymes were proteins, because enzyme properties could easily be destroyed, as proteins could be denatured, by gentle heating. But, in the 1920's, the German biochemist Richard Willstätter reported that certain purified enzyme solutions, from which he believed he had eliminated all protein, showed marked catalytic effects. He concluded from this that enzymes were not proteins but were relatively simple chemicals, which might, indeed, utilize a protein as a "carrier molecule." Most biochemists went along with Willstätter, who was a Nobel Prize winner and had great prestige.

However, the Cornell University biochemist James Batcheller Sumner produced strong evidence against this theory almost as soon as it was advanced. From jackbeans (the white seeds of a tropical American plant), Sumner isolated crystals that, in solution, showed the properties of an enzyme called "urease." This enzyme catalyzed the breakdown of urea to carbon dioxide and ammonia. Sumner's crystals showed definite protein properties, and he could find no way to separate the protein from the enzyme activity. Anything that denatured the protein also destroyed the enzyme. All this seemed to show that what he had was an enzyme in pure and crystalline form and that enzyme was a protein.

Willstätter's greater fame for a time minimized Sumner's discovery. But, in 1930, the chemist John Howard Northrop and his co-workers at the Rockefeller Institute clinched Sumner's case. They crystallized a number of enzymes, including pepsin, and found all to be proteins. Northrop, furthermore, showed that these crystals were pure proteins and retained their catalytic activity even when dissolved and diluted to the point where the ordinary chemical tests, such as those used by Willstätter, could no longer detect the presence of protein.

Enzymes were thus established to be "protein catalysts." By now nearly a hundred enzymes have been crystallized, and all without exception are proteins.

For their work, Sumner and Northrop shared in the Nobel Prize in chemistry in 1946.

Enzymes are remarkable as catalysts in two respects—efficiency and specificity. There is an enzyme known as catalase, for instance, which catalyzes the breakdown of hydrogen peroxide to water and oxygen. Now the breakdown of hydrogen peroxide in solution can also be catalyzed by iron filings or manganese dioxide. However, weight for weight, catalase speeds up the rate of breakdown far more than any inorganic catalyst can. Each molecule of catalase can bring about the breakdown of 44,000 molecules of hydrogen peroxide per second at 0° C. The result is that an enzyme need be present only in small concentration to perform its function.

For this same reason, it takes but small quantities of substances ("poisons"), capable of interfering with the workings of a key enzyme, to put an end to life. Heavy metals, when administered in such forms as mercuric chloride or barium nitrate, react with thiol groups, which are essential to the working of many enzymes. The action of those enzymes stops, and the organism is poisoned. Compounds such as potassium cyanide or hydrogen cyanide place their cyanide group (–CN) in combination with the iron atom of other key enzymes and bring death quickly and, it is to be hoped, painlessly, for hydrogen cyanide is the gas used for execution in the gas chambers of some of our Western states.

Carbon monoxide is an exception among the common poisons. It does not act on enzymes primarily, but ties up the hemoglobin molecule (a protein but not an enzyme), which ordinarily carries oxygen from lungs to cells but cannot do so with carbon monoxide hanging on to it. Animals that do not use hemoglobin are not harmed by carbon monoxide.

Enzymes, with catalase a good example, are highly specific; catalase breaks down hydrogen peroxide and nothing else, whereas inorganic catalysts, such as iron filing and manganese dioxide, may break down hydrogen peroxide, but will also catalyze numerous other reactions.

What accounts for the remarkable specificity of enzymes? Lunge's and Langmuir's theories about the behavior of a catalyst as a middle-man suggested an answer. Suppose we consider that an enzyme forms a temporary combination with the "substrate"—the substance whose reaction it catalyzes. The form, or configuration, of the particular enzyme may therefore play a highly important role. Plainly, each enzyme must present a very complicated surface, for it has a number of different side chains sticking out of the peptide backbone. Some of these side chains have a negative charge, some positive, some no charge. Some are bulky, some small. One can imagine that each enzyme may have a surface that just fits a particular substrate. In other words, it fits the substrate as a key fits a lock. Therefore, it will combine readily with that substance, but only clumsily or not at all with others. This would explain the high specificity of enzymes; each has a surface made to order, so to speak, for combining with a particular compound. That being the case, no wonder that proteins are built of so many different units and are constructed by living tissue in such great variety.

This view of enzyme action was borne out by the discovery that the presence of a substance similar in structure to a given substrate would slow down or inhibit the substrate's enzyme-catalyzed reaction. The best-known case involves an enzyme called succinic acid dehydrogenase, which catalyzes the removal of two hydrogen atoms from succinic acid. That reaction will not proceed if a substance called malonic acid, which is very similar to succinic acid, is present. The structures of succinic acid and malonic acid are:

The only difference between these two molecules is that succinic acid has one more CH_2 group at the left. Presumably the malonic acid, because of its structural similarity to succinic acid, can attach itself to the surface of the enzyme. Once it has pre-empted the spot on the surface to which the succinic acid would attach itself, it remains jammed there, so to speak, and the enzyme is out of action. The malonic acid "poisons" the enzyme, so far as its normal function is concerned. This sort of thing is called "competitive inhibition."

The most positive evidence in favor of the enzyme-substrate-complex theory has come from spectrographic analysis. Presumably, if an enzyme combines with its substrate, there should be a change in the absorption spectrum: the combination's absorption of light should be different from that of the enzyme or the substrate alone. In 1936, the British biochemists David Keilin and Thaddeus Mann detected a change of color in a solution of the enzyme peroxidase after its substrate, hydrogen peroxide, was added. The American biophysicist Britton Chance made a spectral analysis and found that there were two progressive changes in the absorption pattern, one following the other. He attributed the first change in pattern to the formation of the enzyme-substrate complex at a certain rate and the second to the decline of this combination as the reaction was completed. In 1964 the Japanese biochemist Kunio Yagi announced the isolation of an enzyme-substrate complex, made up of a loose union of the enzyme D-amino acid oxidase and its substrate alanine.

Now the question arises: Is the entire enzyme molecule necessary for catalysis, or would some part of it be sufficient? This is an important question from a practical as well as a theoretical standpoint. Enzymes are in wide use today; they have been put to work in the manufacture of drugs, citric acid, and many other chemicals. If the entire enzyme molecule is not essential and some small fragment of it would do the job, perhaps this active portion could be synthesized, so that the processes would not have to depend on the use of living cells, such as yeasts, molds, and bacteria.

Some promising advances toward this goal have been made. For instance, Northrop found that when a few acetyl groups (CH_3CO) were added to the side chains of the amino acid tyrosine in the pepsin molecule, the enzyme lost some of its activity. There was no loss, however, when acetyl groups were added to the lysine side chains in pepsin. Tyrosine, therefore, must contribute to pepsin's activity while lysine obviously did not. This was the first indication that an enzyme might possess portions not essential to its activity.

Recently the "active region" of another digestive enzyme was pinpointed with more precision. This enzyme is chymotrypsin. The pancreas first secretes it in an inactive form called "chymotrypsinogen." This inactive molecule is converted into the active one by the splitting of a single peptide link (accomplished by the digestive enzyme trypsin).

That is to say, it looks as if the uncovering of a single amino acid endows chymotrypsin with its activity. Now it turns out that the attachment of a molecule known as DFP (diisopropylfluorophosphate) to chymotrypsin stops the enzyme's activity. Presumably, the DFP attaches itself to the key amino acid. Thanks to its tagging by DFP, that amino acid had been identified as serine. In fact, DFP has also been found to attach itself to serine in other digestive enzymes. In each case the serine is in the same position in a sequence of four amino acids: glycine-aspartic acid-serine-glycine.

It turns out that a peptide consisting of those four amino acids alone will not display catalytic activity. In some way, the rest of the enzyme molecule plays a role, too. We can think of the four-acid sequence—the active center—as analogous to the cutting edge of a knife, which is useless without a handle.

Nor need the active center, or cutting edge, necessarily exist all in one piece in the amino acid chain. Consider the enzyme ribonuclease. Now that the exact order of its 124 amino acids is known, it has become possible to devise methods for deliberately altering this or that amino acid in the chain and noting the effect of the change on the enzyme's action. It was discovered that three amino acids, in particular, were necessary for action, but that they were widely separated. They were a histidine in position 12, a lysine in position 41, and another histidine in position 119.

This separation, of course, existed only in the chain viewed as a long string. In the working molecule, the chain was coiled into a specific three-dimensional configuration, held in place by four cystine molecules, stretching across the loops. In such a molecule, the three necessary amino acids are brought together into a close-knit unit.

The matter of an active center was made even more specific in the case of lysozyme, an enzyme found in many places, including tears and nasal mucus. It brings about the dissolution of bacterial cells by catalyzing the breakdown of key bonds in some of the substances that make up the bacterial cell wall. It is as though it causes the wall to crack and the cell contents to leak away.

Lysozyme was the first enzyme whose structure was completely analyzed (in 1965) in three dimensions. Once this was done, it could be shown that the molecule of the bacterial cell wall that was subject to lysozyme's action fitted neatly along a cleft in the enzyme structure. The key bond was found to lie between an oxygen atom in the side-chain of glutamic acid (position #35) and another oxygen atom in the side-chain of aspartic acid (position #52). The two positions were brought together by the folding of the amino-acid chain with just enough separation that the molecule to be attacked could fit in between. The chemical reaction necessary for breaking the bond could easily take place under

554

those circumstances—and it is in this fashion that lysozyme is specifically organized to do its work.

Then, too, it happens sometimes that the cutting edge of the enzyme molecule is not a group of amino acids at all, but an atom combination of an entirely different nature. A few such cases will be mentioned later in the book.

We cannot tamper with the cutting edge, but could we modify the handle without impairing the usefulness of the tool? The existence of different varieties of a such protein as insulin, for instance, encourages us to believe that we might. Insulin is a hormone, not an enzyme, but its function is highly specific. At a certain position in the G-chain of insulin there is a three-amino-acid sequence which differs in different animals: in cattle it is alamine-serine-valine; in swine, threonine-serine-isoleucine; in sheep, alanine-glycine-valine; in horses, threonine-glycine-isoleucine; and so on. Yet any of these insulins can be substituted for any other and still perform the same function.

What is more, a protein molecule can sometimes be cut down drastically without any serious effect on its activity (as the handle of a knife or an ax might be shortened without much loss in effectiveness). A case in point is the hormone called ACTH (adrenocorticotropic hormone). This is a peptide chain made up of thirty-nine amino acids, the order of which has now been fully determined. Up to fifteen of the amino acids have been removed from the C-terminal end without destroying the hormone's activity. On the other hand, the removal of one or two amino acids from the N-terminal end (the cutting edge, so to speak) kills activity at once.

The same sort of thing has been done to an enzyme called "papain," from the fruit and sap of the papaya tree. Its enzymatic action is similar to that of pepsin. Removal of eighty of the pepsin molecule's 180 amino acids from the N-terminal end does not reduce its activity to any detectable extent.

So it is at least conceivable that enzymes may yet be simplified to the point where they will fall within the region of practical synthesis. Synthetic enzymes, in the form of fairly simple organic compounds, may then be made on a large scale for various purposes. This would be a form of "chemical miniaturization."

Metabolism

An organism, such as the human body, is a chemical plant of great diversity. It breathes in oxygen and drinks water. It takes in as food carbohydrates, fats, proteins, minerals, and other raw materials. It eliminates various indigestible materials plus bacteria and the products of the putrefaction they bring about. It also excretes carbon dioxide via the lungs, gives up water both by way of the lungs and the sweat glands, and excretes urine, which carries off a number of compounds in solution, the chief of these being urea. These chemical reactions determine the body's metabolism.

By examining the raw materials that enter the body and the waste products that leave it, we can tell a few things about what goes on within the body. For instance, since protein supplies most of the nitrogen entering the body, we know that urea (NH_2CONH_2) must be a product of the metabolism of proteins. But between protein and urea lies a long, devious, complicated road. Each enzyme of the body catalyzes only a specific small reaction, rearranging perhaps no more than two or three atoms. Every major conversion in the body involves a multitude of steps and many enzymes. Even an apparently simple organism such as the tiny bacterium must make use of many thousands of separate enzymes and reactions.

All this may seem needlessly complex, but it is the very essence of life. The vast complex of reactions in tissues can be controlled delicately by increasing or decreasing the production of appropriate enzymes. The enzymes control body chemistry as the intricate movements of fingers on the strings control the playing of a violin, and without this intricacy the body could not perform its manifold functions.

To trace the course of the myriads of reactions that make up the body's metabolism is to follow the outline of life. The attempt to follow it in detail, to make sense of the intermeshing of countless reactions all taking place at once, may indeed seem a formidable and even hopeless undertaking. Formidable it is, but not hopeless.

The chemists' study of metabolism began modestly with an effort to find out how yeast cells converted sugar to ethyl alcohol. In 1905, two British chemists, Arthur Harden and W. J. Young, suggested that this process involved the formation of sugars bearing phosphate groups. They were the first to note that phosphorus played an important role in metabolism (and phosphorus has been looming larger and larger ever since). Harden and Young even found in living tissue a sugar-phosphate ester consisting of the sugar fructose with two phosphate groups (PO_3H_2) attached. This "fructose diphosphate" (still sometimes known as "Harden-Young ester") was the first "metabolic intermediate" to be identified definitely, the first compound, that is, recognized to be formed momentarily, in the process of passing from the compounds as taken into the body to the compounds eliminated by it. Harden and Young had thus founded the study of "intermediary metabolism," which concentrates on the nature of such intermediates and the reactions involving them. For this work and for further work on the enzymes involved in the conversion of sugar to alcohol by yeast (see Chapter 14), Harden shared the Nobel Prize in chemistry in 1929.

What began by involving only the yeast cell became of far broader importance when the German chemist Otto Fritz Meyerhof demonstrated in 1918 that animal cells, such as those of muscle, broke down sugar in much the same way as yeast did. The chief difference was that in animal cells the breakdown did not proceed so far in this particular route of metabolism. Instead of converting the six-carbon glucose molecule all the way down to the two-carbon ethyl alcohol (CH_3CH_2-OH), they broke it down only as far as the three-carbon lactic acid ($CH_3CHOHCOOH$).

Meyerhof's work made clear for the first time a general principle that has since become commonly accepted: that, with only minor differences, metabolism follows the same routes in all creatures, from the simplest to the most complex. For his studies on the lactic acid in muscle, Meyerhof shared the Nobel Prize in physiology and medicine in 1922 with the English physiologist Archibald Vivian Hill. The latter had tackled muscle from the standpoint of its heat production and had come to conclusions quite similar to those obtained from Meyerhof's chemical attack.

The details of the individual steps involved in the transition from sugar to lactic acid were evolved between 1937 and 1941 by Carl Ferdinand Cori and his wife Gerty Theresa Cori, working at Washington University in St. Louis. They used tissue extracts and purified enzymes to bring about changes in various sugar-phosphate esters, then put all the changes together like a jigsaw puzzle. The scheme of step-by-step changes that they presented has stood with little modification to this day, and the Coris were awarded a share in the Nobel Prize in physiology and medicine in 1947.

In the path from sugar to lactic acid, a certain amount of energy is produced and is utilized by the cells. The yeast cell lives on it when it is fermenting sugar, and so, when necessary, does the muscle cell. It is important to remember that this energy is obtained without the use of oxygen from the air. Thus, a muscle is capable of working even when it must expend more energy than can be replaced by reactions involving the oxygen brought to it at a relatively slow rate by the blood. As the lactic acid accumulates, however, the muscle grows weary, and eventually it must rest until oxygen breaks up the lactic acid.

Next comes the question: In what form is the energy from the sugar-to-lactic-acid breakdown supplied to the cells, and how do they use it? The German-born American chemist Fritz Albert Lipmann found an answer in researches beginning in 1941. He showed that certain phosphate compounds formed in the course of carbohydrate metabolism store unusual amounts of energy in the bond that connects the phosphate group to the rest of the molecule. This "high-energy phosphate bond" is transferred to energy carriers present in all cells. The best known of these carriers is "adenosine triphosphate" (ATP). The ATP molecule and certain similar compounds represent the small currency of the body's energy. They store the energy in neat, conveniently sized, readily negotiable packets. When the phosphate bond is hydrolyzed off, the energy is available to be converted into chemical energy for the building of proteins from amino acids, or into electrical energy for the transmission of a nerve impulse, or into kinetic energy via the contraction of muscle, and so on. Although the quantity of ATP in the body is small at any one time, there is always enough (while life persists), for as fast as the ATP molecules are used up, new ones are formed.

For his key discovery, Lipmann shared the Nobel Prize in physiology and medicine in 1953.

The mammalian body cannot convert lactic acid to ethyl alcohol (as yeast can); instead, by another route of metabolism, it bypasses ethyl alcohol and breaks down lactic acid all the way to carbon dioxide (CO_2) and water. In so doing, it consumes oxygen and produces a great deal more energy than is produced by the nonoxygen-requiring conversion of glucose to lactic acid.

The fact that consumption of oxygen is involved offers a convenient means of tracing a metabolic process—that is, finding out what intermediate products are created along the route. Let us say that at a given step in a sequence of reactions a certain substance (e.g., succinic acid) is suspected to be the intermediate substrate. We can mix this with living tissue (or in many cases with a single enzyme) and measure the rate at which the mixture consumes oxygen. If it shows a rapid uptake of oxygen, we can be confident that this particular substance can indeed further the process.

The German biochemist Otto Heinrich Warburg devised the key instrument used to measure the rate of uptake of oxygen. Called the "Warburg manometer," it consists of a small flask (where the substrate and the tissue or enzyme are mixed) connected to one end of a thin U-tube, the other end of which is open. A colored fluid fills the lower part of the U. As the mixture of enzyme and substrate absorbs oxygen from the air in the flask, a slight vacuum is created there, and the colored liquid in the U-tube rises on the side of the U connected to the flask. The rate at which the liquid rises can be used to calculate the rate of oxygen uptake.

Warburg's experiments on the uptake of oxygen by tissues won him the Nobel Prize in physiology and medicine in 1931.

Warburg and another German biochemist, Heinrich Wieland, identified the reactions that yield energy during the breakdown of lactic acid. In the course of the series of reactions, pairs of hydrogen atoms are removed from intermediate substances by means of enzymes called "dehydrogenases." These hydrogen atoms then combine with oxygen, with the catalytic help of enzymes called "cytochromes." In the late 1920's, Warburg and Wieland argued strenuously over which of these reactions was the important one, Warburg contending that it was the uptake of oxygen and Wieland that it was the removal of hydrogen. Eventually, David Keilin showed that both steps were essential.

The German biochemist Hans Adolf Krebs went on to work out the complete sequence of reactions and intermediate products from lactic acid to carbon dioxide and water. This is called the Krebs cycle, or the citric-acid cycle, citric acid being one of the key products formed along the way. For this achievement, completed in 1940, Krebs received a share in the Nobel Prize in physiology and medicine in 1953 (with Lipmann).

The Krebs cycle produces the lion's share of energy for those organisms that make use of molecular oxygen in respiration (which means all organisms except a few types of anaerobic bacteria that depend for energy on chemical reactions not involving oxygen). At different points in the Krebs cycle, a compound will lose two hydrogen atoms, which are eventually combined with oxygen to form water. This "eventually" hides a good deal of detail. The two hydrogen atoms are passed from one variety of cytochrome molecule to another, until the final one, "cytochrome oxidase," passes it on to molecular oxygen. Along the line of cytochromes, molecules of ATP are formed and the body is supplied with its chemical "small change" of energy. All told, for every turn of the Krebs cycle, eighteen molecules of ATP are formed. Exactly where in the chain the ATP is formed, and exactly how, is still not certain. The entire process, because it involves oxygen and the piling-up of phosphate groups to form the ATP, is called "oxidative phosphorylation," and this is a key reaction of living tissue. Any serious interference with it (as when one swallows potassium cyanide) brings death in minutes.

All the substances and all the enzymes that take part in oxidative phosphorylation are contained in tiny granules within the cytoplasm. These were first detected in 1898 by the German biologist C. Benda, who did not at that time, of course, understand their importance. He called them "mitochondria" ("threads of cartilage," which he wrongly thought they were), and the name stuck.

The average mitochondrion is football-shaped, about 1/10,000 of an inch long and 1/25,000 of an inch thick. An average cell might contain anywhere from several hundred to a thousand mitochondria. Very large cells may contain a couple of hundred thousand, while anaerobic bacteria contain none. After World War II, electron-microscopic investigation showed the mitochondrion to have a complex structure of its own, for all its tiny size. The mitochondrion has a double membrane, the outer one smooth and the inner one elaborately wrinkled to present a large surface. Along the inner surface of the mitochrondrion are several thousand tiny structures called "elementary particles." It is these that seem to represent the actual sites of oxidative phosphorylation.

Meanwhile biochemists also made headway in solving the metabolism of fats. It was known that the fat molecules were carbon chains, that they could be hydrolyzed to "fatty acids" (most commonly sixteen or eighteen carbon atoms long), and that the molecules were broken down two carbons at a time. In 1947, Fritz Lipmann discovered a rather complex compound, which played a part in "acetylation"—that is, transfer of a two-carbon fragment from one compound to another. He called the compound "coenzyme A" (the A standing for acetylation). Three years later the German biochemist Feodor Lynen found that coenzyme A was deeply involved in the breakdown of fats. Once it attached itself to a fatty acid, there followed a series of four steps which ended in lopping off the two carbons at the end of the chain to which the coenzyme A was attached. Then another coenzyme A molecule would attach itself to what was left of the fatty acid, chop off two more atoms, and so on. This is celled the "fatty-acid oxidation cycle." This and other work won Lynen a share in the 1964 Nobel Prize in physiology and medicine.

The breakdown of proteins obviously must be, in general, more complicated than that of carbohydrates or fats, because some twenty different amino acids are involved. In some cases it turns out to be rather

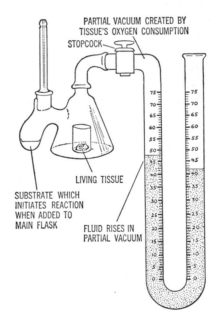

Warburg manometer.

simple: one minor change in an amino acid may convert it into a compound that can enter the citric-acid cycle (as the two-carbon fragments from fatty acids can). But mainly amino acids are decomposed by complex routes.

We can now go back to the conversion of protein into urea—the question that we considered at the start. This conversion happens to be comparatively simple.

A group of atoms which is essentially the urea molecule forms part of a side chain of the amino acid arginine. This group can be chopped off by an enzyme called "arginase," and it leaves behind a kind of truncated amino acid, called "ornithine." In 1932, Krebs and a co-worker, K. Henseleit, while studying the formation of urea by rat-liver tissue, discovered that when they added arginine to the tissue, it produced a flood of urea—much more urea, in fact, than the splitting of every molecule of arginine they had added could have produced. Krebs and Henseleit decided that the arginine molecules must be acting as agents that produced urea over and over again. In other words, after an arginine molecule had its urea combination chopped off by arginase, the ornithine that was left picked up amine groups from other amino acids (plus carbon dioxide from the body) and formed arginine again. So the arginine molecule was repeatedly split, re-formed, split again, and so on, each time yielding a molecule of urea. This is called the "urea cycle," the "ornithine cycle," or the "Krebs-Henseleit cycle."

After the removal of nitrogen, by way of arginine, the remaining "carbon skeletons" of the amino acids can be broken down by various routes to carbon dioxide and water, producing energy.

Tracers

The investigations of metabolism by all these devices still left biochemists in the position of being on the outside looking in, so to speak. They could work out general cycles, but to find out what was really going on in the living animal they needed some means of tracing, in fine detail, the course of events through the stages of metabolism—to follow the fate of particular molecules, as it were. Actually, techniques for doing this had been discovered early in the century, but the chemists were rather slow in making full use of them.

The first to pioneer along these lines was a German biochemist named Franz Knoop. In 1904, he conceived the idea of feeding labeled fat molecules to dogs to see what happened to the molecules. He labeled them by attaching a benzene ring at one end of the chain; he used the benzene ring because mammals possess no enzymes that can break it down. Knoop expected that what the benzene ring carried with it when it showed up in the urine might tell something about how the fat molecule broke down in the body—and he was right. The benzene ring invariably turned up with a two-carbon side chain attached. From this he deduced that the body must split off the fat molecule's carbon atoms two at a time. (As we have seen, more than 40 years later the work with coenzyme A confirmed his deduction.)

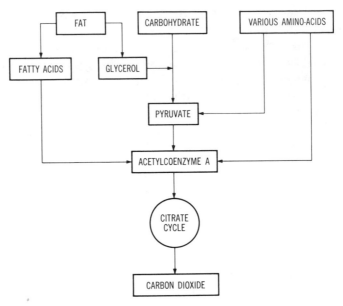

The overall scheme of metabolism of carbohydrates, fats, and proteins.

The carbon chains in ordinary fats all contain an even number of carbon atoms. What if you used a fat whose chain had an odd number of carbon atoms? In that case, if the atoms were chopped off two at a time, you should end up with just one carbon atom attached to the benzene ring. Knoop fed this kind of fat molecule to dogs and did indeed end up with that result.

Knoop had employed the first "tracer" in biochemistry. In 1913, the Hungarian chemist Georg von Hevesy and his co-worker, the German chemist Friedrich Adolf Paneth hit upon another way to tag molecules: radioactive isotopes. They began with radioactive lead, and their first biochemical experiment was to measure how much lead, in the form of a lead-salt solution, a plant would take up. The amount was certainly too small to be measured by any available chemical method, but if radiolead was used, it could easily be measured by its radioactivity. Hevesy and Paneth fed the radioactively tagged lead-salt solution to plants, and at periodic intervals they would burn a plant and measure the radioactivity of its ash. In this way, they were able to determine the rate of absorption of lead by plant cells..

But the benzene ring and lead were very "unphysiological" substances to use as tags. They might easily upset the normal chemistry of living cells. It would be much better to use as tags atoms that actually took part in the body's ordinary metabolism—such atoms as oxygen, nitrogen, carbon, hydrogen, phosphorus.

Once the Joliot-Curies had demonstrated artificial radioactivity in 1934, Hevesy took this direction at once and began using phosphates containing radioactive phosphorus. With these he measured phosphate uptake in plants. Unfortunately, the radioisotopes of some of the key elements in living tissue—notably, nitrogen and oxygen—are not usable, because they are very short-lived, having a half-life of only a few minutes at most. But the most important elements do have *stable* isotopes that can be used as tags. These isotopes are carbon 13, nitrogen 15, oxygen 18, and hydrogen 2. Ordinarily, they occur in very small amounts (about 1 per cent or less); consequently, by "enriching" natural hydrogen, say, in hydrogen 2, it can be made to serve as a distinguishing tag in a hydrogen-containing molecule fed to the body. The presence of the heavy hydrogen in any compound can be detected by means of the mass spectrograph, which separates it by virtue of its extra weight. Thus, the fate of the tagged hydrogen can be traced through the body.

Hydrogen, in fact, served as the first physiological tracer. It became available for this purpose when Harold Urey isolated hydrogen 2 (deuterium) in 1931. One of the first things brought to light by the use of deuterium as a tracer was that hydrogen atoms in the body were much less fixed to their compounds than had been thought. It turned out that they shuttled back and forth from one compound to another, exchanging places on the oxygen atoms of sugar molecules, water molecules, and so on. Since one ordinary hydrogen atom cannot be told from another, this shuttling had not been detected before the deuterium atoms disclosed it. What the discovery implied was that hydrogen atoms hopped about throughout the body, and, if deuterium atoms were attached to oxygen, they would spread through the body regardless of whether or not the compounds involved underwent over-all chemical change. Consequently, the investigator must make sure that a deuterium atom found in a compound got there by some definite enzyme-catalyzed reaction and not just by the shuttling, or exchange, process. Fortunately, hydrogen atoms attached to carbon do not exchange, so deuterium found along carbon chains has metabolic significance.

The roving habits of atoms were further emphasized in 1937 when the German-born American biochemist Rudolf Schoenheimer and his associates began to use nitrogen 15. They fed rats on amino acids tagged with nitrogen 15, killed the rats after a set period, and analyzed the tissues to see which compounds carried nitrogen 15. Here again, exchange was found to be important. After one tagged amino acid had entered the body, almost all the amino acids were shortly found to carry nitrogen 15. In 1942, Schoenheimer published a book entitled *The Dynamic State of Body Constituents*. That title describes the new look in biochemistry that the isotopic tracers brought about. A restless traffic in atoms goes on ceaselessly, quite aside from actual chemical changes.

Little by little the use of tracers filled in the details of the metabolic routes. It corroborated the general pattern of such things as sugar breakdown, the citric-acid cycle, and the urea cycle. It resulted in the addition of new intermediates, in the establishment of alternate routes of reaction, and so on.

When, thanks to the nuclear reactor, over a hundred different radioactive isotopes became available in quantity after World War II, tracer work went into high gear. Ordinary compounds could be bombarded by neutrons in a reactor and come out loaded with radioactive isotopes. Almost every biochemical laboratory in the United States (I might almost say in the world, for the United States soon made isotopes available to other countries for scientific use) started research programs involving radioactive tracers.

The stable tracers were now joined by radioactive hydrogen (tritium), radiophosphorus (phosphorus 32), radiosulfur (sulfur 35), radiopotassium (potassium 42), radiosodium, radioiodine, radioiron, radiocopper, and most important of all, radiocarbon (carbon 14). Carbon 14 was discovered in 1940 by the American chemists Martin David Kamen and Samuel Ruben, and, to their surprise, it turned out to have a half-life of more than 5,000 years—unexpectedly long for a radioisotope among the light elements.

Carbon 14 solved problems that had defied chemists for years and against which they had seemed to be able to make no headway at all. One of the riddles to which it gave the beginning of an answer was the production of the substance known as "cholesterol." Cholesterol's formula, worked out by many years of painstaking investigation by men such as Wieland (who received the 1927 Nobel Prize in chemistry for his work on compounds related to cholesterol), had been found to be:

The function of cholesterol in the body is not yet completely understood, but the substance is clearly of central importance. Cholesterol is found in large quantity in the fatty sheaths around nerves, in the adrenal glands, and in combination with certain proteins. An excess of it can cause gallstones and atherosclerosis. Most significant of all, cholesterol is the prototype of the whole family of "steroids," the steroid nucleus being the four-ring combination you see in the formula. The steroids are a group of solid, fatlike substances, which include the sex hormones and the adrenocortical hormones. All of them undoubtedly are formed from cholesterol. But how is cholesterol itself synthesized in the body?

Until tracers came to their help, biochemists had not the foggiest notion. The first to tackle the question with a tracer were Rudolf Schoenheimer and his co-worker David Rittenberg. They gave rats heavy water to drink and found that its deuterium turned up in the cholesterol molecules. This in itself was not significant, because the deuterium could have got there merely by exchanges. But, in 1942 (after Schoenheimer tragically had committed suicide), Rittenberg and another coworker, the German-American biochemist, Konrad Emil Bloch, discovered a more definite clue. They fed rats acetate ion (a simple two-carbon group, CH_3COO-) with the deuterium tracer attached to the carbon atom in the CH_3 group. The deuterium again showed up in cholesterol molecules, and this time it could not have arrived there by exchange; it must have been incorporated in the molecule as part of the CH_3 group.

Two-carbon groups (of which the acetate ion is one version) seem to represent a general crossroads of metabolism. Such groups, then, might very well serve as the pool of material for building cholesterol. But just how did they form the molecule?

In 1950, when carbon 14 had become available, Bloch repeated the experiment, this time labeling the two carbons of the acetate ion, each with a different tag. He marked the carbon of the CH_3 group with the stable tracer carbon 13, and he labeled the carbon of the COO– group with radioactive carbon 14. Then, after feeding the compound to a rat, he analyzed its cholesterol to see where the two tagged carbons would appear in the molecule. The analysis was a task that called for delicate chemical artistry, and Bloch and a number of other experimenters worked at it for years, identifying the source of one after another of the cholesterol carbon atoms. The pattern that developed eventually suggested that the acetate groups probably first formed a substance called "squalene," a rather scarce thirty-carbon compound in the body to which no one had ever dreamed of paying serious attention before. Now it appeared to be a way station on the road to cholesterol, and biochemists have begun to study it with intense interest. For this work, Bloch shared the 1964 Nobel Prize in physiology and medicine with Lynen.

In much the same way as they tackled the synthesis of cholesterol, biochemists have gone after the construction of the porphyrin ring of heme, a key structure in hemoglobin and in many enzymes. David Shemin of Columbia Universiy fed ducks the amino acid glycine, labeled in various ways. Glycine (NH_2CH_2COOH) has two carbon atoms. When he tagged the CH_2 carbon with carbon 14, that carbon showed up in the porphyrin extracted from the ducks' blood. When he labeled the COOH carbon, the radioactive tracer did not appear in the porphyrin. In short, the CH_2 group entered into the synthesis of porphyrin but the COOH group did not.

Shemin, working with Rittenberg, found that the incorporation of glycine's atoms into porphyrin could take place just as well in red blood cells in the test tube as it could in living animals. This simplified matters, gave more clear-cut results, and avoided sacrificing or inconveniencing the animals.

He then labeled glycine's nitrogen with nitrogen 15 and its CH_2 carbon with carbon 14, then mixed the glycine with duck blood. Later, he carefully took apart the porphyrin produced and found that all four nitrogen atoms in the porphyrin molecule came from the glycine. So did an adjacent carbon atom in each of the four small pyrrole rings (see the formula on page 461), and also the four carbon atoms that serve as bridges between the pyrrole rings. This left twelve other carbon atoms in the porphyrin ring itself and fourteen in the various side-chains. These were shown to arise from acetate ion, some from the CH_3 carbon and some from the COO– carbon.

From the distribution of the tracer atoms it was possible to deduce the manner in which the acetate and glycine entered into the porphyrin. First they formed a one-pyrrole ring; then two such rings combined, and finally two two-ring combinations joined to form the four-ring porphyrin structure.

In 1952, a compound called "porphobilinogen" was isolated in pure form, as a result of an independent line of research by the English chemist R. G. Westall. This compound occurs in the urine of persons with defects in porphyrin metabolism, so it was suspected of having something to do with porphyrins. Its structure turned out to be just about identical with the one-pyrrole-ring structure that Shemin and his co-workers had postulated as one of the early steps in porphyrin synthesis. Porphobilinogen was a key way station.

It was next shown that "delta-aminolevulinic acid," a substance with a structure like that of a porphobilinogen molecule split in half, could supply all the atoms necessary for incorporation into the porphyrin ring by the blood cells. The most plausible conclusion is that the cells first form delta-aminolevulinic acid from glycine and acetate (eliminating the COOH group of glycine as carbon dioxide in the process), that two molecules of delta-aminolevulinic acid then combine to form porphobilinogen (a one-pyrrole ring), and that the latter in turn combines first into a two-pyrrole ring and finally into the four-pyrrole ring of porphyrin.

Photosynthesis

Of all the triumphs of tracer research, perhaps the greatest has been the tracing of the complex series of steps that builds green plants—on which all life on this planet depends.

The animal kingdom could not exist if animals could feed only on one another, any more than a community of people could grow rich solely by taking in one another's washing or a man could lift himself by yanking upward on his belt buckle. A lion that eats a zebra or a man who eats a steak is consuming precious substance that was obtained at great pains and with considerable attrition from the plant world. The second law of thermodynamics tells us that at each stage of the cycle something is lost. No animal stores all of the carbohydrate, fat, and protein contained in the food it eats, nor can it make use of all the energy available in the food. Inevitably a large part, indeed most, of the energy is wasted in unusable heat. At each level of eating, then, some chemical energy is frittered away. Thus, if all animals were strictly carnivorous, the whole animal kingdom would die off in a very few generations. In fact, it would never have come into being in the first place.

The fortunate fact is that most animals are herbivorous. They feed on the grass of the field, on the leaves of trees, on seeds, nuts, and fruit, or on the seaweed and microscopic green plant cells that fill the upper layers of the oceans. Only a minority of animals can be supported in the luxury of being carnivorous.

As for the plants themselves, they would be in no better plight were they not supplied with an external source of energy. They build carbohydrates, fats, and proteins from simple molecules, such as carbon dioxide and water. This synthesis calls for an input of energy, and the plants get it from the most copious possible source: sunlight. Green plants convert the energy of sunlight into the chemical energy of complex compounds, and that chemical energy supports all life forms (except for certain bacteria). This was first clearly pointed out in 1845 by the German physicist Julius Robert von Mayer, who was one of those who pioneered the law of conservation of energy and who was therefore particularly aware of the problem of energy balance. The process by which green plants make use of sunlight is called "photosynthesis," from Greek words meaning "put together by light."

The first attempt at a scientific investigation of plant growth was made early in the seventeenth century by the Flemish chemist Jan Baptista Van Helmont. He grew a small willow tree in a tub containing a weighed amount of soil, and he found, to everyone's surprise, that although the tree grew large, the soil weighed just as much as before. It had been taken for granted that plants derived their substance from the soil. (Actually plants do take some minerals and ions from the soil, but not in any easily weighable amount.) If they did not get it there, where did they get it from? Van Helmont decided that plants must manufacture their substance from water, with which he had supplied the soil liberally. He was only partly right.

A century later the English physiologist Stephen Hales showed that plants built their substance in great part from a material more ethereal than water, namely, air. Half a century later, the Dutch physician Jan Ingen-Housz identified the nourishing ingredient in air as carbon dioxide. He also demonstrated that a plant did not absorb carbon dioxide in the dark; it needed light (the "photo" of photosynthesis). Meanwhile Priestley, the discoverer of oxygen, had learned that green plants gave off oxygen. And, in 1804, the Swiss chemist Nicholas Théodore de Saussure proved that water was incorporated in plant tissue, as Van Helmont had suggested.

The next important contribution came in the 1850's, when the French mining engineer Jean Baptiste Boussingault grew plants in soil completely free of organic matter. He showed, in this way, that plants could obtain their carbon from atmospheric carbon dioxide only. On the other hand, plants would not grow in soil free of nitrogen compounds, and this showed they derived their nitrogen from the soil and that atmospheric nitrogen was not utilized (except, as it turned out, by certain bacteria). From Boussingault's time, it became apparent that the service of soil as direct nourishment for the plant was confined to certain inorganic salts, such as nitrates and phosphates. It was these ingredients that organic fertilizers (such as manure) added to soil. Chemists began to advocate the addition of chemical fertilizers, which served the purpose excellently and which eliminated noisome odors as well as decreasing the dangers of infection and disease, much of which could be traced to the farm's manure pile.

Thus, the skeleton of the process of photosynthesis was established. In sunlight, a plant took up carbon dioxide and combined it with water to form its tissues, giving off "left-over" oxygen in the process. Hence, it became plain that green plants not only provided food but also renewed the earth's oxygen supply. Were it not for this, within a matter of centuries the oxygen would fall to a low level and the atmosphere would be loaded with enough carbon dioxide to asphyxiate animal life.

The scale on which the earth's green plants manufacture organic matter and release oxygen is enormous. The Russian-American biochemist Eugene I. Rabinowitch, a leading investigator of photosynthesis, estimates that each year the green plants of the earth combine a total of 150 billion tons of carbon (from carbon dioxide) with 25 billion tons of hydrogen (from water) and liberate 400 billion tons of oxygen. Of this gigantic performance, the plants of the forests and fields on land account for only 10 per cent; for 90 per cent we have to thank the one-celled plants and seaweed of the oceans.

We still have only the skeleton of the process. What about the details? Well, in 1817, Pierre Joseph Pelletier and Joseph Bienaimé Caventou of France, who were later to be the discoverers of quinine, strychnine, caffeine, and several other specialized plant products, isolated the most important plant product of all—the one that gives the green color to green plants. They called the compound "chlorophyll," from Greek words meaning "green leaf." Then, in 1865, the German botanist Julius von Sachs showed that chlorophyll was not distributed generally through plant cells (though leaves appear uniformly green), but was localized in small subcellular bodies, later called "chloroplasts."

It became clear that photosynthesis took place within the chloroplasts and that chlorophyll was essential to the process. Chlorophyll was not enough, however. Chlorophyll by itself, however carefully extracted, could not catalyze the photosynthetic reaction in a test-tube.

Chloroplasts generally are considerably larger than mitochondria. Some one-celled plants possess only one large chloroplast per cell. Most plant cells, however, contain many smaller chloroplasts, each from two to three times as long and as thick as the typical mitochondrion.

The structure of the chloroplast seems to be even more complex than that of the mitochondrion. The interior of the chloroplast is made up of many thin membranes stretching across from wall to wall. These are the "lamellae." In most types of chloroplasts, these lamellae thicken and darken in places to produce "grana," and it is within the grana that the chlorophyll molecules are found.

If the lamellae within the grana are studied under the electron microscope, they in turn seem to be made up of tiny units, just barely visible, that look like the neatly laid tiles of a bathroom floor. Each of these objects may be a photosynthesizing unit containing 250 to 300 chlorophyll molecules.

The chloroplasts are more difficult than mitochondria to isolate intact. It was not until 1954 that the Polish-American biochemist Daniel I. Arnon, working with disrupted spinach-leaf cells, could obtain chloroplasts completely intact and was able to carry through the complete photosynthetic reaction.

The chloroplast contains not only chlorophyll, but a full complement of enzymes and associated substances, all properly and intricately arranged. It even contains cytochromes by which the energy of sunlight, trapped by chlorophyll, can be converted into ATP through oxidative phosphorylation.

Meanwhile, though, what about the structure of chlorophyll, the most characteristic substance of the chloroplasts? For decades, chemists had tackled this key substance with every tool at their command, but it yielded only slowly. Finally, in 1906, Richard Willstätter of Germany (who was later to rediscover chromatography and to insist, incorrectly, that enzymes were not proteins) identified a central component of the chlorophyll molecule. It was the metal magnesium. (Willstätter received the Nobel Prize in chemistry in 1915 for this discovery and other work on plant pigments.) Willstätter and Hans Fischer went on to work on the structure of the molecule—a task that took a full generation to complete. By the 1930's, it had been determined that chlorophyll had a porphyrin ring structure basically like that of heme (a molecule which Fischer had deciphered). Where heme had an iron atom at the center of the porphyrin ring, chlorophyll had a magnesium atom.

If there were any doubt on this point, it was removed by R. B. Woodward. That master synthesist, who had put together quinine in 1945, strychnine in 1947, and cholesterol in 1951, now capped his previous efforts by putting together a molecule in 1960 that matched the formula worked out by Willstätter and Fischer, and, behold, it had all the properties of chlorophyll isolated from green leaves. Woodward received the 1965 Nobel Prize for chemistry as a result.

Exactly what reaction in the plant did chlorophyll catalyze? All that was known, up to the 1930's, was that carbon dioxide and water went in and oxygen came out. Investigation was made more difficult by the fact that isolated chlorophyll could not be made to bring about photosynthesis. Only intact plant cells or, at best, intact chloroplasts, would do, which meant that the system under study was very complex.

As a first guess, biochemists assumed that the plant cells synthesized glucose ($C_6H_{12}O_6$) from the carbon dioxide and water and then went on to build from this the various plant substances, adding nitrogen, sulfur, phosphorus, and other inorganic elements from the soil.

On paper, it seemed as if glucose might be formed by a series of steps which first combined the carbon atom of carbon dioxide with water (releasing the oxygen atoms of CO_2), and then polymerized the combination, CH_2O (formaldehyde), into glucose. Six molecules of formaldehyde would make one molecule of glucose.

This synthesis of glucose from formaldehyde could indeed be performed in the laboratory, in a tedious sort of way. Presumably, the plant might possess enzymes that speeded the reactions. To be sure, formaldehyde is a very poisonous compound, but the chemists assumed that the formaldehyde was turned into glucose so quickly that at no time did the plant contain more than a very small amount of it. This formaldehyde theory, first proposed in 1870 by Baeyer (the synthesizer of indigo), lasted for two generations, simply because there was nothing better to take its place.

A fresh attack on the problem began in 1938, when Ruben and Kamen undertook to probe the chemistry of the green leaf with tracers. By the use of oxygen 18, the uncommon stable isotope of oxygen, they made one clear-cut finding. It turned out that when the water given a plant was labeled with oxygen 18, the oxygen released by the plant carried this tag, but the oxygen did not carry the tag when only the carbon dioxide supplied to the plant was labeled. In short, the experiment showed that the oxygen given off by plants came from the water molecule and not from the carbon dioxide molecule, as had been mistakenly assumed in the formaldehyde theory.

Ruben and his associates tried to follow the fate of the carbon atoms in the plant by labeling the carbon dioxide with the radioactive isotope carbon 11 (the only radiocarbon known at the time). But this attempt failed. For one thing, carbon 11 has a half-life of only 20.5 minutes. For another, they had no available method at the time for separating individual compounds in the plant cell quickly and thoroughly enough.

But, in the early 1940's, the necessary tools came to hand. Ruben and Kamen discovered carbon 14, the long-lived radioisotope, which made it possible to trace carbon through a series of leisurely reactions. And the development of paper chromatography provided a means of separating complex mixtures easily and cleanly. (In fact, radioactive isotopes allowed a neat refinement of paper chromatography: the radioactive spots on the paper, representing the presence of the tracer, would produce dark spots on a photographic film laid under it, so that the chromatogram would take its own picture—a technique called "autoradiography.")

After World War II, another group, headed by the American biochemist Melvin Calvin, picked up the ball. They exposed microscopic one-celled plants ("chlorella") to carbon dioxide containing carbon 14 for short periods, in order to allow the photosynthesis to progress only through its earliest stages. Then they mashed the plant cells, separated their substances on a chromatogram, and made an autoradiograph.

They found that even when the cells had been exposed to the tagged carbon dioxide for only a minute and a half, the radioactive carbon atoms turned up in as many as fifteen different substances in the cell. By cutting down the exposure time, they reduced the number of substances in which radiocarbon was incorporated, and eventually they decided that the first, or almost the first, compound in which the cell incorporated the carbon-dioxide carbon was "glyceryl phosphate." (At no time did they detect any formaldehyde, so the venerable formaldehyde theory passed quietly out of the picture.)

Glyceryl phosphate is a three-carbon compound. Evidently it must be formed by a roundabout route, for no one-carbon or two-carbon precursor could be found. Two other phosphate-containing compounds were located that took up tagged carbon within a very short time. Both were varieties of sugars: "ribulose diphosphate" (a five-carbon compound) and "sedoheptulose phosphate" (a seven-carbon compound). The investigators identified enzymes that catalyzed reactions involving such sugars, studied those reactions, and worked out the travels of the carbon-dioxide molecule. The scheme that best fits all their data is the following.

First, carbon dioxide is added to the five-carbon ribulose diphosphate, making a six-carbon compound. This quickly splits in two, creating the three-carbon glyceryl phosphate. A series of reactions involving sedoheptulose phosphate and other compounds then puts two glyceryl phosphates together to form the six-carbon glucose phosphate. Meanwhile ribulose diphosphate is regenerated and is ready to take on another carbon dioxide molecule. You can imagine six such cycles turning. At each turn, each cycle supplies one carbon atom (from the carbon dioxide), and out of these a molecule of glucose phosphate is built. Another turn of the six cycles produces another molecule of glucose phosphate, and so on.

This is the reverse of the citric-acid cycle, from an energy standpoint. Whereas the citric-acid cycle converts the fragments of carbohydrate breakdown to carbon dioxide, the ribulose-diphosphate cycle builds up carbohydrates from carbon dioxide. The citric-acid cycle delivers energy to the organism; the ribulose-diphosphate cycle, conversely, has to consume energy.

Here the earlier results of Ruben and Kamen fit in. The energy of sunlight is used, thanks to the catalytic action of chlorophyll, to split a molecule of water into hydrogen and oxygen, a process called "photolysis" (from Greek words meaning "loosening by light"). This is the way that the radiant energy of sunlight is converted into chemical energy for the hydrogen and oxygen molecules contain more chemical energy than did the water molecule from which they came.

In other circumstances it takes a great deal of energy to break up water molecules into hydrogen—for instance, heating the water to something like 2,000 degrees or sending a strong electric current through it. But chlorophyll does the trick easily at ordinary temperatures. All it needs is the relatively weak energy of visible light. The plant uses the light-energy that it absorbs with an efficiency of at least 30 per cent; some investigators believe its efficiency may approach 100 per cent under ideal conditions. If man could harness energy as efficiently as the plants do, he would have much less to worry about with regard to his supplies of food and energy.

After the water molecules have been split, half of the hydrogen atoms find their way into the ribulose-diphosphate cycle, and half of the oxygen atoms are liberated into the air. The rest of the hydrogens and oxygens recombine into water. In doing so, they release the excess of energy that was given to them when sunlight split the water molecules, and this energy is transferred to high-energy phosphate compounds such as ATP. The energy stored in these compounds is then used to power the ribulose-diphosphate cycle. For his work in deciphering the reactions involved in photosynthesis, Calvin received the Nobel Prize in chemistry in 1961.

To be sure, there are some forms of life that gain energy without chlorophyll. About 1880, "chemosynthetic bacteria" were discovered;

bacteria that trapped carbon dioxide in the dark and did not liberate oxygen. Some oxidized sulfur compounds to gain energy; some oxidized iron compounds; and some indulged in still other chemical vagaries.

Then, too, some bacteria have chlorophyll-like compounds ("bacteriochlorophyll"), which enable them to convert carbon dioxide to organic compounds at the expense of light energy—even, in some cases, in the near infrared, where ordinary chlorophyll will not work. However, only chlorophyll itself can bring about the splitting of water and the conservation of the large energy store so gained; bacteriochlorophyll must make do with less energetic devices.

All methods of fundamental energy gain, other than that which uses sunlight by way of chlorophyll, are essentially dead-end, and no creature more complicated than a bacterium has successfully made use of them. For the rest of life (and even for most bacteria), chlorophyll and photosynthesis, directly or indirectly, are the basis of life.

CHAPTER 12

The Cell

Chromosomes

It is an odd paradox that man until recent times knew very little about his own body. In fact, it was only some 300 years ago that he learned about the circulation of the blood, and only within the last fifty years or so has he discovered the functions of many of the organs.

Prehistoric man, from cutting up animals for cooking and from embalming his own dead in preparation for afterlife, was aware of the existence of the large organs, such as the brain, liver, heart, lungs, stomach, intestines, and kidneys. This awareness was intensified through the frequent use of the appearance of the internal organs of a ritually sacrificed animal (particularly the appearance of its liver) in foretelling the future or estimating the extent of divine favor or disfavor. Egyptian papyri, dealing validly with surgical technique and presupposing some familiarity with body structure can be dated earlier than 2000 B.C.

The ancient Greeks went so far as to dissect animals and an occasional human cadaver with the deliberate purpose of learning something about "anatomy" (from Greek words meaning "to cut up"). Some delicate work was done. Alcmaeon of Croton, about 500 B.C., first described the optic nerve and the Eustachian tube. Two centuries later, in Alexandria, Egypt (then the world center of science), a school of Greek anatomy started brilliantly with Herophilus and his pupil Erasistratus. They investigated the parts of the brain, distinguishing the cerebrum and cerebellum, and studied the nerves and blood vessels as well.

Ancient anatomy reached its peak with Galen, a Greek physician who practiced in Rome in the latter half of the second century. Galen worked up theories of bodily functions that were accepted as gospel for 1,500 years afterward. But his notions about the human body were full of curious errors. This is understandable, for the ancients obtained most of their information from dissecting animals. Inhibitions of one kind or another made men uneasy about dissecting the human body.

In their denunciations of the pagan Greeks, early Christian writers accused them of having practiced heartless vivisections on human beings. But this comes under the heading of polemical literature; not only is it doubtful that the Greeks did human vivisections, but obviously they did not even dissect enough dead bodies to learn much about the human anatomy. In any case, the Church's disapproval of dissection virtually put a stop to anatomical studies throughout the Middle Ages. As this period of history approached its end, anatomy began to revive in Italy. In 1316, an Italian anatomist, Mondino de Luzzi, wrote the first book to be devoted entirely to anatomy, and he is therefore known as the "Restorer of Anatomy."

The interest in naturalistic art during the Renaissance also fostered anatomical research. In the fifteenth century, Leonardo da Vinci performed some dissections by means of which he revealed new facts of anatomy, picturing them with the power of artistic genius. He showed the double curve of the spine and the sinuses that hollow the bones of the face and forehead. He used his studies to derive theories of physiology more advanced than Galen's. But Leonardo, though he was a genius in science as well as in art, had little influence on scientific thought in his time. Either from neurotic disinclination or from sober caution, he did not publish any of his scientfic work, but kept it hidden in coded notebooks. It was left for later generations to discover his scientific achievements when his notebooks were finally published.

The French physician Jean Fernel was the first modern to take up dissection as an important part of a physician's duties. He published a book on the subject in 1542. However, his work was almost completely overshadowed by a much greater work published in the following year. This was the famous De Humani Corporis Fabrica (Concerning the Structure of the Human Body) of Andreas Vesalius, a Belgian who did most of his work in Italy. On the theory that the proper study of mankind was man, Vesalius dissected the appropriate subject and corrected many of Galen's errors. The drawings of the human anatomy in his book (which are reputed to have been made by Jan Stevenzoon van Calcar, a pupil of the artist Titian) are so beautiful and accurate that they are still republished today and will always stand as classics. Vesalius can be called the father of modern anatomy. His Fabrica was as revolutionary in its way as Copernicus's De Revolutionibus Orbium Coelestium, published in the very same year.

Just as the revolution initiated by Copernicus was brought to fruition by Galileo, so the one initiated by Vesalius came to a head in the crucial discoveries of William Harvey. Harvey was an English physician and experimentalist, of the same generation as Galileo and William Gilbert, the experimenter with magnetism. His particular interest was that vital body juice—the blood. What did it do in the body, anyway?

It was known that there were two sets of blood vessels: the veins and the arteries. (Praxagoras of Cos, a Greek physician of the third century B.C., had given the latter the name "artery" from Greek words meaning "I carry air," because these vessels were found to be empty in dead bodies. Galen had later shown that in life they carried blood.) It was also known that the heartbeat drove the blood in some sort of motion, for when an artery was cut, the blood gushed out in pulses that synchronized with the heartbeat.

Galen had proposed that the blood seesawed to and fro in the blood vessels, traveling first in one direction through the body and then in the other. This theory required him to explain why the back-and-forth movement of the blood was not blocked by the wall between the two halves of the heart; Galen answered simply that the wall was riddled with invisibly small holes that let the blood through.

Harvey took a closer look at the heart. He found that each half was divided into two chambers, separated by a one-way valve that allowed blood to flow from the upper chamber ("auricle") to the lower ("ventricle"), but not vice versa. In other words, blood entering one of the auricles could be pumped into its corresponding ventricle and from there into blood vessels issuing from it, but there could be no flow in the opposite direction.

Harvey then performed some simple but beautifully clear-cut experiments to determine the direction of flow in the blood vessels. He would tie off an artery or a vein in a living animal to see on which side of this blockage the pressure within the blood vessel would build up. He found that when he stopped the flow in an artery, the vessel always bulged on the side between the heart and the block. This meant that the blood in arteries must flow in the direction away from the heart. When he tied a vein, the bulge was always on the other side of the block; therefore, the blood flow in veins must be toward the heart. Further evidence in favor of this one-way flow in veins rests in the fact that the larger veins contain valves that prevent blood from moving away from the heart. This had been discovered by Harvey's teacher, the Italian anatomist Hieronymus Fabrizzi (better known by his Latinized name, Fabricius). Fabricius, however, under the load of Galenic tradition, refused to draw the inevitable conclusion and left the glory to his English student.

Harvey went on to apply quantitative measurements to the blood flow (the first time anyone had applied mathematics to a biological problem). His measurements showed that the heart pumped out blood at such a rate that in twenty minutes its output equaled the total amount of blood contained in the body. It did not seem reasonable to suppose that

576

the body could manufacture new blood, or consume the old, at any such rate. The logical conclusion, therefore, was that the blood must be recycled through the body. Since it flowed away from the heart in the arteries and toward the heart in the veins, Harvey decided that the blood was pumped by the heart into the arteries, then passed from them into the veins, then flowed back to the heart, then was pumped into the arteries again, and so on. In other words, it circulated continuously in one direction through the heart-and-blood-vessel system.

Earlier anatomists, including Leonardo da Vinci, had hinted at such an idea, but Harvey was the first to state and investigate the theory in detail. He set forth his reasoning and experiments in a small, badly printed book entitled *De Motus Cordis* (Concerning the Motion of the Heart), which was published in 1628 and has stood ever since as one of the great classics of science.

The main question left unanswered by Harvey's work was: How did the blood pass from the arteries into the veins? Harvey said there must be connecting vessels of some sort, though they were too small to be seen. This was reminiscent of Galen's theory about small holes in the heart wall, but whereas Galen's holes in the heart were never found and do not exist, Harvey's connecting vessels were confirmed as soon as a microscope became available. In 1661, just four years after Harvey's death, an Italian physician named Marcello Malpighi examined the lung tissues of a frog with a primitive microscope, and, sure enough, there were tiny blood vessels connecting the arteries with the veins. Malpighi named them "capillaries," from a Latin word meaning "hair-like."

The use of the microscope made it possible to see other minute structures as well. The Dutch naturalist Jan Swammerdam discovered the red blood corpuscles, while the Dutch anatomist Regnier de Graaf discovered tiny "ovarian follicles" in animal ovaries. Small creatures, such as insects, could be studied in detail.

Work in such fine detail encouraged the careful comparison of structures in one species with structures in others. The English botanist Nehemiah Grew was the first "comparative anatomist" of note. In 1675, he published his studies comparing the trunk structure of various trees, and in 1681 studies comparing the stomachs of various animals.

The coming of the microscope introduced biologists, in fact, to a more basic level of organization of living things: a level at which all ordinary structures could be reduced to a common denominator. In 1665, the English scientist Robert Hooke, using a compound microscope of his own design, discovered that cork, the bark of a tree, was built of extremely tiny compartments, like a superfine sponge. He called these holes "cells," likening them to small rooms, such as the cells in a monastery. Other microscopists then found similar "cells," but full of fluid, in living tissue.

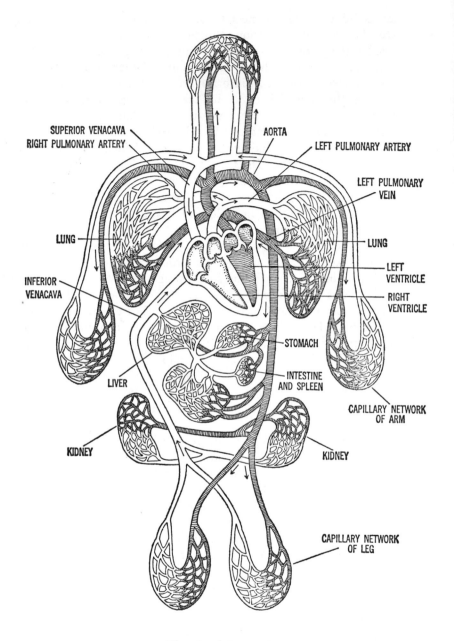

The circulatory system.

Over the next century and a half it gradually dawned on biologists that all living matter was made up of cells and that each cell was an independent unit of life. Some forms of life—certain microorganisms—consisted of only a single cell; the larger organisms were composed of many cooperating cells. One of the earliest to propose this view was the French physiologist René Joachim Henri Dutrochet. His report, published in 1824, went unnoticed, however, and the cell theory gained prominence only after Matthias Jakob Schleiden and Theodor Schwann of Germany independently formulated it in 1838 and 1839.

The colloidal fluid filling certain cells was named "protoplasm" ("first form") by the Czech physiologist Jan Evangelista Purkinje in 1839, and the German botanist Hugo von Mohl extended the term to signify the contents of all cells. The German anatomist Max Johann Sigismund Schultze emphasized the importance of protoplasm as the "physical basis of life" and demonstrated the essential similarity of protoplasm in all cells, both plant and animal, and in both very simple and very complex creatures.

The cell theory is to biology about what the atomic theory is to chemistry and physics. Its importance in the dynamics of life was established when, around 1860, the German pathologist Rudolf Virchow asserted, in a succinct Latin phrase, that all cells arose from cells. He showed that the cells in diseased tissue were produced by the division of originally normal cells.

By that time it had become clear that every living organism, even the largest, began life as a single cell. One of the earliest microscopists, Johann Ham, an assistant of Leeuwenhoek, had discovered in seminal fluid tiny bodies that were later named "spermatozoa" (from Greek

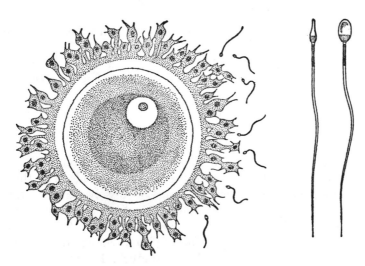

Human egg and sperm cells.

words meaning "animal seed"). Much later, in 1827, the German physiologist Karl Ernst von Baer had identified the ovum, or egg cell, of mammals. Biologists came to realize that the union of an egg and a spermatozoon formed a fertilized ovum from which the animal eventually developed by repeated divisions and redivisions.

The big question was: How did cells divide? The answer lay in a small globule of comparatively dense material within the cell, making up about a tenth its volume, first reported by Robert Brown (the discoverer of Brownian motion) in 1831 and named the "nucleus." (To distinguish it from the nucleus of the atom, I shall refer to it from now on as the "cell nucleus.")

If a one-celled organism was divided into two parts, one of which contained the intact cell nucleus, the part containing the cell nucleus was able to grow and divide, but the other part could not. (Later it was also learned that the red blood cells of mammals, lacking nuclei, are short-lived and have no capacity for either growth or division. For that reason, they are not considered true cells and are usually called "corpuscles.")

Unfortunately, further study of the cell nucleus and the mechanism of division was thwarted for a long time by the fact that the cell was more or less transparent, so that its substructures could not be seen. Then the situation was improved by the discovery that certain dyes would stain parts of the cell and not others. A dye called "hematoxylin" (obtained from logwood) stained the cell nucleus black and brought it out prominently against the background of the cell. After Perkin and other chemists began to produce synthetic dyes, biologists found themselves with a variety of dyes from which to choose.

In 1879, the German biologist Walther Flemming found that with certain red dyes he could stain a particular material in the cell nucleus which was distributed through it as small granules. He called this material "chromatin" (from the Greek word for "color"). By examining this material, Flemming was able to follow some of the changes in the process of cell division. To be sure, the stain killed the cell, but in a slice of tissue he would catch various cells at different stages of cell division. They served as still pictures, which he put together to form a kind of "moving picture" of the progress of cell division.

In 1882, Flemming published an important book in which he described the process in detail. At the start of cell division, the chromatin material gathered itself together in the form of threads. The thin membrane enclosing the cell nucleus seemed to dissolve, and at the same time a tiny object just outside it divided in two. Flemming called this object the "aster," from a Greek word for "star," because radiating threads gave it a starlike appearance. After dividing, the two parts of the aster traveled to opposite sides of the cell. Its trailing threads apparently entangled the threads of chromatin, which had meanwhile lined up in the center of the cell, and the aster pulled half the chromatin threads to one side of the cell, half to the other. As a result, the cell pinched in at the middle and split into two cells. A cell nucleus developed in each, and the chromatin material that the nuclear membrane enclosed broke up into granules again.

Flemming called the process of cell division "mitosis," from the Greek word for "thread," because of the prominent part played in it by the chromatin threads. In 1888, the German anatomist Wilhelm von Waldeyer gave the chromatin thread the name "chromosome" (from the Greek for "colored body"), and that name has stuck. It should be mentioned, though, that chromosomes, despite their name, are colorless in their unstained natural state, and of course are then quite difficult to make out against the very similar background. (Nevertheless, even so, they had dimly been seen in flower cells as early as 1848 by the German amateur botanist Wilhelm Friedrich Benedict Hofmeister.)

Continued observation of stained cells showed that the cells of each species of plant or animal had a fixed and characteristic number of chromosomes. Before a cell divides in two during mitosis, the number

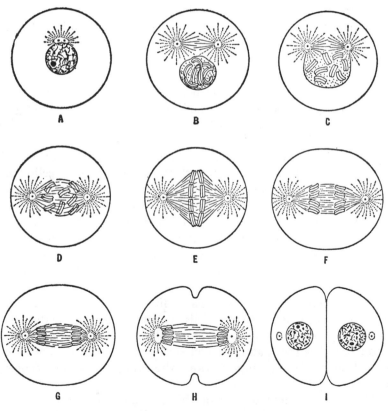

Division of a cell by mitosis.

of chromosomes is doubled, so that each of the two daughter cells after the division has the same number as the original mother cell.

The Belgian embryologist Eduard van Beneden discovered in 1885 that the chromosomes did *not* double in number when egg and sperm cells were being formed. Consequently each egg and each sperm cell had only half the number of chromosomes that ordinary cells of the organism possessed. (The cell division that produces sperm and egg cells therefore is called "meiosis," from a Greek word meaning "to make less.") When an egg and a sperm cell combined, however, the combination (the fertilized ovum) had a complete set of chromosomes, half contributed by the mother through the egg cell and half by the father through the sperm cell. This complete set was passed on by ordinary mitosis to all the cells that made up the body of the organism developing from the fertilized egg.

Even though the use of dyes makes the chromosomes visible, they do not make it easy to see one individual chromosome among the rest. Generally, they look like a tangle of stubby spaghetti. Thus, it was long thought that each human cell contained 24 pairs of chromosomes. It was not until 1956 that a more painstaking count of these cells (certainly earnestly studied) showed 23 pairs to be the correct count.

Fortunately, this problem no longer exists. A technique has been devised whereby treatment with a low-concentration salt-solution, in the proper manner, swells the cells and disperses the chromosomes. They can then be photographed, and that photograph can be cut into sections, each containing a separate chromosome. If these chromosomes are matched into pairs and then arranged in the order of decreasing length, the result is a "karyotype," a picture of the chromosome-content of the cell, consecutively numbered.

The karyotype offers a subtle tool in medical diagnosis, for separation of the chromosomes is not always perfect. In the process of cell division, a chromosome may be damaged or even broken. Sometimes the separation may not be even, so that one of the daughter cells gets an extra chromosome while the other is missing one. Such abnormalities are sure to damage the working of the cell, often to such an extent that the cell cannot function. (This is what keeps the process of mitosis so seemingly accurate—not that it really is as accurate as it seems, but that the mistakes are buried.)

Such imperfections are particularly dire when they take place in the process of meiosis, for then egg cells or sperm cells are produced with imperfections in the chromosome-complement. If an organism can develop at all from such an imperfect start (and usually it cannot), every cell in its body has the imperfection: the result is a serious congenital disease.

The most frequent disease of this type involves severe mental retardation. It is called "Down's syndrome" (because it was first described in 1866 by the English physician John Langdon Haydon Down), and it occurs once in every thousand births. It is more commonly known as "mongolism," because one of the symptoms is a slant to the eyelids that is reminiscent of the epicanthic fold of the peoples of eastern Asia. Since the syndrome has no more to do with the Asians, however, than with others, this is a poor name.

It was not until 1959 that the cause of Down's syndrome was discovered. In that year, three French geneticists—Jerome Jean Lejeune, M. Gautier, and P. Turpin—counted the chromosomes in cells from three cases and found that each had 47 chromosomes instead of 46. It turned out that the error was in the possession of *three* members of chromosome-pair #21. Then, in 1967, the mirror-image example of the disease was located. A mentally retarded three-year-old girl was found to have a single chromosome-21. She was the first discovered case of a living human being with a missing chromosome.

Cases of this sort involving other chromosomes seem less common but are turning up. Patients with a particular type of leukemia show a tiny extra chromosome-fragment in their cells. This is called the "Philadelphia chromosome" because it was first located in a patient hospitalized in that city. Broken chromosomes, in general, turn up with greater-than-normal frequency in certain not-very-common diseases.

Genes

In the 1860's an Austrian monk named Gregor Johann Mendel, who was too occupied with the affairs of his monastery to pay attention to the biologist's excitement about cell division, was quietly carrying through some experiments in his garden which were destined eventually to make sense out of chromosomes. Abbé Mendel was an amateur botanist, and he became particularly interested in the results of cross-breeding pea plants of varying characteristics. His great stroke of intuition was to study one clearly defined characteristic at a time.

He would cross plants with different seed colors (green or yellow), or smooth-seeded peas with wrinkle-seeded ones, or long-stemmed plants with short-stemmed ones, and then would follow the results in the offspring of the succeeding generations. Mendel kept a careful statistical record of his results, and his conclusions can be summarized essentially as follows:

1. Each characteristic was governed by "factors" that (in the cases that Mendel studied) could exist in one of two forms. One version of the factor for seed color, for instance, would cause the seeds to be green; the other form would make them yellow. (For convenience, let us use the present-day terms. The factors are now called "genes," a term put forward in 1909 by the Danish biologist Wilhelm Ludwig Johannsen from

a Greek word meaning "to give birth to," and the different forms of a gene controlling a given characteristic are called "alleles." Thus the seed-color gene possessed two alleles, one for green seeds, the other for yellow seeds.)

2. Every plant had a pair of genes for each characteristic, one contributed by each parent. The plant transmitted one of its pair to a germ cell, so that when the germ cells of two plants united by pollination, the offspring had two genes for the characteristic once more. The two genes might be either identical or alleles.

3. When the two parent plants contributed alleles of a particular gene to the offspring, one allele might overwhelm the effect of the other. For instance, if a plant producing yellow seeds was crossed with one producing green seeds, all the members of the next generation would produce yellow seeds. The yellow allele of the seed-color gene was "dominant," the green allele "recessive."

4. Nevertheless, the recessive allele was not destroyed. The green allele, in the case just cited, was still present, even though it produced no detectable effect. If two plants containing mixed genes (i.e., each with one yellow and one green allele) were crossed, some of the offspring might have two green alleles in the fertilized ovum; in that case those particular offspring would produce green seeds, and the offspring of such parents in turn would also produce green seeds. Mendel pointed out that there were four possible ways of combining alleles from a pair of hybrid parents, each possessing one yellow and one green allele. A yellow allele from the first parent might combine with a yellow allele from the second; a yellow allele from the first might combine with a green allele from the second; a green allele from the first might combine with a yellow allele from the second; and a green allele from the first might combine with a green allele from the second. Of the four combinations, only the last would result in a plant that would produce green seeds. Assuming that all four combinations were equally probable, one fourth of the plants of the new generation should produce green seeds. Mendel found that this was indeed so.

5. Mendel also found that characteristics of different kinds—for instance, seed color and flower color—were inherited independently of each other. That is, red flowers were as apt to go with yellow seeds as with green seeds. The same was true of white flowers.

Mendel performed these experiments in the early 1860's, wrote them up carefully, and sent a copy of his paper to Karl Wilhelm von Nägeli, a Swiss botanist of great reputation. Von Nägeli's reaction was negative. Von Nägeli had, apparently, a predilection for all-encompassing theories (his own theoretical work was semimystical and turgid in expression), and he saw little merit in the mere counting of pea plants as a way to truth. Besides, Mendel was an unknown amateur.

It seems that Mendel allowed himself to be discouraged by von Nägeli's comments, for he turned to his monastery duties, grew fat (too fat to bend over in the garden), and abandoned his researches. He did, however, publish his paper in 1866 in a provincial Austrian journal, where it attracted no further attention for a generation.

But other scientists were slowly moving toward the same conclusions to which (unknown to them) Mendel had already come. One of the routes by which they arrived at an interest in genetics was the study of "mutations," that is, of freak animals, or monsters, which had always been regarded as bad omens. (The word "monster" came from a Latin word meaning "warning.") In 1791, a Massachusetts farmer named Seth Wright took a more practical view of a sport that turned up in his flock of sheep. A lamb was born with abnormally short legs, and it occurred to the shrewd Yankee that short-legged sheep could not escape over the low stone walls around his farm. He therefore deliberately bred a line of short-legged sheep from his not unfortunate accident.

This practical demonstration stimulated others to look for useful mutations. By the end of the nineteenth century the American horticulturist Luther Burbank was making a successful career of breeding hundreds of new varieties of plants which were improvements over the old in one respect or another, not only by mutations, but by judicious crossing and grafting.

Meanwhile botanists tried to find an explanation of mutation. And in what is perhaps the most startling coincidence in the history of science, no fewer than three men, independently, and in the very same year, came to precisely the same conclusions that Mendel had reached a generation earlier. They were Hugo De Vries of Holland, Karl Erich Correns of Germany, and Erich von Tschermak of Austria. None of them knew of each other's or Mendel's work. All three were ready to publish in 1900. All three, in a final check of previous publications in the field, came across Mendel's paper, to their own vast surprise. All three did publish in 1900, each citing Mendel's paper, giving Mendel full credit for the discovery, and advancing his own work only as confirmation.

A number of biologists immediately saw a connection between Mendel's genes and the chromosomes that could be seen under the microscope. The first to draw a parallel was an American cytologist named Walter S. Sutton, in 1904. He pointed out that chromosomes, like genes, came in pairs, one of which was inherited from the father and one from the mother. The only trouble with this analogy was that the number of chromosomes in the cells of any organism was far smaller than the number of inherited characteristics. Man, for instance, has only 23 pairs of chromosomes and yet certainly possesses thousands of inheritable characteristics. Biologists therefore had to conclude that chromosomes were not genes. Each must be a collection of genes.

had to conclude that chromosomes were not genes. Each must be a collection of genes.

In short order, biologists discovered an excellent tool for studying specific genes. It was not a physical instrument but a new kind of laboratory animal. In 1906, the Columbia University zoologist Thomas Hunt Morgan, who was at first skeptical of Mendel's theories, conceived the idea of using fruit flies (*Drosophila melanogaster*) for research in genetics. (The term "genetics" was coined in 1902 by the British biologist William Bateson.)

Fruit flies had considerable advantages over pea plants (or any ordinary laboratory animal) for studying the inheritance of genes. They bred quickly and prolifically, could easily be raised by the hundreds on very little food, had scores of inheritable characteristics which could be observed readily, and had a comparatively simple chromosomal setup— only four pairs of chromosomes per cell.

With the fruit fly, Morgan and his coworkers discovered an important fact about the mechanism of inheritance of sex. They found that the female fruit fly has four perfectly matched pairs of chromosomes so that all the egg cells, receiving one of each pair, are identical so far as chromosome make-up is concerned. However, in the male one of each of the four pairs consists of a normal chromosome, called the "X chromosome," and a stunted one, which was named the "Y chromosome." Therefore when sperm cells are formed, half have an X chromosome and half a Y chromosome. When a sperm cell with the X chromosome fertilizes an egg cell, the fertilized egg, with four matched pairs, naturally becomes a female. On the other hand, a sperm cell with a Y chromosome produces a male. Since both alternatives are equally probable, the number of males and females in the typical species of living things is roughly equal. (In some creatures, notably various birds, it is the female that has a Y chromosome.)

This chromosomal difference explains why some disorders or mutations show up only in the male. If a defective gene occurs on one of a pair of X chromosomes, the other member of the pair is still likely to be normal and can salvage the situation. But in the male, a defect on the X chromosome paired with the Y chromosome generally cannot be compensated for, because the latter carries very few genes. Therefore the defect shows up.

The most notorious example of such a "sex-linked disease" is "hemophilia," a condition in which blood clots only with difficulty, if at all. Individuals with hemophilia run the constant risk of bleeding to death from slight causes or of suffering agonies from internal bleeding. A woman who carries a gene that will produce hemophilia on one of her X chromosomes, is very likely to have a normal gene at the same position in the other X chromosome. She will therefore not show the disease. She will, however, be a "carrier." Of the egg cells she forms, half will have the normal X chromosome and half the hemophiliac X chromosome. If the egg with the abnormal X chromosome is fertilized by sperm with an X chromosome from a normal male, the resulting child will be a girl who will not be hemophiliac but who will again be a carrier; if it is fertilized by sperm with a Y chromosome from a normal male, the hemophiliac gene in the ovum will not be counteracted by anything in the Y chromosome, and the result is a boy with hemophilia. By the laws of chance, half the sons of hemophilia-carriers will be hemophiliacs; half the daughters will be, in their turn, carriers.

The most eminent hemophilia-carrier in history was Queen Victoria of England. Only one of her four sons (the oldest, Leopold) was hemophiliac. Edward VII—from whom later British monarchs descended—escaped, so there is no hemophilia now in the British royal family. However, two of Victoria's daughters were carriers. One had a daughter (also a carrier) who married Czar Nicholas II of Russia. As a result, their only son was a hemophiliac; this helped alter the history of Russia and the world, for it was through his influence on the hemophiliac that the monk Gregory Rasputin gained power in Russia and helped bring on the discontent that eventually led to revolution. The other daughter of Victoria had a daughter (also a carrier) who married into the royal house of Spain, producing hemophilia there. Because of its presence among the Spanish Bourbons and the Russian Romanoffs, hemophilia was sometimes called the "royal disease," but it has no particular connection with royalty, except for Victoria's misfortune.

A lesser sex-linked disorder is color-blindness, which is far more common among men than among women. Actually, the absence of one X chromosome may produce sufficient weakness among men generally as to account for the fact that where women are protected against death from childbirth infections they tend to live some three to seven years longer, on the average, than men. That twenty-third complete pair makes women the sounder biological organism, in a way.

The X and Y chromosomes (or "sex chromosomes") are arbitrarily placed at the end of the karyotype, even though the X chromosome is among the longest. Apparently chromosome abnormalities, such as those involved in Down's syndrome, are more common among the sex chromosomes than among the others. This may not be because the sex chromosomes are most likely to be involved in abnormal mitoses, but perhaps because sex-chromosome abnormalities are less likely to be fatal, so that more young manage to be born with them.

The type of sex-chromosome abnormality that has drawn the most attention, is one in which a male ends up with an extra Y chromosome in his cells, so that he is XYY, so to speak. It turns out that XYY males are difficult to handle. They are tall, strong, and bright, but are characterized by a tendency to rage and violence. Richard Speck, who killed eight nurses in Chicago in 1966, is supposed to have been an XYY. A murderer was acquitted in Australia in October 1968 on the grounds that he was an XYY and therefore not responsible for his action. Nearly 4 per cent of the male inmates in a certain Scottish prison have turned out to be XYY, and there are some estimates that XYY combinations may occur in as many as 1 man in every 3,000.

There seems to be some reason for considering it desirable to run a chromosome check on everyone and certainly on every newborn child. As is the case of other procedures, simple in theory but tedious in practice, attempts are being made to computerize such a process.

Research on fruit flies showed that traits were not necessarily inherited independently, as Mendel had thought. It happened that the seven characteristics of pea plants that he had studied were governed by genes on separate chromosones. Morgan found that where two genes governing two different characteristics were located on the same

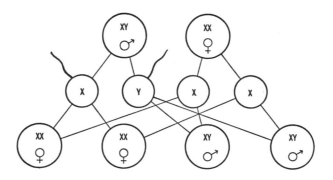

Combinations of X and Y chromosomes.

chromosome, those characteristics were generally inherited together (just as a passenger in the front seat of a car and one in the back seat travel together).

This genetic linkage is not, however, unchangeable. Just as a passenger can change cars, so a piece of one chromosome occasionally switches to another, swapping places with a piece from the other. Such "crossing over" may occur during the division of a cell. As a result, linked traits are separated and reshuffled in a new linkage. For instance, there is a variety of fruit fly with scarlet eyes and curly wings. When it is mated with a white-eyed, miniature-winged fruit fly, the offspring will generally be either red-eyed and curly-winged or white-eyed and miniature-winged. But the mating may sometimes produce a white-eyed, curly-winged fly or a red-eyed, miniature-winged one as a result of crossing over. The new form will persist in succeeding generations unless another crossing over takes place.

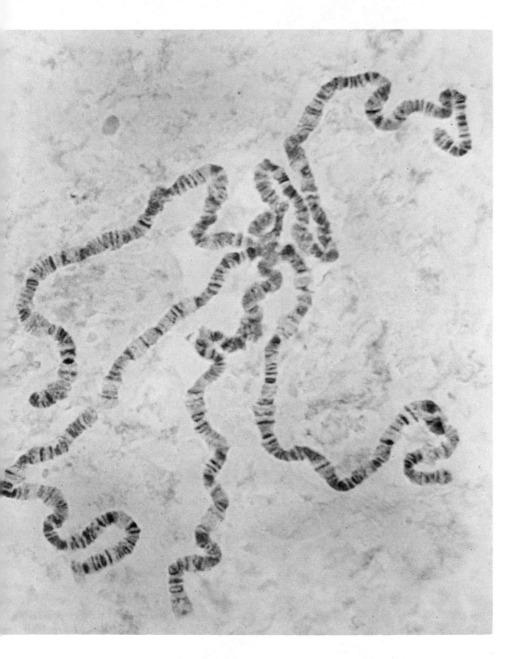

Normal chromosomes of *Drosophila.*

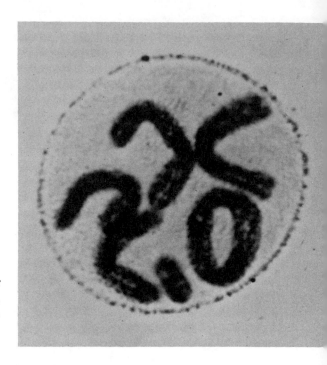

Chromosomes damaged by radiation. Some are broken, and one is coiled into a ring.

Radiation damage caused these chromosomes to divide into three groups instead of the normal two.

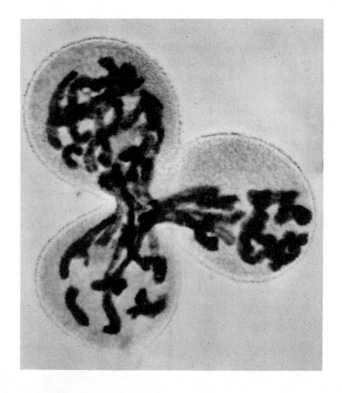

Mutations in fruit flies, shown here in the form of shriveled wings. The mutations were produced by exposure of the male parent to radiation.

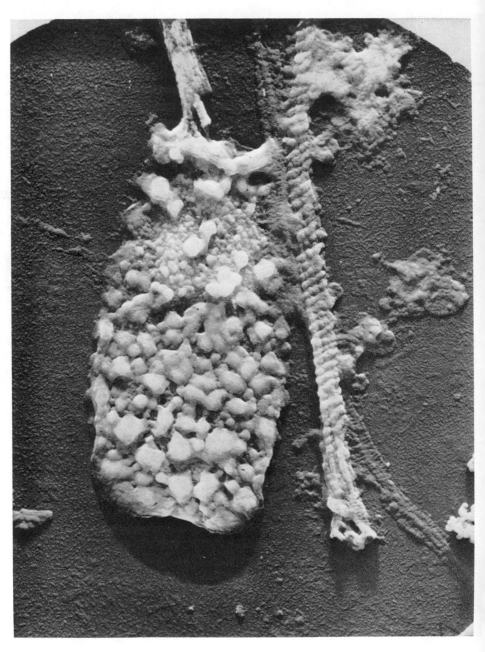

Sperm cell of a bull.

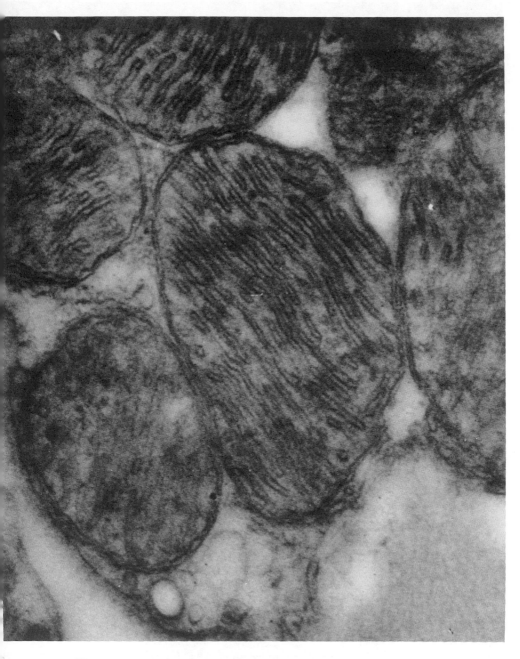

Heart cell in the right ventricle of a rat, magnified more than 100,000 times by the electron microscope.

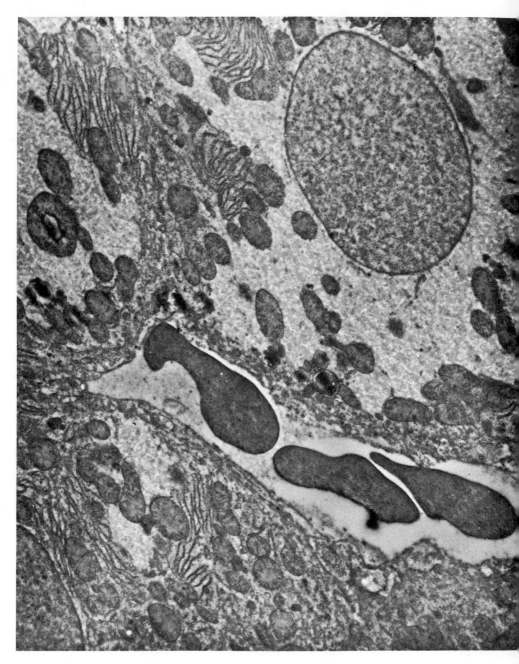

Liver cell, magnified about 10,000 times.

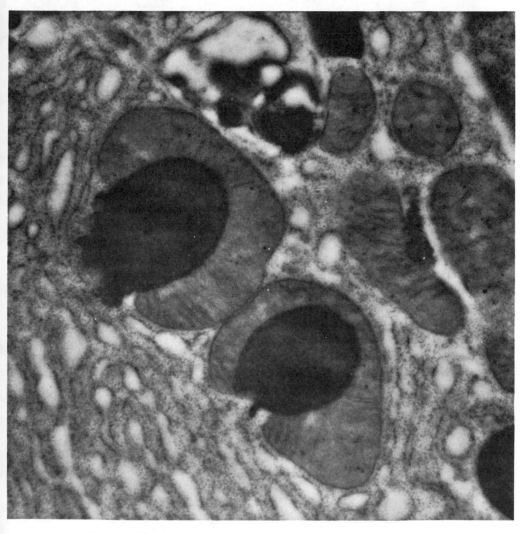

Mitochondria, sometimes called "powerhouses of the cell" because they carry out energy-yielding chemical reactions. The mitochondria are the black bodies in the center of the picture.

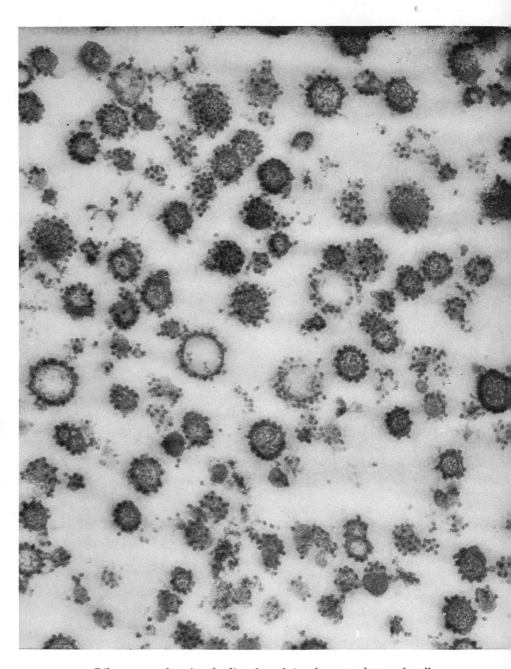

Ribosomes, the tiny bodies found in the cytoplasm of cells. These were separated from pancreas cells by a centrifuge and magnified about 100,000 times under the electron microscope.

Now picture a chromosome with a gene for red eyes at one end and a gene for curly wings at the other end. Let us say that in the middle of the chromosome's length there are two adjacent genes governing two other characteristics. Obviously, the probability of a break occuring at that particular point, separating these two genes, is smaller than the probability of a break coming at one of the many points along the length of the chromosome that would separate the genes at the opposite ends. By noting the frequency of separation of given pairs of linked characteristics by crossing over, Morgan and his co-workers, notably Alfred Henry Sturtevant, were able to deduce the relative locations of the genes in question, and in this way they worked out chromosome "maps" of gene locations for the fruit fly. The location, so determined, is the "locus" of a gene.

From such chromosome maps and from a study of giant chromosomes many times the ordinary size, found in the salivary glands of the fruit fly, it has been established that the insect has a minimum of 10,000 genes in a chromosome pair. This means the individual gene must have a molecular weight of 60 million. Working from this, man's somewhat larger chromosomes may contain from 20,000 to 90,000 genes per chromosome pair, or up to 2 million altogether.

For his work on the genetics of fruit flies, Morgan received the Nobel Prize in medicine and physiology in 1933.

Increasing knowledge of genes raises hopes that the genetic endowment of individual humans might someday be analyzed and modified: either preventing seriously anomalous conditions from developing, or correcting them if they slip by. Such "genetic engineering" would require human chromosome-maps, clearly a tremendously larger job than in the case of the fruit fly. The task was made somewhat simpler in a startling way in 1967, when Howard Green of New York University formed hybrid cells containing both mouse and human chromosomes. Relatively few human chromosomes persisted after several cell-divisions, and the effects due to their activity was more easily pinpointed.

Another step in the direction of gene-knowledge and gene-manipulation came in 1969, when the American biochemist Jonathan Beckwith and his co-workers, isolated an individual gene for the first time in history. It was from an intestinal bacterium, and it controlled an aspect of sugar metabolism.

Every once in a while, with a frequency which can be calculated, a sudden change occurs in a gene. The mutation shows itself by some new and unexpected physical characteristic, such as the short legs of Farmer Wright's lamb. Mutations in nature are comparatively rare. In 1926, the geneticist Hermann Joseph Muller, who had been a member of Morgan's research team, discovered a way to increase the rate of mutations artificially in fruit flies so that the inheritance of such changes could be studied more easily. He found that X-rays would do the trick; presumably, they damaged the genes. The study of mutations made possible by Muller's discovery won him the Nobel Prize in medicine and physiology in 1946.

As it happens, Muller's researches have given rise to some rather disquieting thoughts concerning the future of the human species. While mutations are an important driving force in evolution, occasionally producing an improvement that enables a species to cope better with its environment, the beneficial mutation is very much the exception. Most mutations—at least 99 per cent of them—are detrimental, some even lethal. Eventually, even those that are only slightly harmful die out, because their bearers do not get along as well and leave fewer descendants than healthy individuals do. But in the meantime a mutation may cause illness and suffering for many generations. Furthermore, new mutations keep cropping up continually, and every species carries a constant load of defective genes.

The great number of different gene varieties—including large quantities of seriously harmful ones—in normal populations was clearly shown by the work of the Russian-American geneticist Theodosius Dobzhansky in the 1930's and 1940's. It is this diversity that makes evolution march on as it does, but it is the number of deleterious genes (the "genetic load") that gives rise to some fears for the future.

Two modern developments seem to be adding steadily to this load. First, the advances in medicine and social care tend to compensate for the handicaps of people with detrimental mutations, at least so far as the ability to reproduce is concerned. Eyeglasses are available to individuals with defective vision; insulin keeps alive sufferers from diabetes (a hereditary disease), and so on. Thus they pass on their defective genes to future generations. The alternatives—allowing defective individuals to die young or sterilizing or imprisoning them—are, of course, unthinkable, except where the handicap is sufficiently great to make the individual less than human, as in idiocy or homicidal paranoia. Undoubtedly, the human species can still bear its load of negatively mutated genes, despite its humanitarian impulses.

A   B   C   D
Crossing over in chromosomes.

But there is less excuse for the second modern hazard—namely, adding to the load by unnecessary exposure to radiation. Genetic research shows incontrovertibly that for the population as a whole even a slight increase in general exposure to radiation means a corresponding slight increase in the mutation rate. And since 1895 mankind has been exposed to types and intensities of radiation of which previous generations knew nothing. Solar radiation, the natural radioactivity of the soil, and cosmic rays have always been with us. Now, however, we use X-rays in medicine and dentistry with abandon; we concentrate radioactive material; we form artificially radioactive isotopes of terrifying radiant potency; we even explode nuclear bombs. All of this increases the background radiation.

No one, of course, suggests that research in nuclear physics be abandoned, or that X-rays never be used by the doctor and dentist. There is, however, a strong recommendation that the danger be recognized and that exposure to radiation be minimized; that, for instance, X-rays be used with discrimination and care and that the sexual organs be routinely shielded during all such use. Another suggested precaution is that each individual keep a record of his total accumulated exposure to X-rays, so that he will have some idea of whether he is in danger of exceeding a reasonable limit.

Of course, the geneticists could not be sure that the principles established by experiments on plants and insects necessarily applied to man. After all, man was neither a pea plant nor a fruit fly. But direct studies of certain characteristics in man showed that human genetics did follow the same rules. The best-known example is the inheritance of blood types.

Blood transfusion is a very old practice, and early physicians occasionally even tried to transfuse animal blood into persons weakened by loss of blood. But transfusions even of human blood often turned out badly, so that laws were sometimes passed forbidding transfusion. In the 1890's, the Austrian pathologist Karl Landsteiner finally discovered that human blood came in different types, some of which were incompatible with each other. He found that sometimes when blood from one person was mixed with a sample of serum (the blood fluid remaining after the red cells and a clotting factor are removed) from another person, the red cells of the first person's whole blood would clump together. Obviously such a mixture would be very bad if it occurred in transfusion, and it might even kill the patient if the clumped cells blocked the blood circulation in key vessels. Landsteiner also found, however, that some bloods could be mixed without causing any deleterious clumping.

By 1902, Landsteiner was able to announce that there were four types of human blood, which he called A, B, AB, and O. Any given individual had blood of just one of these types. Of course, a particular type could be transferred without danger from one person to another having the same type. In addition, O blood could safely be transfused to a person possessing any of the other three types, and either A blood or B blood could be given to an AB patient. But red-cell clumping ("agglutination") would result when AB blood was transfused to an A or B individual, when A and B were mixed, or when an O individual received a transfusion of any blood other than O. (Nowadays, because of possible serum reactions, in good practice patients are given only blood of their own type.)

In 1930, Landsteiner (who by then had become a United States citizen) received the Nobel Prize in medicine and physiology.

Geneticists have established that these blood types (and all the others since discovered, including the Rh variations) are inherited in a strictly Mendelian manner. It seems that there are three gene alleles, responsible respectively for A, B, and O blood. If both parents have O-type blood, all the children of that union will have O-type blood. If one parent is O-type and the other A-type, all the children may show A-type blood, for the A allele is dominant over the O. The B allele likewise is dominant over the O allele. The B allele and A allele, however, show no dominance with respect to each other, and an individual possessing both alleles has AB-type blood.

The Mendelian rules work out so strictly that blood groups can be (and are) used to test paternity. If an O-type mother has a B-type child, the child's father must be B-type, for that B allele must have come from somewhere. If the woman's husband happens to be A or O, it is clear that she has been unfaithful (or there has been a baby mix-up at the hospital). If an O-type woman with a B-type child accuses an A or O man of being the parent, she is either mistaken or lying. On the other hand, while blood type can sometimes prove a negative, it can never prove a positive. If the woman's husband, or the man accused, is indeed a B-type, the case remains unproved. Any B-type man, or any AB-type man, could have been the father.

The applicability of the Mendelian rules of inheritance to human beings has also been borne out by the existence of sex-linked traits. Color blindness and hemophilia (a hereditary failure of the blood to clot) are found almost exclusively in males, and they are inherited in precisely the manner that sex-linked characteristics are inherited in the fruit fly.

Naturally, the thought will arise that by forbidding people with such afflictions to have children, the disorder can be wiped out. By directing proper mating, the human breed might even be improved, as breeds of cattle have been. This is by no means a new idea. The ancient Spartans believed this and tried to put it into practice 2,500 years ago. In modern times, the notion was revived by an English scientist, Francis Galton (a cousin of Charles Darwin). In 1883, he coined the word "eugenics" to describe his scheme. (The derivation of the word is from the Greek and means "good birth.")

Galton was not aware, in his time, of the findings of Mendel. He did not understand that characteristics might seem to be absent, yet be carried as recessives. He did not understand that groups of characteristics would be inherited intact and that it might be difficult to get rid of an undesirable one without also getting rid of a desirable one. Nor was he aware that mutations would reintroduce undesirable characteristics in every generation.

Human genetics is an enormously complicated subject that is not likely to be completely or neatly worked out in the foreseeable future. Because man breeds neither as frequently nor as prolifically as the fruit fly; because his matings cannot be subjected to laboratory control for experimental purposes; because he has many more chromosomes and many more inherited characteristics than the fruit fly; because the human characteristics in which we are most interested, such as creative genius, intelligence, and moral strength, are extremely complex, involving the interplay of numerous genes plus environmental influences —for all these reasons, geneticists cannot deal with human genetics with the same confidence with which they study fruit-fly genetics.

Eugenics remains a dream, therefore, made hazy and insubstantial by lack of knowledge. Those who are today most articulate in favor of elaborate eugenic programs tend to be racists or eccentrics.

Just how does a gene bring the physical characteristic for which it is responsible into being? What is the mechanism whereby it gives rise to yellow seeds in pea plants, or curled wings in fruit flies, or blue eyes in human beings?

Biologists are now certain that genes exert their effects by way of enzymes. One of the clearest cases in point involves the color of eyes, hair, and skin. The color (blue or brown, yellow or black, pink or brown, or shades in between) is determined by the amount of pigment, called "melanin" (from the Greek word for "black"), that is present in the eye's iris, the hair, or the skin. Now melanin is formed from an amino acid, tyrosine, by way of a number of steps, most of which have now been worked out. A number of enzymes are involved, and the amount of melanin formed will depend upon the quantity of these enzymes. For instance, one of the enzymes, which catalyzes the first two steps, is tyrosinase. Presumably some particular gene controls the production of tyrosinase by the cells. In that way, it will control the coloring of the skin, hair, and eyes. And since the gene is transmitted from generation to generation, children will naturally resemble their parents in coloring. If a mutation happens to produce a defective gene that cannot form tyrosinase, there will be no melanin, and the individual will be an "albino." The absence of a single enzyme (and hence the deficiency of a single gene) will thus suffice to bring about a major change in personal characteristics.

Granted that an organism's characteristics are controlled by its enzyme make-up, which in turn is controlled by genes, the next question is: How do the genes work? Unfortunately, even the fruit fly is much too complex an organism to trace out the matter in detail. But, in 1941, the American biologists George Wells Beadle and Edward Lawrie Tatum began such a study with a simple organism which they found admirably suited to this purpose. It is the common pink bread mold (scientific name, *Neurospora crassa*).

Neurospora is not very demanding in its diet. It will grow very well on sugar plus inorganic compounds that supply nitrogen, sulfur, and various minerals. Aside from sugar, the only organic substance that has to be supplied to it is a vitamin called "biotin."

At a certain stage in its life cycle, the mold produces eight spores, all identical in genetic constitution. Each spore contains seven chromosomes; as in the sex cell of a higher organism, its chromosomes come singly, not in pairs. Consequently, if one of its chromosomes is changed, the effect can be observed, because there is no normal partner present to mask the effect. Beadle and Tatum therefore were able to create mutations in *Neurospora* by exposing the mold to X-rays and then could follow the specific effects in the behavior of the spores.

If, after the mold had received a dose of radiation, the spores still thrived on the usual medium of nutrients, clearly no mutation had taken place, at least so far as the organism's nutritional requirements for growth were concerned. If the spores would not grow on the usual medium, the experimenters proceeded to determine whether they were alive or dead, by feeding them a complete medium containing all the vitamins, amino acids, and other items they might possibly need. If the spores grew on this, the conclusion was that the X-rays had produced a mutation that had changed *Neurospora's* nutritional requirements. Apparently it now needed at least one new item in its diet. To find out what that was, the experimenters tried the spores on one diet after another, each time with some items of the complete medium missing. They might omit all the amino acids, or all the various vitamins, or all but one or two amino acids or one or two vitamins. In this way they narrowed down the requirements until they identified just what it was that the spore now needed in its diet before the mutation.

It turned out sometimes that the mutated spore required the amino acid arginine. The normal, "wild strain" had been able to manufacture its own arginine from sugar and ammonium salts. Now, thanks to the genetic change, it could no longer synthesize arginine, and unless this amino acid was supplied in its diet, it could not make protein and therefore could not grow.

The clearest way to account for such a situation was to suppose that the X-rays had disrupted a gene responsible for the formation of an enzyme necessary for manufacturing arginine. For lack of the normal gene, *Neurospora* could no longer make the enzyme. No enzyme, no arginine.

Beadle and his co-workers went on to use this sort of information to study the relation of genes to the chemistry of metabolism. There was a way to show, for instance, that more than one gene was involved in the making of arginine. For simplicity's sake, let us say there are two—gene A and gene B—responsible for the formation of two different enzymes, both of which are necessary for the synthesis of arginine. Then a mutation of either gene A or gene B will rob *Neurospora* of the ability to make the amino acid. Suppose we irradiate two batches of *Neurospora* and produce an arginineless strain in each one. If we are lucky, one mutant may have a defective A gene and a normal B gene, the other a normal A and defective B. To see if that has happened, let us cross the two mutants at the sexual stage of their life cycle. If the two strains do indeed differ in this way, the recombination of chromosomes may produce some spores whose A and B genes are both normal. In other words, from two mutants that are incapable of making arginine, we will get some offspring that *can* make it. Sure enough, exactly that sort of thing happened when the experiments were performed.

It was possible to explore the metabolism of *Neurospora* in finer detail than this. For instance, here were three different mutant strains incapable of making arginine on an ordinary medium. One would grow only if it was supplied with arginine itself. The second would grow if it received either arginine or a very similar compound called citrulline. The third could grow on arginine or citrulline or still another similar compound called ornithine.

What conclusion would you draw from all this? Well, we can guess that these three substances are steps in a sequence of which arginine is the final product. Each requires an enzyme. First ornithine is formed from some simpler compound with the help of an enzyme, then another enzyme converts ornithine to citrulline, and finally a third enzyme converts citrulline to arginine. (Actually, chemical analysis shows that each of the three is slightly more complex than the one before.) Now a *Neurospora* mutant that lacks the enzyme for making ornithine but possesses the other enzymes can get along if it is supplied with ornithine, for from it the spore can make citrulline and then the essential arginine. Of course, it can also grow on citrulline, from which it can make arginine, and on arginine itself. By the same token, we can reason that the second mutant strain lacks the enzyme needed to convert ornithine to citrulline. This strain therefore must be provided with citrulline, from which it can make arginine, or with arginine itself. Finally, we can conclude that the mutant that will grow only on arginine has lost the enzyme (and gene) responsible for converting citrulline to arginine.

By analyzing the behavior of the various mutant strains they were able to isolate, Beadle and his co-workers founded the science of "chemical genetics." They worked out the course of synthesis of many important compounds by organisms. Beadle proposed what has become known as the "one-gene-one-enzyme theory"—that is, that every gene governs the formation of a single enzyme—a suggestion that is now generally accepted by geneticists. For their pioneering work, Beadle and Tatum shared in the Nobel Prize in medicine and physiology in 1958.

Beadle's discoveries put biochemists on the *qui vive* for evidence of gene-controlled changes in proteins, particularly in human mutants, of course. A case turned up, unexpectedly, in connection with the disease called "sickle-cell anemia."

This disease had first been reported in 1910 by a Chicago physician named James Bryan Herrick. Examining a sample of blood from a Negro teenage patient under the microscope, he found that the red cells, normally round, had odd, bent shapes, many of them resembling the crescent shape of a sickle. Other physicians began to notice the same peculiar phenomenon, almost always in Negro patients. Eventually investigators decided that sickle-cell anemia was a hereditary disease. It followed the Mendelian laws of inheritance. Apparently there is a sickle-cell gene that, when inherited in double dose from both parents, produces these distorted red cells. Such cells are unable to carry oxygen properly and are exceptionally short-lived, so there is a shortage of red cells in the blood. Those who inherit the double dose tend to die of the disease in childhood. On the other hand, when a person has only one sickle-cell gene, from one of his parents, the disease does not appear. Sickling of his red cells shows up only when the person is deprived of oxygen to an unusual degree, as at high altitudes. Such people are considered to have the "sickle-cell trait," but not the disease.

It was found that about 9 per cent of the Negroes in America have the trait, and 0.25 per cent have the disease. In some localities in Central Africa as much as a quarter of the Negro population shows the trait. Apparently the sickle-cell gene arose as a mutation in Africa and has been inherited ever since by individuals of African descent. If the disease is fatal, why has the defective gene not died out? Studies in Africa during the 1950's turned up the answer. It seems that people with the sickle-cell trait tend to have greater immunity to malaria than do normal individuals. The sickle cells are somehow inhospitable to the malarial parasite. It is estimated that in areas infested with malaria, children with the trait have a 25 per cent better chance of surviving to child-bearing age than those without the trait have. Hence, possessing a single dose of the sickle-cell gene (but not the anemia-causing double dose) confers an advantage. The two opposing tendencies—promotion of the defective gene by the protective effect of the single dose and elimination of the gene by its fatal effect in double dose—tend to produce an equilibrium which maintains the gene at a certain level in the population.

In regions where malaria is not an acute problem, the gene does tend to die out. In America, the incidence of sickle-cell genes among Negroes may have started as high as 25 per cent. Even allowing for a reduction to an estimated 15 per cent by admixture with non-Negro individuals, the present incidence of only 9 per cent shows that the gene is dwindling away. In all probability it will continue to do so. If Africa is freed of malaria, it will presumably dwindle there, too.

The biochemical significance of the sickle-cell gene suddenly came into prominence in 1949 when Linus Pauling and his co-workers at Cal Tech (where Beadle also was working) showed that the gene affected the hemoglobin of the red blood cells. Persons with a double dose of the sickle-cell gene were unable to make normal hemoglobin. Pauling proved this by means of the technique called "electrophoresis," a method that uses an electric current to separate proteins by virtue of differences in the net electric charge on the various protein molecules. (The electrophoretic technique was developed by the Swedish chemist Arne Wilhelm Kaurin Tiselius, who received the Nobel Prize in chemistry in 1948 for this valuable contribution.) Pauling, by electrophoretic analysis, found that patients with sickle-cell anemia had an abnormal hemoglobin (named "hemoglobin S"), which could be separated from normal hemoglobin. The normal kind was given the name hemoglobin A (for "adult") to distinguish it from a hemoglobin in fetuses, called hemoglobin F.

Since 1949, biochemists have discovered a number of other abnormal hemoglobins besides the sickle-cell one, and they are lettered from hemoglobin C to hemoglobin M. Apparently, the gene responsible for the manufacture of hemoglobin has been mutated into many defective alleles, each giving rise to a hemoglobin that is inferior for carrying out the functions of the molecule in ordinary circumstances but perhaps helpful in some unusual condition. Thus, just as hemoglobin S in a single dose improves resistance to malaria, so hemoglobin C in a single dose improves the ability of the body to get along on marginal quantities of iron.

Since the various abnormal hemoglobins differ in electric charge, they must differ somehow in the arrangement of amino acids in the peptide chain, for the amino-acid make-up is responsible for the charge pattern of the molecule. The differences must be very small, because the abnormal hemoglobins all function as hemoglobin after a fashion. The hope of locating the difference in a huge molecule of some 600 amino acids was correspondingly small. Nevertheless, the German-American biochemist Vernon Martin Ingram and co-workers tackled the problem of the chemistry of the abnormal hemoglobins.

They first broke down hemoglobin A, hemoglobin S, and hemoglobin C into peptides of various sizes by digesting them with a protein-splitting enzyme. Then they separated the fragments of each hemoglobin by "paper electrophoresis"—that is, using the electric current to convey the molecules along a moistened piece of filter paper instead of through a solution. (We can think of this as a kind of electrified paper chromatography.) When the investigators had done this with each of the three hemoglobins, they found that the only difference among them was that a single peptide turned up in a different place in each case.

They proceeded to break down and analyze this peptide. Eventually they learned that it was composed of nine amino acids and that the arrangement of these nine was exactly the same in all three hemoglobins except at one position. The respective arrangements were:

Hemoglobin A: His-Val-Leu-Leu-Thr-Pro-Glu-Glu-Lys
Hemoglobin S: His-Val-Leu-Leu-Thr-Pro-Val-Glu-Lys
Hemoglobin C: His-Val-Leu-Leu-Thr-Pro-Lys-Glu-Lys

As far as could be told, the only difference among the three hemoglobins lay in that single amino acid in the seventh position in the peptide: it was glumatic acid in hemoglobin A, valine in hemoglobin S, and lysine in hemoglobin C. Since glumatic acid gives rise to a negative charge, lysine to a positive charge, and valine to no charge at all, it is not surprising that the three proteins behave differently in electrophoresis. Their charge pattern is different.

But why should so slight a change in the molecule result in so drastic a change in the red cell? Well, the normal red cell is one-third hemoglobin A. The hemoglobin A molecules are packed so tightly in the cell that they just barely have room for free movement. In short, they are on the point of precipitating out of solution. Part of the influence that determines whether a protein is to precipitate out or not is the nature of its charge. If all the proteins have the same net charge, they repel one another and keep from precipitating. The greater the charge (i.e., the repulsion), the less likely the proteins are to precipitate. In hemoglobin S the intermolecular repulsion may be slightly less than in hemoglobin A, and hemoglobin S is correspondingly less soluble and more likely to precipitate. When a sickle-cell is paired with a normal gene, the latter may form enough hemoglobin A to keep the hemoglobin S in solution, though it is a near squeak. But when both of the genes are sickle-cell mutants, they will produce only hemoglobin S. This molecule cannot remain in solution. It precipitates out into crystals, and they distort and weaken the red cell.

This theory would explain why the change of just one amino acid in each half of a molecule made up of nearly 600 is sufficient to produce a serious disease and the near-certainty of an early death.

Albinism and sickle-cell anemia are not the only human defects that have been traced to the absence of a single enzyme or the mutation of a single gene. There is phenylketonuria, a hereditary defect of metabolism, which often causes mental retardation. It results from the lack of an enzyme needed to convert the amino acid phenylalanine to tyrosine. There is galactosemia, a disorder causing eye cataracts and damage to the brain and liver, which has been traced to the absence of an enzyme required to convert a galactose phosphate to a glucose phosphate. There is a defect, involving the lack of one or another of the enzymes that control the breakdown of glycogen (a kind of starch) and its conversion to glucose, which results in abnormal accumulations of glycogen in the liver and elsewhere and usually leads to early death. These are examples of "inborn errors of metabolism," a congenital lack of the capacity to form some more or less vital enzyme found in normal human beings. This concept was first introduced to medicine by the English physician Archibald Edward Garrod in 1908, but it lay disregarded for a generation until, in the mid-1930's, the English geneticist John Burdon Sanderson Haldane brought the matter to the attention of scientists once more.

Such disorders are generally governed by a recessive allele of the gene that produces the enzyme involved. When only one of a pair of genes is defective, the normal one can carry on, and the individual is usually capable of leading a normal life (as in the case of possessor of the sickle-cell trait). Trouble generally comes only when two parents happen to have the same unfortunate gene and have the further bad luck of combining those two in a fertilized egg. Their child then is the victim of a kind of Russian roulette. Probably all of us carry our load of abnormal, defective, even dangerous genes, usually masked by normal genes. You can understand why the human geneticists are so concerned about radiation or anything else that may increase the mutation-rate and add to the load.

Nucleic Acids

The really remarkable thing about heredity is not these spectacular, comparatively rare aberrations, but the fact that by and large inheritance runs so strictly true to form. Generation after generation, millennium after millennium, the genes go on reproducing themselves in exactly the same form and generating exactly the same enzymes, with only an occasional accidental variation of the blueprint. They rarely fail by so much as the introduction of a single wrong amino acid in a large protein molecule. How do they manage to make true copies of themselves over and over again with such astounding faithfulness?

A gene is built of two major components. Perhaps half of it is protein, but the other part is not. To this nonprotein portion we must now direct our attention.

In 1869, a Swiss biochemist named Friedrich Miescher, while

breaking down the protein of cells with pepsin, discovered that the pepsin did not break up the cell nucleus. The nucleus shrank some, but it remained intact. By chemical analysis, Miescher then found that the cell nucleus consisted largely of a phosphorus-containing substance that did not at all resemble protein in its properties. He called the substance "nuclein." It was renamed "nucleic acid" twenty years later when the substance was found to be strongly acid.

Miescher devoted himself to a study of this new material and eventually discovered that sperm cells (which have very little material outside the cell nucleus) were particularly rich in nucleic acid. Meanwhile, the German chemist Felix Hoppe-Seyler, in whose laboratories Miescher had made his first discovery, and who had personally confirmed the young man's work before allowing it to be published, isolated nucleic acid from yeast cells. This seemed different in properties from Miescher's material, so Miescher's variety was named "thymus nucleic acid" (because it could be obtained with particular ease from the thymus gland of animals), and Hoppe-Seyler's, naturally, was called "yeast nucleic acid." Since thymus nucleic acid was at first derived only from animal cells and yeast nucleic acid only from plant cells, it was thought for a while that this might represent a general chemical distinction between animals and plants.

The German biochemist Albrecht Kossel, another pupil of Hoppe-Seyler, was the first to make a systematic investigation of the structure of the nucleic-acid molecule. By careful hydrolysis, he isolated from it a series of nitrogen-containing compounds, which he named "adenine," "guanine," "cytosine," and "thymine." Their formulas are now known to be:

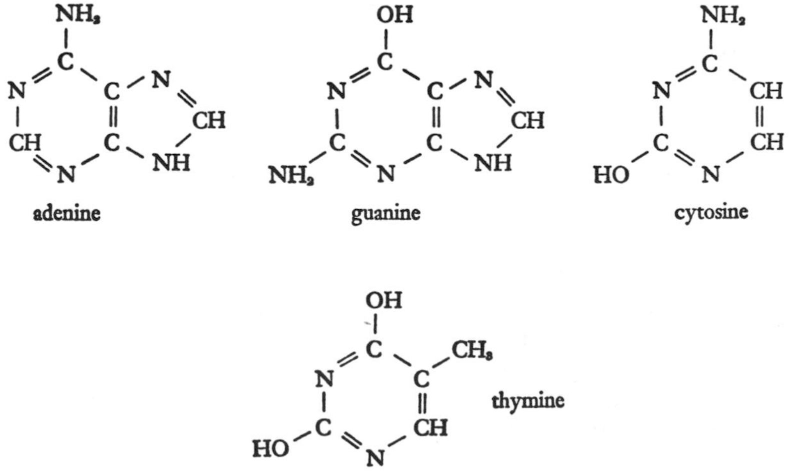

adenine

guanine

cytosine

thymine

The double-ring formation in the first two compounds is called the "purine ring," and the single ring in the other two is the "pyrimidine ring." Therefore adenine and guanine are referred to as purines, and cytosine and thymine are pyrimidines.

For these researches, which started a very fruitful train of discoveries, Kossel received the Nobel Prize in medicine and physiology in 1910.

In 1911, the Russian-born American biochemist Phoebus Aaron Theodore Levene, a pupil of Kossel, carried the investigation a stage further. Kossel had discovered in 1891 that nucleic acids contained carbohydrate, but now Levene showed that the nucleic acids contained five-carbon sugar molecules. (This was, at the time, an unusual finding. The best-known sugars, such as glucose, contain six carbons.) Levene followed this by showing that the two varieties of nucleic acid differed in the nature of the five-carbon sugar. Yeast nucleic acid contained "ribose," while thymus nucleic acid contained a sugar very much like ribose except for the absence of one oxygen atom, so this sugar was called "deoxyribose." Their formulas are:

ribose   deoxyribose

In consequence, the two varieties of nucleic acid came to be called "ribonucleic acid" (RNA) and "deoxyribonucleic acid" (DNA).

Besides the difference in their sugars, the two nucleic acids also differ in one of the pyrimidines. RNA has "uracil" in place of thymine. Uracil is very like thymine, however, as you can see from the formula:

uracil

By 1934, Levene was able to show that the nucleic acids could be broken down to fragments which contained a purine or a pyrimidine, either the ribose or the deoxyribose sugar, and a phosphate group. This combination is called a "nucleotide." Levene proposed that the nucleic-acid molecule was built up of nucleotides as a protein is built up of amino acids. His quantitative studies suggested to him that the molecule consisted of just four nucleotide units, one containing adenine, one guanine, one cytosine, and one either thymine (in DNA) or uracil (in RNA). It turned out, however, that what Levene had isolated were not nucleic-acid molecules but pieces of them, and by the middle 1950's biochemists found that the molecular weights of nucleic acids ran as high as six million. Nucleic acids are thus certainly equal and very likely superior to proteins in molecular size.

The exact manner in which nucleotides are built up and interconnected was confirmed by the British biochemist Alexander Robertus Todd, who built up a variety of nucleotides out of simpler fragments and carefully bound nucleotides together under conditions that allowed only one variety of bonding. He received the Nobel Prize in chemistry in 1957 for this work.

As a result, the general structure of the nucleic acid could be seen to be somewhat like the general structure of protein. The protein molecule is made up of a polypeptide backbone out of which jut the side chains of the individual amino acids. In nucleic acids, the sugar portion of one nucleotide is bonded to the sugar portion of the next by means of a phosphate group attached to both. There is thus a "sugar-phosphate backbone" running the length of the molecule. From this, there extended purines and pyrimidines, one to each nucleotide.

By the use of cell-staining techniques, investigators began to pin down the location of nucleic acids in the cell. The German chemist Robert Feulgen, employing a red dye that stained DNA but not RNA, found that DNA was located in the cell nucleus, specifically in the chromosomes. He detected it not only in animal cells but also in plant cells. In addition, by staining RNA he showed that this nucleic acid, too, occurred in both plant and animal cells. In short, the nucleic acids were universal materials existing in all living cells.

The Swedish biochemist Torbjörn Caspersson studied the subject further by removing one of the nucleic acids (by means of an enzyme which reduced it to soluble fragments that could be washed out of the cell) and concentrating on the other. He would photograph the cell in ultraviolet light; since a nucleic acid absorbs ultraviolet much more strongly than do other cell materials, the location of the DNA or the RNA—whichever he had left in the cell—showed up clearly. DNA showed up only in the chromosomes. RNA made its appearance mainly in certain particles in the cytoplasm. Some RNA also showed up in the "nucleolus," a structure within the nucleus. (In 1948, the Rockefeller Institute biochemist Alfred Ezra Mirsky showed that small quantities of RNA are present even in the chromosomes, while Ruth Sager showed that DNA can occur in the cytoplasm, notably in the chloroplasts of plants.)

Caspersson's pictures disclosed that the DNA lay in localized bands in the chromosomes. Was it possible that DNA molecules were none other than the genes, which up to this time had had a rather vague and formless existence?

Through the 1940's, biochemists pursued this lead with growing excitement. They found it particularly significant that the amount of DNA in the cells of an organism was always rigidly constant, except that the sperm and egg cells had only half this amount, which would be expected, since they had only half the chromosome supply of normal cells. The amount of RNA and of the protein in chromosomes might vary all over the lot, but the quantity of DNA remained fixed. This certainly seemed to indicate a close connection between DNA and genes.

There were, of course, a number of things that argued against the idea. For instance, what about the protein in chromosomes? Proteins of various kinds were associated with the nucleic acid, forming a combination called "nucleoprotein." Considering the complexity of proteins, and their great and specific importance in other capacities in the body, should not the protein be the important part of the molecule? It would seem that the nucleic acid might well be no more than an adjunct—at most a working portion of the molecule, like the heme in hemoglobin.

But the proteins (known as protamine and histone) most commonly found in isolated nucleoprotein turned out to be rather simple, as proteins went. Meanwhile DNA was steadily being found to be more and more complex. The tail was beginning to wag the dog.

At this point some remarkable evidence, which seemed to show that the tail *was* the dog, was disclosed. It involved the pneumococcus, the well-known pneumonia microbe. Bacteriologists had long studied two different strains of pneumococci grown in the laboratory—one with a smooth coat made of a complex carbohydrate, the other lacking this coat and therefore rough in appearance. Apparently the rough strain lacked some enzyme needed to make the carbohydrate capsule. But an English bacteriologist named Fred Griffith had discovered that, if killed bacteria of the smooth variety were mixed with live ones of the rough strain and then injected into a mouse, the tissues of the infected mouse would eventually contain live pneumococci of the smooth variety! How could this happen? The dead pneumococci had certainly not been brought to life. Something must have transformed the rough pneumococci so that they were now capable of making the smooth coat. What was that something? Evidently it was a factor of some kind contributed by the dead bacteria of the smooth strain.

In 1944, three American biochemists, Oswald Theodore Avery, Colin Munro Macleod, and Maclyn McCarty, identified the transforming principle. It was DNA. When they isolated pure DNA from the smooth strain and gave it to rough pneumococci, that alone sufficed to transform the rough strain to a smooth.

Investigators went on to isolate other transforming principles, involving other bacteria and other properties, and in every case the principle turned out to be a variety of DNA. The only plausible conclusion was that DNA could act like a gene. In fact, various lines of research, particularly with viruses (see Chapter 13), showed that the protein associated with DNA is almost superfluous from a genetic point of view: DNA can produce genetic effects all by itself, either in the chromosome or—in the case of nonchromosomal inheritance—in cytoplasmic bodies such as the chloroplasts.

If DNA is the key to heredity, it must have a complex structure, because it has to carry an elaborate pattern, or code of instructions (the "genetic code"), for the synthesis of specific enzymes. Assuming that it is made up of the four kinds of nucleotide, they cannot be strung in a regular arrangement, such as 1, 2, 3, 4, 1, 2, 3, 4, 1, 2, 3, 4. . . . Such a molecule would be far too simple to carry a blueprint for enzymes. In fact, the American biochemist Erwin Chargaff and his co-workers found definite evidence in 1948 that the composition of nucleic acids was more complicated than had been thought. Their analysis showed that the various purines and pyrimidines were not present in equal amounts, and that the proportions varied in different nucleic acids.

Everything seemed to show that the four purines and pyrimidines were distributed along the DNA backbone as randomly as the amino acid side chains were distributed along the peptide backbone. Yet some regularities did seem to exist. In any given DNA molecule, the total number of purines seemed to be equal, always, to the total number of pyrimidines. In addition, the number of adenines (one purine) was always equal to the number of thymines (one pyrimidine), while the number of guanines (the other purine) was always equal to the number of cytosines (the other pyrimidine).

We could symbolize adenine as A, guanine as G, thymine as T, and cytosine as C. The purines would then be A + G and the pyrimidines T + C. The findings concerning any given DNA molecule could then be summarized as:

$$A = T$$
$$G = C$$
$$A + G = T + C$$

More general regularities also emerged. As far back as 1938, Astbury had pointed out that nucleic acids scattered X-rays in diffraction patterns, a good sign of the existence of structural regularities in the molecule. The New Zealand-born British biochemist Maurice Hugh Frederick

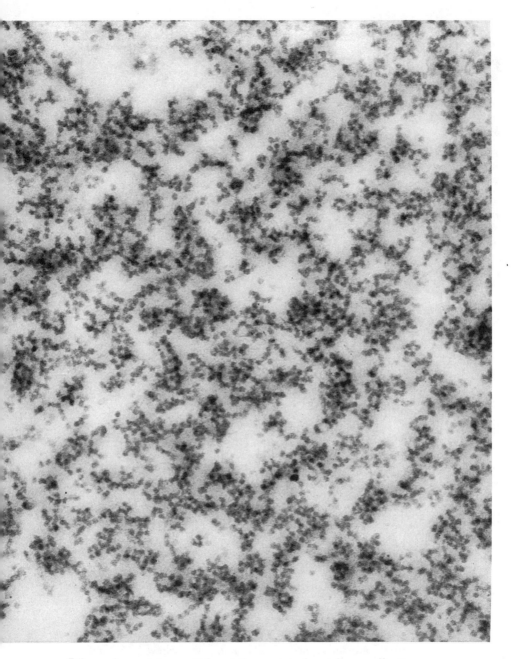

Ribonucleoprotein particles (ribosomes) from liver cells in a guinea pig. These particles are the main sites of the synthesis of proteins in the cell.

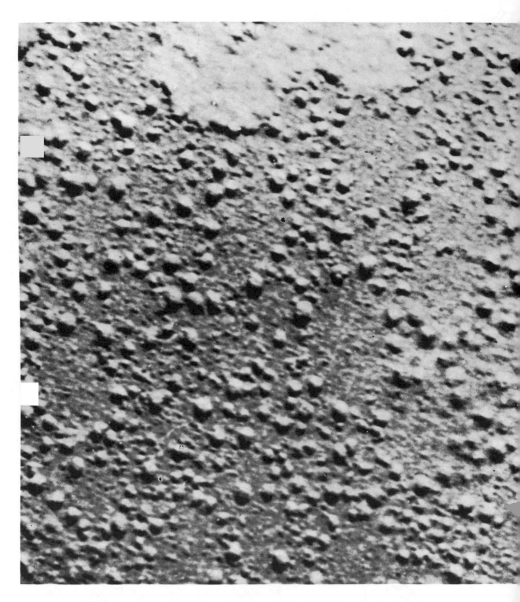

DNA-protein complex photographed with the electron microscope. The spherical bodies, isolated from the germ cells of a sea animal and magnified 77,500 times, are believed to consist of DNA in combination with protein.

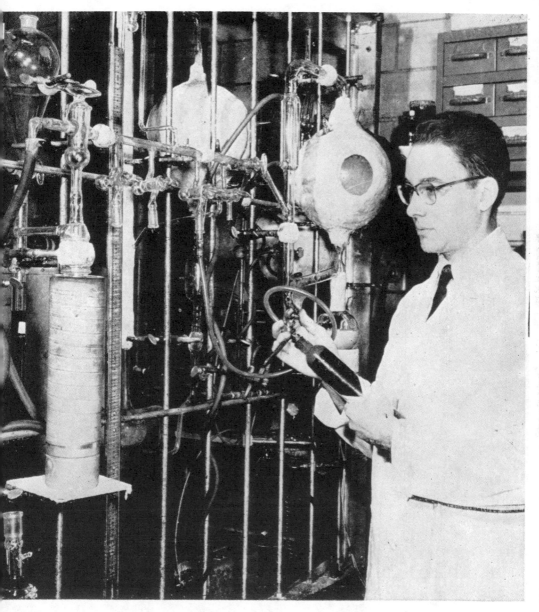

Stanley Miller's historic experiment creating amino acids in the laboratory.

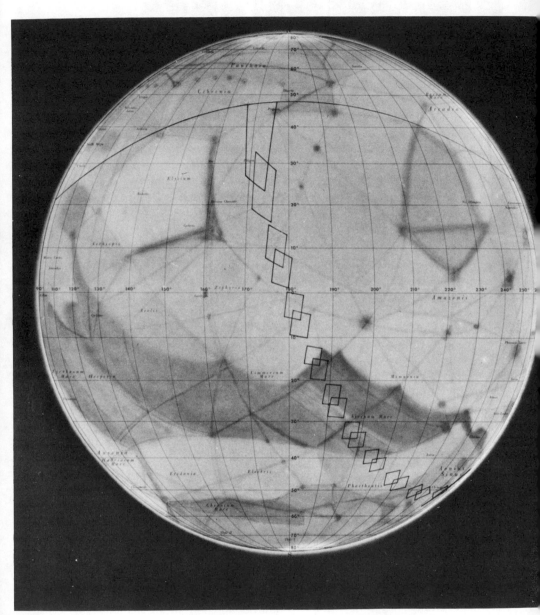

This is a map of Mariner IV's programmed photographic coverage of the planet Mars as computed by program officials two days after the historic event of July 14, 1965.

Wilkins calculated that these regularities repeated themselves at intervals considerably greater than the distance from nucleotide to nucleotide. One logical conclusion was that the nucleic-acid molecule took the form of a helix, with the coils of the helix forming the repetitive unit noted by the X-rays. This thought seemed the more attractive because Linus Pauling was at that time demonstrating the helical structure of certain protein molecules.

In 1953, the English physicist Francis Harry Compton Crick and his co-worker, the American biochemist (and one-time Quiz Kid) James Dewey Watson, put all the information together and came up with a revolutionary model of the nucleic-acid molecule—a model that represented it not merely as a helix but (and this was the key point) as a double helix—two sugar-phosphate backbones winding like a double-railed spiral staircase up the same vertical axis. From each sugar-phosphate chain, purines and peptides extended inward toward each other, meeting as though to form the steps of this double-railed spiral staircase.

Just how might the purines and pyrimidines be arrayed along these parallel chains? To make a good uniform fit, a double-ring purine on one side should always face a single-ring pyrimidine on the other, to make a three-ring width altogether. Two pyrimidines could not stretch far enough to cover the space; while two purines would be too crowded. Furthermore, an adenine from one chain would always face a thymine on the other, and a guanine on one chain would always face a cytosine on the other. In this way, one could explain the finding that $A = T$, $G = C$, and $A + T = G + C$.

This "Watson-Crick model" of nucleic-acid structure has proved to be extraordinarily fruitful and Wilkins, Crick, and Watson shared the 1962 Nobel Prize in medicine and physiology as a result.

The Watson-Crick model makes it possible, for instance, to explain just how a chromosome may duplicate itself in the process of cell division. Consider the chromosome as a string of DNA molecules. The molecules can first divide by a separation of the two helices making up the double helix; the two chains unwind themselves from each other, so to speak. This can be done because opposing purines and pyrimidines are held by hydrogen bonds, weak enough to be easily broken. Each chain is a half-molecule that can bring about the synthesis of its own missing complement. Where it has a thymine, it attaches an adenine; where it has a cytosine, it attaches a guanine; and so on. All the raw materials for making the units, and the necessary enzymes, are on hand in the cell. The half-molecule simply plays the role of a "template," or mold, for putting the units together in the proper order. The units eventually will fall into the appropriate places and stay there because that is the most stable arrangement.

To summarize, then, each half-molecule guides the formation of its own complement, held to itself by hydrogen bonds. In this way, it rebuilds the complete, double-helix DNA molecule, and the two half-molecules into which the original molecule divided thus form two molecules where only one existed before. Such a process, carried out by all the DNA's down the length of a chromosome, will create two chromosomes which are exactly alike and perfect copies of the original mother chromosome. Occasionally something may go wrong: the impact of a subatomic particle or of energetic radiation, or the intervention of certain chemicals, may introduce an imperfection somewhere or other in the new chromosome. The result is a mutation.

Evidence in favor of this mechanism of replication has been piling up. Tracer studies, employing heavy nitrogen to label chromosomes and following the fate of the labeled material during cell division, have tended to bear out the theory. In addition, some of the important enzymes involved in replication have been identified.

In 1955, the Spanish-American biochemist Severo Ochoa isolated from a bacterium (*Aztobacter vinelandii*) an enzyme which proved capable of catalyzing the formation of RNA from nucleotides. In 1956, a former pupil of Ochoa's, Arthur Kornberg, isolated another enzyme (from the bacterium *Escherichia coli*), which could catalyze the formation of DNA from nucleotides. Ochoa proceeded to synthesize RNA-like molecules from nucleotides, and Kornberg did the same for DNA. (The two men shared the Nobel Prize in medicine and physiology in 1959.) Kornberg also showed that his enzyme, given a bit of natural DNA to serve as a template, could catalyze the formation of a molecule which seemed to be identical with natural DNA. In 1965, Sol Spiegelman of the University of Illinois used RNA from a living virus (the simplest class of living things) and produced additional molecules of that sort. Since these additional molecules showed the essential properties of the virus, this was the closest approach yet to producing test-tube life. In 1967, Kornberg and others did the same, using DNA from a living virus as template.

The amount of DNA which is associated with the simplest manifestations of life is small—a single molecule in a virus—and can be made smaller. In 1967, Spiegelman allowed the nucleic acid of a virus to replicate and selected samples after shorter and shorter intervals for further replication. In this way, he selected molecules that completed the job unusually quickly—because they were smaller than average. In the end he had reduced the virus size to one-sixth normal and multiplied replication-speed fifteenfold.

Although it is DNA that replicated in cells, many of the simpler (subcellular) viruses contain RNA only. RNA molecules in double strands replicate in such viruses. The RNA in cells is single-stranded and does not replicate.

Nevertheless, a single-stranded structure and replication are not mutually exclusive. The American biophysicist Robert Louis Sinsheimer

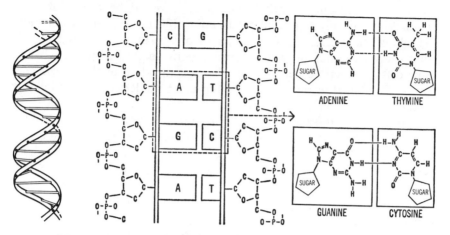

Model of the nucleic-acid molecule. The drawing at the left shows the double helix; in the center a portion of it is shown in detail (omitting the hydrogen atoms); at the right is a detail of the nucleotide combinations.

discovered a strain of virus that contained DNA made up of a single strand. That DNA molecule had to replicate itself; but how could that be done with but a single strand? The answer was not difficult. The single strand brought about the production of its own complement and the complement then brought about the production of the "complement to the complement," that is, a replica of the original strand.

It is clear that the single-strand arrangement is less efficient than the double-strand arrangement (which is probably why the former exists only in certain very simple viruses and the latter in all other living creatures). For one thing, a single strand must replicate itself in two successive steps, whereas the double strand does so in a single step. Second, it now seems that only one strand of the DNA molecule is the important working structure; the cutting edge of the molecule, so to speak. Its complement may be thought of as a protecting scabbard for that cutting edge. The double strand represents the cutting edge protected within the scabbard except when actually in use; the single strand is the cutting edge always exposed and continually subjected to blunting by accident.

Replication, however, merely keeps a DNA molecule in being. How does it accomplish its work, that of bringing about the synthesis of a specific enzyme—that is, of a specific protein molecule? To form a protein, the DNA molecule has to direct the placement of amino acids in a certain specific order in a molecule made up of hundreds or thousands of units. For each position it must choose the correct amino acid from some 20 different amino acids. If there were 20 corresponding units in the DNA molecule, it would be easy. But DNA is made up of only four different building blocks—the four nucleotides. Thinking about this, the astronomer George Gamow suggested in 1954 that the nucleotides, in various combinations, might be used as what we now call a "genetic code" (just as the dot and dash of the Morse code can be combined in various ways to represent the letters of the alphabet, numerals, and so on).

If you take the four different nucleotides (A, G, C, T), two at a time, there are 4×4, or 16 possible combinations (AA, AG, AC, AT, GA, GG, GC, GT, CA, CG, CC, CT, TA, TG, TC, and TT). This is still not enough. If you take them three at a time, there are $4 \times 4 \times 4$, or 64 different combinations, more than enough. (The reader may amuse himself trying to list the different combinations and see if he can find a sixty-fifth.)

It seemed as though each different "nucleotide triplet" or "codon" represented a particular amino acid. In view of the great number of different codons possible, it could well be that two or even three different codons represented one particular amino acid. In this case, the genetic code would be what cryptographers call "degenerate."

This left two chief questions: Which codon (or codons) corresponded to which amino acid? And how did the codon information (which was securely locked in the nucleus where, alone, the DNA was to be found) reach the sites of enzyme formation that were in the cytoplasm?

To take the second problem first, suspicion soon fell upon RNA as the substance serving as go-between. The French biochemists François Jacob and Jacques Lucien Monod were the first to suggest this. The structure of such RNA would have to be very like DNA with such differences as existed not affecting the genetic code. RNA had ribose in place of deoxyribose (one extra oxygen atom per nucleotide) and uracil in place of thymine (one missing methyl group, CH_3, per nucleotide). Furthermore, RNA was present chiefly in the cytoplasm, but also, to a small extent, in the chromosomes themselves.

It was not hard to see, and then demonstrate, what was happening. Every once in a while, when the two coiled strands of the DNA molecule unwound, one of those strands (always the same one, the cutting edge) replicates its structure, not on nucleotides that form a DNA molecule, but on nucleotides that form an RNA molecule. In this case, the adenine of the DNA strand does not attach thymine nucleotides to itself, but uracil nucleotides instead. The resulting RNA molecule, carrying the genetic code imprinted on its nucleotide pattern, can then leave the nucleus and enter the cytoplasm.

Since it carries the DNA "message," it has been named "messenger-RNA," or more simply, "mRNA."

The Rumanian-American biochemist George Emil Palade, thanks

to careful work with the electron microscope, demonstrated, in 1956, the site of enzyme manufacture in the cytoplasm to be tiny particles, about two millionths of a centimeter in diameter. They were rich in RNA and were therefore named "ribosomes." There are as many as 15,000 ribosomes in a bacterial cell, perhaps ten times as many in a mammalian cell. They are the smallest of the subcellular particles or "organelles." It was soon determined that the messenger-RNA—carrying the genetic code on its structure—made its way to the ribosomes and layered itself onto one or more of them, and that the ribosomes were the site of protein synthesis.

The next step was taken by the American biochemist Mahlon Bush Hoagland, who had also been active in working out the notion of mRNA showed that in the cytoplasm were a variety of small RNA molecules, which might be called "soluble-RNA" or "sRNA," because their small size enabled them to dissolve freely in the cytoplasmic fluid.

At one end of each sRNA molecule was a particular triplet of nucleotides that just fitted a complementary triplet somewhere on the mRNA chain. That is, if the sRNA triplet were AGC, it would fit tightly to a UCG triplet on the mRNA and only there. At the other end of the sRNA molecule was a spot where it would combine with one particular amino acid and none other. On each sRNA molecule, the triplet at one end meant a particular amino acid on the other. Therefore, a complementary triplet on the mRNA meant that only a certain sRNA molecule carrying a certain amino acid molecule would affix itself there. A large number of sRNA molecules would affix themselves one after the other, right down the line, to the triplets making up the mRNA structure (triplets that had been molded right on the DNA molecule of a particular gene). All the amino acids properly lined up could then easily be hooked together to form an enzyme molecule.

Because the information from the messenger-RNA is, in this way, transferred to the protein molecule of the enzyme, sRNA has come to be called "transfer-RNA," and it is this name which is now well established.

In 1964, the molecule of alanine-transfer-RNA (the transfer-RNA that attaches itself to the amino acid, alanine) was completely analyzed by a team headed by the American biochemist, Robert W. Holley. This was done by the Sanger-method of breaking down the molecule into small fragments by appropriate enzymes, then analyzing the fragments and deducing how they must fit together. The alanine-transfer-RNA, the first naturally occurring nucleic acid to be completely analyzed, was found to be made up of a chain of 77 nucleotides. These include not only the four nucleotides generally found in RNA (A, G, C, and U) but also several of seven others closely allied to them.

It had been supposed at first that the single chain of a transfer-RNA would be bent like a hairpin at the middle and the two ends would twine about each other in a double helix. The structure of alanine-transfer-RNA did not lend itself to this. Instead, it seemed to consist of three loops, so that it looked rather like a lopsided three-leaf clover. In subsequent years, other transfer-RNA molecules were analyzed in detail, and all seemed to have the same three-leaf-clover structure. For his work, Holley received a share of the 1968 Nobel Prize for medicine and physiology.

In this way, the structure of a gene controls the synthesis of a specific enzyme. Much, of course, remains to be worked out, for genes do not simply organize the production of enzymes at top speed at all times. The gene may be working efficiently now, slowly at another time, and not at all at still another time. Some cells manufacture protein at great rates, with an ultimate capacity of combining some 15 million amino acids per chromosome per minute, some only slowly, some scarcely at all, yet all the cells in a given organism have the same genic organization. Then, too, each type of cell in the body is highly specialized, with characteristic functions and chemical behavior of its own. An individual cell may synthesize a given protein rapidly at one time, slowly at another. And again all have the same genic organization all the time.

It is clear that cells have methods for blocking and unblocking the DNA molecules of the chromosomes. Through the pattern of blocking and unblocking, different cells with identical gene-patterns can produce different combinations of proteins, while a particular cell with an unchanging gene-pattern can produce different combinations from time to time.

In 1961, Jacob and Monod suggested that each gene has its own repressor, coded by a "regulator gene." This repressor—depending on its geometry, which can be altered by delicate changes in circumstances within the cell—will block or release the gene. In 1967, such a repressor was isolated and found to be a small protein. Jacob and Monod, together with a co-worker, André Michael Lwoff, received the 1965 Nobel Prize for medicine and physiology as a result.

Nor is the flow of information entirely one-way, from gene to enzyme. There is "feedback" as well. Thus, there is a gene that brings about the formation of an enzyme that catalyzes a reaction that converts the amino acid, threonine, to another amino acid, isoleucine. Isoleucine, by its presence, somehow serves to activate the repressor, which begins to shut down the very gene that produces the particular enzyme that led to that presence. In other words, as isoleucine concentration goes up, less is formed; if the concentration declines, the gene is unblocked, and more isoleucine is formed. The chemical machinery within the cell—genes, repressors, enzymes, end-products—is enormously complex and intricately interrelated. The complete unraveling of the pattern is not likely to take place rapidly.

he genetic code. In the left-hand column are the initials of the four RNA bases (uracil, cytosine, adenine, ïanine) representing the first "letter" of the codon triplet; the second letter is represented by the initials ross the top, while the third but less important letter appears in the final column. For example, tyrosine yr) is coded for by either UAU or UAC. Amino acids coded by each codon are shown abbreviated as llows: Phe — phenylalanine; Leu — leucine; Ileu — isoleucine; Met — methionine; Val — valine; Ser — ïine; Pro — proline; Thr — threonine; Ala — alanine; Tyr — tyrosine; His — histidine; Glun — glutamine; pn — asparagine; Lys — lysine; asp — aspastic acid; Glu — glutamic acid; Cys — cysteine; Tryp — ïptophan; Arg — arginine; Gly — glycine.

rst position	Second position				Third position
	U	C	A	G	
U	Phe	Ser	Tyr	Cys	U
	Phe	Ser	Tyr	Cys	C
	Leu	Ser	(normal "full stop")	"full stop"	A
	Leu	Ser	(less common "full stop")	Tryp	G
C	Leu	Pro	His	Arg	U
	Leu	Pro	His	Arg	C
	Leu	Pro	Glun	Arg	A
	Leu	Pro	Glun	Arg	G
A	Ileu	Thr	Aspn	Ser	U
	Ileu	Thr	Aspn	Ser	C
	Ileu?	Thr	Lys	Arg	A
	Met ("capital letter")	Thr	Lys	Arg	G
G	Val	Ala	Asp	Gly	U
	Val	Ala	Asp	Gly	C
	Val	Ala	Glu	Gly	A
	Val ("capital letter")	Ala	Glu	Gly	G

But meanwhile, what of the other question: Which codon goes along with which amino acid? The beginning of an answer came in 1961, thanks to the work of the American biochemists Marshall Warren Nirenberg and J. Heinrich Matthaei. They began by making use of a synthetic nucleic acid, built up according to Ochoa's system from uracil nucleotides only. This "polyuridylic acid" was made up of a long chain of . . . UUUUUUUU . . . and could only possess one codon, UUU.

Nirenberg and Matthaei added this polyuridylic acid to a system that contained various amino acids, enzymes, ribosomes, and all the other components necessary to synthesize proteins. Out of the mixture tumbled a protein made up only of the amino acid phenylalanine. This meant that UUU was equivalent to phenylalanine. The first item in the "codon dictionary" was worked out.

The next step was to prepare a nucleotide made out of a preponderance of uridine nucleotides with a small quantity of adenine nucleotides added. This meant that along with the UUU codon, an occasional UUA, or AUU, or UAU codon might appear. Ochoa and Nirenberg showed that, in such a case, the protein formed was mainly phenylalanine, but also contained an occasional leucine, isoleucine, and tyrosine, three other amino acids.

Slowly, by methods such as these, the dictionary was extended. It was found that the code is indeed degenerate and that GAU and GAC might each stand for aspartic acid, for instance, and the GUU, GAU, GUC, GUA, and GUG, all stand for glycine. In addition, there was some punctuation. The codon AUG not only stood for the amino acid, methionine, but apparently signified the beginning of a chain. It was a "capital letter," so to speak. Then, too, UAA and UAG signaled the end of a chain: they were "periods."

By 1967, the dictionary was complete. Nirenberg and his collaborator, the Indian-American chemist Har Gobind Khorana, were awarded shares (along with Holley) in the 1968 Nobel Prize for medicine and physiology.

The possible implications of a true understanding of the genetic code and of the manner in which it is modified from tissue to tissue and cell to cell are staggering. Probably nothing more exciting has happened in the life sciences in a century.

The Origin of Life

Once we get down to the nucleic-acid molecules, we are as close to the basis of life as we can get. Here, surely, is the prime substance of life itself. Without DNA, living organisms could not reproduce, and life as we know it could not have started. All the substances of living matter—enzymes and all the others, whose production is catalyzed by enzymes—depend in the last analysis on DNA. How, then, did DNA, and life, start?

This is a question that science has always hesitated to ask, because the origin of life has been bound up with religious beliefs even more strongly than has the origin of the earth and the universe. It is still dealt with only hesitantly and apologetically. But in recent years a book entitled The Origin of Life, by the Russian biochemist Aleksandr Ivanovich Oparin, has brought the subject very much to the fore. The book was published in the Soviet Union in 1924 and in English translation in 1936. In it the problem of life's origin for the first time was dealt with in detail from a completely materialistic point of view. Since the Soviet Union is not inhibited by the religious scruples to which the Western nations feel bound, this, perhaps, is not surprising.

Most of man's early cultures developed myths telling of the creation of the first human beings (and sometimes of other forms of life as well) by gods or demons. However, the formation of life itself was rarely thought of as being entirely a divine prerogative. At least the lower forms of life might arise spontaneously from nonliving material without supernatural intervention. Insects and worms might, for instance, arise from decaying meat, frogs from mud, mice from rotting wheat. This idea was based on actual observation, for decaying meat, to take most obvious example, did indeed suddenly give rise to maggots. It was only natural to assume that the maggots were formed from the meat.

Aristotle believed in the existence of "spontaneous generation." So did the great theologians of the Middle Ages, such as Thomas Aquinas. So did William Harvey and Isaac Newton. After all, the evidence of one's own eyes was hard to refute.

The first to put this belief to the test of experimentation was the Italian physician Francesco Redi. In 1668, he decided to check on whether maggots really formed out of decaying meat. He put pieces of meat in a series of jars and then covered some of them with fine gauze and left others uncovered. Maggots developed only in the meat in the uncovered jars, to which flies had had free access. Redi concluded that the maggots had arisen from microscopically small eggs laid on the meat by the flies. Without flies and their eggs, he insisted, meat could never produce maggots, however long it decayed and putrefied.

Experimenters who followed Redi confirmed this, and the belief that visible organisms arose from dead matter died. But when microbes were discovered, shortly after Redi's time, many scientists decided that these forms of life at least must come from dead matter. Even in gauze-covered jars, meat would soon begin to swarm with microorganisms. For two centuries after Redi's experiments, belief in the possibility of the spontaneous generation of microorganisms remained very much alive.

It was another Italian, the naturalist Lazzaro Spallanzani, who first cast serious doubt on this notion. In 1765, he set out two sets of vessels containing a broth. One he left open to the air. The other, which he had boiled to kill any organisms already present, he sealed up to keep out any organisms that might be floating in the air. The broth in the first vessels soon teemed with microorganisms, but the boiled and sealed-up broth remained sterile. This proved to Spallanzani's satisfaction that even microscopic life could not arise from inanimate matter. He even isolated a single bacterium and witnessed its division into two bacteria.

The proponents of spontaneous generation were not convinced. They maintained that boiling destroyed some "vital principle" and that this was why no microscopic life developed in Spallanzani's boiled, sealed flasks. It remained for Pasteur to settle the question, in 1862, seemingly once and for all. He devised a flask with a long swan neck in the shape of a horizontal S. With the opening unstoppered, air could percolate into the flask, but dust particles and microorganisms could not, for the curved neck would serve as a trap, like the drain trap under a sink. Pasteur put some broth in the flask, attached the S-shaped neck, boiled the broth until it steamed (to kill any microorganisms in the neck as well as in the broth), and waited for developments. The broth remained sterile. There was no vital principle in air. Pasteur's demonstration apparently laid the theory of spontaneous generation to rest permanently.

All this left a germ of embarrassment for scientists. How had life arisen, after all, if not through divine creation or through spontaneous generation?

Toward the end of the nineteenth century some theorists went to the other extreme and made life eternal. The most popular theory was advanced by Svante Arrhenius (the chemist who had developed the concept of ionization). In 1907, he published a book entitled *Worlds in the Making,* picturing a universe in which life had always existed and migrated across space, continually colonizing new planets. It traveled in the form of spores that escaped from the atmosphere of a planet by random movement and then were driven through space by the pressure of light from the sun.

Such light pressure is by no means to be sneered at as a possible driving force. The existence of radiation pressure had been predicted in the first place by Maxwell, on theoretical grounds, and, in 1899, it had been demonstrated experimentally by the Russian physicist Peter Nicolaevich Lebedev.

Arrhenius' views held, then, that spores traveled on and on through interstellar space, driven by light radiation this way and that, until they died or fell on some planet, where they would spring into active life and compete with life forms already present, or inoculate the planet with life if it was uninhabited but habitable.

At first blush, this theory looks attractive. Bacterial spores, protected by a thick coat, are very resistant to cold and dehydration and might conceivably last a long time in the vacuum of space. Also, they are of just the proper size to be more affected by the outward pressure of a

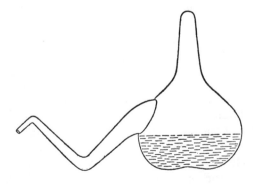

Pasteur's flask for the experiment on spontaneous generation.

sun's radiation than by the inward pull of its gravity. But Arrhenius's suggestion fell before the onslaught of ultraviolet light. In 1910, experimenters showed that ultraviolet light quickly killed bacterial spores, and in interplanetary space the sun's ultraviolet light is intense—not to speak of other destructive radiations, such as cosmic rays, solar X-rays, and zones of charged particles like the Van Allen belts around the earth. Conceivably, there may be spores somewhere that are resistant to radiation, but spores made of protein and nucleic acid, as we know them, could not make the grade. To be sure, some particularly resistant microorganisms were exposed to the radiation of outer space on board the "Gemini 9" capsule in 1966 and survived six hours of harsh unfiltered sunlight. But we are talking of exposures, not of hours, but of months and years.

Such a highly hydrogenated atmosphere we might call "Atmosphere I." Through photodissociation, this would slowly turn into an atmosphere of carbon dioxide and nitrogen (see page 000) or "Atmosphere II." After that an ozone layer would form in the upper atmosphere, and spontaneous change would halt. Could it be, then, that life could form in one or the other of the early atmospheres?

Consider Atmosphere II, for instance. Carbon dioxide is soluble in water, and while earth was bathed in Atmosphere II, the ocean would be a vast collection of seltzer. The ultraviolet radiation from the sun at sea level would be much more intense than it is today while Atmosphere II was in its last stages of formation and before the ozone layer was completely in place. What's more, the Earth's soil would have had a larger supply of radioactive atoms then than now. Under such conditions, could organic matter have sprung into existence?

H. C. Urey felt life started in Atmosphere I. In 1952, Stanley Lloyd Miller, then a graduate student in Urey's laboratories, circulated water, plus ammonia, methane and hydrogen, past an electric discharge (to simulate the ultraviolet radiation of the sun). At the end of a week he analyzed his solution by paper chromatography and found that, in addition to the simple substances without nitrogen atoms, he also had glycine and alanine, the two simplest of the amino acids, plus some indication of one or two more complicated ones.

Miller's experiment was significant in several ways. In the first place, these compounds had formed quickly and in surprisingly large quantities. One-sixth of the methane with which he had started had gone into the formation of more complex organic compounds, yet the experiment had only been in operation for a week.

Then, too, the kind of organic molecules formed in Miller's experiments were just those present in living tissue. The path taken by the simple molecules, as they grew more complex, seemed pointed directly toward life. This pointing-toward-life continued consistently in later, more elaborate experiments. At no time were molecules formed in significant quantity that seemed to point in an unfamiliar nonlife direction.

Thus, P. H. Abelson followed Miller's work by trying a variety of similar experiments with starting materials made up of different gases in different combinations. It turned out that as long as he began with molecules that included atoms of carbon, hydrogen, oxygen, and nitrogen, amino acids of the kind normally found in proteins were formed. Nor were electric discharges the only source of energy that would work. In 1959, two German scientists, Wilhelm Groth and H. von Weyssenhoff, designed an experiment in which ultraviolet light could be used instead, and they also got amino acids.

If there was any doubt that the direction-toward-life was the line of least resistance, there was the fact that in the late 1960's, more and more complicated molecules, representing the first stages of that direction, were found in gas clouds of outer space (see Chapter 2). It may be, then, that at the time the earth was formed out of clouds of dust and gas, the first stages of building up complex molecules had already taken place.

The earth, at its first formation, may have had a supply of amino acids. Evidence in favor of that came in 1970. The Ceylon-born biochemist Cyril Ponnamperuma studied a meteorite that had fallen in Australia on September 28, 1969. Careful analyses showed the presence of small traces of five amino acids: glycine, alanine, glutamic acid, valine, and proline. There was no optical activity in these amino acids, so they were formed not by life processes (and hence their presence was not the result of earthly contamination) but by the nonliving chemical processes of the type that took place in Miller's flask.

Could chemists in the laboratory progress beyond the amino acid stage? One way of doing so would be to start with larger samples of raw materials and subject them to energy for longer periods. This would produce increasing numbers of more and more complicated products, but the mixtures of these products would become increasingly complex and would be increasingly difficult to analyze.

Instead, chemists began with later stages. The products formed in earlier experiments would be used as new raw materials. Thus, one of Miller's products was hydrogen cyanide. The Spanish-American biochemist Juan Oro added hydrogen cyanide to the starting mixture in 1961. He obtained a richer mixture of amino acids and even a few short peptides. He also formed purines, in particular adenine, a vital component of nucleic acids. In 1962, Oro used formaldehyde as one of his raw materials and produced ribose and deoxyribose, also components of nucleic acids.

In 1963, Ponnamperuma also performed experiments similar to those of Miller, using electron beams as a source of energy, and found that adenine was formed. Together with Ruth Mariner and Carl Sagan, he went on to add adenine to a ribose solution and under ultraviolet light, "adenosine," a molecule formed of adenine and ribose linked together, was formed. If phosphate was also present, it, too, was hooked on to form the adenine nucleotide. Indeed, three phosphate groups could be added to form adenosine triphosphate (ATP), which, as was explained in Chapter 11, is essential to the energy-handling mechanisms of living tissue. In 1965, he formed a "dinucleotide," two nucleotides bound together. Additional products can be built up if substances such as cyanamide ($CNNH_2$) and ethane (CH_3CH_3)—substances which may well have been present in the primordial era—are added to the mixtures employed by various experimenters in this field. There is no question, then, but that normal chemical and physical changes in the primordial ocean and atmosphere could have acted in such a way as to build up proteins and nucleic acids.

Any compound that formed in the lifeless ocean would tend to endure and accumulate. There were no organisms, either large or small, to consume them or cause them to decay. Moreover, in the primeval atmosphere there was no free oxygen to oxidize and break down the molecules. The only important factors tending to break down complex molecules would have been the very ultraviolet and radioactive energies that built them up. But ocean currents might have carried much of the material to a safe haven at mid-levels in the sea, away from the ultraviolet-irradiated surface and the radioactive bottom. Indeed, Ponnamperuma and his co-workers have estimated that fully 1 per cent of the primordial ocean may have been made up of these built-up organic compounds. If so, this would represent a mass of over a million billion tons. This is certainly an ample quantity for natural forces to play with, and in such a huge mass even substances of most unlikely complexity are bound to be built up in not too long a period (particularly considering a couple of billions of years are available for the purpose).

There is no logical barrier, then, to supposing that out of the simple compounds in the primordial ocean and atmosphere there appeared, with time, higher and higher concentrations of the more complicated amino acids, as well as simple sugars; that amino acids combined to form peptides; that purines, pyrimidines, sugar, and phosphate combined to form nucleotides; and that, gradually, over the ages, proteins and nucleic acids were created. Then, eventually, must have come the key step—the formation, through chance combinations, of a nucleic acid molecule, capable of inducing replication. That moment marked the beginning of life.

Thus a period of "chemical evolution" preceded the evolution of life itself.

A single living molecule, it seems, might well have been sufficient to get life under way and give rise to the whole world of widely varying living things, as a single fertilized cell can give rise to an enormously complex organism. In the organic "soup" that constituted the ocean at that time, the first living molecule could have replicated billions and billions of molecules like itself in short order. Occasional mutations would create slightly changed forms of the molecule, and those that were in some way more efficient than the others would multiply at the expense of their neighbors and replace the old forms. If one group was more efficient in warm water and another group in cold water, two varieties would arise, each restricted to the environment it fitted best. In this fashion, the course of "organic evolution" would be set in motion.

Even if several living molecules came into existence independently at the beginning, it is very likely that the most efficient one would have outbred the others, so that all life today may very well be descended from a single original molecule. In spite of the great present diversity of living things, all have the same basic ground plan. Their cells all carry out metabolism in pretty much the same way. Furthermore, it seems particularly significant that the proteins of all living things are composed of L-amino acids rather than amino acids of the D type. It may be that the original nucleoprotein from which all life is descended happened to be built from L-amino acids by chance, and since D could not be associated with L in any stable chain, what began as chance persisted by replication into grand universality. (This is not to imply that D-amino acids are totally absent in nature. They occur in the cell walls of some bacteria and in some antibiotic compounds. These, however, are quite exceptional cases.)

Of course, the step from a living molecule to the kind of life we know today, is still an enormous one. Except for the viruses, all life is organized into cells, and a cell, however small it may seem by human standards, is enormously complex in its chemical structure and interrelationships. How did that start?

The question of the origin of cells was illuminated by the researches of the American biochemist Sidney W. Fox. It seemed to him that the early earth must have been quite hot and that the energy of heat alone could be sufficient to form complex compounds out of simple ones. To test this, Fox in 1958 heated a mixture of amino acids and found they formed long chains that resembled those in protein molecules. These "proteinoids" were digested by enzymes that digested ordinary proteins and could be used as food by bacteria.

Most startling of all, when Fox dissolved the proteinoids in hot water and let the solution cool, he found they would cling together in little "microspheres" about the size of small bacteria. These microspheres were not alive by the usual standards, but they behaved as cells do, in some respects at least (they are surrounded by a kind of membrane, for instance). By adding certain chemicals to the solution, Fox could make the microspheres swell or shrink, much as ordinary cells do. They can produce buds, which sometimes seem to grow larger and then break off. Microspheres can separate, divide in two, or cling together in chains.

Perhaps in primordial times, such tiny not-quite-living aggregates of materials formed in several varieties. Some were particularly rich in DNA and were very good at replicating, though only moderately successful at storing energy. Other aggregates could handle energy well but replicated only limpingly. Eventually, collections of such aggregates might have cooperated, each supplying the deficiencies of the other, to form the modern cell, which was much more efficient than any of its parts alone. The modern cell still has the nucleus—rich in DNA but unable of itself to handle oxygen—and numerous mitochondria—which handle oxygen with remarkable efficiency but cannot reproduce in the absence of nuclei. (That mitochondria may once have been independent entities is indicated by the fact that they still possess small quantities of DNA.)

Throughout the existence of Atmospheres I and II, primitive lifeforms could only exist at the cost of breaking down complex chemical substances into simpler ones and storing the energy evolved. The complex substances were rebuilt by the action of the ultraviolet radiation of the sun. Once Atmosphere II was completely formed and the ozone layer was in place, the danger of starvation set in, for the ultraviolet supply was cut off.

By then, though, some mitochondria-like aggregate was formed which contained chlorophyll—the ancestor of the modern chloroplast. In 1966, the Canadian biochemists G. W. Hodson and B. L. Baker began with pyrrole and paraformaldehyde (both of which can be formed from still simpler substances in Miller-type experiments) and demonstrated the formation of porphyrin rings, the basic structure of chlorophyll, after merely three hours gentle heating.

Even the inefficient use of visible light by the first primitive chlorophyll-containing aggregates must have been much preferable to the procedure of nonchlorophyll systems at the time when the ozone layer was forming. Visible light could easily penetrate the ozone layer, and the lower energy of visible light (compared with ultraviolet) was enough to activate the chlorophyll system.

The first chlorophyll-using organisms may have been no more complicated than individual chloroplasts today. There are, in fact, two thousand species of a group of one-celled photosynthesizing organisms called "blue-green algae" (they are not all blue-green, but the first ones studied were). These are very simple cells, rather bacteria-like in structure, except that they contain chlorophyll and bacteria do not. Blue-green algae may be the simplest descendants of the original chloroplast, while bacteria may be the descendants of chloroplasts that lost their chlorophyll and took to parasitism or to foraging on dead tissue and its components.

As chloroplasts multiplied in the ancient seas, carbon dioxide was gradually consumed and molecular oxygen took its place. The present Atmosphere III was formed. Plant cells grew steadily more efficient, each one containing numerous chloroplasts. At the same time, elaborate cells without chlorophyll could not exist on the previous basis, for the plant cells cleared the oceans of the food supply and no more was formed except within those cells. However, cells without chlorophyll but with elaborate mitochondrial equipment that could handle complex molecules with great efficiency and store the energy of their breakdown, could live by ingesting the plant cells and stripping the molecules the latter had painstakingly built up. Thus originated the animal cell of today. Eventually, organisms grew complex enough to begin to leave the fossil record (plant and animal) that we have today.

Meanwhile, the earth environment had changed fundamentally, from the standpoint of creation of new life. Life could no longer originate and develop from purely chemical evolution. For one thing, the forms of energy that had brought it into being in the first place— ultraviolet and radioactive energy—were effectively gone. For another, the well-established forms of life would quickly consume any organic molecules that arose spontaneously. For both these reasons there is virtually no chance of any new and independent breakthrough from nonlife into life (barring some future intervention by man, if he learns to turn the trick). Spontaneous generation today is so highly improbable that it can be regarded as essentially impossible.

Life in Other Worlds

If we accept the view that life arose simply from the workings of physical and chemical laws, it follows that in all likelihood life is not confined to the earth. What are the possibilities of life elsewhere in the universe?

Beginning close to home, we can eliminate the moon pretty definitely, because of its lack of detectable air or water. There has been some speculation that very primitive forms of life may exist here and there in deep crannies, where traces of water and air might conceivably be present. The British astronomer V. A. Firsoff has argued in his book, *Strange World of the Moon*, that water may lie below the moon's surface and that gases absorbed on the dust-covered surface may give rise to a very shallow "atmosphere" that might support life of a sort. Urey maintained that meteoric bombardment of a primordial earth may have splashed water to the moon, giving it temporary lakes and streams. Lunar Orbiter photographs of the moon have indeed shown markings that seem for all the world like tracks of ancient rivers.

However, the rocks brought back from the moon by successive expeditions to our satellite beginning in 1969, have proven absolutely without water or organic material. To be sure, they are surface rocks only, and the story may be different once samples from several feet below are obtained. Nevertheless, what evidence we have so far seems to indicate there may be no life of any sort, however simple, on the moon.

Venus might seem a somewhat better candidate for life, judging by its mass alone (for it is quite close to the earth in size) and by the indisputable fact that it possesses an atmosphere even denser than ours. In 1959, the bright star Regulus was eclipsed by Venus. From studies of the manner in which the light of Regulus penetrated Venus' atmosphere, new knowledge concerning its depth and density was obtained. The American astronomer Donald Howard Menzel was able to announce that the diameter of the solid body of Venus was 3783.5 miles, seventy miles less than the best previous estimate.

The atmosphere of Venus seems to be mainly carbon dioxide. Venus has no detactable free oxygen, but this lack is not a fatal barrier to life. As I explained above, the earth very likely supported life for ages before it possessed free oxygen in its atmosphere (although, to be sure, that life was undoubtedly very primitive).

Until recently no sign of vapor was found in Venus' air, and that *was* a fatal objection to the possibility of life there. But astronomers were uncomfortable about the failure to find water on Venus, because it left them with no completely satisfactory explanation of the planet's cloud cover. Then, in 1959, Charles B. Moore of the Arthur D. Little Company laboratory went up in a balloon and at an altitude of fifteen miles in the stratosphere, above most of the earth's interfering atmosphere, took pictures of Venus with a telescope. Its spectrum in the infrared showed that Venus' upper atmosphere contained as much water vapor as does the earth's.

However, the possibilities of life on Venus took another and, apparently, final nose dive, because of its surface temperature. Radio waves coming from Venus suggested that its surface temperature might be far above the boiling point of water. There was some hope, however, that the radio waves indicating this temperature might arise somewhere in Venus' upper atmosphere and that the planet's actual surface might have a bearable temperature.

This hope, however, was dashed in December 1962, when the American Venus probe, Mariner II, passed close by Venus' position in space and scanned its surface for radio-wave radiation. The results, when analyzed, clearly showed that Venus' surface was far too hot for it to retain an ocean. All its water supply was in its clouds. Venus' temperature—which may be as high as 500° C., judging by a Soviet probe that actually transmitted surface temperatures in December 1970—seems effectively to eliminate the possibility of life as we know it. Mercury, which is closer to the sun than Venus, smaller, as hot, and lacking an atmosphere, is also eliminated.

Mars is without doubt a more hopeful possibility. It has a thin atmosphere of carbon dioxide and nitrogen, and it seemed to have enough water to show thin ice caps (probably an inch or so thick at most), which form and melt with the seasons. The Martian temperatures are low, but not too low for life and probably no worse at any time than Antarctica. The temperature may even reach an occasional balmy 80° F. at Mar's equator during the height of the summer day.

The possibility of life on Mars has excited the world for nearly a century. In 1877, the Italian astronomer Giovanni Virginio Schiaparelli detected fine, straight lines on the surface of the planet. He named them "canali." That is the Italian word for "channels," but it was mistranslated into English as "canals," whereupon people jumped to the conclusion that the lines were artificial waterways, perhaps built to bring water from the ice caps to other parts of the planet for irrigation. The American astronomer Percival Lowell vigorously championed this interpretation of the markings, and the Sunday supplements (and science-fiction stories) went to town on the "evidence" of intelligent life on Mars.

Actually, there was considerable doubt that the markings exist at all. Many astronomers have never seen them, despite earnest attempts; others have seen them only in flashes. Many believe that they may be optical illusions, arising from strained efforts to see something just at the limit of vision. In any case, no astronomer believes that Mars could support any form of advanced life.

Yet there remained the chance that Mars might support simple forms of life. There are, for instance, green patches on Mars' surface that change with the seasons, expanding in the hemisphere that is experiencing summer and contracting in the hemisphere that is experiencing winter. Could this be a sign of a cover of a simple form of plant life? Laboratory experiments have shown that some lichens and microorganisms, though adapted to the earth's environment rather than Mars', could live and grow at temperatures and in an atmosphere that are believed to simulate the Martian environment.

Such hopes began to wither as the result of the results of the "Mariner IV" Martian probe, launched on November 28, 1964. In 1965, it passed within 6,000 miles of Mars and sent back photographs of its surface. This revealed no canals but showed, for the first time, that the Martian surface was heavily cratered, rather like the moon's. It was deduced from this that not only was the Martian atmosphere thin and desiccated now, but it might always have been.

Additional information of the Martian surface gained from an even more sophisticated "flyby" in 1969 made the situation seem still worse.

The atmosphere was even thinner than had been believed, and the temperature lower. The temperature at the Martian south pole seemed to be no higher than 150° K., and the "icecaps," on which so much hope for a water supply had been based, may well be solid carbon-dioxide. While there can be no certainty until an actual Martian landing has been made, it seems rather likely now that life as we know it is not present on Mars.

As for planetary bodies beyond Mars (or satellites or planetoids), it would seem that conditions are harsher still. To be sure, Carl Sagan has suggested that the atmosphere of Jupiter would produce so strong a greenhouse effect as to give rise to moderate temperatures that might support life of a sort. This seemed less far-fetched when it turned out that Jupiter was radiating three times as much energy as it receives from the sun, so that some other energy source (planetary contraction, perhaps) may be involved and temperatures higher than expected may exist. This, however, remains strongly speculative, and lacking a planetary probe in Jupiter's neighborhood, nothing more can be said.

We can conclude, then, that as far as the solar system is concerned, the earth and only the earth seems to be an abode of life. But the solar system isn't all there is. What are the possibilities of life elsewhere in the universe?

The total number of stars in the known universe is estimated to be at least 1,000,000,000,000,000,000,000 (a billion trillion). Our own Galaxy contains well in excess of 100,000,000,000 stars. If all the stars developed by the same sort of process as the one that is believed to have created our own solar system (i.e., the condensing of a large cloud of dust and gas), then it is likely that no star is solitary, but each is part of a local system containing more than one body. We know that there are many double stars, revolving around a common center, and it is estimated that at least one out of three stars belongs to a system containing two or more stars.

What we really want, though, is a multiple system in which a number of members are too small to be self-luminous and are planets rather than stars. Though we have no means (so far) of detecting directly any planet beyond our own solar system, even for the nearest stars, we can gather indirect evidence. This has been done at the Sproul Observatory of Swarthmore College under the guidance of the Dutch-American astronomer Peter Van de Kamp.

In 1943, small irregularities of one of the stars of the double-star system, 61 Cygni, showed that a third component, too small to be self-luminous, must exist. This third component, 61 Cygni C, had to be about eight times the mass of Jupiter and therefore (assuming the same density) about twice the diameter. In 1960, a planet of similar size was located circling about the small star Lalande 21185 (located, at least, in the sense that its existence was the most logical way of accounting for irregularities in the star's motion). In 1963, a close study of Barnard's star indicated the presence of a planet there, too, one that was only one and one-half times the mass of Jupiter.

Barnard's star is second closest to ourselves, Lalande 21185 third closest, and 61 Cygni twelfth closest. That three planetary systems should exist in close proximity to ourselves would be extremely unlikely, unless planetary systems were very common generally. Naturally, at the vast distances of the stars, only the largest planets could be detected and even then with difficulty. Where super-Jovian planets exist, it seems quite reasonable (and even inevitable) to suppose that smaller planets also exist.

But even assuming that all or most stars have planetary systems and that many of the planets are earthlike in size, we must know what criteria must such planets fulfill to be habitable. The American space scientist Stephen H. Dole has made a particular study of this problem in his book *Habitable Planets for Man*, published in 1964, and has reached certain conclusions, admittedly speculative, but reasonable.

He points out, in the first place, that a star must be of a certain size in order to possess a habitable planet. The larger the star, the shorter-lived it is, and, if it is larger than a certain size, it will not live long enough to allow a planet to go through the long stage of chemical evolution prior to the development of complex life forms. A star that is too small cannot warm a planet sufficiently, unless that planet is so close that it will suffer damaging tidal effects. Dole concludes that only stars of spectral classes F2 to K1 are suitable for the nurturing of planets that are comfortably habitable for mankind: planets that he can colonize (if travel between the stars ever becomes practicable) without undue effort. There are, Dole estimates, 17 billion such stars in our Galaxy.

Such a star might be capable of possessing a habitable planet and yet might not possess one. Dole estimates the probabilities that a star of suitable size might have a planet of the right mass and at the right distance, with an appropriate period of rotation and an appropriately regular orbit, and by making what seem to him to be reasonable estimates, he concludes that there are likely to be 600 million habitable planets in our Galaxy alone, each of them already containing some form of life.

If these habitable planets are spread more or less evenly throughout the Galaxy, Dole estimates that there is one habitable planet per 80,000 cubic light-years. This means that the nearest habitable planet to ourselves may be some twenty-seven light-years away and that, within one-hundred light-years of ourselves, there may be a total of fifty habitable planets.

Dole lists fourteen stars within twenty-two light-years of ourselves that might possess habitable planets and weighs the probabilities that this might be true in each case. He concludes that the greatest likelihood of habitable planets is to be found in the stars closest to us, the two sun-like stars of the Alpha Centauri system, Alpha Centauri A and Alpha Centauri B. These two companion stars, taken together, have, Dole estimates, 1 chance in 10 of possessing habitable planets. The total probability for all 14 neighboring stars is about 2 chances in 5.

If life is the consequence of the chemical reactions described in the previous section, its development should prove inevitable on any earth-like planet. Of course, a planet may possess life and yet not possess intelligent life. We have no way of making even an intelligent guess as to the likelihood of the development of intelligence on a planet, and Dole, for instance, is careful to make none. After all, our own earth, the only habitable planet we really know we can study, existed for at least 2 billion years with a load of life, but not intelligent life.

It is possible that the porpoises and some of their relatives are intelligent, but, as sea creatures, they lack limbs and could not develop the use of fire; consequently, their intelligence, if it exists, could not be bent in the direction of a developed technology. If land life alone is considered, then it is only for about a million years or so that the earth has been able to boast a living creature with intelligence greater than that of an ape.

Still, this means that the earth has possessed intelligent life for 1/2000 of the time it has possessed life of any kind (as a rough guess). If we can say that of all life-bearing planets, 1 out of 2,000 bears intelligent life, then that would still mean that out of the 640 million habitable planets Dole speaks of, there may be 320,000 intelligences. We may well be far from alone in the universe.

Until recently this sort of possibility was considered seriously only in science-fiction stories. Those of my readers who happen to be aware that I have written a few science-fiction stories in my time and who may put down my remarks here to overenthusiasm may be assured that today many astronomers accept the high probability of intelligent life on many planets.

In fact, United States scientists took the possibilities seriously enough to set up an enterprise, under the leadership of Frank D. Drake, called Project Ozma (deriving its name from one of the Oz books for children) to listen for possible radio signals from other worlds. The idea is to look for some pattern in radio waves coming in from space. If they detect signals in an ordered pattern, as opposed to the random, formless broadcasts from radio stars or excited matter in space, it may be assumed that such signals will represent messages from some extra-terrestrial intelligence. Of course, even if such messages were received, communication with the distant intelligence would still be a problem. The messages would have been many years on the way, and a reply also would take many years to reach the distant broadcasters, since the nearest potentially habitable planet is 4⅓ light-years away.

The sections of the heavens listened to at one time or another in the course of Project Ozma included the directions in which lie Epsilon Eridani, Tau Ceti, Omicron-2 Eridani, Epsilon Indi, Alpha Centauri, 70 Ophiuchi, and 61 Cygni. After two months of negative results, however, the project was suspended.

The notion, however, is by no means dead. Because the universe is so vast, evidence of faraway life cannot yet be detected, and may never be detected. But precisely because the universe is so vast, such faraway life *must* exist, even intelligent life, perhaps in many millions of varieties.

And perhaps if we never find them, they may yet, in time to come, find us.

CHAPTER 13

The Microorganisms

Bacteria

Before the seventeenth century, the smallest known living creatures were tiny insects. It was taken for granted, of course, that no smaller organisms existed. Living beings might be made invisible by a supernatural agency (all cultures believed that in one way or another), but no one supposed that there were creatures in nature too small to be seen.

Had man suspected such a thing, he might have come much sooner to the deliberate use of magnifying devices. Even the Greeks and Romans knew that glass objects of certain shapes would focus sunlight on a point and would magnify objects seen through it. A hollow glass sphere filled with water would do so, for instance. Ptolemy discussed the optics of burning glasses, and Arabic writers such as Alhazen, about 1000 A.D., extended his observations.

It was Robert Grosseteste, an English bishop, philosopher, and keen amateur scientist, who, early in the thirteenth century, first suggested a peacetime use for this weapon. He pointed out that lenses (so named because they were shaped like lentils) might be useful in magnifying objects too small to see conveniently. His pupil, Roger Bacon, acted on this suggestion and devised spectacles to improve poor vision.

At first only convex lenses, to correct farsightedness, were made. Concave lenses, to correct nearsightedness, were not developed until about 1400. The invention of printing brought more and more demand for spectacles, and by the sixteenth century spectaclemaking was a skilled profession. It became a particular specialty in the Netherlands.

(Bifocals, serving both for far and near vision, were invented by Benjamin Franklin in 1760. In 1827, the British astronomer George Biddell Airy designed the first lenses to correct astigmatism, from which he suffered himself. And, around 1888, a French physician introduced the idea of contact lenses, which may some day make ordinary spectacles more or less obsolete.)

To get back to the Dutch spectaclemakers. In 1608, so the story goes, an apprentice to a spectaclemaker named Hans Lippershey, whiling away an idle hour, amused himself by looking at objects through two lenses held one behind the other. He was amazed to find that when he held them a certain distance apart, far-off objects appeared close at hand. The apprentice promptly told his master about it, and Lippershey proceeded to build the first "telescope," placing the two lenses in a tube to hold them at the proper spacing. Prince Maurice of Nassau, commander of the Dutch forces in rebellion against Spain, saw the military value of the instrument and endeavored to keep it secret.

He reckoned without Galileo, however. Hearing rumors of the invention of a far-seeing glass, Galileo, knowing no more than that it was made with lenses, soon discovered the principle and built his own telescope; his was completed within six months after Lippershey's.

By rearranging the lenses of his telescope, Galileo found that he could magnify close objects, so that it was in effect a "microscope." Over the next decades several scientists built microscopes. An Italian naturalist named Francesco Stelluti studied insect anatomy with one; Malpighi discovered the capillaries; and Hooke discovered the cells in cork.

But the importance of the microscope was not really appreciated until Anton van Leeuwenhoek, a merchant in the city of Delft, took it up (see page 507). Some of van Leeuwenhoek's lenses could enlarge up to 200 times.

Van Leeuwenhoek looked at all sorts of objects quite indiscriminately, describing what he saw in lengthy detail in letters to the Royal Society in London. It was rather a triumph for the democracy of science that the tradesman was elected a fellow of the gentlemanly Royal Society. Before he died, the Queen of England and Peter the Great, Czar of all the Russias, visited the humble microscope maker of Delft.

Through his lenses van Leeuwenhoek discovered sperm cells, red blood cells, and actually saw blood moving through capillaries in the tail of a tadpole. More important, he was the first to see living creatures too small to be seen by the unaided eye. He discovered these "animalcules" in stagnant water in 1675. He also resolved the tiny cells of yeast, and, at the limit of his lenses' magnifying power, he finally, in 1676, came upon "germs," which today we know as bacteria.

Microscopes improved only slowly, and it took a century and a half before objects the size of germs could be studied with ease. For instance, it was not until 1830 that the English optician Joseph Jackson Lister devised an "achromatic microscope," which eliminated the rings of color that limited the sharpness of the image. Lister found that red blood corpuscles (first detected as featureless blobs by the Dutch physician, Jan Swammerdam, in 1658) were biconcave disks—like tiny doughnuts with dents instead of a hole. The achromatic microscope was a great advance, and in 1878 the German physicist Ernst Abbé began a series of improvements that resulted in what might be called the modern optical microscope.

The members of the new world of microscopic life gradually received names. Van Leeuwenhoek's "animalcules" actually were animals, feeding on small particles and moving about by means of small whips (flagellae) or hairlike cilia or advancing streams of protoplasm (pseudopods). These animals were given the name "protozoa" (Greek for "first animals"), and the German zoologist Karl Theodor Ernst Siebold identified them as single-celled creatures.

"Germs" were something else: much smaller than protozoa and much simpler. Although some could move about, most lay quiescent and merely grew and multiplied. Except for their lack of chlorophyll, they showed none of the properties associated with animals. For that reason they were usually classified among the fungi—plants that lack chlorophyll and live on organic matter. Nowadays most biologists tend to consider them as neither plant nor animal but put them in a class by themselves. "Germ" is a misleading name for them. The same term may apply to the living part of a seed (e.g., the "wheat germ"), or to sex cells ("germ cells"), or to embryonic organs ("germ layers"), or, in fact, to any small object possessing the potentiality of life.

The Danish microscopist Otto Frederik Müller managed to see the little creatures well enough in 1773 to distinguish two types: "bacilli" (from a Latin word meaning "little rods") and "spirilla" (for their spiral shape). With the advent of achromatic microscopes, the Austrian surgeon Theodor Billroth saw still smaller varieties to which he applied the term "coccus" (from the Greek word for "berry"). It was the German botanist Ferdinand Julius Cohn who finally coined the name "bacterium" (also from a Latin word meaning "little rod").

Pasteur popularized the general term "microbe" ("small life") for all forms of microscopic life—plant, animal, and bacterial. But this word was soon adopted for the bacteria, just then coming into notoriety. Today the general term for microscopic forms of life is "microorganism."

It was Pasteur who first definitely connected microorganisms with disease, thus founding the modern science of "bacteriology" or, to use a more general term, "microbiology." This came about through Pasteur's concern with something that seemed an industrial problem rather than a medical one. In the 1860's, the French silk industry was being ruined by a disease of the silkworms. Pasteur, having already rescued France's wine makers, was put to work on this problem, too. Again making inspired use of the microscope, as he had in studying asymmetric crystals and varieties of yeast cells, Pasteur found microorganisms infecting the sick silkworms and the mulberry leaves on which they fed. He recommended that all infected worms and leaves be destroyed and a fresh start be made with the uninfected worms and leaves that remained. This drastic step was taken, and it worked.

Pasteur did more with these researches than merely to revive the silk industry. He generalized his conclusions and enunciated the "germ theory of disease"—without question one of the greatest single medical discoveries ever made (and it was made, not by a physician, but by a chemist, as chemists delight in pointing out).

Before Pasteur, doctors had been able to do little more for their patients than recommend rest, good food, fresh air, and clean surroundings, and, occasionally, handle a few types of emergency. This much had been advocated by the Greek physician Hippocrates of Cos (the "father of medicine") as long ago as 400 B.C. It was Hippocrates who introduced the rational view of medicine, turning away from the arrows of Apollo and demonic possession to proclaim that even epilepsy, called the "sacred disease," was not the result of being affected by some god's influence, but was a mere physical disorder to be treated as such. The lesson was never entirely forgotten by later generations.

However, medicine progressed surprisingly little in the next two millennia. Doctors could lance boils, set broken bones, and prescribe a few specific remedies that were simply products of folk wisdom: such drugs as quinine from the bark of the cinchona tree (originally chewed by the Peruvian Indians to cure themselves of malaria) and digitalis from the plant called foxglove (an old herbwomen's remedy to stimu-

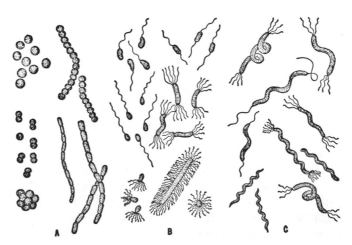

Types of bacteria: cocci (A), bacilli (B), and spirilla (C).
Each type has a number of varieties.

late the heart). Aside from these few treatments (and the smallpox vaccine, which I shall discuss later), many of the medicines and treatments dispensed by physicians after Hippocrates tended to heighten the death rate rather than lower it.

One of the interesting advances made in the first two and a half centuries of the Age of Science was the invention, in 1819, of the stethoscope by the French physician, René Théophile Hyacinthe Laennec. In its original form, it was little more than a wooden tube designed to help the doctor hear and interpret the sounds of the beating heart. Improvements since then have made it as characteristic and inevitable an accompaniment of the physician as the slide rule is of an engineer.

It is not surprising, then, that up to the nineteenth century even the most civilized countries were periodically swept by plagues, some of which had a profound effect on history. The plague in Athens that killed Pericles, at the time of the Peloponnesian War, was the first step in the ultimate ruin of Greece. Rome's downfall probably began with the plagues that fell upon the empire during the reign of Marcus Aurelius. The Black Death of the fourteenth century is estimated to have killed off a fourth of the population of Europe; this plague and gunpowder combined to destroy the social structure of the Middle Ages.

To be sure, plagues did not end when Pasteur discovered that infectious diseases were caused and spread by microorganisms. In India cholera is still endemic, and other underdeveloped countries suffer severely from epidemics. Disease has remained a major hazard of wartime. Virulent new organisms arise from time to time and sweep over the world; indeed, the influenza pandemic of 1918 killed an estimated 15 million people, a larger number of people than died in any other plague in the history of mankind, and nearly twice as many as were killed in the then just-completed World War I.

Nevertheless, Pasteur's discovery was a great turning point. The death rate in Europe and the United States began to fall markedly, and life expectancy steadily rose. Thanks to the scientific study of disease and its treatment, which began with Pasteur, men and women in the more advanced regions of the world can now expect to live an average of 70 years, whereas before Pasteur the average was only 40 years under the most favorable conditions and perhaps only 25 years under unfavorable conditions. Since World War II, life expectancy has been zooming upward even in the less advanced regions of the world.

Even before Pasteur advanced the germ theory in 1865, a Viennese physician named Ignaz Philipp Semmelweiss had made the first effective attack on bacteria, without, of course, knowing what he was fighting. He was working in the maternity ward of one of Vienna's hospitals, where 12 per cent or more of the new mothers died of something called "puerperal" fever (in plain English, "childbed fever"). Semmelweiss noted uneasily that women who bore their babies at home with only the services of ignorant midwives practically never got puerperal fever. His suspicions were further aroused by the death of a doctor in the hospital with symptoms that strongly resembled those of puerperal fever, after the doctor had cut himself while dissecting a cadaver. Were the doctors and students who came in from the dissection wards somehow transmitting this disease to the women whose delivery they attended? Semmelweiss insisted that the doctors wash their hands in a solution of chlorinated lime. Within a year the death rate in the maternity wards fell from 12 per cent to 1.5 per cent.

But the veteran doctors were livid. Resentful of the implication that they had been murderers, and humiliated by all the hand-washing, they drove Semmelweiss out of the hospital. (In this they were helped by the fact that he was a Hungarian and Hungary was in revolt against the Austrian rulers.) Semmelweiss went to Budapest, where he reduced the maternal death rate, while in Vienna the hospitals reverted to death traps for another decade or so. But Semmelweiss himself died of puerperal fever from an accidental infection (at the age of 47) in 1865—just too soon to see the scientific vindication of his suspicions about the transmission of disease. That was the year that Pasteur discovered microorganisms in the diseased silkworms, and an English surgeon named Joseph Lister (the son of the inventor of the achromatic microscope) independently introduced the chemical attack upon germs.

Lister resorted to the drastic substance phenol (carbolic acid). He used it first in dressings for a patient with a compound fracture. Up to that time, any serious wound almost invariably led to infection. Of course, Lister's phenol killed the tissues around the wound, but it did kill the bacteria. The patient made a remarkably untroubled recovery.

Lister followed up this success with the practice of spraying the operating room with phenol. It must have been hard on those who had to breathe it, but it began to save lives. As in Semmelweiss's case, there was opposition, but Pasteur's experiments had created a rationale for antisepsis, and Lister easily won the day.

Pasteur himself had somewhat harder going in France (unlike Lister, he lacked the union label of the M.D.), but he prevailed on surgeons to boil their instruments and steam their bandages. Sterilization with steam à la Pasteur replaced Lister's unpleasant phenol spray. Milder antiseptics, which could kill bacteria without unduly damaging tissue, were sought and found. The French physician Casimir Joseph Davaine reported on the antiseptic properties of iodine in 1873, and "tincture of iodine" (i.e., iodine dissolved in a mixture of alcohol and water) is now in common use in the home. It and similar products are automatically applied to every scratch. The number of infections prevented in this way is undoubtedly enormous.

In fact, the search for protection against infection leaned more and more in the direction of preventing germ entry ("asepsis") rather than of destroying them after they had gained a foothold, as was implied in antisepsis. The American surgeon William Stewart Halstead introduced the practice of using sterilized rubber gloves during operations in 1890; by 1900, the British physician William Hunter had added the gauze mask to protect the patient against the germ content of the physician's breath.

Meanwhile the German physician Robert Koch had begun to identify the specific bacteria responsible for various diseases. To do this he introduced a vital improvement in the nature of culture media, i.e., the food supply in which bacteria were grown. Where Pasteur used liquid media, Koch introduced solid media. He planted isolated samples on gelatin (for which agar, a gelatinlike substance obtained from seaweed, was substituted later). If a single bacterium was deposited (with a fine needle) in a spot on this medium, a pure colony would grow around the spot, because on the solid surface of the agar the bacteria lacked the ability to move or drift away from the original parent, as they would have done in a liquid. An assistant of Koch, Julius Richard Petri, introduced the use of shallow glass dishes with covers, to protect the cultures from contamination by bacterial spores floating in air; such "Petri dishes" have been used for the purpose ever since.

In this way, individual bacteria would give rise to colonies which could then be cultured separately and tested to see what disease they would produce in an experimental animal. The technique not only made it possible to identify a given infection but also permitted experiments with various possible treatments to kill specific bacteria.

With his new techniques, Koch isolated a bacillus that caused anthrax and, in 1882, another that caused tuberculosis. In 1884, he also isolated the bacterium that caused cholera. Others followed in Koch's path. In 1883, for instance, the German pathologist Edwin Klebs isolated the bacterium that caused diphtheria. In 1905, Koch received the Nobel Prize in medicine and physiology.

Once bacteria had been identified, the next task was to find drugs that would kill a bacterium without killing the patient as well. To such a search, the German physician and bacteriologist Paul Ehrlich, who had worked with Koch, now addressed himself. He thought of the task as looking for a "magic bullet" which would not harm the body but strike only the bacteria.

Ehrlich was interested in dyes that stained bacteria. This had an important relationship to cell research. The cell in its natural state is colorless and transparent so that little detail within it could be seen. Early microscopists had tried to use dyes to color the cells, but it was only after Perkin's discovery of aniline dyes (see Chapter 10) that the technique became practical. Though Ehrlich was not the first to use synthetic dyes in staining, he worked out the techniques in detail in the late 1870's, and it was this that led the way to Flemming's study of mitosis and Feulgen's study of DNA in the chromosomes (see Chapter 12).

But Ehrlich had other game in mind, too. He turned to these dyes as possible bactericides. A stain that reacted with bacteria more strongly than with other cells might well kill the bacteria, even when it was injected into the blood in a concentration low enough not to harm the cells of the patient. By 1907, Ehrlich had discovered a dye, called "Trypan red," which would stain trypanosomes, the organisms responsible for the dreaded African sleeping sickness, transmitted via the tsetse fly. Trypan red, when injected in the blood in proper doses, could kill trypanosomes without killing the patient.

Ehrlich was not satisfied: he wanted a surer kill of the microorganisms. Assuming that the toxic part of the trypan-red molecule was the "azo" combination—that is, a pair of nitrogen atoms $(-N=N-)$ —he wondered what a similar combination of arsenic atoms $(-As=AS-)$ might accomplish. Arsenic is chemically similar to nitrogen but much more toxic. Ehrlich began to test arsenic compounds one after the other almost indiscriminately, numbering them methodically as he went. In 1909, a Japanese student of Ehrlich's, Sahachiro Hata, tested compound 606, which had failed against the trypanosomes, on the bacterium that caused syphilis. It proved deadly against this microbe (called a "spirochete" because it is spiral-shaped).

At once Ehrlich realized he had stumbled on something more important than a cure for trypanosomiasis, which after all was a limited disease confined to the tropics. Syphilis had been a hidden scourge of Europe for more than 400 years, ever since Columbus' time. (Columbus' men are supposed to have brought it back from the Caribbean Indians; in return, Europe donated smallpox to the Indians.) Not only was there no cure for syphilis, but prudishness had clothed the disease in a curtain of silence that let it spread unchecked.

Ehrlich devoted the rest of his life (he died in 1915) to the attempt to combat syphilis with compound 606, or, as he called it, "Salvarsan" —"safe arsenic." (Its chemical name is arsphenamine.) It could cure the disease, but its use was not without risk, and Ehrlich had to bully hospitals into using it correctly.

With Ehrlich, a new phase of chemotherapy came into being. Pharmacology, the study of the action of chemicals other than foods (that is "drugs") upon organisms, finally came into its own as a twentieth-century adjunct of medicine. Arsphenamine was the first synthetic drug, as opposed to the plant remedies such as quinine or the mineral remedies of Paracelsus and those who imitated him.

Naturally, the hope at once arose that every disease might be fought with a little tailored antidote all its own. But for a quarter of a century after Ehrlich's discovery the concocters of new drugs had little luck. About the only success of any sort was the synthesis by German chemists of "plasmochin" in 1924 and "atabrine" in 1930; they could be used as substitutes for quinine against malaria. (They were very helpful to Western troops in jungle areas during World War II, when the Japanese held Java, the source of the world supply of quinine, which, like rubber, had moved from South America to Southeast Asia.)

In 1932 came a breakthrough. A German chemist named Gerhard Domagk had been injecting various dyes into infected mice. He tried a new red dye called "Prontosil" on mice infected with the deadly hemolytic streptococcus. The mice survived! He used it on his own daughter, who was dying of streptococcal blood poisoning. She survived also. Within three years Prontosil had gained worldwide renown as a drug that could stop the strep infection in man.

Oddly, Prontosil did not kill streptococci in the test tube—only in the body. At the Pasteur Institute in Paris, Jacques Trefouël and his co-workers decided that the body must change Prontosil into some other substance that took effect on the bacteria. They proceeded to break down Prontosil to the effective fragment, named "sulfanilamide." This compound had been synthesized in 1908, reported perfunctorily, and forgotten. Sulfanilamide's structure is:

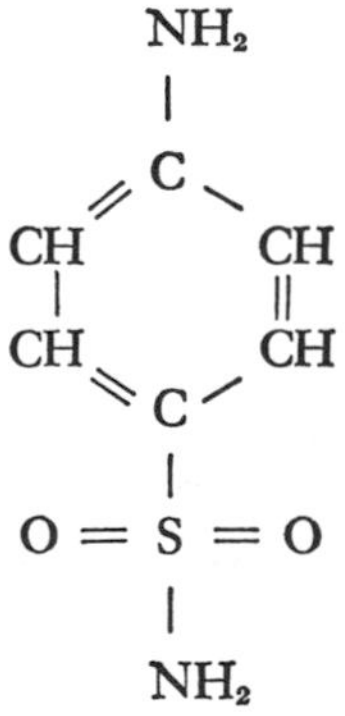

$$NH_2$$

$$
\begin{array}{ccc}
 & C & \\
CH & & CH \\
CH & & CH \\
 & C & \\
\end{array}
$$

$$O = S = O$$

$$NH_2$$

It was the first of the "wonder drugs." One after another bacterium fell before it. Chemists found that by substituting various groups for one of the hydrogen atoms on the sulfur-containing group, they could obtain a series of compounds, each of which had slightly different antibacterial properties. "Sulfapyridine" was introduced in 1937, "sulfathiazole" in 1939, and "sulfadiazine" in 1941. Physicians now could choose from a whole platoon of "sulfa drugs" for various infections. In the medically advanced countries, the death rates from bacterial diseases, notably, pneumococcal pneumonia, dropped dramatically.

Domagk was awarded the Nobel Prize in medicine and physiology in 1939. When he wrote the usual letter of acceptance, he was promptly arrested by the Gestapo; the Nazi government, for peculiar reasons of its own, refused to have anything to do with the Nobel prizes. Domagk felt it the better part of valor to refuse the prize. After World War II, when he was at last free to accept the honor, Domagk went to Stockholm to receive it officially.

The sulfa drugs had only a brief period of glory, for they were soon put in the shade by the discovery of a far more potent kind of antibacterial weapon—the antibiotics.

All living matter (including man) eventually returns to the soil to decay and decompose. With the dead matter and the wastes of living creatures go the germs of the many diseases that infect those creatures. Why is it, then, that the soil is usually so remarkably clean of infectious germs? Very few of them (the anthrax bacillus is one of the few) survive in the soil. A number of years ago bacteriologists began to suspect that the soil harbored microorganisms or substances that destroyed bacteria. As early as 1877, for instance, Pasteur had noticed that some bacteria died in the presence of others. And if this is so, the soil offers a large variety of organisms within which to search for death to others of their kind. It is estimated that each acre of soil contains about 2,000 pounds of molds, 1,000 pounds of bacteria, 200 pounds of protozoa, 100 pounds of algae, and 100 pounds of yeast.

One of those who conducted a deliberate search for bactericides in the soil was the French-American microbiologist René Jules Dubos. In 1939, he isolated from a soil microorganism called *Bacillus brevis* a substance, "tyrothricin," from which he isolated two bacteria-killing compounds that he named "gramicidin" and "tyrocidin." They turned out to be peptides containing D-amino acids—the mirror images of the ordinary L-amino acids that make up most natural proteins.

Gramicidin and tyrocidin were the first antibiotics produced as such. But an antibiotic which was to prove immeasurably more important had been discovered—and merely noted in a scientific paper—twelve years earlier.

The British bacteriologist Alexander Fleming one morning found that some cultures of staphylococcus (the common pus-forming bacterium), which he had left on a bench, were contaminated with something that had killed the bacteria. There were little clear circles where the staphylococci had been destroyed in the culture dishes. Fleming, being interested in antisepsis (he had discovered that an enzyme in tears, called "lysozome," had antiseptic properties), at once investigated to see what had killed the bacteria, and he discovered that it was a common bread mold, *Penicillium notatum*. Some substance, which he named "penicillin," produced by the mold was lethal to germs. Fleming dutifully published his results in 1929, but no one paid much attention at the time.

Ten years later the British biochemist Howard Walter Florey and his German-born associate Ernst Boris Chain became intrigued by the almost forgotten discovery and set out to try to isolate the antibacterial substance. By 1941, they had obtained an extract which proved effective clinically against a number of "gram-positive" bacteria (bacteria that retain a dye developed in 1884 by the Danish bacteriologist Hans Christian Joachim Gram).

Because wartime Britain was in no position to produce the drug, Florey went to the United States and helped to launch a program which developed methods of purifying penicillin and speeding up its production by the mold. In 1943, five hundred cases were treated with penicillin and, by the war's end, large-scale production and use of penicillin were under way. Not only did penicillin pretty much supplant the sulfa drugs, but it became (and still is) one of the most important drugs in the entire practice of medicine. It is effective against a wide range of infections, including pneumonia, gonorrhea, syphilis, puerperal fever, scarlet fever, and meningitis. (The range of effectivity is called the "antibiotic spectrum.") Furthermore, it has practically no toxicity or undesirable side effects, except in penicillin-sensitive individuals.

In 1945, Fleming, Florey, and Chain shared the Nobel Prize in medicine and physiology.

Penicillin set off an almost unbelievably elaborate hunt for other antibiotics. (The word was coined in 1942 by the Rutgers University bacteriologist Selman Abraham Waksman.)

In 1943, Waksman isolated from a soil mold of the genus *Streptomyces* the antibiotic known as "streptomycin." Streptomycin hit the "gram-negative" bacteria (those that easily lose the Gram stain). Its greatest triumph was against the tubercle bacillus. But streptomycin, unlike penicillin, is rather toxic, and it must be used with caution.

For the discovery of streptomycin, Waksman received the Nobel Prize in medicine and physiology in 1952.

Another antibiotic, chloramphenicol, was isolated from molds of the genus *Streptomyces* in 1947. It attacks not only gram-positive and gram-negative bacteria but also certain smaller organisms, notably, those causing typhus fever and psittacosis ("parrot fever"). But its toxicity calls for care in its use.

Then came a whole series of "broad-spectrum" antibiotics, found after painstaking examination of many thousands of soil samples—Aureomycin, Terramycin, Achromycin, and so on. The first of these, Aureomycin, was isolated by Benjamin Minge Duggar and his co-workers in 1944 and was placed on the market in 1948. These antibiotics are called "tetracyclines," because in each case the molecule is composed of four rings side by side. They are effective against a wide infectious diseases have fallen to cheeringly low levels. (Of course human beings left alive by the continuing mastery of man over infectious disease have a much greater chance of succumbing to a metabolic disorder. Thus, in the last eighty years, the incidence of diabetes, the most common such disorder, has increased tenfold.)

The chief disappointment in the development of chemotherapy has been the speedy rise of resistant strains of bacteria. In 1939, for instance, all cases of meningitis and pneumococcal pneumonia showed a favorable response to the administration of sulfa drugs. Twenty years later, only half the cases did. The various antibiotics also became less effective with time. It is not that the bacteria "learn" to resist but that resistant mutants among them flourish and multiply when the "normal" strains are killed off. This danger is greatest in hospitals, where antibiotics are used constantly and where the patients naturally have below-normal resistance to infection. Certain new strains of staphylococci resist antibiotics with particular stubbornness. This "hospital staph" is now a serious worry in maternity wards, for instance, and attained headline fame in 1961, when an attack of pneumonia, sparked by such resistant bacteria, nearly ended the life of screen star Elizabeth Taylor.

Fortunately, where one antibiotic fails another may still attack a resistant strain. New antibiotics, and synthetic modifications of the old, may hold the line in the contest against mutations. The ideal thing would be to find an antibiotic to which no mutants are immune. Then there would be no survivors of that particular bacterium to multiply. A number of such candidates have been produced. For instance, a modified penicillin, called "Staphcyllin," was developed in 1960. It is partly synthetic, and because its structure is strange to bacteria, its molecule is not split and its activity ruined by enzymes such as "penicillinase" (first discovery by Chain), which resistant strains use against ordinary penicillin. Consequently, Staphcyllin is death to otherwise resistant strains; it was used to save Miss Taylor's life, for instance. Yet strains of staphylococcus, resistant to synthetic penicillins, have also turned up. Presumably, the merry-go-round will go on forever.

Additional allies against resistant strains are various other new antibiotics and modified versions of old ones. One can only hope that the stubborn versatility of chemical science will manage to keep the upper hand over the stubborn versatility of the disease germs.

The same problem of the development of resistant strains arises in man's battle with his larger enemies, the insects, which not only compete dangerously for food, but which also spread disease. The modern chemical defenses against insects arose in 1939, with the development by the Swiss chemist Paul Müller of the chemical, "dichlorodiphenyltrichloroethane," better known by its initials, "DDT." Müller was awarded the Nobel prize in medicine and physiology for this feat in 1948.

By then DDT had come into large-scale use and, already, resistant strains of houseflies had developed. Newer "insecticides" (or, to use a more general term that will cover chemicals used against rats or against weeds, "pesticides") must continually be developed.

In addition, there are critics of the overchemicalization of man's battle against other forms of life. There are some who are concerned lest mankind make it possible for an increasingly large segment of the population to remain alive only through the grace of chemistry; they fear that if ever man's technological organization falters, even temporarily, great carnage will result as populations fall prey to the infections and diseases to which they lack natural resistance.

As for the pesticides, the American science-writer, Rachel Louise Carson, published a book, *Silent Spring*, in 1962 that dramatically brought to the fore the possibility that by indiscriminate use of chemicals, mankind might kill harmless and even useful species along with those he was actually attempting to destroy. Furthermore, Miss Carson maintained that to destroy living things without due consideration might lead to a serious upsetting of the intricate system whereby one species depended on another and, in the end, hurt man more than it helps him. The study of this interlinking of species is termed "ecology," and there is no question but that Miss Carson's book encourages a new hard look at this branch of biology.

The answer, of course, is not to abandon technology and give up all attempts to control insects (the price in disease and starvation would be too high) but to find methods that are more specific and less damaging to the ecological structure generally.

For instance, insects have their enemies. Those enemies, whether insect-parasites or insect-eaters, might be encouraged. Sounds and odors might be used to repel insects or to lure them to their death. Insects might be sterilized through radiation. In each case, every effort should be made to zero in on the insect being fought.

One hopeful line of attack, led by the American biologist Carroll Milton Williams, is to make use of the insects' own hormones. Insects molt periodically and pass through two or three well-defined stages: larva, pupa, and adult. The transitions are complex and are controlled by hormones. Thus, one called "juvenile hormone" prevents formation of the adult stage until an appropriate time. By isolating and applying the juvenile hormone, the adult stage is held back to the point where the insect is killed. Each insect has its own juvenile hormone and is affected only by its own. A particular juvenile hormone might thus be used to attack one particular species of insect without affecting any other organism in the world. Guided by the structure of the hormone, biologists may even prepare synthetic substitutes that will be much cheaper and do the job as well.

In short, the answer to the fact that scientific advance may sometimes have damaging side-effects, is not to abandon scientific advance, but to substitute still more advance—intelligently and cautiously applied.

As to how the chemotherapeutic agents work, the best guess seems to be that each drug inhibits some key enzyme in the microorganism in a competitive way. This is best established in the case of the sulfa drugs. They are very similar to "para aminobenzoic acid" (generally written *p*-aminobenzoic acid), which has this structure:

$$
\begin{array}{c}
\text{NH}_2 \\
| \\
\underset{\displaystyle \;}{\text{C}} \\
\text{CH} \qquad \text{CH} \\
| \qquad\quad || \\
\text{CH} \qquad \text{CH} \\
\text{C} \\
| \\
\text{C} = \text{O} \\
| \\
\text{OH}
\end{array}
$$

P-aminobenzoic acid is necessary for the synthesis of "folic acid," a key substance in the metabolism of bacteria as well as other cells. A bacterium that picks up a sulfanilamide molecule instead of *p*-amino-benzoic acid can no longer produce folic acid, because the enzyme needed for the process is put out of action. Consequently, the bacterium ceases to grow and multiply. The cells of the human patient, on the other hand, are not disturbed; they obtain folic acid from food and do not have to synthesize it. There are no enzymes in human cells to be inhibited by moderate concentrations of the sulfa drugs in this fashion.

Even where a bacterium and the human cell possess similar enzymes, there are other ways of attacking the bacterium selectively. The bacterial enzyme may be more sensitive to a given drug than the human enzyme is, so that a certain dose will kill the bacterium without seriously disturbing the human cells. Or a drug of the proper design may be able to penetrate the bacterial cell membrane but not the human cell membrane. Penicillin, for instance, interferes with the manufacture of cell walls, which bacteria possess but animals cells do not.

Do the antibiotics also work by competitive inhibition of enzymes? Here the answer is less clear. But there is good ground for believing that at least some of them do.

Gramicidin and tyrocidin, as I mentioned earlier, contain the "unnatural" D-amino acids. Perhaps these jam up the enzymes that form compounds from the natural L-amino acids. Another peptide antibiotic, bacitracin, contains ornithine; thus may inhibit enzymes from making use of arginine, which ornithine resembles. There is a similar situation in streptomycin: its molecule contains an odd variety of sugar which may interfere with some enzyme acting on one of the normal sugars of living cells. Again, chloramphenicol resembles the amino acid phenylalanine; likewise, part of the penicillin molecule resembles the amino acid cysteine. In both of these cases the possibility of competitive inhibition is strong.

The clearest evidence of competitive action by an antibiotic turned up so far involves "puromycin," a substance produced by a *Streptomyces* mold. This compound has a structure much like that of nucleotides (the building units of nucleic acids), and Michael Yarmolinsky and his co-workers at Johns Hopkins University have shown that puromycin, competing with transfer-RNA, interferes with the synthesis of proteins. Again, streptomycin interferes with transfer-RNA, forcing the misreading of the genetic code and the formation of useless protein. Unfortunately, this form of interference makes it toxic to other cells besides bacteria, because it prevents their normal production of necessary proteins. Thus puromycin is too dangerous a drug to use, and streptomycin is nearly so.

Viruses

To most people it may seem mystifying that the "wonder drugs" have had so much success against the bacterial diseases and so little success against the virus diseases. Since viruses, after all, can cause disease only if they reproduce themselves, why should it not be possible to jam the virus' machinery just as we jam the bacterium's machinery? The answer is quite simple, and indeed obvious, once you realize how a virus reproduces itself. As a complete parasite, incapable of multiplying anywhere except inside a living cell, the virus has very little, if any, metabolic machinery of its own. To make copies of itself, it depends entirely on materials supplied by the cell it invades. This, however, it can do with great efficiency. One virus within a cell can become 200 in 25 minutes. And it is therefore difficult to deprive it of those materials or jam the machinery without destroying the cell itself.

Biologists discovered the viruses only recently, after a series of encounters with increasingly simple forms of life. Perhaps as good a place as any to start this story is the discovery of the cause of malaria.

Malaria has, year in and year out, probably killed more people in the world than any other infectious ailment, since until recently about 10 per cent of the world's population suffered from the disease, with 3,000,000 deaths a year caused by it. Until 1880, it was thought to be caused by the bad air (*mala aria* in Italian) of swampy regions. Then a French bacteriologist, Charles Louis Alphonse Laveran, discovered that the red blood cells of malaria-stricken individuals were infested with parasitic protozoa of the genus *Plasmodium*. (For this discovery, Laveran was awarded the Nobel Prize in medicine and physiology in 1907.)

In the early 1890's a British physician named Patrick Manson, who had conducted a missionary hospital in Hong Kong, pointed out that swampy regions harbored mosquitoes as well as dank air, and he suggested that mosquitoes might have something to do with the spread of malaria. A British physician in India, Ronald Ross, pursued this idea, and he was able to show that the malarial parasite did indeed pass part of its life cycle in mosquitoes of the genus *Anopheles*. The mosquito picked up the parasite in sucking the blood of an infected person and then would pass it on to any person it bit.

For his work, bringing to light for the first time the transmission of a disease by an insect "vector," Ross received the Nobel Prize in medicine and physiology in 1902. It was a crucial discovery of modern medicine, for it showed that a disease might be stamped out by killing off the insect carrier. Drain the swamps that breed mosquitoes; eliminate stagnant water; destroy the mosquitoes with insecticides, and you can stop the disease. Since World War II, large areas of the world have been freed of malaria in just this way, and the total number of deaths from malaria has declined by at least one third from its maximum.

Malaria was the first infectious disease traced to a nonbacterial microorganism (a protozoan in this case). Very shortly afterward, another nonbacterial disease was tracked down to a similar cause. It was the deadly yellow fever, which as late as 1898, during an epidemic in Rio de Janeiro, killed nearly 95 per cent of those it struck. In 1899, when an epidemic of yellow fever broke out in Cuba, a United States board of inquiry, headed by the bacteriologist Walter Reed, went to Cuba to investigate the causes of the disease.

Reed suspected a mosquito vector, such as had just been exposed as the transmitter of malaria. He first established that the disease could not be transmitted by direct contact between the patients and doctors or by way of the patient's clothing or bedding. Then some of the doctors deliberately let themselves be bitten by mosquitoes that had previously bitten a man sick with yellow fever. They got the disease, and one of the courageous investigators, Jesse William Lazear, died. But the culprit was identified as the *Aedes aegypti* mosquito. The epidemic in Cuba was checked, and yellow fever is no longer a serious disease in the medically advanced parts of the world.

As a third example of a nonbacterial disease, there is typhus fever. This infection is endemic in North Africa and was brought into Europe via Spain during the long struggle of the Spaniards against the Moors of North Africa. Commonly known as "plague," it is very contagious and has devastated nations. In World War I the Austrian armies were driven out of Serbia by the typhus when the Serbian army itself was unequal to the task. The ravages of typhus in Poland and Russia during that war and its aftermath (some three million persons died of the disease) did as much as military action to ruin those nations.

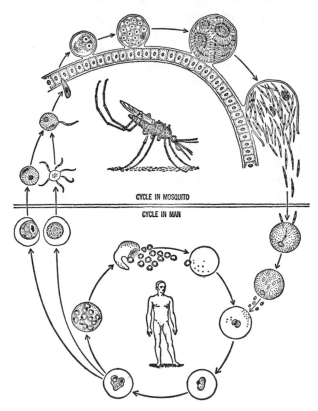

CYCLE IN MOSQUITO

CYCLE IN MAN

Life cycle of the malarial microorganism.

At the turn of the twentieth century the French bacteriologist Charles Nicolle then in charge of the Pasteur Institute in Tunis, noticed that although typhus was rife in the city, no one caught it in the hospital. The doctors and nurses were in daily contact with typhus-ridden patients and the hospital was crowded, yet there was no spread of the disease there. Nicolle considered what happened when a patient came into the hospital, and it struck him that the most significant change was a thorough washing of the patient and removal of his lice-infested clothing. Nicolle decided that the body louse must be the vector of typhus. He proved the correctness of his guess by experiments. He received the Nobel Prize in medicine and physiology in 1928 for his discovery. Thanks to his finding, and the discovery of DDT, typhus fever did not repeat its deadly carnage in World War II. In January 1944, DDT was brought into play against the body louse. The population of Naples was sprayed en masse and the lice died. For the first time in history, a winter epidemic of typhus (when the multiplicity of clothes, not removed very often, made louse-infestation almost certain and almost universal) was stopped in its tracks. A similar epidemic was stopped in Japan in late 1945 after the American occupation. World War II became almost unique among history's wars by possessing the dubious merit of killing fewer people by disease than by guns and bombs.

Typhus, like yellow fever, is caused by an agent smaller than a bacterium, and we must now enter the strange and wonderful realm populated by subbacterial organisms.

To get some idea of the dimensions of objects in this world, let us look at them in order of decreasing size. The human ovum is about 100 microns (100 millionths of a meter) in diameter, and it is just barely visible to the naked eye. The paramecium, a large protozoan which in bright light can be seen moving about in a drop of water, is about the same size. An ordinary human cell is only 1/10 as large (about 10 microns in diameter), and it is quite invisible without a microscope. Smaller still is the red blood corpuscle—some 7 microns in maximum diameter. The bacteria, starting with species as large as ordinary cells, drop down to a tinier level: the average rod-shaped bacterium is only 2 microns long, and the smallest bacteria are spheres perhaps no more than 4/10 of a micron in diameter. They can barely be seen in ordinary microscopes.

At this level, organisms apparently have reached the smallest possible volume into which all the metabolic machinery necessary for an independent life can be crowded. Any smaller organism cannot be a self-sufficient cell and must live as a parasite. It must shed most of the enzymatic machinery as excess baggage, so to speak. It is unable to grow or multiply on any artificial supply of food, however ample; hence it cannot be cultured, as bacteria can, in the test tube. The only place it can grow is in a living cell, which supplies the enzymes that it lacks. Such a parasite grows and multiplies, naturally, at the expense of the host cell.

The first subbacteria were discovered by a young American pathologist named Howard Taylor Ricketts. In 1909, he was studying a disease called Rocky Mountain spotted fever, which is spread by ticks (blood-sucking arthropods, related to the spiders rather than to insects). Within the cells infected hosts he found "inclusion bodies" that turned out to be very tiny organisms, now called "rickettsia" in his honor. Ricketts and others soon found that typhus also was a rickettsial disease. In the process of establishing a proof of this fact, Ricketts himself caught typhus, and he died in 1910 at the age of thirty-nine.

The rickettsia are still big enough to be attacked by antibiotics such as chloramphenicol and the tetracyclines. They range from about four-fifths of a micron to one-fifth of a micron in diameter. Apparently they possess enough metabolic machinery of their own to differ from the host cells in their reaction to drugs. Antibiotic therapy has therefore considerably reduced the danger of rickettsial diseases.

At the lowest end of the scale, finally, come the viruses. They overlap the rickettsia in size; in fact, there is no actual dividing line between rickettsia and viruses. But the smallest viruses are small indeed. The virus of yellow fever, for instance, is only 1/50 of a micron in diameter. The viruses are much too small to be detected in a cell or to be seen

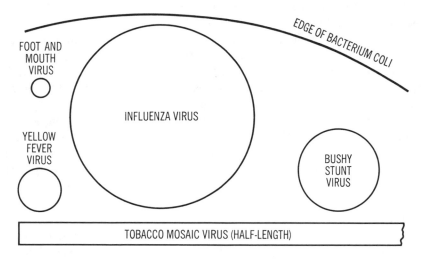

Relative sizes of simple substances and proteins and of
various particles and bacteria. (An inch and a half on this
scale = 1/10,000 of a millimeter in life.)

under any optical microscope. The average virus is only 1/1,000 the size of the average bacterium.

A virus is stripped practically clean of metabolic machinery. It depends almost entirely upon the enzyme equipment of the host cell. Some of the largest viruses are affected by certain antibiotics, but against the run-of-the-mill viruses drugs are helpless.

The existence of viruses was suspected many decades before they were finally seen. Pasteur, in his studies of hydrophobia, could find no organism in the body that could reasonably be suspected of causing the disease. Rather than decide that his germ theory of disease was wrong, Pasteur suggested that the germ in this case was simply too small to be seen. He was right.

In 1892, a Russian bacteriologist, Dmitri Ivanovski, was studying "tobacco mosaic disease," a disease that gave the leaves of the tobacco plant a mottled appearance. He found that the juice of infected leaves could transmit the disease when placed on the leaves of healthy plants. In an effort to trap the germs, he passed the juice through porcelain filters with holes so fine that not even the smallest bacterium could pass through. Yet the filtered juice still infected tobacco plants. Ivanovski decided that his filters must be defective and were actually letting bacteria through.

A Dutch bacteriologist, Martinus Willem Beijerinck, repeated the experiment in 1897, and he came to the decision that the agent of the disease was small enough to pass through the filter. Since he could see nothing in the clear, infective fluid under any microscope, and was unable to grow anything from it in a test-tube culture, he thought the infective agent might be a small molecule, perhaps about the size of a sugar molecule. Beijerinck called the infective agent a "filtrable virus" (virus being a Latin word meaning "poison").

In the same year a German bacteriologist, Friedrich August Johannes Löffler, found that the agent causing hoof-and-mouth disease in cattle could also pass through a filter. And, in 1901, Walter Reed, in the course of his yellow-fever researches, found that the infective agent of that disease also was a filtrable virus. In 1914, the German bacteriologist Walther Kruse demonstrated the common cold to be virus-produced.

By 1931, some forty diseases (including measles, mumps, chickenpox, influenza, smallpox, poliomyelitis, and hydrophobia) were known to be caused by viruses, but the nature of viruses was still a mystery. Then an English bacteriologist, William Joseph Elford, finally began to trap some in filters and to prove that at least they were material particles of some kind. He used fine collodion membranes, graded to keep out smaller and smaller particles, and he worked his way down to membranes fine enough to remove the infectious agent from a liquid. From the fineness of the membrane that could filter out the agent of a given disease, he was able to judge the size of that virus. He found that Beijerinck had been wrong: even the smallest virus was larger than most molecules. The largest viruses approached the rickettsia in size.

For some years afterward, biologists debated whether viruses were living or dead particles. Their ability to multiply and transmit disease certainly suggested that they were alive. But in 1935 the American biochemist Wendell Meredith Stanley produced a piece of evidence which seemed to speak forcefully in favor of "dead." He mashed up tobacco leaves heavily infected with the tobacco mosaic virus and set out to isolate the virus in as pure and concentrated a form as he could get, using protein-separation techniques for the purpose. Stanley succeeded beyond his expectations, for he obtained the virus in crystalline form! His preparation was just as crystalline as a crystallized molecule, yet the virus evidently was still intact; when he redissolved it in liquid, it was just as infectious as before.

For his crystallization of the virus, Stanley shared the 1946 Nobel Prize in chemistry with Summer and Northrop, the crystallizers of enzymes (see Chapter 11).

Still, for twenty years after Stanley's feat, the only viruses that could be crystallized were the very simple "plant viruses" (those infesting plant cells). Not until 1955 was the first "animal virus" crystallized. In that year, Carlton E. Schwerdt and Frederick L. Schaffer crystallized the poliomyelitis virus.

The fact that viruses could be crystallized seemed to many, including Stanley himself, to be proof that they were merely dead protein. Nothing living had ever been crystallized, and life and crystallinity just seemed to be mutually contradictory. Life was flexible, changeable, dynamic; a crystal was rigid, fixed, strictly ordered.

Yet the fact remained that viruses were infective, thay they could grow and multiply even after having been crystallized. And growth and reproduction had always been considered the essence of life.

The turning point came when two British biochemists, Frederick Charles Bawden and Norman W. Pirie, showed that the tobacco mosaic virus contained ribonucleic acid! Not much, to be sure; the virus was 94 per cent protein and only 6 per cent RNA. But it was nonetheless definitely a nucleoprotein. Furthermore, all other viruses proved to be nucleoprotein, containing RNA or DNA or both.

The difference between being nucleoprotein and being merely protein is practically the difference between being alive and dead. Viruses turned out to be composed of the same stuff as genes, and the genes are the very essence of life. The larger viruses give every appearance of being chromosomes on the loose, so to speak. Some contain as many as 75 genes, each of which controls the formation of some aspect of its structure—a fiber here, a folding there. By producing mutations in the nucleic acid, one gene or another may be made defective, and through this means, its function and even its location can be determined. The total gene analysis (both structural and functional) of a virus is within reach, though of course this represents but a small step toward a similar total analysis for cellular organisms, with their much more elaborate genic equipment.

We can picture viruses in the cell as raiders that, pushing aside the supervising genes, take over the chemistry of the cell in their own interests, often causing the death of the cell or of the entire host organism in the process. Sometimes, a virus may even replace a gene, or series of genes, with its own, introducing new characteristics that can be passed along to daughter cells. This phenomenon is called "transduction."

If the genes carry the "living" properties of a cell, then viruses are living things. Of course, a lot depends on how one defines life. I, myself, think it fair to consider any nucleoprotein molecule that is capable of replication to be living. By that definition, viruses are as alive as elephants and human beings.

No amount of indirect evidence of the existence of viruses is as good as seeing one, of course. Apparently the first man to lay eyes on a virus was a Scottish physician named John Brown Buist. In 1887, he reported that in the fluid from a vaccination blister he had managed to make out some tiny dots under the microscope. Presumably they were the cowpox virus, the largest known virus.

To get a good look—or any look at all—at a typical virus, something better than an ordinary microscope was needed. The something better was finally invented in the late 1930's: the electron microscope, which could reach magnifications as high as 100,000 and resolve objects as small as 1/1,000 of a micron in diameter.

The electron microscope has its drawbacks. The object has to be placed in a vacuum, and the inevitable dehydration may change its shape. An object such as a cell must be sliced extremely thin. The image is only two-dimensional; furthermore, the electrons tend to go right through a biological material, so that it does not stand out against the background.

In 1944, the American astronomer and physicist Robley Cook Williams and the electron microscopist Ralph Walter Graystone Wyckoff jointly worked out an ingenious solution of these last difficulties. It occurred to Williams, as an astronomer, that just as the craters and mountains of the moon are brought into relief by shadows when the sun's light falls on them obliquely, so viruses might be seen in three dimensions in the electron microscope if they could somehow be made to cast shadows. The solution the experimenters hit upon was to blow vaporized metal obliquely across the virus particles set up on the stage of the microscope. The metal stream left a clear space—a "shadow"—behind each virus particle. The length of the shadow indicated the height of the blocking particle. And the metal, condensing as a thin film, also defined the virus particles sharply against the background.

The shadow pictures of various viruses then disclosed their shapes. The cowpox virus was found to be shaped something like a barrel. It turned out to be about 0.25 of a micron thick—about the size of the smallest rickettsia. The tobacco mosaic virus proved to be a thin rod 0.28 micron long by 0.015 micron thick. The smallest viruses, such as those of poliomyelitis, yellow fever, and hoof-and-mouth disease, were tiny spheres ranging in diameter from 0.025 down to 0.020 micron. This is considerably smaller than the estimated size of a single human gene. The weight of these viruses is only about 100 times that of an average protein molecule. The brome grass mosaic virus, the smallest yet characterized, has a molecular weight of 4.5 million. It is only one-tenth the size of the tobacco mosaic virus and may, perhaps, bear off the prize as "smallest living thing."

In 1959, the Finnish cytologist Alvar P. Wilska designed an electron microscope using comparatively low-speed electrons. Because they are less penetrating than high-speed electrons, they can define some of the internal detail in the structure of viruses. And in 1961, the French cytologist Gaston DuPouy devised a way of placing bacteria in air-filled capsules and taking electron microscope views of living cells in this way. In the absence of metal-shadowing, however, detail was lacking.

Virologists have actually begun to take viruses apart and put them together again. For instance, at the University of California, the German-American biochemist, Heinz Fraenkel-Conrat, working with Robley Williams, found that gentle chemical treatment broke down the protein of the tobacco-mosaic virus into some 2,200 fragments, consisting of peptide chains made up of 158 amino acids apiece, and individual molecular weights of 18,000. The exact amino acid constitution of these virus-protein units was completely worked out in 1960. When such units are dissolved, they tend to coalesce once more into the long, hollow rod (in which form they exist in the intact virus). The units are held together by calcium and magnesium atoms.

In general, virus-protein units make up geometric patterns when they combine. Those of tobacco mosaic virus, just discussed, form segments of a helix. The sixty subunits of the protein of the poliomyelitis virus are arranged in 12 pentagons. The twenty subunits of the *Tipula* iridescent virus are aranged in a regular twenty-sided solid, an icosahedron.

The protein of the virus is hollow. The protein helix of tobacco mosaic virus, for instance, is made up of 130 turns of the peptide chain, producing a long, straight cavity within. Inside the protein cavity is the nucleic acid portion of the virus. This may be DNA or RNA, but, in either case, it is made up of about 6,000 nucleotides although Sol Spiegelman has detected an RNA molecule with as few as 470 nucleotides that is capable of replication.

Fraenkel-Conrat separated the nucleic acid and protein portions of tobacco mosaic viruses and tried to find out whether each portion alone could infect a cell. It developed that separately they could not, as far as he could tell. But when he mixed the protein and nucleic acid together again, as much as 50 per cent of the original infectiousness of the virus sample could eventually be restored!

What had happened? The separated virus protein and nucleic acid had seemed dead, to all intents and purposes; yet, mixed together again, some at least of the material seemed to come to life. The public press hailed Fraenkel-Conrat's experiment as the creation of a living organism from nonliving matter. The stories were mistaken, as we shall see in a moment.

Apparently, some recombination of protein and nucleic acid had taken place. Each, it seemed, had a role to play in infection. What were the respective roles of the protein and the nucleic acid, and which was more important?

Fraenkel-Conrat performed a neat experiment that answered the question. He mixed the protein part of one strain of the virus with the nucleic-acid portion of another strain. The two parts combined to form an infectious virus with a mixture of properties! In virulence (i.e., the degree of its power to infect tobacco plants) it was the same as the strain of virus that had contributed the protein; in the particular disease produced (i.e., the nature of the mosaic pattern on the leaf), it was identical with the strain of virus that had supplied the nucleic acid.

This finding fitted well with what virologists already suspected about the respective functions of the protein and the nucleic acid. It seems that when a virus attacks a cell, its protein shell, or coat, serves to attach itself to the cell and to break open an entrance into the cell. Its nucleic acid then invades the cell and engineers the production of virus particles.

After Fraenkel-Conrat's hybrid virus had infected a tobacco leaf, the new generation of virus that it bred in the leaf's cells turned out to be not a hybrid but just a replica of the strain that had contributed the nucleic acid. It copied that strain in degree of infectiousness as well as in the pattern of disease produced. In other words, the nucleic acid had dictated the construction of the new virus' protein coat. It had produced the protein of its own strain, not that of the strain with which it had been combined in the hybrid.

This reinforced the evidence that the nucleic acid is the "live" part of a virus, or, for that matter, of any nucleoprotein. Actually, Fraenkel-Conrat found in further experiments that pure virus nucleic acid alone could produce a little infection in a tobacco leaf—about 0.1 per cent as much as the intact virus. Apparently once in a while the nucleic acid somehow managed to breach an entrance into a cell all by itself.

So putting virus nucleic acid and protein together to form a virus is not creating life from nonlife; the life is already there, in the shape of the nucleic acid. The protein merely serves to protect the nucleic acid against the action of hydrolyzing enzymes ("nucleases") in the environment and to help it go about the business of infection and reproduction more efficiently. We might compare the nucleic-acid fraction to a man and the protein fraction to an automobile. The combination makes easy work of traveling from one place to another. The automobile by itself could never make the trip. The man could make it on foot (and occasionally does), but the automobile is a big help.

The clearest and most detailed information about the mechanism by which viruses infect a cell has come from studies of the viruses called bacteriophages, first discovered by the English bacteriologist Frederick William Twort in 1915 and, independently, by the Canadian bacteriologist Félix Hubert d' Hérelle in 1917. Oddly enough, these viruses are germs that prey on germs—namely, bacteria. D'Hérelle gave them the name "bacteriophage," from Greek words meaning "bacteria-eater."

The bacteriophages are beautifully convenient things to study, because they can be cultured with their hosts in a test tube. The process of infection and multiplication goes about as follows.

A typical bacteriophage (usually called "phage" by the workers with the beast) is shaped like a tiny tadpole, with a blunt head and a tail. Under the electron microscope investigators have been able to see that the phage first lays hold of the surface of a bacterium with its tail. The best guess as to how it does this is that the pattern of electric charge on the tip of the tail (determined by charged amino acids) just fits the charge pattern on certain portions of the bacterium's surface. The configurations of the opposite, and attracting, charges on the tail and on the bacterial surface match so neatly that they come together with something like the click of perfectly meshing gear teeth. Once the virus has attached itself to its victim by the tip of its tail, it cuts a tiny opening in the cell wall, perhaps by means of an enzyme that cleaves the molecules at that point. As far as the electron-microscope pictures show, nothing whatever is happening. The phage, or at least its visible shell, remains attached to the outside of the bacterium. Inside the bacterial cell there is no visible activity. But, within half an hour the cell bursts open and hundreds of full-grown viruses pour out.

Evidently only the protein shell of the attacking virus stays outside the cell. The nucleic acid within the virus' shell must pour into the bacterium through the hole in its wall made by the protein. That the invading material is just nucleic acid, without any detectable admixture

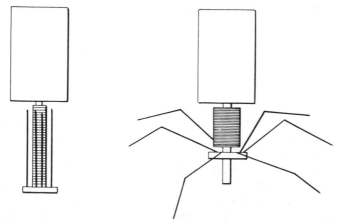

Model of T-2 bacteriophage, a tadpole-shaped virus that preys on other germs, in "untriggered" (*left*) and "triggered" (*right*) forms.

of protein, was proved by the American bacteriologist Alfred Day Hershey by means of radioactive tracers. He tagged phages with radioactive phosphorus and radioactive sulfur atoms (by growing them in bacteria that had incorporated these radioisotopes from their nutritive medium). Now phosphorus occurs both in proteins and in nucleic acids, but sulfur will turn up only in proteins, because there is no sulfur in a nucleic acid. Therefore if a phage labeled with both tracers invaded a bacterium and its progeny turned up with radiophosphorus but no radiosulfur, the experiment would indicate that the parent virus' nucleic acid had entered the cell but its protein had not. The absence of radiosulfur would suggest that all the protein in the virus progeny was supplied by the host bacterium. The experiment, in fact, turned out just this way: the new viruses contained radiophosphorus (contributed by the parent) but no radiosulfur.

Once more, the dominant role of nucleic acid in the living process was demonstrated. Apparently, only the phage's nucleic acid went into the bacterium, and there it superintended the construction of new viruses —protein and all—from the material in the cell. Indeed, the potato spindle tuber virus, an unusually small one, seems to be nucleic acid without a protein coat at all.

On the other hand, it may be that even nucleic acid is not altogether vital to the production of a virus effect. In 1967, it was found that a sheep disease called "scrapie" is caused by particles with a molecular weight of 700,000—considerably smaller than that of any known virus and, even more important, lacking in nucleic acid. The particle may be an independent "repressor" that alters the gene action of a cell in such a way as to bring about its own formation. Not only are the cell's enzymes thus used for the invader's own purpose, but even the cell's genes. The importance to man lies in the fact that the human disease multiple sclerosis may be related to scrapie.

Immunity

Viruses are man's most formidable living enemy(except man himself). By virtue of their intimate association with the body's own cells, viruses have been all but invulnerable to attack by drugs or any other artificial weapon. And yet man has been able to hold his own against them, even under the most unfavorable conditions. The human organism is endowed with impressive natural defenses against disease.

Consider the Black Death, the great plague of the fourteenth century. It attacked a Europe living in appalling filth, without any modern conception of cleanliness and hygiene, without plumbing, without any form of reasonable medical treatment—a crowded and helpless population. To be sure, people could flee from the infected villages, but the

Louis Pasteur working in his laboratory (1885).

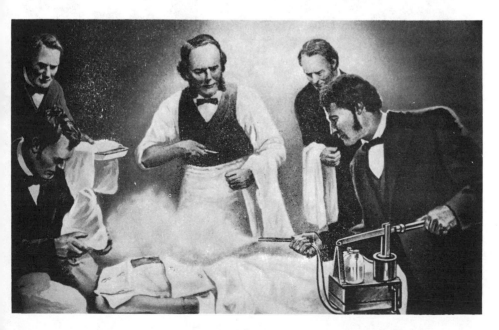

Joseph Lister directing an assistant who is spraying a surgical patient with carbolic acid as a disinfectant.

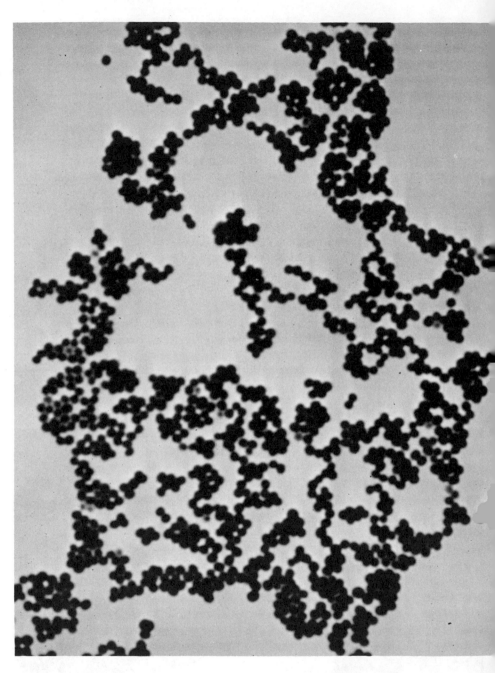

Staphylococcus bacteria, magnified about 2,000 times. The picture clearly shows why they were named from the Greek word for "a bunch of grapes."

666

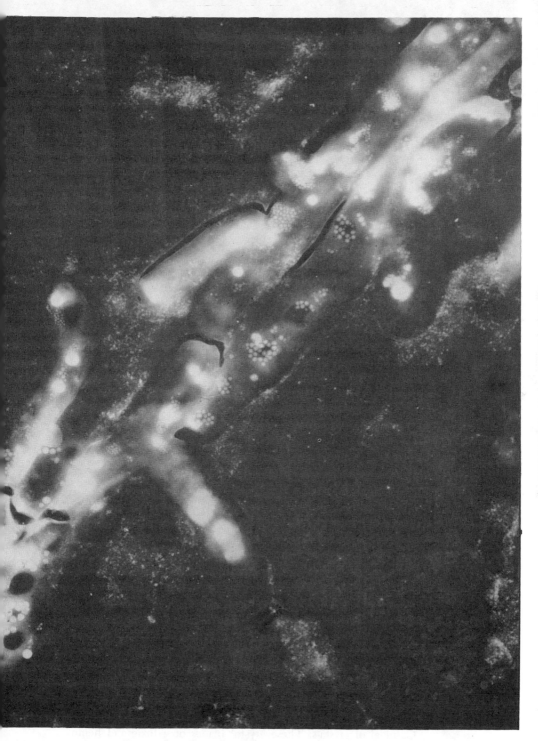

The rod-shaped bacillus of tuberculosis, photographed with the electron microscope.

Carnival of crystals. These crystal blades of a toxic and hemolytic lipoid were isolated from cultures of *Pseudomonas pseudomallei*. The lipoid, of low molecular weight, killed mice. The photograph, made through a polarizing microscope at about × 75, was taken at the Naval Biological Laboratory, University of California, Oakland, where experiments are being conducted with bacterial endotoxins.

The tobacco mosaic virus.

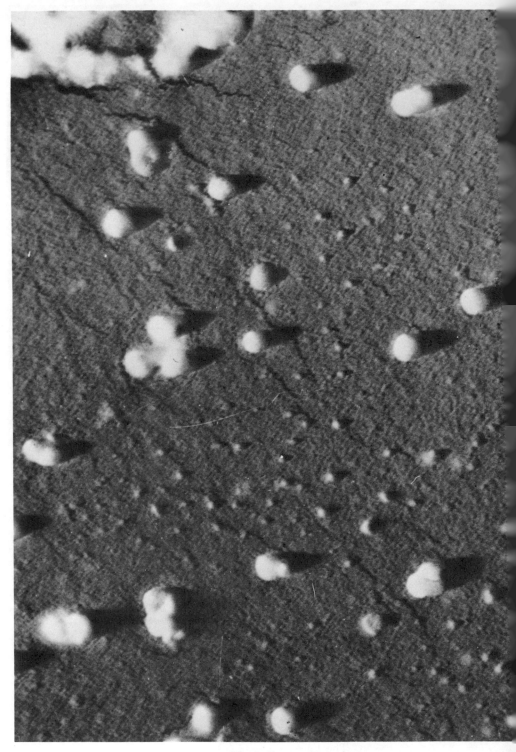

The influenza virus.

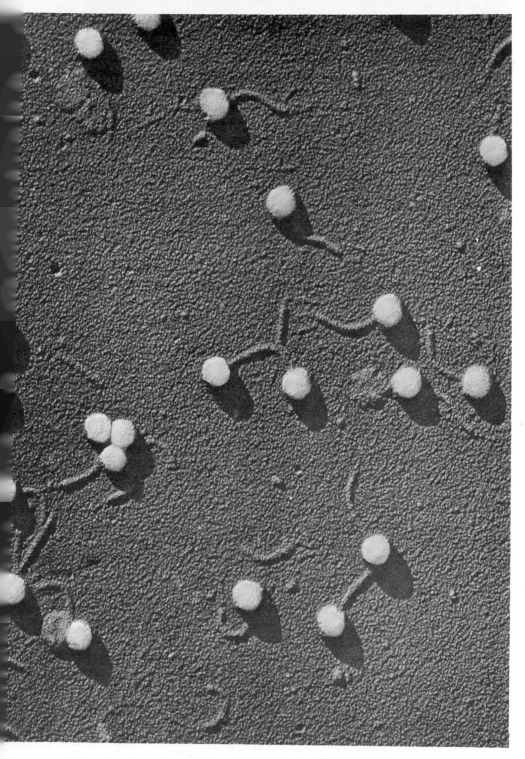

Bacteriophages. The tails by which they attached themselves to bacteria can clearly be seen in this electron micrograph.

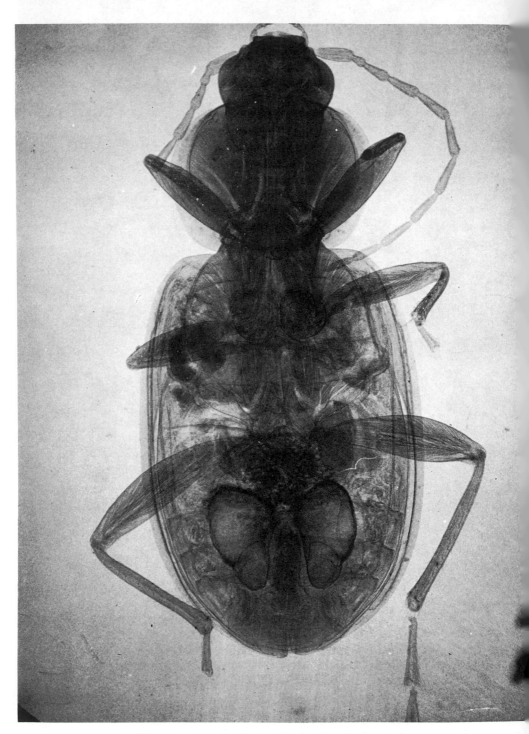

An X-ray micrograph of a beetle showing its internal organs and even muscles. The photograph was made in the Cavendish Physics Laboratory at the University of Cambridge.

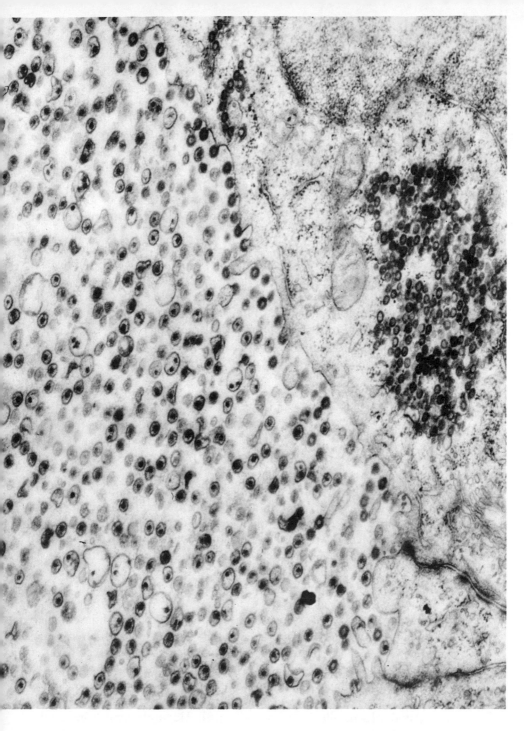

Cancer-producing particles, the so-called "milk factor," which may consist of viruses, are shown here in a mouse breast tumor. They are the group of small, black bodies at the right of the picture. This electron micrograph magnifies them 50,000 times.

fugitive sick only spread the epidemics faster and farther. Notwithstanding all this, three fourths of the population successfully resisted the infections. Under the circumstances, the marvel is not that one out of four died; the marvel is that three out of four survived.

There is clearly such a thing as natural resistance to any given disease. Of a number of people exposed to a serious contagious disease, some will have a relatively mild case, some will be very sick, some will die. There is also such a thing as complete immunity—sometimes inborn, sometimes acquired. A single attack of measles, mumps, or chickenpox, for instance, will usually make a person immune to that particular disease for the rest of his life.

All three of these diseases, as it happens, are caused by viruses. Yet they are comparatively minor infections, seldom fatal. Measles usually produces only mild symptoms, at least in a child. How does the body fight off these viruses, and then fortify itself so that the virus it has defeated never troubles it again? The answer to that question forms a thrilling episode in modern medical science, and for the beginning of the story we must go back to the conquest of smallpox.

Up to the end of the eighteenth century, smallpox was a particularly dreaded disease, not only because it was often fatal but also because those who recovered were permanently disfigured. A light case would leave the skin pitted; a severe attack could destroy all traces of beauty and almost of humanity. A very large proportion of the population bore the marks of smallpox on their faces. And those who had not yet caught it lived in fear of when it might strike.

In the seventeenth century, people in Turkey began to infect themselves deliberately with mild forms of smallpox, with the hope of making themselves immune to severe attack. They would have themselves scratched with the serum from blisters of a person who had a mild case. Sometimes they developed only a light infection; sometimes they suffered the very disfigurement or death they had sought to avoid. It was risky business, but it is a measure of the horror of the disease that people were willing to risk the horror itself in order to escape from it.

In 1718, the famous beauty Lady Mary Wortley Montagu learned about this practice when she went to Turkey with her husband, sent there briefly as the British ambassador, and she had her own children inoculated. They escaped without harm. But the idea did not catch on in England, perhaps partly because Lady Montagu was considered a notorious eccentric. A similar case, across the ocean, was that of Zabdiel Boylston, an American physician. During a smallpox epidemic in Boston, he inoculated 241 people, of whom six died. He underwent considerable criticism for this.

Certain country folk in Gloucestershire had their own idea about how to avoid smallpox. They believed that a case of cowpox, a disease that attacked cows and sometimes people, would make a person immune to both cowpox and smallpox. This was wonderful, if true, for cowpox produced hardly any blisters and left hardly any marks. A Gloucestershire doctor, Edward Jenner, decided that there might be some truth in this folk "superstition." Milkmaids, he noticed, were particularly prone to catch cowpox and apparently also particularly prone not to be pock-marked by smallpox. (Perhaps the eighteenth-century vogue of romanticizing the beautiful milkmaid was based on the fact that milkmaids, having clear complexions, were indeed beautiful in a pock-marked world.)

Was it possible that cowpox and smallpox were so alike that a defense formed by the body against cowpox would also protect against smallpox? Very cautiously Dr. Jenner began to test this notion (probably experimenting on his own family first). In 1796, he decided to chance the supreme test. First he inoculated an eight-year-old boy named James Phipps with cowpox, using fluid from a cowpox blister on a milkmaid's hand. Two months later came the crucial and desperate part of the test. Jenner deliberately inoculated young James with smallpox itself.

The boy did not catch the disease. He was immune.

Jenner called the process "vaccination," from *vaccinia*, the Latin name for cowpox. Vaccination spread through Europe like wildfire. It is one of the rare cases of a revolution in medicine that was adopted easily and almost at once—a true measure of the deadly fear inspired by smallpox and the eagerness of the public to try anything that promised escape. Even the medical profession put up only weak opposition to vaccination—though its leaders put up such stumbling blocks as they could. When Jenner was proposed for election to the Royal College of Physicians in London in 1813, he was refused admission, on the ground that he was not sufficiently up on Hippocrates and Galen.

Today smallpox has practically been wiped out in civilized countries, though its terrors as a disease are still as strong as ever. A report of a single case in any large city is sufficient to send virtually the entire population running to doctors' offices for renewed vaccination.

Attempts to discover similar inoculations for other severe diseases got nowhere for more than a century and a half. It was Pasteur who made the next big step forward. He discovered, more or less by accident, that he could change a severe disease into a mild one by weakening the microbe that produced it.

Pasteur was working with a bacterium that caused cholera in chickens. He concentrated a preparation so virulent that a little injected under the skin of a chicken would kill it within a day. On one occasion he used a culture that had been standing for a week. This time the chickens became only slightly sick and recovered. Pasteur decided that the culture was spoiled and prepared a virulent new batch. But his fresh culture failed to kill the chickens that had recovered from the dose of "spoiled" bacteria. Clearly, the infection with the weakened bacteria had equipped the chickens with a defense against the fully potent ones.

In a sense, Pasteur had produced an artificial "cowpox" for this particular "smallpox." He recognized the philosophical debt he owed to Jenner by calling his procedure vaccination, too, although it had nothing to do with vaccinia. Since then, the term has been used quite generally to mean inoculations against any disease, and the preparation used for the purpose is called a "vaccine."

Pasteur developed other methods of weakening (or "attenuating") disease agents. For instance, he found that culturing anthrax bacteria at a high temperature produced a weakened strain that would immunize animals against the disease. Until then, anthrax had been so hopelessly fatal and contagious that as soon as one member of a herd came down with it, the whole herd had to be slaughtered and burned.

Pasteur's most famous victory, however, was over the virus disease called hydrophobia, or "rabies" (from a Latin word meaning "to rave," because the disease attacked the nervous system and produced symptoms akin to madness). A person bitten by a rabid dog would, after an incubation period of a month or two, be seized by violent symptoms and almost invariably die an agonizing death.

Pasteur could find no visible microbe as the agent of the disease (of course, he knew nothing of viruses), so he had to use living animals to cultivate it. He would inject the infectious fluid into the brain of a rabbit, let it incubate, mash up the rabbit's spinal cord, inject the extract into the brain of another rabbit, and so on. Pasteur attenuated his preparations by aging and testing them continuously until the extract could no longer cause the disease in a rabbit. He then injected the attenuated virus into a dog, which survived. After a time, he infected the dog with hydrophobia in full strength and found the animal immune.

In 1885, Pasteur got his chance to try the cure on a human being. A nine-year-old boy, Joseph Meister, who had been severely bitten by a rabid dog, was brought to him. With considerable hesitation and anxiety, Pasteur treated the boy with inoculations of successively less and less attenuated virus, hoping to build up resistance before the incubation period had elapsed. He succeeded. At least, the boy survived. (Meister became the gatekeeper of the Pasteur Institute, and in 1940 he committed suicide when the Nazi army in Paris ordered him to open Pasteur's crypt.)

In 1890, a German army doctor named Emil von Behring, working in Koch's laboratory, tried another idea. Why take the risk of injecting the microbe itself, even in attenuated form, into a human being? Assuming that the disease agent caused the body to manufacture some defensive substance, would it not serve just as well to infect an animal with the agent, extract the defense substance that it produced, and inject that substance into the human patient?

Von Behring found that this scheme did indeed work. The defensive substance turned up in the blood serum, and von Behring called it "antitoxin." He caused animals to produce antitoxins against tetanus and diphtheria. His first use of the diphtheria antitoxin on a child with the disease was so dramatically successful that the treatment was adopted immediately and proceeded to cut the death rate from diphtheria drastically.

Paul Ehrlich (who later was to discover the "magic bullet" for syphilis) worked with von Behring, and it was probably he who calculated the appropriate antitoxin dosages. Later he broke with von Behring (Ehrlich was an irascible individual who found it easy to break with anyone), and alone he went on to work out the rationale of serum therapy in detail. Von Behring received the Nobel Prize in medicine and physiology in 1901, the first year in which it was awarded. Ehrlich also was awarded that Nobel Prize, sharing it with a Russian biologist in 1908.

The immunity conferred by an antitoxin lasts only as long as the antitoxin remains in the blood. But the French bacteriologist Gaston Ramon found that by treating the toxin of diphtheria or tetanus with formaldehyde or heat he was able to change its structure in such a way that the new substance (called "toxoid") could safely be injected in a human patient. The antitoxin then made by the patient himself lasts longer than that from an animal; furthermore, new doses of the toxoid can be injected when necessary to renew immunity. After toxoid was introduced in 1925, diphtheria lost most of its terrors.

Serum reactions were also used to detect the presence of disease. The best-known example of this is the "Wasserman test," introduced by the German bacteriologist August von Wasserman, in 1906, for the detection of syphilis. This was based on techniques first developed by a Belgian bacteriologist, Jules Bordet, who worked with serum fractions that came to be called "complement." For his work, Bordet received the Nobel Prize in medicine and physiology in 1919.

Pasteur's laborious wrestle with the virus of rabies showed the difficulty of dealing with viruses. Bacteria can be cultured, manipulated, and attenuated on artificial media in the test tube. Viruses cannot; they can be grown only in living tissue. In the case of smallpox, the living hosts for the experimental material (the cowpox virus) were cows and milkmaids. In the case of rabies, Pasteur used rabbits. But living animals are, at best, an awkward, expensive, and time-consuming type of medium for culturing microorganisms.

In the first quarter of this century the French biologist Alexis Carrel won considerable fame with a feat which was to prove immensely valuable to medical research—keeping bits of tissue alive in the test tube. Carrel had become interested in this sort of thing through his work as a surgeon. He had developed new methods of transplanting animals' blood vessels and organs, for which he received the Nobel Prize in medicine and physiology in 1912. Naturally, he had to keep the excised organ alive while he was getting ready to transplant it. He worked out a way to nourish it, which consisted in perfusing the tissue with blood and supplying the various extracts and ions. As an incidental dividend, Carrel, with the help of Charles Augustus Lindbergh, developed a crude "mechanical heart" to pump the blood through the tissue.

Carrel's devices were good enough to keep a piece of embryonic chicken heart alive for thirty-four years—much longer than a chicken's lifetime. Carrel even tried to use his tissue cultures to grow viruses—and he succeeded in a way. The only trouble was that bacteria also grew in the tissues, and in order to keep the virus pure, such tedious aseptic precautions had to be taken that it was easier to use animals.

The chick-embryo idea, however, was in the right ball park, so to speak. Better than just a piece of tissue would be the whole thing—the chick embryo itself. A chick embryo is a self-contained organism, protected by the egg shell, equipped with its own natural defenses against bacteria, and cheap and easy to come by in quantity. And, in 1931, the pathologist Ernest William Goodpasture and his co-workers at Vanderbilt University succeeded in transplanting a virus into a chick embryo. For the first time, pure viruses could be cultured almost as easily as bacteria.

The first great medical victory by means of the culture of viruses in fertile eggs came in 1937. At the Rockefeller Institute, bacteriologists were still hunting for further protection against the yellow-fever virus. It was impossible to eradicate the mosquito completely, after all, and infected monkeys maintained a constantly threatening reservoir of the disease in the tropics. The South-African bacteriologist Max Theiler at the Institute set out to produce an attenuated yellow-fever virus. He passed the virus through 200 mice and 100 chick embryos until he had a mutant that caused only mild symptoms yet gave rise to complete immunity against yellow fever. For this achievement Theiler received the 1951 Nobel Prize in medicine and physiology.

When all is said and done, nothing can beat culture in glassware for speed, control of the conditions, and efficiency. In the late 1940's John Franklin Enders, Thomas Huckle Weller, and Frederick Chapman Robbins at the Harvard Medical School went back to Carrel's approach. (He had died in 1944 and was not to see their success.) This time they had a new and powerful weapon against bacteria contaminating the tissue culture—the antibiotics. They added penicillin and streptomycin to the supply of blood that kept the tissues alive, and they found that they could grow viruses without trouble. On impulse, they tried the poliomyelitis virus. To their delight, it flourished in this medium. It was the breakthrough that was to conquer polio, and the three men received the Nobel Prize in medicine and physiology in 1954.

The poliomyelitis virus could now be bred in the test tube, instead of solely in monkeys (which are expensive and temperamental laboratory subjects). Large-scale experimentation with the virus became possible. Thanks to the tissue-culture technique, Jonas Edward Salk of the University of Pittsburgh was able to experiment with chemical treatment of the virus, to learn that polio viruses killed by formaldehyde could still produce immune reactions in the body, and to develop his now-famous Salk vaccine.

Polio's sizable death rate, its dreaded paralysis, its partiality for children (so that it has the alternate name of "infantile paralysis"), the fact that it seems to be a modern scourge with no epidemics on record prior to 1840, and particularly the interest attracted to the disease by its eminent victim, Franklin Delano Roosevelt, made its conquest one of the most celebrated victories over a disease in all human history. Probably no medical announcement ever received such a Hollywood-premiere-type reception as did the report, in 1955, of the evaluating committee that found the Salk vaccine effective. Of course, the event merited such a celebration—more than do most of the performances that arouse people to throw ticker tape and trample one another. But science does not thrive on furor or wild publicity. The rush to respond to the public pressure for the vaccine apparently resulted in a few defective, disease-producing samples of the vaccine slipping through, and the subsequent counterfuror set back the vaccination program against the disease.

The setback was, however, made up, and the Salk vaccine was found effective and, properly prepared, safe. In 1957, the Polish-American microbiologist Albert Bruce Sabin went a step further. He did not make use of dead virus (which, when not entirely dead, could be dangerous) but of strains of living virus, incapable of producing the disease itself, but capable of bringing about the production of appropriate antibodies. Such a "Sabin vaccine" could be taken by mouth, moreover, and did not require the hypodermic. The Sabin vaccine gained popularity first in the Soviet Union and then in the east European countries; but by 1960 it came into use in the United States as well, and the fear of poliomyelitis has lifted.

What does a vaccine do, exactly? The answer to this question may some day give us the chemical key to immunity.

For more than half a century biologists have known the body's main defenses against infection as "antibodies." (Of course, there are also the white blood cells called "phagocytes," which devour bacteria. This was discovered in 1883 by the Russian biologist Ilya Ilitch Mechnikov, who later succeeded Pasteur as the head of the Pasteur Institute in Paris and shared the 1908 Nobel Prize in medicine and physiology with Ehrlich. But phagocytes are no help against viruses and seem not to be involved in the immunity process we are considering.) A virus, or indeed almost any foreign substance entering into the body's chemistry, is called an "antigen." The antibody is a substance manufactured by the body to fight the specific antigen. It puts the antigen out of action by combining with it.

Long before the chemists actually ran down an antibody, they were pretty sure the antibodies must be proteins. For one thing, the best-known antigens were proteins, and presumably it would take a protein to catch a protein. Only a protein could have the subtlety of structure necessary to single out and combine with a particular antigen.

Early in the 1920's Landsteiner (the discoverer of blood groups) carried out a series of experiments which clearly showed that antibodies were very specific indeed. The substances he used to generate antibodies were not antigens but much simpler compounds whose structure was well known. They were arsenic-containing compounds called "arsanilic acids." In combination with a simple protein, such as the albumin of egg white, an arsanilic acid acted as an antigen: when injected into an animal, it gave rise to an antibody in the blood serum. Furthermore, this antibody was specific for the arsanilic acid; the blood serum of the animal would clump only the arsanilic-albumin combination, not albumin alone. Indeed, sometimes the antibody could be made to react with just an arsanilic acid, not combined with albumin. Landsteiner also showed that very small changes in the structure of the arsanilic acid would be reflected in the antibody. An antibody evoked by one variety of arsanilic acid would not react with a slightly altered variety.

Landsteiner coined the name "haptens" (from a Greek word meaning "to bind") for compounds, such as the arsanilic acids, that can give rise to antibodies when they are combined with protein. Presumably, each natural antigen has a specific region in its molecule that acts as a hapten. On that theory, a germ or virus that can serve as a vaccine is one that has had its structure changed sufficiently to reduce its ability to damage cells but still has its hapten group intact, so that it can cause the formation of a specific antibody.

It would be interesting to learn the chemical nature of the natural haptens. If that could be determined, it might be possible to use a hapten, perhaps in combination with some harmless protein, to serve as a vaccine giving rise to antibodies for a specific antigen. That would avoid the necessity of resorting to toxins or attenuated viruses, which always carries some small risk.

Just how an antigen evokes an antibody has not been determined. Ehrlich believed that the body normally contains a small supply of all the antibodies it may need, and that when an invading antigen reacts with the appropriate antibody, this stimulates the body to produce an extra supply of that particular antibody. Some immunologists still adhere to that theory or to modifications of it. Yet it seems highly unlikely that the body is prepared with specific antibodies for all the possible antigens, including unnatural substances such as the arsanilic acids.

The alternate suggestion is that the body has some generalized protein molecule which can be molded to fit any antigen. The antigen, then, acts as a template to shape the specific antibody formed in response to it. Pauling proposed such a theory in 1940. He suggested that the specific antibodies are varying versions of the same basic molecule, merely folded in different ways. In other words, the antibody is molded to fit its antigen as a glove fits the hand.

By 1969, however, the advance of protein analysis had made it possible for a team under Gerald M. Edelman to work out the amino-acid structure of a typical antibody made up of well over a thousand amino acids. No doubt this will pave the way for determining the manner of working of such molecules with considerably more subtlety than had hitherto been possible.

The very specificity of antibodies is a disadvantage in some ways. Suppose a virus mutates so that its protein has a slightly different structure. The old antibody for the virus often will not fit the new structure. It follows that immunity against one strain of virus is no safeguard against another strain. The virus of influenza and of the common cold are particularly prone to minor mutations, and that is one reason why we are plagued by frequent recurrences of these diseases. Influenza, in particular, will occasionally develop a mutant of extraordinary virulence, which may then sweep a surprised and nonimmune world. This happened in 1918 and, with much less fatal result, in the "Asian flu" pandemic of 1957.

A still more annoying effect of the body's oversharp efficiency in forming antibodies is its tendency to produce them even against a harmless protein that happens to enter the body. The body then becomes "sensitized" to that protein, and it may react violently to any later incursion of the originally innocent protein. The reaction may take the form of itching, tears, production of mucus in the nose and throat, asthma, and so on. Such "allergic reactions" are evoked by the pollen of certain plants (causing hay fever), by certain foods, by the fur or dandruff of animals, and so on. An allergic reaction may be acute enough to cause serious disablement, or even death. The discovery of such "anaphylactic shock" won for the French physiologist Charles Robert Richet the Nobel Prize in medicine and physiology in 1913.

In a sense, every human being is more or less allergic to every other human being. A transplant, or graft, from one individual to another will not take, because the receiver's body treats the transplanted tissue as foreign protein and manufactures antibodies against it. The person-to-person graft that will work best is from one identical twin to the other. Since their identical heredity gives them exactly the same proteins, they can exchange tissues or even a whole organ, such as a kidney.

The first successful kidney transplant took place in December 1954 in Boston, from one identical twin to another. The receiver died in 1962 at the age of thirty of coronary artery disease. Since then, hundreds of individuals have lived for months and even years with kidneys transplanted from *other* than identical twins.

Attempts at transplanting other organs, such as the lungs or the liver, have been made, but that which most caught the public fancy was the heart transplant. The first reasonably successful heart-transplants were conducted in December 1967 by the South African surgeon Christiaan Barnard. The fortunate receiver—Philip Blaiberg, a retired South African dentist—lived on for many months on someone else's heart.

For a while after that, heart transplants became the rage, but the furor by late 1969 had died down. Few receivers lived very long, for the problems of tissue rejection seemed mountainous, despite massive attempts to solve the reluctance of the body to incorporate any tissue but its own.

The Australian bacteriologist Macfarlane Burnet had suggested that embryonic tissues might be "immunized" to foreign tissues and that the free-living animal might then tolerate grafts of that tissue. The British biologist Peter Medawar demonstrated this to be so, using mouse embryos. The two men shared in the 1960 Nobel Prize in medicine and physiology as a result.

In 1962, a French-Australian immunologist, Jacques Francis Albert Pierre Miller, working in England, went even further and discovered what may be the reason for this ability to work with embryos in order to make future toleration possible. He discovered that the thymus gland (a piece of tissue which until then had had no known use) was the tissue capable of forming antibodies. If the thymus gland was removed from mice at birth, those mice died after three or four months out of sheer incapacity to protect themselves against the environment. If the thymus was allowed to remain in the mice for three weeks, it already had time to bring about the development of antibody-producing cells in the body, and the thymus gland might then be removed without harm. Embryos in which the thymus has not yet done its work may be so treated as to "learn" to tolerate foreign tissue; the day may yet come when, by the way of the thymus, we may improve tissue toleration, when that is desirable, perhaps even in adults.

And yet, even if the problem of tissue rejection were surmounted, there would remain serious problems. After all, every person who receives a living organ, must receive it from someone who is giving it up, and the question arises as to when the prospective donor may be considered dead enough to yield up his organs.

In that respect it might prove better if mechanical organs were prepared which would involve neither tissue rejection nor knotty ethical issues. Artificial kidneys became practical in the 1940's, and it is possible for patients without natural kidney function to visit a hospital once or twice a week and have their blood cleansed of wastes. It makes for a restricted life even for those fortunate enough to be serviced, but it is preferable to death.

In the 1940's, researchers found that allergic reactions are brought about by the liberation of small quantities of a substance called "histamine" into the blood stream. This led to the successful search for neutralizing "antihistamines," which can relieve the allergic symptoms but, of course, do not remove the allergy. The first successful antihistamine was produced at the Pasteur Institute in Paris in 1937 by the Swiss-born chemist Daniel Bovet, who for this and subsequent researches in chemotherapy was awarded the Nobel Prize in physiology and medicine in 1957.

Noting that sniffling and other allergic symptoms were much like those of the common cold, pharmaceutical firms decided that what worked for one ought to work for the other, and in 1949 and 1950 they flooded the country with antihistamine tablets. (The tablets turned out to do little or nothing for colds, and their vogue diminished.)

In 1937, thanks to the protein-isolating techniques of electrophoresis, biologists finally tracked down the physical location of antibodies in the blood. The antibodies were located in the blood fraction called "gamma globulin."

Physicians have long been aware that some children are unable to form antibodies and therefore are easy prey to infection. In 1951, doctors at the Walter Reed Hospital in Washington made an electrophoretic analysis of the plasma of an eight-year-old boy suffering from a serious septicemia ("blood poisoning"), and to their astonishment they discovered that his blood had no gamma globulin at all. Other cases were quickly discovered. Investigators established that this lack is due to an inborn defect of metabolism which deprives the person of the ability to make gamma globulin; it is called "agammaglobulinemia." Such persons cannot develop immunity to bacteria. They can now be kept alive, however, by antibiotics. Surprisingly enough, they *are* able to become immune to virus infections, such as measles and chickenpox, after having the disease once. Apparently, antibodies are not the body's only defense against viruses.

In 1957, a group of British bacteriologists, headed by Alick Isaacs, showed that cells, under the stimulus of a virus invasion, liberated a protein that had broad antiviral properties. It countered not only the virus involved in the immediate infection but others as well. This protein, named "interferon," is produced more quickly than antibodies are and may explain the antivirus defenses of those with agammaglobulinemia. Apparently its production is stimulated by the presence of RNA in the double-stranded variety found in viruses. Interferon seems to direct the synthesis of a messenger-RNA that produces an antivirus protein that inhibits production of virus protein but not of other forms of protein. Interferon seems to be as potent as antibiotics and doesn't activate resistance. It is, however, fairly species-specific. Only interferon from humans and from other primates will work on human beings.

Cancer

As the danger of infectious diseases diminishes, the incidence of other types of disease increases. Many people who a century ago would have died young of tuberculosis or diphtheria or pneumonia or typhus now live long enough to die of heart disease or cancer. That is one reason why heart disease and cancer have become, respectively, the number one and the number two killers in the Western world. Cancer, in fact, has succeeded plague and smallpox as the great fear of man. It is a nightmare hanging over all of us, ready to strike anyone without warning or mercy. Three hundred thousand Americans die of it each year while 10,000 new cases are recorded each week. The incidence has risen 50 per cent since 1900.

Cancer is actually a group of many diseases (about 200 types are known), affecting various parts of the body in various fashions. But the primary disorder is always the same: disorganization and uncontrolled growth of the affected tissues. The name cancer (the Latin word for "crab") comes from the fact that Hippocrates and Galen fancied the disease spreading its ravages through diseased veins like the crooked, outstretched claws of a crab.

"Tumor" (from the Latin word meaning "grow") is by no means synonymous with cancer; it applies to harmless growths such as warts and moles ("benign tumors") as well as to cancers ("malignant tumors"). The cancers are variously named according to the tissues affected. Cancers of the skin or the intestinal linings (the most common malignancies) are called "carcinomas" (from the Greek word for "crab"); cancers of the connective tissues are "sarcomas"; of the liver, "hepatoma"; of glands generally, "adenomas"; of the white blood cells, "leukemia"; and so on.

Rudolf Virchow of Germany, the first to study cancer tissue under the microscope, believed that cancer was caused by the irritations and shocks of the outer environment. This is a natural thought, for it is just those parts of the body most exposed to the outer world that are most subject to cancer. But when the germ theory of disease became popular, pathologists began to look for some microbe as the cause of cancer. Virchow, a staunch opponent of the germ theory of disease, stubbornly insisted on the irritation theory. (He quit pathology for archaeology and politics when it turned out that the germ theory of disease was going to win out. Few scientists in history have gone down with the ship of mistaken beliefs in quite so drastic a fashion.)

If Virchow was stubborn for the wrong reason, he may have been so in the right cause. There has been increasing evidence that some environments are particularly conducive to cancer. In the eighteenth century, chimney sweeps were found to be more prone to cancer of the scrotum than other people were. After the coal-tar dyes were developed, workers in the dye industries showed an above-average incidence of cancers of the skin or bladder. It seemed that something in soot and in the aniline dyes must be capable of causing cancer. Then, in 1915, two Japanese scientists, K. Yamagiwa and K. Ichikawa, discovered that a certain coal-tar fraction could produce cancer in rabbits when it was applied to the rabbits' ears for long periods. In 1930, two British chemists induced cancer in animals with a synthetic chemical called "dibenzanthracene" (a hydrocarbon with a molecule made up of five benzene rings). This does not occur in coal tar, but three years later it was discovered that "benzpyrene" (also containing five benzene rings but in a different arrangement), a chemical that *does* occur in coal tar, can cause cancer.

Quite a number of "carcinogens" (cancer-producers) have now been identified. Many are hydrocarbons made up of numerous benzene rings, like the first two discovered. Some are molecules related to the aniline dyes. In fact, one of the chief concerns about using artificial dyes in foods is the possibility that in the long run such dyes may be carcinogenic.

Many biologists believe that man has introduced a number of new cancer-producing factors into his environment within the last two or three centuries. There is the increased use of coal; there is the burning of oil on a large scale, particularly in gasoline engines; there is the growing use of synthetic chemicals in food, cosmetics, and so on. The most dramatic of the suspects, of course, is cigarette-smoking, which, statistically at least, seems to be accompanied by a relatively high rate of incidence of lung cancer.

One environmental factor that is certainly carcinogenic is energetic radiation, and man has been exposed to such radiation in increasing measure since 1895.

On November 5, 1895, the German physicist Wilhelm Konrad Roentgen performed an experiment to study the luminescence produced by cathode rays. The better to see the effect, he darkened the room. His cathode-ray tube was enclosed in a black cardboard box. When he turned on the cathode-ray tube, he was startled to catch a flash of light from something across the room. The flash came from a sheet of paper coated with barium platinocyanide, a luminescent chemical. Was it possible that radiation from the closed box had made it glow? Roentgen turned off his cathode-ray tube, and the glow stopped. He turned it on again—the glow returned. He took the paper into the next room, and it still glowed. Clearly, the cathode-ray tube was producing some form of radiation which could penetrate cardboard and walls.

Roentgen, having no idea what kind of radiation this might be, called it simply "X-rays." Other scientists tried to change the name to "Roentgen rays," but this was so hard for anyone but Germans to pronounce that "X-rays" stuck. (We now know that the speeding electrons making up the cathode rays are strongly decelerated on striking a metal barrier. The kinetic energy lost is converted into radiation that is called "*Bremsstrahlung*"—German for "braking radiation." X-rays are an example of such radiation.)

The X-rays revolutionized physics. They captured the imagination of physicists, started a typhoon of experiments, led within a few months to the discovery of radioactivity, and opened up the inner world of the atom. When the award of Nobel Prizes began in 1901, Roentgen was the first to receive the prize in physics.

The hard X-radiation also started something else—exposure of human beings to intensities of energetic radiation such as man had never experienced before. Four days after the news of Roentgen's discovery reached the United States, X-rays were used to locate a bullet in a patient's leg. They were a wonderful means of exploring the interior of the body. X-rays pass easily through the soft tissues (consisting chiefly of elements of low atomic weight) and tend to be stopped by elements of higher atomic weight, such as make up the bones (composed largely of phosphorus and calcium). On a photographic plate placed behind the body, bone shows up as a cloudy white, in contrast to the black areas where X-rays have come through in greater intensity because they have been much less absorbed by the soft tissues. A lead bullet shows up as pure white; it stops the X-rays completely.

X-rays are obviously useful for showing bone fractures, calcified joints, cavities in the teeth, foreign objects in the body, and so on. But it is also a simple matter to outline the soft tissues by introducing an insoluble salt of a heavy element. Barium sulfate, when swallowed, will outline the stomach or intestines. An iodine compound injected into the blood will travel to the kidneys and the ureter and outline those organs, for iodine has a high atomic weight and therefore is opaque to X-rays.

Even before X-rays were discovered, a Danish physician, Niels Ryberg Finsen, had found that high-energy radiation could kill microorganisms; he used ultraviolet light to destroy the bacteria causing lupus vulgaris, a skin disease. (For this he was awarded the Nobel Prize in physiology and medicine in 1903.) The X-rays turned out to be far more deadly. They could kill the fungus of ringworm. They could damage or destroy human cells, and were eventually used to kill cancer cells beyond reach of the surgeon's knife.

What was also discovered—the hard way—was that high-energy

radiation could *cause* cancer. At least one hundred of the early workers with X-rays and radioactive materials died of cancer, the first death taking place in 1902. As a matter of fact, both Marie Curie and her daughter, Irène Joliot-Curie, died of leukemia, and it is easy to believe that radiation was a contributing cause in both cases. In 1928, a British physician, George William Marshall Findlay, found that even ultraviolet radiation was energetic enough to cause skin cancer in mice.

It is certainly reasonable to suspect that man's increasing exposure to energetic radiation (in the form of medical X-rays and so on) may be responsible for part of the increased incidence of cancer. And the future will tell whether the accumulation in our bones of traces of strontium 90 from fallout will increase the incidence of bone cancer and leukemia.

What can all the various carcinogens—chemicals, radiation, and so on—possibly have in common? One reasonable thought is that all of them may cause genetic mutations, and that cancer may be the result of mutations in body cells.

Suppose that some gene is changed so that it no longer can produce a key enzyme needed in the process that controls the growth of cells. When a cell with such a defective gene divides, it will pass on the defect. With the control mechanism not functioning, further division of these cells may continue indefinitely, without regard to the needs of the body as a whole or even to the needs of the tissue involved (for example, the specialization of cells in an organ). The tissue is disorganized. It is, so to speak, a case of anarchy in the body.

That energetic radiation can produce mutations is well established. What about the chemical carcinogens? Well, mutation by chemicals also has been demonstrated. The "nitrogen mustards" are a good example. These compounds, like the "mustard gas" of World War I, produce burns and blisters on the skin resembling those caused by X-rays. They can also damage the chromosomes and increase the mutation rate. Moreover, a number of other chemicals have been found to imitate energetic radiation in the same way.

The chemicals that can induce mutations are called "mutagens." Not all mutagens have been shown to be carcinogens, and not all carcinogens have been shown to be mutagens. But there are enough cases of compounds that are both carcinogenic and mutagenic to arouse suspicion that the coincidence is not accidental.

Meanwhile, the notion that microorganisms may have something to do with cancer is far from dead. With the discovery of viruses, this suggestion of the Pasteur era was revived. In 1903, the French bacteriologist Amédée Borrel suggested that cancer might be a virus disease, and, in 1908, two Danes, Wilhelm Ellerman and Olaf Bang, showed that fowl leukemia was indeed caused by a virus. However, leukemia was not at the time recognized as a form of cancer, and the issue hung fire. In 1909, however, the American physician Francis Peyton Rous ground up a chicken tumor, filtered it, and injected the clear filtrate into other chickens. Some of them developed tumors. The finer the filter, the fewer the tumors. This certainly looked as if particles of some kind were responsible for the initiation of tumors, and it seemed that these particles were the size of viruses.

The "tumor viruses" have had a rocky history. At first, the tumors pinned down to viruses turned out to be uniformly benign; for instance, viruses were shown to cause such things as rabbits' papillomas (similar to warts). In 1936, John Joseph Bittner, working in the famous mouse-breeding laboratory at Bar Harbor, Maine, came on something more exciting. Maude Slye of the same laboratory had bred strains of mice that seemed to have an inborn resistance to cancer and other strains that seemed cancer-prone. The mice of some strains rarely developed cancer; those of other strains almost invariably did, after reaching maturity. Bittner tried the experiment of switching mothers on the newborn mice so that they would suckle at the opposite strain. He discovered that when baby mice of a "cancer-resistant" strain suckled at mothers of a "cancer-prone" strain, they usually developed cancer. On the other hand, supposedly cancer-prone baby mice that were fed by cancer-resistant mothers did not develop cancer. Bittner concluded that the cancer cause, whatever it was, was not inborn but was transmitted in the mothers' milk. He called it the "milk factor."

Naturally, Bittner's milk factor was suspected to be a virus. Eventually the Columbia University biochemist Samuel Graff identified the factor as a particle containing nucleic acids. Other tumor viruses, causing certain types of mouse tumors and animal leukemias, have been found, and all of them contain nucleic acids. No viruses have been detected in connection with human cancers, but research on human cancer is obviously limited.

Now the mutation and virus theories of cancer begin to converge. Perhaps the seeming contradiction between the two notions is not a contradiction after all. Viruses and genes have a very important thing in common: the key to the behavior of both lies in their nucleic acids. Indeed, in 1959, G. A. di Mayorca and co-workers at the Sloan-Kettering Institute and the National Institutes of Health isolated DNA from a mouse-tumor virus and found that the DNA alone could induce cancers in mice just as effectively as the virus did.

Thus the difference between the mutation theory and the virus theory boils down to whether the cancer-causing nucleic acid arises by a mutation in a gene within the cell or is introduced by a virus invasion from outside the cell. These ideas are not mutually exclusive; cancer may come about in both ways.

It was not until 1966, however, that the virus hypothesis was deemed fruitful enough to be worth a Nobel Prize. Fortunately, Peyton Rous, who had made his discovery fifty-five years, before was still alive and received a share of the 1966 Nobel Prize for medicine and physiology. (He lived on to 1970, dying at the age of ninety, active in research nearly to the end.)

What goes wrong in the metabolic machinery when cells grow unrestrainedly? This question has as yet received no answer. But strong suspicion rests on some of the hormones, especially the sex hormones.

For one thing, the sex hormones are known to stimulate rapid, localized growth in the body (as in the breasts of an adolescent girl). For another, the tissues of sexual organs—the breasts, cervix, and ovaries in a woman, the testes and prostate in a man—are particularly prone to cancer. Strongest of all is the chemical evidence. In 1933, the German biochemist Heinrich Wieland (who had won the Nobel Prize in chemistry in 1927 for his work with bile acids) managed to convert a bile acid into a complex hydrocarbon called "methylcholanthrene," a powerful carcinogen. Now methylcholanthrene (like the bile acids) has the four-ring structure of a steroid, and it so happens that all the sex hormones are steroids. Could a misshapen sex-hormone molecule act as a carcinogen? Or might even a correctly shaped hormone be mistaken for a carcinogen, so to speak, by a distorted gene pattern in a cell, and so stimulate uncontrolled growth? It is anyone's guess, but these are interesting speculations.

Curiously enough, changing the supply of sex hormones sometimes checks cancerous growth. For instance, castration, to reduce the body's manufacture of male sex hormone, or the administration of neutralizing female sex hormone, has a mitigating effect on cancer of the prostate. As a treatment, this is scarcely something to shout about, and it is a measure of the desperation regarding cancer that such devices are resorted to.

The main line of attack against cancer still is surgery. And its limitations are still what they have always been: sometimes the cancer cannot be cut out without killing the patient; often the knife frees bits of malignant tissue (since the disorganized cancer tissue has a tendency to fragment), which are then carried by the blood stream to other parts of the body where they take root and grow.

The use of energetic radiation to kill the cancer tissue also has its drawbacks. Artificial radioactivity has added new weapons to the traditional X-rays and radium. One of them is cobalt 60, which yields high-energy gamma rays and is much less expensive than radium; another is a solution of radioactive iodine (the "atomic cocktail"), which concentrates in the thyroid gland and thus attacks a thyroid cancer. But the body's tolerance of radiation is limited, and there is always the danger that the radiation will start more cancers than it stops.

Still, surgery and radiation are the best we have, and they have saved, or at least prolonged, many a life. They will perforce be man's main reliance against cancer until biologists find what they are seeking: a "magic bullet" that, without harming normal cells, will search out the cancer cells and either destroy them or stop their wild division in its tracks.

A great deal of work is going on along two principal routes. One is to find out everything possible about how cells divide. The other is to learn in the greatest possible detail exactly how cells conduct metabolism, with the hope of finding some decisive difference between cancer cells and normal cells. Differences have been found, but they are pretty minor—so far.

Meanwhile a stupendous sifting of chemicals by trial and error is being carried out. About 50,000 new drugs a year are tested. For a time the nitrogen mustards looked hopeful, on the theory that they would mimic radiation in killing cancer cells. Some of the drugs of this type do seem to help against certain types of cancer, at least to the extent of prolonging life, but they are obviously only a stopgap.

More hope lies in the direction of the nucleic acids themselves. There must be some difference between the nucleic acids in cancer cells and those in normal cells. The object, then, is to find a way to interfere with the chemical workings of one and not the other. Then again, perhaps the disorganized cancer cells are less efficient than normal cells in manufacturing nucleic acids. If so, throwing a few grams of sand into the machinery might cripple the less efficient cancer cells without seriously disturbing the more efficient normal cells.

For instance, one substance that is vital to the production of nucleic acid is folic acid. It plays a major role in the formation of the purines and pyrimidines, the building blocks for nucleic acid. Now a compound resembling folic acid might, by competitive inhibition, slow things up just enough to prevent cancer cells from making nucleic acid while allowing normal cells to produce it at an adequate rate. And, of course, without nucleic acid the cancer cells could not multiply. There are, in fact, "folic-acid antagonists" of this sort. One of them, called "amethopterin," has shown some effect against leukemia.

There is a still more direct attack. Why not inject competitive substitutes for the purines and pyrimidines themselves? The most hopeful candidate is "6-mercaptopurine." This compound is just like adenine except that it has an $-SH$ group in place of adenine's $-NH_2$.

Even the possibility of treatment of just one of the cancer group of diseases is not to be ignored. The malignant cells of certain types of leukemias require an outside source of the substance asparagine, something healthy cells can manufacture for themselves. Treatment with the enzyme asparaginase, which catalyzes the breakdown of asparagine, reduces the body's supply and starves the malignant cells while the normals manage to survive.

The world-wide research attack upon cancer is keen, resourceful, and, in comparison with other biological research, handsomely financed. Treatment has reached the point where one out of three cancer victims will survive and live out a normal life span. But a cure will not be found easily, for the secret of cancer is as subtle as the secret of life itself.

CHAPTER 14

The Body

Food

Perhaps the first great advance in medical science was the recognition by physicians that good health called for a simple, balanced diet. The Greek philosophers recommended moderation in eating and drinking, not only for philosophical reasons but also because those who followed this rule were more comfortable and lived longer. That was a good start, but biologists eventually learned that moderation alone was not enough. Even if one has the good fortune to avoid eating too little and the good sense to avoid eating too much, he will still do poorly if his diet happens to be shy of certain essential ingredients, as is actually the case for large numbers of people in some parts of the world.

The human body is rather specialized (as organisms go) in its dietary needs. A plant can live on just carbon dioxide, water, and certain inorganic ions. Some of the microorganisms likewise get along without any organic food; they are called "autotrophic" ("self-feeding"), which means that they can grow in environments in which there is no other living thing. The bread mold *Neurospora* begins to get a little more complicated: in addition to inorganic substances it has to have sugar and the vitamin biotin. And as the forms of life become more and more complex, they seem to become more and more dependent on their diet to supply the organic building blocks necessary for building living tissue. The reason is simply that they have lost some of the enzymes that primitive organisms possess. A green plant has a complete supply of enzymes for making all the necessary amino acids, proteins, fats, and carbohydrates from inorganic materials. *Neurospora* has all the enzymes except one or more of those needed to make sugar and biotin. By the time we get to man, we find that he lacks the enzymes required to make many of the amino acids, the vitamins, and various other necessities, and he must get these ready-made in his food.

This may seem a kind of degeneration—a growing dependence on the environment which puts the organism at a disadvantage. Not so. If the environment supplies the building blocks, why carry the elaborate enzymatic machinery needed to make them? By dispensing with this machinery, the cell can use its energy and space for more refined and specialized purposes.

It was the English physician William Prout (the same Prout who was a century ahead of his time in suggesting that all the elements were built from hydrogen) who first suggested that the organic foods could be divided into three types of substances, later named carbohydrates, fats, and proteins.

The chemists and biologists of the nineteenth century, notably Justus von Liebig of Germany, gradually worked out the nutritive properties of these foods. Protein, they found, is the most essential, and the organism could get along on it alone. The body cannot make protein from carbohydrate and fat, because those substances have no nitrogen, but it could make the necessary carbohydrates and fats from materials supplied by protein. Since protein is comparatively scarce in the environment, however, it would be wasteful to live on an all-protein diet—like stoking the fire with furniture when firewood is available.

Under favorable circumstances the human body's daily requirement of proteins, by the way, is surprisingly low. The Food and Nutrition Board of the National Research Council in its 1958 chart of recommendations suggested that the minimum for adults is one gram of protein per kilogram of body weight per day, which amounts to a little more than two ounces for the average grown man. About two quarts of milk can supply that amount. Children and pregnant or nursing mothers need somewhat more protein.

Of course, a lot depends on what proteins you choose. Nineteenth-century experimenters tried to find out whether the population could get along, in times of famine, on gelatin—a protein material obtained by heating bones, tendons, and other otherwise inedible parts of animals. But the French physiologist François Magendie demonstrated that dogs lost weight and died when gelatin was their sole source of protein. This does not mean there is anything wrong with gelatin as a food, but it simply does not supply all the necessary building blocks when it is the only protein in the diet. The key to the usefulness of a protein lies in the efficiency with which the body can use the nitrogen it supplies. In 1854 the English agriculturists John Bennet Lawes and Joseph Henry Gilbert fed pigs protein in two forms—lentil meal and barley meal. They found that the pigs retained much more of the nitrogen in barley than of that in lentils. These were the first "nitrogen balance" experiments.

A growing organism gradually accumulates nitrogen from the food it ingests ("positive nitrogen balance"). If it is starving or suffering a wasting disease, and gelatin is the sole source of protein, the body continues to starve or waste away, from a nitrogen-balance standpoint (a situation called "negative nitrogen balance"). It keeps losing more nitrogen than it takes in, regardless of how much gelatin it is fed.

Why so? The nineteenth-century chemists eventually discovered that gelatin is an unusually simple protein. It lacks tryptophan and other amino acids present in most proteins. Without these building blocks, the body cannot build the proteins it needs for its own substance. Therefore, unless it gets other protein in its food as well, the amino acids that do occur in the gelatin are useless and have to be excreted. It is as if housebuilders found themselves with plenty of lumber but no nails. Not only could they not build the house, but the lumber would just be in the way and eventually would have to be disposed of. Attempts were made in the 1890's to make gelatin a more efficient article of diet by adding some of those amino acids in which it was deficient, but without success. Better luck was obtained with proteins not as drastically limited as gelatin.

In 1906, the English biochemists Frederick Gowland Hopkins and E. G. Willcock fed mice a diet in which the only protein was "zein," found in corn. They knew that this protein had very little of the amino acid tryptophan. The mice died in about fourteen days. The experimenters then tried mice on zein plus tryptophan. This time the mice survived twice as long. It was the first hard evidence that amino acids, rather than protein, might be the essential components of the diet. (Although the mice still died prematurely, this was probably due mainly to a lack of certain vitamins not known at the time.)

In the 1930's, the American nutritionist William Cumming Rose got to the bottom of the amino-acid problem. By that time the major vitamins were known, so he could supply the animals with those needs and focus on the amino acids. Rose fed rats a mixture of amino acids instead of protein. The rats did not live long on this diet. But when he fed rats on the milk protein casein, they did well. Apparently there was something in casein—some undiscovered amino acid, in all probability —which was not present in the amino-acid mixture he was using. Rose broke down the casein and tried adding various of its molecular fragments to his amino-acid mixture. In this way he tracked down the amino acid "threonine," the last of the major amino acids to be discovered. When he added the threonine from casein to his amino-acid mixture, the rats grew satisfactorily, without any intact protein in the diet.

Rose proceeded to remove the amino acids from their diet one at a time. By this method he eventually identified ten amino acids as indispensable items in the diet of the rat: lysine, tryptophan, histidine, phenylalanine, leucine, isoleucine, threonine, methionine, valine, and arginine. If supplied with ample quantities of these, the rat could manufacture all it needed of the others, such as glycine, proline, aspartic acid, alanine, and so on.

In the 1940's, Rose turned his attention to man's requirements of amino acids. He persuaded graduate students to submit to controlled diets in which a mixture of amino acids was the only source of nitrogen. By 1949, he was able to announce that the adult male required only eight amino acids in the diet: phenylalanine, leucine, isoleucine, methionine, valine, lysine, tryptophan, and threonine. Since arginine and histidine, indispensable to the rat, are dispensable in the human diet, it would seem that in this respect man is less specialized than the rat, or, indeed, than any other mammal that has been tested in detail.

Potentially a person could get along on the eight dietarily essential amino acids; given enough of these, he could make not only all the other amino acids he needs but also all the carbohydrates and fats. Actually a diet made up only of amino acids would be much too expensive, to say nothing of its flatness and monotony. But it is enormously helpful to have a complete blueprint of our amino-acid needs so that we can reinforce natural proteins when necessary for maximum efficiency in absorbing and utilizing nitrogen.

Vitamins

Food fads and superstitions unhappily still delude too many people—and spawn too many cure-everything best sellers—even in these enlightened times. In fact, it is perhaps because these times are enlightened that food faddism is possible. Through most of man's history, his food consisted of whatever could be produced in the vicinity, of which there usually was not very much. It was eat what there was to eat or starve; no one could afford to be picky, and without pickiness there can be no food faddism.

Modern transportation has made it possible to ship food from any part of the earth to any other, particularly since the use of large-scale refrigeration has arisen. This reduced the threat of famine, which, before modern times, was invariably local, with neighboring provinces loaded with food that could not be transported to the famine area.

Home storage of a variety of foods became possible as early man learned to preserve foods by drying, salting, increasing the sugar content, fermenting, and so on. It became possible to preserve food in states closer to the original when methods of storing cooked food in vacuum were developed. (The cooking kills microorganisms and the vacuum prevents others from growing and reproducing.) Vacuum storage was first made practical by a French chef, François Appert, who developed the technique in response to a prize offered by Napoleon I for a way of preserving food for his armies. Appert made use of glass jars, but nowadays, tin-lined steel cans (inappropriately called "tin cans" or just "tins") are used for the purpose. Since World War II, fresh-frozen food has become popular and the growing number of home freezers has further increased the general availability and variety of fresh foods. Each broadening of food availability has increased the practicality of food-faddism.

All this is not to say that a shrewd choice of food may not be useful. There are certain cases in which specific foods will definitely cure a particular disease. In every instance, these are "deficiency diseases," diseases produced by the lack in the diet of some substance essential to the body's chemical machinery. These arise almost invariably when a person is deprived of a normal, balanced diet—one containing a wide variety of foods.

To be sure, the value of a balanced and variegated diet was understood by a number of medical practitioners of the nineteenth century and before, when the chemistry of food was still a mystery. A famous example is that of Florence Nightingale, the heroic English nurse of the Crimean War who pioneered the adequate feeding of soldiers, as well as decent medical care. And yet "dietetics" (the systematic study of diet) had to await the end of the century and the discovery of trace substances in food, essential to life.

The ancient world was well acquainted with scurvy, a disease in which the capillaries become increasingly fragile, gums bleed and teeth loosen, wounds heal with difficulty if at all, and the patient grows weak and eventually dies. It was particularly prevalent in besieged cities and on long ocean voyages. (Magellan's crew suffered more from scurvy than from general undernourishment.) Ships on long voyages, lacking refrigeration, had to carry nonspoilable food, which meant hardtack and salt pork. Nevertheless, physicians for many centuries failed to connect scurvy with diet.

In 1536, while the French explorer Jacques Cartier was wintering in Canada, 110 of his men were stricken with scurvy. The native Indians knew and suggested a remedy: drinking water in which pine needles had been soaked. Cartier's men in desperation followed this seemingly childish suggestion. It cured them of their scurvy.

Two centuries later, in 1747, the Scottish physician James Lind took note of several incidents of this kind and experimented with fresh fruits and vegetables as a cure. Trying his treatments on scurvy-ridden sailors, he found that oranges and lemons brought about improvement most quickly. Captain Cook, on a voyage of exploration across the Pacific from 1772 to 1775, kept his crew scurvy-free by enforcing the regular eating of sauerkraut. Nevertheless, it was not until 1795 that the brass hats of the British Navy were sufficiently impressed by Lind's experiments (and by the fact that a scurvy-ridden flotilla could lose a naval engagement with scarcely a fight) to order daily rations of lime juice for British sailors. (They have been called "limeys" ever since, and the Thames area in London where the crates of limes were stored is still called "Limehouse.") Thanks to the lime juice, scurvy disappeared from the British Navy.

A century later, in 1891, Admiral Takaki of the Japanese Navy similarly introduced a broader diet into the rice monotony of his ships. The scourge of a disease known as "beri-beri" came to an end in the Japanese Navy as a result.

In spite of occasional dietary victories of this kind (which no one could explain), nineteenth-century biologists refused to believe that a disease could be cured by diet, particularly after Pasteur's germ theory of disease came into its own. In 1896, however, a Dutch physician named Christiaan Eijkman convinced them almost against his own will.

Eijkman was sent to the Dutch East Indies to investigate beri-beri, which was endemic in those regions (and which, even today, when medicine knows its cause and cure, still kills 100,000 people a year). Takaki had stopped beri-beri by dietary measures, but the West, apparently, placed no stock in what might have seemed merely the mystic lore of the Orient.

Supposing that beri-beri was a germ disease, Eijkman took along some chickens as experimental animals in which to establish the germ. A highly fortunate piece of skulduggery upset his plans. Without warning, most of his chickens came down with a paralytic disease from which some died, but after about four months those still surviving regained their health. Eijkman, mystified by failing to find any germ responsible for the attack, finally investigated the chickens' diet. He discovered that the person originally in charge of feeding the chickens had economized (and no doubt profited) by using scraps of leftover food, mostly polished rice, from the wards of the military hospital. It happened that after a few months a new cook had arrived and taken over the feeding of the chickens; he had put a stop to the petty graft and supplied the animals with the usual chicken feed, containing unhulled rice. It was then that the chickens had recovered.

Eijkman experimented. He put chickens on a polished-rice diet, and they fell sick. Back on the unhulled rice, they recovered. It was the first case of a deliberately produced dietary-deficiency disease. Eijkman decided that this "polyneuritis" that afflicted fowls was similar in symptoms to human beri-beri. Did human beings get beri-beri because they ate only polished rice?

For human consumption, rice was stripped of its hulls mainly so that it would keep better, for the rice germ removed with the hulls contains oils that go rancid easily. Eijkman and a co-worker, Gerrit Grijns, set out to see what it was in rice hulls that prevented beri-beri. They succeeded in dissolving the crucial factor out of the hulls with water, and they found that it would pass through membranes which would not pass proteins. Evidently the substance in question must be a fairly small molecule. They could not, however, identify it.

Meanwhile other investigators were coming across other mysterious factors that seemed to be essential for life. In 1905, a Dutch nutritionist, C. A. Pekelharing, found that all his mice died within a month on an artificial diet which seemed ample as far as fats, carbohydrates, and proteins were concerned. But mice did fine when he added a few drops of milk to this diet. And in England the biochemist Frederick Hopkins, who was demonstrating the importance of amino acids in the diet, carried out a series of experiments in which he, too, showed that something in the casein of milk would support growth if added to an artificial diet. This something was soluble in water. Even better than casein as the dietary supplement was a small amount of a yeast extract.

For their pioneer work in establishing that trace substances in the diet were essential to life, Eijkman and Hopkins shared the Nobel Prize in medicine and physiology in 1929.

The next task was to isolate these vital trace factors in food. By 1912, three Japanese biochemists, Umetaro Suzuki, T. Shimamura, and S. Ohdake, had extracted from rice hulls a compound which was very potent in combating beri-beri. Doses of five to ten milligrams sufficed to effect a cure in fowl. In the same year the Polish-born biochemist Casimir Funk (then working in England and later to come to the United States) prepared the same compound from yeast.

Because the compound proved to be an amine (that is, one containing the amine group, NH_2), Funk called it a "vitamine," Latin for "life amine." He made the guess that beri-beri, scurvy, pellagra, and rickets all arose from deficiencies of "vitamines." Funk's guess was correct as far as his identification of these diseases as dietary-deficiency diseases was concerned. But it turned out that not all "vitamines" were amines.

In 1913, two American biochemists, Elmer Vernon McCollum and Marguerite Davis, discovered another trace factor vital to health in butter and egg yolk. This one was soluble in fatty substances instead of water. McCollum called it "fat-soluble A," to contrast it with "water-soluble B," which was the name he applied to the antiberi-beri factor. In the absence of chemical information as to the nature of the factors, this seemed fair enough, and it started the custom of naming them by letters. In 1920, the British biochemist Jack Cecil Drummond changed the names to "vitamin A" and "vitamin B," dropping the final e of "vitamine" as a gesture toward taking "amine" out of the name. He also suggested that the antiscurvy factor was still a third such substance, which he named "vitamin C."

Vitamin A was quickly identified as a food factor required to prevent the development of abnormal dryness of the membranes around the eye, called "xerophthalmia," from Greek words meaning "dry eyes." In 1920,

McCollum and his associates found that a substance in cod-liver oil, which was effective in curing both xerophthalmia and a bone disease called "rickets," could be so treated as to cure rickets only. They decided the anti-rickets factor must represent a fourth vitamin, which they named vitamin D. Vitamins D and A are fat-soluble; C and B are water-soluble.

By 1930, it had become clear that vitamin B was not a simple substance but a mixture of compounds with different properties. The food factor that cured beri-beri was named vitamin B_1, a second factor was called vitamin B_2, and so on. Some of the reports of new factors turned out to be false alarms, so that one does not hear of B_3, B_4, or B_5 any longer. However, the numbers worked their way up to B_{14}. The whole group of vitamins (all water-soluble) is frequently referred to as the "B-vitamin complex."

New letters also were added. Of these, vitamins E and K (both fat-soluble) remain as veritable vitamins, but "vitamin F" turned out to be not a vitamin and "vitamin H" turned out to be one of the B-complex vitamins.

Nowadays, with their chemistry identified, the letters of even the true vitamins are going by the board, and most of them are known by their chemical names, though the fat-soluble vitamins, for some reason, have held on to their letter designations more tenaciously than the water-soluble ones.

It was not easy to work out the chemical composition and structure of the vitamins, for these substances occur only in minute amounts. For instance, a ton of rice hulls contains only about five grams (a little less than a fifth of an ounce) of vitamin B_1. Not until 1926 did anyone extract enough of the reasonably pure vitamin to analyze it chemically. Two Dutch biochemists, Barend Coenraad Petrus Jansen and William Frederick Donath, worked up a composition for vitamin B, from a tiny sample, but it turned out to be wrong. In 1932, Ohdake tried again on a slightly larger sample and got it almost right. He was the first to detect a sulfur atom in a vitamin molecule.

Finally in 1934 Robert R. Williams, then director of chemistry at the Bell Telephone Laboratories, climaxed 20 years of research by painstakingly separating vitamin B_1 from tons of rice hulls until he had enough to work out a complete structural formula. The formula follows:

$$
\begin{array}{c}
\text{CH}_3 \\
\end{array}
\quad
\begin{array}{c}
\text{CH}_2 - \text{CH}_2 - \text{OH} \\
\end{array}
$$

CH$_3$ — C $=$ N — C — NH$_2$ — C $=$ C — CH$_2$ — CH$_2$ — OH — S — N $=$ CH — C — CH$_2$ — N$^{(+)}$ — CH

Since the most unexpected feature of the molecule was the atom of sulfur ("theion" in Greek), the vitamin was named "thiamine."

Vitamin C was a different sort of problem. Citrus fruits furnish a comparatively rich source of this material, but one difficulty was finding an experimental animal that did not make its own vitamin C. Most mammals, aside from man and the other primates, have retained the capacity to form this vitamin. Without a cheap and simple experimental animal that would develop scurvy, it was difficult to follow the location of vitamin C among the various fractions into which the fruit juice was broken down chemically.

In 1918 the American biochemists B. Cohen and Lafayette Benedict Mendel solved this problem by discovering that guinea pigs could not form the vitamin. In fact, guinea pigs developed scurvy much more easily than men did. But another difficulty remained. Vitamin C was found to be very unstable (it is the most unstable of the vitamins), so it was easily lost in chemical procedures to isolate it. A number of research workers ardently pursued the vitamin without success.

As it happened, Vitamin C was finally isolated by someone who was not particularly looking for it. In 1928, the Hungarian-born biochemist Albert Szent-Györgi, then working in London in Hopkins' laboratory and interested mainly in finding out how tissues made use of oxygen, isolated from cabbages a substance which helped transfer hydrogen atoms from one compound to another. Shortly afterward Charles Glen King and his co-workers at the University of Pittsburgh, who *were* looking for vitamin C, prepared some of the substance from cabbages and found that it was strongly protective against scurvy. Furthermore, they found it identical with crystals they had obtained from lemon juice. King determined its structure in 1933, and it turned out to be a sugar molecule of six carbons, belonging to the L-series instead of the D-series:

$$O = C \underset{\displaystyle C = C}{\overset{\displaystyle O}{\diagup \diagdown}} CH - CH - CH_2OH$$

It was named "ascorbic acid" (from Greek words meaning "no scurvy").

As for vitamin A, the first hint as to its structure came from the observation that the foods rich in vitamin A were often yellow or orange (butter, egg yolk, carrots, fish-liver oil, and so on). The substance largely responsible for this color was found to be a hydrocarbon named "carotene," and in 1929 the British biochemist Thomas Moore demonstrated that rats fed on diets containing carotene stored vitamin A in the liver. The vitamin itself was not colored yellow, so the deduction was that though carotene was not itself vitamin A, the liver converted it into something which was vitamin A. (Carotene is now considered an example of a "provitamin.")

In 1937, the American chemists Harry Nicholls Holmes and Ruth Elizabeth Corbet isolated vitamin A as crystals from fish-liver oil. It turned out to be a 20-carbon compound—half of the carotene molecule with a hydroxyl group added:

$$CH_3 \quad CH_3$$
$$C$$
$$CH_2 \quad C - CH = CH - C = CH - CH = CH - C = CH - CH_2 - OH$$
$$CH_2 \quad C - CH_3$$
$$CH_2$$

The chemists hunting for vitamin D found their best chemical clue by means of sunlight. As early as 1921, the McCollum group (who first demonstrated the existence of the vitamin) showed that rats did not develop rickets on a diet lacking vitamin D if they were exposed to sunlight. Biochemists guessed that the energy of sunlight converted some provitamin in the body into vitamin D. Since vitamin D was fat-soluble, they went searching for the provitamin in the fatty substances of food.

By breaking down fats into fractions and exposing each fragment separately to sunlight, they determined that the provitamin that sunlight converted into vitamin D was a steroid. What steroid? They tested cholesterol, the most common steroid of the body, and that was not it. Then, in 1926, the British biochemists Otto Rosenheim and T. A. Webster found that sunlight would convert a closely related sterol, "ergosterol" (so named from the fact that it was first isolated from ergot-infested rye), into vitamin D. The German chemist Adolf Windaus discovered this independently at about the same time. For this and other work in steroids, Windaus received the Nobel Prize in chemistry in 1928.

The difficulty in producing vitamin D from ergosterol rested on the fact that ergosterol did not occur in animals. Eventually the human provitamin was identified as "7-dehydrocholesterol," which differs from cholesterol only in having two hydrogen atoms fewer in its molecule. The vitamin D formed from it has this formula:

Vitamin D in one of its forms is called "calciferol," from Latin

701

words meaning "calcium-carrying," because it is essential to the proper laying down of bone structure.

Not all the vitamins show their absence by producing an acute disease. In 1922, Herbert McLean Evans and K. J. Scott at the University of California implicated a vitamin as a cause of sterility in animals. Evans and his group did not succeed in isolating this one, vitamin E, until 1936. It was then given the name "tocopherol" (from Greek words meaning "to bear children").

Unfortunately, whether human beings need vitamin E, or how much, is not yet known. Obviously, dietary experiments designed to bring about sterility cannot be tried on human subjects. And even in animals, the fact that they can be made sterile by withholding vitamin E does not necessarily mean that natural sterility arises in this way.

In the 1930's, the Danish biochemist Carl Peter Henrik Dam discovered by experiments on chickens that a vitamin was involved in the clotting of blood. He named it *Koagulationsvitamine,* and this was eventually shortened to vitamin K. Edward Doisy and his associates at St. Louis University then isolated vitamin K and determined its structure. Dam and Doisy shared the Nobel Prize in medicine and physiology in 1943.

Vitamin K is not a major vitamin nor a nutritional problem. Normally a more than adequate supply of this vitamin is manufactured by the bacteria in the intestines. In fact, they make so much of it that the feces may be richer in vitamin K than the food is. Newborn infants are the most likely to run a danger of poor blood clotting and consequent hemorrhage because of vitamin-K deficiency. In the hygienic modern hospital it takes infants three days to accumulate a reasonable supply of intestinal bacteria, and they are protected by injections of the vitamin into themselves directly or into the mother shortly before birth. In the old days, the infants picked up the bacteria almost at once, and though they might die of various infections and disease, they were at least safe from the dangers of hemorrhage.

In fact, one might wonder whether organisms could live at all in the complete absence of intestinal bacteria, or whether the symbiosis had not become too intimate to abandon. However, germ-free animals have been grown from birth under completely sterile conditions and have even been allowed to reproduce under such conditions. Mice have been carried through twelve generations in this fashion. Experiments of this sort have been conducted at the University of Notre Dame since 1928.

During the late 1930's and early 1940's, biochemists identified several additional B vitamins, which now go under the names of biotin, pantothenic acid, pyridoxine, folic acid, and cyanocobalamine. These vitamins are all made by intestinal bacteria; moreover, they are present so universally in foodstuffs that no cases of deficiency diseases have appeared. In fact, investigators have had to feed animals an artificial diet deliberately excluding them, and even to add "antivitamins" to neutralize those made by the intestinal bacteria, in order to see what the deficiency symptoms are. (Antivitamins are substances similar to the vitamin in structure. They immobilize the enzyme making use of the vitamin by means of competitive inhibition.)

The determination of the structure of each of the various vitamins was usually followed speedily (or even preceded) by synthesis of the vitamin. For instance, Williams and his group synthesized thiamine in 1937, three years after they had deduced its structure. The Polish-born Swiss biochemist Tadeus Reichstein and his group synthesized ascorbic acid in 1933, somewhat before the structure was completely determined by King. Vitamin A, for another example, was synthesized in 1936 (again somewhat before the structure was completely determined) by two different groups of chemists.

The use of synthetic vitamins has made it possible to fortify food (milk was first vitamin-fortified as early as 1924) and to prepare vitamin mixtures at reasonable prices and sell them over the drugstore counter. The need for vitamin pills varies with individual cases. Of all the vitamins, the one most apt to be deficient in supply is vitamin D. Young children in northern climates, where sunlight is weak in winter time, run the danger of rickets, so they may require irradiated foods or vitamin supplements. But the dosage of vitamin D (and of vitamin A) should be carefully controlled, because an overdose of these vitamins can be harmful. As for the B vitamins, anyone eating an ordinary, rounded diet does not need to take pills for them. The same is true of vitamin C, which in any case should not present a problem, for there are few people who do not enjoy orange juice or who do not drink it regularly in these vitamin-conscious times.

On the whole, the wholesale use of vitamin pills, while redounding chiefly to the profit of drug houses, usually does people no harm and may be partly responsible for the fact that the current generation of Americans is taller and heavier than previous generations.

Biochemists naturally were curious to find out how the vitamins, present in the body in such tiny quantities, exerted such important effects on the body chemistry. The obvious guess was that they had something to do with enzymes, also present in small quantities.

The answer finally came from detailed studies of the chemistry of enzymes. Protein chemists had known for a long time that some proteins were not made up solely of amino acids, and that nonamino-acid prosthetic groups might exist, such as the heme in hemoglobin (see Chapter 10). In general, these prosthetic groups tended to be tightly bound to the rest of the molecule. With enzymes, however, there were in some cases nonamino-acid portions that were quite loosely bound and might be removed with little trouble.

This was first discovered in 1904 by Arthur Harden (who was soon to discover phosphorus-containing intermediates; see Chapter 11). Harden worked with a yeast extract capable of bringing about the fermentation of sugar. He placed it in a bag made of a semipermeable membrane and placed that bag in fresh water. Small molecules could penetrate the membrane, but the large protein molecule could not. After this "dialysis" had progressed for a while, Harden found that the activity of the extract was lost. Neither the fluid within nor that outside the bag would ferment sugar. If the two fluids were combined, activity was regained.

Apparently, the enzyme was made up not only of a large protein molecule, but also of a "coenzyme" molecule, small enough to pass through the pores of the membrane. The coenzyme was essential to enzyme activity (it was the "cutting edge," so to speak).

Chemists at once tackled the problem of determining the structure of this coenzyme (and of similar adjuncts to other enzymes). The German-Swedish chemist Hans Karl August Simon von Euler-Chelpin was the first to make real progress in this respect. As a result, he and Harden shared the Nobel Prize in chemistry in 1929.

The coenzyme of the yeast enzyme studied by Harden proved to consist of a combination of an adenine molecule, two ribose molecules, two phosphate groups, and a molecule of "nicotinamide." Now this last was an unusual kind of thing to find in living tissue, and interest naturally centered on the nicotinamide. (It is called "nicotinamide" because it contains an amide group, $CONH_2$, and can be formed easily from nicotinic acid. Nicotinic acid is structurally related to the tobacco alkaloid "nicotine," but they are utterly different in properties; for one thing, nicotinic acid is necessary to life, whereas nicotine is a deadly poison.) The formulas of nicotinamide and nicotinic acid are:

nicotinic acid   nicotinamide

Once the formula of Harden's coenzyme was worked out, it was promptly renamed "diphosphopyridine nucleotide" (DPN)—"nucleotide" from the characteristic arrangement of the adenine, ribose, and phosphate, similar to that of the nucleotides making up nucleic acid, and "pyridine" from the name given to the combination of atoms making up the ring in the nicotinamide formula.

Soon a very similar coenzyme was found, differing from DPN only in the fact that it contained three phosphate groups rather than two. This, naturally, was named "triphosphopyridine nucleotide" (TPN).

Both DPN and TPN proved to be coenzymes for a number of enzymes in the body, all serving to transfer hydrogen atoms from one molecule to another. (Such enzymes are called "dehydrogenases.") It was the coenzyme that did the actual job of hydrogen transfer; the enzyme proper in each case selected the particular substrate on which the operation was to be performed. The enzyme and the coenzyme each had a vital function, and if either was deficient in supply, the release of energy from foodstuffs via hydrogen transfer would slow to a limp.

What was immediately striking about all this was that the nicotinamide group represented the only part of the enzyme the body cannot manufacture itself. It can make all the protein it needs and all the ingredients of DPN and TPN except the nicotinamide; that it must find ready-made (or at least in the form of nicotinic acid) in the diet. If not, then the manufacture of DPN and TPN stops and all the hydrogen-transfer reactions they control slow down.

Was nicotinamide or nicotinic acid a vitamin? As it happened, Funk (who coined the word "vitamine") had isolated nicotinic acid from rice hulls. Nicotinic acid was not the substance that cured beri-beri, so he had ignored it. But on the strength of nicotinic acid's appearance in connection with coenzymes, the University of Wisconsin biochemist Conrad Arnold Elvehjem and his co-workers tried it on another deficiency disease.

In the 1920's, the American physician Joseph Goldberger had studied pellagra (sometimes called Italian leprosy), a disease endemic in the Mediterranean area and almost epidemic in the southern United States in the early part of this century. Pellagra's most noticeable symptoms are a dry, scaly skin, diarrhea, and an inflamed tongue; it sometimes leads to mental disorders. Goldberger noticed that the disease struck people who lived on a limited diet (e.g., mainly cornmeal) and spared families that owned a milch cow. He began to experiment with artificial diets, feeding them to animals and inmates of jails (where pellagra seemed to blossom). He succeeded in producing "blacktongue" (a disease analogous to pellagra) in dogs, and in curing this disease with a yeast extract. He found he could cure jail inmates of pellagra by adding milk to their diet. Goldberger decided that a vitamin must be involved, and he named it the P-P ("pellagra-preventive") factor.

It was pellagra, then, that Elvehjem chose for the test of nicotinic acid. He fed a tiny dose to a dog with blacktongue, and the dog responded with a remarkable improvement. A few more doses cured him. Nicotinic acid was a vitamin, all right; it was the P-P factor.

The American Medical Association, worried that the public might get the impression there were vitamins in tobacco, urged that the vitamin not be called nicotinic acid and suggested instead the name "niacin" (an abbreviation of *nic*otine *ac*id) or "niacinamide." Niacin has caught on fairly well.

Gradually, it became clear that the various vitamins were merely portions of coenzymes, each consisting of a molecular group an animal or a human being cannot make for itself. In 1932, Warburg had found a yellow coenzyme that catalyzed the transfer of hydrogen atoms. The Austrian chemist Richard Kuhn and his associates shortly afterward isolated vitamin B_2, which proved to be yellow, and worked out its structure:

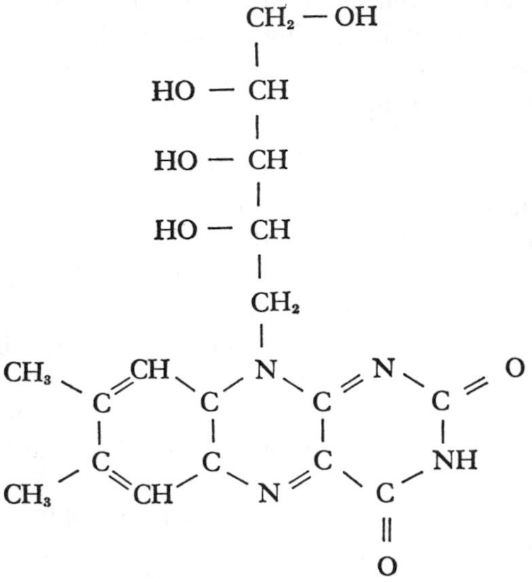

$$CH_2 - OH$$
$$|$$
$$HO - CH$$
$$|$$
$$HO - CH$$
$$|$$
$$HO - CH$$
$$|$$
$$CH_2$$

The carbon chain attached to the middle ring is like a molecule called "ribitol," so vitamin B_2 was named "riboflavin," "flavin" coming from a Latin word meaning "yellow." Since examination of its spectrum showed that riboflavin was very similar in color to Warburg's yellow coenzyme, Kuhn tested the coenzyme for riboflavin activity in 1935 and found such activity to be there. In the same year the Swedish biochemist Hugo Theorell worked out the structure of Warburg's yellow coenzyme and showed that it was riboflavin with a phosphate group added. (In 1954, a second and more complicated coenzyme also was shown to have riboflavin as part of its molecule.)

Kuhn was awarded the 1938 Nobel Prize in chemistry, and Theorell received the 1955 Nobel Prize in medicine and physiology. Kuhn, however, was unfortunate enough to be selected for his prize shortly after Austria had been absorbed by Nazi Germany, and (like Gerhard Domagk) he was compelled to refuse it.

Riboflavin was synthesized, independently, by the Swiss chemist Paul Karrer. For this and other work on vitamins, Karrer was awarded a share of the 1937 Nobel Prize in chemistry. (He shared it with the English chemist Walter Norman Haworth, who had worked on the structure of carbohydrate molecules.)

In 1937, the German biochemists K. Lohmann and P. Schuster discovered an important coenzyme that contained thiamine as part of its structure. Through the 1940's other connections were found between B vitamins and coenzymes. Pyridoxine, pantothenic acid, folic acid, biotin—each in turn was found to be tied to one or more groups of enzymes.

The vitamins beautifully illustrate the economy of the human body's chemical machinery. The human cell can dispense with making them because they serve only one special function, and the cell can take the reasonable risk of finding the necessary supply in the diet. There are many other vital substances that the body needs only in trace amounts but must make for itself. ATP, for instance, is formed from much the same building blocks that make up the indispensable nucleic acids. It is inconceivable that any organism could lose any enzyme necessary for nucleic-acid synthesis and remain alive, for nucleic acid is needed in such quantities that the organism dare not trust to the diet for its supply of the necessary building blocks. And to be able to make nucleic acid automatically implies the ability to make ATP. Consequently, no organism is known that is incapable of manufacturing its own ATP, and in all probability no such organism will ever be found.

To make such special products as vitamins would be like setting up a special machine next to an assembly line to turn out nuts and bolts for the automobiles. The nuts and bolts can be obtained more efficiently from a parts supplier, without any loss to the apparatus for assembling the automobiles; by the same token the organism can obtain vitamins in its diet, with a saving in space and material.

The vitamins illustrate another important fact of life. As far as is known, all living cells require the B vitamins. The coenzymes are an essential part of the cell machinery of every cell alive—plant, animal, or bacterial. Whether the cell gets the B vitamins from its diet or makes them itself, it must have them if it is to live and grow. This universal need for a particular group of substances is an impressive piece of evidence for the essential unity of all life and its descent (possibly) from a single original scrap of life formed in the primeval ocean.

While the roles of the B vitamins are now well known, the chemical functions of the other vitamins have proved rather hard nuts to crack. The only one on which any real advance has been made is vitamin A.

In 1925, the American physiologists L. S. Fridericia and E. Holm found that rats fed on a diet deficient in vitamin A had difficulty performing tasks in dim light. An analysis of their retinas showed that they were deficient in a substance called "visual purple."

There are two kinds of cell in the retina of the eye—"rods" and "cones." The rods specialize in vision in dim light, and they contain the visual purple. A shortage of visual purple therefore hampers only vision in dim light, and it results in what is known as "night blindness."

In 1938, the Harvard biologist George Wald began to work out the chemistry of vision in dim light. He showed that light causes visual purple, or "rhodopsin," to separate into two components: the protein "opsin" and a nonprotein called "retinene." Retinene proved to be very similar in structure to vitamin A.

The retinene always recombines with the opsin to form rhodopsin in the dark. But during its separation from opsin in the light, a small percentage of it breaks down, because it is unstable. However, the supply of retinene is replenished from vitamin A, which is converted to retinene by the removal of two hydrogen atoms with the aid of enzymes. Thus vitamin A acts as a stable reserve for retinene. If vitamin A is lacking in the diet, eventually the retinene supply and the amount of visual purple decline, and night blindness is the result. For his work in this field, Wald shared in the 1967 Nobel Prize for medicine and physiology.

Vitamin A must have other functions as well, for a deficiency causes dryness of the mucous membranes and other symptoms which cannot very well be traced to troubles in the retina of the eye. But the other functions are still unknown.

The same has to be said about the chemical functions of vitamins C, D, E, and K. In 1970, Linus Pauling created a stir by maintaining that massive doses of vitamin C would reduce the incidence of colds. The public stripped druggists' shelves of the vitamin at once, and no doubt the contention will forthwith receive a thorough testing.

Minerals

It is natural to suppose that the materials making up anything as wonderful as living tissue must themselves be something pretty exotic. Wonderful the proteins and nucleic acids certainly are, but it is a little humbling to realize that the elements making up the human body are as common as dirt, and the whole lot could be bought for a few dollars. (It used to be cents, but inflation has raised the price of everything.)

In the early nineteenth century, when chemists were beginning to analyze organic compounds, it became quite clear that living tissue was made up, in the main, of carbon, hydrogen, oxygen, and nitrogen. These four elements alone constituted about 96 per cent of the weight of the human body. Then there was also a little sulfur in the body. If you burned off these five elements, you were left with a bit of white ash, mostly the residue from the bones. The ash was a collection of minerals.

It was not surprising to find common salt, sodium chloride, in the ash. After all, salt is not a mere condiment to improve the taste of food—as dispensable as, say, basil, rosemary, or thyme. It is a matter of life and death. You need only taste blood to realize that salt is a basic component of the body. Herbivorous animals, which presumably lack sophistication as far as the delicacies of food preparation are concerned, will undergo much danger and privation to reach a "salt lick," where they can make up the lack of salt in their diet of grass and leaves.

As early as the mid-eighteenth century, the Swedish chemist Johann Gottlieb Gahn had shown that bones were made up largely of calcium phosphate, and an Italian scientist, V. Menghini, had established that the blood contained iron. In 1847, Justus von Liebig found potassium and magnesium in the tissues. By the mid-nineteenth century, then, the mineral constituents of the body were known to include calcium, phosphorus, sodium, potassium, chlorine, magnesium, and iron. Furthermore, these were as active in life processes as any of the elements usually associated with organic compounds.

The case of iron is the clearest. If it is lacking in the diet, the blood becomes deficient in hemoglobin and transports less oxygen from the lungs to the cells. The condition is known as "iron-deficiency anemia." The patient is pale for lack of the red pigment and tired for lack of oxygen.

In 1882 the English physician Sidney Ringer found that a frog heart could be kept alive and beating outside its body in a solution (called "Ringer's solution" to this day) containing, among other things, sodium, potassium, and calcium in about the proportions found in the frog's blood. Each was essential for functioning of muscle. An excess of calcium caused the muscle to lock in permanent contraction ("calcium rigor"), whereas an excess of potassium caused it to unlock in permanent relaxation ("potassium inhibition"). Calcium, moreover, was vital to blood clotting. In its absence blood would not clot, and no other element could substitute for calcium in this respect.

Of all the minerals, phosphorus was eventually discovered to perform the most varied and crucial functions in the chemical machinery of life (see Chapter 12).

Calcium, a major component of bone, makes up 2 per cent of the body; phosphorus, 1 per cent. The other minerals I have mentioned come in smaller proportions, down to iron, which makes up only 0.004 per cent of the body. (That still leaves the average adult male 1/10 of an ounce of iron in his tissues.) But we are not at the end of the list; there are other minerals that, though present in tissue only in barely detectable quantities, are yet essential to life.

The mere presence of an element is not necessarily significant; it may be just an impurity. In our food we take in at least traces of every element in our environment, and some small amount of each finds its way into our tissues. But elements such as silicon and aluminum, for instance, contribute nothing. On the other hand, zinc is vital. How does one distinguish an essential mineral from an accidental impurity?

The best way is to show that some necessary enzyme contains the trace element as an essential component. (Why an enzyme? Because in no other way can any trace component possibly play an important role.) In 1939, David Keilin and T. Mann of England showed that zinc was an integral part of the enzyme carbonic anhydrase. Now carbonic anhydrase is essential to the body's handling of carbon dioxide, and the proper handling of that important waste material in turn is essential to life. It follows in theory that zinc is indispensable to life, and experiment shows that it actually is. Rats fed on a diet low in zinc stop growing, lose hair, suffer scaliness of the skin, and die prematurely for lack of zinc as surely as for lack of a vitamin.

In the same way it has been shown that copper, manganese, cobalt, and molybdenum are essential to animal life. Their absence from the diet gives rise to deficiency diseases. Molybdenum, the latest of the essential trace elements to be identified (in 1954), is a constituent of an enzyme called "xanthine oxidase." The importance of molybdenum was first noticed in the 1940's in connection with plants, when soil scientists found that plants would not grow well in soils deficient in the element. It seems that molybdenum is a component of certain enzymes in soil microorganisms that catalyze the conversion of the nitrogen of the air into nitrogen-containing compounds. Plants depend on this help from microorganisms because they cannot themselves take nitrogen from the air. (This is only one of an enormous number of examples of the close interdependence of all life on our planet. The living world is a long and intricate chain which may suffer hardship or even disaster if any link is broken.)

MINERALS NECESSARY TO LIFE

Sodium (Na)	Zinc (Zn)
Potassium (K)	Copper (Cu)
Calcium (Ca)	Manganese (Mn)
Phosphorus (P)	Cobalt (Co)
Chlorine (Cl)	Molybdenum (Mo)
Magnesium (Mg)	Iodine (I)
Iron (Fe)	

Not all "trace elements" are universally essential. Boron seems to be essential in traces to plant life, but not, apparently, to animals. Certain tunicates gather vanadium from sea water and use it in their oxygen-transporting compound, but few, if any, other animals require vanadium for any reason.

It is now realized that there are trace-element deserts, just as there are waterless deserts; the two usually go together but not always. In Australia soil scientists have found that an ounce of molybdenum in the form of some appropriate compound spread over sixteen acres of molybdenum-deficient land results in a considerable increase in fertility. Nor is this a problem of exotic lands only. A survey of American farm-land in 1960 showed areas of boron deficiency in 41 states, of zinc deficiency in twenty-nine states, and of molybdenum deficiency in twenty-one states. The dosage of trace elements is crucial. Too much is as bad as too little, for some substances that are essential for life in small quantities (e.g., copper) become poisonous in large quantities.

This, of course, carries to its logical extreme the much older custom of using "fertilizers" for soil. Until modern times, fertilization was through the use of animal excreta, manure or guano, which restored nitrogen and phosphorus to the soil. While this worked, it was accompanied by foul odors and by the ever-present possibility of infection. The substitution of chemical fertilizers, clean and odor-free, was through the work of Justus von Liebig in the early nineteenth century.

One of the most dramatic episodes in the discovery of mineral deficiencies has to do with cobalt. It involves the once incurably fatal disease called "pernicious anemia."

In the early 1920's, the University of Rochester pathologist George Hoyt Whipple was experimenting on the replenishment of hemoglobin by means of various food substances. He would bleed dogs to induce anemia and then feed them various diets to see which would permit them to replace the lost hemoglobin most rapidly. He did this not because he was interested in pernicious anemia, or in any kind of anemia, but because he was investigating bile pigments, compounds produced by the body from hemoglobin. Whipple discovered that the food that enabled the dogs to make hemoglobin most quickly was liver.

In 1926, two Boston physicians, George Richards Minot and William Parry Murphy, considered Whipple's results, decided to try liver as a treatment for pernicious-anemia patients. The treatment worked. The incurable disease was cured, so long as the patients ate liver as an important portion of their diet. Whipple, Minot, and Murphy shared the Nobel Prize in physiology and medicine in 1934.

Unfortunately liver, although it is a great delicacy when properly cooked, then chopped, and lovingly mixed with such things as eggs, onions, and chicken fat, becomes wearing as a steady diet. (After a while, a patient might be tempted to think pernicious anemia was preferable.) Biochemists began to search for the curative substance in liver, and by 1930 Edwin Joseph Cohn and his co-workers at the Harvard

Medical School had prepared a concentrate a hundred times as potent as liver itself. To isolate the active factor, however, further purification was needed. Fortunately, chemists at the Merck Laboratories discovered in the 1940's that the concentrate from liver could accelerate the growth of certain bacteria. This provided an easy test of the potency of any preparation from it, so the biochemists could proceed to break down the concentrate into fractions and test them in quick succession. Because the bacteria reacted to the liver substance in much the same way that they reacted to, say, thiamine or riboflavin, the investigators now suspected strongly that the factor they were hunting for was a B vitamin. They called it "vitamin B_{12}."

By 1948, using bacterial response and chromatography, Ernest Lester Smith in England and Karl August Folkers at Merck succeeded in isolating pure samples of vitamin B_{12}. The vitamin proved to be a red substance, and both scientists thought it resembled the color of certain cobalt compounds. It was known by this time that a deficiency of cobalt caused severe anemia in cattle and sheep. Both Smith and Folkers burned samples of vitamin B_{12}, analyzed the ash, and found that it did indeed contain cobalt. The compound has now been named "cyanocobalamine." So far it is the only cobalt-containing compound that has been found in living tissue.

By breaking it up and examining the fragments, chemists quickly decided that vitamin B_{12} was an extremely complicated compound, and they worked out an empirical formula of $C_{63}H_{88}O_{14}N_{14}PCo$. Then a British chemist, Dorothy Crowfoot Hodgkin, determined its over-all structure by means of X-rays. The diffraction pattern given by crystals of the compound allowed her to build up a picture of the "electron densities" along the molecule, that is, those regions where the probability of finding an electron is high and those where it is low. If lines are drawn through regions of equal probability, a kind of skeletal picture is built up of the shape of the molecule as a whole.

This is not as easy as it sounds. Complicated organic molecules can produce an X-ray scattering truly formidable in its complexity. The mathematical operations required to translate that scattering into electron densities are tedious in the extreme. By 1944, electronic computers had been called in to help work out the structural formula of penicillin. Vitamin B_{12} was much more complicated and Miss Hodgkin had to use a more advanced computer—the National Bureau of Standards Western Automatic Computer (SWAC)—and do some heavy spade-work. It eventually earned for her, however, the 1964 Nobel Prize for chemistry.

The molecule of vitamin B_{12}, or cyanocobalamine, turned out to be a lopsided porphyrin ring, with one of the carbon bridges connecting two of the smaller pyrrole rings missing, and with complicated side chains on the pyrrole rings. It resembled the somewhat simpler heme molecule, with this key difference: where heme had an iron atom at the center of the porphyrin ring, cyanocobalamine had a cobalt atom.

Cyanocobalamine is active in very small quantities when injected into the blood of pernicious-anemia patients. The body can get along on only 1/1,000 as much of this substance as it needs of the other B vitamins. Any diet, therefore, ought to have enough cyanocobalamine for our needs. Even if it did not, the bacteria in the intestines manufacture quite a bit of it. Why, then, should anyone ever have pernicious anemia?

Apparently, the sufferers from this disease are simply unable to absorb enough of the vitamin into the body through the intestinal walls. Their feces are actually rich in the vitamin (for want of which they are dying). From feedings of liver, providing a particularly abundant supply, such a patient manages to absorb enough cyanocobalamine to stay alive. But he needs 100 times as much of the vitamin if he takes it by mouth as he does when it is injected directly into the blood.

Something must be wrong with the patient's intestinal apparatus, preventing the passage of the vitamin through the walls of the intestines. It has been known since 1929, thanks to the researches of the American physician William Bosworth Castle, that the answer lies somehow in the gastric juice. Castle called the necessary component of gastric juice "intrinsic factor." And in 1954 investigators found a product, from the stomach linings of animals, which assists the absorption of the vitamin and proved to be Castle's intrinsic factor. Apparently this substance is missing in pernicious-anemia patients. When a small amount of it is mixed with cyanocobalamine, the patient has no difficulty in absorbing the vitamin through the intestines. Just how this intrinsic factor helps absorption is still not known.

Getting back to the trace elements. . . . The first one discovered was not a metal; it was iodine, an element with properties like those of chlorine. This story begins with the thyroid gland.

In 1896, a German biochemist, Eugen Baumann, discovered that the thyroid was distinguished by containing iodine, practically absent from all other tissues. In 1905, a physician named David Marine, who had just set up practice in Cleveland, was amazed to find how widely prevalent goiter was in that area. Goiter is a conspicuous disease, sometimes producing grotesque enlargement of the thyroid and causing its victims to become either dull and listless, or nervous, overactive, and pop-eyed. For the development of surgical techniques in the treatment of abnormal thyroids with resulting relief from goitrous conditions, the Swiss physician Emil Theodor Kocher earned the 1909 Nobel Prize in medicine and physiology.

But Marine wondered whether the enlarged thyroid might not be the result of a deficiency of iodine, the one element in which the thyroid specialized, and whether goiter might not be treated more safely and expeditiously by chemicals rather than by the knife. Iodine deficiency and the prevalence of goiter in the Cleveland area might well go hand in hand, at that, for Cleveland, being inland, might lack the iodine that was so plentiful in the soil near the ocean and in the seafood that is an important article of diet there.

The doctor experimented on animals and after ten years felt sure enough of his ground to try feeding iodine-containing compounds to goiter patients. He was probably not too surprised to find that it worked. Marine then suggested that iodine-containing compounds be added to table salt and to the water supply of inland cities where the soil was poor in iodine. There was strong opposition to his proposal, however, and it took another ten years to get water iodination and iodized salt generally accepted. Once the iodine supplements became routine, simple goiter declined in importance as one of mankind's woes.

Today American researchers (and the public) are engaged in studies and discussion of a similar health question—the fluoridation of water to prevent tooth decay. This issue is still a matter of bitter controversy in the nonscientific and political arena; so far the opposition has been far more stubborn and successful than in the case of iodine. Perhaps one reason is that cavities in the teeth do not seem nearly as serious as the disfigurement of goiter.

In the early decades of this century dentists noticed that people in certain areas in the United States (e.g., some localities in Arkansas) tended to have darkened teeth—a mottling of the enamel. Eventually this was traced to a higher-than-average content of fluorine compounds ("fluorides") in the natural drinking water of those areas. With the attention of researchers directed to fluoride in the water, another interesting discovery turned up. Where the fluoride content of the water was above average, the population had an unusually low rate of tooth decay. For instance, the town of Galesburg in Illinois, with fluoride in its water, had only one-third as many cavities per youngster as the nearby town of Quincy, whose water contained practically no fluoride.

Tooth decay is no laughing matter, as anyone with a toothache will readily agree. It costs the people of the United States more than a billion and a half dollars a year in dental bills, and by the age of thirty-five two thirds of all Americans have lost at least some of their teeth. Dental researchers succeeded in getting support for large-scale studies to find out whether fluoridation of water would be safe and would really help to prevent tooth decay. They found that one part per million of fluoride in the drinking water, at an estimated cost of 5 to 10 cents per person per year, did not mottle teeth and yet showed an effect in decay prevention. They therefore adopted one part per million as a standard for testing the results of fluoridation of community water supplies.

The effect is, primarily, on those whose teeth are being formed; that is, on children. The presence of fluoride in the drinking water ensures the incorporation of tiny quantities of fluoride into the tooth structure; it is this, apparently, that makes the tooth mineral unpalatable to bacteria. (The use of small quantities of fluoride in pill-form or in toothpaste has also shown some protective effect against tooth decay.)

The dental profession is now convinced, on the basis of a quarter of a century of research, that for a few pennies per person per year, tooth decay can be reduced by about two thirds, with a saving of at least a billion dollars a year in dental costs and a relief of pain and of dental handicaps that cannot be measured in money. The nation's dental and medical organizations, the United States Public Health Service, and state health agencies recommend fluoridation of public water supplies. And yet, in the realm of politics fluoridation has lost a majority of its battles. A group called the National Committee Against Fluoridation has aroused community after community to vote down fluoridation and even to repeal it in some localities where it had been adopted.

Two chief arguments have been employed by the opponents with the greatest effect. One is that fluorine compounds are poisonous. So they are, but not in the doses used for fluoridation! The other is that fluoridation is compulsory medication, infringing the individual's freedom. That may be so, but it is questionable whether the individual in any society should have the freedom to expose others to preventable sickness. If compulsory medication is evil, then we have a quarrel not only with fluoridation but also with chlorination, iodination, and, for that matter, with all the forms of inoculation, including vaccination against smallpox, that are compulsory in most civilized countries today.

Hormones

Enzymes, vitamins, trace elements—how potently these sparse substances decide life-or-death issues for the organism! But there is a fourth group of substances that, in a way, are even more potent. They conduct the whole performance; they are like a master switch that awakens a city to activity, or the throttle that controls an engine, or the red cape that excites the bull.

At the turn of the century two English physiologists, William Maddock Bayliss and Ernest Henry Starling, became intrigued by a striking little performance in the digestive tract. The gland behind the stomach known as the pancreas releases its digestive fluid into the upper intestines at just the moment when food leaves the stomach and enters the intestine. How does it get the message? What tells the pancreas that the right moment has arrived? The obvious guess was that the information must be transmitted via the nervous system, which was then the only known means of communication in the body. Presumably, the entry of food into the intestines from the stomach stimulated nerve endings that relayed the message to the pancreas by way of the brain or the spinal cord.

To test this theory, Bayliss and Starling cut every nerve to the pancreas. Their maneuver failed! The pancreas still secreted juice at precisely the right moment.

The puzzled experimenters went hunting for an alternate signaling system. In 1902, they tracked down a "chemical messenger." It was a substance secreted by the walls of the intestine. When they injected this into an animal's blood, it stimulated the secretion of pancreatic juice even though the animal was not eating. Bayliss and Starling concluded that, in the normal course of events, food entering the intestines stimulates their linings to secrete the substance, which then travels via the bloodstream to the pancreas and triggers the gland to start giving forth pancreatic juice. The two investigators named the substance secreted by the intestines "secretin," and they called it a "hormone," from a Greek word meaning "rouse to activity." Secretin is now known to be a small protein molecule.

Several years earlier, physiologists had discovered that an extract of the adrenals (two small organs just above the kidneys) could raise blood pressure if injected into the body. The Japanese chemist Jokichi Takamine, working in the United States, isolated the responsible substance in 1901 and named it "adrenalin." (This later became a trade name; the chemists' name for it now is "epinephrine.") Its structure proved to resemble that of the amino acid tyrosine, from which it is derived in the body.

Plainly, adrenalin, too, was a hormone. As the years went on, the physiologists found that a number of other "glands" in the body secreted hormones. (The word "gland" comes from the Greek word for acorn, and it was originally applied to any small lump of tissue in the body. But it became customary to give the name gland to any tissue that secreted a fluid, even large organs such as the liver and the mammaries. Small organs that did not secrete fluids gradually lost this name, so that the "lymph glands," for instance, were renamed the "lymph nodes." Even so, when lymph nodes in the throat or the armpit become enlarged during infections, physicians and mothers alike still refer to them as "enlarged glands.")

Many of the glands, such as those along the alimentary canal, the sweat glands, and the salivary glands, discharge their fluids through ducts. Some, however, are "ductless"; they release substances directly into the blood stream, which then circulates the secretions through the body. It is the secretions of these ductless or "endocrine" glands that contain hormones. The study of hormones is for this reason termed "endocrinology."

Naturally, biologists are most interested in hormones that control functions of the mammalian body and, in particular, that of man. However, I should like at least to mention the fact that there are "plant hormones" that control and accelerate plant growth, "insect hormones" that control pigmentation and molting, and so on.)

When biochemists found that iodine was concentrated in the thyroid gland, they made the reasonable guess that the element was part of a hormone. In 1915, Edward Calvin Kendall of the Mayo Foundation in Minnesota isolated from the thyroid an iodine-containing amino acid which behaved like a hormone, and he named it "thyroxine." Each molecule of thyroxine contained four atoms of iodine. Like adrenalin, thyroxine has a strong family resemblance to tyrosine and is manufactured from it in the body. (Many years later, in 1952, the biochemist Rosalind Pitt-Rivers and her associates isolated another thyroid hormone—"tri-iodothyronine," so named because its molecule contains three atoms of iodine rather than four. It is less stable than thyroxine but three to five times as active.)

The thyroid hormones control the over-all rate of metabolism in the body: they arouse all the cells to activity. People with an underactive thyroid are sluggish, torpid, and after a time may become mentally retarded, because the various cells are running in low gear. Conversely, people with an overactive thyroid are nervous and jittery, because their cells are racing. Either an underactive or an overactive thyroid can produce goiter.

The thyroid controls the body's "basal metabolism," that is, its rate of consumption of oxygen at complete rest in comfortable environmental conditions—the "idling rate," so to speak. If a person's basal metabolism is above or below the norm, suspicion falls upon the thyroid gland. Measurement of the basal metabolism is a tedious affair, for the subject must fast for a period in advance and lie still for half an hour while the rate is measured, to say nothing of an even longer period beforehand. Instead of going through this troublesome procedure, why not go straight to the horse's mouth—that is, measure the amount of rate-controlling hormone that the thyroid is producing? In recent years researchers have developed a method of measuring the amount of "protein-bound iodine" (PBI) in the bloodstream; this indicates the rate of thyroid-hormone production and so has provided a simple, quick blood test to replace the basal-metabolism determination.

The best-known hormone is insulin, the first protein whose structure was fully worked out (see Chapter 11). Its discovery was the culmination of a long chain of events.

Diabetes is the name of a whole group of diseases, all characterized by unusual thirst and, in consequence, an unusual output of urine. It is the most common of the inborn errors of metabolism. There are 1.5 million diabetics in the United States, 80 per cent of whom are over forty-five. It is one of the few diseases to which the female is more subject than the male; women diabetics outnumber men four to three.

The name comes from a Greek word meaning "syphon" (apparently the coiner pictured water syphoning endlessly through the body). The most serious form of the disease is "diabetes mellitus." "Mellitus" comes from the Greek word for "honey," and it refers to the fact that in advanced stages of certain cases of the disease the urine has a sweet taste. (This may have been determined directly by some heroic physician, but the first indication of this was rather indirect. Diabetic urine tended to gather flies.) In 1815, the French chemist Michel Eugène Chevreul was able to show the sweetness was due to the presence of the simple sugar glucose. This waste of glucose plainly indicates that the body is not utilizing its food efficiently. In fact, the diabetic patient, despite an increase in appetite, may steadily lose weight as the disease advances. Up to a generation ago there was no helpful treatment for the disease.

In the nineteenth century, the German physiologists Joseph von Mering and Oscar Minkowski found that removal of the pancreas gland from a dog produced a condition just like human diabetes. After Bayliss and Starling discovered the hormone secretin, it began to appear that a hormone of the pancreas might be involved in diabetes. But the only known secretion from the pancreas was the digestive juice. Where did the hormone come from? A significant clue turned up. When the duct of the pancreas was tied off, so that it could not pour out its digestive secretions, the major part of the gland shriveled, but the groups of cells known as the "islets of Langerhans" (after the German physician Paul Langerhans, who had discovered them in 1869) remained intact.

In 1916, a Scottish physician, Albert Sharpey-Schafer, suggested, therefore, that the islets must be producing the antidiabetes hormone. He named the assumed hormone "insulin," from the Latin word for "island."

Attempts to extract the hormone from the pancreas at first failed miserably. As we now know, insulin is a protein, and the protein-splitting enzymes of the pancreas destroyed it even while the chemists were trying to isolate it. In 1921, the Canadian physician Frederick Grant Banting and the physiologist Charles Herbert Best (working in the laboratories of John James Rickard MacLeod at the University of Toronto) tried a new approach. First they tied off the duct of the pancreas. The enzyme-producing portion of the gland shriveled, the production of protein-splitting enzymes stopped, and the scientists were then able to extract the intact hormone from the islets. It proved indeed effective in countering diabetes, and it is estimated that in the next fifty years it saved the lives of some 20 to 30 million diabetics. Banting called the hormone "isletin," but the older and more Latinized form proposed by Sharpey-Schafer won out. Insulin it became and still is.

In 1923, Banting and, for some reason, MacLeod (whose only service to the discovery of insulin was to allow the use of his laboratory over the summer while he was on vacation) received the Nobel Prize in physiology and medicine.

The effect of insulin within the body shows most clearly in connection with the level of glucose concentration in the blood. Ordinarily the body stores most of its glucose in the liver, in the form of a kind of starch called "glycogen" (discovered in 1856 by the French physiologist Claude Bernard), keeping only a small quantity of glucose in the blood stream to serve the immediate energy needs of the cells. If the glucose concentration in the blood rises too high, this stimulates the pancreas to increase its production of insulin, which pours into the blood stream and brings about a lowering of the glucose level. On the other hand, when the glucose level falls too low, the lowered concentration inhibits the production of insulin by the pancreas, so that the sugar level rises. Thus a balance is achieved. The production of insulin lowers the level of glucose, which lowers the production of insulin, which raises the level of glucose, which raises the production of insulin, which lowers the level of glucose—and so on. This is an example of what is called "feedback." The thermostat that controls the heating of a house works in the same fashion.

Feedback is probably the customary device by which the body maintains a constant internal environment. Another example involves the hormone produced by the parathyroid glands, four small bodies embedded in the thyroid gland. The hormone "parathormone" was finally purified in 1960 by the American biochemists Lyman Creighton Craig and Howard Rasmussen after five years of work.

The molecule of parathormone is somewhat larger than that of insulin, being made up of eighty-three amino acids and possessing a molecular weight of 9,500. The action of the hormone is to increase calcium absorption in the intestine and decrease calcium loss through the kidneys. Whenever calcium concentration in the blood falls slightly below normal, secretion of the hormone is stimulated. With more calcium coming in and less going out, the blood level soon rises; this rise inhibits the secretion of the hormone. This interplay between calcium concentration in the blood and parathyroid hormone flow keeps the calcium level close to the needed level at all times. (And a good thing, too, for even a small departure of the calcium concentration from the proper level can lead to death. Thus, removal of the parathyroids is fatal. At one time, doctors, in their anxiety to snip away sections of thyroid to relieve goiter, thought nothing of tossing away the much smaller and less prominent parathyroids. The death of the patient taught them better.)

At some times, the action of feedback is refined by the existence of two hormones working in opposite directions. In 1961, for instance, D. Harold Copp, at the University of British Columbia, demonstrated the presence of a thyroid hormone he called "calcitonin," which acted to depress the level of calcium in the blood by encouraging the deposition of its ions in bone. With parathormone pulling in one direction and calcitonin in the other, the feedback produced by calcium levels in the blood can be all the more delicately controlled. (The calcitonin molecule is made up of a single polypeptide chain that is 32 amino acids long.)

Then, too, in the case of blood-sugar concentration, where insulin is involved, a second hormone, also secreted by the islets of Langerhans, cooperates. The islets are made up of two distinct kinds of cells, "alpha" and "beta." The beta cells produce insulin, while the alpha cells produce "glucagon." The existence of glucagon was first suspected in 1923 and it was crystallized in 1955. Its molecule is made up of a single chain of twenty-nine acids, and, by 1958, its structure had been completely worked out.

Glucagon opposes the effect of insulin, so the two hormonal forces push in opposite directions, and the balance shifts very slightly this way and that under the stimulus of the glucose concentration in blood. Secretions from the pituitary gland (which I shall discuss shortly) also have a countering effect on insulin activity. For the discovery of this, the Argentinian physiologist Bernardo Alberto Houssay shared in the 1947 Nobel Prize for medicine and physiology.

Now the trouble in diabetes is that the islets have lost the ability to turn out enough insulin. The glucose concentration in the blood therefore drifts upward. When the level rises to about 50 per cent higher than normal, it crosses the "renal threshold"—that is, glucose spills over into the urine. In a way this loss of glucose into the urine is the lesser of two evils, for if the glucose concentration were allowed to build up any higher, the resulting rise in viscosity of the blood would cause undue heartstrain. (The heart is designed to pump blood, not molasses.)

The classic way of checking for the presence of diabetes is to test the urine for sugar. For instance, a few drops of urine can be heated with "Benedict's solution" (named for the American chemist Francis Gano Benedict). The solution contains copper sulfate, which gives it a deep blue color. If glucose is not present in the urine, the solution remains blue. If glucose is present, the copper sulfate is converted to cuprous oxide. Cuprous oxide is a brick-red, insoluble substance. A reddish precipitate at the bottom of the test tube therefore is an unmistakable sign of sugar in the urine, which usually means diabetes.

Nowadays an even simpler method is available. Small paper strips about two inches long are impregnated with two enzymes, glucose dehydrogenase and peroxidase, plus an organic substance called "orthotolidine." The yellowish strip is dipped into a sample of the patient's urine and then exposed to the air. If glucose is present, it combines with oxygen from the air with the catalytic help of the glucose dehydrogenase. In the process, hydrogen peroxide is formed. The peroxidase in the paper then causes the hydrogen peroxide to combine with the orthotolidine to form a deep blue compound. In short, if the yellowish paper is dipped into urine and turns blue, diabetes can be strongly suspected.

Once glucose begins to appear in the urine, diabetes mellitus is fairly far along in its course. It is better to catch the disease earlier by checking the glucose level in the blood before it crosses the renal threshold. The "glucose tolerance test," now in general use, measures the rate of fall of the glucose level in the blood after it has been raised by feeding the person glucose. Normally, the pancreas responds with a flood of insulin. In a healthy person the sugar level will drop to normal within two hours. If the level stays high for three hours or more, it shows a sluggish insulin response, and the person is likely to be in the early stages of diabetes.

It is possible that insulin has something to do with controlling appetite.

To begin with, we are all born with what some physiologists call an "appestat," which regulates appetite as a thermostat regulates a furnace. If the appestat is set too high, the individual finds himself continually taking in more calories than he expends, unless he exerts a strenuous self-control which sooner or later wears him out.

In the early 1940's, a physiologist, Stephen Walter Ranson, showed that animals grew obese after destruction of a portion of the hypothalamus (located in the lower part of the brain). This seems to fix the location of the appestat. What controls its operation? "Hunger pangs" spring to mind. An empty stomach contracts in waves, and the entry of food ends the contractions. Perhaps it is these contractions that signal to the appestat. Not so; surgical removal of the stomach has never interfered with appetite control.

The Harvard physiologist Jean Mayer has advanced a more subtle suggestion. He believes that the appestat responds to the level of glucose in the blood. After food has been digested, the glucose level in the blood slowly drops. When it falls below a certain level, the appestat is turned on. If, in response to the consequent urgings of the appetite, the person eats, the glucose level in his blood momentarily rises, and the appestat is turned off.

The hormones I have discussed so far are all either proteins (as insulin, glucagon, secretin, parathormone) or modified amino acids (as thyroxine, triiodothyronine, adrenalin). We come now to an altogether different group—the steroid hormones.

The story of these begins in 1927, when two German physiologists,

Bernhard Zondek and Selmar Aschheim, discovered that extracts of the urine of pregnant women, when injected into female mice or rats, aroused them to sexual heat. (Their discovery led to the first early test for pregnancy.) It was clear at once that they had found a hormone, specifically, a "sex hormone."

Within two years pure samples of the hormone were isolated by Adolf Butenandt in Germany and by Edward Adelbert Doisy at St. Louis University. It was named "estrone," from "estrus," the term for sexual heat in females. Its structure was quickly found to be that of a steroid, with the four-ring structure of cholesterol. For his part in the discovery of sex hormones, Butenandt was awarded the Nobel Prize for chemistry in 1939. He, like Domagk and Kuhn, was forced to reject it and could only accept the honor in 1949 after the destruction of the Nazi tyranny.

Estrone is now one of a group of known female sex hormones, called "estrogens" ("giving rise to estrus"). In 1931, Butenandt isolated the first male sex hormone, or "androgen" ("giving rise to maleness"). He called it "androsterone."

It is the production of sex hormones that governs the changes that take place during adolescence: the development of facial hair in the male and of enlarged breasts in the female, for instance. The complex menstrual cycle in females depends on the interplay of several estrogens.

The female sex hormones are produced in large part in the ovaries, the male sex hormones in the testes.

The sex hormones are not the only steroid hormones. The first nonsexual chemical messenger of the steroid type was discovered in the adrenals. These, as a matter of fact, are double glands, consisting of an inner gland called the adrenal "medulla" (the Latin word for "marrow") and an outer gland called the adrenal "cortex" (the Latin word for "bark"). It is the medulla that produces adrenalin. In 1929, investigators found that extracts from the cortex could keep animals alive after their adrenal glands had been removed—a 100 per cent fatal operation. Naturally, a search immediately began for "cortical hormones."

The search had a practical medical reason behind it. The well-known affliction called "Addison's disease" (first described by the English physician Thomas Addison in 1855) had symptoms like those resulting from the removal of the adrenals. Clearly the disease must be caused by a failure in hormone production by the adrenal cortex. Perhaps injections of cortical hormones might deal with Addison's disease as insulin dealt with diabetes.

Two men were outstanding in this search. One was Tadeus Reichstein (who was later to synthesize vitamin C); the other was Edward Kendall (who had first discovered the thyroid hormone nearly twenty years before). By the late 1930's, the researchers had isolated more than two dozen different compounds from the adrenal cortex. At least four showed hormonal activity. Kendall named the substances Compounds A, B, E, F, and so on. All the cortical hormones proved to be steroids.

Now the adrenals are very tiny glands, and it would take the glands of countless numbers of animals to provide enough cortical extracts for general use. Apparently, the only reasonable solution was to try to synthesize the hormones.

A false rumor drove cortical-hormone research forward under full steam during World War II. It was reported that the Germans were buying up adrenal glands in Argentine slaughterhouses to manufacture cortical hormones that improved the efficiency of their airplane pilots in high-altitude flight. There was nothing to it, but the rumor had the

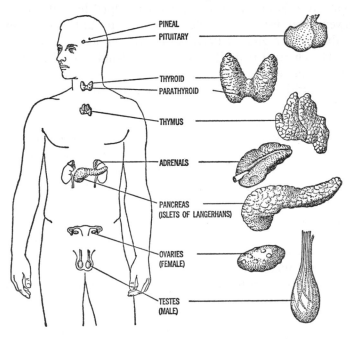

The endocrine glands.

effect of stimulating the United States Government to place a high priority on research into methods for the synthesis of the cortical hormones; the priority was even higher than that given to the synthesis of penicillin or the antimalarials.

Compound A was synthesized by Kendall in 1944, and by the following year Merck & Co. had begun to produce it in substantial amounts. It proved of little value for Addison's disease, to the disappointment of all. After prodigious labor the Merck biochemist Lewis H. Sarrett then synthesized, by a process involving thirty-seven steps, Compound E, which was later to become known as "cortisone."

The synthesis of Compound E created little immediate stir in medical circles. The war was over; the rumor of cortical magic worked on German pilots had proved untrue; and Compound A had fizzled. Then, in an entirely unexpected quarter, Compound E suddenly came to life.

For twenty years, the Mayo Clinic physician Philip Showalter Hench had been studying rheumatoid arthritis, a painful, sometimes paralytic disease. Hench suspected that the body possessed natural mechanisms for countering this disease, because the arthritis was often relieved during pregnancy or during attacks of jaundice. He could not think of any biochemical factor that jaundice and pregnancy held in common. He tried injections of bile pigments (involved in jaundice) and sex hormones (involved in pregnancy) but neither helped his arthritic patients.

However, various bits of evidence pointed toward cortical hormones as a possible answer, and, in 1949, with cortisone available in reasonable quantity, Hench tried that. It worked! It did not cure the disease, any more than insulin cures diabetes, but it seemed to relieve the symptoms, and to an arthritic that alone is manna from heaven. What was more, cortisone later proved to be helpful as a treatment for Addison's disease, where Compound A had failed.

For their work on the cortical hormones, Kendall, Hench, and Reichstein shared the Nobel Prize in medicine and physiology in 1950.

Unfortunately, the influences of the cortical hormones on the body's workings are so multiplex that there are always side effects, sometimes serious. Physicians are reluctant to use cortical-hormone therapy unless the need is clear and urgent. Synthetic substances related to cortical hormones (some with a fluorine atom inserted in the molecule) are being used in an attempt to avoid the worst of the side effects, but nothing approaching a reasonable ideal has yet been found. One of the most active of the cortical hormones discovered so far is "aldosterone," isolated in 1953 by Reichstein and his co-workers.

What controls all the varied and powerful hormones? All of them (including a number I have not mentioned), can exert more or less drastic effects in the body. Yet they are tuned together so harmoniously that they keep the body functioning smoothly without a break in the rhythm. Seemingly, there must be a conductor somewhere that directs their cooperation.

The nearest thing to an answer is the pituitary, a small gland suspended from the bottom of the brain (but not part of it). The name of the gland arose from an ancient notion that its function was to secrete phlegm, the Latin word for which is *pituita* (also the source of the word "spit"). Because this notion is false, scientists have renamed the gland the "hypophysis" (from Greek words meaning "growing under" —i.e., under the brain), but pituitary is still the more common term.

The gland has three parts: the anterior lobe, the posterior lobe, and, in some organisms, a small bridge connecting the two. The anterior lobe is the most important, for it produces at least six hormones (all small-molecule proteins), which seem to act specifically upon other ductless glands. In other words, the anterior pituitary can be viewed as the orchestra leader that keeps the other glands playing in time and in tune. (It is interesting that the pituitary is located just about in the center of the skull, as if deliberately placed in a spot of maximum security.)

One of the pituitary's messengers is the "thyroid-stimulating hormone" (TSH). It stimulates the thyroid on a feedback basis. That is, it causes the thyroid to produce thyroid hormone; the rise in concentration of thyroid hormone in the blood in turn inhibits the formation of TSH by the pituitary; the fall of TSH in the blood in its turn reduces the thyroid's production; that stimulates the production of TSH by the pituitary, and so the cycle maintains a balance.

In the same way, the "adrenal-cortical-stimulating hormone," or "adrenocorticotropic hormone" (ACTH), maintains the level of cortical hormones. If extra ACTH is injected into the body, it will raise the level of these hormones, and thus it can serve the same purpose as the injection of cortisone itself. ACTH has therefore been used to treat rheumatoid arthritis.

Research into the structure of ACTH has proceeded with vigor because of this tie-in with arthritis. By the early 1950's, its molecular weight had been determined as 20,000, but it was easily broken down into smaller fragments ("corticotropins"), which possessed full activity. One of them, made up of a chain of thirty-nine amino acids, has had its structure worked out completely, and even shorter chains have been found effective.

ACTH has the ability of influencing the skin pigmentation of animals, and even man is affected. In diseases involving overproduction of ACTH, human skin darkens. It is known that in lower animals, particularly the amphibians, special skin-darkening hormones exist. A hormone of this sort was finally detected among the pituitary products in the human being in 1955. It is called "melanocyte-stimulating hormone" (melanocytes being the cells that produce skin pigment) and is usually abbreviated as "MSH."

The molecule of MSH has been largely worked out; it is interesting to note that MSH and ACTH share a seven amino-acid sequence in common. The indication that structure is allied to function (as, indeed, it must be) is unmistakable.

While on the subject of pigmentation, it might be well to mention the pineal gland, a conical body attached, like the pituitary, to the base of the brain and so-named because of its resemblance to a pine cone in shape. The pineal gland has seemed glandular in nature, but no hormone could be located until the late 1950's. Then the discoverers of MSH, working with 200,000 beef pineals, finally isolated a tiny quantity of substance that, on injection, lightened the skin of a tadpole. The hormone, named "melatonin," does not, however, appear to have any effect on human melanocytes.

The list of pituitary hormones is not yet complete. A couple of pituitary hormones, ICSH ("interstitial cell-stimulating hormone") and FSH ("follicle-stimulating hormone") control the growth of tissues involved in reproduction. There is also the "lactogenic hormone," which stimulates milk production.

Lactogenic hormone stimulates other postpregnancy activities. Young female rats injected with the hormone busy themselves with nest-building even though they have not given birth. On the other hand, mice whose pituitaries have been removed shortly before giving birth to young exhibit little interest in the baby mice. The newspapers at once termed lactogenic hormone the "mother-love hormone."

These pituitary hormones, associated with sexual tissues, are lumped together as the "gonadotropins." Another substance of this type is produced by the placenta (the organ that serves to transfer nourishment from the mother's blood to the blood of the developing infant and to transfer wastes in the opposite direction). The placental hormone is called "human chorionic gonadotropin" and is abbreviated "HCG." As early as two to four weeks after the beginning of pregnancy, HCG is produced in appreciable quantities and makes its appearance in the urine. When extracts of the urine of a pregnant woman are injected into mice, frogs, or rabbits, recognizable effects can be detected. Pregnancy can be determined in this way at a very early stage.

The most spectacular of the anterior pituitary hormones is the "somatotropic hormone" (STH), more popularly known as the "growth hormone." Its effect is a general one stimulating growth of the whole body. A child who cannot produce a sufficient supply of the hormone will become a dwarf; one who produces too much will turn into a circus giant. If the disorder that results in an oversupply of the growth hormone does not occur until after the person has matured (i.e., when his bones have been fully formed and hardened), only the extremities, such as the hands, feet, and chin, grow grotesquely large—a condition known as "acromegaly" (Greek for "large extremities"). It is this growth hormone that Li (who first determined its structure in 1966) synthesized in 1970.

As to how the hormones work, in chemical terms, investigators have so far made little headway.

It seems certain that the hormones do not act as enzymes. At least, no hormone has been found to catalyze a specific reaction directly. The next alternative is to suppose that a hormone, if not itself an enzyme, acts upon an enzyme—that it either promotes or inhibits an enzyme's activity. Insulin, the most thoroughly investigated of all the hormones, does seem to be definitely connected with an enzyme called "gluco-kinase," which is essential for the conversion of glucose to glycogen. This enzyme is inhibited by extracts from the anterior pituitary and the adrenal cortex, and insulin can nullify that inhibition. Thus, insulin in the blood may serve to activate the enzyme and so speed up the conversion of glucose to glycogen. That would help to explain how insulin lowers the glucose concentration in the blood.

Yet the presence or absence of insulin affects metabolism at so many points that it is hard to see how this one action could bring about all the abnormalities that exist in the body chemistry of a diabetic. (The same is true for other hormones.) Some biochemists have therefore tended to look for grosser and more wholesale effects.

There is the suggestion that insulin somehow acts as an agent to get glucose into the cell. On this theory, a diabetic has a high glucose level in his blood for the simple reason that the sugar cannot get into his cells and therefore he cannot use it. (In explaining the insatiable appetite of a diabetic, Mayer, as I have already mentioned, suggested that glucose in the blood has difficulty in entering the cells of the appestat.)

If insulin assists glucose in entering the cell, then it must act on the cell membrane in some way. How? Cell membranes are composed of protein and fatty substances. We can speculate that insulin, as a protein molecule, may somehow change the arrangement of amino-acid side chains in the protein of the membrane and thus open doors for glucose (and possibly many other substances).

If we are willing to be satisfied with generalities of this kind, we can go on to suppose that the other hormones also act on the cell membranes, each in its own fashion because each has its own specific amino-acid arrangement. Similarly, steroid hormones, as fatty substances, may act on the fatty molecules of the membrane, either opening or closing the door to certain substances. Clearly, by helping a given material to enter the cell or preventing it from doing so, a hormone could exert a drastic effect on what goes on in the cell. It could supply one enzyme with plenty of substrate to work on and deprive another of material, thus controlling what the cell produces. Assuming that a single hormone may decide the entrance or nonentrance of several different substances, we can see how the presence or absence of a hormone could profoundly influence metabolism, as in fact it does in the case of insulin.

The picture drawn above is attractive, but it is vague. Biochemists would much prefer to know the exact reactions that take place at the cell membrane under the influence of a hormone. The beginning of such knowledge came with the discovery in 1960 of a nucleotide like adenylic acid except that the phosphate group was attached to two different places in the sugar molecule. Its discoverers, Earl W. Sutherland and T. W. Rall, called it "cyclic AMP." It was "cyclic" because the doubly attached phosphate-group formed a circle of atoms, and the "AMP" stood for "*a*denine *mono*phosphate," an alternate name for adenylic acid.

Once discovered, cyclic AMP was found to be widely spread in tissue, and to have a pronounced effect on the activity of many different enzymes and cell processes. Cyclic AMP is produced from the universally occurring ATP by means of an enzyme named "adenyl cyclase," which is located at the surface of cells. There may be several such enzymes, each geared for activity in the presence of a particular hormone. In other words, the surface activity of hormones serves to activate an adenyl cyclase that leads to the production of cyclic AMP, which alters the enzyme activity within the cell, producing many changes.

Undoubtedly, the details are enormously complex, and compounds other than cyclic AMP may be involved—but it is a beginning.

Death

The advances made by modern medicine in the battle against infection, against cancer, against nutritional disorders, have increased the probability that any given individual will live long enough to experience old age. Half the people born in this generation can be expected to reach the age of seventy (barring a nuclear war or some other prime catastrophe).

The rarity of survival to old age in earlier eras no doubt accounts in part for the extravagant respect paid to longevity in those times. The *Iliad*, for instance, makes much of "old" Priam and "old" Nestor. Nestor is described as having survived three generations of men, but at a time when the average length of life could not have been more than twenty to twenty-five, Nestor need not have been older than seventy to have survived three generations. That is old, yes, but not extraordinary by present standards. Because Nestor's antiquity made such an impression on people in Homer's time, later mythologists supposed that he must have been something like 200 years old.

To take another example at random, Shakespeare's *Richard II* opens with the rolling words: "Old John of Gaunt, time-honored Lancaster." John's own contemporaries, according to the chroniclers of the time, also considered him an old man. It comes as a slight shock to realize that John of Gaunt lived only to the age of fifty-nine. An interesting example from our own history is that of Abraham Lincoln. Whether because of his beard, or his sad, lined face, or songs of the time that referred to him as "Father Abraham," most people think of him as an old man at the time of his death. One could only wish that he had lived to be one. He was assassinated at the age of fifty-six.

All this is not to say that really old age was unknown in the days before modern medicine. In ancient Greece Sophocles, the playwright, lived to be 90, and Isocrates, the orator, to 98. Flavius Cassiodorus of fifth-century Rome died at 95. Enrico Dandolo, the twelfth-century Doge of Venice, lived to be 97. Titian, the Renaissance painter, survived to 99. In the era of Louis XV, the Duc de Richelieu, grandnephew of the famous cardinal, lived 92 years, and the French writer Bernard Le Bovier de Fontenelle managed to arrive at just 100 years.

This emphasizes the point that although the average life expectancy in medically advanced societies has risen greatly, the maximum life span has not. We expect very few men, even today, to attain or exceed the lifetime of an Isocrates or a Fontenelle. Nor do we expect modern non-agenarians to be able to participate in the business of life with any greater vigor. Sophocles was writing great plays in his nineties, and Isocrates was composing great orations. Titian painted to the last year of his life, Dandolo was the indomitable leader of a Venetian war against the Byzantine Empire at the age of 96. (Among comparably vigorous oldsters of our day, the best example I can think of is George Bernard Shaw, who lived to 94 and the English mathematician and philosopher Bertrand Russell who was still active in his ninety-eighth year, when he died.

Although a far larger proportion of our population reaches the age of 60 than ever before, beyond that age life expectancy has improved very little over the past. The Metropolitan Life Insurance Company estimates that the life expectancy of a sixty-year-old American male in 1931 was just about the same as it was a century and a half earlier—that is, 14.3 years against the estimated earlier figure of 14.8. For the average American woman the corresponding figures were 15.8 and 16.1. Since 1931, the advent of antibiotics has raised the expectancy at 60 for both sexes by two and a half years. But on the whole, despite all that medicine and science have done, old age overtakes a person at about the same rate and in the same way as it always has. Man has not yet found a way to stave off the gradual weakening and eventual breakdown of the human machine.

As in other forms of machinery, it is the moving parts that go first. The circulatory system—the pulsing heart and arteries—is man's Achilles' heel in the long run. His progress in conquering premature death has raised disorders of this system to the rank of the number one killer. Circulatory diseases are responsible for just over half the deaths in the United States, and of these diseases, a single one, atherosclerosis, accounts for one death out of four.

Atherosclerosis (from Greek words meaning "mealy hardness") is characterized by grainlike fatty deposits along the inner surface of the arteries, which force the heart to work harder to drive blood through the vessels at a normal pace. The blood pressure rises, and the consequent increase in strain on the small blood vessels may burst them. If this happens in the brain (a particularly vulnerable area) there is a cerebral hemorrhage, or "stroke." Sometimes the bursting of a vessel is so minor that it occasions only a trifling and temporary discomfort or even goes unnoticed, but a massive collapse of vessels will bring on paralysis or a quick death.

The roughening and narrowing of the arteries introduces another hazard. Because of the increased friction of the blood scraping along the roughened inner surface of the vessels, blood clots are more likely to form, and the narrowing of the vessels heightens the chances that a clot will completely block the blood flow. In the coronary artery, feeding the heart muscle itself, a block ("coronary thrombosis") can produce almost instant death.

Just what causes the formation of deposits on the artery wall is a matter of much debate among medical scientists. Cholesterol certainly seems to be involved, but how it is involved is still far from clear. The plasma of human blood contains "lipoproteins," which consist of cholesterol and other fatty substances bound to certain proteins. Some of the fractions making up lipoprotein maintain a constant concentration in the blood—in health and in disease, before and after eating, and so on. Others fluctuate, rising after meals. Still others are particularly high in obese individuals. One fraction, rich in cholesterol, is particularly high in overweight people and in those with atherosclerosis.

Atherosclerosis tends to go along with a high blood-fat content, and so does obesity. Overweight people are more prone to atherosclerosis than are thin people. Diabetics also have high blood-fat levels, and they are more prone to atherosclerosis than are normal individuals. And, to round out the picture, the incidence of diabetes among the stout is considerably higher than among the thin.

It is thus no accident that those who live to a great age are so often scrawny, little fellows. Large, fat men may be jolly, but they do not keep the sexton waiting unduly, as a rule. (Of course, there are always exceptions, and one can point to men such as Winston Churchill and Herbert Hoover, who passed their ninetieth birthdays, although they were never noted for leanness.)

The key question, at the moment, is whether atherosclerosis can be fostered or prevented by the diet. Animal fats, such as those in milk, eggs, and butter, are particularly high in cholesterol; plant fats are particularly low in it. Moreover, the fatty acids of plant fats are mainly of the unsaturated type, which has been reported to counter the deposition of cholesterol. Despite the fact that investigations of these matters have yielded no conclusive results one way or the other, people have been flocking to "low-cholesterol diets," in the hope of staving off thickening of the artery walls. No doubt this will do no harm.

Of course, the cholesterol in the blood is not derived from the cholesterol of the diet. The body can and does make its own cholesterol with great ease, and even though you live on a diet that is completely free of cholesterol, you will still have a generous supply of cholesterol in your blood lipoproteins. It therefore seems reasonable to suppose that what matters is not the mere presence of cholesterol but the individual's tendency to deposit it where it will do the most harm. It may be that there is a hereditary tendency to manufacture excessive amounts of cholesterol. Biochemists are seeking drugs that will inhibit cholesterol formation, in the hope that such drugs may forestall the development of atherosclerosis in those who are prone to the disease.

But even those who escape atherosclerosis grow old. Old age is a disease of universal incidence. Nothing can stop the creeping enfeeblement, the increasing brittleness of the bones, the weakening of the muscles, the stiffening of the joints, the slowing of reflexes, the dimming of sight, the declining agility of the mind. The rate at which this happens is somewhat slower in some than in others, but, fast or slow, the process is inexorable.

Perhaps mankind ought not complain too loudly about this. If old age and death must come, they arrive unusually slowly. In general, the life span of mammals correlates with size. The smallest mammal, the shrew, may live 1½ years and a rat may live 4 or 5. A rabbit may live up to 15 years, a dog up to 18, a pig up to 20, a horse up to 40, and an elephant up to 70. To be sure, the smaller the animal the more rapidly it lives—the faster its heartbeat, for instance. A shrew with a heartbeat of 1,000 per minute can be matched against an elephant with a heartbeat of 20 per minute.

In fact, mammals in general seem to live, at best, as long as it takes their hearts to count a billion. To this general rule, man himself is the most astonishing exception. Man is considerably smaller than a horse and far smaller than an elephant, yet no mammal can live as long as he can. Even if we discount tales of vast ages from various backwoods where accurate records have not been kept, there are reasonably convincing data for life spans of up to 115 years. The only vertebrates to outdo this, without question, are certain large, slow-moving tortoises.

Man's heartbeat of about seventy-two per minute is just what is to be expected of a mammal of his size. In seventy years, which is the average life expectancy of man in the technologically advanced areas of the world, the human heart has beaten 2.5 billion times; at 115 years it has beaten about 4 billion times. Even man's nearest relatives, the great apes, cannot match this, even closely. The gorilla, considerably larger than a man, is in extreme old age at fifty.

There is no question but that man's heart outperforms all other hearts in existence. (The tortoise's heart may last longer but it lives nowhere near as intensely.) Why man should be so long-lived is not known, but man, being what he is, is far more interested in asking why he does not live still longer.

What is old age, anyway? So far, there are only speculations. Some have suggested that the body's resistance to infection slowly decreases with age (at a rate depending on heredity). Others speculate that "clinkers" of one kind or another accumulate in the cells (again at a rate that varies from individual to individual). These supposed side products of normal cellular reactions, which the cell can neither destroy nor get rid of, slowly build up in the cell as the years pass, until they eventually interfere with the cell's metabolism so seriously that it ceases to function. When enough cells are put out of action, so the theory goes, the body dies. A variation of this notion holds that the protein molecules themselves become clinkers, because cross links develop between them so that they become stiff and brittle and finally bring the cell machinery grinding to a halt.

If this is so, then "failure" is built into the cell machinery. Carrel's ability to keep a piece of embryonic tissue alive for decades had made it seem that cells themselves might be immortal: it was only the organization into combinations of trillions of individual cells that brought death. The organization failed, not the cells.

Not so, apparently. It is now thought that Carrel may (unwittingly) have introduced fresh cells into his preparation in the process of feeding the tissue. Attempts to work with isolated cells or groups of cells in which the introduction of fresh cells was rigorously excluded, seems to show that the cells inevitably age—presumably through irreversible changes in the key cell-components.

And yet there is man's extraordinarily long life-span. Can it be that human tissue has developed methods of reversing or inhibiting cellular aging effects, that are more efficient than those in any other mammal? Again, birds tend to live markedly longer than mammals of the same size despite the fact that bird-metabolism is even more rapid than mammal-metabolism—again, superior ability of old age reversal or inhibition.

If old age can be staved off more by some organisms than by others, there seems no reason to suppose that man cannot learn the method and improve upon it. Might not old age then be curable, and might not mankind develop the ability to enjoy an enormously extended lifespan—or even immortality?

General optimism in this respect is to be found among some people. Medical miracles in the past would seem to herald unlimited miracles in the future. And if that is so, what a shame to live in a generation that will just miss a cure for cancer, or for arthritis, or for old age!

In the late 1960's, therefore, a movement grew to freeze human bodies at the moment of death, in order that the cellular machinery might remain as intact as possible, until the happy day when whatever it was that marked the death of the frozen individual, could be cured. He would then be revived and made healthy, young, and happy.

To be sure, there is no sign at the present moment that any dead body can be restored to life, or that any frozen body—even if alive at the moment of freezing—can be thawed to life. Nor do the proponents of this procedure ("cryonics") give much attention to the complications that might arise in the flood of dead bodies returned to life—the personal hankering for immortality governs all.

Actually, it makes little sense to freeze intact bodies, even if all possible revival could be done. It is wasteful. Biologists have so far had much more luck with the developing of whole organisms from groups of specialized cells. Skin cells or liver cells, after all, have the same genetic equipment that other cells have, and that the original fertilized ovum had in the first place. The cells are specialized because the various genes are inhibited or activated to varying extents; but might not the genes be de-inhibited or de-activated, and might they not then make their cell into the equivalent of a fertilized ovum and develop an organism all over again—the same organism, genetically speaking, as the one of which they had formed part? Surely, this procedure (called "cloning") offers more hope for a kind of preservation of the personality (if not the memory). Instead of freezing an entire body, chop off the little toe and freeze that.

But do we really want immortality—either through cryonics, through cloning, or through simple reversal of the aging phenomenon in each individual? There are few human beings who wouldn't eagerly accept an immortality reasonably free of aches, pains, and the effects of age—but suppose we were all immortal?

Clearly, if there were few or no deaths on earth, there would have to be few or no births. It would mean a society without babies. Presumably that is not fatal; a society self-centered enough to cling to immortality would not stop at eliminating babies altogether.

But will that do? It would be a society composed of the same brains, thinking the same thoughts, circling the same ruts in the same way, endlessly. It must be remembered that babies possess not only young brains but *new* brains. Each baby (barring identical multiple births) has genetic equipment unlike that of any human individual who ever lived. Thanks to babies, there are constantly fresh genetic combinations injected into mankind, so that the way is open toward improvement and development.

It would be wise to lower the level of the birth-rate, but ought we to wipe it out entirely? It would be pleasant to eliminate the pains and discomforts of old age, but ought we to create a species consisting of the old, the tired, the bored, the same, and never allow for the new and the better?

Perhaps the prospect of immortality is worse than the prospect of death.

CHAPTER 15

The Species

Varieties of Life

Man's knowledge of his own body is incomplete without a knowledge of his relationship to the rest of life on the earth.

In primitive cultures, the relationship was often considered to be close indeed. Many tribes regarded certain animals as their ancestors or blood brothers, and made it a crime to kill or eat them, except under certain ritualistic circumstances. This veneration of animals as gods or near-gods is called "totemism" (from an American Indian word), and there are signs of it in cultures that are not so primitive. The animal-headed gods of Egypt were a hangover of totemism, and so, perhaps, is the modern Hindu veneration of cows and monkeys.

On the other hand, Western culture, as exemplified in Greek and Hebrew ideas, very early made a sharp distinction between man and the "lower animals." Thus, the Bible emphasizes that man was produced by a special act of creation in the image of God, "after our likeness" (Genesis 1:26). Yet the Bible attests, nevertheless, to man's remarkably keen interest in the lower animals. Genesis mentions that Adam, in his idyllic early days in the Garden of Eden, was given the task of naming "every beast of the field, and every fowl of the air."

Offhand, that seems not to difficult a task—something that one could do in perhaps an hour or two. The scriptural chroniclers put "two of every sort" of animal in Noah's Ark, whose dimensions were 450 by 75 by 45 feet (if we take the cubit to be 18 inches). The Greek natural philosophers thought of the living world in similarly limited terms: Aristotle could list only about 500 kinds of animals, and his pupil Theophrastus, the most eminent botanist of ancient Greece, listed only about 500 different plants.

Such a list might make some sense if one thought of an elephant as always an elephant, a camel as just a camel, or a flea as simply a flea. Things began to get a little more complicated when naturalists realized that animals had to be differentiated on the basis of whether they could breed with each other. The Indian elephant could not interbreed with the African elephant; therefore, they had to be considered different "species" of elephant. The Arabian camel (one hump) and the Bactrian camel (two humps) also are separate species. As for the flea, the small biting insects (all resembling the common flea) are divided into 500 different species!

Through the centuries, as naturalists counted new varieties of creatures in the field, in the air, and in the sea, and as new areas of the world came into view through exploration, the number of identified species of animals and plants grew astronomically. By 1800 it had reached 70,000. Today more than 1.25 million different species, ⅔ animal and ⅓ plant, are known, and no biologist supposes that the count is complete.

The living world would be exceedingly confusing if we were unable to classify this enormous variety of creatures according to some scheme of relationships. One can begin by grouping together the cat, the tiger, the lion, the panther, the leopard, the jaguar, and other catlike animals in the "cat family"; likewise, the dog, the wolf, the fox, the jackal, and the coyote form a "dog family," and so on. On the basis of obvious general criteria one can go on to classify some animals as meat-eaters and others as plant-eaters. The ancients also set up general classifications based on habitat, and so they considered all animals that lived in the sea to be fishes and all that flew in the air to be birds. But this made the whale a fish and the bat a bird. Actually, in a fundamental sense the whale and the bat are more like each other than the one is like a fish or the other like a bird. Both bear live young. Moreover, the whale has air-breathing lungs, rather than the gills of a fish, and the bat has hair instead of the feathers of a bird. Both are classed with the mammals, which give birth to living babies (instead of laying eggs) and feed them on mother's milk.

One of the earliest attempts to make a systematic classification was that of an Englishman named John Ray (or Wray), who in the seventeenth century classified all the known species of plants (about 18,600), and later the species of animals, according to systems which seemed to him logical. For instance, he divided flowering plants into two main groups, on the basis of whether the seed contained one embryonic leaf or two. The tiny embryonic leaf or pair of leaves had the name "cotyledon," from the Greek word for a kind of cup ("kotyle"), because it lay in a cuplike hollow in the seed. Ray therefore named the two types respectively "monocotyledonous" and "dicotyledonous." The classification (similar, by the way, to one set up 2,000 years earlier by Theophrastus) proved so useful that it is still in effect today. The difference between one embryonic leaf and two in itself is unimportant, but there are a number of important ways in which all monocotyledonous plants differ from all dicotyledonous ones. The difference in the embryonic leaves is just a handy tag which is symptomatic of many general differences. (In the same way, the distinction between feathers and hair is minor in itself but is a handy marker for the vast array of differences that separates birds from mammals.)

Although Ray and others contributed some useful ideas, the real founder of the science of classification, or "taxonomy" (from a Greek word meaning "arrangement"), was a Swedish botanist best known by his Latinized name of Carolus Linnaeus, who did the job so well that the main features of his scheme still stand today. Linnaeus set forth his system in 1737 in a book entitled *Systema Naturae*. He grouped species resembling one another into a "genus" (from a Greek word meaning "race" or "sort"), put related genera in turn into an "order," and grouped similar orders in a "class." Each species was given a double name, made up of the name of the genus and of the species itself. (This is much like the system in the telephone book, which lists Smith, John; Smith, William, and so on.) Thus the members of the genus of cats are *Felis domesticus* (the pussycat), *Felis leo* (the lion), *Felis tigris* (the tiger), *Felis pardus* (the leopard), and so on. The genus to which the dog belong includes *Canis familiaris* (the dog), *Canis lupus* (the European gray wolf), *Canis occidentalis* (the American timber wolf), and so on. The two species of camel are *Camelus bactrianus* (the Bactrian camel) and *Camelus dromedarius* (the Arabian camel).

Around 1800, the French naturalist Georges Leopold Cuvier went beyond "classes" and added a more general category called the "phylum" (from a Greek word for "tribe"). A phylum includes all animals with the same general body plane (a concept that was emphasized and made clear by none other than the great German poet Johann Wolfgang von Goethe). For instance, the mammals, birds, reptiles, amphibia, and fishes are placed in one phylum because all have backbones, a maximum of four limbs, and red blood containing hemoglobin. Insects, spiders, lobsters, and centipedes are placed in another phylum; clams, oysters, and mussels in still another, and so on. In the 1820's, the Swiss botanist Augustin Pyramus de Candolle similarly improved Linnaeus' classification of plants. Instead of grouping species together according to external appearance, he laid more weight on internal structure and functioning.

The tree of life now is arranged as I shall describe in the following paragraphs, going from the most general divisions to the more specific.

We start with the "kingdoms"—plant, animal, and in-between (that is, those microorganisms, such as the bacteria, that cannot be classed definitely as plant or animal in nature). The German biologist Ernst Heinrich Haeckel suggested, in 1866, that this in-between group be called "Protista," a name which is coming into increasing use among biologists even though the world of life is still divided exclusively, in popular parlance, into "animal and vegetable."

The plant kingdom, according to one system of classification, is divided into two subkingdoms. In the first subkingdom, called Thallophyta, are placed all the plants that do not have roots, stems, or leaves —that is, the algae (one-celled green plants and various seaweeds), which contain chlorophyll, and the fungi (the one-celled molds plus such organisms as mushrooms), which do not. The members of the second subkingdom, the Embryophyta, are divided into two main phyla—the Bryophyta (the various mosses) and the Tracheophyta (plants with systems of tubes for the circulation of sap), which includes all the species that we ordinarily think of as plants.

This last great phylum is made up of three main classes: the Filicineae, the Gymnospermae, and the Angiospermae. In the first class are the ferns, which reproduce by means of spores. The gymnosperms, forming seeds on the surface of the seed-bearing organs, include the various evergreen cone-bearing trees. The angiosperms, with the seeds enclosed in ovules, make up the vast majority of the familiar plants.

As for the animal kingdom, I shall list only the more important phyla.

The Protozoa ("first animals") are, of course, the one-celled animals. Next there are the Porifera, animals consisting of colonies of cells within a pore-bearing skeleton; these are the sponges. The individual cells show signs of specialization but retain a certain independence, for after all are separated by straining through a silk cloth, they may aggregate to form a new sponge.

(In general, as the animal phyla grow more specialized, individual cells and tissues grow less "independent." Simple creatures can regrow to entire organisms even though badly mutilated, a process called "regeneration." More complex ones can regrow limbs. By the time we reach man, however, the capacity for regeneration has sunk quite low. We can regrow a lost fingernail, but not a lost finger.)

The first phylum whose members can be considered truly multicelled animals is the Coelenterata (meaning "hollow gut"). These animals have the basic shape of a cup and consist of two layers of cells— the ectoderm ("outer skin") and the endoderm ("inner skin"). The most common examples of this phylum are the jellyfish and the sea anemones.

All the rest of the animal phyla have a third layer of cells—the mesoderm ("middle skin"). From these three layers, first recognized in

737

1845 by the German physiologists Johannes Peter Müller and Robert Remak, are formed the many organs of even the most complex animals, including man.

The mesoderm arises during the development of the embryo, and the manner in which it arises divides the animals involved into two "superphyla." Those in which the mesoderm forms at the junction of the ectoderm and the endoderm make up the Annelid superphylum; those in which the mesoderm arises in the endoderm alone are the Echinoderm superphylum.

Let us consider the Annelid superphylum first. Its simplest phylum is Platyhelminthes (Greek for "flat worms"). This includes not only the parasitic tapeworm but also free-living forms. The flatworms have contractile fibers that can be considered primitive muscles, and they also possess a head, a tail, special reproductive organs, and the beginnings of excretory organs. In addition, the flatworms display bilateral symmetry: that is, they have left and right sides that are mirror images of each other. They move headfirst, and their sense organs and rudimentary nerves are concentrated in the head area, so that the flatworm can be said to possess the first step toward a brain.

Next comes the phylum Nematoda (Greek for "thread worm"), whose most familiar member is the hookworm. These creatures possess a primitive blood stream—a fluid within the mesoderm that bathes all the cells and conveys food and oxygen to them. This allows the nematodes, in contrast to animals such as the flat tapeworm, to have bulk, for the fluid can bring nourishment to interior cells. The nematodes also possess a gut with two openings, one for the entry of food, the other (the anus) for ejection of wastes.

The next two phyla in this superphylum have hard external "skeletons"—that is, shells (which are found in some of the simpler phyla, too). These two groups are the Brachiopoda, which have calcium carbonate shells on top and bottom and are popularly called "lampshells," and the Mollusca (Latin for "soft"), whose soft bodies are enclosed in shells originating from the right and left sides instead of the top and bottom. The most familiar molluscs are the clams, oysters, and snails.

A particularly important phylum in the Annelid superphylum is Annelida. These are worms, but with a difference: they are composed of segments, each of which can be looked upon as a kind of organism in itself. Each segment has its own nerves branching off the main nerve stem, its own blood vessels, its own tubules for carrying off wastes, its own muscles, and so on. In the most familiar annelid, the earthworm, the segments are marked off by little constrictions of flesh which look like little rings around the animal; in fact, Annelida is from a Latin word meaning "little ring."

Segmentation apparently endows an animal with superior efficiency, for all the most successful species of the animal kingdom, including man, are segmented. (Of the nonsegmented animals, the most complex and successful is the squid.) If you wonder how the human body is segmented, think of the vertebrae and the ribs; each vertebra of the backbone and each rib represents a separate segment of the body, with its own nerves, muscles, and blood vessels.

The annelids, lacking a skeleton, are soft and relatively defenseless. The phylum Arthropoda ("jointed feet"), however, combines segmentation with a skeleton, the skeleton being as segmented as the rest of the body. The skeleton is not only more maneuverable for being jointed; it is also light and tough, being made of a polysaccharide called "chitin" rather than of heavy, inflexible limestone or calcium carbonate. On the whole, the Arthropoda, which includes the lobsters, spiders, centipedes, and insects, is the most successful phylum in existence. At least the phylum contains more species than all the other phyla put together.

This accounts for the main phyla in the Annelid superphylum. The other superphylum, the Echinoderm, contains only two important phyla. One is Echinodermata ("spiny skin"), which includes such creatures as the starfish and the sea urchin. The echinoderms differ from other mesoderm-containing phyla in possessing radial symmetry and having no clearly defined head and tail (though in early life echinoderms do show bilateral symmetry, which they lose as they mature).

The second important phylum of the Echinoderm superphylum is important indeed, for it is the one to which man himself belongs.

The general characteristic that distinguishes the members of this phylum (which embraces man, ostrich, snake, frog, mackerel, and a varied host of other animals) is an internal skeleton. No animal outside this phylum possesses one. The particular mark of such a skeleton is the backbone. In fact, the backbone is so important a feature that in common parlance all animals are loosely divided into vertebrates and invertebrates. Actually, there is an in-between group which has a rod of cartilage called a "notochord" ("backcord") in the place of the backbone. The notochord, first discovered by Von Baer, who had also discovered the mammalian ovum, seems to represent a rudimentary backbone; in fact, it makes its appearance even in mammals during the development of the embryo. So the animals with notochords (various wormlike, sluglike, and mollusclike creatures) are classed with the vertebrates. The whole phylum was named Chordata in 1880, by the English zoologist Francis Maitland Balfour; it is divided into four subphyla, three of which have only a notochord. The fourth, with a true backbone and general internal skeleton, is Vertebrata.

The vertebrates in existence today form two superclasses: the Pisces ("fishes") and the Tetrapoda ("four-footed" animals).

The Pisces group is made up of three classes: (1) the Agnatha ("jawless") fishes, which have true skeletons but no limbs or jaws—the best-known representative, the lamprey, possessing a rasping set of files in a round suckerlike mouth; (2) the Chondrichthyes ("cartilage fish"), with a skeleton of cartilage instead of bone, sharks being the most familiar example; and (3) the Osteichthyes, or "bony fishes."

The tetrapods, or four-footed animals, all of which breathe by means of lungs, make up four classes. The simplest are the Amphibia ("double life")—for example, the frogs and toads. The double life means that in their immature youth (e.g., as tadpoles) they have no limbs and breathe by means of gills; then as adults they develop four feet and lungs. The amphibians, like fishes, lay their eggs in the water.

The second class are the Reptilia (from a Latin word meaning "creeping"). They include the snakes, lizards, alligators, and turtles. They breathe with lungs from birth, and hatch their eggs (enclosed in a hard shell) on land. The most advanced reptiles have essentially four-chambered hearts, whereas the amphibian's heart has three chambers and the fish's heart only two.

The final two groups of tetrapods are the Aves (birds) and the Mammalia (mammals). All are warm-blooded: that is, their bodies possess devices which maintain an even internal temperature regardless of the temperature outside (within reasonable limits). Since the internal temperature is usually higher than the external, these animals require insulation. As aids to this end, the birds are equipped with feathers and the mammals with hair, both serving to trap a layer of insulating air next to the skin. The birds lay eggs like those of reptiles. The mammals, of course, bring forth their young already "hatched" and supply them with milk produced by mammary glands (*mammae* in Latin).

In the nineteenth century zoologists heard reports of a great curiosity so amazing that they refused to believe it. The Australians had found a creature that had hair and produced milk (through mammary glands that lacked nipples), yet laid eggs! Even when the zoologists were shown specimens of the animal (not alive, unfortunately, because it is not easy to keep it alive away from its natural habitat), they were inclined to brand it a clumsy fraud. The beast was a land-and-water animal that looked a good deal like a duck: it had a bill and webbed feet. Eventually the "duckbilled platypus" had to be recognized as a genuine phenomenon and a new kind of mammal. Another egg-laying mammal,

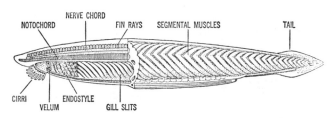

Amphioxus, a primitive, fishlike chordate with a notochord.

the echidua has since been found in Australia and New Guinea. Nor is it only in the laying of eggs that these mammals show themselves to be still close to the reptile. They are only imperfectly warm-blooded; on cold days their internal temperature may drop as much as 10° C.

The mammals are now divided into three subclasses. The egg-laying mammals form the first class, Prototheria (Greek for "first beasts"). The embryo in the egg is actually well developed by the time the egg is laid, and it hatches out not long afterward. The second subclass of mammals, Metatheria ("mid-beasts"), includes the opossums and kangaroos. Their young, though born alive, are in a very undeveloped form and will die in short order unless they manage to reach the mother's protective pouch and stay at the mammary nipples until they are strong enough to move about. These animals are called "marsupials" (from *marsupium*, Latin for pouch).

Finally, at the top of the mammalian hierarchy, we come to the subclass Eutheria ("true beasts"). Their distinguishing feature is the placenta, a blood-suffused tissue that enables the mother to supply the embryo with food and oxygen and carry off its wastes, so that she can develop the offspring for a long period inside her body (nine months in the case of the human being, two years in the case of elephants and whales). The eutherians are usually referred to as "placental mammals."

The placental mammals are divided into well over a dozen orders, of which the following are examples:

Insectivora ("insect-eating")—shrews, moles, and others.

Chiroptera ("hand-wings")—the bats.

Carnivora ("meat-eating")—the cat family, the dog family, weasels, bears, seals, and so on, but not including man.

Rodentia ("gnawing")—mice, rats, rabbits, squirrels, guinea pigs, beavers, porcupines, and so on.

Edentata ("toothless")—the sloths and armadillos, which have teeth, and anteaters, which do not.

Artiodactyla ("even toes")—hoofed animals with an even number of toes on each foot, such as cattle, sheep, goats, swine, deer, antelopes, camels, giraffes, and so on.

Perissodactyla ("odd toes")—horses, donkeys, zebras, rhinoceroses, and tapirs.

Proboscidea ("long nose")—the elephants, of course.

Odontoceti ("toothed whales")—the sperm whale and others with teeth.

Mysticeti ("mustached whales")—the right whale, the blue whale, and others that filter their small sea food through fringes of whalebone that look like a colossal mustache inside the mouth.

Primates ("first")—man, apes, monkeys, and some other creatures with which man may be surprised to find himself associated.

The primates are characterized by hands and sometimes feet that are equipped for grasping, with opposable thumbs and big toes. The digits are topped with flattened nails rather than with sharp claws or enclosing hoofs. The brain is enlarged, and the sense of vision is more important than the sense of smell. There are many other, less obvious, anatomical criteria.

The primates are divided into nine families. Some have so few primate characteristics that it is hard to think of them as primates, but so they must be classed. One is the family Tupaiidae, which includes the insect-eating tree-shrews! Then there are the lemurs—nocturnal, tree-living creatures with foxlike muzzles and a rather squirrely appearance, found particularly in Madagascar.

The families closest to man are, of course, the monkeys and apes. There are three families of monkeys (a word possibly derived from the Latin *homunculus*, meaning "little man").

The two monkey families in the Americas, known as the "New-World monkeys," are the Cebidae (e.g., the organ-grinder's monkey) and the Callithricidae (e.g., the marmoset). The third, the "Old World" family, are the Cercopithecidae; they include the various baboons.

The apes all belong to one family, called Pongidae. They are native to the Eastern Hemisphere. Their most noticeable outward differences from the monkeys are, of course, their larger size and their lack of tails. The apes fall into four types: the gibbon, smallest, hairiest, longest-armed, and most primitive of the family; the orangutan, larger, but also a tree-liver like the gibbon; the gorilla, rather larger than a man, mainly ground-dwelling, and a native of Africa; and the chimpanzee, also a dweller in Africa, rather smaller than a man and the most intelligent primate next to man himself.

As for our own family, Hominidae, it consists today of only one genus and, as a matter of fact, only one species. Linnaeus named the species *Homo sapiens* ("man the wise"), and no one has dared change the name, despite provocation.

Evolution

It is almost impossible to run down the roster of living things, as we have just done, without ending with a strong impression that there has been a slow development of life from the very simple to the complex. The phyla can be arranged so that each seems to add something to the one before. Within each phylum, the various classes can be arranged likewise, and within each class the orders.

Furthermore, the species often seem to melt together, as if they were still evolving along their slightly separate roads from common ancestors not very far in the past. Some species are so close together that under special circumstances they will interbreed, as in the case of the horse and donkey, which, by appropriate cooperation, can produce the mule. Cattle can interbreed with buffaloes, and lions with tigers. There are also intermediate species, so to speak—creatures that link together two larger groups of animals. The cheetah is a cat with a smattering of doggish characteristics, and the hyena is a dog with some cattish characteristics. The platypus is a mammal only halfway removed from a reptile. There is a creature called "peripatus," which seems half worm, half centipede. The dividing lines become particularly thin when we look at certain animals in their youthful stages. The infant frog seems to be a fish, and there is a primitive chordate called "balanoglossus," discovered in 1825, which as a youngster is so like a young echinoderm that at first it was so classified.

We can trace practically a re-enactment of the passage through the phyla, even in the development of a human being from the fertilized egg. The study of this development ("embryology") began in the modern sense with Harvey, the discoverer of the circulation of the blood. In 1759, the German physiologist Kaspar Friedrich Wolff demonstrated that the change in the egg was really a development; that is, specialized tissues grew out of unspecialized precursors by progressive alteration rather than (as many had previously thought) through the mere growth of tiny, already specialized structures existing in the egg to begin with.

In the course of this development, the egg starts as a single cell (a kind of protozoon), then becomes a small colony of cells (as in a sponge), each of which at first is capable of separating and starting life on its own, as happens when identical twins develop. The developing embryo passes through a two-layered stage (like a coelenterate), then adds a third layer (like an echinoderm), and so it continues to add complexities in roughly the order that the higher and higher species do. The human embryo has at some stage in its development the notochord of a primitive chordate, later gill pouches reminiscent of a fish, and still later the tail and body hair of a lower mammal.

From Aristotle on, many men speculated on the possibility that organisms had evolved from one another. But as Christianity grew in power, such speculations were discouraged. The first chapter of Genesis in the Bible stated flatly that each living thing was created "after his kind," and, taken literally, this had to mean that the species were "immutable" and had had the same form from the very beginning. Even Linnaeus, who must have been struck by the apparent kinships among living things, insisted firmly on the immutability of species.

The literal story of Creation, strong as its hold was on the minds of men, eventually had to yield to the evidence of the "fossils" (from the Latin word meaning "to dig"). As long ago as 1669, the Danish scientist Nicolaus Steno had pointed out that lower layers ("strata") of rock had to be older than the upper strata. At any reasonable rate of rock formation, it became more and more evident that lower strata had to be *much* older than upper strata. Petrified remnants of once living things were often found buried so deeply under layers of rock that they had to be immensely older than the few thousand years that had elapsed since the creation as described in the Bible. The fossil evidence also pointed to vast changes in the structure of the earth. As long ago as the sixth century B.C., the Greek philosopher Xenophanes of Colophon had noted fossil sea shells in the mountains and had surmised that those mountains had been under water long ages before.

Believers in the literal words of the Bible could and did maintain that the fossils resembled once-living organisms only through accident, or that they had been created deceitfully by the Devil. Such views were most unconvincing, and a more plausible suggestion was made that the fossils were remnants of creatures drowned in the Flood. Sea shells on mountain tops would certainly be evidence for that, since the Biblical account of the Deluge states that water covered all the mountains.

But on close inspection, many of the fossil organisms proved to be different from any living species. John Ray, the early classifier, wondered if they might represent extinct species. A Swiss naturalist named Charles Bonnet went further. In 1770, he suggested that fossils were indeed remnants of extinct species which had been destroyed in ancient geological catastrophes going back to long before the Flood.

It was an English land surveyor named William Smith, however, who laid a scientific foundation for the study of fossils ("paleontology"). While working on excavations for a canal in 1791, he was impressed by the fact that the rock through which the canal was being cut was divided into strata, and that each stratum contained its own characteristic fossils. It now became possible to put fossils in a chronological order, depending on their place in the series of successive layers, and to associate each fossil with a particular type of rock stratum which would represent a certain period in geological history.

About 1800, Cuvier (the man who invented the notion of the phylum) classified fossils according to the Linnaean system and extended comparative anatomy into the distant past. Although many fossils represented species and genera not found among living creatures, all fitted neatly into one or another of the known phyla and so made up an integral part of the scheme of life. In 1801, for instance, Cuvier studied a long-fingered fossil of a type first discovered twenty years earlier, and demonstrated it to be the remains of a leathery-winged flying creature like nothing now existing—at least like nothing now existing *exactly*. He was able to show from the bone structure that these "pterodactyls" ("wing-fingers"), as he called them, were nevertheless reptiles, clearly related to the snakes, lizards, alligators, and turtles of today.

744

Fossil of a crinoid, or sea lily, a primitive animal of the echinoderm superphylum. This specimen was found in Indiana.

A palm-leaf fossil, found in Colorado.

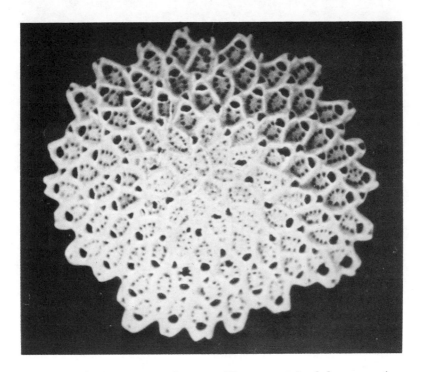

Fossil of a bryozoan, a tiny, mosslike water animal, here magnified about twenty times. It was brought up from an oil drill-hole on Cape Hatteras.

Fossil of a foraminifer, also found in a Cape Hatteras drill-hole. Chalk and some limestones are composed mainly of the shells of these microscopic, one-celled animals. Notable examples are the White Cliffs of Dover and the stones used in the construction of the pyramids.

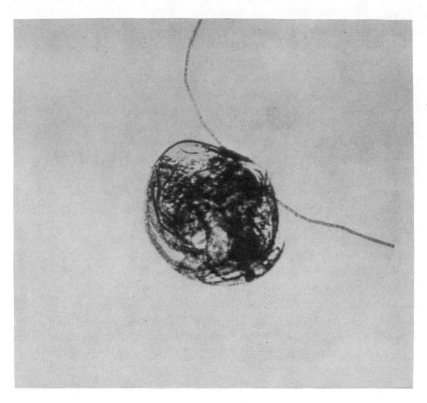

A 3,000-year-old crustacean. It was found in a frozen deposit and revived on being thawed, according to Soviet paleontologists.

An ancient ant delicately preserved in amber.

Model of an extinct lungfish found in Germany. The lung-fishes, living in tidal waters, were forerunners of the land animals.

Cast of a coelacanth. This ancient fish was found still living in deep water near Madagascar.

Fossil of a trilobite, a long-extinct sea arthropod of the Paleozoic era. The name "trilobite" comes from the fact that it was
divided into three lobes. This specimen was found in Vermont.

Fossil of a coral, an ancient coelenterate that lived in the sea. This was found in Indiana.

A fossilized fish.

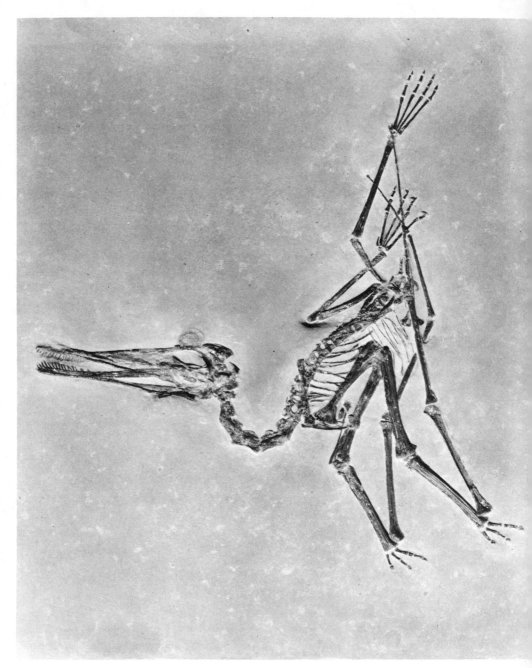

Skeleton of a pterodactyl, an extinct flying reptile.

Furthermore, the deeper the stratum in which the fossil was to be found, and therefore the older the fossil, the simpler and less highly developed it seemed. Not only that, but fossils sometimes represented intermediate forms connecting two groups of creatures which, as far as living forms were concerned, seemed entirely separate. A particularly startling example, discovered after Cuvier's time, is a very primitive bird called Archaeopteryx (Greek for "ancient wing"). This now-extinct creature had wings and feathers, but it also had a lizardlike, feather-fringed tail and a beak that contained reptilian teeth! In these and other respects it was clearly midway between a reptile and a bird.

Cuvier still supposed that terrestrial catastrophes, rather than evolution, had been responsible for the disappearance of the extinct forms of life, but in the 1830's Charles Lyell's new view of fossils and geological history in his history-making work *The Principles of Geology* killed "catastrophism" deader than a doornail (see Chapter 3). Some reasonable theory of evolution became a necessity, if any sense at all was to be made of the paleontological evidence.

If animals had evolved from one form to another, what had caused them to do so? This was the main stumbling block in the efforts to explain the varieties of life. The first to attempt an explanation was the French naturalist Jean Baptiste de Lamarck. In 1809, he published a book, entitled *Zoological Philosophy*, in which he suggested that the environment caused organisms to acquire small changes which were then passed on to their descendants. Lamarck illustrated his idea with the giraffe (a newly discovered sensation of the time). Suppose that a primitive, antelopelike creature that fed on tree leaves ran out of food within easy reach, and had to stretch its neck as far as it could to get more food. By habitual stretching of its neck, tongue, and legs, it would gradually lengthen those appendages. It would then pass on these

Archaeopteryx.

developed characteristics to its offspring, which in turn would stretch further and pass on a still longer neck to their descendants, and so on. Little by little, by generation after generation of stretching, the primitive antelope would evolve into a giraffe.

Lamarck's notion of the "inheritance of acquired characteristics" quickly ran afoul of difficulties. How had the giraffe developed its blotched coat, for instance? Surely no action on its part, deliberate or otherwise, could have effected this change. Furthermore, a skeptical experimenter, the German biologist August Friedrich Leopold Weismann, cut off the tails of mice for generation after generation and reported that the last generation grew tails not one whit shorter than the first. (He might have saved himself the trouble by considering the case of the circumcision of Jewish males, which after more than a hundred generations had produced no shriveling of the foreskin.)

By 1883, Weismann had observed that the germ cells, which were eventually to produce sperm or ova, separated from the remainder of the embryo at an early stage and remained relatively unspecialized. From this, and from his experiments with rat tails, Weismann deduced the notion of the "continuity of the germ plasm." The germ plasm (that is, the protoplasm making up the germ cells) had, he felt, an independent existence, continuous across the generations, with the remainder of the organism but a temporary housing, so to speak, built up and destroyed in each generation. The germ plasm guided the characteristics of the body and was not itself affected by the body. In all this, he was at the extreme opposite to Lamarck and was also wrong, although, on the whole, the actual situation seemed closer to the Weismann view than to that of Lamarck.

Despite its rejection by most biologists, Lamarckism lingered on into the twentieth century and even had a strong but apparently temporary revival in the form of Lysenkoism (hereditary modification of plants by certain treatments) in the Soviet Union. (Trofim Denisovich Lysenko, the exponent of this belief, was powerful under Stalin, retained much influence under Khrushchev, but underwent an eclipse when Khrushchev fell from power in 1964.) Modern geneticists do not exclude the possibility that the action of the environment may bring about certain transmittable changes in simple organisms, but the Lamarckian idea as such was demolished by the discovery of genes and the laws of heredity.

In 1831, a young Englishman named Charles Darwin, a dilettante and sportsman who had spent a more or less idle youth and was restlessly looking for something to do to overcome his boredom, was persuaded by a ship captain and a Cambridge professor to sign on as naturalist on a ship setting off on a five-year voyage around the world. The expedition was to study continental coastlines and make observations of flora and fauna along the way. Darwin, aged twenty-two, made the voyage of the "Beagle" the most important sea voyage in the history of science.

As the ship sailed slowly down the east coast of South America and then up its west coast, Darwin painstakingly collected information on the various forms of plant and animal life. His most striking discovery came in a group of islands in the Pacific, about 650 miles west of Ecuador, called the Galapagos Islands because of giant tortoises living on them (Galapagos coming from the Spanish word for tortoise). What most attracted Darwin's attention during his five-week stay was the variety of finches on the islands; they are known as "Darwin's finches" to this day. He found the birds divided into at least fourteen different species, distinguished from one another mainly by differences in the size and shape of their bills. These particular species did not exist anywhere else in the world, but they resembled an apparently close relative on the South American mainland.

What accounted for the special character of the finches on these islands? Why did they differ from ordinary finches, and why were they themselves divided into no fewer than fourteen species? Darwin decided that the most reasonable theory was that all of them were descended from the mainland type of finch and had differentiated during long isolation on the islands. The differentiation had resulted from varying methods of obtaining food. Three of the Galapagos species still fed on seeds, as the mainland finch did, but each ate a different kind of seed and varied correspondingly in size, one species being rather large, one medium, and one small. Two other species fed on cacti; most of the others fed on insects.

The problem of the changes in the finches' eating habits and physical characteristics preyed on Darwin's mind for many years. In 1838 he began to get a glimmering of the answer from reading a book that had been published forty years before by an English clergyman named Thomas Robert Malthus. It was entitled *An Essay on the Principle of Population*; in it Malthus maintained that a population always outgrew its food supply, so that eventually starvation, disease, or war cut it back. It was in this book that Darwin came across the phrase "the struggle for existence," which his theories later made famous. Thinking of his finches, Darwin at once realized that competition for food would act as a mechanism favoring the more efficient individuals. When the finches that had colonized the Galapagos multiplied to the point of outrunning the seed supply, only the stronger birds or those particularly adept at obtaining seeds or those able to get new kinds of food would survive. A bird that happened to be equipped with slight variations of the finch characteristics, which enabled it to eat bigger seeds or tougher seeds or, better still, insects, would find an untapped food supply. A bird with a slightly thinner and longer bill could reach food that others could not, or one with an unusually massive bill could use otherwise unusable food. Such birds, and their descendants, would gain in numbers at the expense of the original variety of finch. Each of the adaptive types would find and fill a new, unoccupied niche in the environment. On the Galapagos Islands, virtually empty of bird life to begin with, all sorts of niches were there for the taking, with no established competitors to bar the way. On the South American mainland, with all the niches occupied, the ancestral finch did well merely to hold its own. It proliferated into no further species.

Darwin suggested that every generation of animals was composed of an array of individuals varying randomly from the average. Some would be slightly larger; some would possess organs of slightly altered shape; some abilities would be a trifle above or below normal. The differences might be minute, but those whose make-up was even slightly better suited to the environment would tend to live slightly longer and have more offspring. Eventually, an accumulation of favorable characteristics might be coupled with an inability to breed with the original type or other variations of it, and thus a new species would be born.

Darwin called this process "natural selection." According to his view, the giraffe got its long neck not by stretching but because some giraffes were born with longer necks than their fellows, and the longer the neck, the more chance a giraffe had of reaching food. By natural selection, the long-necked species won out. Natural selection explained the giraffe's blotched coat just as easily: an animal with blotches on its skin would blend against the sun-spotted vegetation and thus have more chance of escaping the attention of a prowling lion.

Darwin's view of the way in which species were formed also made clear why it was often so difficult to make clear-cut distinctions between species or between genera. The evolution of species is a continuous process, and, of course, takes a very long time. There must be any number of species with members which are even now slowly drifting apart into separate species.

Darwin spent many years collecting evidence and working out his theory. He realized that it would shake the foundations of biology and man's thinking about his own place in the scheme of things, and he wanted to be sure of his ground in every possible respect. Darwin started collecting notes on the subject and thinking about it in 1834, even before he read Malthus, and in 1858 he was still working on a book dealing with the subject. His friends (including Lyell, the geologist) knew what he was working on; several had read his preliminary drafts. They urged him to hurry, lest he be anticipated. Darwin would not (or could not) hurry, and he *was* anticipated.

The man who anticipated him was Alfred Russel Wallace, fourteen years younger than Darwin. Wallace's life paralleled that of Darwin. He, too, went on an around-the-world scientific expedition as a young man. In the East Indies, he noticed that the plants and animals in the eastern islands were completely different from those in the western islands. A sharp line could be drawn between the two types of life forms; it ran between Borneo and Celebes, for instance, and between the small islands of Bali and Lombok farther to the south. The line is still called "Wallace's Line." (Wallace went on, later in his life, to divide the earth into six large regions, characterized by differing varieties of animals, a division that, with minor modifications, is still considered valid today.)

Now the mammals in the eastern islands and in Australia were distinctly more primitive than those in the western islands and Asia, or indeed in the rest of the world. It looked as if Australia and the eastern islands had split off from Asia at some early time when only primitive mammals existed, and the placental mammals had developed later only in Asia. New Zealand must have been isolated even longer, for it lacked mammals altogether and was inhabited by primitive flightless birds, of which the best-known survivor today is the kiwi.

How had the higher mammals in Asia arisen? Wallace first began puzzling over this in 1855, and in 1858 he, too, came across Malthus' book and from it, he, too, drew the conclusions Darwin had drawn. But Wallace did not spend fourteen years writing his conclusions. Once the idea was clear in his mind, he sat down and wrote a paper on it in two days. Wallace decided to send his manuscripts to some well-known competent biologist for criticism and review, and he chose Charles Darwin.

When Darwin received the manuscript, he was thunderstruck. It expressed his own thoughts in almost his own terms. At once he passed Wallace's paper to other important scientists and offered to collaborate with Wallace on reports summarizing their joint conclusions. Their reports appeared in the *Journal of the Linnaean Society* in 1858.

The next year Darwin's book was finally published. Its full title is *On the Origin of Species by Means of Natural Selection, or the Preservation of Favoured Races in the Struggle for Life*. We know it simply as *The Origin of Species*.

The theory of evolution has been modified and sharpened since Darwin's time, through knowledge of the mechanism of inheritance, of genes, and of mutations (see Chapter 12). It was not until 1930, indeed, that the English statistician and geneticist Ronald Aylmer Fisher succeeded in showing that Mendelian genetics provided the necessary mechanism for evolution by natural selection. Only then did evolutionary theory gain its modern guise. Nevertheless, Darwin's basic conception of evolution by natural selection has stood firm, and indeed the evolutionary idea has been extended to every field of science— physical, biological, and social.

The announcement of the Darwinian theory naturally blew up a storm. At first, a number of scientists held out against the notion. The most important of these was the English zoologist Richard Owen, who was the successor of Cuvier as an expert on fossils and their classification. Owen stooped to rather unmanly depths in his fight against Darwinism. He not only urged others into the fray while remaining hidden himself, but even wrote anonymously against the theory and quoted himself as an authority.

The English naturalist Philip Henry Gosse tried to wriggle out of the dilemma by suggesting that the earth had been created by God complete with fossils to test man's faith. To most people, however, the suggestion that God would play juvenile tricks on mankind seemed more blasphemous than anything Darwin had suggested.

Its counterattacks blunted, opposition within the scientific world gradually subsided, and, within the generation, nearly disappeared. The opponents outside science, however, carried on the fight much longer and much more intensively. The Fundamentalists (literal interpreters of the Bible) were outraged by the implication that man might be a mere descendant from an apelike ancestor. Benjamin Disraeli (later to be Prime Minister of Great Britain) created an immortal phrase by remarking acidly: "The question now placed before society is this, 'Is man an ape or an angel?' I am on the side of the angels." Churchmen, rallying to the angels' defense, carried the attack to Darwin.

Darwin himself was not equipped by temperament to enter violently into the controversy, but he had an able champion in the eminent biologist Thomas Henry Huxley. As "Darwin's bulldog," Huxley fought the battle tirelessly in the lecture halls of England. He won his most telling victory almost at the very beginning of the struggle, in the famous debate with Samuel Wilberforce, a bishop of the Anglican Church, a mathematician, and so accomplished and glib a speaker that he was familiarly known as "Soapy Sam."

Bishop Wilberforce, after apparently having won the audience, turned at last to his solemn, humorless adversary. As the report of the debate quotes him, Wilberforce "begged to know whether it was through his grandfather or his grandmother that [Huxley] claimed his descent from a monkey."

While the audience roared with glee, Huxley rose slowly to his feet and answered: "If, then, the question is put to me, would I rather have a miserable ape for a grandfather, or a man highly endowed by nature and possessing great means and influence, and yet who employs those faculties and that influence for the mere purpose of introducing ridicule into a grave scientific discussion—I unhesitatingly affirm my preference for the ape."

Huxley's smashing return apparently not only crushed Wilberforce but also put the Fundamentalists on the defensive. In fact, so clear was the victory of the Darwinian viewpoint that, when Darwin died in 1882, he was buried, with widespread veneration, in Westminster

Abbey, where lie England's greats. In addition, the town of Darwin in northern Australia was named in his honor.

Another powerful proponent of evolutionary ideas was the English philosopher Herbert Spencer, who popularized the phrase "survival of the fittest" and the word "evolution"—a word Darwin himself rarely used. Spencer tried to apply the theory of evolution to the development of human societies (he is considered the founder of the science of sociology). His arguments, contrary to his intention, were later misused to support war and racism.

The last open battle against evolution took place in 1925; it ended with the antievolutionists winning the battle and losing the war.

The Tennessee legislature had passed a law forbidding teachers in publicly supported schools of the state from teaching that man had evolved from lower forms of life. To challenge the law's constitutionality, scientists and educators persuaded a young high-school biology teacher named John T. Scopes to tell his class about Darwinism. Scopes was thereupon charged with violating the law and brought to trial in Dayton, Tennessee, where he taught. The world gave fascinated attention to his trial.

The local population and the judge were solidly on the side of antievolution. William Jennings Bryan, the famous orator, three times unsuccessful candidate for the Presidency, and outstanding Fundamentalist, served as one of the prosecuting attorneys. Scopes had as his defenders the noted criminal lawyer Clarence Darrow and associated attorneys.

The trial was for the most part disappointing, for the judge refused to allow the defense to place scientists on the stand to testify to the evidence behind the Darwinian theory, and restricted testimony to the question whether Scopes had or had not discussed evolution. But the issues nevertheless emerged in the courtroom when Bryan, over the protests of his fellow prosecutors, volunteered to submit to cross-examination on the Fundamentalist position. Darrow promptly showed that Bryan was ignorant of modern developments in science and had only a stereotyped Sunday-school acquaintance with religion and the Bible.

Scopes was found guilty and fined $100. (The conviction was later reversed on technical grounds by the Tennessee Supreme Court.) But the Fundamentalist position (and the state of Tennessee) had stood in so ridiculous a light in the eyes of the educated world that the antievolutionists have not made any serious stand since then—at least not in broad daylight.

In fact, if any confirmation of Darwinism were needed, it has turned up in examples of natural selection that have taken place before the eyes of mankind (now that mankind knows what to watch for). A notable example occurred in Darwin's native land.

In England, it seems, the peppered moth exists in two varieties, a light and a dark. In Darwin's time, the white variety was predominant because it was less prominently visible against the light lichen-covered bark of the trees it frequented. It was saved by this "protective coloration" more often than the clearly visible, dark variety from those animals who would feed on it. In modern, industrialized England, however, soot has killed the lichen cover and blackened the tree bark. Now it is the dark variety that is less visible against the bark and therefore protected. It is the dark variety that is now predominant—through the action of natural selection.

A study of the fossil record has enabled paleontologists to divide the history of the earth into a series of "eras." These were roughed out and named by various nineteenth-century British geologists, including Lyell himself, Adam Sedgwick, and Roderick Impey Murchison. The eras start, as is now known, some 500 or 600 million years ago with the first fossils (when all the phyla except Chordata were already established). The first fossils do not, of course, represent the first life. For the most part, it is only the hard portions of a creature that fossilize, so the clear fossil record contains only animals that possessed shells or bones. Even the simplest and oldest of these creatures are already far advanced and must have a long evolutionary background. One evidence of that is that in 1965 fossil remains of small clamlike creatures were discovered that seem to be about 720 million years old.

Paleontologists can now do far better than that. It stands to reason that simple one-celled life must extend much farther back in time than anything with a shell, and indeed, signs of blue-green algae and of bacteria have been found in rocks that were a billion years old and more. In 1965, the American paleontologist Elso Sterrenberg Barghoorn discovered minute bacteriumlike objects ("microfossils") in rocks over 3 billion years old. They are so small, their structure must be studied by electron microscope.

It would seem then that chemical evolution, moving toward the origin of life, began almost as soon as the earth took on its present shape some 4.6 billion years ago. Within 1.5 billion years, chemical evolution had reached the stage where systems complicated enough to be called living had formed. About 2.5 billion years ago, blue-green algae may have been in existence, and the process of photosynthesis began the slow change from a nitrogen–carbon dioxide atmosphere into a nitrogen–oxygen atmosphere. About a billion years ago, the one-celled life of the seas must have been quite diversified and included distinctly animal protozoa, who would then have been the most complicated forms of life in existence—monarchs of the world.

For 2 billion years after blue-green algae came into existence, the oxygen content must have been very slowly increasing. As the most recent billion years of earth's history began to unfold, the oxygen concentration may have been 1 or 2 per cent of the atmosphere, enough to supply a rich source of energy for animal cells beyond anything that had earlier existed. Evolutionary change spurted in the direction of increased complication, and by 600 million years ago there could begin the rich fossil record of elaborate organisms.

The earliest rocks with elaborate fossils are said to belong to the Cambrian age, and the entire 4-billion-year history of our planet that preceded it has been, until recently, dismissed as the "pre-Cambrian age." Now that the traces of life have unmistakably been found in it, the more appropriate name of "Cryptozoic eon" (Greek for "hidden life") is used, while the last 600 million years make up the "Phanerozoic eon" ("visible life").

The Cryptozoic eon is even divided into two sections: the earlier "Archeozoic era" ("ancient life"), to which the first traces of unicellular life belong, and the later "Proterozoic era" ("early life.")

The division between the Cryptozoic eon and the Phanerozoic eon is extraordinarily sharp. At one moment in time, so to speak, there are no fossils at all above the microscopic level, and at the next there are elaborate organisms of a dozen different basic types. Such a sharp division is called an "unconformity," and an unconformity leads invariably to speculations about possible catastrophes. It seems there should have been a more gradual appearance of fossils, and what may have happened is that geological events of some extremely harsh variety wiped out the earlier record.

One fascinating (if highly speculative) suggestion was put forward in 1967 by Walter S. Olson. He suggested that about a billion years ago or somewhat less, our moon was captured by the earth—that prior to that the earth was without a satellite. (This possibility is seriously considered by astronomers as a way of explaining certain anomalies of the earth-moon system.) When first captured, the moon was considerably closer than it is now. The enormous tides suddenly set in motion would have broken up the superficial layers of rock on earth and, so to speak, erased any fossil record, which began again—apparently full-born—only after the tides had receded in power to the point where the rocky surface was left relatively untouched.

The broad divisions of the Phanerozoic eon are the Paleozoic era ("ancient life"), the Mesozoic ("middle life"), and the Cenozoic ("new life"). According to modern methods of geological dating, the Paleozoic era covered a span of perhaps 350 million years, the Mesozoic 150 million years, and the Cenozoic the last 50 million years of the earth's history.

Each era is in turn subdivided into ages. The Paleozoic begins, as stated above, with the Cambrian age (named for a location in Wales—actually an ancient tribe that occupied it—where these strata were first uncovered). During the Cambrian period shellfish were the most elaborate form of life. This was the era of the "trilobites," primitive arthropods of which the modern king crab is the closest living relative. The king crab, because it has survived with few evolutionary changes over long ages, is an example of what is sometimes rather dramatically called a "living fossil."

The next age is the Ordovician (named for another Welsh tribe). This was the age, between 450 and 500 million years ago, when the chordates made their first appearance in the form of "graptolites," small animals living in colonies and now extinct. They are possibly related to the balanoglossus, which, like the graptolites, belongs to the "hemichordata," the most primitive subphylum of the chordate phylum.

Then came the Silurian (named for still another Welsh tribe) and the Devonian (from Devonshire). The Devonian age, between 350 and 400 million years ago, witnessed the rise of fish to dominance in the ocean, a position they still hold. In that age, however, came also the colonization of the dry land by life forms. It is hard to realize, but true, that during perhaps three quarters or more of its history, life was confined to the waters and the land remained dead and barren. Considering the difficulties represented by the lack of water, by extremes of temperature, by the full force of gravity unmitigated by the buoyancy of water, it must be understood that the spread to land of life forms that evolved to meet the conditions of the ocean represented the greatest single victory won by life over the inanimate environment.

The move toward the land probably began when competition for food in the crowded sea drove some organisms into shallow tidal waters, until then unoccupied because the bottom was exposed for hours at a time at low tide. As more and more species crowded into the tidewaters, relief from competition could be attained only by moving farther and farther up the shore, until eventually some mutant organisms were able to establish themselves on dry land.

The first life forms to manage the transition were plants. This took place about 400 million years ago. The pioneers belonged to a now extinct plant group called "psilopsids"—the first multicellular plants. (The name comes from the Greek word for "bare" because the stems were bare of leaves, a sign of the primitive nature of these plants.) More complex plants developed and by 350 million years ago, the land was covered with forest. Once plant life had begun to grow on dry land, animal life could follow suit. Within a few million years the land was occupied by arthropods, molluscs, and worms. All these first land animals were small, because heavier animals, without an internal skeleton, would have collapsed under the force of gravity. (In the ocean, of course, buoyancy largely negated gravity, which was not therefore a factor. Even today the largest animals live in the sea.) The first land creatures to gain much mobility were the insects; thanks to their development of wings, they were able to counteract the force of gravity, which held other animals to a slow crawl.

Finally, 100 million years after the first invasion of the land, there came a new invasion by creatures that could afford to be bulky despite gravity because they had a bracing of bone within. The new colonizers from the sea were bony fishes belonging to the subclass Crossopterygii ("fringed fins"). Some of their fellow members had migrated to the uncrowded sea deeps, including the coelacanth, which biologists discover in 1939 to be still in existence (much to their astonishment).

The fishy invasion of land began as a result of competition for oxygen in brackish stretches of fresh water. With oxygen available in unlimited quantities in the atmosphere, those fish best survived who could most effectively gulp air when the oxygen content of water fell below the survival point. Devices for storing such gulped air had survival value, and fish developed pouches in their alimentary canals in which swallowed air could be kept. These pouches developed into simple lungs in some cases. Descendants of these early fish include the "lungfishes," a few species of which still exist in Africa and Australia. These live in stagnant water where ordinary fishes would suffocate, and they can even survive summer droughts when their habitat dries up. Even fish who live in the sea where the oxygen supply is no problem, show signs of their descent from the early-lunged creatures, for they still possess air-filled pouches, used not for respiration but for buoyancy.

Some of the lung-possessing fishes, however, carried the matter to the logical extreme and began living, for shorter or longer stretches, out of the water altogether. These crossopterygian species with the strongest fins could do so most successfully for, in the absence of water buoyancy, they had to prop themselves up against the pull of gravity. By the end of the Devonian age some of the primitive-lunged crossopterygians found themselves standing on the dry land, propped up shakily on four stubby legs.

After the Devonian came the Carboniferous ("coal-bearing") age, so named by Lyell because it was the period of the vast, swampy forests that, some 300 million years ago, represented what was perhaps the lushest vegetation in earth's history; eventually, they were buried and became this planet's nearly endless coal beds. This was the age of the amphibians; the crossopterygians by then were spending their entire adult lives on land. Next came the Permian age (named for a district in the Urals, for the study of which Murchison made the long trip from England). The first reptiles now made their appearance. They ushered in the Mesozoic era, in which reptiles were to dominate the earth so thoroughly that it has become known as the age of the reptiles.

The Mesozoic is divided into three ages—the Triassic (it was found in three strata), the Jurassic (from the Jura mountains in France), and the Cretaceous ("chalk-forming"). In the Triassic arose the dinosaurs (Greek for "terrible lizards"). The dinosaurs reached their peak form

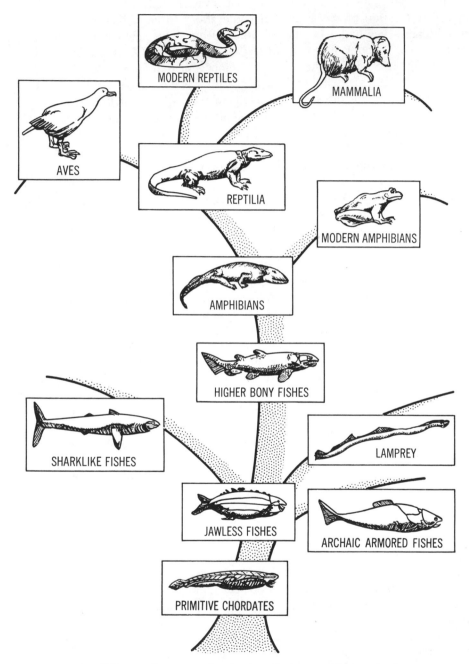

A philogenetic tree, showing evolutionary lines of the vertebrates.

in the Cretaceous, when *Tyrannosaurus rex* thundered over the land—the largest carnivorous land animal in the history of our planet.

It was during the Jurassic that the earliest mammals and birds developed, each from a separate group of reptiles. For millions of years these creatures remained inconspicuous and unsuccessful. With the end of the Cretaceous, however, the gigantic reptiles began to disappear (for some unknown reason, so that the cause of the "great dying" remains the most tantalizing problem in paleontology), and the mammals and birds came into their own. The Cenozoic era that followed became the age of mammals; it brought in placental mammals and the world we know.

The unity of present life is demonstrated in part by the fact that all organisms are composed of proteins built from the same amino acids. The same kind of evidence has recently established our unity with the past as well. The new science of "paleobiochemistry" (the biochemistry of ancient forms of life) was opened in the late 1950's, when it was shown that certain 300-million-year-old fossils contained remnants of proteins consisting of precisely the same amino acids that make up proteins today—glycine, alanine, valine, leucine, glutamic acid, and aspartic acid. Not one of the ancient amino acids differed from present ones. In addition, traces of carbohydrates, cellulose, fats, and porphyrins were located, with (again) nothing that would be unknown or unexpected today.

From our knowledge of biochemistry we can deduce some of the biochemical changes that may have played a part in the evolution of animals.

Let us take the excretion of nitrogenous wastes. Apparently, the simplest way to get rid of nitrogen is to excrete it in the form of the small ammonia molecule (NH_3), which can easily pass through cell membranes into the blood. Ammonia happens to be extremely poisonous; if its concentration in the blood exceeds one part in a million, the organism will die. For a sea animal, this is no great problem; it can discharge the ammonia into the ocean continuously through its gills. But for a land animal, however, ammonia excretion is out of the question. To discharge ammonia as quickly as it is formed would require such an excretion of urine that the animal would quickly be dehydrated and die. Therefore a land organism must produce its nitrogenous wastes in a less toxic form than ammonia. The answer is urea. This substance can be carried in the blood in concentrations up to one part in a thousand without serious danger.

Now fish eliminate nitrogenous wastes as ammonia, and so do tadpoles. But when a tadpole matures to a frog, it begins to eliminate nitrogenous wastes as urea. This change in the chemistry of the organism is every bit as crucial for the change-over from life in the water to life on land as is the visible change from gills to lungs.

Such a biochemical change must have taken place when the cross-opterygians invaded the land and became amphibians. Thus, there is every reason to believe that biochemical evolution played as great a part in the development of organisms as "morphological" evolution (that is, changes in form and structure).

Another biochemical change was necessary before the great step from amphibian to reptile could be taken. If the embryo in a reptile's egg excreted urea, it would build up to toxic concentrations in the limited quantity of water within the egg. The change that took care of this problem was the formation of uric acid instead of urea. Uric acid (a purine molecule resembling the adenine and guanine that occur in nucleic acids) is insoluble in water; it is therefore precipitated in the form of small granules and thus cannot enter the cells. This changeover from urea excretion to uric-acid excretion was as essential in the development of reptiles as was the changeover, for instance, from a three-chambered heart to an essentially four-chambered one.

In adult life, reptiles continue eliminating nitrogenous wastes as uric acid. They have no urine in the liquid sense. Instead, the uric acid is eliminated as a semisolid mass through the same body opening that serves for the elimination of feces. This single body opening is called the "cloaca" (Latin for "sewer").

Birds and egg-laying mammals, which lay eggs of the reptilian type, preserve the uric-acid mechanism and the cloaca. In fact, the egg-laying mammals are often called "monotremes" (from Greek words meaning "one hole").

Placental mammals, on the other hand, can easily wash away the embryo's nitrogenous wastes, for the embryo is connected, indirectly, to the mother's circulatory system. Mammalian embryos, therefore, manage well with urea. It is transferred to the mother's blood stream and passes out through the mother's kidneys.

An adult mammal has to excrete substantial amounts of urine to get rid of its urea. This calls for two separate openings: an anus to eliminate the indigestible solid residues of food and a urethral opening for the liquid urine.

The account just given of nitrogen excretion demonstrates that although life is basically a unity there are systematic minor variations from species to species. Furthermore, these variations seem to be greater as the species considered are farther removed from each other in the evolutionary sense.

Consider, for instance, that antibodies can be built up in animal blood to some foreign protein or proteins as, for example, those in human blood. Such "antisera," if isolated, will react strongly with human blood, coagulating it, but will not react in this fashion with the blood of other species. (This is the basis of the tests indicating that bloodstains are of human origin—or not—which sometimes lend drama to murder investigations.) Interestingly, antisera that will react with human blood will respond weakly with chimpanzee blood, while antisera that will react strongly with chicken blood will react weakly with duck blood, and so on. Antibody specificity thus can be used to indicate close relationships among life forms.

Such tests indicate, not surprisingly, the presence of minor differences in the complex protein molecule—differences small enough in closely related species to allow some overlapping in antiserum reactions.

When biochemists developed techniques for determining the precise amino acid structure of proteins, in the 1950's, this method of arranging species according to protein structure was vastly sharpened.

In 1965, even more detailed studies were reported on the hemoglobin molecules of various types of primates, including man. Of the two kinds of peptide chains in hemoglobin, one, the "alpha chain," varied little from primate to primate. The other, the "beta chain," varied considerably. Between a particular primate and man there were only six differences in the amino acids and the alpha chain, but twenty-three in those of the beta chains. Judging by differences in the hemoglobin molecules, it is believed man diverged from the other apes about 75 million years ago, or just about the time the ancestral horses and donkeys diverged.

Still broader distinctions can be made by comparing molecules of "cytochrome c," an iron-containing protein molecule made up of about 105 amino acids and found in the cells of every oxygen-breathing species —plant, animal, or bacterial. By analyzing the cytochrome-c molecules from different species it was found that the molecules in man differed from those of the rhesus monkey in only one amino acid in the entire chain. Between the cytochrome-c of man and that of a kangaroo, there were 10 differences in amino acid; between those of man and a tuna fish, 21 differences; between those of man and a yeast cell, some 40 difference.

With the aid of computer analysis, biochemists have decided it takes on the average some 7 million years for a change in one amino-acid residue to establish itself, and estimates can be made of the time in the past when one type of organism diverged from another. It was about 2.5 billion years ago, judging from cytochrome-c analysis, that higher organisms diverged from bacteria (that is, it was about that long ago that a living creature was last alive who might be considered a common ancestor). Similarly, it was about 1.5 billion years ago that plants and animals had a common ancestor, and 1 billion years ago that insects and vertebrates had a common ancestor.

If mutations in the DNA chain, leading to changes in amino acid pattern, were established by random factors only, it might be supposed that the rate of evolution would continue at an approximately constant rate. Yet there are occasions when evolution seems to progress more rapidly than at others—when there is a sudden flowering of new species or a sudden spate of deaths of old ones. The period at the end of the Cretaceous, when all the dinosaurs then living—together with other groups of organisms—utterly died out over a relatively short period of time (while other groups of organisms lived on undisturbed), is a case in point. It may be that the rate of mutations is greater at some periods in earth's history than at others, and these more frequent mutations may establish an extraordinary number of new species or render an extraordinary number of old ones unviable. (Or else some of the new species may prove more efficient than the old and compete them to death.)

One environmental factor which encourages the production of mutations is energetic radiation, and earth is constantly bombarded by energetic radiation from all directions at all times. The atmosphere absorbs most of it, but even the atmosphere is helpless to ward off cosmic radiation. Can it be that cosmic radiation is greater at some period than at others?

A difference can be postulated in each of two different ways. Cosmic radiation is diverted to some extent by earth's magnetic field. However, the magnetic field varies in intensity, and there are periods, at varying intervals, when it sinks to zero intensity. Bruce Heezen suggested in 1966 that these periods when the magnetic field, in the process of reversal, goes through a time of zero intensity, may also be periods when unusual amounts of cosmic radiation reach the surface of the earth, bringing about a jump in mutation-rate. This is a sobering thought in view of the fact that the earth seems to be heading toward such a peroid of zero intensity.

Then, too, what about the occurrence of supernovas in earth's vicinity —close enough to the solar system, that is, to produce a distinct increase in the intensity of bombardment by cosmic rays of the earth's surface? Two American astronomers, K. D. Terry and Wallace H. Tucker, have speculated on that possibility. Could a combination of these two effects, taking place with fortuitous simultaneity—a nearby supernova just when earth's magnetic field had temporarily faded—account for the sudden dying of the dinosaurs? Well, perhaps, but as yet there is no firm evidence.

The Descent of Man

James Ussher, a seventeenth-century Irish archbishop, dated the creation of man precisely in the year 4004 B.C.

Before Darwin, few men dared to question the Biblical interpretation of man's early history. The earliest reasonably definite date to which the events recorded in the Bible can be referred is the reign of Saul, the first king of Israel, who is believed to have become king about 1025 B.C.

Bishop Ussher and other Biblical scholars who worked back through the chronology of the Bible came to the conclusion that man and the universe could not be more than a few thousand years old.

The documented history of man, as recorded by Greek historians, began only about 700 B.C. Beyond this hard core of history, dim oral traditions went back to the Trojan War, about 1200 B.C., and more dimly still to a pre-Greek civilization on the island of Crete under a King Minos. Nothing beyond documentation—the writings of historians in known languages, with all the partiality and distortion that might involve—was known to moderns concerning the everyday life of ancient times prior to the eighteenth century. Then, in 1738, the cities of Pompeii and Herculaneum, buried in an eruption of Vesuvius in 79 A.D., began to be excavated. For the first time, historians grew aware of what could be done by digging, and the science of archaeology got its start.

At the beginning of the nineteenth century, archaeologists began to get their first real glimpses of human civilizations that came before the periods described by the Greek and Hebrew historians. In 1799, during General Bonaparte's invasion of Egypt, an officer in his army, named Boussard, discovered an inscribed stone in the town of Rosetta, on one of the mouths of the Nile. The slab of black basalt had three inscriptions, one in Greek, one in an ancient form of Egyptian picture writing called "hieroglyphic" ("sacred writing"), and one in a simplified form of Egyptian writing called "demotic" ("of the people").

The inscription in Greek was a routine decree of the time of Ptolemy V, dated the equivalent of March 27, 196 B.C. Plainly, it must be a translation of the same decree given in the other two languages on the slab (compare the no-smoking signs and other official notices that often appear in three languages in public places, especially airports, today). Archaeologists were overjoyed: at last they had a "pony" with which to decipher the previously undecipherable Egyptian scripts. Important work was done in "cracking the code" by Thomas Young, the man who had earlier established the wave theory of light (see Chapter 7), but it fell to the lot of a French student of antiquities, Jean François Champollion, to solve the "Rosetta stone" completely. He ventured the guess that Coptic, a still-remembered language of certain Christian sects in Egypt, could be used as a guide to the ancient Egyptian language. By 1821, he had cracked the hieroglyphs and the demotic script and opened the way to reading all the inscriptions found in the ruins of ancient Egypt.

An almost identical find later broke the undeciphered writing of ancient Mesopotamia. On a high cliff near the ruined village of Behistun in western Iran, scholars found an inscription that had been carved about 520 B.C. at the order of the Persian emperor Darius I. It announced the manner in which he had come to the throne after defeating a usurper;

to make sure that everyone could read it, Darius had had it carved in three languages—Persian, Sumerian, and Babylonian. The Sumerian and Babylonian writings were based on pictographs formed as long ago as 3100 B.C. by indenting clay with a stylus; these had developed into a "cuneiform" ("wedge-shaped") script, which remained in use until the first century A.D.

An English army officer, Henry Creswicke Rawlinson, climbed the cliff, transcribed the entire inscription, and, by 1846, after ten years of work, managed to work out a complete translation, using local dialects as his guide where necessary. The deciphering of the cuneiform scripts made it possible to read the history of the ancient civilizations between the Tigris and the Euphrates.

Expedition after expedition was sent to Egypt and Mesopotamia to look for tablets and the remains of the ancient civilizations. In 1854 a Turkish scholar, Hurmuzd Rassam, discovered the remnants of a library of clay tablets in the ruins of Nineveh, the capital of ancient Assyria—a library that had been collected by the last great Assyrian king, Ashurbanipal, about 650 B.C. In 1873 the English Assyriologist George Smith discovered clay tablets giving legendary accounts of a flood so like the story of Noah that it became clear that much of the first part of the book of Genesis was based on Babylonian legend. Presumably, the Jews picked up the legends during their Babylonian captivity in the time of Nebuchadrezzar, a century after the time of Ashurbanipal. In 1877, a French expedition to Iraq uncovered the remains of the culture preceding the Babylonian—that of the aforementioned Sumerians. This carried the history of the region back to earliest Egyptian times. And in 1921, remains of a totally unexpected civilization were discovered along the Indus Valley in what is now Pakistan. It had flourished between 2500 and 2000 B.C.

Yet Egypt and Mesopotamia were not quite in the same league with Greece when it came to dramatic finds on the origins of modern Western culture. Perhaps the most exciting moment in the history of archaeology came in 1873 when a German ex-grocer's boy found the most famous of all legendary cities.

Heinrich Schliemann as a boy developed a mania for Homer. Although most historians regarded the *Iliad* as mythology, Schliemann lived and dreamed of the Trojan War. He decided that he must find Troy, and by nearly superhuman exertions he raised himself from grocer's boy to millionaire so that he could finance the quest. In 1868, at the age of forty-six, he set forth. He persuaded the Turkish government to give him permission to dig in Asia Minor, and, following only the meager geographical clues afforded by Homer's accounts, he finally settled upon a mound near the village of Hissarlik. He browbeat the local population into helping him dig into the mound. Excavating in a completely amateurish, destructive and unscientific manner, he began to uncover a series of buried ancient cities, each built on the ruins of the other. And then, at last, success: he uncovered Troy—or at least a city he proclaimed as Troy. Actually, the particular ruins he named Troy are now known to be far older than Homer's Troy, but Schliemann had proved that Homer's tales were not mere legends.

Inexpressibly excited by his triumph, Schliemann went on to mainland Greece and began to dig at the site of Mycenae, a ruined village which Homer had described as the once powerful city of Agamemnon, leader of the Greeks in the Trojan War. Again he uncovered an astounding find—the ruins of a city with gigantic walls which we now know to date back to 1500 B.C.

Schliemann's successes prompted the British archaeologist Arthur John Evans to start digging on the island of Crete, described in Greek legends as the site of a powerful early civilization under a King Minos. Evans, exploring the island in the 1890's, laid bare a brilliant, lavishly ornamented "Minoan" civilization that stretched back many centuries before the time of Homer's Greece. Here, too, written tablets were found. They were in two different scripts, one of which, called "Linear B," was finally deciphered in the 1950's and shown to be a form of Greek through a remarkable feat of cryptography and linguistic analysis by a young English architect named Michael Vestris.

As other early civilizations were uncovered—the Hittites and the Mittanni in Asia Minor, the Indus civilization in India, and so on—it became obvious that the history recorded by Greece's Herodotus and the Hebrews' Old Testament represented comparatively advanced stages of human civilization. Man's earliest cities were at least several thousand years old, and his prehistoric existence in less civilized modes of life must stretch many thousands of years farther into the past.

Anthropologists find it convenient to divide cultural history into three major periods: the Stone Age, the Bronze Age, and the Iron Age (a division first suggested by the Roman poet and philosopher, Lucretius, and introduced to modern science by the Danish paleontologist Christian Jurgenson Thomsen in 1834). Before the Stone Age, there may have been a "Bone Age," when pointed horns, chisellike teeth, and clublike thigh bones served men at a time when the working of relatively intractable rock had not yet been perfected.

The Bronze and Iron Ages are, of course, very recent; as soon as we delve into the time before written history, we are back in the Stone Age. What we call civilization (from the Latin word for "city") began perhaps around 6000 B.C., when man first turned from hunting to agriculture, learned to domesticate animals, invented pottery and new types of tools, and started to develop permanent communities and a settled way of life. Because the archaeological remains from this period of transition are marked by advanced stone tools formed in new ways, it is called the New Stone Age, or the "Neolithic" period.

This Neolithic Revolution seems to have started in the Near East, at the crossroads of Europe, Asia, and Africa (where later the Bronze and Iron Ages also were to originate). From there, it appears, the revolution slowly spread in widening waves to the rest of the world. It did not reach western Europe and India until 3000 B.C., northern Europe and eastern Asia until 2000 B.C., and central Africa and Japan until perhaps 1000 B.C. or later. Southern Africa and Australia remained in the Old Stone Age until the eighteenth and nineteenth centuries. Most of America also was still in the hunting phase when the Europeans arrived in the sixteenth century, although a well-developed civilization, possibly originated by the Mayas, had developed in Central America and Peru as early as the first centuries of the Christian era.

Evidences of man's pre-Neolithic cultures began to come to light in Europe at the end of the eighteenth century. In 1797 an Englishman named John Frere dug up in Suffolk some crudely fashioned flint tools too primitive to have been made by Neolithic man. They were found thirteen feet underground, which, allowing for the normal rate of sedimentation, testified to great age. In the same stratum with the tools were bones of extinct animals. More and more signs of the great antiquity of tool-making man were discovered, notably by two nineteenth-century French archeologists, Jacques Boucher de Perthes and Edouard Armand Lartet. Lartet, for instance, found a mammoth tooth on which some early man had scratched an excellent drawing of the mammoth, obviously from living models. The mammoth was a hairy species of elephant that disappeared from the earth well before the beginning of the New Stone Age.

Archaeologists launched upon an active search for early stone tools. They found that these could be assigned to a relatively short Middle Stone Age ("Mesolithic") and a long Old Stone Age ("Paleolithic"). The Paleolithic was divided into Lower, Middle, and Upper periods. The earliest objects that could be considered tools ("eoliths," or "dawn stones") seemed to date back nearly a million years!

What sort of creature had made the Old Stone Age tools? It turned out that Paleolithic man, at least in his late stages, was far more than a hunting animal. In 1879, a Spanish nobleman, the Marquis de Sautuola, explored some caves that had been discovered a few years earlier—after having been blocked off by rock slides since prehistoric times—at Altamira in northern Spain near the city of Santander. While he dug into the floor of a cave, his five-year-old daughter, who had come along to watch papa dig, suddenly cried: "Toros! Toros!" ("Bulls! Bulls!"). The father looked up, and there on the walls of the cave were drawings of various animals, in vivid color and vigorous detail.

Anthropologists found it hard to believe that these sophisticated drawings could have been made by primitive man. But some of the pictured animals were plainly extinct types. The French archaeologist Henri Edouard Prosper Breuil found similar art in caves in southern France. All the evidence finally forced archaeologists to agree with Breuil's firmly expressed views and to conclude that the artists must have lived in the late Paleolithic, say about 10,000 B.C.

Something was already known about the physical appearance of these Paleolithic men. In 1868, workmen excavating a roadbed for a railroad had uncovered the skeletons of five human beings in the so-called Cro-Magnon caves in southwest France. The skeletons were unquestionably *Homo sapiens*, yet some of them, and similar skeletons soon found elsewhere, seemed to be up to 35,000 or 40,000 years old, according to the geological evidence. They were given the name "Cro-Magnon man." Taller than the average modern man and equipped with a large braincase, Cro-Magnon man is pictured by artists as a handsome, stalwart fellow, modern enough, it would certainly appear, to be able to interbreed with present-day human beings.

Mankind, traced thus far back, was not a planet-wide species as it is now. Prior to 20,000 B.C. or so, he was confined to the great "world island" of Africa, Asia, and Europe. It was only later that hunting bands began to migrate across narrow ocean passages into the Americas, Indonesia, and Australia. It was not until 400 B.C., and later, that daring Polynesian navigators crossed wide stretches of the Pacific, without compasses, and in what were little more than canoes, to colonize the islands of the Pacific. Finally, it was not until the twentieth century that the foot of man rested on Antarctica.

But if we are to trace the fortunes of prehistoric man at a time when he was confined to only part of the earth's land area, there must be some manner of dating events, at least roughly. A variety of ingenious methods have been used.

Archaeologists have, for instance, used tree rings for the purpose, a technique ("dendrochronology") introduced in 1914 by the American astronomer Andrew Ellicott Douglass. Tree rings are widely separated in wet summers when much new wood is laid down, and closely spaced in dry summers. The pattern over the centuries is quite distinctive. A piece of wood forming part of a primitive abode can have its ring pattern matched with the one place of the scheme where it will fit, and, in this way, its date can be obtained.

A similar system can be applied to layers of sediment or "varves" laid down summer after summer by melting glaciers in such places as Scandinavia. Warm summers will leave thick layers, cool summers thin ones, and again there is a distinctive pattern. In Sweden, events can be traced back 18,000 years in this way.

An even more startling technique is that developed in 1946 by the American chemist Willard Frank Libby. Libby's work had its origin in the 1939 discovery by the American physicist Serge Korff that cosmic ray bombardment of the atmosphere produced neutrons. Nitrogen reacts with these neutrons, producing radioactive carbon 14 in nine reactions out of every ten, and radioactive hydrogen 3 in the tenth reaction.

As a result, the atmosphere would always contain small traces of carbon 14 (and even smaller traces of hydrogen 3). Libby reasoned that radioactive carbon 14 created in the atmosphere by cosmic rays would enter all living tissue via carbon dioxide, first absorbed by plants and then passed on to animals. As long as a plant or animal lived, it would continue to receive radiocarbon and maintain it at a constant level in its tissues. But when the organism died and ceased to take in carbon, the radiocarbon in its tissues would begin to diminish by radioactive breakdown, at a rate determined by its 5,600-year half-life. Therefore, any piece of preserved bone, any bit of charcoal from an ancient camp-fire, or organic remains of any kind could be dated by measuring the amount of radiocarbon left. The method is reasonably accurate for objects up to 30,000 years old, and this covers archaeological history from the ancient civilizations back to the beginnings of Cro-Magnon man. For developing this technique of "archaeometry," Libby was awarded the Nobel Prize for chemistry in 1960.

Cro-Magnon was not the first early man dug up by the archaeologists. In 1857, in the Neanderthal valley of the German Rhineland, a digger discovered part of a skull and some long bones that looked human in the main but only crudely human. The skull had a sharply sloping forehead and very heavy brow ridges. Some archaeologists maintained that they were the remains of a human being whose bones had been deformed by disease, but as the years passed other such skeletons were found, and a detailed and consistent picture of Neanderthal man was developed. Neanderthal was a short, squat, stooping biped, the men averaging a little taller than five feet, the women somewhat shorter. The skull was roomy enough for a brain nearly as large as modern man's. Anthropological artists picture the creature as barrel-chested, hairy, beetle-browed, chinless, and brutish in expression, a picture originated by the French paleontologist Marcellin Boule, who was the first to describe a nearly complete Neanderthal skeleton in 1911. Actually, he was probably not as subhuman as pictured. Modern examination of the skeleton described by Boule show it to have belonged to a badly arthritic creature. A normal skeleton gives rise to a far more human image. In fact, give a Neanderthal man a shave and a haircut, dress him in well-fitted clothes, and he could probably walk down New York's Fifth Avenue without getting much notice.

Traces of Neanderthal man were eventually found not only in Europe but also in northern Africa, in Russia and Siberia, in Palestine, and in Iraq. About a hundred different skeletons have now been located at some forty different sites, and men of this sort may still have been alive as recently as 30,000 years ago. Skeletal remains somewhat resembling Neanderthal man were discovered in still more widely separated places; these were Rhodesian man, dug up in northern Rhodesia in southern Africa in 1921, and Solo man, found on the banks of the Solo River in Java in 1931. They were considered separate species of the genus *Homo*, and so the three types were named *Homo neanderthalensis*, *Homo rhodesiensis*, and *Homo solensis*. But some anthropologists and evolutionists maintain that all three should be placed in the same species as *Homo sapiens*, as "varieties" or "subspecies" of man. There were men that we call *sapiens* living at the same time as Neanderthal, and intermediate forms have been found which suggest that there may have been interbreeding between them. If Neanderthal and his cousins can be classed as *sapiens*, then our species is perhaps 200,000 years old. Another apparent relative of Neanderthal, "Heidelberg man" (of which only a jawbone was discovered in 1907 and nothing additional since then), is much older, and if we include him in our species, the history of *Homo sapiens* can be pushed even further back. Indeed, at Swanscombe in England archaeologists have found skull fragments which seem to be definitely *sapiens* and even older than Neanderthal. One *Homo sapiens* remnant, discovered just west of Budapest in 1966, may be 500,000 years old.

Darwin's *Origin of Species* launched a great hunt for man's distinctly subhuman ancestors—what the popular press came to call the "missing link" between man and his presumably apelike forerunners. This hunt, in the very nature of things, could not be an easy one. Primates are quite intelligent, and few allow themselves to be trapped in situations that lead to fossilization. It has been estimated that the chance of finding a primate skeleton by random search is only one in a quadrillion.

In the 1880's, a Dutch paleontologist, Marie Eugène François Thomas Dubois, got it into his head that the ancestors of man might be found in the East Indies (modern Indonesia), where great apes still flourished (and where he could work conveniently because those islands then belonged to the Netherlands). Surprisingly enough, Dubois, working in Java, the most populous of the Indonesian islands, did turn up a creature somewhere between an ape and a man! After three years of hunting, he found the top of a skull which was larger than an ape's but smaller than any recognized as human. The next year he found a similarly intermediate thighbone. Dubois named his "Java man" *Pithecanthropus erectus* ("erect apeman"). Half a century later, in the 1930's, another Dutchman, Gustav H. R. von Koenigswald, discovered more bones of *Pithecanthropus*, and they composed a clear picture of a small-brained, very beetling-browed creature with a distant resemblance to Neanderthal.

Meanwhile other diggers had found, in a cave near Peking, skulls, jaws, and teeth of a primitive man they called "Peking man." Once this discovery was made, it came to be realized that such teeth had been located earlier—in a Peking drugstore, where they were kept for medicinal purposes. The first intact skull was located in December 1929, and Peking man was eventually recognized as markedly similar to "Java man." It lived perhaps half a million years ago, used fire, and had tools of bone and stone. Eventually, fragments from forty-five individuals were accumulated, but they disappeared in 1941 during an attempted evacuation of the fossils in the face of the advancing Japanese.

Peking man was named "Sinanthropus pekinensis" ("China man of Peking"), but closer examination of more and more of these comparatively small-brained "hominids" ("manlike" creatures) made it seem that it was poor practice to place Peking man and Java man in separate genera. The German-American biologist Ernst Walter Mayr felt it wrong to place them in a separate genus from modern man, so that Peking man and Java man are now considered two varieties of the species *Homo erectus*.

It is unlikely that mankind originated in Java, despite the existence

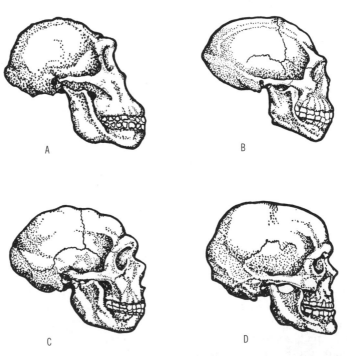

Reconstructed skulls of (A) Zinjanthropus, (B) Pithecanthropus, (C) Neanderthal, and (D) Cro-Magnon.

there of a small-brained hominid. For a while the vast continent of Asia, early inhabited by "Peking man," was suspected of being the birthplace of man, but as the twentieth century progressed, attention focused more and more firmly on Africa, which, after all, is the continent richest in primate life generally, and of the higher primates particularly.

The first significant African finds were made by two English scientists, Raymond Dart and Robert Broom. One spring day in 1924, workers blasting in a limestone quarry near Taungs in South Africa picked up a small skull that looked nearly human. They sent it to Dart, an anatomist working in Johannesburg. Dart immediately identified it as a being between an ape and a man, and he called it *Australopithecus africanus* ("southern ape of Africa"). When his paper announcing the find was published in London, anthropologists thought he had blundered, mistaking a chimpanzee for an apeman. But Broom, an ardent fossil-hunter who had long been convinced that man originated in Africa, rushed to Johannesburg and proclaimed *Australopithecus* the closest thing to a missing link that had yet been discovered.

Through the following decades Dart, Broom, and several anthropologists searched for and found many more bones and teeth of the South African apeman, as well as clubs that he used to kill game, the bones of animals that he killed, and caves in which he lived. *Australopithecus* was a short, small-brained creature with a snoutlike face, in many ways less human than Java man. But he had more human brows and more human teeth than *Pithecanthropus*. He walked erect, used tools, and probably had a primitive form of speech. In short, he was an African variety of hominid living at least half a million years ago and definitely more primitive than *Homo erectus*.

There were no clear grounds for suspecting priority between the African and Asian varieties of *Homo erectus* at first, but the balance swung definitely and massively toward Africa with the work of the Kenya-born Englishman Louis Seymour Bazett Leakey and his wife Mary. With patience and persistence, the Leakeys combed likely areas in eastern Africa for early fossil hominids. The most promising was Olduvai Gorge, in what is now Tanzania, and there, on July 17, 1959, Mary Leakey crowned a more than quarter-century search by discovering fragments of a skull that, when pieced together, proved to encase the smallest brain of any hominid yet discovered. Other features showed this hominid, however, to be closer to man than to ape, for he walked upright and the remains were surrounded by small tools formed out of pebbles. The Leakeys named their find *Zinjanthropus* ("East African man," using the Arabic word for East Africa).

Zinjanthropus does not seem to be in the direct line of ancestry of modern man. Still older fossils, some 2 million years old, may qualify. These, given the name of *Homo habilis* ("nimble man"), were 4½-foot-tall creatures who already had hands with opposable thumbs which were nimble enough (hence the name) to make them utterly manlike in his respect.

Prior to *Homo habilis*, we approach fossils that are too primitive to be called hominids and begin to come nearer to the common ancestor of man and the other apes. There is "Ramapithecus," of which an upper jaw was located in northern India in the early 1930's by G. Edward Lewis. The upper jaw was distinctly closer to the human than is that of any living primate other than man himself; it was perhaps 3 million years old. In 1962, Leakey discovered an allied species which isotope studies showed to be 14 million years old.

In 1948, Leakey had discovered a still older fossil (perhaps 25 million years old), which was named "Proconsul." (This name, meaning "before Consul" honored Consul, a chimpanzee in the London Zoo.) Proconsul seems to be the common ancestor of the larger great apes, the gorilla, chimpanzee, and orangutan. Farther back, then, there must be a common ancestor of Proconsul and *Ramapithecus* (and of the primitive ape that was ancestral to the smallest modern ape, the gibbons). Such a creature, the first of all the apelike creatures, would date back perhaps 40 million years.

For many years anthropologists were greatly puzzled by a fossil that did look like a missing link, but of a curious and incredible kind. In 1911, near a place called Piltdown Common in Sussex, England, workmen building a road found an ancient, broken skull in a gravel bed. The skull came to the attention of a lawyer named Charles Dawson, and he took it to a paleontologist, Arthur Smith Woodward, at the British Museum. The skull was high-browed, with only slight brow ridges; it looked more modern than Neanderthal. Dawson and Woodward went searching in the gravel pit for other parts of the skeleton. One day Dawson, in Woodward's presence, came across a jawbone in about the place where the skull fragments had been found. It had the same reddish-brown hue as the other fragments, and therefore appeared to have come from the same head. But the jawbone, in contrast to the human upper skull, was like that of an ape! Equally strange, the teeth in the jaw, though apelike, were ground down as human teeth are by chewing.

Woodward decided that this half-ape, half-man might be an early creature with a well-developed brain and a backward jaw. He presented the find to the world as the "Piltdown man," or *Eoanthropus dawsoni* ("Dawson's dawn man").

Piltdown man became more and more of an anomaly as anthropologists found that in all other fossil finds that included the jaw, jawbone development did keep pace with skull development. Finally, in the early 1950's three British scientists, Kenneth Oakley, Wilfrid Le Gros Clark, and Joseph Sidney Weiner, decided to investigate the possibility of fraud. It *was* a fraud. The jawbone, that of a modern ape, had been planted.

778

Tracks of dinosaur. The living example of youthful *Homo sapiens* was planted to indicate the size of the tracks.

Tyrannosaurus rex, reconstructed from fossilized bones and displayed in the Cretaceous Dinosaur Hall of the American Museum of Natural History in New York City. This big carnivore preyed on other dinosaurs with vegetarian diets.

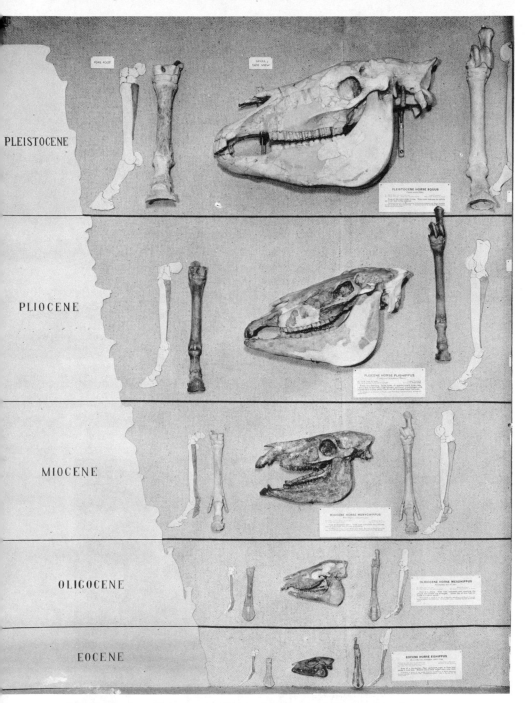

The evolution of the horse, illustrated by the skull and foot bones.

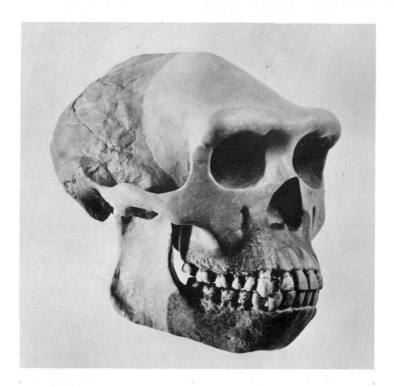

Skull of Pithecanthropus, as reconstructed by Franz
Weidenreich.

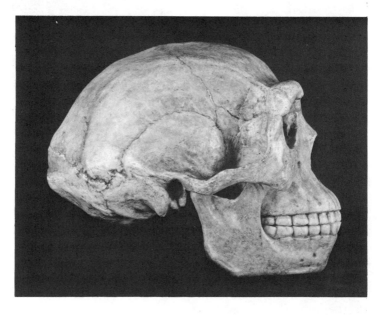

Skull of Sinanthropus.

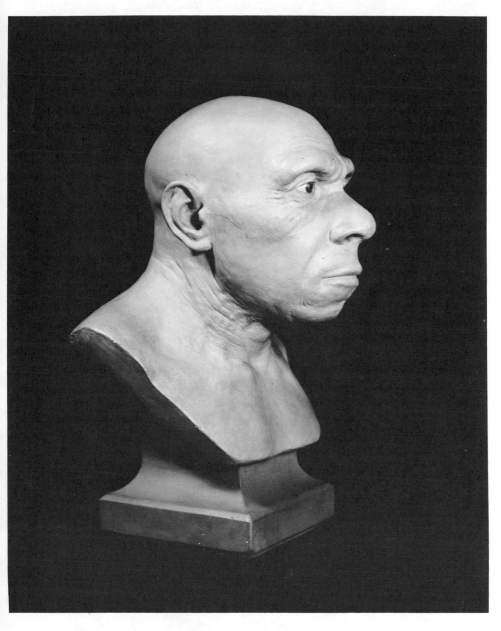

Neanderthal man, according to a restoration by J. H. McGregor.

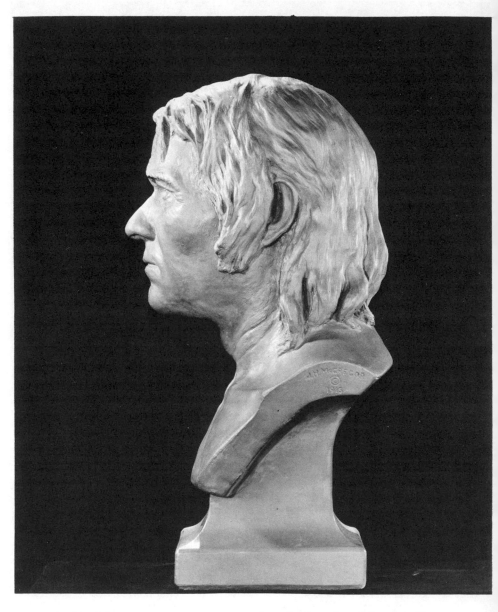

Cro-Magnon man, as reconstructed at the American Museum of
Natural History.

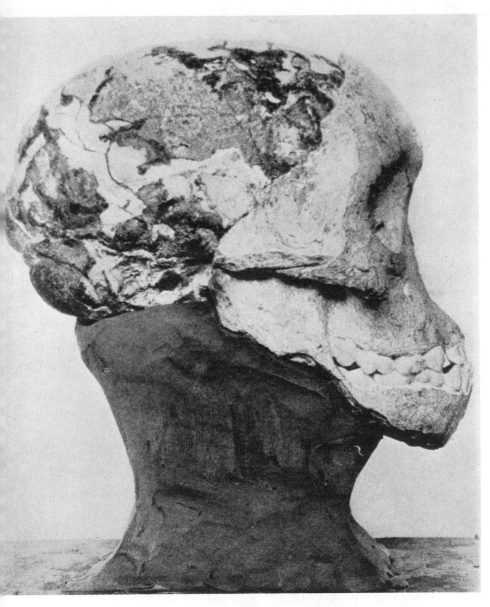

Australopithecus, the South African ape-man, reconstructed by Raymond Dart.

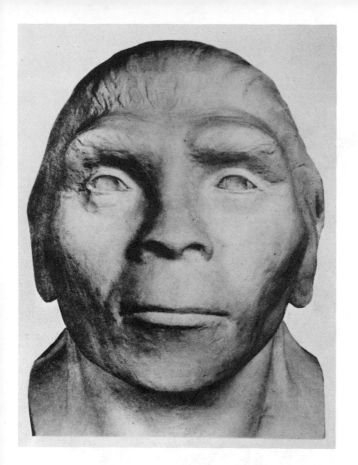

Sinanthropus woman, as reconstructed by Franz Weidenreich and Lucile Swan.
Below is a restored skull.

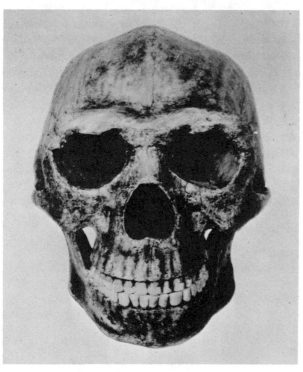

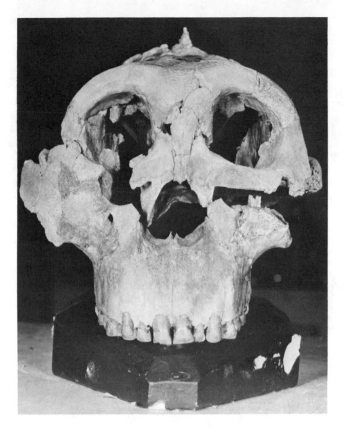

Skull of Australopithecus (*Zinjanthropus*) *BOESEI*, found by Mrs. M. Leakey in a gorge in Tanganyika.

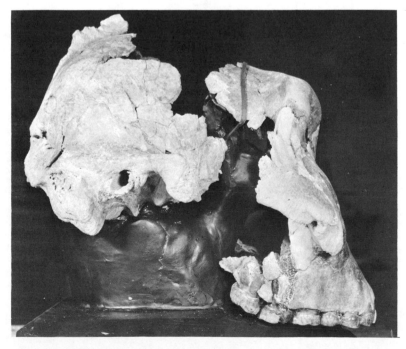

Zinjanthropus—side view of the partly reassembled skull.

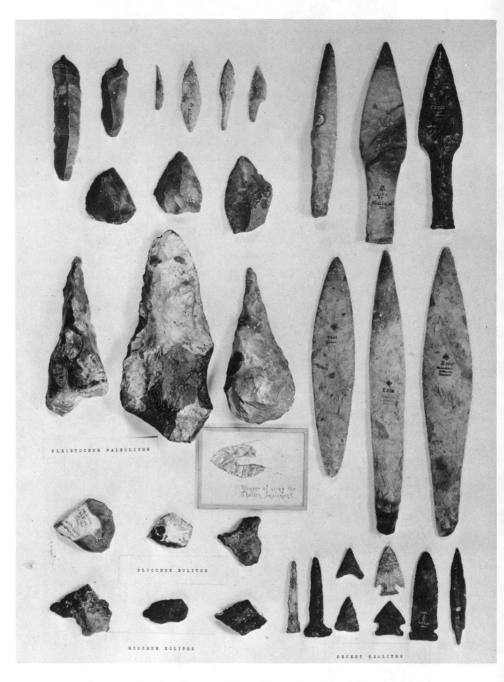

PLEISTOCENE PALEOLITHS

Manner of using the
Chellean Implement

PLIOCENE EOLITHS

MIOCENE EOLITHS

RECENT NEOLITHS

Stone tools of early man. The oldest, from the Miocene period,
are at the lower left; the most recent, at the lower right.

Another odd story of primate relics had a happier ending. In 1935, von Koenigswald had come across a huge but manlike fossil-tooth for sale in a Hong Kong pharmacy. The Chinese pharmacist considered it a "dragon tooth" of valuable medicinal properties. Von Koenigswald ransacked other Chinese pharmacies and had four such molars before World War II temporarily ended his activities.

The manlike nature of the teeth made it seem that gigantic human beings, possibly nine feet high, once roamed the earth. There was a tendency to accept this, perhaps, because the Bible says (Genesis 6:4), "There were giants in the earth in those days."

Between 1956 and 1968, however, four jawbones were discovered into which such teeth would fit. The creature "Gigantopithecus" was seen to be the largest primate ever known to exist, but was distinctly an ape and not a hominid, for all its human-appearing teeth. Very likely it was a gorilla-like creature, standing nine feet tall when upright and weighing six hundred pounds. It may have existed contemporaneously with *Homo erectus* and possessed the same feeding habits (hence the similarity in teeth). It has, of course, been extinct for at least a million years and could not possibly have been responsible for that Biblical verse.

It is important to emphasize that the net result of human evolution has been the production today of a single species. That is, while there may have been a number of species of hominids, one only has survived. All men today, regardless of differences in appearances, are *Homo sapiens,* and the difference between blacks and whites is approximately that between horses of different coloring.

Still, ever since the dawn of civilization, men have been more or less acutely conscious of racial differences, and usually they have viewed other races with the emotions generally evoked by strangers, ranging from curiosity to contempt to hatred. But seldom has racism had such tragic and long-persisting results as the modern conflict between white men and Negroes. (White men are often referred to as "Caucasians," a term first used, in 1775, by the German anthropologist Johann Friedrich Blumenbach, who was under the mistaken impression that the Caucasus contained the most perfect representatives of the group. Blumenbach also classified Negroes as "Ethiopians" and east Asians as "Mongolians," terms that are still sometimes used.)

The racist conflict between white and Negro, between Caucasian and Ethiopian, so to speak, entered its worst phase in the fifteenth century, when Portuguese expeditions down the west coast of Africa began a profitable business of carrying off Negroes into slavery. As the trade grew and nations built their economies on slave labor, rationalizations to justify the Negroes' enslavement were invoked in the name of the Scriptures, of social morality, and even of science.

According to the slave-holders' interpretation of the Bible—an interpretation believed by many to this day—Negroes were descendants of Ham and, as such, an inferior tribe subject to Noah's curse . . . "a servant of servants shall he be unto his brethren" (Genesis 9:25). Actually, the curse was laid upon Ham's son, Canaan, and on his descendants the "Canaanites," who were reduced to servitude by the Israelites when the latter conquered the land of Canaan. No doubt the words in Genesis 9:25 represent a comment after the fact, written by the Hebrew writers of the Bible to justify the enslavement of the Canaanites. In any case, the point of the matter is that the reference is to the Canaanites only, and the Canaanites were certainly white men. It was a twisted interpretation of the Bible that the slave-holders used, with telling effect in centuries past, to defend their subjugation of the Negro.

The "scientific" racists of more recent times took their stand on even shakier ground. They argued that the Negro was inferior to the white man because he obviously represented a lower stage of evolution. Were not his dark skin and wide nose, for instance, reminiscent of the ape? Unfortunately for their case, this line of reasoning actually leads to the opposite conclusion. The Negro is the least hairy of all the groups of mankind; in this respect and in the fact that his hair is crisp and woolly, rather than long and straight, the Negro is farther from the ape than the white man is! The same can be said of the Negro's thick lips; they resemble those of an ape less than do the white man's thin lips.

The fact of the matter is that to attempt to rank the various groups of *Homo sapiens* on the evolutionary ladder is to try to do fine work with blunt tools. Humanity consists of but one species, and so far the variations that have developed in response to natural selection are quite superficial.

The dark skin of dwellers in the earth's tropical and subtropical regions has obvious value in preventing sunburn. The fair skin of northern Europeans is useful to absorb as much ultraviolet radiation as possible from the comparatively feeble sunlight in order that enough vitamin D be formed from the sterols in the skin. The narrowed eyes of the Eskimo and the Mongol have survival value in lands where the glare from snow or desert sands is intense. The high-bridged nose and narrow nasal passages of the European serve to warm the cold air of the northern winter. And so on.

Since the tendency of *Homo sapiens* has been to make our planet one world, no basic differences in the human constitution have developed in the past, and they are even less likely to develop in the future. Interbreeding is steadily evening out man's inheritance. The American Negro is one of the best cases in point. Despite social barriers against intermarriage, nearly four fifths of the Negroes in the United States, it is estimated, have some white ancestry. By the end of the twentieth century probably there will be no "pure-blooded" Negroes in North America.

Anthropologists nevertheless are keenly interested in race, primarily as a guide to the migrations of early man. It is not easy to identify specific races. Skin color, for instance, is a poor guide; the Australian aborigine and the African Negro are both dark in color but are no more closely related to each other than either is to the European. Nor is the shape of the head—"dolichocephalic" (long) versus "brachycephalic" (wide), terms introduced in 1840 by the Swedish anatomist Anders Adolf Retzius—much better despite the classifications of Europeans into subgroups on this basis. The ratio of head length to head width multiplied by a hundred ("cephalic index," or, if skull measurements were substituted, "cranial index") served to divide Europeans into "Nordics," "Alpines," and "Mediterraneans." The differences, however, from one group to another are small, and the spread within a group is wide. In addition, the shape of the skull is affected by environmental factors such as vitamin deficiencies, the type of cradle in which the infant slept, and so on.

But the anthropologists have found an excellent marker for race in blood groups. The Boston University biochemist William Clouser Boyd was prominent in this connection. He pointed out that blood groups are inherited in a simple and known fashion, are unaltered by the environment, and show up in distinctly different distributions in the various races.

The American Indian is a particularly good example. Some tribes are almost entirely O; others are O but with a heavy admixture of A; virtually no Indians have B or AB blood. An American Indian testing as a B or AB is almost certain to possess some European ancestry. The Australian aborigines are likewise high in O and A, with B virtually nonexistent. But they are distinguished from the American Indian in being high in the more recently discovered blood group M and low in blood group N, while the American Indian is high in N and low in M.

In Europe and Asia, where the population is more mixed, the differences between peoples are smaller, yet still distinct. For instance, in London 70 per cent of the population has O blood, 26 per cent A, and 5 per cent B. In the city of Kharkov, Russia, on the other hand, the corresponding distribution is 60, 25, and 15. In general, the percentage of B increases as one travels eastward in Europe, reaching a peak of 40 per cent in central Asia.

Now the blood-type genes show the not-yet-entirely-erased marks of past migrations. The infiltration of the B gene into Europe may be a dim mark of the invasion by the Huns in the fifth century and by the Mongols in the thirteenth. Similar blood studies in the Far

East seem to indicate a comparatively recent infiltration of the A gene into Japan from the southwest and of the B gene into Australia from the north.

A particularly interesting, and unexpected, echo of early human migrations in Europe showed up in Spain. It came out in a study of Rh blood distribution. (The Rh blood groups are so named from the reaction of the blood to antisera developed against the red cells of a rhesus monkey. There are at least eight alleles of the responsible gene; seven are called "Rh positive," and the eighth, recessive to all the others, is called "Rh negative" because it shows its effect only when a person has received the allele from both parents.) In the United States about 85 per cent of the population is Rh positive, 15 per cent Rh negative. The same proportion holds in most of the European peoples. But curiously, the Basques of northern Spain stand apart, with something like 60 per cent Rh negative to 40 per cent Rh positive. And the Basques are also notable in having a language unrelated to any other European language.

The conclusion that can be drawn from this is that the Basques are a remnant of a prehistoric invasion of Europe by an Rh-negative people. Presumably a later wave of invasions by Rh-positive tribes penned them up in their mountainous refuge in the western corner of the continent, where they remain the only sizable group of survivors of the "early Europeans." The small residue of Rh-negative genes in the rest of Europe and in the American descendants of the European colonizers may represent a legacy from those early Europeans.

The peoples of Asia, the African Negroes, the American Indians, and the Australian aborigines are almost entirely Rh positive.

Man's Future

Attempting to foretell the future of the human race is a risky proposition that had better be left to mystics and science-fiction writers (though, to be sure, I am a science-fiction writer myself, among other things). But of one thing we can be fairly sure. Provided there are no worldwide catastrophes, such as a full-scale nuclear war, or a massive attack from outer space, or a pandemic of a deadly new disease, the human population will increase rapidly. It is now nearly three times as large as it was only a century and a half ago. Some estimates are that the total number of human beings who have lived over a period of 600,000 years comes to 77 billion. If so, then 4 per cent of all the human beings who have ever lived are living at this moment. And the world population is still growing at a tremendous rate—indeed, at a faster rate than ever before.

Since we have no censuses of ancient populations, we must estimate them roughly on the basis of what we know about the conditions of human life. Ecologists have estimated that the preagricultural food supply—obtainable by hunting, fishing, collecting wild fruit and nuts, and so on—could not have supported a world population of more than 20 million, and in all likelihood the actual population during the Paleolithic era was only a third or half of this at most. This means that as late as 6000 B.C. it could not have numbered more than 6 to 10 million people—roughly the population of a single present city such as Tokyo or New York. (When America was discovered, the food-gathering Indians occupying what is now the United States probably numbered not much more than 250,000, which is like imagining the population of Dayton, Ohio, spread out across the continent.)

The first big jump in world population came with the Neolithic Revolution and agriculture. The British biologist Julian Sorrell Huxley (grandson of the Huxley who was "Darwin's bulldog") estimates that the population began to increase at a rate which doubled its numbers every 1,700 years or so. By the opening of the Bronze Age, the world population may have been about 25 million; by the beginning of the Iron Age, 70 million; by the start of the Christian era, 150 million, with one third crowded into the Roman Empire, another third into the Chinese Empire, and the rest scattered. By 1600, the earth's population totaled perhaps 500 million, considerably less than the present population of China alone.

At that point the smooth rate of growth ended and the population began to explode. World explorers opened up some 18 million square miles of almost empty land on new continents to colonization by the Europeans. The eighteenth-century Industrial Revolution accelerated the production of food and of people. Even backward China and India shared in the population explosion. The doubling of the world's population now took place not in a period of nearly two millennia but in less than two centuries. The population expanded from 500 million in 1600 to 900 million in 1800. Since then it has grown at an ever faster rate. By 1900, it had reached 1.6 billion. In the first 70 years of the twentieth century, it has climbed to 3.6 billion, despite two world wars.

Currently, the world population is increasing at the rate of 220,000 each day, or 70 million each year. This is an increase at the rate of 2.0 per cent each year (as compared with an estimated increase of only 0.3 per cent per year in 1650). At this rate, the population of the earth will double in about 35 years and in some regions, such as Latin America, the doubling will take place in a shorter time. There is every reason to fear that by the year 2000 the earth's population will be over 6 billion.

At the moment students of the population explosion are leaning strongly toward the Malthusian view, which has been unpopular ever since it was advanced in 1798. As I said earlier, Thomas Robert Malthus maintained in *An Essay on the Principle of Population* that population always tends to grow faster than the food supply, with the inevitable result of periodic famines and wars. Despite his predictions, the world population has grown apace without any serious setbacks in the past century and a half. But for this postponement of catastrophe we can be grateful, in large measure, that large areas of the earth were still open for the expansion of food production. Now we are running out of tillable new lands. A majority of the world's population is underfed, and mankind must make mighty efforts to wipe out this chronic undernourishment. To be sure, the sea can be more rationally exploited and its food yield multiplied. The use of chemical fertilizers must yet be introduced to wide areas. Proper use of pesticides will reduce the loss of food to insect depredation in areas where such loss has not yet been countered. There are also ways of encouraging growth directly. Plant hormones such as "gibberellin" (studied by Japanese biochemists before World War II and coming to Western attention in the 1950's) could accelerate plant growth, while small quantities of antibiotics added to animal feed will accelerate animal growth (perhaps by suppressing intestinal bacteria that otherwise compete for the food supply passing through the intestines and by suppressing mild but debilitating infections). Nevertheless, with new mouths to feed multiplying as fast as they are, it will take Herculean efforts merely to keep the world's population up to the present none-too-good mark in which some 300 million children under five, the world over, are undernourished to the point of suffering permanent brain-damage.

Even so common (and, till recently, disregarded) a resource as fresh water is beginning to feel the pinch. Fresh water is now being used at the rate of nearly 2 trillion gallons a day the world over; although total rainfall, which at the moment is the main source of fresh water, is fifty times this quantity, only a fraction of the rainfall is easily recoverable. And in the United States where fresh water is used at a total rate of 350 billion gallons a day at a larger per capita rate than in the world generally, some 10 per cent of the total rainfall is being consumed one way or another.

The result is that the world's lakes and rivers are being quarreled over more intensely than ever. (The quarrels of Syria and Israel over the Jordan, and of Arizona and California over the Colorado River, are cases in point.) Wells are being dug ever deeper, and in many parts of the world the ground-water level is sinking dangerously. Attempts to conserve fresh water have included the use of cetyl alcohol as a cover for lakes and reservoirs in such regions as Australia, Israel, and East Africa. Cetyl alcohol spreads out into a film one molecule thick, cutting down on water evaporation without polluting the water. (Of course, water pollution by sewage and by industrial wastes is an added strain on the diminishing fresh-water surplus.)

Eventually, it seems, it will be necessary to obtain fresh water from the oceans, which, for the foreseeable future, offer an unlimited supply. The most promising methods of desalting sea water include distillation and freezing. In addition, experiments are proceeding with membranes that will selectively permit water molecules to pass, but not the various ions. Such is the importance of this problem that the Soviet Union and the United States are discussing a joint attack on it, at a time when cooperation between these two competing nations is, in other respects, exceedingly difficult to arrange.

But let us be as optimistic as we can and admit no reasonable limits to human ingenuity. Let us suppose that by miracles of technology we raise the productivity of the earth tenfold; suppose that we mine the metals of the ocean, bring up gushers of oil in the Sahara, find coal in Antarctica, harness the energy of sunlight, develop fusion power. Then what? If the rate of increase of the human population continues unchecked at its present rate, all our science and technical invention will still leave us struggling uphill like Sisyphus.

If you are not certain whether to accept this pessimistic appraisal, let us consider the powers of a geometric progression. It has been estimated that the total quantity of living matter on earth is now equal to 2×10^{19} grams. If so, the total mass of humanity is about $1/100,000$ of the mass of all life.

If the earth's population continues to double every 35 years (as it is now doing) then by 2570 A.D. it will have increased 100,000-fold. It may prove extremely difficult to increase the mass of life as a whole which the earth can support (though one species can always multiply at the expense of others). In that case, by 2570 A.D. the mass of humanity would comprise all of life and we would be reduced to cannibalism if some were to survive.

Even if we could imagine artificial production of foodstuffs out of the inorganic world via yeast culture, hydroponics (the growth of plants in solutions of chemicals), and so on, no conceivable advance could match the inexorable number increase involved in this doubling-every-thirty-five years. At that rate, by 2600 A.D., it would reach 630,000 billion! Our planet would have standing room only, for there would be only two and a half square feet per person on the entire land surface, including Greenland and Antarctica. In fact, if the human species could be imagined as continuing to multiply further at the same rate, by 3550 A.D. the total mass of human tissue would be equal to the mass of the earth.

If there are those who see a way out in emigration to other planets, they may find food for thought in the fact that, assuming there were 1,000 billion other inhabitable planets in the universe and people could be transported to any of them at will, at the present rate of increase of human numbers every one of those planets would be crowded literally to standing room only by 5000 A.D. By 7000 A.D. the mass of humanity would be equal to the mass of the known universe!

Obviously, the human race cannot increase at the present rate for very long, regardless of what is done with respect to the supply of food, water, minerals, and energy. I do not say "will not" or "dare not" or "should not"; I say quite flatly "cannot."

Indeed, it is not mere numbers that will set limit to our growth if it continues at our present rate. It is not only that there are more men, women, and children each minute, but that each individual uses (on the average) more of earth's unrenewable resources, expends more energy, and produces more waste and pollution each minute. Where population doubles every thirty-five years at the moment, energy-utilization is increasing at such a rate that in thirty-five years it will have increased not twice but sevenfold.

The blind urge to waste and poison faster and faster each year is driving us to destruction even more rapidly, then, than mere multiplication alone. For instance, smoke from burning coal and oil is freely dumped into the air by home and factory, as is the gaseous chemical refuse from industrial plants. Automobiles by the hundreds of millions discharge fumes of gasoline and of its breakdown and oxidation products, to say nothing of carbon-monoxide and lead compounds. Oxides of sulfur and nitrogen (produced either directly or through later oxidation by ultra-violet light from the sun), together with other substances, can corrode metals, weather construction materials, embrittle rubber, damage crops, cause and exacerbate respiratory diseases, and even serve as one of the causes of lung cancer.

When atmospheric conditions are such that the air over a city remains stagnant for a period of time, the pollutants collect, seriously contaminating the air and encouraging the formation of a smoky fog ("smog") that was first publicized in Los Angeles but had long existed in many cities and now exists in more. At its worst, it can take thousands of lives among those who, out of age or illness, cannot tolerate the added stress placed on their lungs. Such disasters took place in Donora, Pennsylvania, in 1948, and in London in 1952.

The fresh waters of the earth are polluted by chemical wastes, and occasionally one of them will come to dramatic notice. Thus, in 1970, it was found that mercury compounds heedlessly dumped into the world's waters were finding their way into sea organisms in sometimes dangerous quantities. At this rate, far from finding the ocean a richer source of food, we may make a good beginning at poisoning it altogether.

Indiscriminate use of long-lingering pesticides results in their incorporation first into plants, then into animals. Because of the poisoning, some birds find it increasingly difficult to form normal eggshells, so that in attacking insects we are bringing perilously close to extinction the peregrine falcon.

Almost every new so-called technological advance, hastened into without due caution by the eagerness to overreach one's competitors and multiply one's profits, can bring about difficulties. Since World War II, synthetic detergents have replaced soaps. Important ingredients of those detergents are various phosphates, which washed into the water supply and greatly accelerated the growth of microorganisms that, however, used up the oxygen supply of the waters—thus leading to the death of other sea-organisms. These deleterious changes in water habitats ("eutrophication") are rapidly aging the Great Lakes, for instance—the shallow Lake Erie in particular—and are shortening their natural lives by millions of years. Thus Lake Erie may become Swamp Erie, while the swampy Everglades may dry up altogether.

Living species are utterly interdependent. There are obvious cases like the interconnection of plants and bees, where the plants are pollinated by the bees and the bees are fed by the plants, and a million other cases less obvious. Every time life is made easier or more difficult for one particular species, dozens of others are affected—sometimes in hard-to-predict ways. The study of this interconnectability of life, ecology, is only now attracting attention, for in many cases mankind, in an effort to achieve some short-term benefit for himself has so altered the ecological structure as to bring about some long-term difficulty. Clearly man must learn to look far more carefully before he leaps.

Even so apparently other-worldly an affair as rocketry must be carefully considered. A single large rocket may inject over a hundred tons of exhaust gases into the atmosphere at levels above sixty miles. Such quantities of material could appreciably change the properties of the thin upper atmosphere and lead to hard-to-predict climatic changes. In 1971, there is the prospect of mass use of gigantic supersonic transport planes (SSTs) to travel through the stratosphere at higher-than-sound velocities. Those who object to their use cite not only the noise factor involved in sonic booms but also the chance of climate-affecting pollution.

Another factor that makes the increase in numbers even worse, is the uneven distribution of mankind over the face of the earth. Everywhere there is a trend toward accumulation within metropolitan areas. In the United States, even while the population goes up and up, certain farming states not only do not share in the explosion but are actually decreasing in population. It is estimated that the urban population of the earth is doubling not every thirty-five years but every eleven years. By 2005 A.D., when the earth's total population will have doubled, the metropolitan population will, at this rate, have increased over ninefold.

This is serious. We are already witnessing a breakdown in the social structure—a breakdown that is most strongly concentrated in just those advanced nations where urbanization is most apparent. Within those nations, it is most concentrated in the cities, especially in their most crowded portions. There is no question but that when living beings are crowded beyond a certain point, many forms of pathological behavior become manifest. This has been found to be true in laboratory experiments on rats, and the newspaper and our own experience should convince us that this is also true for human beings.

It would seem obvious, then, that *if present trends continue unchanged*, the world's social and technological structure will have broken down well within the next half-century, with incalculable consequences. Mankind, in sheer madness, may even resort to the ultimate catastrophe of thermonuclear warfare.

But *will* present trends continue?

Clearly, changing them will require a massive effort and will mean that man must change long-cherished beliefs. For most of man's history, he has lived in a world in which life was brief and many children died while still infants. If the tribal population were not to die out, women had to bear as many babies as they could. For this reason, motherhood was deified and every trend that might lower the birthrate was stamped out. The status of women was lowered so that they might be nothing but baby-making and baby-rearing machines. Sexual mores were so controlled that only those actions were approved of that led to conception; everything else was considered perverted and sinful.

But now we live in a crowded world. If we are to avoid catastrophe, motherhood must become a privilege sparingly doled out. Our views on sex and on its connection with childbirth must be changed.

Again, the problems of the world—the really serious problems—are global in nature. The dangers posed by overpopulation, overpollution, the disappearance of resources, the risk of nuclear war, affect every nation, and there can be no real solutions unless all nations cooperate. What this means is that a nation can no longer go its own way, heedless of the others; nations can no longer act on the assumption that there is such a thing as a "national security" whereby something good can happen to them if something bad happens to someone else. In short, an effective world government is necessary.

Mankind is moving in this direction (very much against its will in many cases) but the question is whether the movement will be quick enough.

I do not wish to make it appear as though there is no hope and as though mankind is in a blind alley from which there is no escape at all. While things look dark and difficult, there may yet be ways out.

One possible source of optimism is an impending revolution in communications. The proliferation of communications satellites may make it possible in the near future for every person to be within reach of every other person. Underdeveloped nations can leapfrog over the earlier communications networks' necessity of involving large capital investments and move directly into a world in which every man has his own television station, so to speak, for receipt and emission of messages.

The world will become so much smaller as to resemble in social structure a kind of neighborhood village. (Indeed, the phrase "global village" has come into use to describe the new situation.) Education can penetrate every corner of the global village with the ubiquity of television. The new generation of every underdeveloped nation may grow up learning about modern agricultural methods, about the proper use of fertilizers and pesticides, and about the techniques of birth control.

There may even be, for the first time in earth's history, a tendency toward decentralization. With ubiquitous television making all parts of the world equally accessible to business conferences and libraries and cultural programs, there will be less need to conglomerate everything into a large, decaying mass.

Who knows, then? Catastrophe seems to have the edge, but the race for salvation is perhaps not quite over.

Assuming that the race for salvation is won; that the population levels off and a slow and humane decrease begins to take place; that an effective and sensible world government is instituted, allowing local diversity but not local murder; that the ecological structure is cared for and the earth systematically preserved—what then?

For one thing, man will probably continue to extend his range. Beginning as a primitive hominid in east Africa—at first perhaps no more widespread or successful than the modern gorilla—he slowly moved outward until by 15,000 years ago he had colonized the entire "world island" (Asia, Africa, and Europe). He then made the leap into the Americas, Australia, and even through the Pacific islands. By the twentieth century, the population remained thin in particularly undesirable areas—such as the Sahara, the Arabian Desert, and Greenland—but no sizable area was utterly uninhabited by man except for Antarctica. Now scientific stations, at least, are permanently established even on that least habitable of continents.

Where next?

One possible answer is the sea. It was in the sea that life originated and where it still flourishes best in terms of sheer quantity. Every kind of land animal, except for the insects, has tried the experiment of returning to the sea for the sake of its relatively unfailing food supply and for the relative equability of the environment. Among mammals, such examples as the otter, the seal, or the whale, indicate progressive stages of re-adaptation to a watery environment.

Can man return to the sea, not by the excessively slow alteration of his body through evolutionary change, but by the rapid help of technological advance? Encased in the metal walls of submarines and bathyscaphes, he has penetrated the ocean to its very deepest floor.

For bare submergence, much less is required. In 1943, the French oceanographer Jacques-Ives Cousteau invented the aqualung. This device brings oxygen to a man's lungs from a cylinder of compressed air worn on his back and makes possible the modern sport of scuba diving ("scuba" is an acryonym for "self-contained underwater-breathing apparatus").

799

This makes it possible for a man to stay underwater for considerable periods in his skin, so to speak, without being encased in ships or even in enclosed suits.

Cousteau also pioneered in the construction of underwater living quarters in which men could remain submerged for even longer periods. In 1964, for instance, two men lived two days in an air-filled tent 432 feet below sea level. (One was Jon Lindbergh, son of the aviator.) At shallower depths, men have remained underwater for many weeks.

Even more dramatic is the fact that beginning in 1961, the biologist Johannes A. Kylstra, at the University of Leyden, began to experiment with actual water-breathing in mammals. The lung and the gill act similarly, after all, except that the gill is adapted to work on lower levels of oxygenation. Kylstra made use of a water solution sufficiently like mammalian blood to avoid damaging lung tissue and then oxygenated it heavily. He found that both mice and dogs could breathe such liquid for extended periods without apparent ill effect.

Hamsters have been kept alive under ordinary water when they were enclosed in a sheet of thin silicone rubber through which oxygen could pass from water to hamster and carbon dioxide from hamster to water. The membrane was virtually an artificial gill. With such advances and still others to be expected, can man look forward to a future in which he can remain under water for indefinite periods and make all the planet's surface—land and sea—his home?

And what of outer space? Need man remain on his home planet, or can he venture to other worlds?

Once the first satellites were launched into orbit in 1957, the thought naturally arose that the dream of space travel, till then celebrated only in science-fiction stories, might become an actuality. It took only three and a half years after the launching of Sputnik I for the first step to be taken.

On April 12, 1961, the Soviet cosmonaut Yuri Alexeyevich Gagarin was launched into orbit and returned safely. Three months later, on August 6, another Soviet cosmonaut, Gherman Stepanovich Titov, flew 17 orbits before landing, spending 24 hours in free flight. On February 20, 1962, the United States put its first man in orbit when the astronaut John Herschel Glenn circled the earth 3 times. Since then dozens of men have left the earth and, in some cases, remained in space for weeks. Included is one woman—the Soviet cosmonaut Valentina V. Tereshkova—who was launched on June 16, 1963, and remained in free flight for 71 hours, making 17 orbits altogether.

Rockets have left the earth carrying two and three men at a time. The first such launching was that of the Soviet cosmonauts Vladimir M. Komarov, Konstantin P. Feokstistov, and Boris G. Yegorov, on October 12, 1964. The Americans launched Virgil I. Grissom and John W. Young in the first multi-manned U. S. rocket on March 23, 1965.

The first man to leave his rocket ship in space was the Soviet cosmonaut Aleksei A. Leonov, who did so on March 18, 1965. This "space walk" was duplicated by the American astronaut Edward H. White on June 3, 1965.

Although most of the space "firsts" through 1965 had been made by the Soviets, the Americans thereafter went into the lead. Manned vehicles maneuvered in space, rendezvoused with each other, docked, and began to move farther and farther out.

The space program, however, did not continue without tragedy. In January 1967, three American astronauts—Grissom, White, and Roger Chaffee—died on the ground in a fire that broke out in their space capsule during routine tests. Then, on April 23, 1967, Komarov died when his parachute fouled during re-entry. He was the first man to die in the course of a space flight.

The American plans to reach the moon by means of three-man vessels (the "Apollo" program) were delayed by the tragedy while the space capsules were redesigned for greater safety, but the plans were not abandoned. The first manned Apollo vehicle, Apollo 7, was launched on October 11, 1968, with its three-man crew under the command of Walter M. Schirra. Apollo 8, launched on December 21, 1968, under the command of Frank Borman, approached the moon, circling it at close quarters. Apollo 10, launched on May 18, 1969, also approached the moon, detached the lunar module, and sent it down to within nine miles of the lunar surface.

Finally, on July 16, 1969, Apollo 11 was launched under the command of Neil A. Armstrong. On July 20, Armstrong was the first human being to stand on the soil of another world.

Since then three other Apollo vehicles have been launched. Three of them, Apollo 12, 14, and 15, completed their missions with outstanding success. Apollo 13 had trouble in space and was forced to return without landing on the moon, but *did* return safely without loss of life.

The Soviet space program has not as yet included manned flights to the moon. However, on September 12, 1970, an unmanned vessel was fired to the moon. It soft-landed safely, gathered up specimens of soil and rock, then safely brought these back to earth. Still later, an automatic Soviet vehicle landed on the moon and moved about under remote control for months, sending back data.

And the future?

The space program has been expensive and has met with growing resistance from scientists who think that too much of it has been public-relations-minded and too little scientific, or who think it obscures other programs of greater scientific importance. It has also met with growing resistance from the general public, which considers it too expensive, particularly in the light of urgent sociological problems on earth.

Nevertheless, the space program will probably continue, if only at a reduced pace; and if mankind can figure out how to spend less of its energies and resources on the suicidal folly of war, the program may even accelerate. There are plans for the establishment of space stations—in effect, large vehicles in more or less permanent orbit about the earth and capable of housing sizable numbers of men and women for extended periods—so that observations and experiments can be conducted that will presumably be of great value. Shuttle vessels, perhaps reusable, will be devised.

It is to be hoped that further trips to the moon will eventually result in the establishment of more or less permanent colonies there that, we may further hope, can exploit lunar resources and become independent of earth's day-to-day help.

But can man penetrate beyond the moon?

In theory, there is no reason why he cannot, but flights to the next nearest world on which he can land, Mars (Venus, though closer, is too hot for a manned landing), will require flights not of days, as in the case of the moon, but of months. And for those months he will have to take a livable environment along with him.

Man has already had some experience along these lines in descending into the ocean depths in submarines and vessels such as the bathyscaphe. As on those voyages, he will go into space in a bubble of air enclosed in a strong metal shell, carrying a full supply of the food, water, and other necessities he will require for the journey. But the take off into space is complicated enormously by the problem of overcoming gravity. In the space ship, a large proportion of the weight and volume must be devoted to the engine and fuel, and the possible "payload" of crew and supplies will at first be small indeed.

The food supply will have to be extremely compact: there will be no room for any indigestible constituents. The condensed, artificial food might consist of lactose, a bland vegetable oil, an appropriate mixture of amino acids, vitamins, minerals, and a dash of flavoring, the whole enclosed in a tiny carton made of edible carbohydrate. A carton containing 180 grams of solid food would suffice for one meal. Three such cartons would supply 3,000 calories. To this a gram of water per calorie (2½ to 3 liters per day per person) would have to be added; some of it might be mixed in the food to make it more palatable, increasing the size of the carton. In addition, the ship would have to carry oxygen for breathing in the amount of about one liter (1,150 grams) of oxygen in liquid form per day per person.

Thus the daily requirement for each person would be 540 grams of dry food, 2,700 grams of water, and 1,150 grams of oxygen. Total, 4,390 grams, or roughly 9½ pounds. Imagine a trip to the moon, then, taking 1 week each way and allowing 2 days on the moon's surface for exploration. Each man on the ship would require about 150 pounds of food, water, and oxygen. This can probably be managed at present levels of technology.

For an expedition to Mars and back, the requirements are vastly greater. Such an expedition might well take 2½ years, allowing for a wait on Mars for a favorable phase of the planetary orbital positions to start the return trip. On the basis I have just described, such a trip would call for about 5 tons of food, water, and oxygen per man. To transport such a supply in a space ship is, under present technological conditions, unthinkable.

The only reasonable solution for a long trip is to make the space ship self-sufficient, in the same sense that the earth, itself a massive "ship" traveling through space, is self-sufficient. The food, water, and air taken along to start with would have to be endlessly reused by recycling the wastes.

Such "closed systems" have already been constructed in theory. The recycling of wastes sounds unpleasant, but this is, after all, the process that maintains life on the earth. Chemical filters on the ship could collect the carbon dioxide and water vapor exhaled by the crew members; urea, salt, and water could be recovered by distillation and other processes from the urine and feces; the dry fecal residue could be sterilized of bacteria by ultraviolet light; and along with the carbon dioxide and water could then be fed to algae growing in tanks. By photosynthesis the algae would convert the carbon dioxide and nitrogenous compounds of the feces to organic food, plus oxygen, for the crew. The only thing that would be required from outside the system is energy for the various processes, including photosynthesis, and this could be supplied by the sun.

It has been estimated that as little as 250 pounds of algae per man could take care of the crew's food and oxygen needs for an indefinite period. Adding the weight of the necessary processing equipment, the total weight of supplies per man would be perhaps 350 pounds, certainly no more than 1,000 pounds. Studies have also been made with systems in which hydrogen-using bacteria are employed. These do not require light, merely hydrogen which can be obtained through the electrolysis of water. The efficiency of such systems is much higher, according to the report, than that of photosynthesizing organisms.

Aside from supply problems, there is that of prolonged weightlessness. Men have survived weeks of continuous weightlessness without permanent harm, but there have been enough minor disturbances to make prolonged weightlessness a disturbing factor. Fortunately, there are ways to counteract it. A slow rotation of the space vehicle, for instance, could produce the sensation of weight by virtue of the centrifugal force, acting like the force of gravity.

More serious and less easily countered are the hazards of high acceleration and sudden deceleration, which space travelers will inevitably encounter in taking off and landing on rocket flights.

The normal force of gravity at the earth's surface is called 1 g. Weightlessness is 0 g. An acceleration (or deceleration) that doubles the body's weight is 2 g, a force tripling the weight is 3 g, and so on.

The body's position during acceleration makes a big difference. If you are accelerated head first (or decelerated feet first), the blood rushes away from your head. At a high enough acceleration (say 6 g for 5 seconds), this means "blackout." On the other hand, if you are accelerated feet first (called "negative acceleration," as opposed to the "positive" headfirst acceleration), the blood rushes to your head. This is more dangerous, because the heightened pressure may burst blood vessels in the eyes or brain. The investigators of acceleration call it "redout." An acceleration of 2½ g for 10 seconds is enough to damage some of the vessels.

By far the easiest to tolerate is "transverse" acceleration—i.e., with the force applied at right angles to the long axis of the body, as in a sitting position. Men have withstood transverse accelerations as high as 10 g for more than 2 minutes in a centrifuge without losing conciousness.

For shorter periods the tolerances are much higher. Astounding records in sustaining high g decelerations were made by Colonel John Paul Stapp and other volunteers on the sled track of the Holloman Air Force Base in New Mexico. On his famous ride of December 10, 1954, Stapp took a deceleration of 25 g for about a second. His sled was brought to a full stop from a speed of more than 600 miles per hour in just 1.4 seconds. This, it was estimated, amounted to driving an automobile into a brick wall at 120 miles per hour! Of course, Stapp was strapped in the sled in a manner to minimize injury. He suffered only bruises, blisters, and painful eye shocks that produced two black eyes.

An astronaut, on takeoff, must absorb (for a short while) as much as 6½ g and, at re-entry, up to 11 g.

Devices such as contour couches, harnesses, and perhaps even immersion in water in a water-filled capsule or space suit will give a sufficient margin of safety against high g forces.

Similar studies and experiments are being made on the radiation hazards, the boredom of long isolation, the strange experience of being in soundless space where night never falls, and other eerie conditions that space fliers will have to endure. All in all, those preparing for man's first venture away from his home planet see no insurmountable obstacles ahead.

CHAPTER 16

The Mind

The Nervous System

Physically speaking, man is a rather unimpressive specimen, as organisms go. He cannot compete in strength with most other animals his size. He walks awkwardly, compared with, say, the cat; he cannot run with the dog or the deer; in vision, hearing, and the sense of smell he is inferior to a number of other animals. His skeleton is ill-suited to his erect posture: man is probably the only animal that develops "low back pain" from his normal posture and activities. When we think of the evolutionary perfection of other organisms—the beautiful efficiency of the fish for swimming or of the bird for flying, the great fecundity and adaptability of the insects, the perfect simplicity and efficiency of the virus—man seems a clumsy and poorly designed creature indeed. As sheer organism, he could scarcely compete with the creatures occupying any specific environmental niche on earth. He has come to dominate the earth only by grace of one rather important specialization—his brain.

A cell is sensitive to a change in its surroundings ("stimulus") and will react appropriately ("response"). Thus, a protozoon will swim toward a drop of sugar solution deposited in the water near it, or away from a drop of acid. Now this direct, automatic sort of response is fine for a single cell, but it would mean chaos for a collection of cells. Any organism made up of a number of cells must have a system that coordinates their responses. Without such a system, it would be like a city of men completely out of communication with one another and acting at cross purposes. So even the coelenterates, the most primitive multicelled animals, have the beginnings of a nervous system. We can see in them the first nerve cells ("neurons")—special cells with fibers that extend from the main cell body and put out extremely delicate branches.

The functioning of nerve cells is so subtle and complex that even at this simple level we are already a little beyond our depth when it comes to explaining just what happens. In some way not yet understood, a change in the environment acts upon the nerve cell. It may be a change in the concentration of some substance, or in the temperature, or in the amount of light, or in the movement of the water, or it may be an actual touch by some object. Whatever the stimulus, it sets the impulse jumps a tiny gap ("synapse") to the next nerve cell; and so it is transmitted from cell to cell. (In well-developed nervous systems, a nerve cell may make thousands of synapses with its neighbors.) In the case of a coelenterate, such as a jellyfish, the impulse is communicated throughout the organism. The jellyfish responds by contracting some part or all of its body. If the stimulus is a contact with a food particle, the organism engulfs the particle by contraction of its tentacles.

All this is strictly automatic, of course, but since it helps the jellyfish, we like to read purpose into the organism's behavior. Indeed, man, as a creature who behaves in a purposeful, motivated way, naturally tends to attribute purpose even to inanimate nature. Scientists call this attitude "teleological," and they try to avoid such a way of thinking and speaking as much as they can. But in describing the results of evolution it is so convenient to speak in terms of development toward more efficient ends that even among scientists all but the most fanatical purists occasionally lapse into teleology. (Readers of this book have noticed, of course, that I have sinned often.) Let us, however, try to avoid teleology in considering the development of the nervous system and the brain. Nature did not design the brain; it came about as the result of a long series of evolutionary accidents, so to speak, which happened to produce helpful features that at each stage gave an advantage to organisms possessing them. In the fight for survival, an animal that was more sensitive to changes in the environment than its competitors, and could respond to them faster, would be favored by

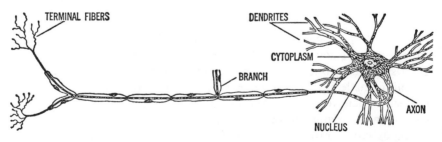

A nerve cell.

natural selection. If, for instance, an animal happened to possess some spot on its body that was exceptionally sensitive to light, the advantage would be so great that evolution of eye spots, and eventually of eyes, would follow almost inevitably.

Specialized groups of cells that amount to rudimentary "sense organs" begin to appear in the Platyhelminthes, or flatworms. Furthermore, the flatworms also show the beginnings of a nervous system that avoids sending nerve impulses indiscriminately throughout the body, but instead speeds them to the critical points of response. The development that accomplishes this is a central nerve cord. The flatworms are the first to develop a "central nervous system."

This is not all. The flatworm's sense organs are localized in its head end, the first part of its body that encounters the environment as it moves along, and so naturally the nerve cord is particularly well developed in the head region. That knob of development is the beginning of a brain.

Gradually the more complex phyla add new features. The sense organs increase in number and sensitivity. The nerve cord and its branches grow more elaborate, developing a widespread system of afferent ("carrying to") nerve cells that bring messages to the cord and efferent ("carrying away") fibers that transmit messages to the organs of response. The knot of nerve cells at the crossroads in the head becomes more and more complicated. Nerve fibers evolve into forms that can carry the impulses faster. In the squid, the most highly developed of the unsegmented animals, this faster transmission is accomplished by a thickening of the nerve fiber. In the segmented animals, the fiber develops a sheath of fatty material ("myelin") which is even more effective in speeding the nerve impulse. In man some nerve fibers can transmit the impulse at 100 meters per second (about 225 miles per hour), compared to only about 1/10 of a mile per hour in some of the invertebrates.

The chordates introduce a radical change in the location of the nerve cord. In them this main nerve trunk (better known as the spinal chord) runs along the back instead of along the belly, as in all lower animals. This may seem a step backward—putting the cord in a more exposed position. But the vertebrates have the cord well protected within the bony spinal column. The backbone, though its first function was protecting the nerve cord, produced amazing dividends, for it served as a girder upon which chordates could hang bulk and weight. From the backbone they can extend ribs that enclose the chest, jawbones that carry teeth for chewing, and long bones that form limbs.

The chordate brain develops from three structures which are already present in simple form in the most primitive vertebrates. These structures, at first mere swellings of nerve tissue, are the "forebrain," "midbrain," and "hindbrain," a division first noted by the Greek anatomist Erasistratus of Chios about 280 B.C. At the head end of the spinal cord, the cord widens smoothly into the hindbrain section known as the "medulla oblongata." On the front side of this section in all but the most primitive chordates is a bulge called the "cerebellum" ("little brain"). Forward of this is the midbrain. In the lower vertebrates the midbrain is concerned chiefly with vision and has a pair of "optic lobes," while the forebrain is concerned with smell and taste and contains "olfactory bulbs." The forebrain, reading from front to rear, is divided into the olfactory-bulb section, the "cerebrum," and the "thalamus," the lower portion of which is the "hypothalamus." (Cerebrum is Latin for "brain"; in man, at least, the cerebrum is the largest and most important part of the organ.) By removing the cerebrum from animals and observing the results, the French anatomist Marie Jean Pierre Flourens was able to demonstrate in 1824 that it was indeed the cerebrum that was responsible for acts of thought and will.

It is the roof of the cerebrum, moreover, the cap called the cerebral cortex, that is the star of the whole show. In fishes and amphibians this is merely a smooth covering (called the "pallium," or cloak). In reptiles a patch of new nerve tissue, called the "neopallium" ("new cloak") appears. This is the real forerunner of things to come. It will eventually take over the supervision of vision and other sensations. In the reptiles the clearing house for visual messages has already moved from the midbrain to the forebrain in part; in birds this move is completed. With the first mammals, the neopallium begins to take charge. It spreads virtually

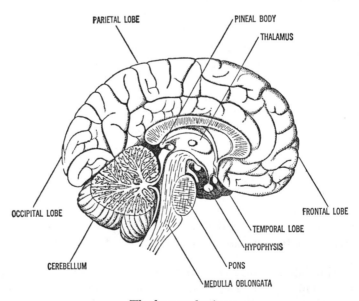

The human brain.

over the entire surface of the cerebrum. At first it remains a smooth coat, but as it goes on growing in the higher mammals, it becomes so much larger in area than the surface of the cerebrum that it is bent into folds, or "convolutions." This folding is responsible for the complexity and capacity of the brain of a higher mammal, notably that of man.

More and more, as one follows this line of species development, the cerebrum comes to dominate the brain. The midbrain fades to almost nothing. In the case of the primates, which gain in the sense of sight at the expense of the sense of smell, the olfactory lobes of the forebrain shrink to mere blobs. By this time the cerebrum has expanded over the thalamus and the cerebellum.

Even the early manlike fossils had considerably larger brains than the most advanced apes. Whereas the brain of the chimpanzee or of the orangutan weighs less than 400 grams (under 14 ounces), and the gorilla, though far larger than a man, has a brain that averages about 540 grams (19 ounces), *Pithecanthropus'* brain apparently weighed about 850 to 1,000 grams (30 to 35 ounces). And these were the "small-brained" hominids. Rhodesian man's brain weighed about 1,300 grams (46 ounces); the brain of Neanderthal and of modern *Homo sapiens* comes to about 1,500 grams (53 ounces or 3.3 pounds). Modern man's mental gain over Neanderthal apparently lies in the fact that a larger proportion of his brain is concentrated in the foreregions, which apparently control the higher aspects of mental function. Neanderthal was a low-brow whose brain bulged in the rear; present-day man, in contrast, is a high-brow whose brain bulges in front.

Modern man's brain is about 1/50 of his total body weight. Each gram of brain weight is in charge, so to speak, of 50 grams of body. In comparison, the chimpanzee's brain is about 1/150 the weight of its body, and the gorilla's about 1/500 of its body. To be sure, some of the smaller primates have an even higher brain/body ratio than man has. (So do the hummingbirds.) A monkey can have a brain that is 1/18 the weight of its body. However, there the mass of the brain is too small in absolute terms for it to be able to pack into itself the necessary complexity for intelligence on the human scale. In short, what is needed, and what man has, is a brain that is both large in the absolute sense, and large in relation to body size.

This is made plain by the fact that two types of mammal have brains that are distinctly larger than the human brain and yet that do not lend those mammals superintelligence. The largest elephants can have brains as massive as 6,000 grams (about thirteen pounds) and the largest whales can have brains that reach a mark of 9,000 grams (or nearly nineteen pounds). The size of the bodies those brains must deal with is, however, enormous. The elephant's brain, despite its size, is only 1/1,000 the weight of its body, and the brain of a large whale may be only 1/10,000 the weight of its body.

In only one direction, however, does man have a possible rival. The dolphins and porpoises, small members of the whale family, show possibilities. Some of these are no heavier than man and yet have brains that are larger (with weights up to 1,700 grams, or 60 ounces) and more extensively convoluted.

It is not safe to conclude from this evidence alone that the dolphin is more intelligent than man, because there is the question of the internal organization of the brain. The dolphin's brain (like that of Neanderthal man) may be oriented more in the direction of what we might consider "lower functions."

The only safe way to tell is to attempt to gauge the intelligence of the dolphin by actual experiment. Some investigators, notably John C. Lilly, seem convinced that dolphin intelligence is indeed comparable to our own, that dolphins and porpoises have a speech pattern as complicated as ours, and that possibly a form of interspecies communication may yet be established.

Even if this is so, there can be no question but that dolphins however intelligent, lost their opportunity to translate that intelligence into control of the environment when they readapted to sea life. It is impossible to make use of fire under water, and it was the discovery of the use of fire that first marked off mankind from all other organisms. More fundamentally, still, rapid locomotion through a medium as viscous as water requires a thoroughly streamlined shape. This has made impossible in the dolphin the development of anything equivalent to the human arm and hand with which the environment could be delicately investigated and manipulated.

In effective intelligence, at least, *Homo sapiens* stands without a peer on earth at present or, so far as we know, in the past.

While considering the difficulty in determining the precise intelligence level of a species such as the dolphin, it might be well to say that no completely satisfactory method exists for measuring the precise intelligence level of individual members of our own species.

In 1904, the French psychologists Alfred Binet and Théodore Simon devised means of testing intelligence by answers given to judiciously chosen questions. Such "intelligence tests" give rise to the expression "intelligence quotient" (or "IQ"), representing the ratio of the mental age, as measured by the test, to the chronological age; this ratio being multiplied by one hundred to remove decimals. The public was made aware of the significance of IQ chiefly through the work of the American psychologist Lewis Madison Terman.

The trouble is that no test has been devised that is not culturally centered. Simple questions about ploughs might stump an intelligent city boy, and simple questions about escalators might stump an equally intelligent farmboy. Both would puzzle an equally intelligent Australian aborigine, who might nevertheless dispose of questions about boomerangs that would leave us gasping.

Another familiar test is aimed at an aspect of the mind even more subtle and elusive than intelligence. This consists of ink-blot patterns first prepared by a Swiss doctor, Hermann Rorschach, between 1911 and 1921. Subjects are asked to convert these ink blots into images; from the type of image a person builds into such a "Rorschach test," conclusions concerning his personality are drawn. Even at best, however, such conclusions are not likely to be truly conclusive.

Oddly enough, many of the ancient philosophers almost completely missed the significance of the organ under man's skull. Aristotle considered the brain merely an air-conditioning device, so to speak, designed to cool the overheated blood. In the generation after Aristotle, Herophilus of Chacedon, working at Alexandria, correctly recognized the brain as the seat of intelligence, but, as usual, Aristotle's errors carried more weight than did the correctness of others.

The ancient and medieval thinkers therefore often tended to place the seat of emotions and personality in organs such as the heart, the liver, and the spleen (*vide* the expressions "broken-hearted," "lily-livered," "vents his spleen").

The first modern investigator of the brain was a seventeenth-century English physician and anatomist named Thomas Willis; he traced the nerves that led to the brain. Later, a French anatomist named Felix Vicq d'Azyr and others roughed out the anatomy of the brain itself. But it was the eighteenth-century Swiss physiologist Albrecht von Haller who made the first crucial discovery about the functioning of the nervous system.

Von Haller found that he could make a muscle contract much more easily by stimulating a nerve than by stimulating the muscle itself. Furthermore, this contraction was involuntary; he could even produce it by stimulating a nerve after the organism had died. Von Haller went on to show that the nerves carried sensations. When he cut the nerves attached to specific tissues, these tissues could no longer react. The physiologist concluded that the brain received sensations by way of nerves and then sent out, again by way of nerves, messages that led to such responses as muscle contraction. He supposed that the nerves all came to a junction at the center of the brain.

In 1811 the Austrian physician Franz Joseph Gall focused attention on the "gray matter" on the surface of the cerebrum (which is distinguished from the "white matter" in that the latter consists merely of the fibers emerging from the nerve-cell bodies, these fibers being white because of their fatty sheaths). Gall suggested that the nerves did not collect at the center of the brain, as von Haller had thought, but that each ran to some definite portion of the gray matter, which he considered the coordinating region of the brain. Gall reasoned that different parts of the cerebral cortex were in charge of collecting sensations from different parts of the body and sending out the messages for responses to specific parts as well.

If a specific part of the cortex was responsible for a specific property of the mind, what was more natural than to suppose that the degree of development of that part would reflect a person's character or mentality? By feeling for bumps on a person's skull one might find out whether this or that portion of the brain was enlarged and so judge whether he was particularly generous or particularly depraved or particularly something else. With this reasoning, some of Gall's followers founded the pseudo science of "phrenology," which had quite a vogue in the nineteenth century and is not exactly dead even today. (Oddly enough, although Gall and his followers emphasized the high forehead and domed head as a sign of intelligence—a view that still influences people today—Gall himself had an unusually small brain, about 15 per cent smaller than the average.)

But the fact that phrenology, as developed by charlatans, is nonsense, does not mean that Gall's original notion of the specialization of functions in particular parts of the cerebral cortex was wrong. Even before specific explorations of the brain were attempted, it was noted that damage to a particular portion of the brain might result in a particular disability. In 1861, the French surgeon Pierre Paul Broca, by assiduous post-mortem study of the brain, was able to show that patients with "aphasia" (the inability to speak, or to understand speech) usually possessed physical damage to a particular area of the left cerebrum, an area called "Broca's convolution" as a result.

Then, in 1870, two German scientists, Gustav Fritsch and Eduard Hitzig, began to map the supervisory functions of the brain by stimulating various parts of it and observing what muscles responded. A half century later, this technique was greatly refined by the Swiss physiologist Walter Rudolf Hess, who was awarded a share of the 1949 Nobel Prize for medicine and physiology in consequence.

It was discovered by such methods that a specific band of the cortex was particularly involved in the stimulation of the various voluntary muscles into movement. This band is therefore called the "motor area." It seems to bear a generally inverted relationship to the body; the uppermost portions of the motor area, toward the top of the cerebrum, stimulate the lowermost portions of the leg; as one progresses downward in the motor area, the muscles higher in the leg are stimulated, then the muscles of the torso, then those of the arm and hand, and finally those of the neck and hand.

Behind the motor area is another section of the cortex that receives many types of sensation and is therefore called the "sensory area." As in the case of the motor area, the regions of the sensory area in the cerebral cortex are divided into sections that seem to bear an inverse relation to the body. Sensations from the foot are at the top of the area, followed successively as we go downward with sensations from the leg, hip, trunk, neck, arm, hand, fingers, and, lowest of all, the tongue. The sections of the sensory area devoted to the lips, tongue, and hand are (as one might expect) larger in proportion to the actual size of those organs than are the sections devoted to other parts of the body.

If, to the motor area and the sensory area, are added those sections of the cerebral cortex primarily devoted to receiving the impressions from the major sense organs, the eye and ear, it still leaves a major portion of the cortex without any clearly assigned and obvious function.

It is this apparent lack of assignment that has given rise to the statement, often encountered, that the human being "uses only one fifth of his brain." That, of course, is not so; the best we can really say is that one fifth of man's brain has an obvious function. We might as well suppose that a construction firm engaged in building a skyscraper is using only one fifth of its employees because that one fifth was actually engaged in raising steel beams, laying down electric cables, transporting equipment, and such. This would ignore the executives, secretaries, filing clerks, supervisors, and others. Analogously, the major portion of the brain is engaged in what we might call white-collar work, in the assembling of sensory data, in its analysis, in deciding what to ignore, what to act upon, and just how to act upon it. The cerebral cortex has distinct "association areas"—some for sound sensations, some for visual sensations, some for others.

When all these association areas are taken into account, there still remains one area of the cerebrum that has no specific and easily definable function. This is the area just behind the forehead, which is called the "prefrontal lobe." Its lack of obvious function is such that it is sometimes called the "silent area." Tumors have made it necessary to remove large areas of the prefrontal lobe without any particular significant effect on the individual; yet surely it is not a useless mass of nerve tissue.

One might even suppose it to be the most important portion of the brain if one considers that in the development of the human nervous system there has been a continual piling up of complication at the forward end. The prefrontal lobe might therefore be the brain area most recently evolved and most significantly human.

In the 1930's, it seemed to a Portuguese surgeon, Antonio Egas Moniz, that where a mentally-ill patient was at the end of his rope, it might be possible to help by taking the drastic step of severing the prefrontal lobes from the rest of the brain. The patient might then be cut off from a portion of the associations he had built up, which were, apparently, affecting him adversely, and make a fresh and better start with the brain he had left. This operation, "prefrontal lobotomy," was first carried out in 1935; in a number of cases it did indeed seem to help. Moniz shared (with W. R. Hess) the Nobel Prize for medicine and physiology in 1949 for his work. Nevertheless, the operation never achieved popularity and is less popular now than ever. Too often, the cure is literally worse than the disease.

The cerebrum is actually divided into two "cerebral hemispheres" connected by a tough bridge of white matter, the "corpus callosum." In effect, the hemispheres are separate organs, unified in action by the nerve fibers that cross the corpus callosum and act to coordinate the two. Despite this, the hemispheres remain potentially independent.

The situation is somewhat analogous to that of our eyes. Our two eyes act as a unit, ordinarily, but if one eye is lost, the other can meet our needs. Similarly, the removal of one of the cerebral hemispheres does not make an experimental animal brainless. The remaining hemisphere learns to carry on.

Ordinarily, each hemisphere is largely responsible for a particular side of the body; the left cerebral hemisphere for the right side, the right cerebral hemisphere for the left side. If both hemispheres are left in place and the corpus callosum is cut, coordination is lost, and the two body halves come under more or less independent control. A literal case of twin brains, so to speak, is set up.

Monkeys can be so treated (with further operation upon the optic nerve to make sure that each eye is connected to only one hemisphere), and when this is done each eye can be separately trained to do particular tasks. A monkey can be trained to select a cross over a circle to indicate, let us say, the presence of food. If only the left eye is kept uncovered during the training period, only the left eye will be useful in this respect. If the right eye is uncovered and the left eye covered, the monkey will have no right-eye memory of his training. He will have to hunt for his food by trial and error. If the two eyes are trained to contradictory tasks and if both are then uncovered, the monkey alternates activities, as the hemispheres politely take their turns.

Naturally, in any such two-in-charge situation, there is always the danger of conflict and confusion. To avoid that, one cerebral hemisphere (almost always the left one in human beings) is dominant, when both are normally connected. Broca's convolution, which controls speech, is in the left hemisphere, for instance. The "gnostic area" which is an over-all association area, a kind of court of highest appeal, is also in the left hemisphere. Since the left cerebral hemisphere controls the motor activity of the right-hand side of the body, it is not surprising that most people are right-handed (though even left-handed people usually have a dominant left cerebral hemisphere). Where clear-cut dominance is not established between left and right, there may ambidexterity, rather than a clear right-handedness or left-handedness, along with some speech difficulties and, perhaps, manual clumsiness.

The cerebrum is not the whole of the brain. There are areas of gray matter embedded below the cerebral cortex. These are called the "basal ganglia"; included is a section called the "thalamus." The thalamus acts as a reception center for various sensations. The more violent of these, such as pain, extreme heat or cold, or rough touch, are filtered out. The milder sensations from the muscles—the gentle touches, the moderate temperatures—are passed on to the sensory area of the cerebral cortex. It is as though mild sensations can be trusted to the cortex, where they can be considered judiciously and where reaction can come after a more or less prolonged interval of consideration. The rough sensations, however, which must be dealt with quickly and for which there is no time for consideration, are handled more or less automatically in the thalamus.

Underneath the thalamus is the "hypothalamus" center for a variety of devices for controlling the body. The body's appestat, mentioned in Chapter 14 as controlling the body's appetite, is located there; so is the control of the body's temperature. It is through the hypothalamus, moreover, that the brain exerts at least some influence over the pituitary gland (see Chapter 14); this is an indication of the manner in which the nervous controls of the body and the chemical controls (the hormones) can be unified into a master supervisory force.

In 1954, the physiologist James Olds discovered another and rather frightening function of the hypothalamus. It contains a region that, when stimulated, apparently gives rise to a strongly pleasurable sensation. An electrode affixed to the "pleasure center" of a rat, so arranged that it can be stimulated by the animal itself, will be stimulated up to 8,000 times an hour for hours or days at a time, to the exclusion of food, sex, and sleep. Evidently, all the desirable things in life are desirable only insofar as they stimulate the pleasure center. To stimulate it directly makes all else unnecessary.

The hypothalamus also contains an area that has to do with the wake-sleep cycle, since damage to parts of it induces a sleeplike state in animals. The exact mechanism by which the hypothalamus performs its function is uncertain. One theory is that it sends signals to the cortex, which sends signals back in response, in mutually stimulating fashion. With continuing wakefulness, the coordination of the two fails, the oscillations become ragged, and the individual becomes sleepy. A violent stimulus (a loud noise, a persistent shake of the shoulder, or, for that matter, a sudden interruption of a steady noise) will arouse one. In the absence of such stimuli, coordination will be restored, eventually, between hypothalamus and cortex, and sleep will end spontaneously;

or perhaps sleep will become so shallow that a perfectly ordinary stimulus, of which the surroundings are always full, will suffice to wake one.

During sleep, dreams—sensory data more or less divorced from reality—will take place. Dreaming is apparently a universal phenomenon; people who report dreamless sleep are merely failing to remember their dreams. The American physiologist W. Dement, studying sleeping subjects in 1952, noticed periods of rapid eye movements that sometimes persisted for minutes ("REM sleep"). During this period his breathing, heartbeat, and blood pressure, rose to waking levels. This takes place about a quarter of the sleeping time. If a sleeper was awakened during these periods, he generally reported having had a dream. Furthermore, if a sleeper was continually disturbed during these periods, he began to suffer psychological distress; the periods of distress were multiplied during succeeding nights as though to make up for the lost dreaming.

It would seem, then, that dreaming has an important function in the working of the brain. It is suggested that dreaming is a device whereby the brain runs over the events of the day to remove the trivial and repetitious that might otherwise clutter it and reduce its efficiency. Sleep is the natural time for such an activity for the brain is then relieved of many of its waking functions. Failure to accomplish this task (because of interruption) may so clog the brain that clearing attempts must be made during waking periods, producing hallucinations (that is, waking dreams, so to speak) and other unpleasant symptoms. One might naturally wonder if this is not a chief function of sleep, since there is very little physical resting in sleep that cannot be duplicated by quiet wakefulness. REM sleep even occurs in infants who spend half their sleeping time at it and who would seem to lack anything about which to dream. It may be that REM sleep helps the development of the nervous system. (It has been observed in mammals other than man, too.)

Below the cerebrum is the smaller cerebellum (also divided into two "cerebellar hemispheres") and the "brain stem," which narrows and leads smoothly into the "spinal cord" that extends about eighteen inches down the hollow center of the spinal column.

The spinal cord consists of gray matter (at the center) and white matter (on the periphery); to it are attached a series of nerves that are largely concerned with the internal organs—the heart, lungs, digestive system, and so on—organs that are more or less under involuntary control.

In general, when the spinal cord is severed, through disease or through injury, that part of the body lying below the severed segment is disconnected, so to speak. It loses sensation and is paralyzed. If the cord is severed in the neck region, death follows, because the chest is paralyzed, and with it the action of the lungs. It is this that makes a "broken neck" fatal, and hanging a feasible form of quick execution. It is the severed cord, rather than a broken bone, that is fatal.

The entire structure of the "central nervous system," consisting of the cerebrum, cerebellum, brain stem, and spinal cord, is carefully co-ordinated. The white matter of the spinal cord is made up of bundles of nerve fibers that run up and down the cord, unifying the whole. Those that conduct impulses downward from the brain are the "descending tracts" and those that conduct them upward to the brain are the "ascending tracts."

In 1964, research specialists at Cleveland's Metropolitan General Hospital reported the isolation from rhesus monkeys of brains which were then kept independently alive for as long as eighteen hours. This offers the possibility of detailed specific study of the brain's metabolism through a comparison of the nutrient medium entering the blood vessels of the isolated brain and of the same medium leaving it.

The next year they were transplanting dogs' heads to the necks of other dogs, hooking them up to the host's blood-supply, and keeping the brains in the transplanted heads alive and working for as long as two days. By 1966, dogs' brains were lowered to temperatures near freezing for six hours and then revived to the point of showing clear indications of normal chemical and electrical activity. Brains are clearly tougher than they might seem to be.

Nerve Action

It is not only the various portions of the central nervous system that are hooked together by nerves, but, clearly, all the body that, in this fashion, is placed under the control of that system. The nerves interlace the muscles, the glands, the skin; they even invade the pulp of the teeth (as we learn to our cost at every toothache.)

The nerves themselves were observed in ancient times but the structure and function were consistently misunderstood. Until modern times, they were felt to be hollow and to function as carriers of a subtle fluid. Rather complicated theories developed by Galen involved three different fluids carried by the veins, the arteries, and the nerves, respectively. The fluid of the nerves, usually referred to as "animal spirits," was the most rarefied of the three. When Galvani discovered that muscles and nerves could be stimulated by an electric discharge, this laid the foundation for a series of studies that eventually showed nerve action to be associated with electricity, a subtle fluid, indeed, more subtle than Galen could have imagined.

Specific work on nerve action began in the early nineteenth century with the German physiologist Johannes Peter Müller, who, among other things, showed that sensory nerves always produced their own sensations regardless of the nature of the stimulus. Thus, the optic nerve registered a flash of light, whether it was stimulated by light itself, or by the mechanical pressure of a punch in the eye. (In the latter case, you "see stars.") This emphasizes that our contact with the world is not a contact with reality at all, but a contact with specialized stimuli that the brain usually interprets in a useful manner, but can interpret in a nonuseful manner.

Study of the nerves was advanced greatly in 1873, when an Italian physiologist, Camillo Golgi, developed a cellular stain involving silver salts that was well-adapted to react with nerve cells, making clear their finest details. He was able to show, in this manner, that nerves were composed of separate and distinct cells, and that the processes of one cell might approach very closely to those of another, but that they did not fuse. There remained the tiny gap of the synapse. In this way, Golgi bore out, observationally, the contentions of a German anatomist, Wilhelm von Waldeyer, to the effect that the entire nervous system consisted of individual nerve cells or neurons (this contention being termed the "neuron theory").

Golgi did not, however, himself support the neuron theory. This proved to be the task of the Spanish neurologist Santiago Ramon y Cajal, who, by 1889, using an improved version of Golgi's stain, worked out the connections of the cells in the gray matter of the brain and spinal cord and fully established the neuron theory. Golgi and Ramon y Cajal, although disputing the fine points of their findings, shared the Nobel Prize for medicine and physiology in 1906.

These nerves form two systems: the "sympathetic" and the "parasympathetic." (The terms date back to semimystical notions of Galen.) Both systems act on almost every internal organ, exerting control by opposing effects. For instance, the sympathetic nerves act to accelerate the heartbeat, the parasympathetic nerves to slow it; the sympathetic nerves slow up secretion of digestive juices, the parasympathetic stimulate such secretions, and so on. Thus, the spinal cord, together with the subcerebral portions of the brain, regulates the workings of the organs in an automatic fashion. This set of involuntary controls was investigated in detail by the British physiologist John Newport Langley in the 1890's, and he named it the "autonomic nervous system."

In the 1830's, the English physiologist Marshall Hall had studied another type of behavior which seemed to have voluntary aspects but proved to be really quite involuntary. When you accidentally touch a hot object with your hand, the hand draws away instantly. If the sensation of heat had to go to the brain, be considered and interpreted there, and evoke the appropriate message to the hand, your hand would be pretty badly scorched by the time it got the message. The unthinking spinal cord disposes of the whole business automatically and much faster. It was Hall who gave the process the name "reflex."

The reflex is brought about by two or more nerves working in coordination, to form a "reflex arc." The simplest possible reflex arc is one consisting of two neurons, a sensory (bringing sensations to a "reflex center" in the central nervous system, usually at some point in the spinal cord) and a motor (carrying instructions for movement from the central nervous system). The two neurons may be connected by one or more "connector neurons." A particular study of such reflex arcs and of their function in the body was made by the English neurologist Charles Scott Sherrington, who won a share in the 1932 Nobel Prize for medicine and physiology in consequence. It was Sherrington who, in 1897, coined the word "synapse."

Reflexes bring about so rapid and certain a response to a particular stimulus that they offer simple methods for checking the general integrity of the nervous system. A familiar example is the "patellar reflex" or, as it is commonly called, the knee jerk. When the legs are crossed, a sudden blow below the knee of the upper leg will cause it to make a quick, kicking motion—a fact first brought into medical prominence in 1875 by the German neurologist Carl Friedrich Otto Westphal. The patellar reflex is not important in itself, but its nonappearance can mean some serious disorder involving the portion of the nervous system in which that reflex arc is to be found.

Sometimes, damage to a portion of the central nervous system brings about the appearance of an abnormal reflex. If the sole of the foot is scratched the normal reflex brings the toes together and bent downward. Certain types of damage to the central nervous system will cause the big toe to bend upward in response to this stimulus, and the little toes to spread apart as they bend down. This is the "Babinski reflex," named for a French neurologist, Joseph François Félix Babinski, who described it in 1896.

In man, reflexes are decidedly subordinate to the conscious will. up your rate of breathing when ordinary reflex action would keep it slow, and so on. The lower phyla of animals, on the other hand, not only are much more strictly controlled by their reflexes but also have them far more highly developed.

One of the best examples is a spider spinning its web. Here the reflexes produce such an elaborate pattern of behavior that it is difficult to think of it as mere reflex action; instead, it is usually called "instinctive" behavior. (Because the word "instinct" is often misused, biologists prefer the term "innate" behavior.) The spider is born with a nerve-wiring system in which the switches have been preset, so to speak. A particular stimulus sets it off on weaving a web, and each act in the process in turn acts as a stimulus determining the next response.

Looking at the spider's intricate web, built with beautiful precision and effectiveness for the function it will serve, it is almost impossible to believe that the thing has been done without purposeful intelligence.

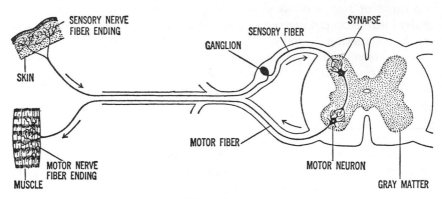

The reflex arc.

Yet the very fact that the complex task is carried through so perfectly and in exactly the same way every time is itself proof that intelligence has nothing to do with it. Conscious intelligence, with the hesitations and weighings of alternatives that are inherent in deliberate thought, will inevitably give rise to imperfections and variations from one construction to another.

With increasing intelligence, animals tend more and more to shed instincts and inborn skills. Thereby they doubtless lose something of value. A spider can build its amazingly complex web perfectly the first time, although it has never seen web-spinning, or even a web, before. Man, on the other hand, is born almost completely unskilled and helpless. A newborn baby can automatically suck on a nipple, wail if it is hungry, and hold on for dear life if about to fall, but it can do very little else. Every parent knows how painfully and with what travail a child comes to learn the simplest forms of suitable behavior. And yet, a spider or an insect, though born with perfection, cannot deviate from it. The spider builds a beautiful web, but if its preordained web should fail, it cannot learn to build another type of web. A boy, on the other hand, reaps great benefits from being unfettered by inborn perfection. He may learn slowly and attain only imperfection at best, but he can attain a variety of imperfection of his own choosing. What man has lost in convenience and security, he has gained in an almost limitless flexibility.

Recent work, however, emphasizes the fact that there is not always a clear division between instinct and learned behavior. It would seem, on casual observation, for instance, that chicks or ducklings, fresh out of the shell, follow their mothers out of instinct. Closer observation shows that this is not so.

The instinct, however, is not to follow their mother, but merely to follow something of a characteristic shape or color or faculty of movement. Whatever object provides this sensation at a certain period of early life is followed by the young creature and is thereafter treated as the mother. This may really be the mother; it almost invariably is, in fact, but it need not be! In other words, following is instinctive, but the "mother" that is followed is learned. (Much of the credit for this discovery goes to the remarkable Austrian naturalist, Konrad Zacharias Lorenz. Lorenz, during the course of studies now some thirty years old, was followed hither and yon by a gaggle of goslings.)

The establishment of a fixed pattern of behavior in response to a particular stimulus encountered at a particular time of life is called "imprinting." The specific time at which imprinting takes place is a "critical period." For chicks the critical period of "mother-imprinting" lies between 13 and 16 hours after hatching. For a puppy there is a critical period between 3 and 7 weeks, during which the stimulations it is usually likely to encounter imprint various aspects of what we consider normal doggish behavior.

Imprinting is the most primitive form of learned behavior, one that is so automatic, takes place in so limited a time, and under so general a set of conditions that it is easily mistaken for instinct.

A logical reason for imprinting is that it allows a certain desirable flexibility. If a chick were born with some instinctive ability of distinguishing its true mother so that it might follow only her, and if the true mother were for any reason absent in the chick's first day of life, the little creature would be helpless. As it is, the question of motherhood is left open for just a few hours and the chick may imprint itself to any hen in the vicinity and thus adopt a foster mother.

As stated earlier, it had been Galvani's experiments just before the opening of the nineteenth century that had first indicated some connection between electricity and the actions of muscle and nerve.

The electrical properties of muscle led to a startling medical application, thanks to the work of the Dutch physiologist Willem Einthoven. In 1903, he developed an extremely delicate galvanometer, one delicate enough to respond to the tiny fluctuations of the electric potential of the beating heart. By 1906, Einthoven was recording the peaks and troughs of this potential (the recording being an "electrocardiogram") and correlating them with various types of heart disorder.

The more subtle electrical properties of nerve impulses were thought to have been initiated and propagated by chemical changes in the nerve. This was elevated from mere speculation to experimental demonstration by the nineteenth-century German physiologist Emil Du Bois-Reymond, who by means of a delicate galvanometer was able to detect tiny electric currents in stimulated nerves.

With modern electronic instruments, researches into the electrical properties of the nerve have been incredibly refined. By placing tiny electrodes at different spots on a nerve fiber and by detecting electrical changes through an oscilloscope, it is possible to measure a nerve impulse's strength, duration, speed of propagation, and so on. For such work, the American physiologists Joseph Erlanger and Herbert Spencer Gasser were awarded the 1944 Nobel Prize for medicine and physiology.

If you apply small electric pulses of increasing strength to a single nerve cell, up to a certain point there is no response whatever. Then suddenly the cell fires: an impulse is initiated and travels along the fiber. The cell has a threshold; it will not react at all to a stimulus below the threshold, and to any stimulus above the threshold it will respond only with an impulse of a certain fixed intensity. The response, in other words, is "all or nothing." And the nature of the impulse elicited by the stimulus seems to be the same in all nerves.

How can such a simple yes-no affair, identical everywhere, lead to the complex sensations of sight, for instance, or to the complex finger responses involved in playing a violin? It seems that a nerve, such as the optic nerve, contains a large number of individual fibers, some of which may be "firing" and others not, and where the "firing" may be in rapid succession or slowly, forming a pattern, possibly a complex one, shifting continuously with changes in the over-all stimulus. (For work in this field, the English physiologist Edgar Douglas Adrian shared, with Sherrington, the 1932 Nobel Prize in medicine and physiology.) Such a changing pattern may be continually "scanned" by the brain and interpreted appropriately. But nothing is known about how the interpretation is made or how the pattern is translated into action such as the contraction of a muscle or secretion by a gland.

The firing of the nerve cell itself apparently depends on the movement of ions across the membrane of the cell. Ordinarily, the inside of the cell has a comparative excess of potassium ions, while outside the cell there is an excess of sodium ions. Somehow the cell holds potassium ions in and keeps sodium ions out so that the concentrations on the two sides of the cell membrane do not equalize. It is now believed that a "sodium pump" of some kind inside the cell keeps pumping out sodium ions as fast as they come in. In any case, there is an electric potential difference of about one tenth of a volt across the cell membrane, with the inside negatively charged with respect to the outside. When the nerve cell is stimulated, the potential difference across the membrane collapses, and this represents the firing of the cell. It takes a couple of thousandths of a second for the potential difference to be reestablished, and during that interval the nerve will not react to another stimulus. This is the "refractory period."

Once the cell fires, the nerve impulse travels down the fiber by a series of firings, each successive section of the fiber exciting the next in turn. The impulse can travel only in the forward direction, because the section that has just fired cannot fire again until after a resting pause.

Research that related, in the fashion just described, nerve action and ion permeability led to the award of the 1963 Nobel Prize for medicine and physiology to two British physiologists, Alan Lloyd Hodgkin and Andrew Fielding Huxley, and to an Australian physiologist, John Carew Eccles.

What happens, though, when the impulse traveling along the length of the nerve fiber comes to a synapse—a gap between one nerve cell and the next? Apparently, the nerve impulse also involves the production of a chemical that can drift across the gap and initiate a nerve impulse in the next nerve cell. In this way, the impulse can travel from cell to cell.

One of the chemicals definitely known to affect the nerves is the hormone adrenalin. It acts upon nerves of the sympathetic system, which slows the activity of the digestive system and accelerates the rate of respiration and the heartbeat. When anger or fear excites the adrenal glands to secrete the hormone, its stimulation of the sympathetic nerves sends a faster surge of blood through the body, carrying more oxygen to the tissues, and by slowing down digestion for the duration it saves energy during the emergency.

The American psychologists and police officers John Augustus Larsen and Leonard Keeler took advantage of this finding in 1921 to devise a machine to detect the changes in blood pressure, pulse rate, breathing rate, and a perspiration brought on by emotion. This device, the "polygraph," detected the emotional effort involved in telling a lie, which always carries with it the fear of detection in any reasonably normal individual and therefore brings adrenalin into play. While not infallible, the polygraph has gained great fame as a "lie detector."

In the normal course, the nerve endings of the sympathetic nervous system themselves secrete a compound very like adrenalin, called "noradrenalin." This chemical serves to carry the nerve impulses across the synapses, transmitting the message by stimulating the nerve endings on the other side of the gap.

In the early 1920's the English physiologist Henry Dale and the German physiologist Otto Loewi (who were to share the Nobel Prize in physiology and medicine in 1930) studied a chemical that performed this function for most of the nerves other than those of the sympathetic system. The chemical is called acetylcholine. It is now believed to be involved not only at the synapses but also in conducting the nerve impulse along the nerve fiber itself. Perhaps acetylcholine acts upon the "sodium pump." At any rate, the substance seems to be formed momentarily in the nerve fiber and to be broken down quickly by an enzyme called "cholinesterase." Anything that inhibits the action of cholinesterase will interfere with this chemical cycle and will stop the transmission of nerve impulses. The deadly substances now known as "nerve gases" are cholinesterase inhibitors. By blocking the conduction of nerve impulses they can stop the heartbeat and produce death within minutes. The application to warfare is obvious. They can be used, less immorally, as insecticides.

A less drastic interference with cholinesterase is that of local anesthetics, which in this way suspend (temporarily) those nerve impulses associated with pain.

Thanks to the electric currents involved in nerve impulses, it is possible to "read" the brain's activity, in a way, though no one has yet been able to translate fully what the brain waves are saying. In 1929, a German psychiatrist, Hans Berger, reported earlier work in which he applied electrodes to various parts of the head and was able to detect rhythmic waves of electrical activity.

Berger gave the most pronounced rhythm the name of "alpha wave." In the alpha wave, the potential varies by about 20 microvolts in a frequency of roughly 10 times a second. The alpha wave is clearest and most obvious when the subject is resting with his eyes closed. When the eyes are open, but are viewing featureless illumination, the alpha wave persists. If, however, the ordinary variegated environment is in view, the alpha view vanishes, or is drowned, by other more prominent rhythms. After a while, if nothing visually new is presented, the alpha wave reappears. Typical names for other types of waves are "beta waves," "delta waves," and "theta waves."

Electroencephalograms ("electrical writings of the brain" or, as they are abbreviated, "EEG") have since been extensively studied, and they show that each individual has his own pattern, varying with excitement and in sleep. Although the electroencephalogram is still far from being a method of "reading thoughts" or tracing the mechanism of the intellect, it does help in the diagnosis of major upsets of brain function, particularly epilepsy. It can also help locate areas of brain damage or brain tumors.

In the 1960's specially designed computers were called into battle. If a particular small environmental change is applied to a subject, it is to be presumed that there will be some response in the brain that will reflect itself in a small alteration in the EEG pattern at the moment when the change is introduced. The brain will be engaged in many other activities, however, and the small alteration in the EEG will not be noticeable. Notwithstanding, if the process is repeated over and over again, a computer can be programed to average out the EEG pattern and find the consistent difference.

By 1964, the American psychologist Manfred Clynes reported analyses fine enough to be able to tell, by a study of the EEG pattern alone, what color a subject was looking at. The English neurophysiologist William Grey Walter similarly reported a brain-signal pattern that seems characteristic of the learning process. It comes when the subject under study has reason to think he is about to be presented with a stimulus that will call for thought or action. Walter calls it the "expectancy wave" and points out that it is absent in children under three and in certain psychotics. The reverse phenomenon, that of bringing about specific actions through direct electrical stimulation of the brain, was also reported in 1965. José Manuel Rodriguez Delgado of Yale, transmitting electrical stimulation by radio signals, caused animals to walk, climb, yawn, sleep, mate, switch emotions, and so on at command. Most spectacularly, a charging bull was made to stop short and trot peacefully away.

Human Behavior

Unlike physical phenomena, such as the motions of planets or the behavior of light, the behavior of living things has never been reduced to rigorous natural laws and perhaps never will be. There are many who insist that the study of human behavior cannot become a true science, in the sense of being able to explain or predict behavior in any given situation on the basis of universal natural laws. Yet life is no exception to the rule of natural law, and it can be argued that living behavior would be fully explainable if all the factors were known. The catch lies in that last phrase. It is unlikely that all the factors will ever be known; they are too many and too complex. Man need not, however, despair of ever being able to understand himself. There is ample room for better knowledge of his own mental complexities, and even if we never reach the end of the road, we may yet hope to travel along it quite a way.

Not only is the subject particularly complex, but the study of the subject has not been progressing for long. Physics came of age in 1600, and chemistry in 1775, but the much more complex study of "experimental psychology" dates only from 1879, when the German physiologist Wilhelm Wundt set up the first laboratory devoted to the scientific study of human behavior. Wundt interested himself primarily in sensation and in the manner in which men perceived the details of the universe about them.

At almost the same time, the study of human behavior in one particular application—that involving man as an industrial cog—arose. In 1881, the American engineer Frederick Winslow Taylor began measuring the time required to do certain jobs, and to work out methods for so organizing the work as to minimize that time. He was the first "efficiency expert" and was (like all efficiency experts who are prone to lose sight of values beyond the stop watch) unpopular with the workers.

But as we study human behavior, step by step, either under controlled conditions in a laboratory or empirically in a factory, it does seem that we are tackling a fine machine with blunt tools.

In the simple organisms we can see direct, automatic responses of the kind called "tropisms" (from a Greek word meaning to "turn"). Plants show "phototropism" (turning toward light), "hydrotropism" (turning toward water, in this case by the roots), and "chemotropism" (turning toward particular chemical substances). Chemotropism is also characteristic of many animals, from protozoa to ants. Certain moths are known to fly toward a scent as far as two miles away. That tropisms are completely automatic is shown by the fact that a phototropic moth will even fly into a candle flame.

The reflexes mentioned earlier in this chapter do not seem to progress far beyond tropisms, and imprinting, also mentioned, represents learning, but in so mechanical a fashion as scarcely to deserve the name. Yet neither reflexes nor imprinting can be regarded as characteristic of the lower animals only; man has his share.

The human infant from the moment of birth will grasp a finger tightly if it touches his palm and will suck at a nipple if that is put to his lips. The importance of such instincts to keep the infant secure from falling and from starvation is obvious.

It seems almost inevitable that the infant is subject also to imprinting. This is not a fit subject for experimentation, of course, but knowledge can be gained through incidental observations. Children who, at the babbling stage, are not exposed to the sounds of actual speech may not develop the ability to speak later, or do so to an abnormally limited extent. Children brought up in impersonal institutions where they are efficiently fed and their physical needs are amply taken care of, but where they are not fondled, cuddled, and dandled, become sad little specimens indeed. Their mental and physical development is greatly retarded and many die for no other reason, apparently, than lack of "mothering"—by which may be meant the lack of adequate stimuli to bring about the imprinting of necessary behavior patterns. Similarly, children who are unduly deprived of the stimuli involved in the company of other children during critical periods in childhood develop personalities that may be seriously distorted in one fashion or another.

Of course, one can argue that reflexes and imprinting are a matter of concern only for infancy. When a man achieves adulthood, he is then a rational being who responds in more than a mechanical fashion. But does he? To put it another way: Does man possess "free will" (as he likes to think), or is his behavior in some respects absolutely determined by the stimulus, as the bull's was in Delgado's experiment described on the previous page.

One can argue for the existence of free will on philosophical or theological grounds, but I know of no one who has ever found a way to demonstrate it experimentally. To demonstrate "determinism," the reverse of free will, is not exactly easy, either. Attempts in that direction, however, have been made. Most notable were those of the Russian physiologist Ivan Petrovich Pavlov.

Pavlov started with a specific interest in the mechanism of digestion. He showed in the 1880's that gastric juice was secreted in the stomach as soon as food was placed on a dog's tongue; the stomach would secrete this juice even if food never reached it. But if the vagus nerve (which runs from the medulla oblongata to various parts of the alimentary canal) was cut near the stomach, the secretions stopped. For his work on the physiology of digestion, Pavlov received the Nobel Prize in physiology and medicine in 1904. But like some other Nobel laureates (notably, Ehrlich and Einstein) Pavlov went on to other discoveries that dwarfed the accomplishments for which he actually received the prize.

He decided to investigate the automatic, or reflex, nature of secretions, and he chose the secretion of saliva as a convenient, easy-to-observe example. The sight or odor of food causes a dog (and a man, for that matter) to salivate. What Pavlov did was to ring a bell every time he placed food before a dog. Eventually, after 20 to 40 associations of this sort, the dog salivated when it heard the bell even though no food was present. An association had been built up. The nerve impulse that carried the sound of the bell to the cerebrum had become equivalent to one representing the sight or odor of food.

In 1903, Pavlov invented the term "conditioned reflex" for this phenomenon; the salivation was a "conditioned response." Willynilly, the dog salivated at the sound of the bell just as it would at the sight of food. Of course, the conditioned response could be wiped out—for instance, by repeatedly denying food to the dog when the bell was rung and subjecting it to a mild electric shock instead. Eventually, the dog would not salivate but instead would wince at the sound of the bell, even though it received no electric shock.

Furthermore, Pavlov was able to force dogs to make subtle decisions by associating food with a circular patch of light and an electric shock with an elliptical patch. The dog could make the distinction, but as the ellipse was made more and more nearly circular, distinction became more difficult. Eventually, the dog, in an agony of indecision, developed what could only be called a "nervous breakdown."

Conditioning experiments have thus become a powerful tool in psychology. Through them, animals sometimes almost "talk" to the experimenter. The technique has made it possible to investigate the learning abilities of various animals, their instincts, their visual abilities, their ability to distinguish colors, and so on. Of all the investigations, not the least remarkable are those of the Austrian naturalist Karl von

Frisch. Von Frisch trained bees to go to dishes placed in certain locations for their food, and he learned that these foragers soon told the other bees in their hive where the food was located. From his experiments von Frisch learned that the bees could distinguish certain colors, including ultraviolet, but excluding red, which they communicated with one another by means of a dance on the honeycombs, that the nature and vigor of the dance told the direction and distance of the food dish from the hive and even how plentiful or scarce the food supply was, and that the bees were able to tell direction from the polarization of light in the sky. Von Frisch's fascinating discoveries about the language of the bees opened up a whole new field of study of animal behavior.

In theory, all learning can be considered to consist of conditioned responses. In learning to type, for instance, you start by watching the typewriter keyboard and gradually substitute certain automatic movements of the fingers for visual selection of the proper key. Thus the thought "k" is accompanied by a specific movement of the middle finger of the right hand; the thought "the" causes the first finger of the left hand, the first finger of the right hand, and the second finger of the left hand, to hit certain spots in that order. These responses involve no conscious thought. Eventually a practiced typist has to stop and think to recall where the letters are. I am myself a rapid and completely mechanical typist, and if I am asked where the letter "f," say, is located on the keyboard, the only way I can answer (short of looking at the keyboard) is to move my fingers in the air as if typing and try to catch one of them in the act of typing "f." Only my fingers know the keyboard; my conscious mind does not.

The same principle may apply to more complex learning, such as reading or playing a violin. Why, after all, does the design CRAYON in black print on this piece of paper automatically evoke a picture of a pigmented stick of wax and a certain sound that represents a word? You do not need to spell out the letters or search your memory or reason out the possible message contained in the design; from repeated conditioning you automatically associate the symbol with the thing itself.

In the early decades of this century the American psychologist John Broadus Watson built a whole theory of human behavior, called "behaviorism," on the basis of conditioning. Watson went so far as to suggest that people had no deliberate control over the way they behaved; it was all determined by conditioning. Although his theory was popular for a time, it never gained wide support among psychologists. In the first place, even if the theory is basically correct—if behavior is dictated solely by conditioning—behaviorism is not very enlightening on those aspects of human behavior that are of most interest to us, such as creative intelligence, artistic ability, and the sense of right and wrong. It would be impossible to identify all the conditioning influences and relate them to the pattern of thought and belief in any measurable way;

828

and something that cannot be measured is not subject to any really scientific study.

In the second place, what does conditioning have to do with a process such as intuition? The mind suddenly puts two previously un-related thoughts or events together, apparently by sheer chance, and creates an entirely new idea or response.

Cats and dogs, in solving tasks (as in finding out how to work a lever in order to open a door) may do so by a process of trial and error. They may move about randomly and wildly until some motion of theirs trips the lever. If they are set to repeating the task a dim memory of the successful movement may lead them to it sooner, and then still sooner at the next attempt, until finally they move to the lever at once. The more intelligent the animal, the fewer attempts will be required to graduate from sheer trial and error to purposive useful action.

By the time we reach man, memory is no longer feeble. The tendency might be to search for a dropped dime by glances randomly directed at the floor, but from past experience he may look in places where he has found the dime before, or look in the direction of the sound, or institute a systematic scanning of the floor. Similarly, if he were in a closed place, he might try to escape by beating and kicking at the walls randomly; but he would also know what a door would look like and would concentrate his efforts on that.

A man can, in short, simplify trial and error by calling on years of experience, and transfer it from thought to action. In seeking a solution, he may do nothing, he may merely act in thought. It is this etherealized trial and error we call reason, and it is not even entirely restricted to the human species.

Apes, whose patterns of behavior are simpler and more mechanical than man's, show some spontaneous insight, which may be called reason. The German psychologist Wolfgang Köhler, trapped in one of the German colonies in Africa by the advent of World War I, discovered some striking illustrations of this in his famous experiments with chimpanzees. In one case a chimp, after trying in vain to reach bananas with a stick that was too short, suddenly picked up another bamboo stick that the experimenter had left lying handy, joined the two sticks together, and so brought the fruit within reach. In another instance a chimp piled one box on another to reach bananas hanging overhead. These acts had not been preceded by any training or experience that might have formed the association for the animal; apparently they were sheer flashes of inspiration.

To Köhler, it seemed that learning involved the entire pattern of a process, rather than individual portions of it. He was one of the founders of the "Gestalt" school of psychology (that being the German word for "pattern").

The power of conditioning has turned out to be greater than had been expected, in fact. For a long time it had been assumed that certain body functions such as heartbeat, blood pressure, and intestinal contractions, were essentially under the control of the autonomic nervous system and therefore beyond conscious control. There were catches, of course. A man adept at yoga can produce effects on his heartbeat by control of chest muscles, but that is no more significant than stopping the blood-flow through a wrist artery by applying thumb-pressure. Again, one can make one's heart beat faster by fantasying a state of anxiety, but that is the conscious manipulation of the autonomic nervous system. Is it possible simply to will the heart to beat faster or the blood pressure to rise without extreme manipulation of either the muscles or the mind?

The American psychologist Neal Elgar Miller and his coworkers, have carried out conditioning experiments where rats were rewarded when they happened to increase their blood pressure for any reason, or where their heartbeat was increased or decreased. Eventually, for the sake of the reward, they learned to perform voluntarily a change effected by the autonomic nervous system—just as they might learn to press a lever, and for the same purpose.

At least one experimental program, using human volunteers (male) who were rewarded by flashes of light revealing photographs of nude girls, demonstrated the volunteers' ability to produce increases or decreases in blood pressure in response. The volunteers did not know what was expected of them in order to produce the flashing light—and the nude— but just found that as time went along, they caught the desired glimpses more often.

There is more subtlety to the autonomic body controls than had once been suspected, too. Since living organisms are subjected to natural rhythms—the ebb and flow of the tides, the somewhat slower alternation of day and night, the still slower swing of the seasons—it is not surprising that they themselves respond rhythmically. Trees shed their leaves in fall and bud in the spring; men grow sleepy at night and rouse themselves at dawn.

What did not come to be fully appreciated until lately, was the complexity and multiplicity of the rhythmic responses, and their automatic nature, which persisted even in the absence of the environmental rhythm.

Thus, the leaves of plants rise and fall in a daylong rhythm to match the coming and going of the sun. This is made apparent by time-lapse photography. Seedlings grown in darkness showed no such cycle, but the potentiality was there. One exposure to light—one only—was enough to convert that potentiality into actuality. The rhythm then began, and it continued even if the light was cut off again. From plant to plant, the exact period of rhythm varied—anywhere from 24 to 26 hours in the absence of light—but it was always about 24 hours, under the regulating effect of the sun. A 20-hour cycle could be established if artificial light were used on a 10-hour-on and 10-hour-off cycle, but as soon as the light was turned off altogether, the about-24-hour rhythm reestablished itself.

This daily rhythm, a kind of "biological clock" that works even in the absence of outside hints, permeates all life. Franz Halberg of the University of Minnesota named it "circadian rhythm," from the Latin "circa dies," meaning "about a day."

Human beings are not immune to such rhythms. Men and women have voluntarily lived for months at a time in caves where they separated themselves from any time-telling mechanism and had no idea whether it was night or day outside. They soon lost all track of time and ate and slept rather erratically. However, they also noted their temperature, pulse, blood pressure, and brain waves, and sent these and other measurements to the surface, where observers kept track of them in connection with time. It turned out that, however time-confused the cave-dwellers were, the bodily rhythm was not. The rhythm remained stubbornly at a period of about a day, with all measurements rising and falling regularly, through all the stay in the cave.

This is by no means only an abstract matter. In nature, the earth's rotation remains steady, and the alternation of day and night remains constant and beyond human interference—but only if you remain in the same spot on earth, or only shift north or south. If you travel east or west for long distances and quite rapidly, however, you change the time of day. You may land in Japan at lunchtime (for Japanese) when your biological clock tells you it is time to go to bed. The jet-age traveler often has difficulty matching his activity to that of the at-home people surrounding him. If he does so—with his pattern of hormone-secretion, for instance, not matching the pattern of his activity—he will be tired and inefficient, suffering from "jet fatigue."

Less dramatically, the ability of an organism to withstand a dose of X-rays or various types of medication, often depends on the setting of the biological clock. It may well be that medical treatment ought to vary with the time of day or, for maximum effect and minimum side-effect, be restricted to one particular time of day.

What keeps the biological clock so well-regulated? Suspicion in this respect has fallen upon the pineal gland. In some reptiles the pineal gland is particularly well-developed and seems to be similar in structure to the eye. In the tuatara, a lizardlike reptile that is the last surviving species of its order and is found only on some small islands off New Zealand, the "pineal eye" is a skin-covered patch on top of its skull, particularly prominent for about six months after birth and definitely sensitive to light.

The pineal gland does not "see" in the ordinary sense of the word, but it may produce some chemical that rises and falls in rhythmic response to the coming and going of light. It thus may regulate the biological clock and do so even after light ceases to be periodic (having learned its chemical lesson by a kind of conditioning).

But then how does the pineal gland work in mammals, where it is no longer located just under the skin at the top of the head but is buried deep in the center of the brain? Can there be something more penetrating than light—something that is rhythmic in the same sense? There are speculations that cosmic rays might be the answer. These have a circadian rhythm of their own, thanks to earth's magnetic field and the solar wind, and perhaps this force is external regulator.

Even if the external regulator is found, is the internal biological clock something that can be identified? Is there some chemical reaction in the body that rises and falls in a circadian rhythm and that controls all the other rhythms? Is there some "master-reaction" that we can tab as *the* biological clock? If so, it has not yet been found.

Simple reasoning from the conditioned reflex finds it difficult to encompass the subtleties of intuition and reason. An attack upon human behavior has been launched by methods that are themselves highly intuitive. These methods can be traced back nearly two centuries to an Austrian physician, Franz Anton Mesmer, who became the sensation of Europe for his experiments with a powerful tool for probing human behavior. He used magnets at first, and then his hands only, obtaining his effects by what he called "animal magnetism" (soon renamed "mesmerism"); he would put a patient into a trance and pronounce the patient cured of his illness. He may well have produced some cures (since some disorders can be treated by suggestion) and gained many ardent followers, including the Marquis de Lafayette, fresh from his American triumph. However, Mesmer, an ardent astrologer and all-round mystic, was investigated skeptically but fairly by a committee, which included Lavoisier and Benjamin Franklin, and was then denounced as a fake and eventually retired in disgrace.

Nevertheless, he had started something. In the 1850's a British surgeon named James Braid revived hypnotism (he was the first to use this term) as a medical device, and other physicians also took it up. Among them was a Viennese doctor named Josef Breuer, who in the 1880's began to use hypnosis specifically for mental and emotional disorders.

Hypnotism (Greek for "putting to sleep") had been known, of course, since ancient times, and had often been used by mystics. But Breuer and others now began to interpret its effects as evidence of the existence of an "unconscious" level of the mind. Motivations of which the individual was unaware were buried there, and they could be brought to light by hypnosis. It was tempting to suppose that these motivations were suppressed from the conscious mind because they were associated with shame or guilt, and that they might account for useless, irrational, or even vicious behavior.

Breuer set out to employ hypnosis to probe the hidden causes of hysteria and other behavior disorders. Working with him was a pupil named Sigmund Freud. For a number of years they treated patients together, putting the patients under light hypnosis and encouraging them to speak. They found that the patients' venting of experiences or impulses buried in the unconscious often acted as a cathartic, relieving their symptoms after they awoke from the hypnosis.

Freud came to the conclusion that practically all of the suppressed memories and motivations were sexual in origin. Sexual impulses tabooed by society and the child's parents were driven underground, but they still strove for expression and generated intense conflicts which were the more damaging for being unrecognized and unadmitted.

In 1894, after breaking with Breuer because the latter disagreed with his concentration on the sexual factor, Freud went on alone to develop his ideas about the causes and treatment of mental disturbances. He dropped hypnosis and urged his patients to babble in a virtually random manner—to say anything that came into their minds. As the patient came to feel that the physician was listening sympathetically without any moral censure, slowly—sometimes very slowly—the individual began to unburden himself, to remember things long repressed and forgotten. Freud called this slow analysis of the "psyche" (Greek for soul or mind) "psychoanalysis."

Freud's involvement with the sexual symbolism of dreams and his description of infantile wishes to substitute for the parent of the same sex in the marital bed (the "Oedipus complex" in the case of boys and the "Electra complex" in girls—named for characters in Greek mythology) horrified some and fascinated others. In the 1920's, after the dislocations of World War I and amid the further dislocations of Prohibition in America and changing mores in many parts of the world, Freud's views struck a sympathetic note, and psychoanalysis attained the status, almost, of a popular fad.

More than half a century after its beginnings, however, psychoanalysis still remains an art rather than a science. Rigorously controlled experiments, such as those conducted in physics and the other "hard" sciences, are, of course, exceedingly difficult in psychiatry. The practitioners must base their conclusions largely on intuition or subjective judgment. Psychiatry (of which psychoanalysis is only one of the techniques) has undoubtedly helped many patients, but it has produced no spectacular cures and has not notably reduced the incidence of mental disease. Nor has it developed any all-embracing and generally accepted theory, comparable to the germ theory of infectious disease. In fact, there are almost as many schools of psychiatry as there are psychiatrists.

Serious mental illness takes various forms, ranging from chronic depression to a complete withdrawal from reality into a world in which some, at least, of the details do not correspond to the way most of us see things. This form of psychosis is usually called "schizophrenia," a term introduced by the Swiss psychiatrist Eugen Bleuler. The word covers such a multitude of disorders that it can no longer be described as a specific disease. About 60 per cent of all the chronic patients in our mental hospitals are diagnosed as schizophrenics.

Until recently, drastic treatments, such as prefrontal lobotomy, or shock therapy using electricity or insulin (the latter technique introduced in 1933 by the Austrian psychiatrist Manfred Sakel), were all that could be offered. Psychiatry and psychoanalysis have been of little avail, except occasionally in the early stages when a physician is still able to communicate with the patient. But some recent discoveries concerning drugs and the chemistry of the brain ("neurochemistry") have introduced an encouraging note.

Even the ancients knew that certain plant juices could induce hallucinations (fantasies of vision, hearing, and so on) and others could bring on happy states. The Delphic priestesses of ancient Greece chewed some plant before they pronounced their cryptic oracles. Indian tribes of the southwestern United States have made a religious ritual of chewing peyote or mescal buttons (which produce hallucinations in color). Perhaps the most dramatic case was that of the Moslem sect in a mountain stronghold in Iran who used "hashish," the juice of hemp leaves, more familiarly known to us as "marijuana." The drug, taken in their religious ceremonies, gave the communicants the illusion that they caught glimpses of the paradise to which their souls would go after death, and they would obey any command of their leader, called "the Old Man of the Mountains," to receive this key to heaven. His commands took the form of ordering them to kill enemy rulers and hostile Moslem government officials; this gave rise to the word "assassin," from "hashishin" (a user of hashish). The sect terrorized the region throughout the twelfth century, until the Mongol invaders in 1226 swarmed into the mountains and killed every last assassin.

The modern counterpart of the euphoric herbs of earlier times (aside from alcohol) is the group of drugs known as the "tranquilizers." As a matter of fact, one of the tranquilizers had been known in India as long ago as 1000 B.C. in the form of a plant called *Rauwolfia serpentinum*. It was from the dried roots of this plant that American chemists in 1952 extracted "reserpine," the first of the currently popular tranquilizing drugs. Several substances with similar effects but simpler chemical structure have since been synthesized.

The tranquilizers are sedatives, but with a difference. They reduce anxiety without appreciably depressing other mental activity. Nevertheless, they do tend to make people sleepy, and they may have other undesirable effects. They were at once found to be immensely helpful in relieving and quieting mental patients, including some schizophrenics. The tranquilizers are not cures for any mental illness, but they suppress certain symptoms that stand in the way of adequate treatment. By reducing the hostilities and rages of patients, and by quieting their fears and anxieties, they reduce the necessity for drastic physical restraints, make it easier for psychiatrists to establish contacts with patients, and increase the patients' chances of release from the hospital.

But where the tranquilizers had their runaway boom was among the public at large, which apparently seized upon them as a panacea to banish all cares.

Reserpine turns out to have a tantalizing resemblance to an important substance in the brain. A portion of its complex molecule is rather similar to the substance called "serotonin." Serotonin was discovered in the blood in 1948, and it has greatly intrigued physiologists ever since. It was found to be present in the hypothalamus region of the human brain and proved to be widespread in the brain and nerve tissues of other animals, including invertebrates.

What is more, various other substances that affect the central nervous system have turned out to resemble serotonin closely. One of them is a compound in toad venom called "bufotenin." Another is mescaline, the active drug in mescal buttons. Most dramatic of all is a substance named "lysergic acid diethylamide" (popularly known as LSD). In 1943, a Swiss chemist named Albert Hofmann happened to absorb some of this compound in the laboratory and was overcome by strange sensations. Indeed, what he seemed to perceive by way of his senses, in no way matched what we would take to be the objective reality of the environment. He suffered what we call hallucinations, and LSD is an example of what we now call a "hallucinogen."

Those who take pleasure in the sensations they experience when under the influence of a hallucinogen, refer to this as "mind-expansion" —apparently indicating that they sense—or think they sense—more of the universe than they would under ordinary conditions. But then, so do drunks once they bring themselves to the stage of "delirium tremens." The comparison is not as unkind as it may seem, for investigations have shown that a small dose of LSD, in some cases, can produce many of the symptoms of schizophrenia!

What can all this mean? Well, serotonin (which is structurally like the amino acid tryptophan) can be broken down by means of an enzyme called "amine oxidase," which occurs in brain cells. Suppose that this enzyme is taken out of action by a competitive substance with a structure like serotonin's—lysergic acid, for example. With the breakdown enzyme removed, serotonin will accumulate in the brain cells, and its level may rise too high. This will upset the serotonin balance in the brain and may bring on the schizophrenic state.

Is it possible that schizophrenia arises from some naturally induced upset of this sort? The manner in which a tendency to schizophrenia is inherited certainly makes it appear that some metabolic disorder (one, moreover, that is gene-controlled) is involved. In 1962, it was found that with a certain course of treatment, the urine of schizophrenics often contained a substance absent from the urine of nonschizophrenics. The substance eventually turned out to be a chemical called "dimethoxy-phenylethylamine," with a structure that lies somewhere between adrenalin and mescaline. In other words, certain schizophrenics seem, through some metabolic error, to form their own hallucinogens and to be, in effect, on a permanent drug-high.

Not everyone reacts identically to a given dose of one drug or another. Obviously, however, it is dangerous to play with the chemical mechanism of the brain. To become a mental cripple is a price surely too high for any amount of "mind-expanding" fun. Nevertheless, the reaction of society to drug use—particularly to that of marijuana, which has not yet been definitely shown to be as harmful as other hallucinogens —tends to be overstrenuous. Many of those who inveigh against the use of drugs of one sort or another are themselves thoroughly addicted to the use of alcohol or tobacco, both of which, in the mass, are responsible for much harm both to the individual and to society. Hypocrisy of this sort tends to decrease the credibility of much of the antidrug movement.

Neurochemistry also offers a hope for understanding that elusive mental property known as "memory." There are, it seems, two varieties of memory: short-term and long-term. If you look up a phone number, it is not difficult to remember it until you have dialed; it is then automatically forgotten and, in all probability, will never be recalled again. A telephone number you use frequently, however, enters the long-term memory category. Even after a lapse of months, you can dredge it up.

Yet even of what we would consider long-term memory items, much is lost. We forget a great deal and even, alas, forget much of vital importance (as every student facing an examination is woefully aware). Yet is it forgotten? Has it really vanished, or is it simply so well-stored that it is difficult to recall—buried, so to speak, under too many extraneous items?

The tapping of such hidden memories has become an almost literal tap. The American-born surgeon Wilder Graves Penfield at McGill University in Montreal, while operating on a patient's brain, accidentally touched a particular spot that caused the patient to hear music. That happened over and over again. The patient could be made to relive an experience in full, while remaining quite conscious of the present. Proper stimulation can apparently reel off memories with great accuracy. The area involved is called the "interpretative cortex." It may be that the accidental tapping of this portion of the cortex gives rise to the phenomenon of *déjà vu* (the feeling that something has happened before) and other manifestations of "extrasensory perception."

But if memory is so detailed, how can the brain find room for it all? It is estimated that, in a lifetime, a brain can store 1,000,000,000-000,000 (a million billion) units of information. To store so much, the units of storage must be of molecular size. There would be room for nothing more.

Suspicion is currently falling on ribonucleic acid (RNA) in which the nerve cell, surprisingly enough, is richer than almost any other type of cell in the body. This is surprising because RNA is involved in the synthesis of protein (see Chapter 12) and is therefore usually found in particularly high quantity in those tissues producing large quantities of protein either because they are actively growing or because they are producing copious quantities of protein-rich secretions. The nerve cell falls into neither classification.

A Swedish neurologist, Holger Hyden, developed techniques that could separate single cells from the brain and then analyze them for RNA content. He took to subjecting rats to conditions where they were forced to learn new skills—that of balancing on a wire for long periods of time, for instance. By 1959, he had discovered that the brain cells of rats that were forced to learn, increased their RNA content up to 12 per cent higher than that of the brain cells of rats allowed to go their normal way.

The RNA molecule is so very large and complex that if each unit of stored memory is marked off by an RNA molecule of distinctive pattern we need not worry about capacity. So many different RNA patterns are available that even a number such as a million billion is insignificant in comparison.

But ought one to consider RNA by itself? RNA molecules are formed according to the pattern of DNA molecules in the chromosomes. Is it that each person carries a vast supply of potential memories—a "memory bank," so to speak—in the DNA molecules he was born with, called upon and activated by actual events with appropriate modifications?

And is RNA the end? The chief function of RNA is to form specific protein-molecules. Is it the protein, rather than the RNA, that is truly related to the memory function?

One way of testing this is to make use of a drug called "puromycin," which interferes with protein formation by way of RNA. The American man-and-wife team Louis Barkhouse Flexner and Josepha Barbar Flexner conditioned mice to solve a maze, then immediately injected puromycin. The mice forgot what they had learned. The RNA molecule was still there, but the key protein-molecule could not be formed. Using puromycin, the Flexners showed that while short-term memory could be erased in this way in rats, long-term memory could not. The proteins for the latter had, presumably, already been formed.

And yet it may be that memory is more subtle and is not to be fully explained on the simple molecular level. There are indications that patterns of neural activity may be involved, too. Much yet remains to do.

Feedback

Man is not a machine and a machine is not man, but as science and technology advance, man and machine seem to be becoming less and less distinguishable from each other.

If you analyze what it is that makes a man, one of the first thoughts that strikes you is that, more than any other living organism, he is a self-regulating system. He is capable of controlling not only himself but also his environment. He copes with changes in the environment, not by yielding but by reacting according to his own desires and standards. Let us see how close a machine can come to this ability.

About the simplest form of self-regulating mechanical device is the controlled valve. Simple versions were devised as early as 50 A.D. by Hero of Alexandria, who used one in a device to dispense liquid automatically. A very elementary version of a safety valve is exemplified in a pressure cooker invented by Denis Papin in 1679. To keep the lid on against the steam pressure, he placed a weight on it, but he used a weight light enough so that the lid could fly off before the pressure rose to the point where the pot would explode. The present-day household pressure cooker or steam boiler has more sophisticated devices for this purpose (such as a plug that will melt when the temperature gets too high); but the principle is the same.

Of course, this is a "one-shot" sort of regulation. But it is easy to think of examples of continuous regulation. A primitive type was a device patented in 1745 by an Englishman, Edmund Lee, to keep a windmill facing squarely to the wind. He devised a "fantail" with small vanes that caught the wind whenever the wind shifted direction; the turning of these vanes operated a set of gears that rotated the windmill itself so that its main vanes were again head-on to the wind in the new quarter. In that position the fantail vanes remained motionless; they turned only when the windmill was not facing the wind.

But the archetype of modern mechanical self-regulators is the "governor" invented by James Watt for his steam engine. To keep the steam output of his engine steady, Watt conceived a device consisting of a vertical shaft with two weights attached to it laterally by hinged rods, allowing the weights to move up and down. The pressure of the steam whirled the shaft. When the steam pressure rose, the shaft whirled faster and the centrifugal force drove the weights upward. In moving up they partly closed a valve, choking off the flow of steam. As the steam pressur fell, the shaft whirled less rapidly, gravity pulled the weights down, and the valve opened. Thus the governor kept the shaft speed, and hence the power delivered, at a uniform level. Each departure from that level set in train a series of events that corrected the deviation. This is called "feedback": the error itself continually sends back information and serves as the measure of the correction required.

A very familiar example of a feedback device is the "thermostat," first used in very crude form by the Dutch inventor Cornelis Drebble in the early seventeenth century. A more sophisticated version, still used today, was invented in principle by a Scottish chemist named Andrew Ure in 1830. Its essential component consists of two strips of different metals laid against each other and soldered together. Since the two metals expand and contract at different rates with changes in temperature, the strip bends. The thermostat is set, say, at 70° F. When the room temperature falls below that, the thermocouple bends in such a fashion as to make a contact that closes an electric circuit and turns on the heating system. When the temperature rises above 70°, the thermocouple bends back enough to break the contact. Thus the heater regulates its own operation through feedback.

It is feedback that similarly controls the workings of the human body. To take one example of many, the glucose level in the blood is controlled by the insulin-producing pancreas, just as the temperature of a house is controlled by the heater. And just as the working of the heater is regulated by the departure of the temperature from the norm, so the secretion of insulin is regulated by the departure of the glucose concentration from the norm. A too-high glucose level turns on the insulin, just as a too-low temperature turns on the heater. Likewise, as a thermostat can be turned up to higher temperature, so an internal change in the body such as the secretion of adrenalin can raise the operation of the human body to a new norm, so to speak.

Self-regulation by living organisms to maintain a constant norm was named "homeostasis" by the American physiologist Walter Bradford

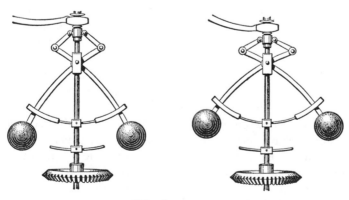

Watt's governor.

Cannon, who was a leader in investigation of the phenomenon in the first decades of the twentieth century.

Most systems, living and nonliving, lag a little in their response to feedback. For instance, after a heater has been turned off, it continues for a time to emit its residual heat; conversely, when it is turned on, it takes a little time to heat up. Therefore, the room temperature does not hold to 70° F. but oscillates around that level; it is always overshooting the mark on one side or the other. This phenomenon, called "hunting," was first studied in the 1830's by George Airy, the Astronomer Royal of England, in connection with devices he had designed to turn telescopes automatically with the motion of the earth.

Hunting is characteristic of most living processes, from control of the glucose level in the blood to conscious behavior. When you reach to pick up an object, the motion of your hand is not a single movement but a series of movements continually adjusted in both speed and direction, with the muscles correcting departures from the proper line of motion, those departures being judged by the eye. The corrections are so automatic that you are not aware of them. But watch an infant, not yet practiced in visual feedback, try to pick up something. It overshoots and undershoots because the muscular corrections are not precise enough. And victims of nerve damage that interferes with the ability to utilize visual feedback go into pathetic oscillations, or wild hunting, whenever they attempt a coordinated muscular movement.

The normal, practiced hand goes smoothly to its target and stops at the right moment because the control center looks ahead and makes corrections in advance. Thus, when you drive a car around a corner you begin to release the steering wheel before you have completed the turn, so that the wheels will be straight by the time you have rounded the corner. In other words, the correction is applied in time to avoid overshooting the mark to any significant degree.

It is the chief role of the cerebellum, evidently, to take care of this adjustment of motion by feedback. It looks ahead, and predicts the position of the arm a few instants ahead, organizing motion accordingly. It keeps the large muscles of the torso in constantly varying tensions to keep you in balance and upright if you are standing. It is hard work to stand and "do nothing"; we all know how tiring just standing can be.

Now this principle can be applied to a machine. Matters can be arranged so that as the system approaches the desired condition, the shrinking margin between its actual state and the desired state will automatically shut off the corrective force before it overshoots. In 1868, a French engineer, Léon Farcot, used this principle to invent an automatic control for a steam-operated ship's rudder. As the rudder approached the desired position, his device automatically closed down the steam valve; by the time the rudder reached the specified position, the steam pressure had been shut off. When the rudder moved away from this position, its motion opened the appropriate valve so that it was pushed back. Farcot called his device a "servomechanism," and in a sense it ushered in the era of "automation" (a term introduced in 1946 by the American engineer D. S. Harder).

Servomechanisms did not really come into their own until the arrival of electronics. The application of electronics made it possible to endow the mechanisms with a sensitivity and swiftness of response even beyond the capabilities of a living organism. Furthermore, radio extended their sphere of action over a considerable distance. The German buzz bomb of World War II was essentially a flying servomechanism, and it introduced the possibility not only of guided missiles but also of self-operated or remotely operated vehicles of all sorts, from subway trains to space ships. Because the military establishments had the keenest interest in these devices, and the most abundant supply of funds, servomechanisms have reached perhaps their highest development in aiming-and-firing mechanisms for guns and rockets. These systems can detect a swiftly moving target hundreds of miles away, instantly calculate its course (taking into account the target's speed of motion, the wind, the temperatures of the various layers of air, and numerous other conditions), and hit the target with pinpoint accuracy, all without any human guidance.

Automation found an ardent theoretician and advocate in the mathematician Norbert Wiener. In the 1940's he and his group at the Massachusetts Institute of Technology worked out some of the fundamental mathematical relationships governing the handling of feedback. He named this branch of study "cybernetics," from the Greek word for "helmsman," which seems appropriate, since the first use of servomechanisms was in connection with a helmsman. (Cybernetics also harks back to Watt's centrifugal governor, for "governor" comes from the Latin word for helmsman.)

Since World War II, automation has progressed fairly rapidly, especially in the United States and the Soviet Union. Oil refineries and factories manufacturing objects such as radio sets and aluminum pistons have been set up on an almost completely automatic basis, taking in raw materials at one end and pouring out finished products at the other, with all the processes handled by self-regulating machines. Automation has even invaded the farm: engineers at the University of Wisconsin announced in 1960 an automated hog-feeding system in which machines would feed each hog the correct amount of the correct type of feed at the correct time.

In a sense, automation marks the beginning of a new Industrial Revolution. Like the first revolution, it may bring great pains of adjustment, not only for workers but also for the economy in general. There is a story of an automobile executive who conducted a union official on a tour of a new automatic factory and observed: "You won't be able to collect union dues from these machines, I'm afraid." The union official shot back: "And you won't be able to sell them automobiles, either."

Automation will not make the human worker obsolete, any more than the steam engine or electricity did. But it will certainly mean a great shift in emphasis. The first Industrial Revolution made it no longer necessary for a man to be a muscle-straining workhorse. The second will make it no longer necessary for him to be a mind-dulled automaton.

Naturally, feedback and servomechanisms have stirred up as much interest among biologists as among engineers. Self-regulating machines can serve as simplified models for studying the workings of the nervous system.

A generation ago the imagination of men was excited—and disturbed—by Karel Čapek's play *R. U. R.* (for "Rossem's Universal Robots," robot coming from the Czech word meaning "to work"). In recent years scientists have begun to experiment with various forms of robots (sometimes called "automata") not as mere mechanical substitutes for men but as tools to explore the nature of living organisms. For instance, L. D. Harmon of the Bell Telephone Laboratories devised a transistorized circuit that, like a neuron, fires electrical pulses when stimulated. Such circuits can be assembled into devices that mimic some of the functions of the eye and the ear. In England the biologist W. Ross Ashby formed a system of circuits that exhibits simple reflex responses. He calls his creature a "homeostat," because it tends to maintain itself in a stable state.

The British neurologist William Grey Walter built a more elaborate system that explores and reacts to its surroundings. His turtlelike object, which he calls a "testudo" (Latin for tortoise"), has a photoelectric cell for an eye, a sensing device to detect touch, and two motors, one to move forward or backward and the other to turn around. In the dark, it crawls about, circling in a wide arc. When it touches an obstacle, it backs off a bit, turns slightly and moves forward again; it will do this until it gets around the obstacle. When its photoelectric eye sees a light, the turning motor shuts off and the testudo advances straight toward the light. But its phototropism is under control; as it gets close to the light the increase in brightness causes it to back away, so that it avoids the mistake of the moth. When its batteries run down, however, the now "hungry" testudo can crawl close enough to the light to make contact with a recharger placed near the light bulb. Once recharged, it is again sensitive enough to back away from the bright area around the light.

The subject of robots brings up the thought of machines that, in general, mimic living systems. In a sense, human beings, as tool-makers, have always mimicked what they have seen in nature about them. The knife is an artificial tusk; the lever is an artificial arm; the wheel developed from the roller, which in turn was inspired by the tree-trunk; and so on.

It is only very recently, however, that the full resources of science have been turned upon the effort to analyze the functioning of living tissues and organs, in order that the manner in which they perform—worked out hit-and-miss over billions of years of evolution—might be imitated in man-made machines. This study is called "bionics," a term—suggested by "*bio*logical electro*nics*" but much broader in scope—coined by the American engineer Jack Steele in 1960.

As one example of what bionics might do, consider the structure of dolphin skin. Dolphins swim at speeds that would require 2.6 horsepower if the water about them were as turbulent as it would be about a vessel of the same size. For some reason, water flows past the dolphin without turbulence, and therefore little power is consumed overcoming water resistance. Apparently this happens because of the nature of dolphin skin. If we can reproduce that effect in vessel walls, the speed of an ocean liner could be increased and its fuel consumption decreased—simultaneously.

Then, too, the American biophysicist Jerome Lettvin studied the frog's retina in detail by inserting tiny platinum electrodes into its optic nerve. It turned out that the retina did not merely transmit a melange of light and dark dots to the brain and leave it to the brain to do all the interpretation. Rather, there were five different types of cells in the retina, each designed for a particular job. One reacted to edges—that is, to sudden changes in the nature of illumination, as at the edge of a tree marked off against the sky. A second reacted to dark curved objects (the insects eaten by the frog). A third reacted to anything moving rapidly (a dangerous creature that might better be avoided). A fourth reacted to dimming light, and a fifth to the watery blue of a pond. In other words, the retinal message went to the brain already analyzed to a considerable degree. If man-made sensors made use of the tricks of the frog's retina, they could be made far more sensitive and versatile than they now are.

If, however, we are to build a machine that will imitate some living device, the most attractive possibility is the imitation of that unique device that interest us most profoundly—the human brain.

Thinking Machines

Can we build a machine that thinks? To try to answer that question we must first define "thinking."

Certainly we can take mathematics as a representation of one form of thinking. And it is a particularly good example for our purposes. For one thing, it is distinctly a human attribute. Some lower organisms are able to distinguish between three objects and four, say, but no species except *Homo sapiens* can perform the simple operation of dividing ¾ by ⅞. Second, mathematics involves a type of reasoning that operates by fixed rules and includes (ideally) no undefined terms or procedures. It can be analyzed in a more definite and more precise way than can the kind of thinking that goes into, say, literary composition or high finance or industrial management or military strategy. As early as 1936, the English mathematician Alan Mathison Turing showed that any problem could be solved mechanically if it could be expressed in the form of a finite number of manipulations that could be performed by the machine. So let us consider machines in relation to mathematics.

Tools to aid mathematical reasoning are undoubtedly as old as mathematics itself. The first tools for the purpose must have been man's own fingers. Mathematics began when man used his fingers to represent numbers and combinations of numbers. It is no accident that the word "digit" stands both for a finger (or toe) and for a numerical integer.

From that, another step leads to the use of other objects in place of fingers—small pebbles, perhaps. There are more pebbles than fingers, and intermediate results can be preserved for future reference in the course of solving the problem. Again, it is no accident that the word "calculate" comes from the Latin word for pebble.

Pebbles or beads lined up in slots or strung on wires formed the "abacus," the first really versatile mathematical tool. With this device it became easy to represent units, tens, hundreds, thousands, and so on. By manipulating the pebbles, or counters, of an abacus, one could quickly carry through an addition such as $576 + 289$. Furthermore, any instrument that can add can also multiply, for multiplication is only repeated addition. And multiplication makes raising to a power possible, because this is only repeated multiplication (e.g., 4^5 is shorthand for $4 \times 4 \times 4 \times 4 \times 4$). Finally, running the instrument backward, so to speak, makes possible the operations of subtraction, division, and extracting a root.

The abacus can be considered the second "digital computer." (The first, of course, was the fingers.)

For thousands of years the abacus remained the most advanced form of calculating tool. It actually dropped out of use in the West after the end of the Roman Empire and was reintroduced by Pope Sylvester II about 1000 A.D., probably from Moorish Spain, where its use had lingered. It was greeted on its return as an eastern novelty, its western ancestry forgotten.

The abacus was not replaced until a numerical notation was introduced which imitated the workings of the abacus. (This notation, the

844

fingers → abacus
logarithms → slide rule
Pascal gears → adding machine → electronic calculator

one familiar to us nowadays as "Arabic numerals," was originated in India some time about 800 A.D., was picked up by the Arabs, and finally introduced to the West about 1200 A.D. by the Italian mathematician Leonardo of Pisa.)

In the new notation, the nine different pebbles in the units row of the abacus were represented by nine different symbols, and those same nine symbols were used for the tens row, hundreds row, and thousands row. Counters differing only in position were replaced by symbols differing only in position, so that in the written number 222, for instance, the first 2 represents 200, the second 20, and third represents two itself; that is, $200 + 20 + 2 = 222$.

This "positional notation" was made possible by recognition of an all-important fact which the ancient users of the abacus had overlooked. Although there are only nine counters in each row of the abacus, there are actually ten possible arrangements. Besides using any number of counters from one to nine in a row, it is also possible to use *no* counter—that is, to leave the place at the counting position empty. This escaped all the great Greek mathematicians and was not recognized until the ninth century, when some unnamed Hindu thought of representing the tenth alternative by a special symbol which the Arabs called "sifr" ("empty") and which has come down to us, in consequence, as "cipher" or, in more corrupt form, "zero." The importance of the zero is recorded in the fact that the manipulation of numbers is still sometimes called "ciphering," and that to solve any hard problem is to "decipher" it.

Another powerful tool grew out of the use of the exponents to express powers of numbers. To express 100 as 10^2, 1,000 as 10^3, 100,000 as 10^5, and so on, is a great convenience in several respects; not only does it simplify the writing of large numbers but it reduces multiplication and division to simple addition or subtraction of the exponents (e.g., $10^2 \times 10^3 = 10^5$) and makes raising to a power or extraction of a root a simple matter of multiplying or dividing exponents (e.g., the cube root of 1,000,000 is $10^{6/3} = 10^2$). Now this is all very well, but very few numbers can be put into simple exponential form. What could be done with a number such as 111? The answer to that question led to the tables of logarithms.

The first to deal with this problem was the seventeenth-century Scottish mathematician John Napier. Obviously, expressing a number such as 111 as a power of 10 involves assigning a fractional exponent to 10 (the exponent is between 2 and 3). In more general terms, the exponent will be fractional whenever the number in question is not a multiple of the base number. Napier worked out a method of calculating the fractional exponents of numbers, and he named these exponents "logarithms." Shortly afterward the English mathematician Henry Briggs simplified the technique and worked out logarithms with 10 as the base.

The "Briggsian logarithms" are less convenient in calculus, but they are the more popular for ordinary computations.

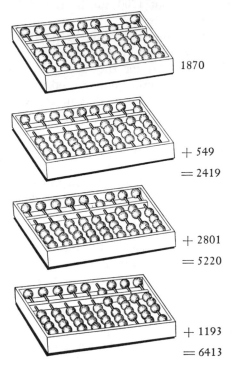

1870

+ 549

= 2419

+ 2801

= 5220

+ 1193

= 6413

Adding with an abacus. Each counter below the bar counts 1; each counter above the bar counts 5. A counter registers when it is pushed to the bar. Thus in the top setting here the right-hand column reads 0; the one to the left of that reads 7 or (5 + 2); the next left reads 8 or (5 + 3); and the next left reads 1: the number shown, then, is 1870. When 549 is added to this, the right column becomes 9 or (9 + 0); the next addition (4 + 7) becomes 1 with 1 to carry, which means that one counter is pushed up in the next column; the third addition is 9 + 5, or 4 with 1 to carry; and the fourth addition is 1 + 1 or 2: the addition gives 2419, as the abacus shows. The simple maneuver of carrying 1 by pushing up a counter in the next column makes it possible to calculate very rapidly; a skilled operator can add faster than an adding machine can, as was shown by an actual test in 1946.

All nonintegral exponents are irrational, that is, they cannot be expressed in the form of an ordinary fraction. They can be expressed only as an indefinitely long decimal expression lacking a repeating pattern. Such a decimal can be calculated, however, to as many places as necessary for the desired precision.

For instance, let us say we wish to multiply 111 by 254. The Briggsian logarithm of 111 to five decimal places is 2.04532, and for 254 it is 2.40483. Adding these logarithms, we get $10^{2.04532} \times 10^{2.40483} = 10^{4.45015}$. That number is approximately 28,194, the actual product of 111×254. If we want to get still closer accuracy, we can use the logarithms to six or more decimal places.

Tables of logarithms simplified computation enormously. In 1622 an English mathematician named William Oughtred made things still easier by devising a "slide rule." Two rulers are marked with a "logarithmic scale," in which the distances between numbers get shorter as the numbers get larger: for example, the first division holds the numbers from 1 to 10; the second division, of the same length, holds the numbers from 10 to 100; the third from 100 to 1,000, and so on. By sliding one rule along the other to an appropriate position, one can read off the result of an operation involving multiplication or division. The slide rule makes computations as easy as addition and subtraction on the abacus, though in both cases, to be sure, one must be skilled in the use of the instrument.

The first step toward a truly automatic calculating machine was taken in 1642 by the French mathematician Blaise Pascal. He invented an adding machine that did away with the need to move the counters separately in each row of the abacus. His machine consisted of a set of wheels connected by gears. When the first wheel—the units wheel— was turned ten notches to its 0 mark, the second wheel turned one notch to the number 1, so that the two wheels together showed the number 10. When the tens wheel reached its 0, the third wheel turned a notch, showing 100, and so on. (The principle is the same as that of the mileage indicator in an automobile.) Pascal is supposed to have had more than fifty such machines constructed; at least five are still in existence.

Pascal's device could add and subtract. In 1674 the German mathematician Gottfried Wilhelm von Leibnitz went a step further and arranged the wheels and gears so that multiplication and division were as automatic and easy as addition and subtraction. In 1850 a United States inventor named D. D. Parmalee patented an important advance which added greatly to the calculator's convenience: in place of moving the wheels by hand, he introduced a set of keys—pushing down a marked key with the finger turned the wheels to the correct number. This is the mechanism of what is now familiar to us as the old-fashioned cash register.

It remained only to electrify the machine (so that motors did the work dictated by the touching of keys), and the Pascal-Leibnitz device graduated into the modern desk computer.

The desk computer, however, represents a dead end, not the wave of the future. The computer that we have in mind when we consider thinking machines is an entirely different affair. Its granddaddy is an idea conceived early in the nineteenth century by an English mathematician named Charles Babbage.

calculating machine

analytical engine

A genius far in advance of his time, Babbage imagined an analytical machine that would be able to perform any mathematical operation, be instructed by punch cards, store numbers in a memory device, compare the results of operations, and so on. He worked for thirty-seven years on his ideas, spending a fortune, both his own and the government's money, putting together awesomely elaborate structures of wheels, cams, levers, and wires, in an age when every part had to be hand-fitted. In the end he failed, and died a bitterly disappointed man, because what he was seeking to achieve could not be accomplished with mere mechanical devices.

The machine had to wait a century for the development of electronics. And electronics in turn suggested the use of a mathematical language much easier for a machine to handle than the decimal system of numerals. It is called the "binary system" and was invented by Leibnitz. To understand the modern computer we must acquaint ourselves with this system.

The binary notation uses only two digits: 0 and 1. It expresses all numbers in terms of powers of 2. Thus, the number one is 2^0, the number two is 2^1, three is $2^1 + 2^0$, four is 2^2, and so on. As in the decimal system, the power is indicated by the position of the symbol. For instance, the number four is represented by 100, read thus: (1×2^2) $+ (0 \times 2^1) + (0 \times 2^0)$, or $4 + 0 + 0 = 4$ in the decimal system.

As an illustration, let us consider the number 6,413. In the decimal system it can be written $(6 \times 10^3) + (4 \times 10^2) + (1 \times 10^1) +$ (3×10^0); remember that any number to the zero power equals 1. Now in the binary system we add numbers in powers of 2, instead of powers of 10, to compose a number. The highest power of 2 that leaves us short of 6,413 is 12; 2^{12} is 4,096. If we now add 2^{11}, or 2,048, we have 6,144, which is 269 short of 6,413. Next, 2^8 adds 256 more, leaving 13; we can then add 2^3, or 8, leaving 5; then 2^2, or 4, leaving 1; and 2^0 is 1. Thus we might write the number 6,413 as $(1 \times 2^{12}) + (1 \times 2^{11}) + (1 \times 2^8)$ $+ (1 \times 2^3) + (1 \times 2^2) + (1 \times 2^0)$. But as in the decimal system, each digit in a number, reading from the left, must represent the next smaller power. Just as in the decimal system we represent the additions of the third, second, first, and zero powers of 10 in stating the number 6,413, so in the binary system we must represent the additions of the powers of 2 from 12 down to 0. In the form of a table this would read:

$$1 \times 2^{12} = 4{,}096$$
$$1 \times 2^{11} = 2{,}048$$
$$0 \times 2^{10} = 0$$
$$0 \times 2^9 = 0$$

$$1 \times 2^8 = 256$$
$$0 \times 2^7 = 0$$
$$0 \times 2^6 = 0$$
$$0 \times 2^5 = 0$$
$$0 \times 2^4 = 0$$
$$1 \times 2^3 = 8$$
$$1 \times 2^2 = 4$$
$$0 \times 2^1 = 0$$
$$1 \times 2^0 = 1$$
$$\overline{6{,}413}$$

Taking the successive multipliers in the column at the left (as we take 6, 4, 1, and 3 as the successive multipliers in the decimal system), we write the number in the binary system as 1100100001101.

This looks pretty cumbersome. It takes 13 digits to write the number 6,413, whereas in the decimal system we need only four. But for a computing machine the system is just about the simplest imaginable. Since there are only two different digits, any operation can be carried out in terms of just the two states of an electrical circuit—on and off. On (i.e., the circuit closed) can represent 1; off (the circuit open) can represent 0. With proper circuitry and the use of diodes the machine can carry out all sorts of mathematical manipulations. For instance, with two switches in parallel it can perform addition: $0 + 0 = 0$ (off plus off equals off), and $0 + 1 = 1$ (off plus on equals on). Similarly, two switches in series can perform multiplication: $0 \times 0 = 0$; $0 \times 1 = 0$; $1 \times 1 = 1$.

So much for the binary system. As for electronics, its contribution to the computing machine is incredible speed. It can perform an arithmetical operation almost instantaneously: some of the electronic computers carry out billions of operations per second! Calculations that would take a man with a pencil a lifetime can be completed by a computer in a matter of days. To cite a typical case, before the coming of the computer an English mathematician named William Shanks spent 15 years calculating the value of π, the ratio of the circumference of a circle to its diameter, and carried it to 707 decimal places (getting the last hundred-odd places wrong). Some years ago, an electronic computer carried out the calculation accurately to 10,000 decimal places, taking only a few days for the job.

In 1925, the American electrical engineer Vannevar Bush and his colleagues constructed a machine capable of solving differential equations. This was the first modern computer, though it still made use of mechanical switches, and represented a successful version of the sort of thing Babbage had tackled a century earlier.

For complete success, however, the switches had to be electronic. This would greatly increase the speed of computers for a flow of electrons can be started, shifted, or stopped in millionths of a second—far faster than mechanical switches, however delicate, could be manipulated.

The first large electronic computer, containing 19,000 vacuum tubes, was built at the University of Pennsylvania by John Presper Eckert and John William Mauchly during World War II. It was called ENIAC, for "Electronic Numerical Integrator and Computer." ENIAC ceased operation in 1955 and was dismantled in 1957, a hopelessly outmoded dotard at twelve years of age, but it left behind an amazingly numerous and sophisticated progeny. Whereas ENIAC weighed 30 tons and took up 1,500 square feet of floor space, the equivalent computer today—using switching units far smaller, faster, and more reliable than the old vacuum tubes—could be built into an object the size of a refrigerator. Modern computers contain half a million switching components and 10 million high-speed memory-elements.

So fast was progress that by 1948, small electronic computers were being produced in quantity; within five years, 2,000 were in use; by 1961, the number was 10,000. By 1970, the number had passed the 100,000 mark.

Computers working directly with numbers are "digital computers." They can work to any desired accuracy but can only answer the specific problem asked. An abacus, as I said earlier, is a primitive example. A computer may work not with numbers but with current-strengths made to vary in fashion analogous to the variables being considered. This is an "analog computer"; a famous example is Remington Rand's "Universal Analog Computer" (UNIVAC), the first large commercial computer, produced by Eckert and Mauchly in 1950 and put to use on television in 1952 to analyze the results of the presidential election as they came in. The slide rule is actually a very simple analog computer. Analog computers are of limited accuracy, but can supply answers, at one stroke, to a whole related family of questions.

Is ENIAC, or any of its more sophisticated descendants, truly a "thinking machine"? Hardly. Essentially it is no more than a very rapid abacus. It slavishly follows instructions that are fed into it.

The basic tool for instructing most computers is the punch card, still much like those Babbage once planned to use. A hole punched in a definite position on such a card can signify a number; a combination of two or more holes in specific positions can signify a letter of the alphabet, an operation of mathematics, a quality—anything one chooses to let it represent. Thus the card can record a person's name, the color of his eyes and hair, his position or income, his marital status, his special talents or educational qualifications. When the card is passed between electric poles, the contacts made through the holes set up a specific electrical pattern. By scanning cards in this way and selecting only those that set up a particular pattern, the machine can pick out of a large

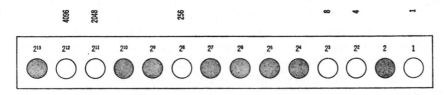

The number 6,413 represented by lights on a computer panel. The unshaded circles are lamps turned on.

population just those individuals, say, who are over six feet tall, have blue eyes, and speak Russian.

Similarly, a punch card carrying a "program" for a computer sets up electrical circuits that cause the machine to put certain numbers through certain operations. In all except the simplest programs, the computer has to store some numbers or data until it is ready to use them in the course of its series of operations. This means it must have a memory device. The memory in a modern computer usually is of the magnetic type: the information is stored in the form of magnetized spots on a spinning drum or, more recently, on magnetic tape. A particular spot may represent the digit 1 when it is magnetized and 0 when it is not magnetized. Each such item of information is a "bit" (for "binary digit"). The spots are magnetized by an electric current in recording the information and also are read by the production of an induced voltage.

It is quite apparent, now, however, that the useful punch card has had its day. Systems are in use whereby instructions can be fed computers (such as UNIVAC III, built in 1960) by means of typed English words. The computer is designed to react properly to certain words and combinations of words, which can be used quite flexibly but according to a definite set of rules. Such "computer languages" have proliferated rapidly; one of the better known ones is FORTRAN (short for "formula translation").

The computer is, of course, the crowning instrument that makes full automation possible. Servomechanisms can only carry out the slave aspects of a task. In any complex process, such as the aiming and firing of a gun or the operation of an automatic factory, computers are necessary to calculate, interpret, coordinate, and supervise—in other words, to serve as mechanical minds to direct the mechanical muscles. And the computer is by no means limited to this sort of function. It makes out your electric bills; it handles reservations for airlines; it keeps the records of checks and banking accounts; it deals with company payrolls, maintains a running record of inventories, in short, handles a great deal of the business of large corporations today.

The speed of a computer, and its immunity to tedium, make it possible to perform tasks too long-enduring (though not necessarily too difficult in principle) for the human brain. The computer can deal with purely random processes like the mutual collisions of thousands of particles by actually conducting mathematical "games" that simulate such processes and following thousands of cases in very short periods, in order to set up the probabilities of various possible events. This is called the "Monte Carlo method," since it is the same principle as that used by human gamblers to try to bring some sort of system into the workings of a roulette wheel.

The mathematicians, physicists, and engineers investigating the possible uses of computers are sure that all this is only a beginning. They have programed computers to play chess, as an exercise in storing and applying information. (Of course, the machine plays only a mediocre game, but very likely it will eventually be possible to program a computer to learn from its opponent's moves.) In laboratory experiments computers have been programed to digest articles from scientific journals, to index such articles, and to translate from one language into another. For instance, with a vocabulary of Russian words and their English equivalents stored in its memory, a computer can scan Russian print, recognize key words, and render a crude translation. A great deal of research is being done on this problem of translating by machine, because the flood of technical literature the world over is so vast that human translators cannot possibly keep up with it.

Not only can a computer "read"; it can also listen. Workers at the Bell Telephone Laboratories have built a machine called "Audrey" which can distinguish words of ordinary speech. Presumably this creates the possibility that computers will some day be able to "understand" spoken instructions. It is already possible with machines to program speech and even song.

In 1965, systems were devised to enable people or machines to communicate with the computer from remote terminals—via a typewriter, for instance. This is called "time-sharing," because many people may be sharing the computer's time, though response is so quick that each user is scarcely aware of the others.

These developments suggest still further possibilities. Computers are fast and far more efficient than a man in performing routine mathematical operations, but they do not begin to approach the flexibility of the human mind. Whereas even a child learning to read has no difficulty in recognizing that a capital B, a small b, and an italic *b* in various type sizes and styles all stand for the same letter, this is far beyond any present machine. To attain any such flexibility in a machine would call for prohibitive size and complexity.

The human brain weighs a little over 3 pounds and works on a practically negligible input of energy. In contrast, ENIAC, which was probably less than one millionth as complex as the brain, weighed 30 tons, and required 150 kilowatts of energy. To be sure, devices such as the transistor and the cryotron, a kind of switch employing superconducting wires at low temperatures, have miniaturized computers and reduced the power requirements. The use of ferrite components to store bits of information has increased the capacity of computer memory as the components have been made smaller. Ferrite components as small as a period on this page now exist.

Tiny tunnel sandwiches, operating under superconductive conditions and called "neuristors," bear the promise of being put together with the compactness and complexity of cells in the human brain. So perhaps we are in sight of duplicating the amazing qualities of that organ, which, in its three pounds, contains some 10 billion nerve cells together with 90 million auxiliary cells.

As computers grow more intricate, subtle, and complex at a breakneck rate, the question naturally arises: Even if computers today cannot truly think, will the day come when one will? They can, already, calculate, remember, associate, compare, recognize. Will one someday reason? The answer, many think (as, for instance, the American mathematician Marvin L. Minsky) is: Yes.

In 1938, a young American mathematician and engineer, Claude Elwood Shannon, pointed out in his master's thesis that deductive logic, in a form known as Boolean algebra, could be handled by means of the binary system. Boolean algebra refers to a system of "symbolic logic" suggested in 1854 by the English mathematician George Boole in a book entitled *An Investigation of the Laws of Thought*. Boole observed that the types of statement employed in deductive logic could be represented by mathematical symbols, and he went on to show how such symbols could be manipulated according to fixed rules to yield appropriate conclusions.

To take a very simple example, consider the following statement: "Both A and B are true." We are to determine the truth or falsity of this statement by a strictly logical exercise, assuming that we know whether A and B, respectively, are true or false. To handle the problem in binary terms, as Shannon suggested, let 0 represent "false" and 1 represent "true." If A and B are both false, then the statement "Both A and B are true" is false. In other words, 0 and 0 yield 0. If A is true but B is false (or vice versa), then the statement again is false. That is, 1 and 0 (or 0 and 1) yield 0. If A is true and B is true, then the statement "Both A and B are true" is true. Symbolically, 1 and 1 yield 1.

Now these three alternatives correspond to the three possible multiplications in the binary system, namely: $0 \times 0 = 0$, $1 \times 0 = 0$, and $1 \times 1 = 1$. Thus the problem in logic posed by the statement "Both A and B are true" can be manipulated by multiplication. A computer (properly programed) therefore can handle this logical problem as easily, and in the same way, as it handles ordinary calculations.

In the case of the statement "Either A or B is true," the problem is handled by addition instead of by multiplication. If neither A nor B is true, then this statement is false. In other words, $0 + 0 = 0$. If A is true and B false, or vice versa, the statement is true; in these cases $1 + 0 = 1$ and $0 + 1 = 1$. If both A and B are true, the statement is certainly true, and $1 + 1 = 10$. (The significant digit in the 10 is the 1; the fact that it is moved over one position is immaterial. In the binary system 10 represents $(1 \times 2^1) + (0 \times 2^0)$, which is equivalent to 2 in the decimal system.)

Boolean algebra has become important in the engineering of communications, and it forms part of what is now known as "information theory."

What remains of thinking, then, that cannot be acquired by the machine? We are left finally with creativity and the ability of the human mind to cope with the unknown: its intuition, judgment, weighing of a situation and the possible consequences—call it what you will. This, too, was put in mathematical form, after a fashion, by the mathematician John von Neumann, who with the economist Oskar Morgenstern wrote *The Theory of Games and Economic Behavior* in the early 1940's.

Von Neumann took certain simple games, such as matching coins and poker, as models for analysis of the typical situation in which one tries to find a winning strategy against an opponent who is himself selecting the best possible course of action for his own purposes. Military campaigns, business competition, many matters of great moment involve decisions of this kind. Even scientific research can be viewed as a game of man against nature, and game theory can be helpful in selecting the optimal strategy of research, assuming that nature stacks the cards in the way that will hamper man most (which, in fact, it often seems to do).

And as machines threaten to become recognizably human, humans become more mechanical. Mechanical organs can replace organic hearts and kidneys. Electronic devices can make up for organic failures so that artificial pacemakers can be implanted in bodies to keep hearts going that would otherwise stop. Artificial hands have been devised that can be controlled by amplified nerve-impulses in the arms to which they are attached. The word "cyborg" ("*cy*bernetic *org*anism") was coined in 1960 to refer to mechanically amplified men.

Attempts to mimic the mind of man are as yet in their infancy. The road, however, is open, and it conjures up thoughts which are exciting but also in some ways frightening. What if man eventually were to produce a mechanical creature, with or without organic parts, equal or superior to himself in all respects, including intelligence and creativity?

854

The original Univac, first of the large electronic computers.

Program on tape (at left) is being fed into Univac 1105, a computer used by the U. S. Census Bureau.

Disk memory unit of the IBM Ramac 305. The access arm
shown can write information bits on a disk or read them off.
Such a unit can hold up to 20 million bits.

Tape memories for a transistorized Univac.

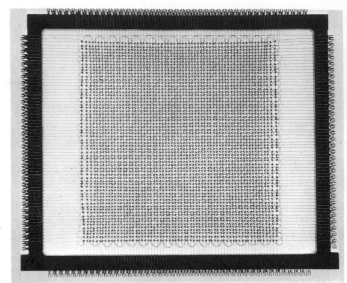

An array of ferrite cores used for the memory of an IBM computer. The complete memory consists of many such matrices stacked above one another.

Reading by machine. This IBM apparatus can read writing or printing on documents. The small "write" head at the lower left first magnetizes the ink of the printed characters, and the "read" head to the right of it then reads the magnetic patterns that identify the characters.

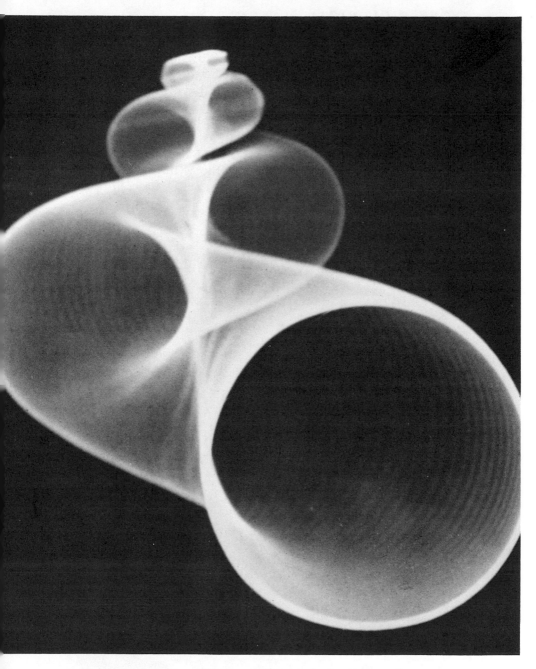

Computer snail, an oscilloscope photograph showing a coil generated by a computer programed by biologists to simulate the
growth of a snail shell. In biological studies of morphology,
computers are helpful in constructing accurate geometries typical
of actual species as well as types that are never found in nature.

859

Would it replace man, as the superior organisms of the earth have replaced or subordinated the less well-adapted in the long history of evolution?

It is a queasy thought: that we represent, for the first time in the history of life on the earth, a species capable of bringing about its own possible replacement. Of course, we have it in our power to prevent such a regrettable denouement by refusing to build machines that are too intelligent. But it is tempting to build them nevertheless. What achievement could be grander than the creation of an object that surpasses the creator? How could we consummate the victory of intelligence over nature more gloriously than by passing on our heritage, in triumph, to a greater intelligence—of our own making?

APPENDIX

Mathematics in Science

Gravitation

As I explained in Chapter 1, Galileo initiated science in its modern sense by introducing the concept of reasoning back from observation and experiment to basic principles. In doing so, he also introduced the essential technique of measuring natural phenomena accurately and abandoned the practice of merely describing them in general terms. In short, he turned from the qualitative description of the universe by the Greek thinkers to a quantitative description.

Although science depends so much on mathematical relationships and manipulations, and could not exist in the Galilean sense without it, I have nevertheless written this book nonmathematically, and have done so deliberately. Mathematics, after all, is a highly specialized tool. To have discussed the developments in science in mathematical terms would have required a prohibitive amount of space, as well as a sophisticated knowledge of mathematics on the part of the reader. But in this appendix I would like to present an example or two of the way in which simple mathematics has been fruitfully applied to science. How better to begin than with Galileo himself?

Galileo (like Leonardo da Vinci nearly a century earlier) suspected that falling objects steadily increased their velocity as they fell. He set out to measure exactly by how much and in what manner the velocity increased.

861

The measurement was anything but easy for Galileo, with the tools he had at his disposal in 1600. To measure a velocity requires the measurement of time. We speak of velocities of 60 miles *an hour*, of 13 feet *a second*. But there were no clocks in Galileo's time that could do more than strike the hour at approximately equal intervals.

Galileo resorted to a crude water clock. He let water trickle slowly from a small spout, assuming, hopefully, that it dripped at a constant rate. This water he caught in a cup, and by the weight of water caught during the interval in which an event took place, Galileo measured the elapsed time. (He also used his pulse beat for the purpose on occasion.)

One difficulty was, however, that a falling object dropped so rapidly that Galileo could not collect enough water, in the interval of falling, to weigh accurately. What he did, then, was to "dilute" the pull of gravity by having a brass ball roll down a groove in an inclined plane. The more nearly horizontal the plane, the more slowly the ball moved. Thus Galileo was able to study falling bodies in whatever degree of "slow motion" he pleased.

Galileo found that a ball rolling on a perfectly horizontal plane moved at constant speed. (This supposes a lack of friction, a condition which could be assumed within the limits of Galileo's crude measurements.) Now a body moving on a horizontal track is moving at right angles to the force of gravity. Under such conditions, the body's velocity is not affected by gravity either way. A ball resting on a horizontal plane remains at rest, as anyone can observe. A ball set to moving on a horizontal plane moves at a constant velocity, as Galileo observed.

Mathematically, then, it can be stated that the velocity v of a body, *in the absence of any external force*, is constant k, or:

$$v = k$$

If k is equal to any number other than zero, the ball is moving at constant velocity. If k is equal to zero, the ball is at rest; thus, rest is a "special case" of constant velocity.

Nearly a century later, when Newton systemized the discoveries of Galileo in connection with falling bodies, this finding became the First Law of Motion (also called the "principle of inertia"). This law can be stated: Every body persists in a state of rest or of uniform motion in a straight line unless compelled by external force to change that state.

When a ball rolls down an inclined plane, however, it is under the continuous pull of gravity. Its velocity then, Galileo found, was not constant but increased with time. Galileo's measurements showed that the velocity increased in proportion to the lapse of time t.

In other words, when a body was under the action of constant external force, its velocity, starting at rest, could be expressed as:

$$v = kt$$

What was the value of k?

That, it was easy to find by experiment, depended on the slope of the inclined plane. The more nearly vertical the plane, the more quickly the rolling ball gained velocity and the higher the value of k. The maximum gain in speed would come when the plane was vertical—in other words, when the ball dropped freely under the undiluted pull of gravity. The symbol g (for "gravity") is used where the undiluted force of gravity is acting, so that the velocity of a ball in free fall, starting from rest, was:

$$v = gt$$

Let us consider the inclined plane in more detail. In the diagram:

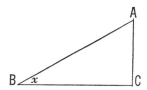

the length of the inclined plane is AB, while its height at the upper end is AC. The ratio of AC to AB is the sine of the angle x, usually abbreviated as sin x.

The value of this ratio—that is, of sin x—can be obtained approximately by constructing triangles with particular angles and actually measuring the height and length involved. Or it can be calculated by mathematical techniques to any degree of precision, and the results can be embodied in a table. By using such a table, we can find, for instance, that sin $10°$ is approximately equal to 0.17365, that sin $45°$ is approximately equal to 0.70711, and so on.

There are two important special cases. Suppose that the "inclined" plane is precisely horizontal. Angle x is then zero and as the height of the inclined plane is zero, the ratio of its height to its length is also zero. In other words, sin $0° = 0$. When the "inclined" plane is precisely vertical, the angle it forms with the ground is a right angle, or $90°$. Its height is then exactly equal to its length, so that the ratio of one to the other is just 1. Consequently, sin $90° = 1$.

Now let us return to the equation showing that the velocity of a ball rolling down an inclined plane is proportional to time:

$$v = kt$$

It can be shown by experiment that the value of k changes with the sine of the angle so that:

$$k = k' \sin x$$

(where k' is used to indicate a constant that is different from k).

(As a matter of fact, the role of the sine in connection with the inclined plane was worked out somewhat before Galileo's time by Simon Stevinus, who also performed the famous experiment of dropping different masses from a height, an experiment traditionally, but wrongly, ascribed to Galileo. Still, if Galileo was not the very first to experiment and measure, he was the first to impress the scientific world, indelibly, with the necessity to experiment and measure, and that is glory enough.)

In the case of a completely vertical inclined plane, sin x becomes sin 90°, which is 1, so that in free fall

$$k = k'$$

It follows that k' is the value of k in free fall under the undiluted pull of gravity, which we have already agreed to symbolize as g. We can substitute g for k' and, for any inclined plane:

$$k = g \sin x.$$

The equation for the velocity of a body rolling down an inclined plane is, therefore:

$$v = (g \sin x) \, t$$

On a horizontal plane with sin $x = 0° = 0$, the equation for velocity becomes:

$$v = 0$$

This is another way of saying that a ball on a horizontal plane, starting from rest, will remain motionless regardless of the passage of time. An object at rest tends to remain at rest, and so on. That is part of the First Law of Motion, and it follows from the inclined plane equation of velocity.

Suppose that a ball does not start from rest but has an initial motion before it begins to fall. Suppose, in other words, you have a ball moving along a horizontal plane at 5 feet per second, and it suddenly finds itself at the upper end of an inclined plane and starts rolling downward.

Experiment shows that its velocity thereafter is 5 feet per second greater, at every moment, than it would have been if it had started rolling down the plane from rest. In other words, the equation for the motion of a ball down an inclined plane can be expressed more completely as follows:

$$v = (g \sin x) \, t + V$$

where V is the original starting velocity. If an object starts at rest, then V is equal to 0 and the equation becomes as we had it before:

$$v = (g \sin x) \, t$$

If we next consider an object with some initial velocity on a horizontal plane, so that angle x is $0°$, the equation becomes:

$$v = (g \sin 0°) + V$$

or, since $\sin 0°$ is 0:

$$v = V$$

Thus the velocity of such an object remains its initial velocity, regardless of the lapse of time. That is the rest of the First Law of Motion, again derived from observed motion on an inclined plane.

The rate at which velocity changes is called "acceleration." If, for instance, the velocity (in feet per second) of a ball rolling down an inclined plane is, at the end of successive seconds, 4, 8, 12, 16 . . . then the acceleration is 4 feet per second per second.

In a free fall, if we use the equation:

$$v = gt$$

each second of fall brings an increase in velocity of g feet per second. Therefore, g represents the acceleration due to gravity.

The value of g can be determined from inclined-plane experiments. By transposing the inclined-plane equation, we get:

$$g = v / (t \sin x)$$

Since v, t, and x can all be measured, g can be calculated, and it turns out to be equal to 32 feet per second per second at the earth's surface. In free fall under normal gravity at earth's surface, then, the velocity of fall is related to time thus:

$$v = 32t$$

This is the solution to Galileo's original problem, namely, determining the rate of fall of a falling body and the manner in which that rate changes.

The next question is: How far does a body fall in a given time? From the equation relating the velocity to time, it is possible to relate distance to time by the process in calculus called "integration." It is not necessary to go into that, however, because the equation can be worked out by experiment, and, in essence, Galileo did this.

He found that a ball rolling down an inclined plane covered a distance proportional to the square of the time. In other words, doubling the time increased the distance fourfold, tripling it increased the distance ninefold, and so on.

For a freely falling body, the equation relating distance d and time is:

$$d = \tfrac{1}{2}gt^2$$

or, since g is equal to 32:

$$d = 16t^2$$

Next, suppose that instead of dropping from rest, an object is thrown horizontally from a position high in the air. Its motion would then be a compound of two motions—a horizontal one and a vertical one.

The horizontal motion, involving no force other than the single original impulse (if we disregard wind, air resistance, and so on), is one of constant velocity, in accordance with the First Law of Motion, and the distance the object covers horizontally is proportional to the time elapsed. The vertical motion, however, covers a distance, as I have just explained, that is proportional to the square of the time elapsed. Prior to Galileo, it had been vaguely believed that a projectile such as a cannon ball traveled in a straight line until the impulse that drove it was somehow exhausted, after which it fell straight down. Galileo, however, made the great advance of *combining* the two motions.

The combination of these two motions (proportional to time horizontally, and proportional to the square of the time vertically) produces a curve called a parabola. If a body is thrown, not horizontally, but upward or downward, the curve of motion is still a parabola.

Such curves of motion, or trajectories, apply, of course, to a projectile such as a cannon ball. The mathematical analysis of trajectories, stemming from Galileo's work, made it possible to calculate where a cannon ball would fall when fired with a given propulsive force and a given angle of elevation of the cannon. Although men had been throwing objects for fun, to get food, to attack, and to defend, for uncounted thousands of years, it was only due to Galileo that for the first time, thanks to experiment and measurement, there was a science of "ballistics." As it happened, then, the very first achievement of modern science proved to have a direct and immediate military application.

It also had an important application in theory. The mathematical analysis of combinations of more than one motion answered several objections to the Copernican theory. It showed an object thrown upward would not be left behind by the moving earth, since the object would have two motions, one imparted to it by the impulse of throwing, and one that it shared along with the moving earth. It also made it reasonable to expect the earth to have two motions at once, rotation about its axis and revolution about the sun—a situation that some of the non-Copernicans insisted was unthinkable.

Isaac Newton extended the Galilean concepts of motion to the heavens and showed that the same set of laws of motion applied to the heavens and the earth alike.

He began by considering that the moon might be falling toward the earth in response to the earth's gravity but never struck the earth's surface because of the horizontal component of its motion. A projectile fired horizontally, as I said, follows a parabolically curved path downward to intersection with earth's surface. But the earth's surface curves downward, too, since the earth is a sphere. A projectile given a sufficiently rapid horizontal motion might curve downward no faster than the earth's surface and would therefore eternally circle the earth.

Now the moon's elliptical motion around the earth can be split into horizontal and vertical components. The vertical component is such that in the space of a second the moon falls a trifle more than 1/20 of an inch toward the earth. In that time it also moves about 3,300 feet in the horizontal direction, just far enough to compensate for the fall and carry it around the earth's curvature.

The question was whether this 1/20 inch fall of the moon was caused by the same gravitational attraction that caused an apple, falling from a tree, to drop 16 feet in the first second of its fall.

Newton visualized the earth's gravitational force as spreading out in all directions like a vast, expanding sphere. The surface area A of a sphere is proportional to the square of its radius r:

$$A = 4\pi r^2$$

He therefore reasoned that the gravitational force, spreading out over the spherical area, must weaken as the square of the radius. The intensity of light and of sound weakened as the square of the distance from the source—why not the force of gravity as well?

The distance from the earth's center to an apple on its surface is roughly 4,000 miles. The distance from the earth's center to the moon is roughly 240,000 miles. Since the distance to the moon was 60 times greater than to the apple, the force of the earth's gravity at the moon must be 60^2, or 3,600, times weaker than at the apple. Divide 16 feet by 3,600, and you come out with roughly 1/20 of an inch. It seemed clear to Newton that the moon did indeed move in the grip of the earth's gravity.

Newton was persuaded further to consider "mass" in relation to gravity. Ordinarily, we measure mass as weight. But weight is only the result of the attraction of the earth's gravitational force. If there were no gravity, an object would be weightless; nevertheless, it would still contain the same amount of matter. Mass, therefore, is independent of weight and should be capable of measurement by a means not involving weight.

Suppose you tried to pull an object on a perfectly frictionless surface in a direction horizontal to the earth's surface, so that there was no resistance from gravity. It would take effort to set the body in motion and to accelerate its motion, because of the body's inertia.

If you measured the applied force accurately, say by pulling on a spring balance attached to the object, you would that the force f required to bring about a given acceleration a would be directly proportional to the mass m. If you doubled the mass, it would take double the force. For a given mass, the force required would be directly proportional to the acceleration desired. Mathematically, this is expressed in the equation:

$$f = ma$$

The equation is known as Newton's Second Law of Motion.

Now, as Galileo had found, the pull of the earth's gravity accelerates all bodies, heavy or light, at precisely the same rate. (Air resistance may slow the fall of very light bodies, but in a vacuum a feather will fall as rapidly as a lump of lead, as can easily be demonstrated.) If the Second Law of Motion is to hold, one must conclude that the earth's gravitational pull on a heavy body must be greater than on a light body, in order to produce the same acceleration. To accelerate a mass that is eight times as great as another, for instance, takes eight times as much force. It follows that the earth's gravitational pull on any body must be exactly proportional to the mass of that body. (That, in fact, is why mass on the earth's surface can be measured quite accurately as weight.)

Newton evolved a Third Law of Motion, too: "For every action there is an equal and opposite reaction." This applies to force. In other words, if the earth pulls at the moon with a certain force, then the moon pulls on the earth with an equal force. If the moon were suddenly doubled in mass, the earth's gravitational force upon it would also be doubled, in accordance with the Second Law; of course, the moon's gravitational force on the earth would then have to be doubled in accordance with the Third Law.

Similarly, if it were the earth rather than the moon that doubled in mass, it would be the moon's gravitational force on the earth that would double, according to the Second Law, and the earth's gravitational force on the moon that would double, in accordance with the Third.

If both the earth and the moon were to double in mass, there would be a doubled doubling, each body doubling its gravitational force twice, for a fourfold increase all told.

Newton could only conclude by this sort of reasoning that the gravitational force between any two bodies in the universe was directly proportional to the product of the masses of the bodies. And, of course, as he had decided earlier, it is inversely proportional to the square of the distance (center to center) between the bodies. This is Newton's Law of Universal Gravitation.

If we let f represent the gravitational force, m_1 and m_2 the masses of the two bodies concerned, and d the distance between them, then the law can be stated:

$$f = \frac{Gm_1m_2}{d^2}$$

G is the "gravitational constant," the determination of which made it possible to "weigh the earth" (see Chapter 3). It was Newton's surmise that G had a fixed value throughout the universe. As time went on, it was found that new planets, undiscovered in Newton's time, tempered their motions to the requirements of Newton's law; even double stars incredibly far away danced in time to Newton's analysis of the universe.

All this came from the new quantitative view of the universe pioneered by Galileo. As you see, much of the mathematics involved was really very simple. Those parts of it I have quoted here are high-school algebra.

In fact, all that was needed to introduce one of the greatest intellectual revolutions of all time was:

1. A simple set of observations any high-school student of physics might make with a little guidance.

2. A simple set of mathematical generalizations.

3. The transcendent genius of Galileo and Newton, who had the insight and originality to make these observations and generalizations for the first time.

Relativity

The laws of motion as worked out by Galileo and Newton depended on the assumption that such a thing as absolute motion existed—that is, motion with reference to something at rest. But everything that we know of in the universe is in motion: the earth, the sun, the Galaxy, the systems of galaxies. Where in the universe, then, can we find absolute rest against which to measure absolute motion?

It was this line of thought that led to the Michelson-Morley experiment, which in turn led to a scientific revolution as great, in some respects, as that initiated by Galileo (see Chapter 7). Here, too, the basic mathematics is rather simple.

The experiment was an attempt to detect the absolute motion of the earth against an "ether" that was supposed to fill all space and to be at rest. The reasoning behind the experiment was as follows.

Suppose that a beam of light is sent out in the direction in which the earth is traveling through the ether, and that at a certain distance in that direction there is a fixed mirror which reflects the light back to the source. Let us symbolize the velocity of light as c, the velocity of the earth through the ether as v, and the distance of the mirror as d. The light starts with the velocity $c + v$: its own velocity plus the earth's velocity. (It is traveling with a tail wind, so to speak.) The time it takes to reach the mirror is d divided by $(c + v)$.

On the return trip, however, the situation is reversed. The reflected light now is bucking the head wind of the earth's velocity, and its net velocity is $c - v$. The time it takes to return to the source is d divided by $(c - v)$.

The total time for the round trip is:

$$\frac{d}{c + v} + \frac{d}{c - v}$$

Combining the terms algebraically, we get:

$$\frac{d(c - v) + d(c + v)}{(c + v)\,(c - v)} = \frac{dc - dv + dc + dv}{c^2 - v^2} = \frac{2dc}{c^2 - v^2}$$

Now suppose that the light beam is sent out to a mirror at the same distance in a direction at right angles to the earth's motion through the ether.

The beam of light is aimed from S (the source) to M (the mirror) over the distance d. However, during the time it takes the light to reach

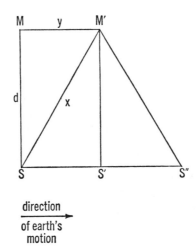

direction
of earth's
motion

the mirror, the earth's motion has carried the mirror from M to M', so that the actual path traveled by the light beam is from S to M'. This distance we call x, and the distance from M to M' we call y.

While the light is moving the distance x at its velocity c, the mirror is moving the distance y at the velocity of the earth's motion v. Since both the light and the mirror arrive at M' simultaneously, the distances traveled must be exactly proportional to the respective velocities. Therefore:

$$\frac{y}{x} = \frac{v}{c}$$

or

$$y = \frac{vx}{c}$$

Now we can solve for the value of x by use of the Pythagorean theorem, which states that the sum of the squares of the sides of a right triangle is equal to the square of the hypotenuse. In the right triangle SMM', then, substituting vx/c for y:

$$x^2 = d^2 + \left(\frac{vx}{c}\right)^2$$

$$x^2 - \left(\frac{vx}{c}\right)^2 = d^2$$

$$x^2 - \frac{v^2x^2}{c^2} = d^2$$

$$\frac{c^2x^2 - v^2x^2}{c^2} = d^2$$

$$(c^2 - v^2)x^2 = d^2c^2$$

$$x^2 = \frac{d^2c^2}{c^2 - v^2}$$

$$x = \frac{dc}{\sqrt{c^2 - v^2}}$$

The light is reflected from the mirror at M' to the source, which meanwhile has traveled on to S'. Since the distance S'S" is equal to SS', the distance M'S" is equal to x. The total path traveled by the light beam is therefore 2x, or 2dc / $\sqrt{c^2 - v^2}$.

The time taken by the light beam to cover this distance at its velocity c is:

$$\frac{2dc}{\sqrt{c^2 - v^2}} \div c = \frac{2d}{\sqrt{c^2 - v^2}}$$

How does this compare with the time that light takes for the round trip in the direction of the earth's motion? Let us divide the time in the parallel case (2dc / (c² − v²)) by the time in the perpendicular case (2d / $\sqrt{c^2 - v^2}$):

$$\frac{2dc}{c^2 - v^2} \div \frac{2d}{\sqrt{c^2 - v^2}} = \frac{2dc}{c^2 - v^2} \times \frac{\sqrt{c^2 - v^2}}{2d} = \frac{c\sqrt{c^2 - v^2}}{c^2 - v^2}$$

Now any number divided by its square root gives the same square root as a quotient, that is, x / $\sqrt{x}$ = $\sqrt{x}$. Conversely, $\sqrt{x}$ / x = 1 / $\sqrt{x}$. So the last equation simplifies to:

$$\frac{c}{\sqrt{c^2 - v^2}}$$

This expression can be further simplified if we multiply both the numerator and the denominator by $\sqrt{1/c^2}$ (which is equal to $1/c$).

$$\frac{c\sqrt{1/c^2}}{\sqrt{c^2-v^2}\sqrt{1/c^2}} = \frac{c/c}{\sqrt{c^2/c^2 - v^2/c^2}} = \frac{1}{\sqrt{1-v^2/c^2}}$$

And there you are. That is the ratio of the time that light should take to travel in the direction of the earth's motion as compared with the time it should take in the direction perpendicular to the earth's motion. For any value of v greater than zero, the expression $1/\sqrt{1-v^2/c^2}$ is greater than 1. Therefore, if the earth is moving through a motionless ether, it should take longer for light to travel in the direction of the earth's motion than in the perpendicular direction. (In fact, the parallel motion should take the maximum time and the perpendicular motion the minimum time.)

Michelson and Morley set up their experiment to try to detect the directional difference in the travel time of light. By trying their beam of light in all directions, and measuring the time of return by their incredibly delicate interferometer, they felt they ought to get differences in apparent velocity. The direction in which they found the velocity of light to be at a minimum should be parallel to the earth's absolute motion, and the direction in which the velocity would be at a maximum should be perpendicular to the earth's motion. From the difference in velocity, the amount (as well as the direction) of the earth's absolute motion could be calculated.

They found no differences at all in the velocity of light with changing direction! To put it another way, the velocity of light was always equal to c, regardless of the motion of the source—a clear contradiction of the Newtonian laws of motion. In attempting to measure the absolute motion of the earth, Michelson and Morley had thus managed to cast doubt not only on the existence of the ether, but on the whole concept of absolute rest and absolute motion, and upon the very basis of the Newtonian system of the universe (see Chapter 7).

The Irish physicist G. F. FitzGerald conceived a way to save the situation. He suggested that all objects decreased in length in the direction in which they were moving by an amount equal to $\sqrt{1-v^2/c^2}$. Thus:

$$L' = L\sqrt{1-v^2/c^2}$$

where L' is the length of a moving body in the direction of its motion and L is what the length would be if it were at rest.

The foreshortening fraction $\sqrt{1-v_2/c^2}$, FitzGerald showed, would just cancel the ratio $1\sqrt{-v^2/c^2}$, which related the maximum and minimum velocities of light in the Michelson-Morley experiment. The ratio would become unity, and the velocity of light would seem to our foreshortened instruments and sense organs to be equal in all directions, regardless of the movement of the source of light through the ether.

Under ordinary conditions the amount of foreshortening is very small. Even if a body were moving at one tenth the velocity of light, or 18,628 miles per second, its length would be foreshortened only slightly, according to the FitzGerald equations. Taking the velocity of light as 1, the equation says:

$$L' = L \sqrt{\left(1 - \frac{0.1}{1}\right)^2}$$
$$L' = L \sqrt{1 - 0.01}$$
$$L' = L \sqrt{0.99}$$

Thus L' turns out to be approximately equal to $0.995L$, a foreshortening of about half of 1 per cent.

For moving bodies, velocities such as this occur only in the realm of the subatomic particles. The foreshortening of an airplane traveling at 2,000 miles per hour is infinitesimal, as you can calculate for yourself.

At what velocity will an object be foreshortened to half its rest length? With L' equal to one-half L, the FitzGerald equation is:

$$L/2 = L\sqrt{1 - v^2/c^2}$$

or, dividing by L:

$$1/2 = \sqrt{1 - v^2/c^2}$$

Squaring both sides of the equation:

$$1/4 = 1 - v^2/c^2$$
$$v^2/c^2 = 3/4$$
$$v = \sqrt{3c/4} = 0.866c$$

Since the velocity of light in a vacuum is 186,282 miles per second, the velocity at which an object is foreshortened to half its length is 0.866 times 186,282, or roughly 161,300 miles per second.

If a body moves at the speed of light, so that v equals c, the Fitz-Gerald equation becomes:

$$L' = L\sqrt{1 - c^2/c^2} = L\sqrt{0} = 0$$

At the speed of light, then, length in the direction of motion becomes zero. It would seem, therefore, that no velocity faster than that of light is possible.

In the decade after FitzGerald had advanced his equation, the electron was discovered, and scientists began to examine the properties of tiny charged particles. Lorentz worked out a theory that the mass of a particle with a given charge was inversely proportional to its radius. In other words, the smaller the volume into which a particle crowded its charge, the greater its mass.

Now if a particle is foreshortened because of its motion, its radius in the direction of motion is reduced in accordance with the FitzGerald equation. Substituting the symbols R and R' for L and L', we write the equation:

$$R' = R\sqrt{1 - v^2/c^2}$$
$$R'/R = \sqrt{1 - v^2/c^2}$$

The mass of a particle is inversely proportional to its radius. Therefore:

$$\frac{R'}{R} = \frac{M}{M'}$$

where M is the mass of the particle at rest and M' is its mass when in motion.

Substituting M / M' for R' / R in the preceding equation, we have:

$$\frac{M}{M'} = \sqrt{1 - v^2/c^2}$$
$$M' = \frac{M}{\sqrt{1 - v^2/c^2}}$$

The Lorentz equation can be handled just as the FitzGerald equation was. It shows, for instance, that for a particle moving at a velocity of 18,628 miles per second (one tenth the speed of light) the mass M' would appear to be 0.5 per cent higher than the rest mass M. At a velocity of 161,300 miles per second the apparent mass of the particle would be twice the rest mass.

Finally, for a particle moving at a velocity equal to that of light, so that v is equal to c, the Lorentz equation becomes:

$$M' = \frac{M}{\sqrt{1 - c^2/c^2}} = \frac{M}{0}$$

Now as the denominator of any fraction with a fixed numerator becomes smaller and smaller ("approaches zero"), the value of the fraction itself becomes larger and larger without limit. In other words, from the equation above, it would seem that the mass of any object traveling at a velocity approaching that of light becomes infinitely large. Again, the velocity of light would seem to be the maximum possible.

All this led Einstein to recast the laws of motion and of gravitation. He considered a universe, in other words, in which the results of the Michelson-Morley experiments were to be expected.

Yet even so we are not quite through. Please note that the Lorentz equation assumes some value for M that is greater than zero. This is true for most of the particles with which we are familiar and for all bodies, from atoms to stars, that are made up of such particles. There are, however, neutrinos and antineutrinos for which M, the mass at rest, or "rest-mass," is equal to zero. This is also true of photons.

Such particles travel at the speed of light in a vacuum, provided they are indeed in a vacuum. The moment they are formed they begin to move at such a velocity without any measurable period of acceleration.

We might wonder how it is possible to speak of the "rest-mass" of a photon or a neutrino, if they are never at rest but can only exist while travelling (in the absence of interfering matter) at a constant speed of 186,280 miles per second. The physicists O. M. Bilaniuk and E. C. G. Sudarshan have therefore suggested that M be spoken of as "proper mass." For a particle with mass greater than zero, the proper mass is equal to the mass measured when the particle is at rest relative to the instruments and observer making the measurement. For a particle with mass equal to zero, the proper mass is obtained by indirect reasoning. Bilaniuk and Sudarshan also suggest that all particles with a proper mass of zero be called "luxons" (from the Latin word for "light") because they travel at light-speed, while particles with a proper mass greater than zero be called "tardyons" because they travel at less than light-speed, or at "subluminal velocities."

In 1962, Bilaniuk and Sudarshan began to speculate on the consequences of faster-than-light velocities ("superluminal velocities"). Any particle travelling with faster-than-light velocities would have an imaginary mass. That is, the mass would be some ordinary value multiplied by the square root of -1.

Suppose, for instance, a particle were going at twice the speed of light, so that in the Lorentz equation $v = 2c$. In that case:

$$M' = \frac{M}{\sqrt{1 - (2c)^2/c^2}} = \frac{M}{\sqrt{1 - 4c^2/c^2}} = \frac{M}{\sqrt{-3}}$$

This works out to the fact that its mass while in motion would be some proper mass (M) divided by $\sqrt{-3}$. But $\sqrt{-3}$ is equal to $3 \times \sqrt{-1}$ and therefore to $1.74\sqrt{-1}$. The proper mass M is therefore equal to $M' \times 1.74 \times \sqrt{-1}$. Since any quantity that includes $\sqrt{-1}$ is called imaginary, we conclude that particles at superluminal velocities must have imaginary proper masses.

Ordinary particles in our ordinary universe always have masses that are zero or positive. An imaginary mass can have no imaginable significance in our universe. Does this mean that faster-than-light particles cannot exist?

Not necessarily. Allowing the existence of imaginary proper-masses, we can make such faster-than-light particles fit all the equations of Einstein's Special Theory of Relativity. Such particles, however, display an apparently paradoxical property: the more slowly they go, the more energy they contain. This is the precise reverse of the situation in our universe and is perhaps the significance of the imaginary mass. A particle with an imaginary mass speeds up when it meets resistance and slows down when it is pushed ahead by a force. As its energy declines, it moves faster and faster, until when it has zero energy it is moving at infinite speed. As its energy increases, it moves slower and slower until, as its energy approaches the infinite, it slows down to approach the speed of light.

Such faster-than-light particles have been given the name of "tachyons" from the Greek word for "speed," by the American physicist Gerald Feinberg.

We may imagine, then, the existence of two kinds of universes. One, our own, is the tardyon-universe, in which all particles go at subluminal velocities and may accelerate to nearly the speed of light as their energy increases. The other is the tachyon-universe, in which all particles go at superluminal velocities and may decelerate to nearly the speed of light as their energy increases. Between is the infinitely narrow "luxon wall" in which there are particles that go at exactly luminal velocities. The luxon wall can be considered as being held by both universes in common.

If a tachyon is energetic enough and therefore moving slowly enough, it might have sufficient energy and remain in one spot for a long enough period of time to give off a detectable burst of photons. (Tachyons would leave a wake of photons even in a vacuum as a kind of Cerenkov radiation.) Scientists are watching for those bursts, but the chance of happening to have an instrument in just the precise place where one of those (possibly very infrequent) bursts appears for a trillionth of a second or less, is not very great.

There are those physicists who maintain that "anything that is not forbidden is compulsory." In other words, any phenomenon that does not actually break a conservation law, *must* at some time or another take place; or, if tachyons do not actually violate special relativity, they *must* exist. Nevertheless, even physicists most convinced of this as a kind of necessary "neatness" about the universe, would be rather pleased (and perhaps relieved) to obtain some evidence for the non-forbidden tachyons. So far, they have not been able to.

One consequence of the Lorentz equation was worked out by Einstein to produce what has become perhaps the most famous scientific equation of all time.

The Lorentz equation can be written in the form:

$$M' = M \, (1 - v^2/c^2)^{-1/2}$$

since in algebraic notation $1/\sqrt{x}$ can be written $x^{-1/2}$. This puts the equation into a form which can be expanded (that is, converted into a series of terms) by a formula discovered by, of all people, Newton. The formula is the binomial theorem.

The number of terms into which the Lorentz equation can be expanded is infinite, but since each term is smaller than the one before, if you take only the first two terms you are approximately correct, the sum of all the remaining terms being small enough to be neglected. The expansion becomes:

$$(1 - v^2/c^2)^{-1/2} = 1 + \frac{\frac{1}{2}v^2}{c^2} \cdot \cdot \cdot \cdot$$

Substituting that in the Lorentz equation, we get:

$$M' = M\left(1 + \frac{\frac{1}{2}v^2}{c^2}\right) = M + \frac{\frac{1}{2}Mv^2}{c^2}$$

Now in classical physics the expression $\frac{1}{2}Mv^2$ represents the energy of a moving body. If we let the symbol e stand for energy, the equation above becomes:

$$M' = M + e/c^2$$

or

$$M' - M = e/c^2$$

The increase in mass due to motion $(M' - M)$ can be represented as m, so:

$$m = e/c^2$$

or

$$e = mc^2$$

It was this equation that for the first time indicated mass to be a form of energy. Einstein went on to show that the equation applied to all mass, not merely to the increase in mass due to motion.

Here again, most of the mathematics involved is only at the high-school level. Yet it presented the world with the beginnings of a view of the universe greater and broader even than that of Newton, and also pointed the way to concrete consequences. It pointed the way to the nuclear reactor and the atom bomb, for instance.

Bibliography

A guide to science would be incomplete without a guide to more reading. I am setting down here a brief selection of books. The list is a miscellaneous one and does not pretend to be a comprehensive collection of the best modern books about science, but I have read most or all of each of them myself and can highly recommend all of them, even my own.

CHAPTER 1 — *What Is Science?*

BERNAL, J. D. *Science in History.* Hawthorn Books, New York, 1965.

CLAGETT, MARSHALL, *Greek Science in Antiquity.* Abelard-Schuman, New York, 1955.

CROMBIE, A. C. *Medieval and Early Modern Science* (2 vols.). Doubleday & Company, New York, 1959.

DAMPIER, SIR WILLIAM CECIL, *A History of Science.* Cambridge University Press, New York, 1958.

DREYER, J. L. E., *A History of Astronomy from Thales to Kepler.* Dover Publications, New York, 1953.

FORBES, R. J. and DIJKSTERHUIS, E. J., *A History of Science and Technology* (2 vols.). Penguin Books, Baltimore, 1963.

MASON, S. F., *Main Currents of Scientific Thought.* Abelard-Schuman, New York, 1953.

TATON, R. (editor), *History of Science* (4 vols.). Basic Books, New York, 1963–1966.

CHAPTER 2 — *The Universe*

ALTER, D., CLEMINSHAW, C. H., and PHILLIPS, J. G., *Pictorial Astronomy* (3rd rev. ed.). Thomas Y. Crowell, New York, 1969.

BIBLIOGRAPHY

Asimov, Isaac, *The Universe* (rev. ed.). Walker & Company, New York, 1971.

Bonnor, William. *The Mystery of the Expanding Universe*. Macmillan Company, New York, 1964.

Burbidge, G. and Burbidge, M., *Quasi-Stellar Objects*, Freeman, San Francisco, 1967.

Flammarion, G. C. *et al.*, *The Flammarion Book of Astronomy*. Simon & Schuster, New York, 1964.

Gamow, George, *The Creation of the Universe*. Viking Press, New York, 1955.

Hoyle, Fred, *Astronomy*. Doubleday & Company, New York, 1962.

Hoyle, Fred, *Frontiers of Astronomy*. New American Library, New York, 1955.

Johnson, Martin, *Astronomy of Stellar Energy and Decay*. Dover Publications, New York, 1959.

Kruse, W. and Dieckvoss, W., *The Stars*. University of Michigan Press, Ann Arbor, 1957.

Ley, Willy, *Watchers of the Skies*, Viking Press, New York, 1966.

McLaughlin, Dean B., *Introduction to Astronomy*. Houghton Mifflin Company, Boston, 1961.

Opik, Ernst J., *The Oscillating Universe*. New American Library, New York, 1960.

Sciama, D. W., *The Unity of the Universe*. Doubleday & Company, New York, 1959.

Shklovskii, I. S. and Sagan, Carl, *Intelligent Life in the Universe*, Holden-Day, San Francisco, 1966.

Smith, F. Graham, *Radio Astronomy*. Penguin Books, Baltimore, 1960.

Struve, Otto and Zebergs, Velta, *Astronomy of the 20th Century*. Macmillan Company, New York, 1962.

Whipple, Fred L., *Earth, Moon and Planets* (rev. ed.). Harvard University Press, Cambridge, Mass., 1963.

Whithrow, G. J., *The Structure and Evolution of the Universe*. Harper & Brothers, New York, 1959.

Chapter 3 — *The Earth*

Adams, Frank Dawson, *The Birth and Development of the Geological Sciences*. Dover Publications, New York, 1938.

Burton, Maurice, *Life in the Deep*. Roy Publishers, New York, 1958.

Gamow, George, *A Planet Called Earth*. Viking Press, New York, 1963.

Gilluly, J., Waters, A. C., and Woodford, A. O., *Principles of Geology*. W. H. Freeman & Company, San Francisco, 1958.

Gutenberg, B. (editor), *Internal Constitution of the Earth*. Dover Publications, New York, 1951.

Hurley, Patrick M., *How Old Is the Earth?* Doubleday & Company, New York, 1959.

Kuenen, P. H., *Realms of Water*. John Wiley & Sons, New York, 1963.

880

MASON, BRIAN, *Principles of Geochemistry*. John Wiley & Sons, New York, 1958.

MOORE, RUTH, *The Earth We Live On*. Alfred A. Knopf, New York, 1956.

SCIENTIFIC AMERICAN (editors), *The Planet Earth*. Simon & Schuster, New York, 1957.

CHAPTER 4 — *The Atmosphere*

BATES, D. R. (editor), *The Earth and Its Atmosphere*. Basic Books, New York, 1957.

GLASSTONE, SAMUEL, *Sourcebook on the Space Sciences*, Van Nostrand, New York, 1965.

LEY, WILLY, *Rockets, Missiles, and Space Travel*. Viking Press, New York, 1957.

LOEBSACK, THEO, *Our Atmosphere*. New American Library, New York, 1961.

NEWELL, HOMER E., JR., *Window in the Sky*. McGraw-Hill Book Company, New York, 1959.

NININGER, H. H., *Out of the Sky*. Dover Publications, New York, 1952.

ORR, CLYDE, JR., *Between Earth and Space*. Collier Books, New York, 1961.

CHAPTER 5 — *The Elements*

ALEXANDER, W. and STREET, A., *Metals in the Service of Man*. Penguin Books, New York, 1954.

ASIMOV, ISAAC, *A Short History of Chemistry*. Doubleday & Company, New York, 1965.

ASIMOV, ISAAC, *The Noble Gases*. Basic Books, New York, 1966.

DAVIS, HELEN MILES, *The Chemical Elements*. Ballantine Books, Boston, 1959.

HOLDEN, ALAN and SINGER, PHYLIS, *Crystals and Crystal Growing*. Doubleday & Company, New York, 1960.

IHDE, AARON, J., *The Development of Modern Chemistry*. Harper & Row, New York, 1964.

JAFFE, BERNARD, *Chemistry Creates a New World*. Thomas Y. Crowell, New York, 1957.

LEICESTER, HENRY M., *The Historical Background of Chemistry*. John Wiley & Sons, New York, 1956.

PAULING, LINUS, *College Chemistry* (3rd ed.). W. H. Freeman & Company, San Francisco, 1964.

SCIENTIFIC AMERICAN (editors), *New Chemistry*. Simon & Schuster, New York, 1957.

WEAVER, E. C. and FOSTER, L. S., *Chemistry for Our Times*. McGraw-Hill Book Company, New York, 1947.

WEEKS, MARY E. and LEICESTER, H. M. *Discovery of the Elements* (7th ed.). Journal of Chemical Education, Easton, Pa., 1968.

CHAPTER 6 — *The Particles*

ALFREN, HANNES, *Worlds Antiworlds*. Freeman, San Francisco, 1966.

ASIMOV, ISAAC, *The Neutrino*. Doubleday & Company, New York, 1966.

FORD, KENNETH W., *The World of Elementary Particles*. Blaisdell Publishing Company, New York, 1963.

FRIEDLANDER, G., KENNEDY, J. W., and MILLER, J. M., *Nuclear and Radiochemistry* (2nd ed.). Wiley & Sons, New York, 1964.

GAMOW, GEORGE, *Mr. Tompkins Explores the Atom*. Cambridge University Press, New York, 1955.

GARDNER, MARTIN, *The Ambidextrous Universe*. Basic Books, New York, 1964.

GLASSTONE, SAMUEL, *Sourcebook on Atomic Energy* (3rd ed.). D. Van Nostrand Company, Princeton, N.J., 1967.

HUGHES, DONALD J., *The Neutron Story*. Doubleday & Company, New York, 1959.

MASSEY, SIR HARRIE, *The New Age in Physics*. Harper & Brothers, New York, 1960.

PARK, DAVID, *Contemporary Physics*. Harcourt, Brace & World, New York, 1964.

SHAMOS, M. H. and MURPHY, G. M. (editors), *Recent Advances in Science*. New York University Press, New York, 1956.

CHAPTER 7 — *The Waves*

BENT, H. A., *The Second Law*. Oxford University Press, New York, 1965.

BERGMANN, P. G., *The Riddle of Gravitation*. Charles Scribner's Sons, New York, 1968.

BLACK, N. H. and LITTLE, E. P., *An Introductory Course in College Physics*. Macmillan Company, New York, 1957.

EDDINGTON, SIR ARTHUR S., *The Nature of the Physical World*. Cambridge University Press, New York, 1953.

EINSTEIN, ALBERT and INFELD, LEOPOLD, *The Evolution of Physics*. Simon & Schuster, New York, 1938.

FREEMAN, IRA M., *Physics Made Simple*. Made Simple Books, New York, 1954.

GARDNER, MARTIN, *Relativity for the Million*. Macmillan Company, New York, 1962.

HOFFMAN, BANESH, *The Strange Story of the Quantum*. Dover Publications, New York, 1959.

ROSSI, BRUNO, *Cosmic Rays*. McGraw-Hill Book Company, New York, 1964.

SHAMOS, MORRIS H., *Great Experiments in Physics*. Henry Holt & Company, New York, 1959.

CHAPTER 8 — *The Machine*

BITTER, FRANCIS, *Magnets*. Doubleday & Company, New York, 1959.

DE CAMP, L. SPRAGUE, *The Ancient Engineers*. Doubleday & Co., New York, 1963.

KOCK, W. E., *Lasers and Holography*. Doubleday & Company, New York, 1969.

LARSEN, EGON, *Transport*. Roy Publishers, New York, 1959.

LEE, E. W., *Magnetism*. Penguin Books, Baltimore, 1963.

LENGYEL, BELA A., *Lasers*. John Wiley & Sons, New York, 1962.

NEAL, HARRY EDWARD, *Communication*. Julius Messner, New York, 1960.

PIERCE, JOHN R., *Electrons, Waves and Messages*. Doubleday & Company, New York, 1956.

PIERCE, JOHN R., *Symbols, Signals and Noise*, Harper & Brothers, New York, 1961.

SINGER, CHARLES, HOLMYARD, E. J., and HALL, A. R. (editors), *A History of Technology* (5 vols.). Oxford University Press, New York, 1954.

TAYLOR, F. SHERWOOD, *A History of Industrial Chemistry*. Abelard-Schuman, New York, 1957.

UPTON, MONROE, *Electronics for Everyone* (2nd rev. ed.). New American Library, New York, 1959.

USHER, ABBOTT PAYSON, *A History of Mechanical Inventions*. Beacon Press, Boston, 1959.

WARSCHAUER, DOUGLAS M., *Semiconductors and Transistors*. McGraw-Hill Book Company, New York, 1959.

CHAPTER 9 — *The Reactor*

ALEXANDER, PETER, *Atomic Radiation and Life*. Penguin Books, New York, 1957.

BISHOP, AMASA S., *Project Sherwood*. Addison-Wesley Publishing Company, Reading, Mass, 1958.

FOWLER, JOHN M., *Fallout: A Study of Superbombs, Strontium 90, and Survival*. Basic Books, New York, 1960.

JUKES, JOHN, *Man-Made Sun*. Abelard-Schuman, New York, 1959.

JUNGK, ROBERT, *Brighter Than a Thousand Suns*. Harcourt, Brace & Company, New York, 1958.

PURCELL, JOHN, *The Best-Kept Secret*. Vanguard Press, New York, 1963.

RIEDMAN, SARAH R., *Men and Women behind the Atom*. Abelard-Schuman, New York, 1958.

SCIENTIFIC AMERICAN (editors), *Atomic Power*. Simon & Schuster, New York, 1955.

WILSON, ROBERT R. and LITTAUER, R., *Accelerators*. Doubleday & Company, New York, 1960.

CHAPTER 10 — *The Molecule*

FIESER, L. F., and M., *Organic Chemistry*. D. C. Heath & Company, Boston, 1956.

GIBBS, F. W., *Organic Chemistry Today*. Penguin Books, Baltimore, 1961.

HUTTON, KENNETH, *Chemistry*. Penguin Books, New York, 1957.

PAULING, LINUS, *The Nature of the Chemical Bond* (3rd ed.). Cornell University Press, Ithaca, N.Y., 1960.

PAULING, LINUS and HAYWARD, R., *The Architecture of Molecules*. W. H. Freeman & Co., San Francisco, 1964.

CHAPTER 11 — *The Proteins*

ASIMOV, ISAAC, *Photosynthesis*. Basic Books, New York, 1969.

BALDWIN, ERNEST, *Dynamic Aspects of Biochemistry* (5th ed.). Cambridge University Press, New York, 1967.

BALDWIN, ERNEST, *The Nature of Biochemistry*. Cambridge University Press, New York, 1962.

HARPER, HAROLD A., *Review of Physiological Chemistry* (8th ed.). Lange Medical Publications, Los Altos, Calif., 1961.

KAMEN, MARTIN D., *Isotopic Tracers in Biology*. Academic Press, New York, 1957.

KARLSON, P., *Introduction to Modern Biochemistry*. Academic Press, New York, 1963.

LEHNINGER, A. L., *Bioenergetics*. Benjamin Company, New York, 1965.

SCIENTIFIC AMERICAN (editors), *The Physics and Chemistry of Life*. Simon & Schuster, New York, 1955.

CHAPTER 12 — *The Cell*

ANFINSEN, CHRISTIAN B., *The Molecular Basis of Evolution*. John Wiley & Sons, New York, 1959.

ASIMOV, ISAAC, *The Genetic Code*. Orion Press, New York, 1962.

ASIMOV, ISAAC, *A Short History of Biology*. Doubleday & Company, New York, 1964.

BOYD, WILLIAM C., *Genetics and the Races of Man*. Little, Brown & Company, Boston, 1950.

BUTLER, J. A. V., *Inside the Living Cell*. Basic Books, New York, 1959.

DOLE, STEPHEN H., *Habitable Planets for Man*. Blaisdell Publishing Company, New York, 1964.

HARTMAN, P. E. and SUSKIND, S. R., *Gene Action*. Prentice-Hall, Englewood Cliffs, 1965.

HUGHES, ARTHUR, *A History of Cytology*. Abelard-Schuman, New York, 1959.

NEEL, J. V. and SCHULL, W. J., *Human Heredity*. University of Chicago Press, Chicago, 1954.

OPARIN, A. I., *The Origin of Life on the Earth*. Academic Press, New York, 1957.

SINGER, CHARLES, *A Short History of Anatomy and Physiology from the Greeks to Harvey*. Dover Publications, New York, 1957.

SULLIVAN, WALTER, *We Are Not Alone*. McGraw-Hill Book Company, New York, 1964.

TAYLOR, GORDON R., *The Science of Life*. McGraw-Hill Book Company, New York, 1963.

WALKER, KENNETH, *Human Physiology*. Penguin Books, New York, 1956.

CHAPTER 13 — *The Microorganism*

BURNET, F. M., *Viruses and Man* (2nd ed.). Penguin Books, Baltimore, 1955.

DE KRUIF, PAUL, *Microbe Hunters*. Harcourt, Brace & Company, New York, 1932.

DUBOS, RENÉ, *Louis Pasteur*. Little, Brown & Company, Boston, 1950.

LUDOVICI, L. J., *The World of the Microscope*. G. P. Putnam's Sons, New York, 1959.

McGRADY, PAT, *The Savage Cell*. Basic Books, New York, 1964.

RIEDMAN, SARAH R., *Shots without Guns*. Rand, McNally & Company, Chicago, 1960.

SMITH, KENNETH M., *Beyond the Microscope*. Penguin Books, Baltimore, 1957.

WILLIAMS, GREER, *Virus Hunters*. Alfred A. Knopf, New York, 1959.

ZINSSER, HANS, *Rats, Lice and History*. Little, Brown & Company, Boston, 1935.

CHAPTER 14 — *The Body*

ASIMOV, ISAAC, *The Human Body*. Houghton Mifflin Company, Boston, 1963.

CARLSON, ANTON J. and JOHNSON, VICTOR, *The Machinery of the Body*. University of Chicago Press, Chicago, 1953.

CHANEY, MARGARET S., *Nutrition*. Houghton Mifflin Company, Boston, 1954.

McCOLLUM, ELMER VERNER, *A History of Nutrition*. Houghton Mifflin Company, Boston, 1957.

WILLIAMS, ROGER J., *Nutrition in a Nutshell*. Doubleday & Company, New York, 1962.

CHAPTER 15 — *The Species*

ASIMOV, ISAAC, *Wellsprings of Life*. Abelard-Schuman, New York, 1960.

BOULE, M. and VALLOIS, H. V., *Fossil Men*. Dryden Press, New York, 1957.

CARRINGTON, RICHARD, *A Biography of the Sea*. Basic Books, New York, 1960.

DARWIN, FRANCIS (editor), *The Life and Letters of Charles Darwin* (2 vols.). Basic Books, New York, 1959.

DE BELL, G., *The Environmental Handbook*. Ballantine Books, New York, 1970.

HANRAHAN, JAMES S. and BUSHNELL, DAVID, *Space Biology*. Basic Books, New York, 1960.

HARRISON, R. J., *Man, the Peculiar Animal*. Penguin Books, New York, 1958.

HOWELLS, WILLIAM, *Mankind in the Making*. Doubleday & Company, New York, 1959.

HUXLEY, T. H., *Man's Place in Nature*. University of Michigan Press, Ann Arbor, 1959.

MEDAWAR, P. B., *The Future of Man*. Basic Books, New York, 1960.

MILNE, L. J. and M. J., *The Biotic World and Man*. Prentice-Hall, New York, 1958.

MONTAGU, ASHLEY, *The Science of Man*. Odyssey Press, New York, 1964.

MOORE, RUTH, *Man, Time, and Fossils* (2nd ed.). Alfred A. Knopf, New York, 1963.

ROMER, A. S., *Man and the Vertebrates* (2 vols.). Penguin Books, New York, 1954.

ROSTAND, JEAN, *Can Man Be Modified?* Basic Books, New York, 1959.

SAX, KARL, *Standing Room Only*. Beacon Press, Boston, 1955.

SIMPSON, GEORGE G., PITTENDRIGH, C. S., and TIFFANY, L. H., *Life: An Introduction to College Biology* (2nd ed.). Harcourt, Brace & Company, New York, 1965.

TINBERGEN, NIKO, *Curious Naturalists*. Basic Books, New York, 1960.

UBBELOHDE, A. R., *Man and Energy*. Penguin Books, Baltimore, 1963.

CHAPTER 16 — *The Mind*

ANSBACHER, H. and R. (editors), *The Individual Psychology of Alfred Adler*. Basic Books, New York, 1956.

ARIETI, SILVANO (editor), *American Handbook of Psychiatry* (2 vols.). Basic Books, New York, 1959.

ASIMOV, ISAAC, *The Human Brain*. Houghton Mifflin Company, Boston, 1964.

BERKELEY, EDMUND C., *Symbolic Logic and Intelligent Machines*. Reinhold Publishing Corporation, New York, 1959.

FREUD, SIGMUND, *Collected Papers* (5 vols.). Basic Books, New York, 1959.

JONES, ERNEST, *The Life and Work of Sigmund Freud* (3 vols.). Basic Books, New York, 1957.

LASSEK, A. M., *The Human Brain*. Charles C Thomas, Springfield, Ill., 1957.

MENNINGER, KARL, *Theory of Psychoanalytic Technique*. Basic Books, New York, 1958.

MURPHY, GARDNER, *Human Potentialities*. Basic Books, New York, 1958.

RAWCLIFFE, D. H., *Illusions and Delusions of the Supernatural and Occult*. Dover Publications, New York, 1959.

SCIENTIFIC AMERICAN (editors), *Automatic Control*. Simon & Schuster, New York, 1955.

SCOTT, JOHN PAUL, *Animal Behavior*. University of Chicago Press, Chicago, 1957.

BIBLIOGRAPHY

THOMPSON, CLARA and MULLAHY, PATRICK, *Psychoanalysis: Evolution and Development.* Grove Press, New York, 1950.

APPENDIX — *Mathematics in Science*

COURANT, RICHARD and ROBBINS, HERBERT, *What Is Mathematics?* Oxford University Press, New York, 1941.

DANTZIG, TOBIAS, *Number, the Language of Science.* Macmillan Company, New York, 1954.

FELIX, LUCIENNE, *The Modern Aspect of Mathematics.* Basic Books, New York, 1960.

FREUND, JOHN E., *A Modern Introduction to Mathematics.* Prentice-Hall, New York, 1956.

KLINE, MORRIS, *Mathematics and the Physical World.* Thomas Y. Crowell Company, New York, 1959.

KLINE, MORRIS, *Mathematics in Western Culture.* Oxford University Press, New York, 1953.

NEWMAN, JAMES R., *The World of Mathematics* (4 vols.). Simon & Schuster, New York, 1956.

STEIN, SHERMAN K., *Mathematics, the Man-Made Universe.* W. H. Freeman & Company, San Francisco, 1963.

VALENS, EVANS G., *The Number of Things.* Dutton & Co., New York, 1964.

General

ASIMOV, ISAAC, *Asimov's Biographical Encyclopedia of Science and Technology.* Doubleday & Company, New York, 1964.

ASIMOV, ISAAC, *Life and Energy.* Doubleday & Company, New York, 1962.

ASIMOV, ISAAC, *Understanding Physics* (3 vol.). Walker & Company, New York, 1966.

ASIMOV, ISAAC, *The Words of Science.* Houghton Mifflin Company, Boston, 1959.

CABLE, E. J. et al., *The Physical Sciences.* Prentice-Hall, New York, 1959.

GAMOW, GEORGE, *Matter, Earth, and Sky.* Prentice-Hall, New York, 1958.

HUTCHINGS, EDWARD, JR. (editor), *Frontiers in Science.* Basic Books, New York, 1958.

SHAPLEY, HARLOW, RAPPORT, SAMUEL, and WRIGHT, HELEN (editors), *A Treasury of Science* (4th ed.). Harper & Brothers, New York, 1958.

SLABAUGH, W. H. and BUTLER, A. B., *College Physical Science.* Prentice-Hall, New York, 1958.

WATSON, JANE WERNER, *The World of Science.* Simon & Schuster, New York, 1958.

Illustration Acknowledgments

888

ILLUSTRATION ACKNOWLEDGMENTS

Page

446 100 KILOTON NUCLEAR CRATER—Lawrence Radiation Laboratory, Livermore, California.

461 SOVIET EXPERIMENTS IN FUSION—SOVFOTO.

462 LOS ALAMOS EXPERIMENTS IN FUSION—Los Alamos Scientific Laboratory.

462 PINCHED PLASMA—Los Alamos Scientific Laboratory.

463 THE LIFE AND DEATH OF A PINCH—Los Alamos Scientific Laboratory.

464 THE BELL SOLAR BATTERY—Bell Telephone Laboratories.

464 A THERMOELECTRIC CELL—United Press International.

465 LASER BEAM—Korad, Union Carbide Corporation.

466 AN ELECTRON MICROSCOPE—RCA Instruments.

489 STRUCTURE OF UREA—Dr. Benjamin Post, Brooklyn Polytechnical Institute.

489 EIGHT-SIDED RING—Dr. Benjamin Post, Brooklyn Polytechnical Institute.

490 SILK FIBERS—Siemens New York, Inc.

491 COLLAGEN FIBERS—The Rockefeller Institute.

492 SECTION OF A MUSCLE FIBER—Dr. David S. Smith, The Rockefeller Institute.

509 THE RUBBER MOLECULE—U.S. Rubber Company.

510 VULCANIZATION OF RUBBER—U.S. Rubber Company.

511 A MODERN PLASTICS PLANT—U.S. Rubber Company.

512 GLASSWARE—U.S. Rubber Company.

589 NORMAL CHROMOSOMES—Franklyn Branley, Ed., *Scientists' Choice*, published by Basic Books, Inc.

590 CHROMOSOMES DAMAGED BY RADIATION—Brookhaven National Laboratory.

590 RADIATION DAMAGE—Brookhaven National Laboratory.

591 MUTATIONS IN FRUIT FLIES—Brookhaven National Laboratory.

592 SPERM CELL—Siemens New York, Inc.

593 HEART CELL—The Rockefeller Institute.

594 LIVER CELL—The Rockefeller Institute.

595 MITOCHONDRIA—The Rockefeller Institute.

596 RIBOSOMES—The Rockefeller Institute.

613 RIBONUCLEOPROTEIN PARTICLES—Dr. J. F. Kirsch, The Rockefeller Institute.

614 DNA-PROTEIN COMPLEX—United Press International.

615 STANLEY MILLER'S HISTORIC EXPERIMENT—United Press International.

616 MARINER IV PHOTOGRAPH—National Aeronautics and Space Administration.

665 LOUIS PASTEUR—The Bettmann Archive.

665 JOSEPH LISTER—The Bettmann Archive.

666 STAPHYLOCOCCUS BACTERIA—National Institutes of Health.

667 THE ROD-SHAPED BACILLUS—Siemens New York, Inc.

668 CARNIVAL OF CRYSTALS—Naval Biological Laboratory, Oakland, California.

669 THE TOBACCO MOSAIC VIRUS—Siemens New York, Inc.

670 THE INFLUENZA VIRUS—National Institutes of Health.

671 BACTERIOPHAGES—Franklyn Branley, Ed., *Scientists' Choice*, published by Basic Books, Inc.

672 AN X-RAY MICROGRAPH OF A BEETLE—Dr. David S. Smith, Cavendish Physics Laboratory, University of Cambridge.

673 CANCER-PRODUCING PARTICLES—Dr. Dan H. Moore, The Rockefeller Institute.

745 FOSSIL OF A CRINOID—American Museum of Natural History.

746 A PALM-LEAF FOSSIL—American Museum of Natural History.

747 FOSSIL OF A BRYOZOAN—United Press International.

747 FOSSIL OF A FORAMINIFER—United Press International.

748 A 3,000-YEAR-OLD CRUSTACEAN—SOVFOTO.

748 AN ANCIENT ANT—American Museum of Natural History.

749 MODEL OF AN EXTINCT LUNG-FISH—American Museum of Natural History.

ILLUSTRATION ACKNOWLEDGMENTS

Name Index

Subject Index

abacus, 844ff.
aberration, chromatic, 69
aberration, light, 343, 344
abnormal hemoglobin, 605, 606
abrasive, 271
absolute motion, 347ff.
absolute rest, 347ff.
absolute space, 348
absolute zero, 261; atomic properties near, 265ff.; temperatures near, 268, 269
absorption spectrum, 539
abstraction, 8
abyss, oceanic, 131
acceleration, 804
accelerators, particle, 301ff.
acetate ion, 565
acetic acid, 475
acetylation, 560
acetylcholine, 823
acetylene, 478
achromatic lenses, 69
achromatic microscope, 640
achromycin, 649
Acrilan, 524
acromegaly, 726
acrylonitrile, 524
ACTH, 555, 725
actinides, 259, 260
actinium, 238
action at a distance, 346
activated carbon, 548
actomyosin, 546
adamantine substances, 417
Addison's disease, 722, 724

adenine, 608
adenosine monophosphate, 728
adenosine triphosphate (ATP), 557
adenyl cyclase, 728
adipic acid, 523, 524
adolescence, 722
Adonis, 209
adrenal glands, 716, 722
adrenalin, 716; nerve action and, 823
adrenocortical hormones, 564
adrenocorticotropic hormone (ACTH), 555, 725
adsorption, 548
afferent nerve cells, 807
agammaglobulinemia, 683
agar, 645
agnatha, 739
air conditioning, 262, 519
airglow, 177, 178
airplane, 405
air resistance, 868
alabamine, 239
alanine, 553, 627
alanine-transfer-RNA, 621
albino, 601
albumen, 528
albuminous substances, 528
alchemy, 226, 227
alcoholism, 500
aldosterone, 724
algae, 737; blue-green, 631
algol, 58; eclipse of, 48
aliphatic compounds, 485
alizarin, 493; synthetic, 494
alkaline earths, 253

alkaloids, 496
alleles, 584
allergic reactions, 681, 683
allotropes, 269
alloys, magnetic, 276
alloy steel, 275
alnico, 276
alpha-amino acids, 529
Alpha Centauri, 637; distance of, 23; magnitude of, 26; spectral classification of, 70
alpha particles, 281; energy of, 314; nature of, 282
alpha rays, 281
alpha waves, 824
alpines, 791
alternating current, 394
aluminum, 277; alloys of, 277, 278
amber, 386
americium, 241
amethopterin, 690
amine group, 529, 698
amine oxidase, 835
amino acids, 483, 523, 529; combinations of, 531; dietary requirement for, 694, 695; essential, 695; fossils and, 765; hemoglobin and, 606; order of, 541ff.; polymerization of, 535; primordial formation of, 615, 627ff.; separation of, 538; three-dimensional structure of, 536
ammonia, 477; masers and, 420; nitrogenous wastes and, 765; planetary atmospheres and, 221; synthetic, 515
ammonium cyanate, 474
amor, 209
ampere, 390
amphibia, 740, 763
amphioxus, 740
amplification, electronic, 414
amplification, sound, 395
amplitude modulation, 412
analog computer, 850
analysis, chemical, 475, 476

anaphylactic shock, 681
anatomy, 574ff.
androgens, 722
andromeda galaxy (nebula), 30, 34, 35; distance of, 31, 42; novae in, 59; radial velocity of, 49; stars in, 32
androsterone, 722
anemia, 711; iron-deficiency, 709; sickle-cell, 604
An Essay on the Principle of Population, 755, 793
anesthetics, 497ff.
angels, 345
angiospermae, 737
angstrom unit, 341
angular momentum, conservation of, 97, 314
aniline, 488
aniline purple, 494
animal(s), 735ff.; land invasion by, 762; one-celled, 641
animal electricity, 389
animal magnetism, 832
animal spirits, 817
animal viruses, 659
annelida, 738
annelid superphylum, 738
annihilation, mutual, 296, 306, 320, 369
anopheles, 654
Antarctica, 152; exploration of, 150, 151
antares, 63
antenna, radio, 412
anterior pituitary, 724
anthrax, 645, 676
antibaryon, 323
antibiotics, 648ff., 656; virus culture and, 678
antibodies, 679ff., 766
antideuteron, 311
antielectron, 295
antiferromagnetism, 183
antigalaxies, 312, 322

913

937

939